Comprehensive Structural Integrity

Comprehensive Structural Integrity

Editors-in-Chief

I. Milne

Structural Integrity Management, Matlock, UK

R. O. Ritchie

University of California, Berkeley, USA

B. Karihaloo

University of Wales, Cardiff, UK

Volume 5

CREEP AND HIGH-TEMPERATURE FAILURE

Volume Editor

A. Saxena

Georgia Institute of Technology, Atlanta, USA

ELSEVIER

PERGAMON

AMSTERDAM – BOSTON – HEIDELBERG – LONDON – NEW YORK – OXFORD –

PARIS – SAN DIEGO – SAN FRANCISCO – SINGAPORE – SYDNEY – TOKYO

UK Elsevier Ltd., The Boulevard, Langford Lane, Kidlington, Oxford, OX5 1GB, UK

USA Elsevier Inc., 525 B Street, Suite 1900, San Diego, CA 92101-4495, USA

First edition 2003

Library of Congress Catalogue Card Number: 2003109071

A catalogue record for this book is available from the British Library

ISBN: 0-08-043749-4 (Set)

This book is printed on acid-free paper

Printed in the United Kingdom

Contents

Preface

Structural integrity of load bearing components is always an important consideration for engineers who design, build, or operate large pieces of machinery. Everyone understands the catastrophic consequences of structural failure and that sometimes it is unavoidable because the factors involved in predicting it are very complex. At elevated temperatures, a consideration of time-dependent creep deformation and damage makes assessments of integrity and failure prediction even more complex than for components operating below the creep temperature regime. Creep is commonly thought to be an important consideration when the operating temperature exceeds $0.35T_M$, where T_M is the melting point in absolute scale (K). Since the early 1950s, developments in approaches and the associated analytical and experimental techniques have contributed immensely to our ability to predict crack initiation, growth, and fracture in the creep regime. This volume comprehensively describes these developments.

Figure 1 shows various elements of a comprehensive integrity analysis for high-temperature components. The models needed include constitutive equations for predicting stress–strain–time–temperature relationships in critical locations of the component, crack initiation models, crack growth models, as well as algorithms for life prediction. The flow chart shows the needed model capabilities and also their interrelationships in comprehensively predicting the integrity of high-temperature components, the design life of a new component, specifying inspection intervals, or in making run–repair–retire type decisions on critical components already in service.

This volume is organized to comprehensively cover the various aspects of the structural integrity analysis of high-temperature components. The first chapter describes the fundamental creep deformation and rupture mechanisms and models at the atomisitic and the mesoscopic levels for predicting the creep rates. Often failures at high temperature occur due to interactions between creep and fatigue and due to high transient and steady-state thermal stresses. The two chapters that follow address the issue of crack initiation under creep, fatigue, and thermal–mechanical loading. These include coverage of models, analysis procedures, constitutive equations, and experimental techniques for generating laboratory data that can be used to evaluate materials and also to conduct structural integrity analysis of components.

A significant number of failures in structural components occur due to pre-existing defects or defects that initiate rapidly from clusters of nonmetallic inclusions or from other imperfections, such as casting, forging, and welding defects. Several defects also lie dormant in the components and pose no threat of fracture. In assessing structural integrity we should be interested in both, the defects that can ultimately cause fracture and those that are benign. Fracture mechanics provides an analytical framework for a comprehensive methodology for predicting the potential for fracture in components by considering: (i) the magnitude and distribution of applied stresses, (ii) the resistance to fracture and crack growth of the component material, and (iii) the size, location, and geometry of defects in the component. This approach is also capable of considering the influence of service environment. Chapters 5.04–5.06 deal with various considerations for predicting crack growth rates at elevated temperature under static as well as cyclic and combined static/cyclic loading. These chapters provide an in-depth coverage of the concepts of time-dependent fracture mechanics, creep crack growth, and creep-fatigue crack growth.

A wide variety of standard test methods exist that help in material and process selection for high-temperature components. Several of these methods have been standardized by international

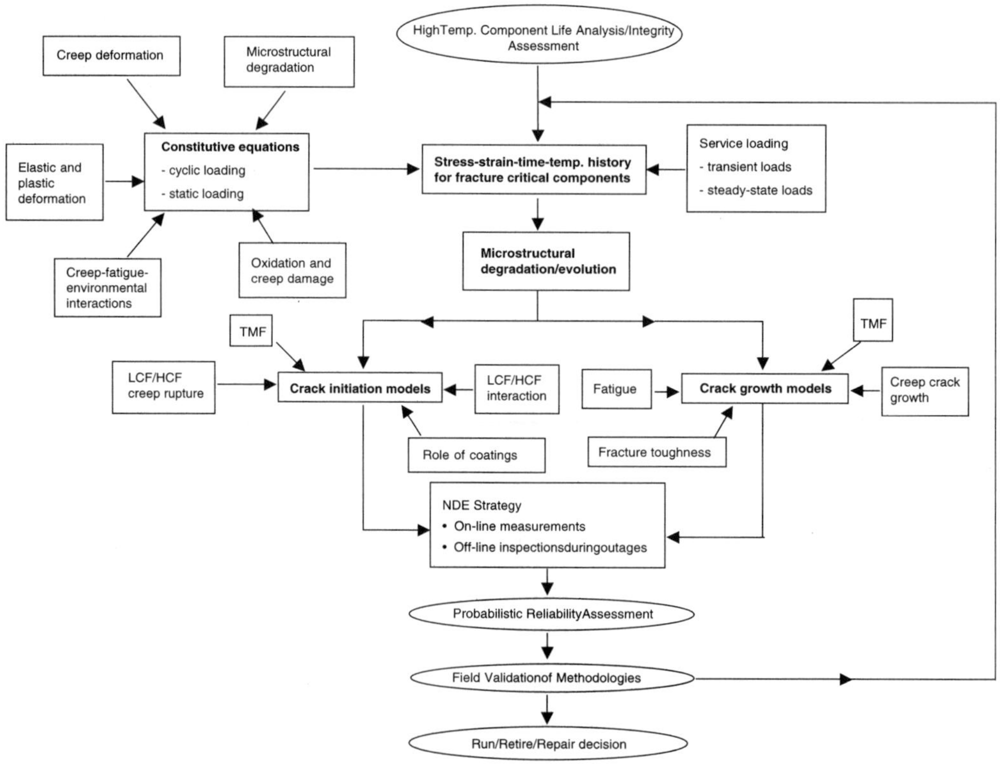

Figure 1 A flow chart of various elements of comprehensive structural integrity analysis of high-temperature components.

standards writing bodies. These methods collectively provide ways for testing and evaluating candidate materials and fabrication processes for critical applications. These more advanced test methods complement the conventional test methods such as tensile strength, hardness, Charpy impact test, or other material screening tests that are often used. In the field of high-temperature material behavior, test methods are available for characterizing creep deformation and rupture behavior, creep-fatigue crack initiation and crack growth behavior, and crack initiation and crack growth behavior under thermal–mechanical loading. A description of these methods is contained within the various chapters that describe the various phenomena.

When design has already been finalized based on other considerations or the component is already in service, engineers are asked to predict how long the component is expected to last. From the knowledge of the magnitude and nature of the operating stresses, the relevant fracture and crack initiation and growth properties and the data from nondestructive inspection for detecting cracks or crack-like defects, it is possible to conduct an analysis to predict the design life, or remaining life of components. The financial incentives are huge for continuing to operate plants with existing equipment as opposed to replacing them with new equipment. However, the additional risk involved in continuing to use critical equipment beyond their initial design life must be balanced by assurance against failure based on sound condition-based assessments and institution of regular inspection intervals.

Very often, the nondestructive inspection techniques that are used during quality assurance or during service to detect cracks or crack-like defects in components are dictated by considerations such as the material, geometry and size of the component, the accessibility of the critical areas of the component for placing transducers, and by economics. Inspections, especially during service, can be expensive because they are often time consuming, involve expensive equipment, and require highly trained professionals to perform them and to interpret results. Conservative inspection criterion and intervals can result in high rejection rates and more frequent inspections than necessary increase the operating cost of conducting business, especially for high-temperature components that are made from expensive materials which often also require expensive processing methods. Alternatively, inadequate inspection criterion and insufficient inspection intervals can cause catastrophic failures with devastating consequences. Structural integrity analysis methods provide an effective analytical tool for estimating realistic inspection intervals and determining realistic inspection criterion making them invaluable.

The final two chapters are focused as creep resistance in ferritic and nonferritic materials. The first of these discusses the use of the concepts and models described in earlier chapters for predicting the life of high-temperature components in steels. This by no means is a trivial task. It is important to carefully understand how the various crack initiation and growth models complement each other in a comprehensive methodology for structural integrity assessment. The final chapter covers a description of high-temperature materials for gas-turbine applications. These include nickel-base superalloys as well as intermetallics and ceremic materials that may see application in the future.

A. Saxena
Georgia Institute of Technology, Atlanta, USA

Contributors to Volume 5

H. K. D. H. Bhadeshia
University of Cambridge, UK

D. J. Gooch
GenExp Ltd., Swindon, UK

H. Harada
National Institute for Materials Science, Tsukuba, Japan

T. Kitamura
Kyoto University, Japan

S. Kubo
Osaka University, Japan

L. Rémy
Centre des Matériaux, UMR CNRS 7633, Ecole des Mines de Paris, France

A. Saxena
Georgia Institute of Technology, Atlanta, USA

R. P. Skelton
Imperial College of Science, Technology and Medicine, London, UK

G. A. Webster
Imperial College, London, UK

5.01
Mechanisms and Models for Creep Deformation and Rupture

H. K. D. H. BHADESHIA
University of Cambridge, UK

NOMENCLATURE

a_0, a_1,... empirical constants
b magnitude of Burgers' vector
B_0 bond order
c concentration
d grain size
D diffusivity
D_{eff} effective diffusion coefficient in coarsening theory
D_{gb} grain boundary diffusion coefficient
D_i diffusion coefficient of component i
D_{P} lattice diffusion coefficient

D_0	pre-exponential term in diffusion coefficient
$\bar{D}$	intrinsic diffusion coefficient
D^*	tracer diffusion coefficient
E_D	noise in output in regression analysis
f_c	Bardeen–Herring correlation factor
G	shear modulus
G^*	activation free energy for diffusion
ΔG_b	free energy of segregation to a grain boundary
ΔG_s	free energy of segregation to a free surface
H^*	activation enthalpy
$^{mg}\Delta H$	excess magnetic enthalpy
J	flux
J_V	vacancy flux
k	Boltzmann constant
ΔK	fraction of kinetic energy attributed to vibration mode
m	mass, or grain size exponent
M_d	term in molecular orbital theory
n	stress exponent
Q	activation energy
r	precipitate radius
R	universal gas constant
S^*	activation entropy
t	time
T	absolute temperature
T_m	absolute melting temperature
v	dislocation velocity
v_k	velocity of Kirkendall markers
V	volume
$\bar{V}_i$	partial molar volume of component i
V_α	molar volume of ferrite
w	weighting coefficient in empirical analysis
x	mole fraction
x_i	ith input variable in regression analysis
y	output variable in regression analysis
z	coordinate
α	ferrite
δ	thickness of grain boundary
$\dot{\varepsilon}$	strain rate
θ	θ parameter or empirical constant
ν	frequency of successful jumps
ν_0	attempt frequency
ρ	density of mobile dislocations
σ	stress
σ_0	threshold stress

5.01.1 INTRODUCTION

The tendency of a material to undergo plastic deformation under the influence of a stress is often characterized using a tensile test. In such a test, the strain is obtained as a function of stress and the onset of permanent strain is taken to correspond to the yield strength of the material. Tensile tests are, in general, conducted under conditions where the measured properties are independent of time. By contrast, the essence of creep deformation is that the plastic strain is time dependent and can occur at stresses which in an ordinary tensile test would be below the yield stress. This chapter begins with a summary of diffusion and the basic constitutive equations for creep, followed by certain useful though more empirical approaches, and models for a variety of aspects of creep-resistant materials.

In a typical uniaxial creep test, the sample is subjected to a constant load such that the stress is below the yield strength, and the engineering strain is monitored as a function of time, as illustrated in Figure 1(a). The application of the load naturally leads to an instantaneous elastic strain. This is followed by a primary creep regime in which the strain rate decreases with time as the material settles down towards the steady-state regime, where there is a balance between dynamic recovery and strain hardening.

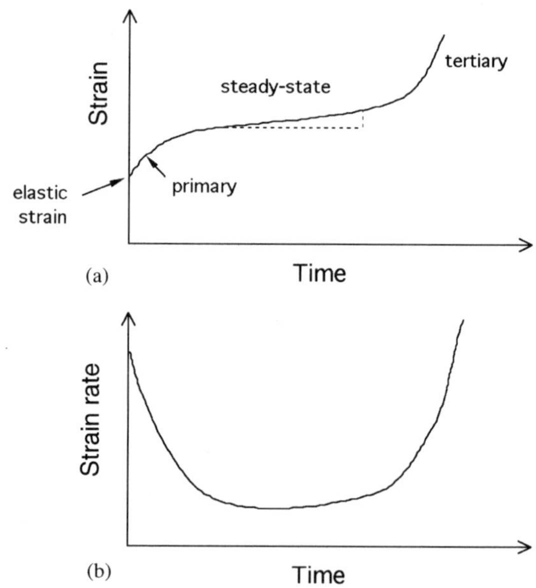

Figure 1 (a) Schematic representation of the stages of creep deformation. (b) Variations in the creep rate.

The creep rate accelerates during tertiary creep. A part of the reason for the increase in strain rate is that the sample may neck so that the true stress increases. However, tertiary creep can occur before the onset of necking, induced by the accumulation of creep damage (e.g., voids).

Creep data can be plotted as in Figure 1(b) in order to identify the minimum creep strain rate. The three stages of creep are also evident in these plots; in the primary region the creep rate decreases with time, whereas in the tertiary region it accelerates.

Steady-state creep is roughly centered at the minimum in the plot of creep rate vs. time (Figure 1(b)). The minimum rate is empirically found to be inversely proportional to the time to rupture. This is often called the Monkman–Grant (1956) relationship in which the product of the minimum creep rate and rupture time is found to be constant for a given creep mechanism.

The extent of each of the stages in the creep curve can vary dramatically with the temperature, stress, and material. Actual creep curves for a variety of ferritic steels are illustrated in Figure 2. The 2Cr steels are not really designed for operation at temperatures as high as 600°C. It is expected, therefore, that there is a rapid onset of tertiary creep. Alternatively, the 12Cr alloy exhibits a prolonged steady-state creep regime, since such alloys are intended for service at the test temperature.

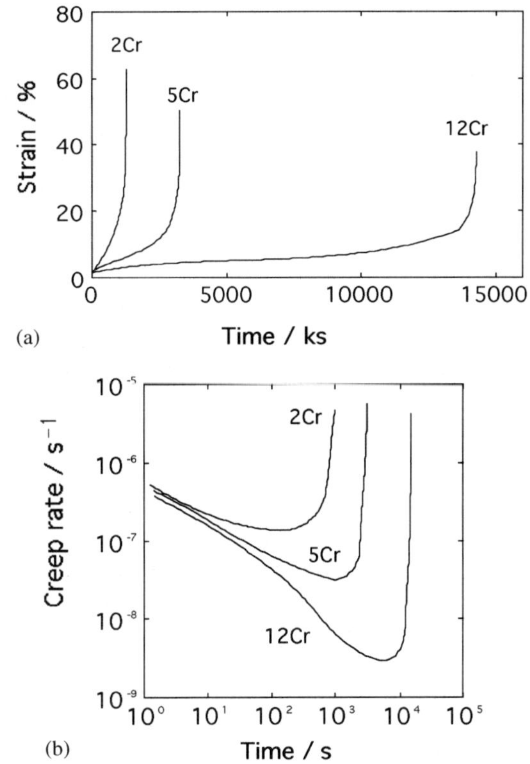

Figure 2 Creep data for a variety of ferritic steels, for tests conducted at 873 K and 108 MPa (Abe and Nakazawa, 1992): (a) actual creep curves and (b) corresponding variations in the creep rate.

5.01.2 DIFFUSION

The mechanism of creep relies on the thermally activated passage of atoms over an activation barrier which is characterized by an activation free energy G^*. The frequency ν_0 with which attempts are made to scale the barrier is multiplied by a Boltzmann factor in order to obtain the frequency ν of successful jumps:

$$\nu = \nu_0 \exp\left(-\frac{G^*}{kT}\right)$$
$$\equiv \underbrace{\nu_0 \exp\left(\frac{S^*}{k}\right)}_{\text{independent of } T} \exp\left(-\frac{H^*}{kT}\right)$$

where k and T are the Boltzmann constant and the absolute temperature, respectively, and H^* and S^* the activation enthalpy and activation entropy, respectively. Since

$$D \propto \nu$$

we find

$$D = D_0 \exp\left(-\frac{H^*}{kT}\right)$$

The activation enthalpy H^* is frequently called the "activation energy" for diffusion and is designated Q. It is necessary to define a variety of diffusion coefficients in order to treat creep deformation in alloys.

Fick's first law assumes that the diffusion flux is related directly to the concentration gradient:

$$J = -D\frac{\mathrm{d}c}{\mathrm{d}z} \tag{1}$$

where the proportionality constant D is the diffusion coefficient and z is the coordinate parallel to the flux. The flux is the rate of transfer of the diffusing substance through a unit of area and the concentration gradient is measured normal to this area. A very large number of measurements have been made within the framework of this law. Deviations from Fick's law are, therefore, treated by making the diffusion coefficient concentration dependent in order to retain the basic proportionality.

Some elements diffuse faster in a solution than others; each component of the solution is associated with its characteristic *intrinsic* diffusion coefficient $\bar{D}_A$. This represents the flux $\bar{J}_A$ of component A in a binary A–B substitutional solution containing a concentration gradient of A (and hence of B).

When the two species in an interdiffusion experiment have unequal intrinsic diffusion coefficients and assuming that diffusion occurs by a vacancy mechanism, there will be a net flow of matter in the direction of the faster diffusing element. If a diffusion couple is constructed with inert markers at the interface, fixed to the laboratory bench, then the specimen will move relative to the markers during the course of diffusion (Figure 3), a phenomenon known as the Kirkendall (1942) effect. This bulk flow of matter contributes to composition change at any point, but is not accounted for in the definition of intrinsic coefficients. The substitution of intrinsic diffusion coefficients into Fick's first law is, therefore, said to define fluxes $\bar{J}_i$ relative to the Kirkendall (or lattice fixed) frame of reference in which $\sum_i \bar{J}_i = -J_V$, where J_V is the flux of vacancies. The Kirkendall frame moves with the velocity of the inert markers relative to the laboratory frame to which the specimen is fixed.

An observer in the laboratory frame sees the change in composition at a point in the sample due both to intrinsic diffusion and the Kirkendall flow of matter. The *chemical* or *interdiffusion* coefficient D is defined relative to the laboratory frame in order to take both of these effects into account, representing the rate at which mixing or unmixing occurs. It is the interdiffusion coefficient that is used most frequently in practical applications, e.g., when dealing with transformation kinetics, homogenization, or tempering reactions. In circumstances where $\bar{D}_A = \bar{D}_B$, there is no Kirkendall effect and $\bar{D}_A = \bar{D}_B = D$; in other words, the laboratory and Kirkendall frames of reference become identical.

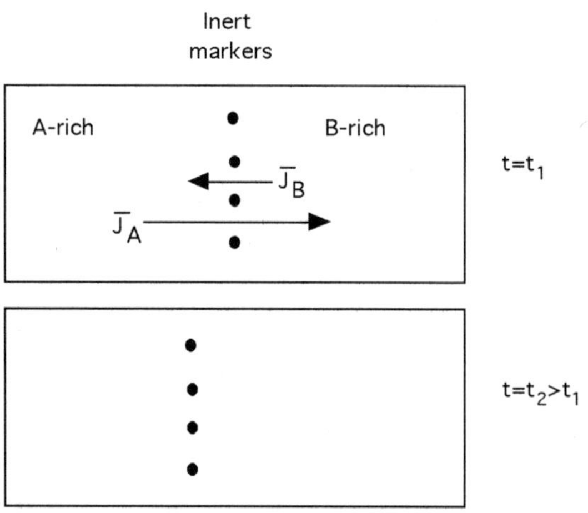

Figure 3 The Kirkendall effect during diffusion by a vacancy mechanism. $\bar{J}$ represent the fluxes of the two elements A and B in the diffusion couple and t is the time: $\bar{D}_A > \bar{D}_B$. The inert markers are fixed to the laboratory frame. Unequal fluxes result in a net transport of matter relative to the markers.

We have thus separated the bulk flow of matter from diffusion due to the random motion of nonuniformly distributed A atoms. Intrinsic coefficients account only for true diffusion, whereas interdiffusion coefficients include the bulk flow of matter.

It is often convenient to define the interdiffusion coefficient D relative to other frames of reference. In the volume-fixed frame, the fluxes J_i are measured across a section such that the total volume on either side of the section remains constant during diffusion. The origin of the coordinate system is at the instantaneous center of volume. In this case,

$$\sum_i^n J_i \bar{V}_i = 0,$$

where $\bar{V}_i$ is the partial molar volume of component i in an n component solution. A coordinate system such as this can only be defined if the partial molar volumes are constant.

The laboratory frame is also known as the number-fixed frame. In this case, the fluxes occur across a section defined such that the total number of atoms on either side of the section remains constant:

$$\sum_i^n J_i = 0.$$

If $\bar{V}_i / \bar{V}_n = 1$ for all i, in other words if the integral molar volume remains constant, then the volume-fixed frame coincides with the number-fixed frame. This may be approximately true in dilute solutions where any changes in the integral molar volume as a function of chemical composition will be small.

The variety of frames of reference are summarized in Table 1. We have seen that the flux described in the laboratory frame includes a component $\bar{J}_i$ due to intrinsic diffusion and the bulk flow of matter due to the Kirkendall effect, i.e., $c_i v_k$ where v_k is the velocity of the Kirkendall markers and c is the local molar concentration per unit volume. It follows that (Darken, 1948)

$$J_A = \bar{J}_A + c_A v_k \tag{2}$$

$$J_B = \bar{J}_B + c_B v_k \tag{3}$$

The velocity of the markers depends on the volume of matter passing through a unit area in a unit of time. The net flux across the markers is $\bar{J}_A + \bar{J}_B$ (moles m^{-2} s^{-1}). Noting that

Table 1 Frames of reference for diffusion. J_V is the flux of vacancies.

Reference frame	*Definition*
Kirkendall (lattice-fixed)	$\sum_i^n \bar{J}_i = -J_V$
Laboratory or number-fixed	$\sum_i^n J_i = 0$
Volume-fixed	$\sum_i^n J_i \bar{V}_i = 0$

$1/(c_A + c_B)$ is simply the volume per mole, we find

$$v_k = -\frac{\text{volume past markers}}{\text{area} \times \text{time}} = -\frac{\bar{J}_A + \bar{J}_B}{c_A + c_B} \quad (4)$$

The negative sign appears because the markers move in a direction opposite to the net flux. Since $c_A/(c_A + c_B)$ is the mole fraction x_A, substitution for the Kirkendall velocity into Equation (2) yields

$$J_A = x_B \bar{J}_A - x_A \bar{J}_B \quad (5)$$

The relationship between the chemical diffusion coefficient and intrinsic coefficients becomes

$$D = x_B \bar{D}_A + x_A \bar{D}_B \quad (6)$$

Notice that the description of interdiffusion in a binary alloy requires just one diffusion coefficient, though some information is omitted in this description unless the Kirkendall velocity is also stated.

The *tracer* diffusion coefficient D^* describes the motion of radioactively labeled isotopes in an otherwise chemically homogeneous solution. When the tracer atoms are of the same species as the nontracer atoms, the tracer diffusion coefficient is called the self-diffusion coefficient. Tracer and self-diffusion coefficients describe purely random motion and do not properly represent diffusion in concentration gradients. In self-diffusion measurements there is a mass difference between the radioactive isotope and the solvent, which causes the two isotopes to vibrate with different frequencies in the lattice. The vibration frequency is proportional to the inverse square root of the mass m of the atom. For two isotopes D_1^* and D_2^* of the same species, it follows that

$$\frac{D_1^* - D_2^*}{D_2^*} \propto \frac{m_1^{-1/2} - m_2^{-1/2}}{m_2^{-1/2}}$$

which can be written as

$$f_c \Delta K = \frac{D_1^*/D_2^* - 1}{(m_2/m_1)^{1/2} - 1} \quad (7)$$

where f_c is the Bardeen–Herring correlation factor which in self-diffusion can be taken to be independent of the isotope. The correlation factor accounts for the fact that the path of a defect such as a vacancy around a solute is not random, because some of its jumps are reversed. ΔK is the fraction of the total kinetic energy attributed to the vibration mode, which resides in the migrating atom (Le Claire, 1966).

By contrast with self-diffusion, an additional virtual force acts on the diffusing species in a concentration gradient, due to the associated chemical potential gradient:

$$D = x_B \bar{D}_A + x_A \bar{D}_B = (x_B D_A^* + x_A D_B^*) \times \text{thermodynamic factor} \quad (8)$$

where the thermodynamic factor depends on the activity coefficients. It is this interdiffusion coefficient which is appropriate when creep is controlled by viscous drag, where the solute atmosphere diffuses with the dislocation by exchanging positions with solvent atoms in the dislocation path (Frost and Ashby, 1982).

The diffusion coefficient for climb-controlled creep in a binary solution is then given by (Burton and Bastow, 1973)

$$D = \frac{D_A^* D_B^*}{x_B D_A^* + x_A D_B^*} \quad (9)$$

Given the diffusion theory outlined above, the thermodynamic effect of solutes on diffusivity can be calculated using commercially available software and databases. However, special considerations are necessary for iron alloys, which represent one of the most important of the class of creep-resistant materials. This is because magnetism has a large role to play. Self-diffusion data for ferrite show anomalies, both in the region of paramagnetic to ferromagnetic transition and in the δ-ferrite temperature range, Figure 4 (Borg and Birchenall, 1960; Buffington *et al.*, 1961; Iijima *et al.*, 1988). The anomaly has been observed also for the diffusion of nickel in ferrite (Hirano *et al.*, 1961) and in a vast range of other measurements including mechanical properties such as creep at elevated temperatures.

To cut a long story (Bhadeshia, 1999c) short, Kučera (1979) proposed a function of the form to explain diffusion in ferrite:

$$D = D_0 \exp\left(\frac{0.35 q\{T\} Q_0}{RT_m}\right) \exp\left(-\frac{Q}{RT}\right) \quad (10)$$

and Jönsson (1992) expressed the function q in terms of thermodynamic parameters alone:

$$q(T) = a_0 \, {}^{mg}\Delta H(T)$$

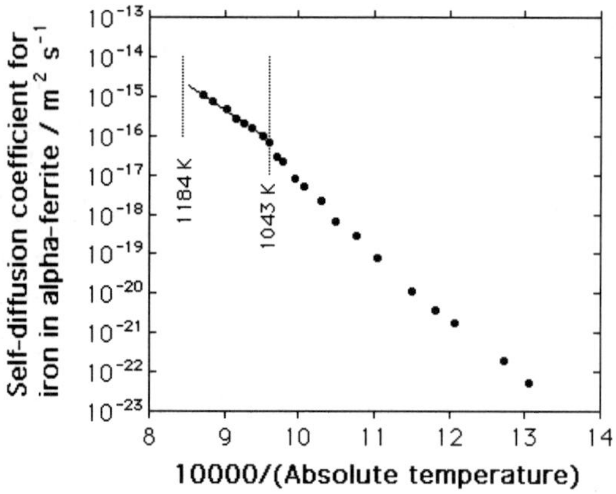

Figure 4 Self-diffusion coefficient of iron in α-ferrite (Iijima *et al.*, 1988). The Curie temperature is 1,043 K and the formation of austenite begins at 1,184 K.

where a_0 is a fitting constant and ${}^{mg}\Delta H(T)$ is the excess magnetic enthalpy at the temperature T. Equation (10) contains the melting temperature whose meaning in the context of diffusion is not clear for an alloy where melting occurs over a range of temperatures. However, there is a well-known empirical relationship between the activation energy for diffusion and the melting temperature of a pure metal:

$$Q_0 = 140T_m \ \text{J mol}^{-1}$$

so that the melting temperature can be eliminated from Equation (10). By fitting the equation to ferritic iron, Jönsson (1992) was able to demonstrate that with appropriate thermodynamic data for ${}^{mg}\Delta H$, he was able to estimate fairly well the tracer diffusion of elements such as cobalt, nickel in ferrite both in the ferromagnetic and paramagnetic regions. The activation energy and D_0 for the paramagnetic state derived by fitting in each case but a_0 was kept fixed, assumed to depend only on the lattice geometry.

To summarize, it is possible, when the thermodynamic data necessary for the estimation of ${}^{mg}\Delta H(T)$ are available, to calculate the effect of solute elements on the tracer diffusion coefficient of iron. It is interesting calculations like these are already being applied in the design of creep-resistant steels (e.g., Hald, 1999).

5.01.2.1 Structure Sensitive Diffusion

Engineering materials are crystalline but imperfect, with defects such as vacancies, dislocations, and grain boundaries. Nonequilibrium concentrations of vacancies tend to rapidly anneal out and hence only have transient effects on diffusion; an exception is materials subjected to irradiation in which case there will be a persistent source of excess vacancies due to displacement damage. Alternatively, dislocations and grain boundaries provide easy diffusion paths through the material. The grain boundary diffusion coefficient (D_{gb}) is much greater than that associated with the perfect structure (D_P). However, the fraction of the sample which can be identified with the grain boundaries is small when compared the material between the boundaries. Consequently, grain boundary or defect diffusion only makes a large contribution to the net diffusion flux at low temperatures where $D_P \ll D_{gb}$. It is easy to show that the effective diffusion coefficient is given by

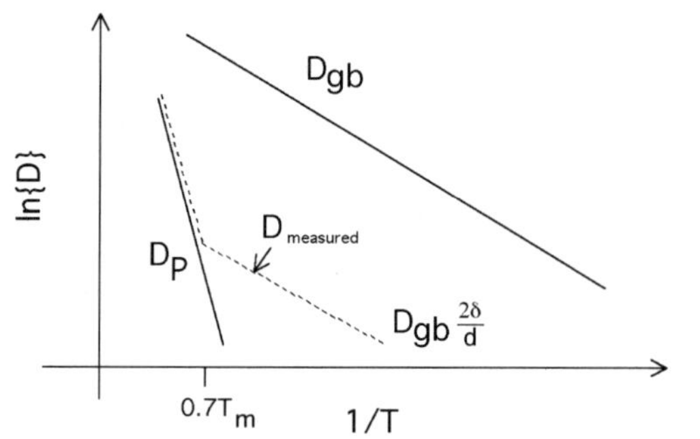

Figure 5 An illustration of structure sensitive diffusion at low homologous temperatures. T_m is the absolute melting temperature.

$$D_{\text{measured}} = D_P + D_{gb}\frac{2\delta}{d} \tag{11}$$

where δ is the grain boundary thickness and d is the grain size. It follows that at high temperatures the diffusion flux is dominated by diffusion through the bulk of the material, whereas at low temperatures it becomes dominated by diffusion through defects (Figure 5).

The flux of matter that leads to the plastic strain is at low temperatures largely through defects, such as the grain boundaries or the cores of dislocations (Figure 5). By contrast, most of the observed effects are due to bulk diffusion when the tests are done at high temperatures ($>0.7T_m$). We now consider some important mechanisms of creep.

5.01.3 CREEP MECHANISMS

The mechanisms of creep can conveniently be summarized using a deformation map (Frost and Ashby, 1982) which identifies domains in a plot of homologous stress vs. homologous temperature, where a particular creep mechanism dominates (Figure 6). The domain boundaries are defined by the locus of points where the strain rate for two or more mechanisms becomes identical.

5.01.3.1 Dislocation Creep

At low strain rates and relatively low temperatures, creep occurs by the glide of slip dislocations. Diffusion-assisted dislocation climb is necessary for continued deformation when the glide process is obstructed, e.g., by the presence of precipitates in the glide plane. An applied stress assists the climb process via a force which tends to push the dislocation onto a parallel plane, such that it can bypass the particle (Figure 7).

5.01.3.2 Diffusion Creep

At a sufficiently high temperature and low strain rate, it is possible to achieve plastic deformation simply by the stress-induced migration of vacancies, almost akin to fluid flow where the strain rate during Newtonian viscous flow is proportional to the stress. In the Nabarro–Herring version of diffusion creep, the flow of matter occurs by diffusion through the lattice (Figure 8), whereas in Coble creep it is dominated by diffusion through the grain boundary structure. Both of these processes require the grains to slide past each other in order to maintain continuity and hence avoid void formation.

Harper–Dorn is a special form of creep in which the constitutive equation is similar to diffusion creep, but the mechanism involves the climb of dislocations at a constant, low dislocation density in a large grain-sized material, so that grain boundary diffusion does not play a major role. Furthermore, Harper–Dorn creep occurs by an intragranular dislocation mechanism, so that grain elongation occurs without a need for grain boundary sliding (Langdon, 2000).

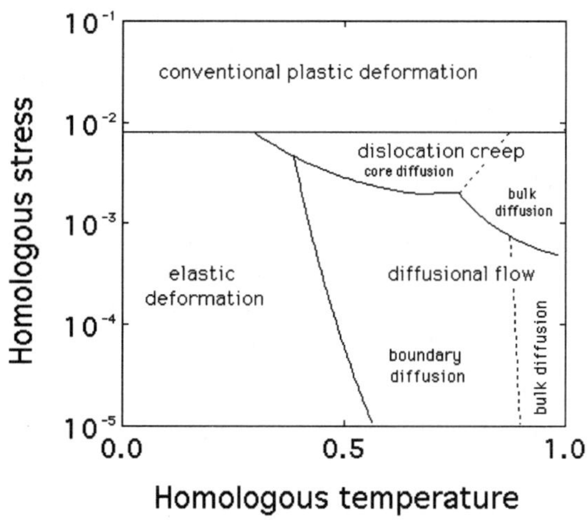

Figure 6 Schematic deformation map as a function of the homologous stress (stress divided by the elastic modulus) and homologous temperature (absolute temperature divided by the absolute melting temperature).

5.01.3.3 Rate Equations

The creep strain rate for metals can be expressed as a function of the applied stress (σ), the absolute temperature T, and the grain size (d):

$$\dot{\varepsilon} = f(\sigma, T, d) \quad (12)$$

or more explicitly as

$$\dot{\varepsilon}_{SS} = a_1\left(\frac{DGb}{kT}\right)\left(\frac{b}{d}\right)^m\left(\frac{\sigma}{G}\right)^n \quad (13)$$

where $\dot{\varepsilon}_{SS}$ is the steady-state creep rate, a_1 is usually an empirical constant, D is an appropriate diffusion coefficient, b is the magnitude of the Burgers vector, and G is the shear modulus. The stress exponent n and the grain size exponent m are dependent on the mechanism of creep.

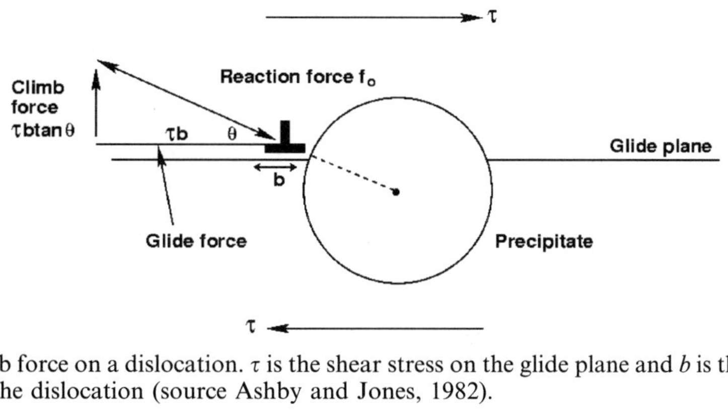

Figure 7 The climb force on a dislocation. τ is the shear stress on the glide plane and b is the magnitude of the Burgers vector of the dislocation (source Ashby and Jones, 1982).

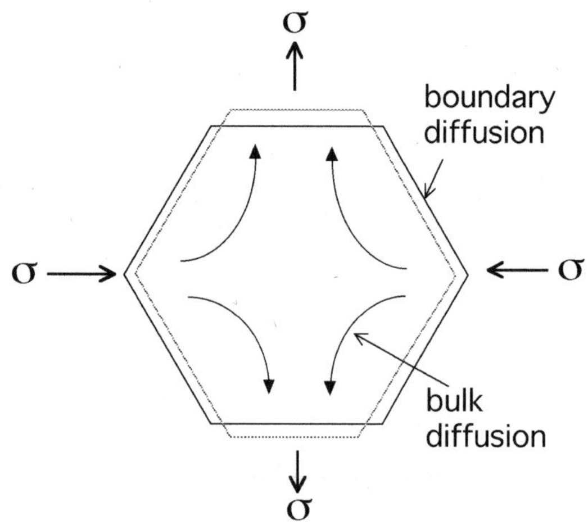

Figure 8 Creep by a diffusion mechanism.

When considering particle-hardened metals, it is necessary to use an effective stress $(\sigma-\sigma_0)$ where σ_0 is a threshold or friction stress (Langdon, 1984):

$$\dot{\varepsilon}_{SS} = a_1\left(\frac{DGb}{kT}\right)\left(\frac{b}{d}\right)^m\left(\frac{\sigma-\sigma_0}{G}\right)^n \qquad (14)$$

Failure to do this can result in artificially large stress exponents.

Figure 9 shows schematic plots of the logarithm of the creep strain rate as a function of the logarithm of the applied stress for pure metals (Langdon, 1984).

Consider first the case for high temperatures close to melting (Figure 9(a)). In samples with a large grain size, $n=5$ for a large, intermediate range of applied stress with creep controlled by the climb of dislocations. Since lattice diffusion is the major component of diffusion flux at high temperatures, the activation energy of the creep deformation is given by that for self-diffusion.

In this regime, the creep rate is independent of grain size. Langdon has argued that this should lead to $n=3$ since, e.g., the creep rate can be written as

$$\dot{\varepsilon} \propto \rho b v \qquad (15)$$

where ρ is the density of mobile dislocations, assumed from deformation theory to be proportional to σ^2, and v is the dislocation velocity which is proportional to σ. Other physical interpretations also lead to this "natural" stress exponent of $n=3$. Therefore, a value of $n=5$ is considered to be an artifact of the analysis of creep data in that the "constant" a_1 may contain a stress dependence.

This power-law behavior with $n=5$ breaks down at high stresses (high strain rates) with $\dot{\varepsilon} \propto \exp(a_2\sigma)$, where a_2 is a constant. This breakdown usually occurs at a point where $\dot{\varepsilon}/D_P \simeq 10^{13}$ m^{-2} (Sherby and Burke, 1967). It occurs when dislocation creep undergoes a transition from climb-controlled to glide-controlled flow (Frost and Ashby, 1982).

There is also a deviation from $n=5$ at low strain rates and high temperatures in a region identified with the Harper–Dorn mechanism (Figure 9(a)). Both the Harper–Dorn creep and dislocation climb mechanisms are, in large grain-sized specimens, intragranular processes and in that sense independent of grain size with $m=0$.

When the grain size is large then at low strain rates, diffusion creep becomes possible,

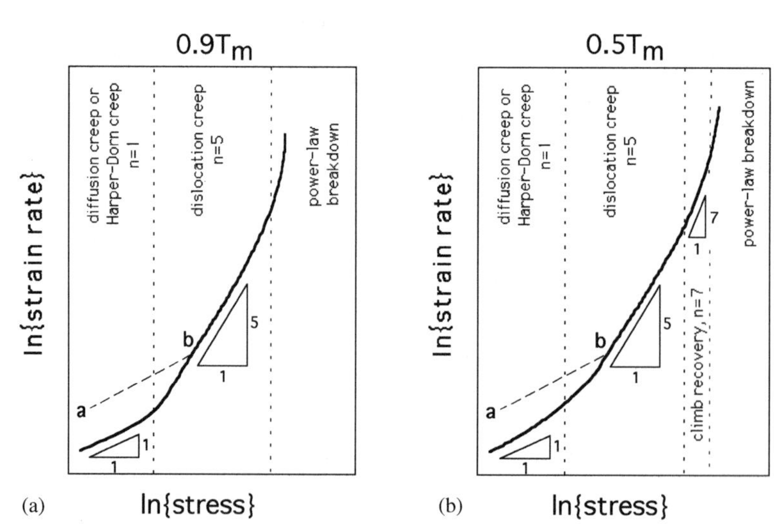

(a) (b)

Figure 9 Schematic logarithmic plots of strain rate vs. stress for pure metals (Langdon, 1985). The dashed line marked ab represents a finer grain size than the solid line: (a) close to the absolute melting temperature and (b) near $0.5T_m$.

with Nabarro–Herring creep controlled by lattice diffusion with $n = 1$ and $m = 2$, whereas in Coble creep which is controlled by grain boundary diffusion, $n = 1$ and $m = 3$. The stress exponent is, of course, expected to be unity, since diffusion creep is essentially viscous flow. In Nabarro–Herring creep, the matter is transported from boundaries which are more parallel to the applied stress to those which are normal to the stress. From stereology, the grain boundary area per unit volume is proportional to d^{-1} and the vacancy chemical potential gradient is also proportional to d^{-1}; since the flux is proportional to both of these quantities, it follows that the strain rate will depend on d^{-2} giving $m = 2$.

Consider now the case of Coble creep. For a grain size d and grain boundary thickness δ, grain elongation leads to an increase in volume along the tensile direction. In the approximation of spherical grains, the rate of volume change along that direction is therefore

$$\frac{\mathrm{d}V}{\mathrm{d}t} \propto (d+\delta)^2 \frac{\mathrm{d}d}{\mathrm{d}t} \quad \text{and} \quad \frac{\mathrm{d}V}{\mathrm{d}t} \propto J$$

where J is the flux through the grain boundary. It follows that

$$\dot{\varepsilon} = \frac{1}{d+\delta}\frac{\mathrm{d}d}{\mathrm{d}t} \propto \frac{1}{(d+\delta)^3} \qquad (16)$$

so in the limit that $\delta \ll d, m = 3$. A derivation (Yamakov *et al.*, 2002) for nanocrystalline metals, where δ is not very different from d, indicates that a better approximation to Equation (16) is

$$\dot{\varepsilon} \propto \left(\frac{2\delta}{d^3} - \frac{\delta}{d^4}\right)$$

a relationship which reduces to the ordinary Coble dependence when δ is relatively small.

Both Nabarro–Herring and Coble creep will be faster when the grain size is smaller, as indicated by the lines ab in Figure 9, with the low-stress $n = 1$ behavior extending to larger stresses.

The situation at low homologous temperatures is illustrated in Figure 9(b). Defect diffusion becomes more prominent relative to lattice diffusion at low temperatures. The additional feature, therefore, is a region of low-temperature climb driven by dislocation pipe diffusion with a stress exponent of $n = 5 + 2 = 7$. The additional factor of 2 in the stress exponent comes about because the intragranular dislocation density is proportional to σ^2 (Robinson and Sherby, 1969).

5.01.4 PARAMETRIC MODELS

The mechanisms discussed above are well established and much more detail can be found in a number of excellent publications (e.g., Frost and Ashby, 1982; Argon, 1982, 1996; Hirth and Lothe, 1982; Evans and Wilshire, 1985). The constitutive relationships can be quite sophisticated with explicit terms accounting for factors such as stacking fault energy, dislocation microstructures, elaborate diffusion theory, etc. Such work has contributed enormously to the understanding of creep deformation and the qualitative design of creep-resistant materials.

However, practical creep-resistant materials tend to be extremely complex, because a large number of design criteria other than creep have to be satisfied when making engineering components. It is important to understand the long-term creep and rupture behavior of such materials for their safe use in safety-critical applications such as power plant and aeroengines. For this reason there exist a large number of empirical or semi-empirical methods, which permit the accurate representation of experimental data and at the same time facilitate the extrapolation of short-term creep data, to varying degrees of success. Many of these methods have been reviewed (Bhadeshia *et al.*, 1998; Evans, 1999). The following sections describe a few of the methods in order to highlight developments in the subject.

Creep strain is a direct measure of remaining life (Figure 10), so the safe extrapolation of short-term creep data is a major goal of structural integrity assessment. A popular method involves the use of a θ-parameter equation (Evans and Wilshire, 1985):

$$\varepsilon = \underbrace{\theta_1[1 - \exp(-\theta_2 t)]}_{\text{decaying rate}} + \underbrace{\theta_3[\exp(\theta_4 t) - 1]}_{\text{accelerating rate}} \qquad (17)$$

where θ_i are obtained by fitting to experimental data, and t is the time at temperature. The first two of these parameters describe the primary or decaying strain rate component, whereas the remaining terms the accelerating regime.

The steady-state creep theory described earlier leads to an equation of the form

$$\dot{\varepsilon} = a_3 \sigma^n \exp\left(-\frac{Q}{RT}\right) \qquad (18)$$

where a_3 is an empirical constant. This can be integrated to find the creep-rupture time t_r as

$$\ln(t_r) = \ln\left(\frac{\varepsilon_r}{a_3}\right) - n\ \ln(\sigma) + \frac{Q}{RT} \qquad (19)$$

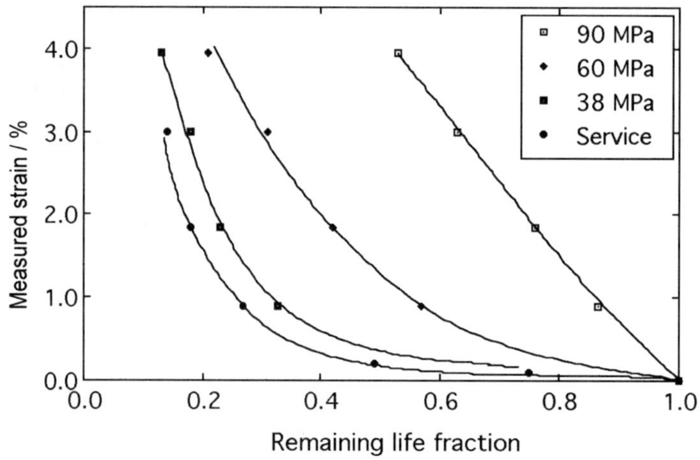

Figure 10 The creep strain as a function of remaining life. The data are for creep samples loaded at three different stresses; the data marked "service" refer to the conditions for a thick-walled pressure vessel (after Parker, 1996).

Nevertheless, it is found, in practice, that slightly better fits are obtained by writing (Cane, 1981; Beech *et al.*, 1995)

$$\ln(t_r) = a_4 + a_5\sigma + a_6T \tag{20}$$

where a_i are fitting constants.

While parametric models like these have been widely and successfully applied, they are not sufficiently general to deal with large number of variables. A more modern technique is the neural network method which thrives in complexity. The method has now been demonstrated to be superior in the extrapolation and representation of creep data (Bhadeshia, 1999a, 1999b, 1999c; Evans, 1999).

5.01.5 NEURAL NETWORKS

A general method of regression which avoids these difficulties is neural network analysis, illustrated at first using the familiar linear regression method. A network representation of linear regression is illustrated in Figure 11(a). The inputs x_i (concentrations) define the *input nodes*, the bainite-start temperature the *output node*. Each input is multiplied by a random weight w_i and the products are summed together with a constant θ to give the output $y = \sum_i w_i x_i + \theta$. The summation is an operation which is hidden at the hidden node. Since the weights and the constant θ were chosen at random, the value of the output will not match with experimental data. The weights are systematically changed until a best-fit description of the output is obtained as a function of the inputs; this operation is known as *training* the network.

The network can be made nonlinear as shown in Figure 11(b). As before, the input data x_j are multiplied by weights ($w_j^{(1)}$), but the sum of all these products forms the argument of a hyperbolic tangent:

$$h = \tanh\left(\sum_j w_j^{(1)} x_j + \theta\right) \quad \text{with } y = w^{(2)}h + \theta^{(2)} \tag{21}$$

where $w^{(2)}$ is a weight and $\theta^{(2)}$ another constant. The strength of the hyperbolic tangent *transfer function* is determined by the weight w_j. The output y is, therefore, a nonlinear function of x_j, the function usually chosen being the hyperbolic tangent because of its flexibility. The exact shape of the hyperbolic tangent can be varied by altering the weights (Figure 12(a)).

A one hidden-unit model may not, however, be sufficiently flexible. Further degrees of nonlinearity can be introduced by combining several of the hyperbolic tangents (Figure 12(b)), permitting the neural network method to capture almost arbitrarily nonlinear relationships. The number of tanh functions per input is the number of hidden units; the structure of a two hidden unit network is shown in Figure 13.

The function for a network with i hidden units is given by

$$y = \sum_i w_i^{(2)} h_i + \theta^{(2)} \tag{22}$$

where

$$h_i = \tanh\left(\sum_j w_{ij}^{(1)} x_j + \theta_i^{(1)}\right) \tag{23}$$

Notice that the complexity of the function is related to the number of hidden units. The availability of a sufficiently complex and

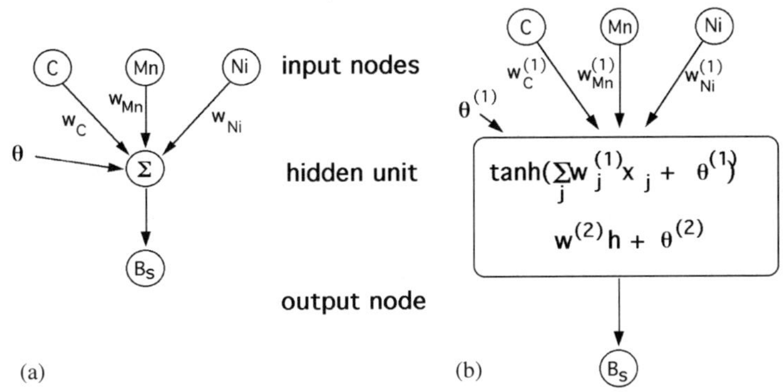

Figure 11 (a) A neural network representation of linear regression. (b) A nonlinear network representation. The inputs are in this case the C, Mn, and Si and the output is a transformation temperature B_S (source Bhadeshia, 1999a, 1999b, 1999c).

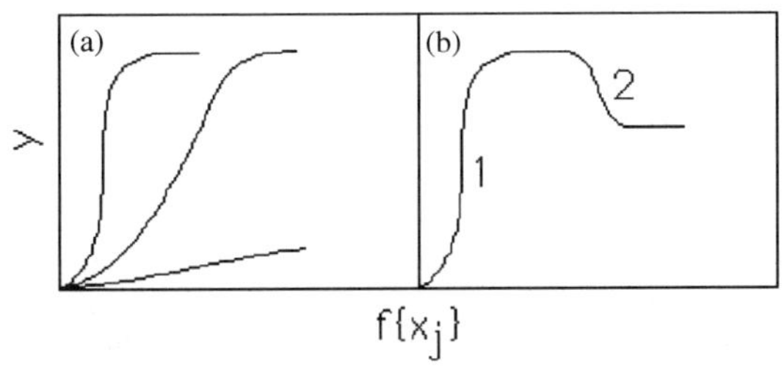

Figure 12 (a) Three different hyperbolic tangent functions, the "strength" of each depends on the weights. (b) A combination of two hyperbolic tangents to produce a more complex model.

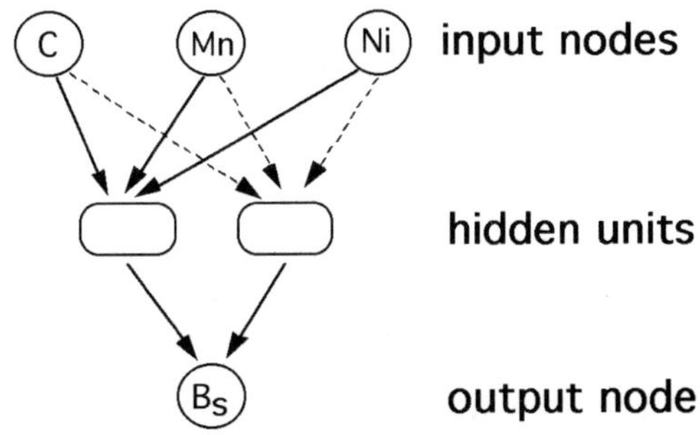

Figure 13 The structure of a two hidden unit neural network. Details have been omitted for simplicity.

flexible function means that the analysis is not as restricted as in linear regression where the form of the equation has to be specified before the analysis.

The neural network can capture interactions between the inputs, because the hidden units are nonlinear. The nature of these interactions is implicit in the values of the weights, but the weights may not always be easy to interpret. For example, there may exist more than just pairwise interactions, in which case the problem becomes difficult to visualize from an examination of the weights. A better method is to actually use the network to make predictions and to see how these depend on various combinations of inputs.

5.01.5.1 Overfitting

A potential difficulty with the use of powerful nonlinear regression methods is the possibility of overfitting data. To avoid this difficulty, the experimental data can be divided into two sets, a *training* dataset and a *test* dataset. The model is produced using only the training data. The test data are then used to check that the model behaves itself when presented with previously unseen data. This is illustrated is Figure 14 which shows three attempts at modeling noisy data for a case where y should vary with x^3. A linear model (Figure 14(a)) is too simple and does not capture the real complexity in the data. An overcomplex function such as that illustrated in Figure 14(c) accurately models the training data but generalizes badly. The optimum model is illustrated in Figure 14(b). The training and test errors are shown schematically in Figure 14(d); not surprisingly, the training error tends to decrease continuously as the model complexity increases. It is the minimum in the test error which enables that model to be chosen which generalizes best to unseen data.

5.01.5.2 Error Estimates

The input parameters are generally assumed in the analysis to be precise and it is normal to calculate an overall error by comparing the predicted values (y_j) of the output against those measured (t_j), e.g.,

$$E_D \propto \sum_j (t_j - y_j)^2 \qquad (24)$$

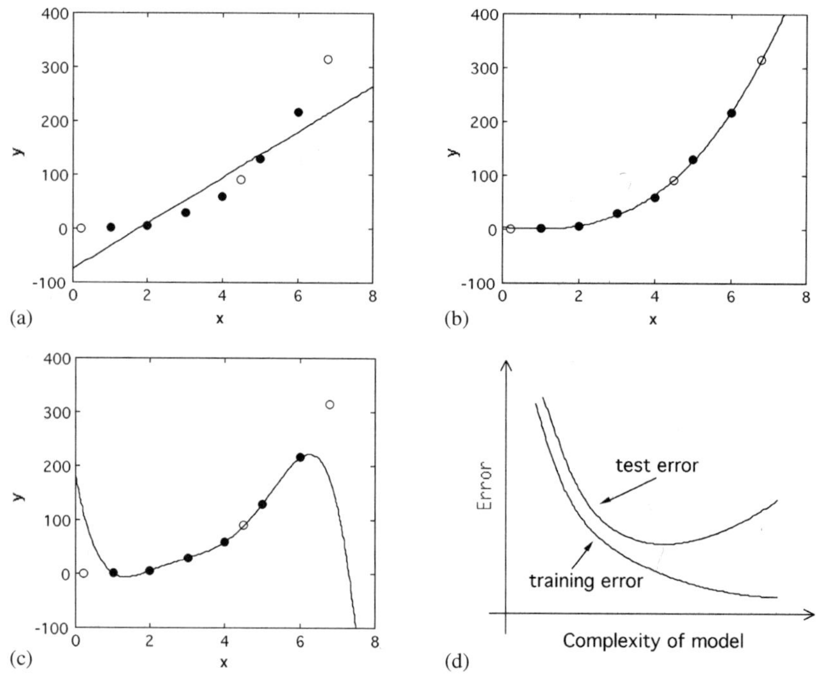

Figure 14 Variations in the test and training errors as a function of model complexity, for noisy data in a case where y should vary with x^3. The filled points were used to create the models (i.e., they represent training data), and the circles constitute the test data: (a) a linear function which is too simple; (b) a cubic polynomial with optimum representation of both the training and test data; (c) a fifth-order polynomial which generalizes poorly; and (d) schematic illustration of the variation in the test and training errors as a function of the model complexity.

E_D is expected to increase if important input variables have been excluded from the analysis. Whereas E_D gives an overall perceived level of noise in the output parameter, it is, on its own, an unsatisfying description of the uncertainties of prediction. Figure 15 illustrates the problem; the practice of using the best-fit function (i.e., the most probable values of the weights) does not adequately describe the uncertainties in regions of the input space where data are sparse (B), or where the data are particularly noisy (A).

MacKay has developed a particularly useful treatment of neural networks in a Bayesian framework (MacKay 1992a, 1992b), which allows the calculation of error bars representing the uncertainty in the fitting parameters. The method recognizes that there are many functions which can be fitted or extrapolated into uncertain regions of the input space, without unduly compromising the fit in adjacent regions which are rich in accurate data. Instead of calculating a unique set of weights, a probability distribution of sets of weights is used to define the fitting uncertainty. The error bars, therefore, become large when data are sparse or locally noisy, as illustrated in Figure 15.

This methodology has proved to be extremely useful in the analysis of creep phenomena and in structural integrity assessment. The application and success of neural networks in creep have been assessed and reviewed (Evans, 1999; Bhadeshia, 1999a, 1999b, 1999c, 2001). We shall now illustrate this with an example on steels.

5.01.6 QUANTITATIVE ESTIMATION OF CREEP-RUPTURE STRENGTH

It has frequently been pointed out that mechanistic studies of creep, useful as they are, have in general been pursued independently of commercial requirement (e.g., Evans and Wilshire, 1985). This is because in industrial alloys, the creep phenomena depend on a very large number of variables, well beyond those implemented in fundamental theories or in simple parametric representations such as that based on the θ-parameter or Larson–Miller plots is also not appropriate. Alternatively, there are no difficulties in dealing with a complex array of variables when applying neural network analysis (Brun *et al.*, 1999; Cole *et al.*, 2000). The number of variables used in the neural network modeling of creep-resistant steels is illustrated in Table 2; there are 37 variables and it is understood from experience that each of these is important in controlling creep properties. Even microstructure is implicitly included in the inputs since it

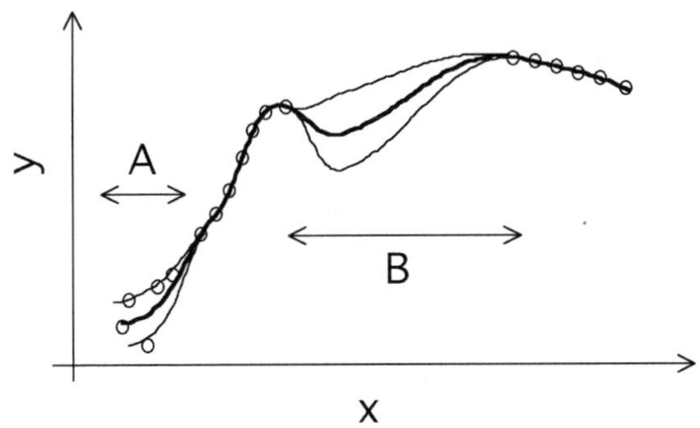

Figure 15 Schematic illustration of the uncertainty in defining a fitting function in regions where data are sparse (B) or where they are noisy (A). The thinner lines represent error bounds due to uncertainties in determining the weights.

Table 2 The minimum set of input parameters required to properly represent the creep-rupture strength of steels. The chemical compositions are all in wt.%.

Steel	*2(0.25)CrMo*	*10CrMoW*
Normalizing temperature (K)	1,203	1,338
Duration (h)	6	2
Cooling rate	water quenched	air cooled
Tempering temperature (K)	908	1,043
Duration (h)	6	4
Cooling rate	air cooled	air cooled
Annealing temperature (K)	873	1,013
Duration (h)	2	4
Cooling rate	air cooled	air cooled
C (wt.%)	0.15	0.12
Si	0.21	0.05
Mn	0.53	0.64
P	0.012	0.016
S	0.012	0.001
Cr	2.4	10.61
Mo	1.01	0.44
W	0.01	1.87
Ni	0.14	0.32
Cu	0.16	0.86
V	0.01	0.21
Nb	0.005	0.01
N	0.0108	0.064
Al	0.018	0.022
B	0.0003	0.0022
Co	0.05	0.015
Ta	0.0003	0.0003
O	0.01	0.01
Re	0.00	0.00

is entirely defined by the full chemical composition and heat treatment. In fact, the list in the table represents the minimum number of variables that must be specified and reported in any experimental work.

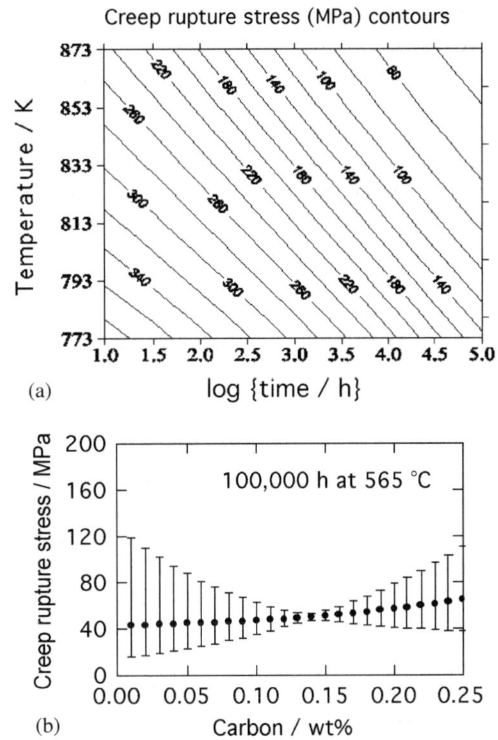

Figure 16 (a) Contour plot of the calculated creep-rupture strength as a function of the temperature and time, for the 2(0.25)Cr1Mo steel listed in Table 2. The error bar on each stress is only $\sim\pm 10\%$ of the quoted value. (b) The calculated creep-rupture strength at 565°C, 1×10^5 h for the same 2(0.25) Cr1Mo steel but as a function of the carbon concentration.

Figure 16(a) shows some typical calculations possible with the model, for the 2(0.25)Cr1Mo steel listed in Table 2. Notice that the region in which there is the sharpest decline in the creep strength with time, i.e., the shortest spacing between the contours, appears at shorter times as the temperature is increased. This is to be expected since coarsening reactions are accelerated at high temperatures and once coarsening becomes significant, the contours again become widely spaced as the inherent strength (Kimura *et al.*, 1997) of the matrix is reached, in this case ~40–80 MPa.

Figure 16(b) illustrates two important points. First, the error bars are clearly a function of the carbon concentration and indicate that there is insufficient knowledge about the creep properties of 2(0.25)Cr1Mo steel for the specified set of parameters, when the carbon concentration falls outside the range 0.1–0.2 wt.%. This is useful because large error bars indicate that new experiments are necessary before the prediction can be taken to heart. The second consideration is that the commercial specification of

2(0.25)Cr1Mo steel only states a maximum carbon concentration of 0.16 wt.%, there is no minimum (Bhadeshia, 1992). This clearly is unsatisfactory.

Calculations like these can now be carried out routinely in the design of creep-resistant steels and nickel-based superalloys, and in the extrapolation of data outside of the range of experiments.

5.01.6.1 Application to Welding Alloys

It has been shown that the models developed for wrought alloys can be applied with little modification to weld metals (Bhadeshia, 1999b; Cole *et al.*, 2000). This is because in the context of creep, the essential difference between a weld metal and wrought alloy is its larger oxygen and nitrogen concentration and the former should not affect creep resistance (the O and N are incorporated into the solidifying metal during welding). Differences in the nitrogen concentration are important; they can be taken into account both in predicting carbonitride formation and in the neural network model where nitrogen is an input.

The microstructure of an as-deposited weld metal is radically different from that of a wrought steel. However, even this is unimportant because the severe tempering heat treatments used following the welding procedure, essentially wipe out the original microstructure and replace it with one which is tempered and similar to that of the steel plate.

Figure 17 shows the very encouraging agreement between the predicted and measured creep rupture lives of alloys designed for the welding of power plant steels, even to times up to 4×10^4 h at temperature.

5.01.6.2 Influence of Microstructure

Although the link between microstructure and the creep properties is well understood for simple systems, the relative importance of the roles of precipitates, solid solution strengthening, microstructure, etc. in determining the long-term creep-rupture properties have not been revealed in the context of applicable alloys.

The neural network method, combined with phase stability calculations, has been used to accomplish a nonlinear factorization of the creep-rupture strength into individual components as illustrated in Figure 18 (Murugananth 2002; Murugananth and Bhadeshia, 2002). It is clear, e.g., that the proportion of the contribution to the 10^5 h creep-rupture strength from

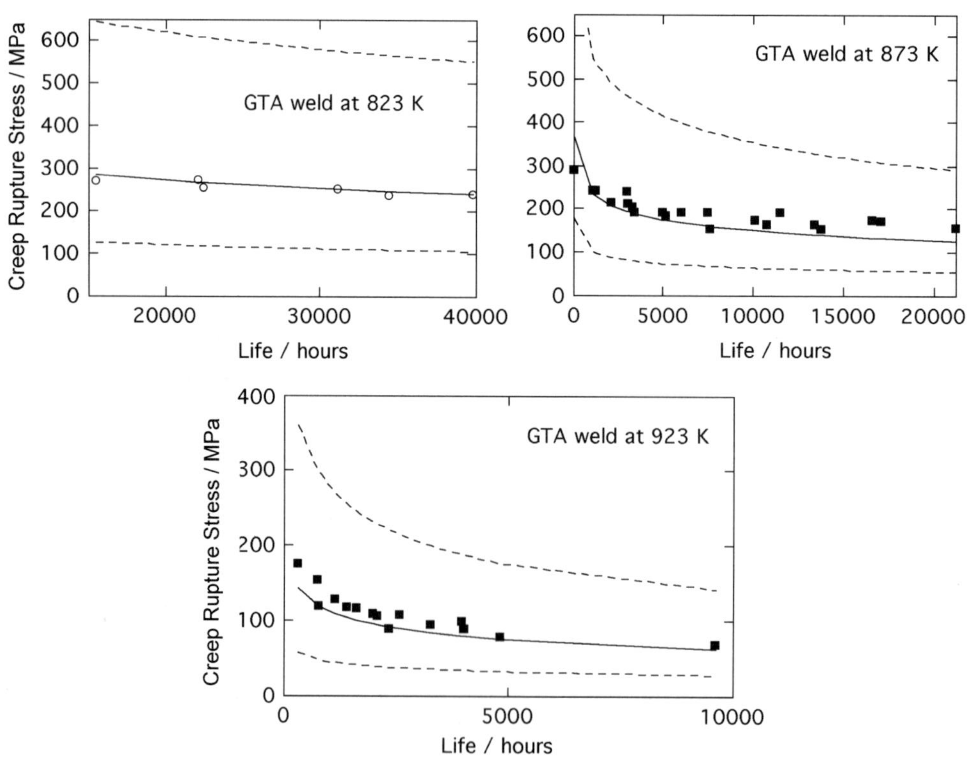

Figure 17 The impressive agreement between the calculated and measured creep-rupture stresses for novel welding alloys (source Cole *et al.*, 2000).

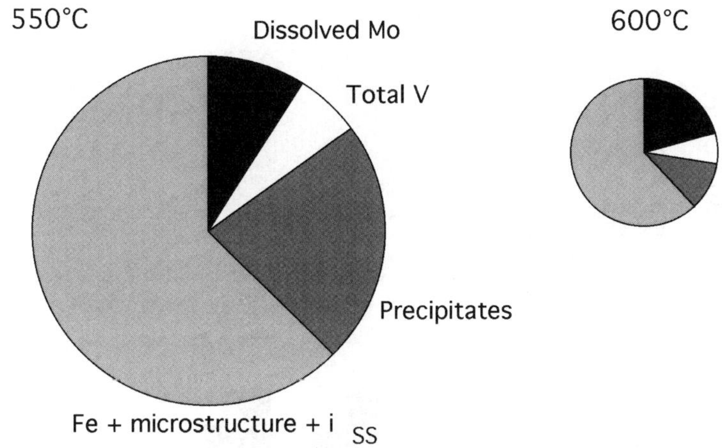

Figure 18 Factorization of the 10^5 h creep-rupture strength of a 2(0.25)Cr1Mo steel at the appropriate temperature. The term i_{SS} represents solid solution strengthening due to elements other than Mo or V. The diameter of each of the pie-charts scales with the creep-rupture strength which is 79 MPa and 30 MPa for 550°C and 600°C, respectively (source Murugananth and Bhadeshia, 2002).

dissolved solutes increases at 600°C when compared with its role at 500°C. The method is now being used to design microstructures and alloys for their long-term properties. Similar methods have been applied to the nickel-based superalloys (Tancret *et al.*, 2000; Fujii *et al.*, 1999).

5.01.7 ELECTRONS, BONDS, CREEP, AND CARBIDES

We now consider some speculative but interesting attempts at modeling creep using the theory of metallic bonds. The atoms in a metal are held together by a special gas of free electrons which allows the ensemble to bond even when different kinds of atoms are put together in solid solution (Cottrell, 1992). There is theory to enable electron densities and total energies to be estimated using only the atomic number and the charge on the electron (Finnis, 1992). The methodology is particularly useful in the prediction of crystal structure, elastic moduli, and surface energies.

There is a limitation in that the technique is computer intensive, so it must deal with small numbers of atoms, which, in practice, means a representative unit cell. It then becomes difficult to handle disordered solid solutions, although some parameters can be derived which may be of use in thermodynamic models of phase diagrams (Ducastelle, 1992).

There have been applications of electron theory to the specific problem of the creep-rupture strength of ferritic steels. Morinaga and co-workers (Morinaga *et al.*, 1994; Murata *et al.*, 1997) have used a molecular orbital method to calculate two parameters, the bond order B_0 and M_d, to suggest trends in creep-resistant steels. The bond order represents the number of electrons in bonding orbitals less than those in antibonding ones (Cottrell, 1995). For a pair of atoms, an antibonding orbital is one where the electron density falls to zero at some point between the adjacent atomic nuclei. A weaker bond thus has a lower bond order; bonding is absent when the bond order is 0. (When two hydrogen atoms, each with a single electron, are brought together, they no longer have separate atomic orbitals. Instead, they have a pair of communal orbitals (bonding and antibonding) each of which can hold two electrons. It follows that for H_2 both the electrons are in the bonding orbitals giving a bond order of 2 and a strong molecule. For He_2, alternatively, the four electrons fill up both the bonding and antibonding orbitals so the bond order is zero, the molecule is not formed; Cottrell, 1995.) In the present context B_0 represents a major contribution to the strength of the bond between a solute and iron atom in the body-centered cubic (b.c.c.) crystal structure. The parameter M_d correlates with electronegativity and the atomic radius of the element concerned. Values of these two parameters for a number of common solutes in ferrite are shown in Figure 19.

The natural question to ask is what does bond strength and electronegativity have to do with creep deformation, which depends on the ability of dislocations to overcome obstacles by thermally activated climb? The answer has to be that there is no physical basis for using either B_0 or M_d in design against creep, but Morinaga claims that there is an empirical *correlation* between the creep-rupture strength and the mean bond order of a commercial steel

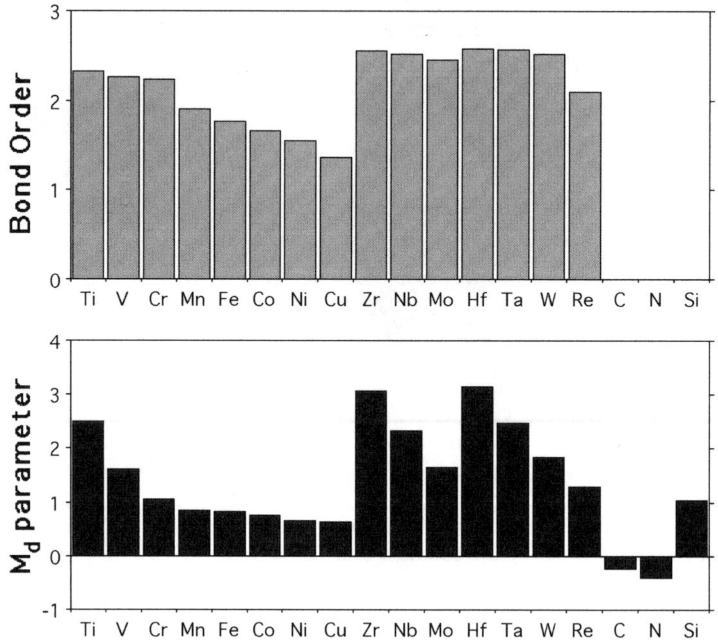

Figure 19 The bond order and M_d parameter for a number of solutes in b.c.c iron (sources Morinaga *et al.*, 1994; Murata *et al.*, 1997).

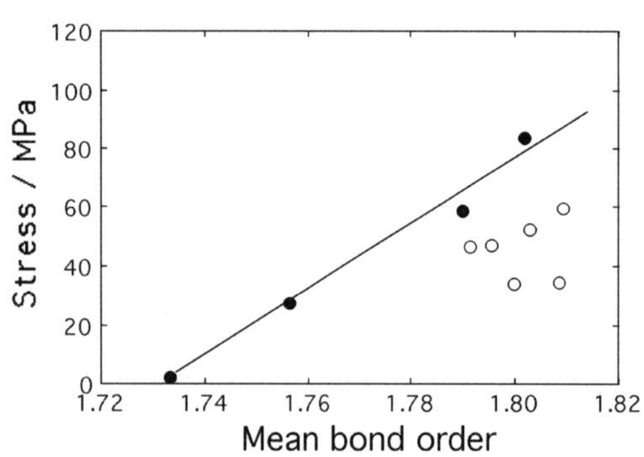

Figure 20 The correlation of the allowable stress at 600°C against the mean bond order of a variety of commercial steels. The open circles represent alloys which may contain some δ-ferrite whereas those which are filled represent δ-ferrite free steels (source Morinaga *et al.*, 1994).

as long as the latter is free from δ-ferrite (Figure 20); furthermore, M_d correlates directly with the propensity to form δ-ferrite in the steels studied. He suggests that the method can be used to optimize steels by maximizing B_0 and minimizing M_d. This was demonstrated by making appropriate additions to the commercial alloy *NF*616 (Table 3) (Morinaga *et al.*, 1994).

It could be argued that the results presented in Table 3 validate the use of the molecular orbital approach in the design of steels. However, a closer study of the data casts doubt on the procedure. For example, tungsten has a significantly larger B_0 than rhenium, and furthermore, it is known that larger concentrations of tungsten can be tolerated in balanced alloys without introducing δ-ferrite or Laves phases (Brun *et al.*, 1999). The addition of rhenium is not justified by the molecular orbital calculations, but as admitted in this chapter, has been made because rhenium is known to improve the creep properties of nickel-based superalloys. Cobalt has a lower B_0 than iron, so its addition is only justified to balance the tendency for chromium at high concentrations to promote δ-ferrite. To summarize, neither the rhenium nor the cobalt modifications of *NF*616 are based on the molecular orbital calculations but rather on intuition and knowledge as reported in the literature. Similar comments can be made about a later paper (Murata *et al.*, 1997) in which the method was used to develop rotor steels. The case that the molecular orbital calculations have predictive power has not, in my opinion, been made. Nor is it obvious why such a method should work when creep rupture is controlled by *defects* and *precipitates* on a scale many millions of times larger than atoms.

5.01.7.1 Carbides

Perhaps a better way to use the concepts of chemical bonding would be to examine the stabilities of the precipitates which strengthen steels. Cottrell (1995) has been able to explain many of the observed trends in the stability, crystal structure, and stoichiometry of the

Table 3 Compositions (wt.%) of steels studied by Morinaga *et al.* All nominally contain 0.07C wt.%, 0.06Si wt.%, 0.45Mn wt.%, 0.5Mo wt.%, 0.2V wt.%, 0.05Nb wt.%, 0.01Al wt.%, 0.004B wt.%, and 0.06N wt.%, so the compositions stated are presumably nominal. The final column gives the time to creep rupture at 700°C for a stress of ~8 MPa.

Alloy	*Cr*	*W*	*Co*	*Re*	*time* (h)
NF616	9.0	1.8	0.0	0.0	406
B1	10.0	1.8	1.5	0.0	919
B2	10.0	1.5	1.5	0.6	1103
B3	11.5	1.8	2.5	0.0	1024
B4	11.5	1.5	2.5	0.6	1263

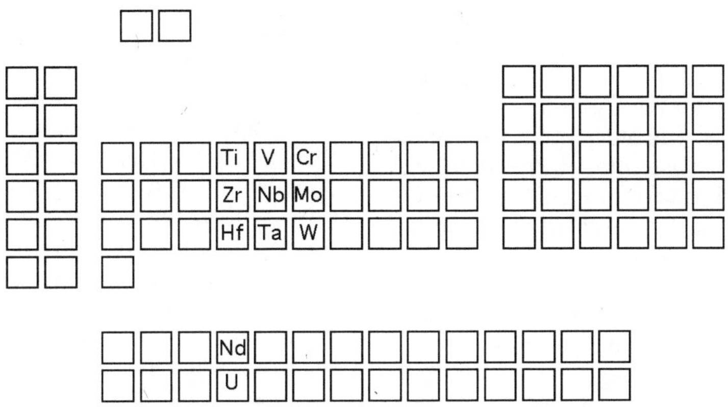

Figure 21 The periodic table showing the positions of strong carbide-forming elements.

carbides of transition metals in terms of chemical bonds. He points out that Ti, Zr, and Hf, which in the periodic table are elements near the beginning of the long periods, form very stable MC carbides but the affinity for carbon diminishes further along the rows of the periodic table (Figure 21). A part of the reason for this is that more electrons have to be accommodated for elements further along the rows, so antibonding states are progressively filled thereby reducing the bond order. This is analogous to the hydrogen–helium example given earlier, but does not completely explain the trend because the maximum bond order occurs with Cr, Mo, and W and we know that carbides of these elements are less stable.

With MC carbides, the metal has to sacrifice four electrons to form the bonds with carbon. Titanium has exactly the right number so that its d-orbitals are left empty on forming TiC. This is not the case with VC, since vanadium has an additional d-electron which forms a V–V bond. The electrons in the two kinds of bonds, V–C and V–V, mutually repel, leading to a reduction in the stability of VC when compared with TiC. This problem becomes larger along the row of the periodic table until MC carbide formation becomes impossible or unlikely.

Although Cottrell has not considered the carbides in the lanthanide or actinide series of elements, it is possible that the same principles should apply there. Both NdC and UC exist. Remarkably, neodymium nitride has already been incorporated into a ferritic creep-resistant steel by Igarashi and Sawaragi (1997) with rather good results. The concentration of neodymium used was only 0.04 wt.% but gave an increase in the creep-rupture life by a factor of ~2 during tests at 650°C. Tantalum behaves in a similar manner by forming stable nitrides. They also tried hafnium but did not recommend it due to a tendency to form coarse particles.

5.01.7.2 Embrittlement and Electron Theory

Impurity-induced embrittlement is a severe problem in the power generation industries, both at the fabrication stage and as we shall see later, during prolonged service at elevated temperatures when the toughness deteriorates.

Modern creep-resistant steels begin with a martensitic microstructure. Martensitic transformation occurs with the disciplined motion of atoms. Such movements cannot be sustained across grain boundaries. It follows that a vestige of the austenite grain boundary remains when the transformation is displacive. These prior boundaries become susceptible to impurity segregation and embrittlement, liquid zinc embrittlement, and any phenomenon able to exploit the disorder associated with these boundaries (Bhadeshia and Svensson, 1993; Iezawa *et al.*, 1993; Ichikawa *et al.*, 1996; Bhadeshia, 1997).

The disorder at the prior austenite grain boundaries is larger than that associated with martensite–martensite boundaries. This is because adjacent martensite plates have coherence with the parent austenite grain and hence a good fit with each other, whereas the orientation relationships between austenite grains can be random. There are considerable practical consequences of this difference. Thus, the extent of segregation is smaller at martensite lath boundaries. Carbides nucleate preferentially at the prior austenite grain boundaries during the tempering of martensite or bainite. The boundaries can be revealed by etching, often with great clarity even though the original grain is no longer present.

The problem of austenite grain boundary embrittlement in martensitic or bainitic steels is not exclusive to high-strength steels used at ambient temperature. It is well established that the fracture toughness of many power plant steels deteriorates during service at elevated temperatures for two reasons. First, the carbide particles, particularly those located at the prior boundaries, coarsen and hence provide easier sites for crack or void nucleation. Second, the elevated temperatures permit the impurities to diffuse relatively rapidly and saturate the boundaries. Figure 22 shows the impact transition curves for both the virgin steel (i.e., at implementation into service) and for an ex-service sample which had experienced 8.8×10^4 h at 813 K (Wignarajah *et al.*, 1990).

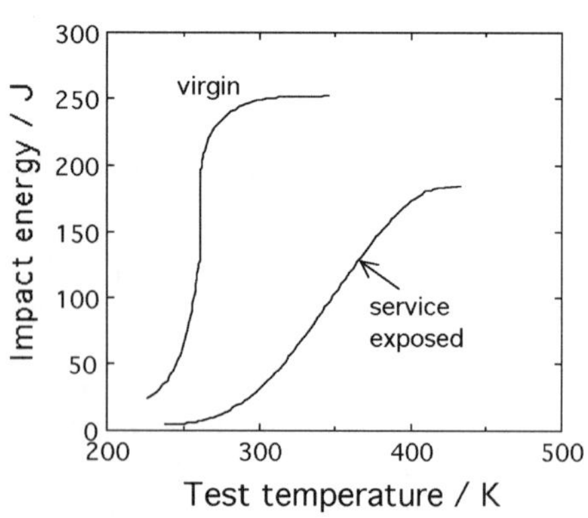

Figure 22 Impact test data on 2(0.25)Cr1Mo steel. Comparison of the impact energies of the steel in its virgin condition, after service exposure at 813 K for 8.8×10^4 h (after Wignarajah *et al.*, 1990).

5.01.7.3 Mechanism of Embrittlement

The Rice and Wang theory (Anderson *et al.*, 1987) defines the impurity effect precisely: embrittlement occurs if the magnitude of the reduction ΔG_b in free energy which occurs when an impurity segregates to a grain boundary is less than when the impurity segregates to a free surface (ΔG_s). This is because the impurity then prefers to be at a free surface than at a boundary. Table 4 lists some values for these terms for a number of elements. Notice that carbon almost has a neutral effect whereas elements such as phosphorus, tin, and antimony are notorious for their embrittlement potency.

It turns out that the values of these segregation energies can be estimated using electron theory. The difference (ΔG_b–ΔG_s) for phosphorus has been confirmed using such calculations with a lower free energy for P in a free surface relative to a grain boundary (Sagert *et al.*, 1998). The work has led to a greater insight into the mechanism of embrittlement. There is a costly structural rearrangement of the iron atoms in the grain boundary when P is introduced, which is partly responsible for the relatively small magnitude of ΔG_b. When molybdenum is added to the grain boundary, the embrittlement process is effectively reversed, not because it affects the embrittlement potency of phosphorus, but because it enhances cohesion at the grain boundary.

5.01.7.4 Avoiding Grain Boundary Embrittlement

The traditional routes for avoiding temper embrittlement involve a reduction in the impurity concentration and the addition of substitutional solutes which remove the impurities (Briant and Banerji, 1987).

A solute causes embrittlement by reducing the cohesion across the boundary plane. Alternatively, those solutes which enhance

Table 4 The free energies of segregation to a grain boundary and to a free surface. These values must naturally depend on factors such as the grain boundary structure.

Segregant	$-\Delta G_b$	$-\Delta G_s$
C	50–75	73–85
Sn	30–35	61–87
P	32–41	76–80
Sb	8–25	83–130
S	50–58	165–190
H	65–68	71–109

Source: Anderson *et al.* (1987)

covalent bonding normal to the boundary must reduce embrittlement. *Ab initio* calculations of the bonding at grain boundaries in iron suggest this mechanism and confirm that phosphorus should embrittle and boron should toughen (Wu *et al.*, 1994).

Boron additions must, therefore, be useful in commercial, martensitic power plant steels, which can be made clean but not free of impurities. The difficulty, however, is that much of the boron enters the $M_{23}C_6$ phase (Hättestrand and Andrén, 1999) and hence less is available to segregate to the austenite grain boundaries. Much further work is needed on this topic.

Another possibility, which has never been investigated, would be to reduce the energy of the prior austenite grain boundaries. This could be done by making the austenite grain structure strongly crystallographically textured. The resulting austenite grain boundaries would have a high degree of fit and hence be less prone to impurity segregation. Whether this approach is worthwhile can, in principle, be predicted from *ab initio* calculations which can be attempted for a variety of different kinds of boundaries (albeit always at special orientations to minimize computations).

5.01.7.5 Summary: Electron Theory

Electron theory has not yet made a direct contribution to the design of creep-resistant steels. The potential is there if appropriate problems can be identified, e.g., in the prediction of the stability of carbides and intermetallic compounds of the kind common in heat-resistant steels. Methods of reducing long-term embrittlement in multicomponent systems is another scenario where first principles calculations have been demonstrated to give valuable insight and qualitative guidance.

5.01.8 STABILITY OF THE MICROSTRUCTURE

Coarsening of precipitation-hardened materials is obviously an important phenomenon in any creep-resistant material (e.g., Bhadeshia, 1999a). In the context of steel, consider a particulate phase θ in a ferritic matrix α. The solute concentration in the ferrite is written $c^{\alpha\theta}$ and that in θ written $c^{\theta\alpha}$ assuming that the two phases are in equilibrium at a flat interface. However, if the interface is curved, e.g., when the particles are in the form of spheres, the equilibrium concentrations become a function of the radius of curvature, given by $c_r^{\alpha\theta}$ and $c_r^{\theta\alpha}$, respectively, where r is the particle radius. This is the Gibbs–Thompson capillarity effect. It accounts for the cost of creating interface as the particle grows.

It can be shown that the concentration $c_r^{\alpha\theta} > c^{\alpha\theta}$, the inequality becoming larger as $r \to 0$. In other words, the solute concentration in the ferrite near a small particle will be greater than that in contact with a large particle, thus setting up a concentration gradient which makes the small particle dissolve and the larger particle grow (Figure 23). This is the process of coarsening, driven by the interfacial energy σ.

The concentration difference $c_r^{\alpha\theta} - c^{\alpha\theta}$ which drives the diffusion flux is given by

$$c_r^{\alpha\theta} - c^{\alpha\theta} = \frac{\sigma V^{\alpha}}{kTr} \times \frac{c^{\alpha\theta}(1 - c^{\alpha\theta})}{c^{\theta\alpha} - c^{\alpha\theta}} \qquad (25)$$

where k is the Boltzmann constant, T the absolute temperature, and V^{α} the molar volume of the ferrite. This flux feeds the growth or dissolution of the particle and hence must match the rate at which solute is absorbed or desorbed at the moving interface:

$$\underbrace{D(c_r^{\alpha\theta} - c^{\alpha\theta})}_{\text{measure of flux}} \propto \underbrace{v(c^{\theta\alpha} - c^{\alpha\theta})}_{\text{rate of solute absorption}}$$

where v is the interfacial velocity and D is the solute diffusivity in the matrix phase. On substituting for the concentration difference and using Fick's first law, it follows that

$$v \propto D\frac{\sigma V^{\alpha}}{kTr} \times \underbrace{\frac{c^{\alpha\theta}(1 - c^{\alpha\theta})}{(c^{\theta\alpha} - c^{\alpha\theta})^2}}_{\text{instability}} \qquad (26)$$

In Equation (26), the equilibrium concentration term on the right-hand side is a thermodynamic term, a larger value of which corresponds to a greater coarsening rate. Figure 24 shows the equilibrium chromium

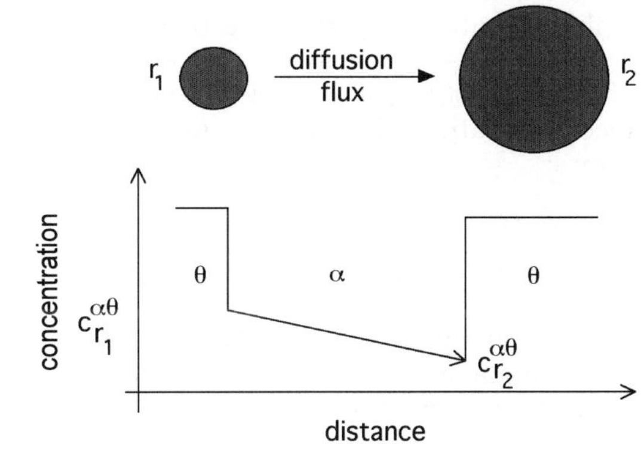

Figure 23 An illustration of the capillarity effect driving coarsening.

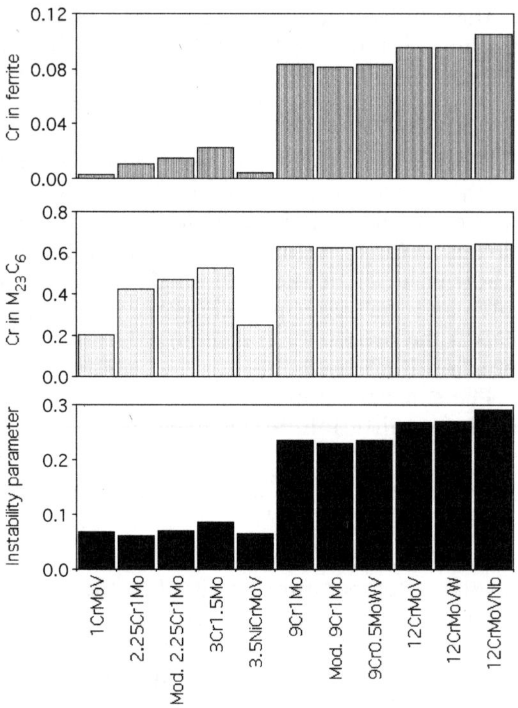

Figure 24 The top two bar charts give the chromium concentrations (mole fractions) in ferrite and in $M_{23}C_6$, respectively. The third bar chart gives a measure of the tendency for $M_{23}C_6$ to coarsen assuming Equation (26).

concentrations in the ferrite and $M_{23}C_6$ phases for a large variety of power plant steels. The concentrations are clearly largest for the more modern martensitic alloys which are Cr rich; such steels are designed for higher temperature service and hence require a greater oxidation resistance. It is evident, however, that the tendency for $M_{23}C_6$ to coarsen should also be greatest in the Cr-rich alloys.

Commercial steels contain many solutes whereas Equation (26) deals with just one. Venugopalan and Kirkaldy (1978) generalized the theory to deal with many solutes, assuming that the effects of the different fluxes could be combined into a single, effective diffusion coefficient D_{eff} by treating the fluxes as a combination of parallel electrical conductances. Equation (26) can be rearranged as follows:

$$\frac{1}{v} \propto \frac{1}{D_i} \times \frac{(c_i^{\theta\alpha} - c_i^{\alpha\theta})^2}{c_i^{\alpha\theta}(1 - c_i^{\alpha\theta})}$$

where i now represents a particular solute. In a multicomponent alloy, the right-hand side is replaced by a summation

$$\frac{1}{v} \propto \frac{1}{D_{\mathrm{eff}}}$$

where

$$\frac{1}{D_{\mathrm{eff}}} = \sum_i \frac{1}{D_i} \times \frac{(c_i^{\theta\alpha} - c_i^{\alpha\theta})^2}{c_i^{\alpha\theta}(1 - c_C^{\alpha\theta})} \tag{27}$$

This multicomponent treatment does not, in fact, make a difference to the ranking of the $M_{23}C_6$ coarsening tendency as presented in Figure 24, because this carbide is chromium rich and because the diffusivities of the substitutional solutes are not very different.

5.01.8.1 Coarsening in Multiphase Microstructures

Abe (1999) has reported interesting data on the coarsening of $M_{23}C_6$ particles in Fe–9Cr–W steels as a function of the tungsten content and of time during creep testing at 600°C (Figure 25(a)). According to Abe, the addition of tungsten reduces the coarsening rate. The data have to be interpreted with care because the addition of tungsten also causes the precipitation of Laves phase (Figure 25(b)), but the fact that the data are from creep test samples should not matter, since it is the trend as a function of W that we are interested in; we shall also assume that the electron microscopy data refer strictly to $M_{23}C_6$ and do not unintentionally include any other phases. Figure 25(c) shows two sets of calculated values of D_{eff}, one assuming that it is only the diffusion of chromium (the major constituent of $M_{23}C_6$ in such alloys) that controls coarsening, and the other using the multicomponent formalism (Equation (27)). As stated earlier, the trends are identical for the two sets of calculations, but in contradiction to the experimental data, they all suggest that the addition of tungsten should *accelerate* the rate at which $M_{23}C_6$ coarsens. The calculations show also that the experimental data cannot be explained by saying that tungsten reduces the overall diffusion coefficient.

A possible explanation is that Laves phase precipitates some time after $M_{23}C_6$, in which case it may be appropriate to consider a constrained equilibrium in which Laves phase is excluded. In fact, when this is done, M_6C is promoted and the situation worsens in the sense that D_{eff} for $M_{23}C_6$ is predicted to increase.

The problem of explaining why the addition of tungsten reduces the coarsening of $M_{23}C_6$ can be explained by taking into account all of the phases that are present. When there is more than one precipitate phase in the microstructure, differences in interfacial energy (σ, Equation (27)) cannot be neglected. Laves phase has a higher interfacial energy so the

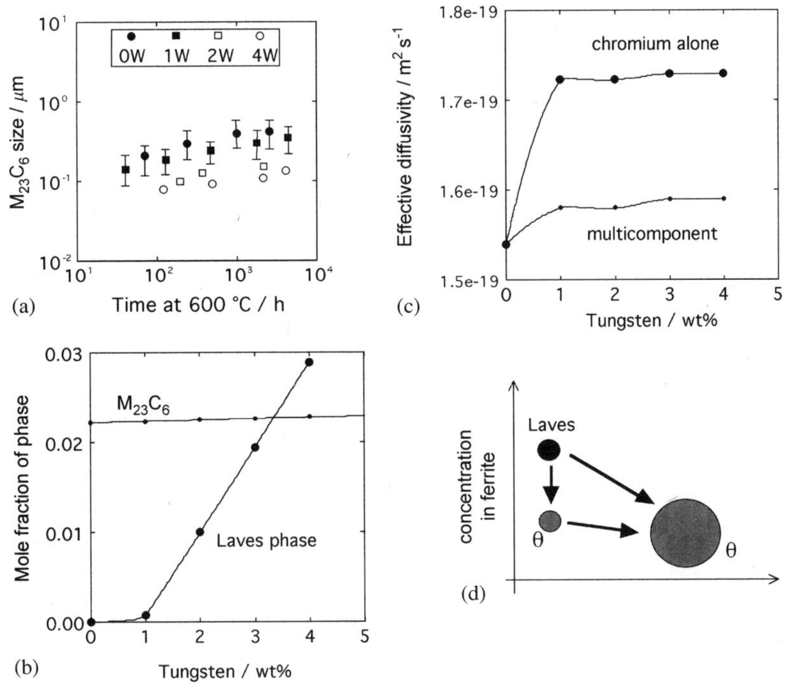

Figure 25 Coarsening data from Fe–0.1C–9Cr–0.5Mn–0.3Si wt.% steel, containing a variety of tungsten additions, tested in creep at 600°C (Abe, 1999). (a) $M_{23}C_6$ size. The error bars have been omitted for the highest W data for clarity. (b) Equilibrium phase fraction calculations for 600°C carried out using MTDATA with the SGTE database, allowing all common carbides and Laves phase and Fe, C, Cr, Mn, Si, W as the components. (c) Values of D_{eff} for $M_{23}C_6$, for two cases, the first where chromium alone is considered and then for multicomponent diffusion. (d) Schematic illustration of why the presence of Laves phase retards the coarsening of $M_{23}C_6$.

concentration in ferrite in equilibrium with Laves will, for the same particle radius, be higher for Laves than for $M_{23}C_6$:

$$c^{\alpha\theta}_{r,\mathrm{Laves}} > c^{\alpha\theta}_{r,}\mathrm{M_{23}C_6}$$

This is illustrated schematically in Figure 25(d) on a concentration–distance plot. In addition to a flux of solute from small Laves phase particles to larger ones, there will also be a flux to both small and large $M_{23}C_6$ particles. The coarsening rate of $M_{23}C_6$ will, therefore, be reduced whereas that of Laves phase accelerated. It is interesting that Abe specifically commented that the coarsening rate of Laves phase was extremely rapid.

Proper calculations of coarsening in multiphase systems will become possible in the near future. The coarsening kinetics do not require an explicit treatment but follow naturally from the precipitation theory which for each phase includes the capillarity effect. Indeed, it is well-known that the separation of precipitation and coarsening is artificial, because capillarity makes the smaller precipitates grow at a slower rate than those which nucleated first and hence are larger, even before the equilibrium fraction is reached (Fujita *et al.*, 2002). Figure 26 shows

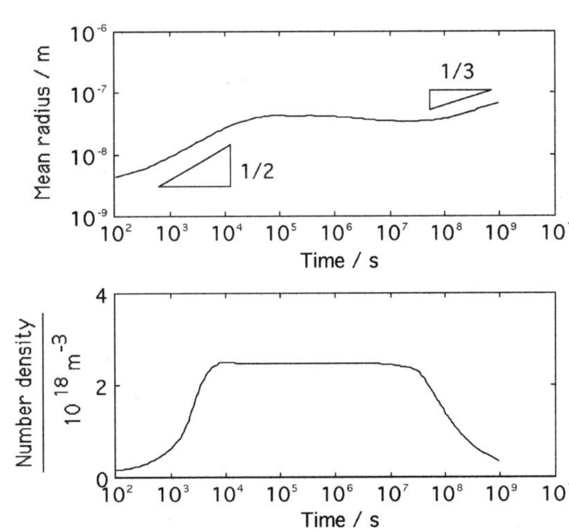

Figure 26 The precipitation and coarsening of niobium carbide precipitates in austenite (Fujita *et al.*, 2002) in Fe–0.05Nb–0.1C wt.% steel, beginning with a supersaturated solid solution at 900°C. The number density of nucleation sites was taken to be $9 \times 10^{19}\,\mathrm{m^{-3}}$, with $\sigma = 0.26\,\mathrm{J\,m^{-2}}$.

the results of calculations of this kind, the mean particle radius at first increasing approximately parabolically with time as all particles grow from solid solution. The mean radius then changes as about $t^{1/3}$ at longer times as the number density of particles decreases, consistent with expectations from coarsening theory. It is important to emphasize that there is no separate treatment of precipitation and coarsening here, both are in an inclusive theory.

5.01.9 SUMMARY

Creep-resistant materials must be reliable over very long periods of time in severe environments. In this chapter we have briefly reviewed the important mechanisms of creep, qualitative understanding of constitutive relationships, and finally empirical methods of vital importance in determining the safety of materials in service.

5.01.10 REFERENCES

F. Abe, 1999, Coarsening behavior of martensite laths in tempered martensitic 9C-W steels during creep deformation. In: "4th International Conference on Recrystallisation and Related Phenomena," eds. T. Sakai and H. G. Suzuki, The Japan Institute of Metals, pp. 289–294.

F. Abe and S. Nakazawa, 1992, Microstructural evolution and creep behaviour of bainitic, martensitic and martensite-ferrite dual phase Cr–2W steels. *Mater. Sci. Technol.*, **8**, 1063–1069.

P. M. Anderson, J.-S. Wang and J. R. Rice, 1987, Thermodynamic and mechanical models of interfacial embrittlement. In: "Innovations in Ultra-high Strength Steel Technology," eds. G. B. Olson, M. Azrin and E. S. Wright, Sagamore Army Materials Research Proceedings, NY, USA, pp. 617–649.

A. S. Argon, 1982, Mechanisms and mechanics of fracture in creeping alloys. In: "Creep and Fracture of Materials and Structures," eds. B. Wilshire and D. R. J. Owen, Pineridge Press, UK, pp. 1–53.

A. S. Argon, 1996, Mechanical properties of single–phase crystalline media: deformation in the presence of diffusion. In: "Physical Metallurgy," eds. R. W. Cahn and P. Haasen, North-Holland, Amsterdam, pp. 1958–2007.

M. F. Ashby and D. Jones, 1982, "Engineering Materials," Pergamon, Oxford.

S. M. Beech, D. J. Gooch and A. Strang, 1995, "3rd International Charles Parsons Conference on Materials Engineering in Turbines and Compressor," Newcastle–upon–Tyne, UK, pp. 277–291.

H. K. D. H. Bhadeshia, 2001, "Bainite in Steels," Institute of Materials, London, 2nd edn., pp. 1–458.

H. K. D. H. Bhadeshia, 1997, Models for the elementary mechanical properties of steel welds. In: "Mathematical Modelling of Weld Phenomena III," eds. H. Cerjak and H. K. D. H. Bhadeshia, The Institute of Materials, London, p. 229.

H. K. D. H. Bhadeshia, 1999a, Neural networks in materials science. *ISIJ Int.*, **39**, 966–979.

H. K. D. H. Bhadeshia, 1999b, Design of creep-resistant steel welds. In: "Trends in Welding Research," eds. S. A. David, T. DebRoy, J. A. Johnson, H. B. Smartt and J. M. Vitek, ASM International, OH, pp. 795–804.

H. K. D. H. Bhadeshia, 1999c, Estimation of the microstructure and properties of ferritic creep-resistant steels. In: "Modelling of Microstructural Evolution in Creep Resistant Materials," eds. A. Strang and M. McLean, The Institute of Materials, London, pp. 15–38.

H. K. D. H. Bhadeshia, 2001, Design of ferritic creep resistant steels. *ISIJ Int.*, **41**, 626–640.

H. K. D. H. Bhadeshia and L.-E. Svensson, 1993, Modelling the evolution of microstructure in steel weld metals. In: "Mathematical Modelling of Weld Phenomena," eds. K. Easterling and H. Cerjak, The Institute of Materials, London, p. 109.

H. K. D. H. Bhadeshia, A. Strang and D. J. Gooch, 1998, Ferritic power plant steels: remanent life assessment and approach to equilibrium. *Int. Mater. Rev.*, **43**, 45–69.

R. J. Borg and C. E. Birchenall, 1960, Self diffusion in alpha iron. *Trans. TMS–AIME*, **218**, 980–984.

C. L. Briant and S. K. Banerji, 1978, Intergranular failure in steel: role of grain boundary composition. *Int. Metals Rev.*, **232**, 164.

F. Brun, T. Yoshida, J. D. Robson, V. Narayan and H. K. D. H. Bhadeshia, 1999, Theoretical design of ferritic creep-resistant steels using neural network, kinetic and thermodynamic models. *Mater. Sci. Technol.*, **15**, 547–555.

F. S. Buffington, K. Hirano and M. Cohen, 1961, Self diffusion in iron. *Acta Metall.*, **9**, 434–439.

B. Burton and B. D. Bastow, 1973, Diffusional creep of binary copper alloys. *Acta Metall.*, **21**, 13.

B. J. Cane, 1981, Interrelationship between creep deformation and creep rupture in 2.25Cr–1Mo steel. *Metal Sci.*, **13**, 287–294.

D. Cole, C. Martin-Moran, A. Sheard, H. K. D. H. Bhadeshia and D. J. C. MacKay, 2000, Modelling the creep rupture strength of ferritic steel welds. *Sci. Technol. Weld. Join.*, **5**, 81–89.

A. H. Cottrell, 1992, Introduction. In: "Electron Theory in Alloy Design," eds. D. G. Pettifor and A. H. Cottrell, The Institute of Materials, London, pp. 1–9.

A. H. Cottrell, 1995, "Chemical Bonding in Transition Metal Carbides," The Institute of Materials, London, pp. 1–97.

L. S. Darken, 1948, Thermodynamics of diffusion. *TMS–AIME*, **175**, 184.

F. Ducastelle, 1992, Order and phase stability in alloys. In: "Electron Theory in Alloy Design," eds. D. G. Pettifor and A. H. Cottrell, The Institute of Materials, London, pp. 122–157.

M. Evans, 1999, Method for improving parametric creep rupture life of 2.25Cr–1Mo steel using artificial neural networks. *Mater. Sci. Technol.*, **15**, 647–658.

R. W. Evans and B. Wilshire, 1985, "Creep of Metals and Alloys," The Institute of Metals, London.

M. W. Finnis, 1992, Modern electron theory. In: "Electron Theory in Alloy Design," eds. D. G. Pettifor and A. H. Cottrell, The Institute of Materials, London, pp. 10–43.

H. J. Frost and M. F. Ashby, 1982, "Deformation Mechanism Maps," Pergamon, Oxford.

H. Fujii, D. J. C. MacKay, H. K. D. H. Bhadeshia, H. Harada and K. Nogi, 1999, Estimation of creep rupture strength of nickel base superalloys. *J. Japan Inst. Metals*, **63**, 905–911.

N. Fujita, H. K. D. H. Bhadeshia and M. Kikuchi, 2002, Modelling M_6C precipitation in niobium-alloyed ferritic stainless steel. *Metall. Mater. Trans. A*, **33**, 3339–3347.

J. Hald, 1999, Thermodynamic modelling of the microstructure of high Cr ferritic creep resistant steels. In: "Modelling of Microstructural Evolution in Creep-resistant Materials," eds. A. Strang and M. McLean, Institute of Materials, London, pp. 1–14.

M. Hättestrand and H.-O. Andrén, 1999, Boron distribution in 9-12Cr chromium steels. *Mater. Sci. Eng. A*, **270**, 33.

K. Hirano, M. Cohen and B. L. Averbach, 1961, Diffusion of nickel into iron. *Acta Metall.*, **9**, 440–445.

J. P. Hirth and J. Lothe, 1982, "Theory of Dislocations," Wiley, New York.

K. Ichikawa, Y. Horii, R. Motomatsu, M. Yamaguchi and N. Yurioka, 1996, Mechanical properties of weld metal of fire resistant steel. *Quarterly J. Japan Weld. Soc.*, **14**, 27.

T. Iezawa, T. Inoue, O. Hirano, T. Okazawa and T. Koseki, 1993, Effect of boron on liquid Zn embrittlement in HAZ of STKT590 steel tube for power transmission tower. *Tetsu to Hagane*, **79**, 96–102.

M. Igarashi and Y. Sawaragi, 1997, Development of 0.1C–11Cr–3W–3Co–V–Nb–Ta–Nd–N ferritic steel for USC boilers. In: "Proc. Int. Conf. on Power Engineering 97," Japan Society of Mechanical Engineers, Tokyo, Japan, pp. 107–112

Y. Iijima, K. Kimura and K. Hirano, 1988, Self diffusion and isotope effect in alpha iron. *Acta Metall.*, **36**, 2811–2820.

B. Jönsson, 1992, On ferromagnetic ordering and lattice diffusion—a simple model. *Z. Metallkunde*, **83**, 349–355.

K. Kimura, H. Kushima, F. Abe, K. Yagi and H. Irie, 1997, Assessment of creep strength properties of 9–12% Cr steels. In: "Advances in Turbine Materials, Design and Manufacturing, 4th Int. Charles Parsons Turbine Conference," eds. A. Strang, W. M. Banks, R. D. Conroy and M. J. Goulette, Institute of Materials, London, pp. 257–269.

E. O. Kirkendall, 1942, *Trans. Am. Inst. Min. Metall. Eng.*, **147**, 104.

J. Kučera, 1979, Analysis of magnetic anomaly in bcc iron. *Czech. J. Phys.*, **B29**, 797–809.

T. G. Langdon, 1985, Dislocations and creep. In: "Dislocations and Properties of Real Materials," Institute of Metals, London, pp. 221–238

T. G. Langdon, 2000, The significance of diffusion creep and Harper–Dorn creep at low stresses. *Key Eng. Mater.*, **171–174**, 205–215.

A. D. Le Claire, 1966, Some comments on the mass effect in diffusion. *Phil. Mag.*, **14**, 1271.

D. J. C. MacKay, 1992a, Bayesian interpolation. *Neural Comp.*, **4**, 415–447.

D. J. C. MacKay, 1992b, Practical bayesian framework of backpropagation networks. *Neural Comp.*, **4**, 448–472.

F. C. Monkman and N. J. Grant, 1956, *Proc. ASTM*, **56**, 593.

M. Morinaga, R. Hashizume and Y Murata, 1994, Electronic approach to the design of ferritic steels for power plants. In: "Materials for Advanced Power Engineering," Kluwer Academic, The Netherlands, Part I, pp. 319–328.

Y. Murata, M. Morinaga and R. Hashizume, 1997, Development of ferritic steels for steam turbine rotors with the aid of a molecular orbital method. In: "Advances in turbine Materials, Design and Manufacturing, 4th Int. Charles Parsons Turbine Conference," eds. A. Strang, W. M. Banks, R. D. Conroy and M. J. Goulette, Institute of Materials, London, pp. 270–282.

M. Murugananth, 2002, Design of welding alloys. Ph.D. thesis, University of Cambridge.

M. Murugananth and H. K. D. H. Bhadeshia, 2002, Components of the creep strength of welds. In: "Mathematical Modelling of Weld Phenomena 6," eds. H. Cerjak and H. K. D. H. Bhadeshia, Maney Publishers, London, pp. 243–260.

J. D. Parker, 1996, Assessment of remanent life using strain monitoring techniques. In: "Creep Resistant Metallic Materials," Vitkovice, Czech Republic pp. 122–130.

S. L. Robinson and O. D. Sherby, 1969, Mechanical behaviour of polycrystalline tungsten at elevated temperatures. *Acta Metall.*, **17**, 109.

L. P. Sagert, G. B. Olson and D. E. Ellis, 1998, Chemical embrittlement of Fe grain boundaries: P and P–Mo couples. *Phil. Mag. B*, **77**, 871–889.

O. D. Sherby and P. M. Burke, 1967, *Prog. Mater. Sci.*, **13**, 325.

F. Tancret, H. K. D. H. Bhadeshia and D. J. C. MacKay, 2000, Design of new creep-resistant nickel-base superalloys for power-plant applications. *Key Eng. Mater.*, **171–174**, 529–536.

D. Venugopalan and J. S. Kirkaldy, 1978, New relations for predicting the mech props of Q & T steels. In: "Hardenability Concepts with Applications to Steels," eds. D. V. Doane and J. S. Kirkaldy, TMS–AIME, Warrendale, Pennsylvania, USA, pp. 249–267.

S. Wignarajah, I. Masumoto and T. Hara, 1990, Evaluation and simulation of the microstructural changes and embrittlement in 2.25Cr–1Mo steel during long term service. *ISIJ Int.*, **30**, 58.

R. Wu, A. J. Freeman and G. B. Olson, 1994, First principles determination of the effects of phosphorus and boron on iron grain boundary cohesion. *Science*, **265**, 376.

V. Yamakov, D. Wolf, S. R. Phillpot and H. Gleiter, 2002, Grain boundary diffusion creep in nanocrystalline palladium by molecular dynamics simulation. *Acta Mater.*, **50**, 61–73.

Comprehensive Structural Integrity
ISBN (set): 0-08-043749-4

Volume 5; (ISBN: 0-08-044154-8); pp. 1–23

5.02
Creep–Fatigue Interactions (Crack Initiation)

R. P. SKELTON
Imperial College of Science, Technology and Medicine, London, UK

5.02.1 INTRODUCTION

In a typical component, the behavior of a fatigue crack after the initiation and short crack growth stage, through to the deep crack growth stage, and on to the possibility of complete penetration depends critically on the section thickness. Some of the causes of such initiation and growth are considered in this chapter—such as thermal shock and other constraints against expansion or contraction—and some ways of simulating propagation behavior in the laboratory are discussed, where a cyclic event in service is identified with a fatigue cycle performed in the laboratory. However, our main purpose is to concentrate on the early stages of a phenomenon that has come to be called the *creep–fatigue interaction*, usually under conditions of constant strain (*q.v.*). This can occur, e.g., in plant where a rapid start-up phase is followed by a period of steady running during which creep processes in the material intervene. The latter can cause internal structural damage that is observed to reduce the cycles to crack initiation and enhance crack growth rates. The review thus covers

(i) associated laboratory-testing techniques,

(ii) mechanistic interpretations of creep–fatigue, and

(iii) illustrative examples and application to practical cases in service.

Finally, a case study (retrospective analysis) of crack initiation and early propagation in a power plant component is described together with a validation of the calculations.

There have of course been reviews in the past of the effect of dwell (i.e., hold) times and the creep–fatigue interaction (Ellison, 1969; Krempl and Wundt, 1971; Esztergar and Ellis, 1971; Conway *et al.*, 1981) but the work considered appears always to accept that data (extrapolated or otherwise) from complete failure of laboratory specimens is the criterion for predicting component behavior in service. Prior to 1967 few investigators cared to comment on likely cycles to crack initiation. However, Edmunds and White (1966) remarked that "cracks which could be seen by the unaided eye were often present at half-life." Coles *et al.* (1967) noted that if a surface crack of length 0.4 mm was defined as initiation, then for a cast CrMoV steel at 550°C with a 30 min tension dwell, the ratio of initiation to failure cycles, N_i/N_f, was 0.17 at a total strain range of 1.44%, increasing to 0.43 at a strain range of 0.52%. This indicated that less of the life was spent in crack propagation at lower strains. Similar values of this parameter, in the range 0.1–0.6, were also reported for several austenitic steels at 600°C (Dawson *et al.*, 1967). In continuous cycling of 1CrMoV steel at 565°C Ellison and Paterson (1976) reported a crack depth of 0.1 mm by 10% of the endurance while Batte *et al.* (1978) identified cracks as 0.2 mm deep at specimen half-life for tests on a 0.5CrMoV steel involving a 0.5 h tension dwell at 550°C. Later reviews (e.g., Tomkins and Wareing, 1977; Batte, 1983; Miller *et al.*, 1984; Priest 1989) indicated that metallurgical considerations and physical mechanisms were beginning to be taken into account in the quest for accurate life prediction.

Throughout this review, which complements those undertaken previously by the author (Skelton, 1982, 1983, 1985, 1988a) and expands on parts of a course on component reliability (Skelton, 1998), the following will always be kept in the foreground for a given component (or laboratory test specimen):

(i) What is the relevant surface strain range?

(ii) What crack size is defined by the user to constitute "initiation" and is the number of fatigue cycles to produce it further reduced by periods of creep?

(iii) Once the fatigue crack has started to propagate, is the cyclic growth rate also accelerated by creep damage?

(iv) Are compressive dwells as damaging as tensile dwells?

A typical example of a fatigue crack surrounded by creep cavity damage in grain boundaries is given in Figure 1. Generally, only traditional smooth-specimen data and their associated results will be considered. Although many elegant studies have been performed using lightly notched specimens of different shapes involving painstaking observations of crack growth, crack coalescence, and relating results to deep-crack growth data (e.g., by means of the ΔJ parameter), it was felt that their inclusion would unnecessarily lengthen and complicate the scope of the present review. These topics are taken up further in Chapter 5.06.

Investigations into creep–fatigue involving long-term testing were at their height during the 1970s and 1980s. Since then, published work has generally involved relatively short-term investigations either to prove a particular point or to demonstrate the applicability of a given model. For reasons of space, it would be impossible to reference all those papers which have the term "creep–fatigue" or even "creep–fatigue interaction" in their titles and so selected publications have been included in the reference list (Carden *et al.*, 1973; Curran, 1976; Pelloux and Stoloff, 1980; Amzallag *et al.*, 1982; Raj, 1985; Solomon *et al.*, 1988; Bensussan and Mascarell, 1990). The proceedings of a major conference held in honor of L. F. Coffin and S. S. Manson, two pioneering workers in the field, contains 72 papers and is indispensable (Solomon *et al.*, 1988). Nor will it prove possible to do justice to the numerous logarithmic plots of "predicted endurance" against "experimental endurance" which have appeared for every model, where agreement within a factor of 2 is considered satisfactory.

This survey begins with the basic phenomena of creep–fatigue interaction and considers typical data and likely mechanisms to explain

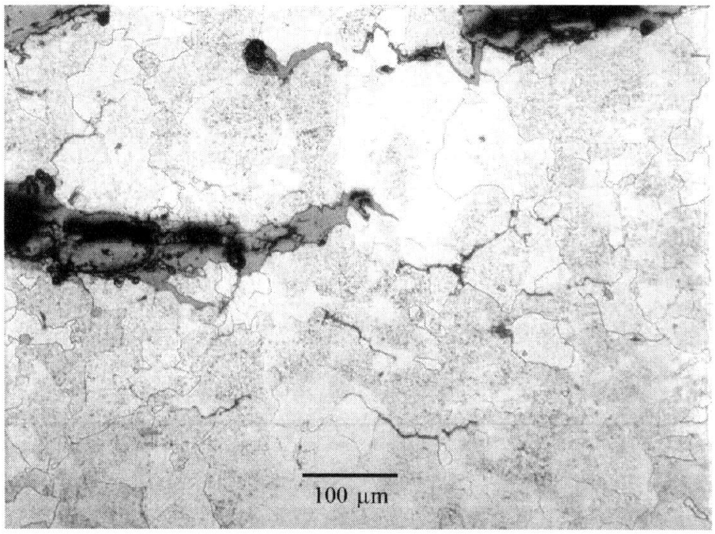

Figure 1 Example of creep–fatigue damage in a ferritic steel at 550°C.

the observations. The approach then becomes increasingly practical in tone and includes illustrative examples for some of the lifetime prediction methods, actual failure examples from industry, and culminates in a fully worked case study constituting a retrospective analysis of a failed component at high temperature. It should therefore be possible for users to undertake their own independent analyses using some of the methods described here.

5.02.1.1 Origin of Thermally Induced Stresses

Owing to the constraints imposed by thermal expansion and conductivity of metals, heating and cooling transients cause internal stresses between tension and compression in many components, causing yield in the surface layers if the temperature changes are severe. Primary stresses arise from pressures, mechanical end loads ("system stresses"), and wholesale constraint, such as a bar prevented from expansion by anchoring both ends. In this case, bulk thermal stresses across the section are independent of the rate of temperature change in the elastic region, becoming modified in the elastic–plastic region only when strain rate/relaxation effects arise in the material. Secondary stresses cannot be measured directly. For example, they may arise from the heating and cooling of a dissimilar metal weld (also independent of heating and cooling rates if yield has not occurred). The most common example is one of thermal shock (Skelton, 1990, 1998) where transient stresses arise due to sudden temperature changes at internal or external surfaces of components. In this case the temperature ramp rate is of major importance in determining the magnitude of such stresses.

In assessing the lifetime of components at elevated temperature, it has proved convenient to recognize several failure modes, each requiring separate treatment. These may be classified as:

(i) inelastic deformation (cyclic plasticity),
(ii) ratchetting,
(iii) creep damage accumulation,
(iv) creep–fatigue crack initiation, and
(v) crack propagation which may also be subdivided into:
(a) creep–fatigue crack growth under transient conditions and
(b) creep crack growth under steady-state conditions.

Many of these topics have been covered elsewhere in Volume 5 of this series. Further interactions may well occur between the modes of types (i)–(iv). Creep damage is generally regarded as a bulk phenomenon while crack initiation ordinarily occurs at the surface (unless it be from an embedded defect). Cyclic plasticity can either occur in the bulk of a component (i.e., across the whole section) or in a contained zone, depending on component thickness and loading type.

5.02.1.2 Simulative Tests

Behavior can be modeled by testing the whole component (sometimes an expensive and unrealistic option) or by replicating one feature of a design ("features" test), and simulating thermomechanical behavior until signs of distress are evident. A further simplifying step is to test smooth specimens in the thermomechanical fatigue mode (TMF, see Section 5.02.5), but this too is a complex technique and investigators commonly test in the isothermal mode at the maximum expected service temperature, on the assumption that this creates the maximum damage and shortest lives. Additionally, simple characterization tests are undertaken to determine material properties used in damage assessment, e.g., cyclic stress–strain and cyclic relaxation behavior. For crack initiation calculations, fatigue endurances to some specified failure criterion are also utilized, as will be discussed. These low cycle fatigue (LCF) tests often incorporate a tension dwell which simulates the steady running period of power plant or aeroengine (Skelton, 1988a, 1988b).

These characterizing tests must be distinguished from tests to determine physical parameters such as Young's modulus, thermal expansion coefficient, and thermal conductivity which are a prerequisite for any stress analysis involving thermal transients. Finally, fundamental experiments are sometimes required which establish a damage parameter, e.g., the extent of grain boundary cavitation or oxidation during fatigue (Levaillant and Pineau, 1982; Levaillant *et al.*, 1988; Rémy and Skelton, 1992). Tests involving a mixture of cycles may then be performed in order to verify a particular model, e.g., on creep–fatigue and oxidation–fatigue interaction.

5.02.1.3 Size Effect and Mechanisms

For purposes of this review, the basic assumption is that significant crack growth cannot be allowed during the lifetime of a component, or indeed in the simulative laboratory specimen, see Sections 5.02.5.2 and 5.02.5.3. The purpose of LCF testing of small specimens is to determine a number of cycles to complete failure, N_f, so that this information can be used for predicting the lifetime. The specimen is taken to represent part of the component and so the term N_f for the small specimen represents crack initiation in a large component. Alternatively, more exacting criteria may be required for critical locations so that knowledge of cycles to crack initiation, N_i, in the specimen itself is desirable. For most ductile materials at high temperature, specimen severance is approached via a gradual tensile load drop with each successive cycle and many workers (e.g., Berling and Slot, 1969; Conway *et al.*, 1981) have adopted a 5% load reduction (measured from some saturation value) as an arbitrary endurance criterion. In some advanced alloys such as Alloy 800 in the range 700–760°C this load drop can take place over very few cycles with little warning (Conway *et al.*, 1981). An additional complication is that in some ferritic and austenitic alloys, crack depths of 0.05 mm have been detected by striation measurements at the end of the cyclic softening and hardening stages, respectively (Skelton, 1993a), which is long before cracking is generally assumed to have taken place in the testpiece.

Disregarding for the moment the effects of cyclic softening or hardening (Skelton, 1993a) the decreasing load range is due to the propagation of a fatigue (or creep–fatigue) crack which has initiated at the smooth specimen surface. It has long been known that the tension and compression load components in a hysteresis loop are unequal. In order to scope the phenomenon it may be stated (Raynor and Skelton, 1985) that a failure criterion of 2% or 5% tensile load drop, respectively, may be taken as indicating 2% or 5% consumption of cross-sectional area by a crack (this laboratory method cannot distinguish the actual shape of the crack). Depending on component size, this could be regarded as early cracking. Conversely, the appearance of a "cusp" near the compressive peak of a hysteresis loop can appear at any time after 25% area consumption and crack behavior thereafter would be considered too "deep" for the present review.

The creep–fatigue interaction and its association with a size effect is well illustrated by the work of Hales (1980) who conducted fully reversed LCF tests with tension dwells up to 1,000 min duration on an austenitic steel at 600°C. Many tests were interrupted part way through their lifetime. This was in order to conduct a detailed metallographic examination by sectioning, polishing, and counting of grain boundary cavities. The first signs of damage were detected after a 1 min hold, so that cracks growing from the surface encountered cavitated boundaries. There was a convincing demonstration of an increasing cavitation level with increased hold time (during which the stress relaxed) such that at the 1,000 min extreme, the specimen failed by bulk intergranular failure. At the other extreme of continuous (no hold) cycling, failure occurred by transgranular crack propagation for which striations were visible on the fracture surface. For an increase in dwell period, therefore, the growing crack interacted more and more with cavitated areas ("mixed mode" where striation spacing was less evident) and the author considered that a true creep–fatigue interaction had taken place.

Hales (1980) considered four criteria of cycles to:

(i) initiate fatigue damage,

(ii) initiate creep damage,
(iii) fatigue failure, and
(iv) creep failure and the order in which they might interchange and overlap according to dwell circumstances.

It is clear from Figure 2 that case (iv) is independent of specimen size since it is a bulk effect whereas case (i) would require a crack depth to be specified for fatigue crack initiation. Equally, if cavities link up late in life owing to the enhancing effect of a crack the failure remains creep dominated and is governed by the failure criterion taken by the load drop (Plumbridge *et al.*, 1982). Examples of a rapid interaction will be given later in this review. Other cases may be considered; Plumbridge (1987a) has extended the possibilities of damage interaction:

(i) transgranular crack initiation and growth,
(ii) transgranular initiation followed by intergranular growth,
(iii) intergranular initiation followed by transgranular growth,
(iv) intergranular initiation and growth,
(v) internal triple point wedge cracks,
(vi) grain boundary voids, and
(vii) initiation from matrix pore or inclusion.

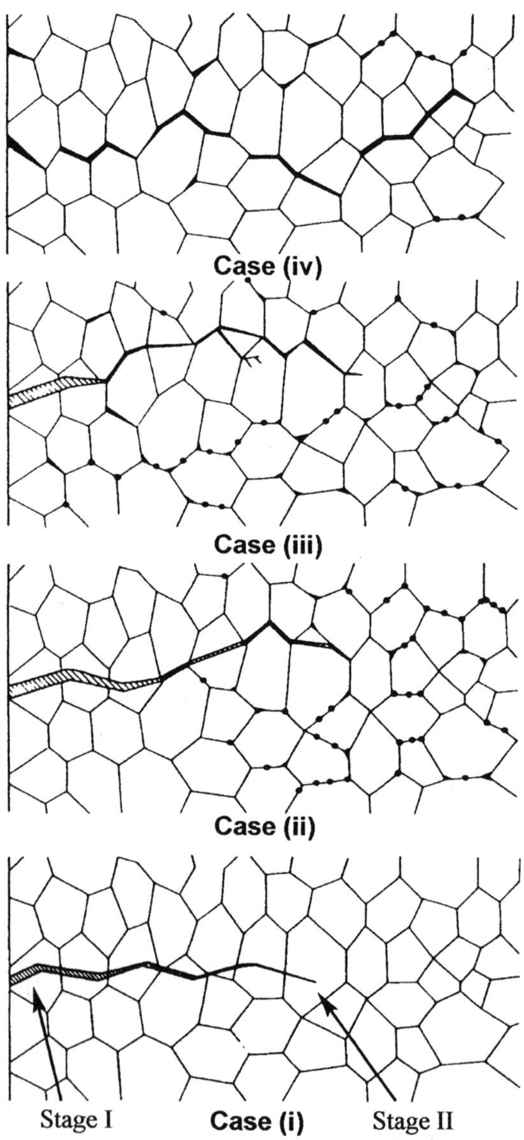

Figure 2 Criteria for creep–fatigue interaction (after Hales, 1980. Reproduced with permission of Blackwell Publishing).

Many instances of the above have been reported in the literature and indeed the crack path can be artificially induced to change by cycle-mixing experiments (e.g., Yamaguchi and Kanazawa, 1979). In Figure 3 is a schematic depiction of a total strain range–endurance diagram after Hales (1980) which indicates the region where "true" creep–fatigue interaction can be said to occur. It may well be linked to cases where linear damage fails to sum to unity or where nonlinear damage rules are invoked, see Sections 5.02.4 and 5.02.9. It may be noted that early work on smooth specimens was generally performed at high strain ranges (>0.7%) with relatively short dwells where the creep–fatigue interaction was likely to occur. In a comprehensive review, Miller *et al.* (1984) divided the standard strain range-endurance plot into four regions, depending on the total strain range, $\Delta\varepsilon_t$:

(i) For $\Delta\varepsilon_t > 3\%$ failure is dominated by fatigue cracking.
(ii) For $3\% > \Delta\varepsilon_t > 0.6\%$ failure is by creep–fatigue interaction.
(iii) For $\Delta\varepsilon_t < 0.6\%$ failure is dominated by creep.
(iv) For $\Delta\varepsilon_t <$ the yield strain, failure is by fatigue if a prior crack exists.

As will be seen later, the modern approach concentrates on the third region, typically involving long dwells and creep–brittle materials, so that the creep effect dominates. In dwell tests on 1CrMoV steel at 565°C at a total strain range of 0.6% Plumbridge *et al.* (1980) considered that there was no interaction since internal cavity growth and cracking from the surface took place independently. Similarly Cailletaud and Levaillant (1984) show for mixed cycling in 316 steel at 1.2% total strain range (500 cycles with 10 min dwell, then 100 cycles with 5 h dwell and finally 55 cycles with 10 min dwell) that enhanced creep damage took place at the crack tip (interaction) whereas a 24 h dwell at the same strain range for 373 cycles produced creep cavity growth throughout the gauge section (no interaction). Even within similar stress relaxation periods there can be differences: Miller *et al.* (1982) have shown that

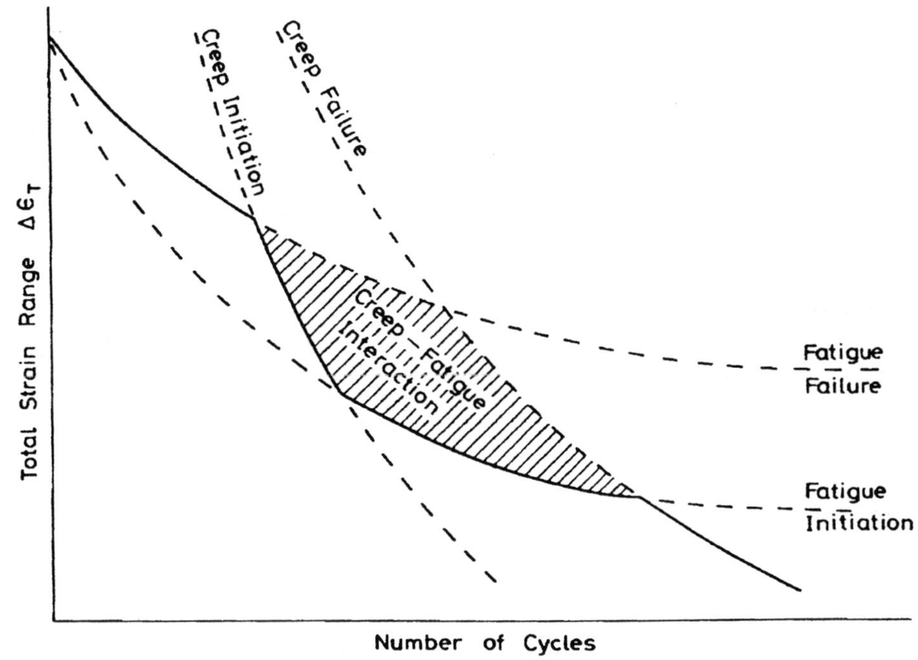

Figure 3 Proposed region of "true" creep–fatigue interaction on an endurance plot (after Hales, 1980. Reproduced with permission of Blackwell Publishing).

for the 1CrMoV steel at 565°C a "constrained" cavity-growth mechanism takes place, while in the 316 steel at 600°C an "unconstrained" mechanism operates, the latter mechanism being more efficient at growing voids.

One aspect of interaction effects must be made clear at the outset. Most tests at elevated temperature are undertaken in air. When dwells are involved, the testing time is much prolonged compared with the equivalent test in continuous cycling. Localized and reversed grain boundary deformation can lead to repeated rupture of the protective oxide film, leading to accelerated oxidation and ease of crack initiation (Coffin, 1963). By implication, the same accelerated processes could occur at a crack tip and so the title of this review should perhaps strictly refer to the creep–oxidation–fatigue interaction. Comments on oxidation effects in fatigue will be made where possible.

5.02.1.4 Practical Problems in Thermal Fatigue

When massive components such as valve chests, turbine casings, and rotors in power plant are subjected to large temperature transients, as during start up and shut down, thermal fatigue cracking may occur at critical locations as has been shown. Similarly, thermal shock in thinner section components such as blades and vanes in aeroengines resulting from repeated takeoff and landing operations may lead to surface crack initiation. It is thus necessary to:

(i) characterize the initiation and

(ii) characterize the growth of such fatigue cracks beneath the surface to assess their likely depth of penetration with known service cycles, allowing also for the effects of creep–fatigue.

5.02.1.4.1 Thick component

The generation of thermal stresses may be examined in more detail as follows (Skelton, 1990). Suppose a thick (up to 150 mm) structure is at a uniformly low temperature and that the temperature suddenly increases at A in Figure 4(a). This is an "upshock" and the surface at A goes into compression as it attempts to expand against the remaining material, yielding along OQ (Figure 4(b)). Heat now flows slowly towards B, and as the temperature gradient reduces, the whole system expands, taking the surface at A into tension at R, i.e., at the original strain value (approximately). If a period of steady running at high temperature ensues, this residual tensile stress relaxes, inducing intergranular cracking in many alloys.

When the temperature suddenly decreases at A, this series of events is reversed. This is a "downshock" and the surface layers now go into tension as contraction is attempted. Peak tension occurs at point S, promoting transgranular cracking since the corresponding

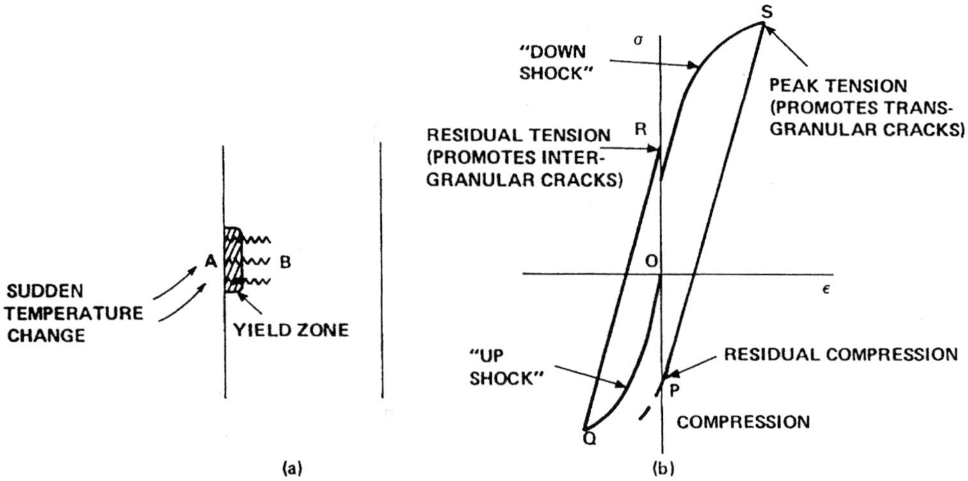

Figure 4 Heating and cooling the surface of a thick structure (a) leads to hysteresis loop (b) (reproduced by permission of Science Reviews).

strain rate is high and the temperature has reduced. Later, when the whole structure cools, the yielded section at A is forced into compression by the surrounding metal. The residual compression stress P (again, approximately at the original strain level) does not necessarily create damage, but when many rapid heating and cooling cycles are repeated from point P, the system shakes down to a closed hysteresis loop PQRS at the surface, leading to the initiation and propagation of (multiple) thermal fatigue cracks due to reversed plasticity.

The sequence described above is most often associated with conventional power plant when required to start up and shut down rapidly (temperature range 25–565°C). However, essentially the same cycle occurs in aeroengine components such as blades and vanes, although

(i) the time sequence for key events in the cycle is very much reduced and

(ii) the section thickness is also very much smaller.

In TMF studies, the dwell at zero strain or thereabouts is often referred to as a "counter-clockwise-diamond" cycle when strain is plotted against temperature, see Chapter 5.03.

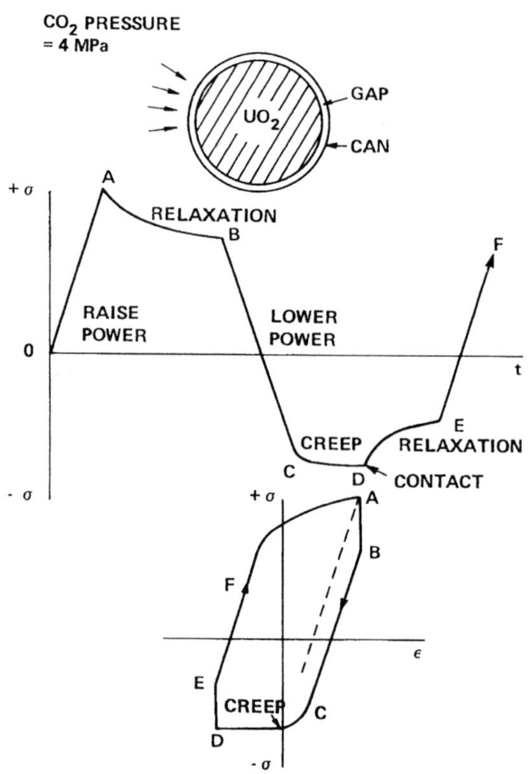

Figure 5 Example of reversed cyclic straining in a thin component (reproduced by permission of Science Reviews).

5.02.1.4.2 Thin section component

In the Advanced Gas Cooled Reactor, fuel element protective cans (0.5 mm thick) are cycled between constant strain limits during power changes (Gittus *et al.*, 1970) and the average temperature of the can is 750°C. Upon increase in power the fuel expands, straining the can at the same rate along OA, see Figure 5. During steady running the can remains in contact with the fuel so that the tension hoop stress relaxes along AB. However, on decreasing power the fuel contracts, leaving the free standing can to creep down (along CD) under the coolant gas pressure, eventually touching the fuel. Further shrinkage thus causes the compressive hoop stress to be relieved by the normal reaction with the can, i.e., relaxation occurs along DE. The next power increase/decrease repeats the cycle and a closed hysteresis loop is described, see Figure 5.

Regarding the size effect, it is immediately apparent from the above examples that complete failure of a smooth laboratory LCF specimen of typical 8 mm diameter is likely to underestimate service cycles in the first case and overestimate them in the latter case.

5.02.1.5 Use of Assessment Codes

When cracks are discovered during routine overhaul of large components the traditional remedy is either to repair (by excavation and welding), to machine out the defected area as in a rotor (Spencer and Timo, 1974; Viswanathan, 1989) or to replace the defected item. The alternative strategy of leaving the component in service until a replacement can be manufactured entails an integrated crack growth calculation which relies on:

(i) knowledge of temperature gradients generated at the area of interest during heating and cooling cycles,

(ii) elasto-plastic stress analyses, and

(iii) relevant high temperature materials data.

This is an example of a remaining or extended-life calculation and for safety, pessimistic (i.e., conservative) materials data should be used to predict future behavior. It may be noted that steps (i)–(iii) may also be applied at the design stage, estimating service cycles to crack initiation followed by integrating crack propagation rates from a notional crack depth to some acceptable final depth, assuming stable growth. In this case also, upper bound crack growth rates should be used, based on reliable laboratory data (see Section 5.02.6). In contrast, a post mortem or retrospective failure analysis would require "best estimate" crack growth data to explain how a particular failure had come about. This may require further testing of service-exposed material.

It thus continues to be important to take account of the specimen size with respect to the component being modeled. To a large extent, the task of the investigating engineer has been simplified by the use of design and assessment codes. The principal procedures for use at high temperature are the ASME Code Case N-47 (ASME, 1991) (now referred to as subsection NH) developed in the USA, Code RCC-MR (Anon, 1985) developed in France and the R5 Assessment Code (Ainsworth, 1996) developed in the UK. They have been compared (Skelton, 1996) for their respective capabilities of predicting crack initiation and growth. In essence, these codes embody over 30 yr work of detailed experimentation and the derivation of complex expressions. They can also help to avoid a detailed literature search since relevant physical and material parameters for an assessment are often tabulated. Care must be taken in applying the codes since safety factors have sometimes been applied to original laboratory data which are not always apparent. The procedures in ASME-N47 and RCC-MR contain much physical property and materials data for use in stress analysis while R5 specifically caters for creep–fatigue crack growth in the short and deep crack regimes, as will become evident.

5.02.2 BACKGROUND TO CREEP–FATIGUE INTERACTION

The two examples in Section 5.02.1.4 show quite clearly that thermal cycling generally involves a limitation on strain. This was reflected in one of the earliest investigations on thermal cycling (Coffin, 1954). Consequently, most of the isothermal simulative tests (with their associated dwells) have been undertaken in the strain-control mode, in sharp contrast to all earlier generations of fatigue test which were performed with load as the control parameter.

5.02.2.1 Fatigue

One of the earliest ideas in fatigue is that a material will fail when the net work expended (accumulated area of hysteresis loops) reaches a critical value. This was taken up by Miner (1945) on some room-temperature tests on aluminum, although the simple damage-accounting procedure had been proposed earlier by Palmgren (1924) in assessing the life of ball bearings and by Langer (1937) for both crack initiation and growth in pressure vessel and piping components. Miner's hypothesis was originally expressed in energy terms but he showed that this could equally well be expressed as:

$$\sum_{i=1} \frac{n}{N_f} = 1 \qquad (1)$$

where n represents the cycles experienced and N_f is the endurance over each cycle of type i. Thus use could be made of a family of endurance curves at different stress levels. Miner defined failure as "the inception of a crack."

5.02.2.2 Creep

In a similar vein, Robinson (1952) by considering stress–rupture plots of turbine alloys in the region 540–680°C proposed, as interpreted by one commentator, "a mathematical accounting system in which operating hours at various stress and temperature levels can be evaluated on the basis of existing creep data, under the assumption that changes from one operating level to another, or rest periods during which the machine returns to no-load or room-temperature conditions, have no effect on creep behavior." In essence:

$$\sum \frac{t}{t_f} = 1 \qquad (2)$$

which Robinson (1952) verified in terms of an equivalent stress or temperature, use being made of a family of stress–rupture plots at different temperatures. Dyson (2000) has examined in some detail the physical justification for the rule and concludes that it is more likely to be obeyed when temperature excursions are made at constant stress rather than when stress excursions are made at constant temperature.

5.02.2.3 Combined Rule

It may be noted that although Miner verified his rule by sequential tests under load control (up to 10^7 cycles and assuming no work hardening), Equations (1) and (2) gradually came to be adopted for application in the LCF regime up to 10^5 cycles:

$$\sum \frac{n}{N_f} + \sum \frac{t}{t_f} = 1 \qquad (3)$$

This came from the realization that assessments may be made from a large body of existing creep–rupture data and relatively simple LCF tests, thus attempting to embrace the intergranular/transgranular failure mechanism of combined tests, see, e.g., Conway *et al.* (1981). One immediate disadvantage (to be discussed later) was that in many practical cases, deformation occurs by repeated stress relaxation rather than by forward creep. There exists very little experimental validation for Miner's law itself for LCF under continuous cycling at elevated temperature, i.e., where reversed plasticity is entailed. Accumulated damage may be dependent on history. Cases involving "major" and "minor" cycling and "incremental step" tests have been reviewed (Skelton *et al.*, 1996). Most investigations during the last 30 yr or so have rather attempted to demonstrate that the creep–fatigue interaction can in fact reduce the fractional damage below unity.

5.02.2.4 Creep–Fatigue

The first tentative steps showing a reduction in endurance with increase in cycle time were taken by Swindeman and Douglas (1959). Working with Inconel at 816°C these authors demonstrated that either a 30 min cycle or relaxation during a 30 min tension dwell (the actuator was clamped during peak displacement) reduced the endurance compared with a standard 2 min cycle. Similar tests were undertaken by Walker (1963). Working on a range of steam turbine alloys at 510°C a simple ranking under 12 h dwell conditions showed a 316/Nb alloy to be inferior to a 1Cr1Mo alloy, see Figure 6. A similar figure was presented in terms of plastic strain range. It was further demonstrated on the ferritic alloys that as the hold time increased from 15 s to 12 h the cycles to failure decreased by a factor of 3 as shown in Figure 7. The increase in hold time also changed the crack initiation and early propagation in the 4.75 mm diameter specimen from a predominantly transgranular to predominantly intergranular mode of failure, thus demonstrating the creep effect.

In the intervening years, many such figures have been produced by other investigators for different alloy systems. Some examples on 316 steel are also given in Figure 7 (Brinkman, 1999, 2000). It should be pointed out that whereas modern systems adopt strain-control on the gauge length, Walker conceded that his tests were conducted under remote displacement conditions. Stress relaxation thus took place under "elastic follow-up" (*q.v.*) but it was nevertheless demonstrated that the calculated increase in plastic strain thereby caused was insufficient to account for the reduction in endurance when referred to data in continuous cycling. The total strain range in Figure 6, e.g., does not include the follow-up strain.

Edmunds and White (1966) were among the first workers to propose a strain-based damage rule, based on their reversed-bend experiments on 2(1/4)Cr1Mo steel at 600°C. Endurances comprising continuous cycling, 30 min and 300 min were produced and the authors attempted to predict these endurances using existing forward creep and specially conducted stress-relaxation data over the same dwell times. The relaxation tests were performed by continuous reloading so that the specimen became longer, rather than being obtained in the push–pull mode, as would be done nowadays. Nevertheless, by summing the relaxation

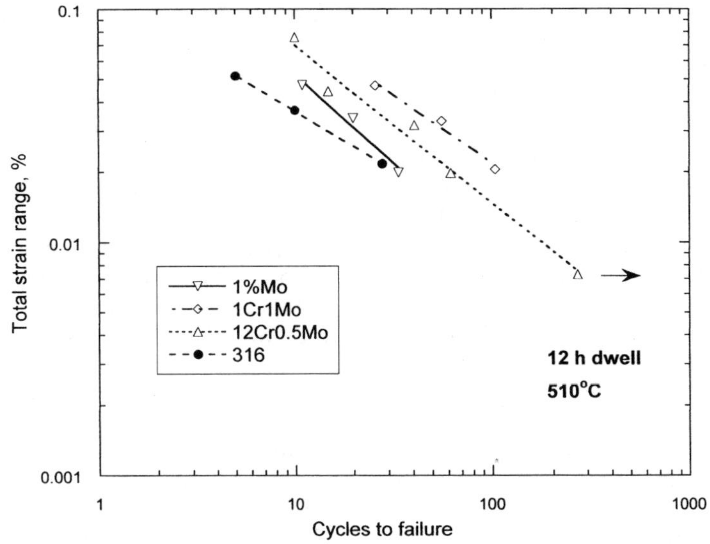

Figure 6 Endurances with dwell (from the data of Walker, 1963).

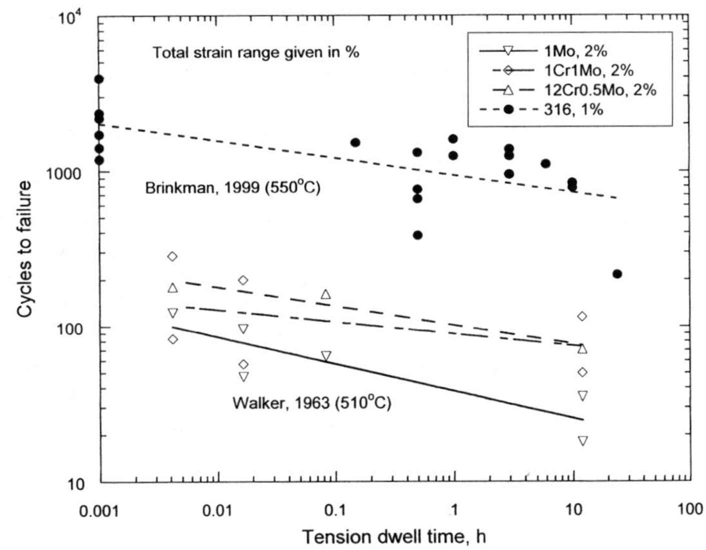

Figure 7 Examples of data indicating reduction in endurance for increasing dwell time.

strains ($\delta\sigma/E$) at the tenth cycle, e.g., where $\delta\sigma$ is the stress drop and E is Young's modulus, and accumulating to an appropriate value of rupture ductility (20%, i.e., 0.2), an endurance of 540 cycles was predicted for the case of a 300 min dwell time. This compared with an observed value of 1,500 cycles for a strain range of 0.3%. (It is significant that use of the prevailing time fraction model gave a more conservative predicted endurance of 290 cycles. As will be seen, this trend has generally been repeated ever since in assessments, even when a more accurate approach is adopted that integrates down the relaxation curves in terms of time or strain, see Section 5.02.4.5.)

In continuous cycling, (multiple) cracking at failure occurred to a depth of 1.4 mm (0.75% strain range). With 30 min dwells, intergranular cracking occurred to a depth of 0.5 mm on the tension face compared with 0.23 mm on the compressive face (Edmunds and White, 1966). Within a few years, similar differences in crack morphology for tension and compression dwells were being reported for austenitic 304 steel at 650°C (Berling and Conway, 1970, see Section 5.02.3.2) and by this time closed-loop control testing equipment in direct stress was employed with extensometers mounted on the specimen gauge length.

5.02.3 EXPERIMENTAL OBSERVATIONS AND INTERPRETATION

5.02.3.1 Environment versus Creep Effect

Testing in oxidizing or inert environments has several purposes. It can demonstrate the accelerating/decelerating effect (compared with an air environment) on directly monitored short crack growth rates (Skelton, 1978) and hence, if total endurance data in that environment are to hand, the cycles to crack initiation may be deduced. Similarly, as hold times are prolonged it has tacitly been assumed that endurances in air are reduced due to the creep–fatigue interaction, i.e., the build up of grain boundary damage. As discussed in Section 5.02.1.3 it could equally be argued that the increased exposure time introduces a creep–fatigue–oxidation interaction whereby the initiation cycles are reduced and crack growth is accelerated by enhanced attack at the tip.

As an example, during tests on a low-alloy steel at 600°C, endurances in argon were twice those in air (White, 1969). Introduction of a hold time of 30 min at zero stress in air (stress-free oxidation) did not impair endurance whilst hold times in tension up to 300 min progressively reduced it. This suggested, therefore, a component of damage under stress either of creep alone or acting together with oxidation. The contribution of each was assessed by testing a 0.5%Mo steel under vacuum (1.3 mPa) and air at 500°C. A 30 min tension dwell under vacuum reduced endurance relative to specimens continuously cycled under vacuum so there was a creep component of damage. However, the specimens continuously cycled in air gave even lower endurances and air specimens with a 30 min dwell time lower still. Thus it was concluded that

(i) both oxidation and creep during a dwell can lower endurance,

(ii) stress-induced oxidation is much more damaging than oxidation alone, and

(iii) in this case oxidation had a more serious effect on endurance than did creep.

Figure 8 shows some results of Maiya (1981) who tested AISI 304 steel in air and under a vacuum of 1.3×10^{-3} mPa (10^{-8} torr). The significant point is that although vacuum endurances were some 4–6 times greater than in air, the strain rate or frequency effect was still present in vacuum, i.e., it can be inferred that internal damage builds up in both cases. Further, when one hour tension hold times were introduced, the vacuum lifetime approached that for the same condition in air.

Stress-free oxidation data are clearly inapplicable to a crack tip. Oxide behavior is sensitive to the total strain range experienced by the underlying metal and continuous cycling studies in a low alloy steel at 550°C have shown (Skelton and Bucklow, 1978) that, at the specimen surface, the oxide growth per cycle increased with increasing total strain range (see Figure 9). It was demonstrated that the effect was due to spalling of the oxide with every cycle, leading to renewed growth. Evidence for this, indicated in Figure 9, was that dwells up to 16 h had little effect on the cyclic oxide growth rate. In attempting to apply these conditions at the crack tip (in terms of metal loss rate) reference was first made to short

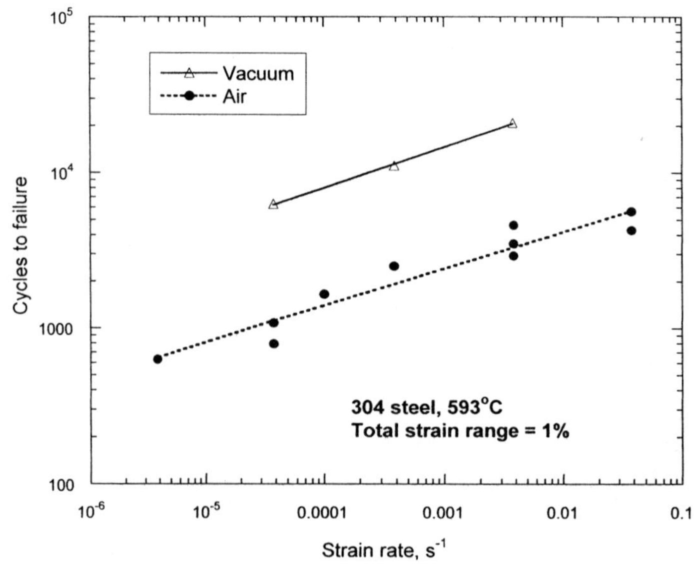

Figure 8 Strain rate effect in an austenitic steel is independent of environment (after Maiya, 1981).

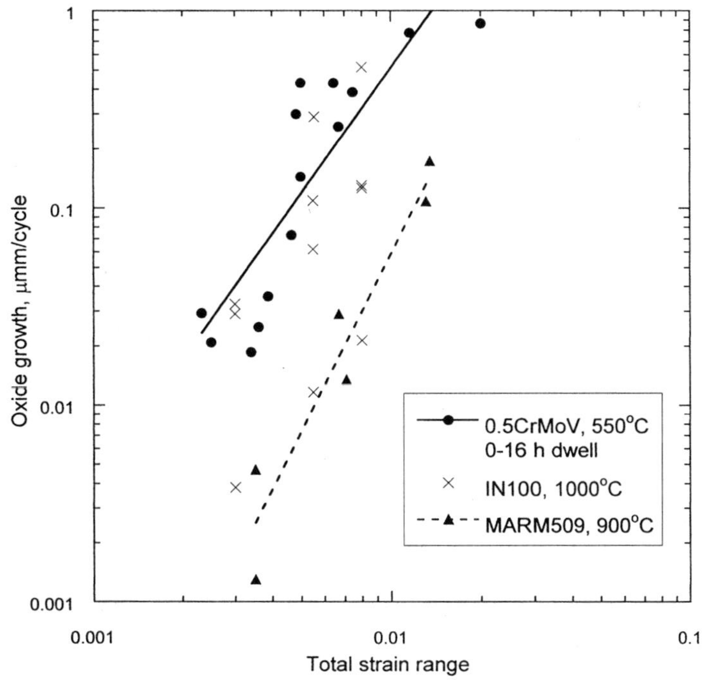

Figure 9 Oxide growth at specimen surface increases with increasing total strain range (see text for sources of data).

crack growth tests in vacuum, as shown in Figure 10. (This type of figure is used to characterize short crack growth and is returned to in Section 5.02.6.) It is noted that growth rates in air are some three times those in vacuum. Thus at a surface strain range of 0.3% the difference in advance rate at the crack tip at unit crack depth (1 mm) in the two environments is $(1\times10^{-3}-3\times10^{-4})=7\times10^{-4}$ mm cycle^{-1}. Crack advance rate is equivalent to metal loss per cycle, and referring to the value at the surface (broken line in Figure 10) we have an indication of the strain magnification in the crack tip region, in this case a factor of 70, i.e., a calculated crack tip strain range of 21%.

Similar enhanced oxidation at the surface has been reported on the superalloys MAR-M509 at 900°C (Reuchet and Rémy, 1983a, 1983b) and IN100 at 1,000°C (Reger and Rémy, 1988). Tests were stopped when the crack depth was in the region 1–1.5 mm. These authors' results have been converted and plotted in Figure 9, and it is seen that the same trend is followed. The authors acknowledged that oxides in these alloys are formed of complex spinels, etc., and may be porous, thus precluding an accurate measure of metal loss. Furthermore as well as the matrix oxidation above, preferential oxidation took place along carbides in the MAR M509 alloy and along grain boundaries or interdendritic areas in IN100. In the former alloy the effect was enhanced by cycling, whereas in the latter alloy it was not.

In a different area, Agatonvic and Taylor (1990) carried out basic "pp" and "cp" (defined later) tests on alloy 800 H at 850°C (deliberately above chosen Code limits) and found that the strain range partitioning method (*q.v.*) of life prediction failed. Metallography indicated that this was associated with heavy oxidation of surface cracks. Life fraction and energy methods also failed. Since the main damage was surface related (oxide-accelerated embrittlement) a correction was applied to the "pp" equation to account for earlier crack initiation via an inverse Arrhenius-type relation whereby the influence of oxidation decreased for crack penetration up to a certain depth. The mechanism was not however dependent on the total strain range.

The fatigue–oxidation effect has been covered in order to illustrate that in some systems it may be the dominant mode (rather than the creep–fatigue interaction). Indeed the Rémy school have, for modeling and lifetime assessment purposes, developed sophisticated calculations for many alloy systems including an oxidation–fatigue model, a creep–oxidation–fatigue models and an oxidation–embrittlement–fatigue damage model. (Rémy *et al.*, 1988; Rémy and Skelton, 1992; Rémy, 1996; see also Section 5.02.5.2 and Chapter 5.04). Several further models of the fatigue–oxidation interaction may also be found in ASTM STP 942 (Solomon *et al.*, 1988).

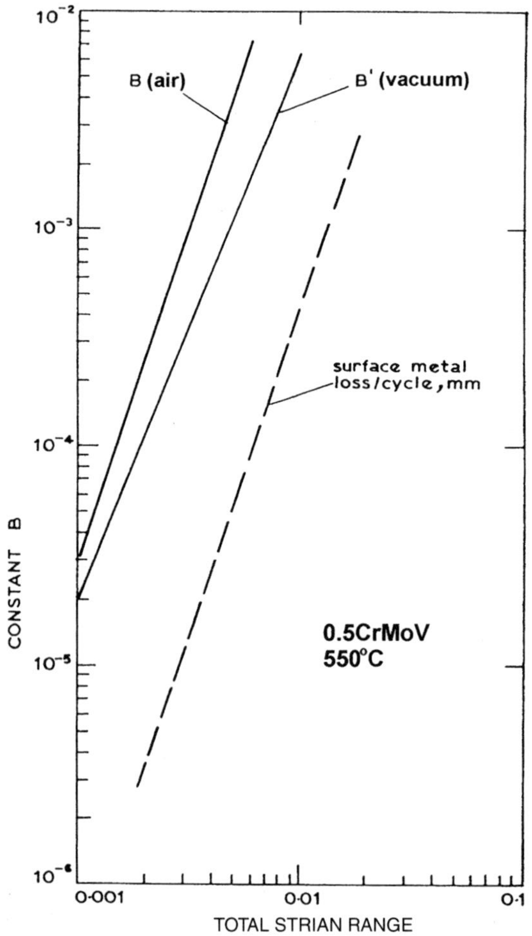

Figure 10 Surface metal loss rate compared with cyclic crack growth rate at 1 mm crack depth (after Skelton and Bucklow, 1978).

5.02.3.2 Effect of Compression Dwell

Amongst researchers and engineers applying the data, few topics in creep–fatigue have aroused as much controversy as the effect of compressive dwells—whether they are more damaging than tension dwells or whether, for example, when applied in combination, there is a healing effect of the compressive dwell. We now attempt to draw together the wide range of findings.

5.02.3.2.1 Austenitic and ferritic alloys

The early work of Berling and Conway (1970) on 304 steel at 649°C has already been cited in Section 5.02.2.4. At 2% strain range (rather large) a 3 h tension dwell reduced the endurance to 0.25 of that in continuous cycling whilst the application of equal 0.5 h dwells at peak tension and peak compression restored this value to ~0.7. In complete contrast, in reviewing data for several alloys, Brinkman (1983) reported that for 2(1/4)Cr1Mo steel at 538°C a compressive dwell only of 6 min reduced the endurance to one tenth of the continuous cycling value. The strain range was 0.3%. The explanation for this behavior was that during the compressive hold period, the surface oxide grew and subsequently spalled during the tension-going direction, leading to easier crack initiation in the underlying metal (Teranishi and McEvily, 1979; Challenger *et al.*, 1981). However, these series of tests have been criticized by Hales (1983a) who noted that the authors used a diametral extensometer to control the strain range and that the output would have been perturbed by gross build up of oxide. It is noteworthy that in another investigation on this steel which employed diametral extensometers (Sandhya *et al.*, 1989), the order of endurance was continuous cycling > compression hold (1 min) > tension hold (1 min) > tension (1 min) + compression (1 min) hold. The temperature was 500°C and the total strain range was 1%.

In a study of a 1CrMoV steel at 530°C Ewald *et al.* (1977) showed that endurances with a 20 min compression dwell were more advantageous than a 20 min tension dwell. The lowest endurance of all was obtained with symmetrical cycles containing a 90 min dwell in both tension and compression. The failure condition was "a distinct drop in stress" on the tension side of the cycle. In some tests on a 1(1/4)CrMoV turbine casting steel at 530°C the following was found (Maile, 1982). In the total strain range region 0.2–2.0% a 20 min compressive dwell gave endurances greater than 20 min tensile dwell values by about a factor 1.5–2. These latter values were identical to symmetrical cycling (10 min each, tension and compression) while 20 min symmetrical cycling reduced endurances only a little further.

Possibly for the reasons outlined above, diametral control on hourglass specimens became less popular but Hecht and Weertman (1998) persisted with this technique on the 2(1/4)Cr1Mo steel, also testing in a vacuum (level unspecified, but insufficient to eliminate oxidation and spalling). Working at a total strain range of 1% throughout and employing a 2 min dwell they found:

(i) At 500°C a compression dwell gave 25% less endurance than a tension dwell, but in vacuum this order was reversed.

(ii) At 600°C a compressive dwell gave about the same endurance as a tension dwell, but in vacuum the compression-dwell endurance exceeded that of the tension dwell by a factor 1.5–2.

As in the tests of Challenger *et al.* (1981) there was intense striated oxide activity growing outwards from the metal surface. For the compressive hold, cracks in the oxide were transmitted into the metal but not for the case of the tension hold. Essentially the same effect for a compressive dwell in a modified 9Cr1Mo steel at 593°C had been reported previously (Kim and Weertman, 1988) where no oxide spalling occurred in any type of test. It is noteworthy in this paper that the authors acknowledged some of the disadvantages of diametral control (e.g., decrease of total strain range by 0.05% during softening, 20% higher stress amplitudes in hourglass specimens for a given axial strain range) compared with conventional axial control.

Tests on forged 2(1/4)Cr1Mo under axial control were undertaken at 565°C at a total strain range of 0.5%. Some specimens were prenotched to a depth of 60 µm (Kschinka and Stubbins, 1989). Continuous cycling tests lay well within the scatter band from various international sources. In air, a 6 min tensile hold, e.g., approximately halved the endurance whereas compressive holds were not quite so damaging. However, a combined 3 min tension and 3 min compression hold reduced the endurance by a further factor of 2. The authors attributed this to the effect of damage inflicted during the early stages of two stress relaxation periods per cycle. The effect of the sharp notches was to reduce all endurances by a further 25%. It was concluded therefore that most of the fatigue life in plain specimens involved crack propagation. Interrupted tests demonstrated surface cracking at 5% of life. Tests were also undertaken in a vacuum of 1.3 mPa (10^{-5} torr). The immediate effect was roughly a fourfold increase in life throughout, except for the case of the compressive hold which exceeded the continuous cycling endurance. This was attributed to crack rewelding and for this reason the combined dwell cycle under vacuum was relatively less damaging. The effect of pre-notching was to reduce all endurances by a factor of 41% compared with the case in air. The authors concluded that a greater part of the cyclic life was spent in crack initiation under vacuum and indicated the accelerating effect of oxidation. Photographs of failed specimens were provided which provided significant evidence for oxide behavior:

(i) With a 6 min tension dwell the oxide appeared to spall off each cycle and build up a thick outer shell.

(ii) With a 6 min compressive dwell the oxide appeared much more adherent and no large diameter increase was observed.

(iii) With symmetrical dwells the oxide behaved similarly to that observed with a compressive dwell.

These observations demonstrate that, as noted above, the strain measuring technique should be chosen with care when assessing the relative contributions of tension and compressive dwell damage. The authors offered a severe critique of the claims of Challenger *et al.* (1981) that compressive dwells are more damaging in this material. In an attempt to resolve the problem, Skelton and Challenger (1984) had earlier undertaken short crack growth tests on 2(1/4)Cr1Mo steel at 525°C. Taking vacuum (10^{-5} torr) data as a reference, the increased propagation rate in air, whether a 0.5 h tension or compression dwell was present or not, was demonstrated to be due to oxidation. Combined tension plus compression dwells gave a slightly greater crack growth rate, but only in annealed material, not normalized and tempered material. A slowing down of the crack in vacuum was demonstrated to be due to rewelding of the crack faces.

Other reports on the compressive dwell effect have been as follows. Asada and Mitsuhashi (1980) found that in 304 steel a factor 2 reduction in endurance occurred for a 10 min compression dwell at 650°C (diametral strain control was used). For 1CrMoV steel at 550°C, a 30 min compressive dwell showed a factor 2 reduction compared with continuous cycling whereas for a 12Cr steel a 10 min dwell reduced endurance by a factor 1.5 (Nam *et al.*, 1992). Significantly in this latter steel, striation spacing was greater than for continuous cycling under equivalent conditions (i.e., the early crack growth rate was greater at the chosen strain range of 2%). A lifetime model was suggested in terms of material behavior in the process zone some 100 µm ahead of the crack tip.

Many more examples could be quoted from the literature but the examples form isolated checks rather than full-scale investigations. As a very rough guide, and bearing in mind the variety of temperatures, strain ranges, and dwell times, it may be stated for austenitic and ferritic steels that a compressive dwell reduces endurances by a factor 0.5–0.6 and that the main effect is due to the elimination of the crack initiation phase, see Section 5.02.5.5. This is reasonable since stress relaxation in compression induces grain boundary sliding which may produce damage initiation sites that are exploited by the subsequent tension-going phase. Similarly, mixed cycling is generally restorative, by the action of sintering the damage previously inflicted in the tension dwell. Further examples based on metallographic examination are given in the following sections where different tension and compression dwell sequences were applied.

5.02.3.2.2 Reverse bend tests

In some reverse bend tests with 30 min dwell on 316 steel at 600°C, striation spacings associated with grain boundary cracks which had initiated at the tensile surface showed signs of a creep–fatigue interaction by sudden changes in spacing and direction, generally associated with triple-point cracks (Dawson *et al.*, 1967). In contrast, on the concurrent compressive dwell side, no intergranular cracks were observed and propagation characteristics were similar to those seen in continuous cycling. In similar tests on a 1%CrMoV steel with a 30 min dwell at 550°C Hill (1967) observed that tension-side induced cracks were both transgranular and intergranular near the surface, becoming wholly intergranular for deeper cracks. On the compression side, however, cracking was transgranular and filled with oxide. In all tests, predominant cracking occurred from the tension side so that the "neutral axis" (region of final failure) was displaced towards the compression dwell face.

As noted, in the four-point reverse bend test, a tensile dwell on one face is accompanied by a compressive dwell on the opposite face, giving an ideal opportunity to compare crack initiation and growth mechanisms. The work of Batte *et al.* (1978) on 0.5CrMoV steel has indicated that:

(i) cracking and damage on the tensile face was intergranular,

(ii) there was a higher density of cracks associated with this face, and

(iii) principal crack propagation to failure took place from this face.

It is thus difficult to understand why the behavior of 2(1/4)Cr1Mo steel should be different, see below. A comparison of the incidence of cracking as a function of strain range and dwell period is given in Figures 11(a) and (b) (Batte *et al.*, 1978). These results were obtained from metallographic sectioning near the surface and show the preponderance of transgranular cracking in the case of compression dwell.

Similarly, reverse bend test on 316 steel at 550°C and 600°C have shown (Wood and Banks, 1990) that, although the average crack depths were slightly greater on the tensile side than on the compression side, there was a great difference between the depths of the major crack to failure, e.g., 1.1, 0.92, and 0.48 mm (tension side) compared with the respective values of 0.47, 0.26, and <0.08 mm (compression side). An example is given in Figure 12(a). Some differences were noted, e.g., at 600°C there were the same number of cracks on the tension and compression side whereas at 550°C there were many more cracks on the tension side. Since cracks were transgranular on the compression side and intergranular on the tension side it was concluded that creep–fatigue in tension was more damaging than creep–fatigue in compression.

In some reverse bend tests on 2(1/4)Cr1Mo steel at 550°C and 0.7% total strain range (Holdsworth, 2001), a 30 min dwell was inserted at the zero strain position. With reference to Figure 4(b) this occurs at a positive tensile stress and so in bending the opposite face remains in compression. Over a 15 mm surface away from the failure crack there was an equispaced array of 10 cracks some 1 mm deep on the tension face. On the compression face there were only five cracks at 0.5 mm deep. For a repeat test using a 16 h dwell period there were 20 secondary cracks 0.75 mm deep on the tensile face and none at all on the compressive face. An example is provided in Figure 12(b). Such evidence makes it highly unlikely that compressive dwells are more damaging for this steel.

5.02.3.2.3 Superalloys and other systems

During the same period, an entirely different view was developing on the behavior of superalloys, where compressive dwells appear to be more damaging. A short review of work current in 1978 has been provided by Miller *et al.* (1984). The effect has been noted in cast IN738, Rene 95, MAR M002 and for cast Rene 80. No satisfactory explanation appeared possible at the time, at least in terms of the phenomena that are available in the simpler and less strong austenitic and ferritic alloys, e.g., visible cavitation and reversible grain boundary sliding. It should be noted that the superalloys have an unusually high yield strain and since associated plastic strain ranges are small, either:

(i) little opportunity exists for the hysteresis loop to shake down towards symmetry in stress or

(ii) reversal of relaxation-induced plastic strains requires a high stress to close the loop.

As a result, high mean stresses are developed in a loop containing a compressive dwell; in the author's laboratory tests on IN738 with a 5 min compressive dwell at 850°C at a total strain range of 0.45% (i.e., ±0.225%) have indicated values of R approaching –3 (where R = minimum stress/maximum stress). An example is given in Figure 13. Consequently, if the failure phenomenon is governed by crack growth, a much greater tension stress range is available to open and close the crack.

For the lesser alloyed Inconel 800, a 10 min dwell time whether in compression only or

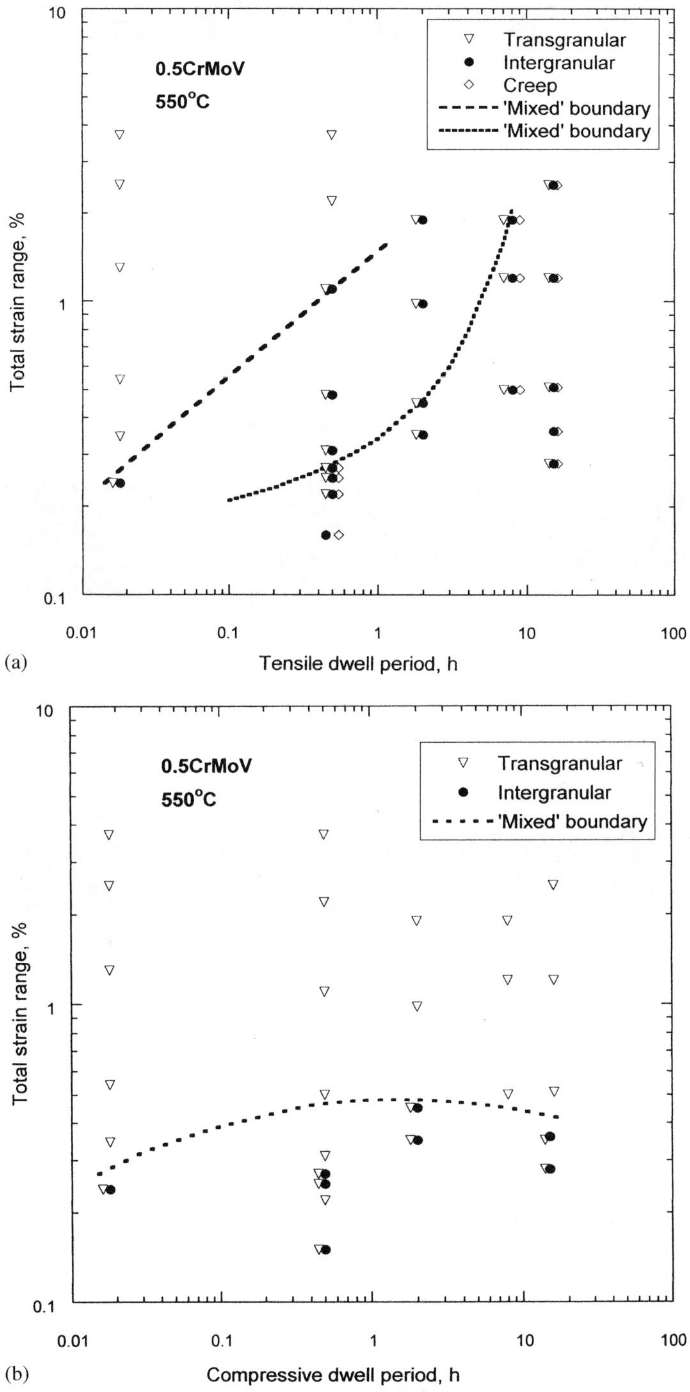

Figure 11 Effect of strain range and dwell period on crack morphology: (a) tension side and (b) compression side in reverse bend experiments (after Batte *et al.*, 1978).

applied both in tension and compression had little effect on the endurance in the range 538–760°C (Jaske *et al.*, 1972). (These tests were in diametral control.) In other tests on the advanced alloy MAR M002 at 750°C, 850°C, and 1,000°C there was little effect of a tension dwell despite the appearance of intergranular cracks (Plumbridge, 1991). A compressive dwell of 5 min at the two lower temperatures led, however, to a reduction in life by a factor of 6, the failure mechanism being transgranular cracking. A balanced tension–compression dwell of 5 min each gave a factor of 2 reduction on reference data. The author tried the ductility-exhaustion and Ostergren (1976) energy method of life prediction for this alloy but both

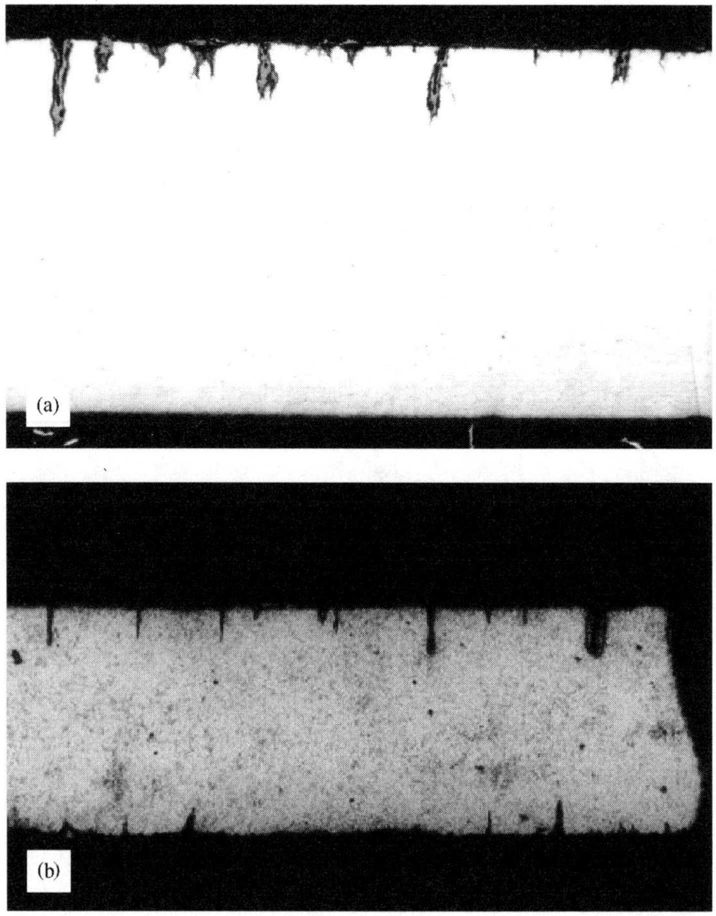

Figure 12 (a) Preponderance of cracking from the tension surface in type 321 austenitic steel plate (AGR roof liner) tested at a total strain range of 0.6%, 650°C with 1 h dwell under reverse bending. Top–bottom thickness of 3.5 mm (reproduced by permission of I. Bretherton; SERCO Assurance). (b) Indicating more incidences of cracking on the tension face of 2(1/4)Cr1MoV steel tests for 1,800 cycles with 0.5 h dwell in reverse bend at a total strain range of 0.7%. Top–bottom thickness of 5 mm (reproduced by permission of S. R. Holdsworth; Alstom-Power).

were unsuccessful. Plumbridge (1987b) has also carried out investigations on a titanium (IMI 829) alloy at 600°C. Compressive dwells of 2 min were most damaging but balanced dwells of 15 min each were less so. Again, tensile dwells produced much intergranular cracking but they had little effect on a life that was fatigue-dominated. Compressive dwells actually produced the greatest number of surface cracks and it was considered that in balanced cycling, environmental effects dominated.

In the Ni-base superalloy IN625 at a temperature of 815°C, a 5 min dwell in compression caused over an order of magnitude reduction in endurance compared with continuous cycling at a fairly low total strain range of 0.4–0.6% (Bui-Quoc *et al.*, 1988). The effect was much greater than tension dwells of 5–30 min. However, in discussion it was pointed out that this alloy becomes very unstable at such elevated temperatures.

Some tests were carried out on 8.5 mm diameter specimens of single crystal SR99 at a strain rate of $4 \times 10^{-3}\,s^{-1}$ at 750°C and 1,050°C with either a tensile or compressive 2 min dwell (Li and Smith, 1996). Associated tests on corner cracking of square cross-section specimens were used to deduce the initiation life by subtraction of cycles, using 30 μm depth for initiation and 1.5 mm for failure. At the higher temperature, tensile dwells were more detrimental; the converse was the case at 750°C. Thus, although crack propagation must perforce be transgranular, creep/oxidation effects persisted. In this class of alloy, competitive initiation existed between internal pores and oxide-induced initiation.

Antill and Smith (1996) also refer to other work where compressive dwells give the shortest lives and balanced dwells have no effect. In their own examination on IN738LC, fully reversed tests were carried out at a strain rate

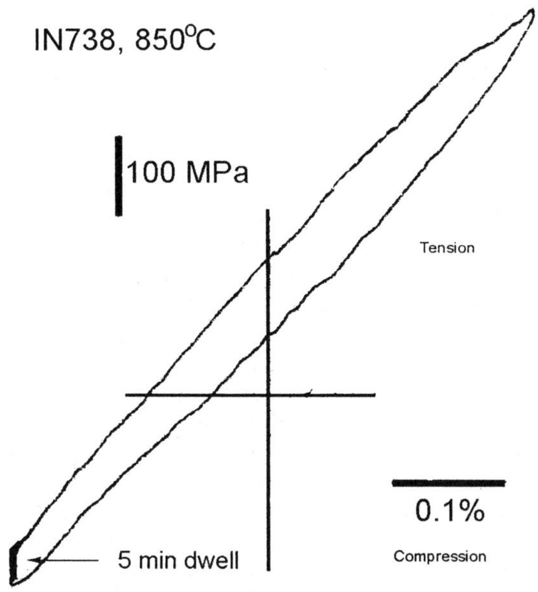

Figure 13 Offset hysteresis loop developed in a superalloy after 16 cycles (reproduced by permission of B. Nath; Innogy plc).

of $10^{-2}\,s^{-1}$. The total strain range was 0.7% with strain-controlled compressive dwell periods of between 120 s and 1,800 s. A 5% load drop criterion was used on an 8 mm diameter specimen. Rapid softening took place over the fist 10–15 cycles. There appeared to be a saturation in endurance for dwell times greater than 120 s (equivalent frequency 0.005 Hz). A tensile mean stress of 200 MPa prevailed for most of the tests. (The relaxed compressive stress was –200 MPa.) Crack initiation was by an intergranular surface oxide intrusion mechanism, but subsequent propagation was transgranular, together with internal damage from cracked carbides. This sometimes led to crack bifurcation. Other literature values gave an order of magnitude reduction in endurance for compressive dwell, independent of total strain range. The effect was attributed to a high mean stress together with creep damage (compressive creep damage was acknowledged but it saturated at 120 s). A "damage diagram" was produced of $1/N_f$ (fatigue attributed to transgranular initiation and propagation) against frequency. The pure creep region had a slope of –1 and a creep–fatigue interaction regime was indicated.

Yamamoto and Nitta (2001) have investigated the deleterious effect of compressive dwell in single crystal superalloys (MDSC-7M) compared with conventional alloys ("mean stress effect" and "oxidation effect"). The longitudinal axis was in the ⟨100⟩ direction. Endurance tests and crack propagation tests were performed at 900°C on hollow specimens of 1.5 mm wall thickness. Continuous cycling, 10 min tension and 10 min compression dwell tests were performed. At 0.7% the compressive dwell endurance was over an order of magnitude less than the continuous cycling value, the tension dwell about an intermediate factor of 3 less. In continuous cycling, crack initiation was internal at a eutectic γ' particle whereas under dwell conditions it was oxide surface-induced. The mean stress gradually approached 200 MPa under the compressive dwell and conversely for the tensile dwell. Crack growth measurements showed that initiation (0.15 mm depth) was particularly easy under dwell conditions (10% of life compared with 40% of life for continuous cycling). Surface coatings (NiCoCrAlY) were found to restore life towards the continuous cycling value but no allowance was made for the size effect, the latter tests being performed on 10 mm diameter specimens.

5.02.3.3 Sequential Tests

Sometimes information on early cracking is sought by performing block-type tests where, e.g., a large period of creep is followed by LCF testing and vice versa. These experiments quite often incur accelerated aging and precipitation effects that affect the follow-on behavior and relevant investigations have been reviewed (Skelton and Horton, 1999).

In tests on two casts of 316 steel at 550°C, Kaneko *et al.* (1992) showed that in slow tensile tests at a strain rate of $2.7 \times 10^{-6}\,s^{-1}$ the reduction in area for virgin material (70% in one cast and 45% in another, higher carbon cast) was reduced by prior fatigue (to 0.25 life) at 1% total strain range at that temperature (to 67% and 35%, respectively). When prior fatigue included an element of creep in the cycle (which increased the exposure times) final reduction in area values were 55% and 20%, respectively. The authors argued that the effect was due to fatigue hardening since fatigue crack initiation was unlikely to affect the outcome.

In creep experiments on 316 steel at 600°C following prior LCF exposure (also at 600°C) to 10% and 40% of endurance at a total strain range of 1.4% (1% plastic strain range) the following effects were observed compared with tests on virgin material (Rezgui, 1982; Rezgui *et al.*, 1981):

(i) an increase in the creep rupture life (by a factor of 2),

(ii) a reduction in secondary creep rate (by a factor of 2) with some suppression of primary creep rates,

(iii) a shortening of the secondary creep stage, and

(iv) a reduction in the strain at the onset of tertiary creep (but little effect on final ductility). A series of sequential tests has been conducted at life fractions in the range 0.1–0.8 on a G91 ferritic steel at 600°C (Yaguchi and Asada, 1996). Where prior loading consisted of time-dependent deformation ("slow–fast" or tension dwell) followed by time-independent loading (symmetrical cycling at $10^{-3}\,s^{-1}$) the damage summation was <1. For the converse type of time-independent prior loading, the damage summed to greater than unity. In the first series, small fatigue cracks had begun to propagate from small cavities throughout the section, subsequently propagated as a true creep–fatigue interaction whereas in the second series intergranular cracking to failure occurred only in the second stage.

Nazmy (1982) demonstrated that the order of sequential tests is important. Specimens of IN738 when crept beforehand at 850°C to fractions of their life developed surface intergranular cracks some 125 µm deep. On subsequent fatigue testing in continuous cycling at $10^{-2}\,s^{-1}$, these were ignored and failure took place by transgranular cracking. Conversely, when the initial period of fatigue produced striated transgranular cracks to a similar depth, these were propagated to failure under creep conditions. Perhaps the story might have been different if lower reversed strain rates in fatigue had been used in both test types. Similar effects have been demonstrated in (1/2)CrMoV steel at 550°C where typically 500 cycles of damage with a tensile dwell of 30 min on a plain specimen, subsequently lightly notched and tested at a frequency of 10^{-2} Hz, produced transgranular cracking (Skelton, 1982). In the course of a normal 30 min dwell LCF test with a small starter notch and a reversed strain rate of $10^{-4}\,s^{-1}$ or thereabouts, intergranular cavitation and cracking was copious (Skelton, 1982).

In 30 min tension dwell tests on a 20Cr25Ni/Nb steel at 650°C it was noted that a prior 25% plastic deformation at room temperature reduced the ductility to 8% compared with a reference value of 60% (Powell *et al.*, 1984). This led to a further reduction on endurance by a factor of 5 in dwell tests compared with similar dwell results for solution-treated material. In the latter material the growing crack interacted with the creep damage whereas in the prestrained material the creep damage was widespread. (The plastic strain range was between 0.3% and 1.0%.) The effect was attributed to an exhaustion of ductility.

Wareing (1983) carried out some step tests on 6 mm diameter type 316 steel at 570°C in order to separate the phases of crack growth. His basic continuous cycling data are plotted in Figure 14, giving continuous cycling endurance at two levels of plastic strain. A 10 h and 1 h tension dwell caused a lifetime reduction at the respective levels. In the first step test at 0.43%, a 1 h dwell was applied for 0.1 of the hold-period life, the remainder being carried out in continuous cycling. The final endurance was identical to the reference (continuous cycling) test. Clearly, the initial creep–fatigue was insufficient to cause accelerated growth though cavitated boundaries. Metallographic examination

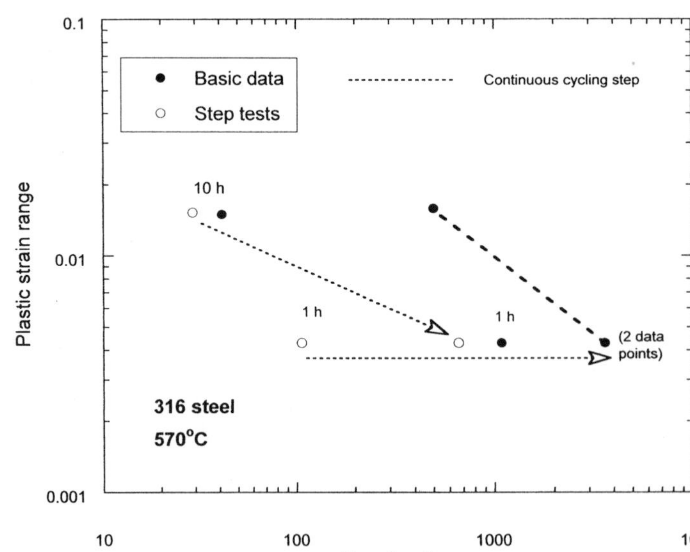

Figure 14 Sequential tests of Wareing (1983) plotted graphically, see text.

revealed predominantly transgranular cracking. In the second step test, a 10 h dwell was applied for 0.7 of the (10 h) hold-period life followed by continuous cycling at the lower strain level, see Figure 14. A rough estimate of the remaining life would be 0.3 of the 1 h dwell-time life which is equal to 300 cycles (i.e., assuming cavitated boundaries). In fact the specimen lasted a further 660 cycles. Metallographic examination in this latter test revealed intergranular crack initiation to a depth of two grains (80 μm) followed by transgranular failure.

Wareing's (1983) basic endurance for a 10 h tension dwell was 41 cycles, as shown in Figure 14. Some idea of the relative cycles for crack propagation can be obtained from a 10 h tension dwell test carried out by the present author on a similar steel at 570°C. The specimen was in fact hollow with a wall thickness of 4.5 mm and an outside diameter of 13.56 mm. A circumferential notch 0.3 mm deep by-passed the crack initiation stage and enabled crack growth to be monitored by a potential drop (p.d.) method. The total strain range was 1.5%, approximating to Wareing's (1983) higher strain level. The results in Figure 15 show that after some cyclic hardening, rapid failure occurred across the whole section during the dwell in the 12th cycle, manifest by a large increase in p.d. reading accompanied by a rapid stress drop. This is an example of unstable crack growth owing to the creep–fatigue interaction at relatively high strain range, see Section 5.02.1.3.

An interesting variation in sequential effects tests is provided by the two-temperature tests at 600°C and 900°C of Reuchet *et al.* (1982) on MAR M509. At the lower temperature there was no oxidation effect whereas an effect was present at the upper temperature. A p.d. change of 0.1% was associated with crack initiation at some 0.2–0.3 mm deep whereas at the "rapid decrease in tensile load" stage the cracks were 1–2 mm deep. Crack initiation at 900°C was affected by preferential oxidation of MC carbides. At temperatures below 650°C crack propagation occurred in matrix dendrites or along carbides in the interdendritic region; above this temperature propagation took place along an inter-dendritic path. These mechanisms illustrate the difficulties experienced by the authors in applying Miner's rule in the step tests. At a plastic strain range of 0.4% the initiation cycles at 600°C and 900°C were 1,905 and 142 cycles, respectively and the associated failure cycles were 3,358 and 442. In the "high/low" sequence the data were scattered about a nonlinear power law which passed through (0.2, 0.2) on the damage diagram, see Sections 5.02.4 and 5.02.9. Alternatively, in the "low/high" sequence, the data were scattered about the linear line. In the propagation phase it was found that, whichever temperature level occurred first:

(i) The main crack at T_1 remained the main crack at T_2.

(ii) Secondary cracking at T_1 evolved at T_2.

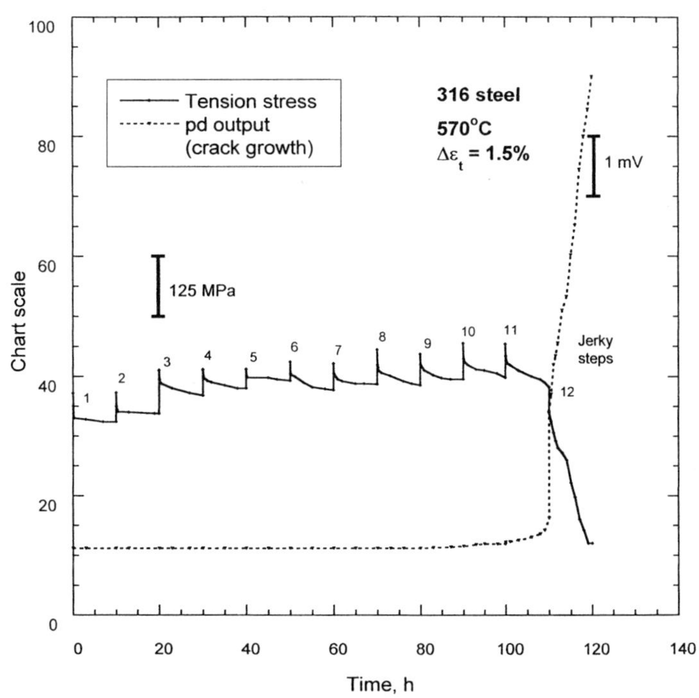

Figure 15 Example of unstable growth in an austenitic steel.

(iii) Final cycling at T_2 produced new cracks.

However, it was observed that there were more newly initiated cracks if the second exposure took place at 900°C, e.g., typical crack densities at 600°C were 3×10^{-2} mm^{-2}, increasing to 2 mm^{-2} at 900°C. Thus, if the cycles undergone during the initiation period at 600°C were low, the initiation period at 900°C would be unaffected owing to a different mechanism (all initiated surface cracks at the higher temperature lying inside the oxidized carbide aggregate). The authors were able to distinguish between initiation criteria for the "designer" (1–2 mm) and the "metallurgist" (0.1 mm), concluding that the more exacting criterion obeyed the linear damage rule if the high temperature cyclic exposure occurred first.

5.02.3.4 Unbalanced Cycling

The term "unbalanced cycling" refers to a type of continuous cycling where the tension- and compression-going phases differ greatly in applied strain rate. The most damaging cycle is for slow loading into tension followed by a quick reversal into compression (slow–fast cycling). This has been demonstrated for AISI 304 steel at 650°C in both air and vacuum (Sidey and Coffin, 1979). The converse (fast–slow) cycle gave endurances greater than when equal (intermediate) strain rates were imposed. Working on AISI 316 steel at 600°C Yamaguchi and Kanazawa (1980) showed that slow–fast cycling between 6.7×10^{-3} s^{-1} and 6.7×10^{-5} s^{-1} gave endurances even less than did a 1 h tensile dwell. Some of their results are shown in Figure 16. Other findings were that a symmetrical tension and compression dwell restored the endurance towards the continuous cycling value, as has been reported by others.

In some tests on 20Cr25Ni/Nb steel at 593°C (Livesey and Wareing, 1983) the standard strain rate adopted by the authors was 10^{-3} s^{-1}. The tension-going strain rate in the slow–fast tests was 2×10^{-6} s^{-1} down to 10^{-7} s^{-1}. A balanced cycle at this slow strain rate did not alter the endurance from the standard rate at a total strain range of about 2%. The slow–fast test gave an endurance less than did a 3.5 h tension dwell test which itself was an order of magnitude less than the fast–fast tests. Microscopical observations and conclusions were:

Fast balanced. Transgranular fatigue crack and striations.

Slow balanced. Intergranular failure with some evidence of cavitation—some sintering proposed.

Tensile dwell. Striations at surface, cavitation in the bulk. Net cavity growth and surface crack interacting with damaged boundaries. It was proposed that unstable growth occurred when half the crack-tip opening displacement (CTOD) was equal to the inter-cavity spacing.

Slow–fast. Cavitation occurred everywhere. A test interrupted at 75% of life showed 10 μm cracks meeting extensively cavitated boundaries. It was proposed that damage was due to the unbalanced nature (not strain rate) since rapid compression sharpened the crack tip.

The general conclusion was that in slow–fast cycling more tensile strain was available for

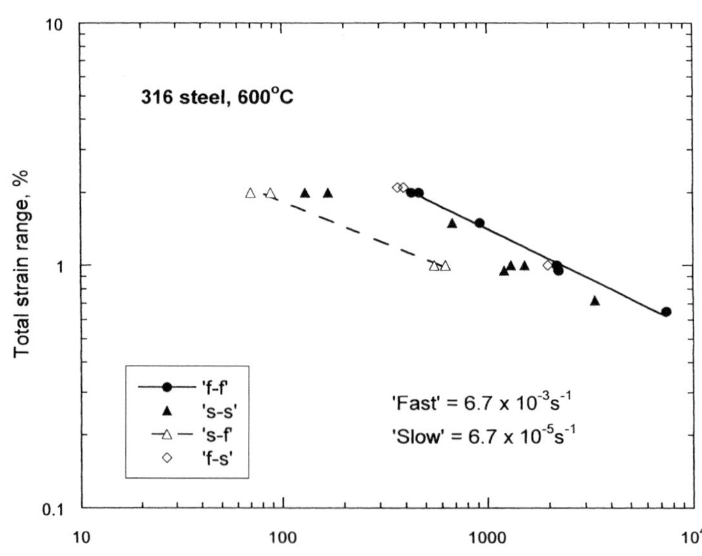

Figure 16 Effect of asymmetrical cycling in an austenitic steel (after Yamaguchi and Kanazawa, 1980).

cavity growth than in the dwell type, hence the creep–fatigue interaction was achieved sooner. The authors noted that, for a given plastic strain, the CTOD was smaller in dwell tests, hence rapid crack growth occurred later than in unbalanced tests. The authors also noted that a model based on constrained/unconstrained cavity growth (Gladwin and Miller, 1982) during the ramp was very conservative.

At 600°C, slow–fast tests ($10^{-5}/10^{-3}\,s^{-1}$) under a vacuum of 0.1 μPa gave a significant reduction in endurance compared with 10 min tension dwell or fast–slow tests (Yaguchi *et al.*, 1992). The results were in line with previous work on 304 steel and 2(1/4)Cr1Mo steel and the authors considered this to be "pure" creep–fatigue behavior, free of environmental effects. The authors developed an "overstress" model (actually related to a back stress manifest on slow unloading) of life prediction and related this to the ratio of intergranular to transgranular damage. A series of slow–fast tests was carried out on a ferritic G91 steel under high vacuum of 0.1 μPa ($\sim 10^{-8}$ torr) at 600°C (Yaguchi and Asada, 1996). The total strain range was between 2% and 0.5%. The strain rate effect persisted such that the endurance curve for 10^{-5}–$10^{-3}\,s^{-1}$ was about an order of magnitude lower than for symmetrical cycling at $10^{-3}\,s^{-1}$. A tension dwell of 600 s with a symmetrical cycle or slow–fast ratio of 10 had an intermediate effect. A compression dwell added to a symmetrical cycle at $10^{-3}\,s^{-1}$ had no effect.

The term "unbalanced cycling" may also be extended to apply to unequal dwell periods. Ellison and Paterson (1976) showed that equal compression-tension dwells in 1CrMoV at 565°C reduced intergranular cracking to a negligible amount. In a continuation of the work (Plumbridge *et al.*, 1980) void nucleation was envisaged as a function of net (microstructural, see Section 5.02.3.5.1) ratcheting which was zero for balanced cycling and *pro rata* for imbalanced types (a 30/3 test gives one quarter of the strain accumulation of a 30/0 test, where the first number gives the tensile dwell in min, the second giving the compressive dwell). In this work a 5% load drop was taken as failure. Interrupted tests (at 10%, 20%, 50%, and 85% of life) for a 30 min tensile dwell showed that the maximum internal crack length was 1 μm, 10 μm, 250 μm, and 570 μm, respectively whereas for 30 min compressive dwell interrupted tests (at 30% and 80% of life) the values were 144 μm and 200 μm, respectively. Surface crack behavior was similar for all tests. Supporting evidence was provided of linking cavities in tension-hold tests. However, the authors considered that creep and fatigue existed as separate processes in these tests, i.e., there was little interaction, unlike the case for austenitic steels.

A "service-type" cycle includes compressive and zero-strain dwell (Granacher and Scholz, 1992). Specimens of low alloy steels and one austenitic steel were tested in the range 525–600°C. They were of 10 mm diameter and testing ceased at 1.5% load drop of stress from a "linear course." This corresponded to a crack (surface) length of 0.5–1 mm. The longest test lasted 3.2×10^4 h. Prior creep reduced the number of creep–fatigue cycles to failure and by the same token prior creep–fatigue reduced the time to failure in a creep test. The Robinson/Miner rule was found to be successful provided that the complex hysteresis loop was partitioned according to the behavior of a "back stress" and cyclic softening (apparent in the long-term tests) was taken into consideration. The earlier work (Granacher and Scholz, 1992) was extended (Scholz *et al.*, 1998) to a three-package cycle which included a dwell at peak tension simulating cold, warm, and hot starts. and included data on a martensitic 12Cr1MoV steel at 550°C. Specimens were 10 mm diameter and the 1.5% drop in stress range at "failure" again corresponded to a crack depth of 0.5–1 mm. Some tests with extended dwell periods and total strain ranges below 0.4% had not failed by 7.1×10^4 h. Time summation was used, modified by an "effective stress," being the difference between the external stress and an internal stress. Such long-term tests are really cyclic creep tests, i.e., for assessment purposes, monotonic creep tests modified by a correction factor depending on the *R* ratio (Granacher *et al.*, 2000).

5.02.3.5 Mechanisms and Predictions from Metallographic Examination

5.02.3.5.1 Observations

In symmetrical continuous cycling, little opportunity is available to elucidate mechanisms from subsequent metallographic examination, cracking being generally transgranular or intergranular according to strain rate. For example, intergranular cracking occurs for symmetrical slow strain-rate cycles, but no wedge-type cracks or cavities appear since in this instance grain boundary sliding is reversible (Yamaguchi and Kanazawa, 1980). Asymmetrical hysteresis loops of whatever nature are more enlightening because the period of tensile stress relaxation, say, or creep produces net cavitation and grain boundary deformation which is not reversed during the fast excursion to the opposite peak;

creep-induced plastic strains are replaced instead by matrix deformation. Therefore, ratcheting occurs over many cycles on a microscale but the specimen as a whole remains within the imposed total strain limits. Such "microstructural ratcheting" (Ellison and Paterson, 1976) indeed forms the basis of the strain range partitioning method (Manson *et al.*, 1971). Damage can be measured as a percentage of cracked or cavitated boundaries or as net grain boundary sliding. To some extent, therefore, this section complements Section 5.02.3.4 on unbalanced cycling. Hales (1980) has found a direct correlation between total length of cavitated boundary and accumulated creep strain in 316 steel at 600°C. Plumbridge (1987a) has ably reviewed the contribution that metallography can make for high temperature lifetime assessment.

Working at relatively high strain ranges between 2% and 4% on a low alloy steel at 560°C, it has been shown (Nam *et al.*, 1988) that an increase in endurance (by a factor of 1.7) occurred when tests were conducted in argon but that the effect disappeared when a tensile dwell of 30 min was introduced. It was hence argued that internal cavitation damage was not affected by oxidation. As the growing crack met damaged material it was demonstrated that unstable growth occurred at depths of about 1.3 mm, 0.4 mm, and 0.15 mm for continuous cycling, 1 min dwell, and 5 min dwell, respectively. Good agreement was obtained with a model incorporating a continuous nucleation of cavities with plastic strain, growth by vacancy diffusion (see later) and the Tomkins/Wareing model (Tomkins and Wareing, 1977) for unstable growth. By SEM examination of 316 stainless steels subjected to LCF, a paper by Nakazawa *et al.* (1992) confirms that as the life reduction factor due to dwell decreased, so did the ratio of intergranular fracture increase. The effect was not observed in one cast of low carbon 316 at 550°C.

In type 304 steel at 800°C Ohtani *et al.* (1996) found that the orientation to the stress axis of most intergranular creep–fatigue cracks was 0° at the surface, 30–35° over the first 300 mm or so of cracking, reverting to 0° internally. There was also a one-to-one relation between surface and internal crack behavior, e.g., as the surface initiation and propagation rates increased with decreasing tension-going strain rate, so did the internal cracks respond (data were inferred from careful crack counting and cycles to failure).

Presenting data on 1(1/4)Cr(1/2)Mo steel, type 304 and 321 and alloy 800 in the range 550–700°C it was shown (Yamaguchi and Nishijima, 1986) that there was a direct correlation of the fraction of intergranular "facets" on the creep–fatigue fracture surface and the ratio N_h/N_f which was effectively independent of material, hold time, and temperature. This was taken as an index of creep damage and the data superimposed on a similar plot of creep rupture ductility against creep-only facets. There was thus a correlation between creep–fatigue life and rupture ductility. Such data support the rule according to the simple Priest–Ellison model (Priest and Ellison, 1981, 1982).

Tests on 316L steel at 600°C with tensile dwells up to 7 h showed a progressive decrease in endurance (Rezgui *et al.*, 1981) whereas dwell tests up to 10 h at 650°C showed less of an effect—in fact, there was a saturation in endurance after 6 h. The authors attributed this to instances of intergranular cracking reverting to the transgranular mode and to the coarsening effect of growing precipitates. An unspecified amount of elastic follow-up took place in these tests. Working on four different casts of similar steels at 570°C and 625°C at dwell times up to 24 h, Wareing (1981) similarly demonstrated this effect, i.e., two casts tended to restore endurances to continuous cycling values after a critical dwell time. Thermal aging effects, coupled with the influence of grain-boundary carbide-precipitation spacing on subsequent crack growth behavior, were held responsible.

Using 7 mm diameter specimens of alloy 800 H at 750°C, Mu *et al.* (1994) performed interrupted tests at 30%, 50%, 80%, and 90% of life. For a balanced cycle, an increase in dwell time progressively decreased endurance. However, employing a "slow–fast" type of cycle which had a constant tension hold time of 1,000 s, an increase in the compressive hold time was found to increase endurance. With a tension dwell, cavitation was present at 30% of life; the density had increased by 60% life, and "w-type" cavities had formed by 80% of life. It was established that compressive strain was indeed important for cavity nucleation by decohesion of the carbide-grain boundary interface or by carbide fracture. Balanced cycling did not permit cavities to open. In all tests a surface crack some 1.5–2 mm long formed, independent of cycle shape, obeying Equation (52) (to be discussed later) with a characterizing value of $B = 1.17 \times 10^{-2}$ for a plastic strain range of 1%. Typically this crack grew much faster after 60 cycles in an asymmetrical tension hold cycle due to the interaction with the internal crack and damage.

At a temperature of 850°C, compressive dwell (6 min and 60 min) tests together with symmetrical tension-compression dwells of

6 min have been made on alloy Inconel 738LC (Yamamoto and Ogata, 2000). The results were augmented by some *in situ* observations *in vacuo* (10^{-5} torr) on miniature specimens. In air the 60 min compressive dwells as well as the balanced type were about four times as more damaging than the tension dwells. Under vacuum, multisite wedge cracks and grain boundary cracks 50–200 μm long initiated after a few cycles of tension dwell, with evidence also of surface cavity damage. With compression dwell there was a single main crack 100 μm long propagating through the grain boundaries at 20% of life. It was clear that accumulation of grain boundary sliding was playing a part and a damage equation was proposed of the form:

$$D_c = [1 + N(C_1 d_{sl} + C_2 d_c)]d_f \quad (4)$$

where D_c is cyclic damage, C_1 and C_2 are constants and the suffixes on the partial damage components, *d*, are due to sliding, creep, and fatigue, respectively. No sliding was observed in continuous cycling (there was a single surface crack at 50% life), but the authors did not comment on the degree of net sliding in balanced dwells and thence the level of damage in the above equation.

Creep–fatigue ($R = -1$) tests with tensile hold time (10 min or 60 min) were conducted on 304 steel at 600°C together with a constant load hold at 147 MPa (Arai *et al.*, 1996). SEM observations were conducted every 20 h and discrete cavities were observed after 24 h. The grain size was 40 μm and sectioning part way through the test confirmed that there were fewer cavities internally and that microcracks had grown to half the grain boundary length. The cavity growth rate (10^{-11} m s^{-1}) was faster in creep–fatigue compared with that in pure creep (5×10^{-13} m s^{-1}). Although the technique could detect increases and decreases in cavity length, it was not possible to determine whether net growth occurred in the dwell itself or by accumulated strain during the (tensile) ramp periods. In creep, the cavities coalesced in the classic manner whereas in creep–fatigue coalescence was assisted by "unstable debonding of the grain boundaries." Cavity growth rates were modeled for pure creep using diffusion-controlled (Hull and Rimmer, 1959) and constrained diffusion-controlled (Needleman and Rice, 1980) mechanisms but a similar model was not undertaken for creep–fatigue.

5.02.3.5.2 Modeling cavitation damage

In the absence of reliable fundamental theory, metallographic assessment of "damage" (away from the major crack) is a powerful adjunct to the phenomenological approach, and the ideas were developed by Levaillant and Pineau (1982) and Levaillant *et al.* (1988) following the earlier work of Hales (1980). The first attempt used the parameter D_s, the fraction of intergranular rupture sites on the fracture surface, which was observed to increase with increasing dwell time in several austenitic steels at 600°C. However, the parameter D_m was preferred, which was the ratio of the length of damaged boundaries per unit area to the average grain boundary length per unit area. The parameter D_m increased with dwell, peaking at 0.15 for dwells up to 200 min and total strain ranges of 2%. Damage was found to increase linearly with cycles and so the "metallurgical" damage per cycle, D_{mc}, is given by

$$D_{mc} = \frac{D_m}{N^*} \quad (5)$$

where the asterisk denotes cycles involved in creep–fatigue, as will be used later. In Section 5.02.6.4 on crack growth we also encounter the empirical damage per cycle, D_c, given by the term $(1/N^* - 1/N)$ where N refers to continuous cycling conditions. Thus by linking observed damage with cycles, the following empirical connection was made (Levaillant and Pineau, 1982):

$$D_c = kD_{mc}^p \quad (6)$$

where typically $k = 2.26$ and $p = 0.923$. Equation (6) was found to be effectively independent of strain level, dwell time and material, at least for the two casts of 316 steel investigated.

Further progress was made by relating D_{mc} (and hence D_c) with the stress drop during relaxation, σ_r, in a manner similar to that of Hales (1980). This was also expressed as a power law:

$$D_c = j\sigma_r^q \quad (7)$$

which again was found to be insensitive to dwell time, strain range and material. The advantage of Equation (7) is that stress drop values and endurances taken from the literature and be linked to the damage per cycle, D_c, though for some austenitic steels Equation (7) breaks down due to aging of the microstructure affecting cavity growth (Levaillant *et al.*, 1988).

In the LCF regime, and for austenitic steel at least, there is thus a good basis for linking metallurgical damage with mechanical response. Further experiments at 650°C have shown (Dabo *et al.*, 1986) that Equation (7) is also independent of temperature and, significantly, the term D_c in compression dwell is found to be four to five times less than in tension. Also, Pineau (1983) showed that the relation between metallurgical damage accumulation and stress relaxation range could be unified for different casts of stainless steel at 600°C provided the early relaxation was included, and not partitioned.

5.02.3.5.3 *Other models*

Most modeling in LCF fatigue is undertaken in a rather different context, which is to predict the shape of the hysteresis loop from given internal state variables and back stresses which enable, e.g., the Bauschinger effect and cyclic hardening or softening to be quantified. Such considerations are outside the present scope. Dyson (2000) notes that a full lifetime model would also take particle aging and dislocation–dislocation interactions into account. Lifetime formulas have been proposed which involve materials-sensitive parameters such as the Burgers vector, shear modulus dislocation cell wall spacing and irreversible slip (Chung *et al.*, 1990) but these have only been applied to 316 steels up to 427°C.

By considering the irreversible action of dislocation pile ups during cycling, Tanaka and Mura (1981) derived a condition for crack initiation when the stored energy of accumulated dislocations reached a critical value. Their expression may be rearranged to give:

$$N_i = \frac{4\pi(1-v)W_i}{G\Delta\gamma^2} \tag{8}$$

where v is Poisson's ratio, G is the shear modulus, $\Delta\gamma$ is the shear strain range, and W_i is the accumulated failure energy/unit volume at failure, a parameter which will be used several times in this review. Thus for $G = 2\times10^5$ MPa, $\Delta\gamma = 0.01$, and assuming $W_i = 1\,\mathrm{J\,mm^{-3}}$, then $N_i = 440$ cycles. The "embryo" crack size was not specified but was of the order of "intrusion/extrusion" dimensions (Tanaka and Mura, 1981).

An alternative approach has been developed (Rie *et al.*, 1988; Nam *et al.*, 1988) which uses the following concepts:

(i) a cavity nucleation rate per unit grain boundary area which is proportional to number of cycles and the plastic strain range (so that the cavity spacing forever decreases),

(ii) cavity growth by the Hull and Rimmer (1959) mechanism, and

(iii) a CTOD due to fatigue or creep–fatigue, leading to rapid crack growth at a critical void spacing.

For cavity growth during repeated stress relaxation the critical cycles to failure were given by an equation of the form:

$$N_f = C\Delta\varepsilon_p^{-3/5}\left[\frac{\exp(-Q_{gb}/RT)}{T}\int_0^t \sigma(t)\,dt\right]^{-2/5} \tag{9}$$

where T is the temperature, R is the gas constant, Q is the activation energy for grain boundary diffusion, and C is a constant involving the critical cavitated area, grain boundary diffusion coefficient, and the atomic volume.

In a later paper (Nam *et al.*, 1996) the cavity nucleation factor was modified and a damage function identified such that the term $D_cN_f^{5/3}$ was a constant. The approach appears to be classical and not apparently material-specific but the authors obtained good agreement for endurances given in many of the papers considered in this review. Atomic diffusion theory was used similarly to predict the amount of reversible grain boundary sliding in slow–fast, fast–slow, and balanced cycling in type 304 steel at 700°C (Tanaka and Iizuka, 1988). At a total strain range of 1% for respective endurances of 103, 330, and 360 cycles, the cycles to initiation, N_i, were 26, 46, and 280, respectively. The level of grain boundary sliding per was $\sim 10^{-9}\,\mathrm{m\,cycle^{-1}}$.

5.02.3.6 Mechanism Maps

Following Solomon and Coffin's work (Solomon and Coffin, 1973) on the frequency effect in A286 at 593°C, mechanism maps embracing the creep–fatigue regime became popular in the late 1970s (e.g., Collins *et al.*, 1979; Sidey and Coffin, 1979) and were found helpful in identifying regimes where certain mechanisms could or could not take place for a particular alloy and they prevented undue extrapolation of endurance data so that "mechanism boundaries" were not crossed. However, the early attempts did not distinguish between crack initiation and crack propagation (Taplin *et al.*, 1984). There are several methods of plotting parameters.

(i) A graph of stress amplitude against homologous temperature, T/T_m, indicated the regions where shearing of γ' precipitates, environment-assisted cavitation, intergranular cavitation, and metallurgical instabilities take place in Inconel 718 (Taplin *et al.*, 1984).

(ii) A similar graph of frequency against homologous temperature indicates where the regions for fatigue-dominant, creep-dominant and creep–fatigue interaction occur for several alloys (Taplin *et al.*, 1984).

(iii) Alternatively, for 304 steel at 650°C (Taplin *et al.*, 1984), a graph of stress amplitude against frequency for a given void spacing provides information on transgranular surface crack formation, environment-assisted cavitation, cavity nucleation at stress-relaxed grain boundaries and (at high stresses) transgranular void formation.

In the creep–fatigue regime, Miller (1982) found it more convenient to plot the total strain range against the tensile dwell period for 1CrMoV steel at 565°C. In this way the number of cycles to failure could also be indicated in separating the creep and fatigue dominated regimes. For our purposes, perhaps the most informative plot is Miller's scheme (Miller, 1983) of plastic strain range against tensile strain rate. The distinct regions are:

(i) creep-dominated (geometric constrained cavity growth),
(ii) creep-dominated (unconstrained cavity growth),
(iii) creep–fatigue interaction, and
(iv) fatigue.

An example is given in Figure 17, where it is seen that contours of constant endurance (and by implication, crack depth) are superimposed. These diagrams are analogous to the uniaxial strain rate–ductility plot considered later. It is notable that in an earlier interpretation (Gladwin and Miller, 1982) only the "tensile" plastic strain range in the hysteresis loop (i.e., in the positive strain quadrant) was used but in this later work the full plastic strain range in the tension-going direction was employed.

In a slightly different approach for stainless steels, Kanazawa *et al.* (1988) plotted a three-dimensional map of strain rate versus temperature versus cyclic 0.2% proof stress and were able to demonstrate

(i) the domains between transgranular fracture, intergranular fracture, and oxidation-dominance and
(ii) on the microstructural level, cell type, planar band, and subgrain dislocation arrays and their association with cyclic hardening or softening.

Similarly, Min (1980) in a graph of tension-going strain rate against the compression-going rate has indicated (by "bar" lengths proportional to the fatigue endurance) the regions where grain boundary sliding damage may or may not be restored.

5.02.4 LINEAR DAMAGE THEORIES OF CREEP–FATIGUE

In Section 5.02.2 the background to linear damage as a "reasonable postulate" was described, making use of already existing creep data and relatively easily determined fatigue data for estimation of life under non-steady conditions. Some of the models are discussed more closely in this section, concentrating on those which have been developed after the mid-1990s, in the R5 procedures, for example (Ainsworth, 1996).

Fatigue damage arises predominantly from cyclic plasticity and is related to the total strain range. Creep damage results from creep strain, either as forward creep or as reversed creep,

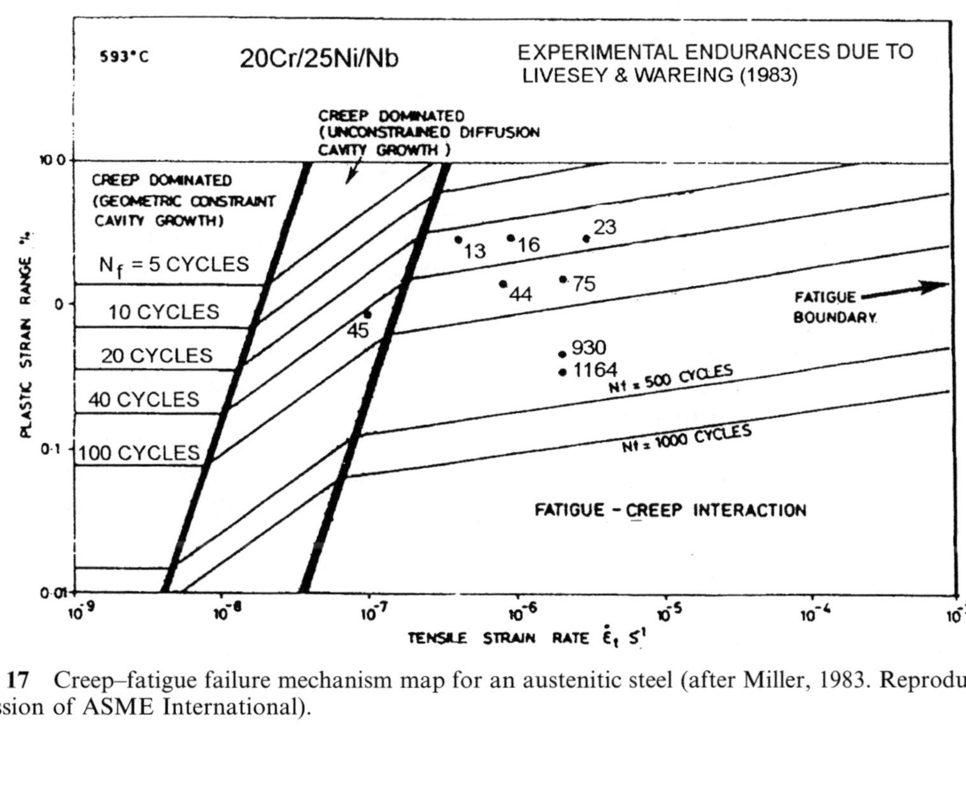

Figure 17 Creep–fatigue failure mechanism map for an austenitic steel (after Miller, 1983. Reproduced with permission of ASME International).

e.g., stress relaxation strain compensated by reversed plastic strain, see Figures 4 and 5. Creep damage is uniformly distributed in the bulk whereas fatigue damage is associated with the initiation of a small surface crack. There is no rigorous mechanistic substantiation for the linear addition of the two damage types to unity—the evidence is empirical and the assumption is common to most of the models. There are, in addition, nonlinear damage accumulation models and oxidation–fatigue models (Bicego *et al.*, 1992; Rémy and Skelton, 1992). Physically, creep strain and the rate at which it is accumulated reflect creep mechanisms giving rise to microstructural damage.

5.02.4.1 Damage Models for Creep–Fatigue Crack Initiation

In design and assessment calculations, the concern is to estimate the number of service cycles to initiate cracking under service creep–fatigue conditions. The assumption is that creep damage (time-dependent) and fatigue damage (cycle-dependent) may be evaluated separately and then added to obtain the total damage due to all service cycles. The simplest damage techniques thus use a linear summation of the separate effects of fatigue and creep, so that Equation (3) may be written in general:

$$\Phi_f + \Phi_c = 1 \tag{10}$$

where the subscripts "f" and "c" in the total damage factor Φ refer to fatigue and creep, respectively. (As will become evident in this review, results are frequently plotted on a damage diagram, see Section 5.02.4.1.10, for example.)

Methods for creep–fatigue lifetime prediction include:

(i) frequency-modified strain range methods,
(ii) strain range partitioning methods,
(iii) double linear damage rule,
(iv) damage parameter techniques,
(v) damage rate approach,
(vi) integrated crack growth methods,
(vii) time fraction methods for creep,
(viii) ductility exhaustion methods for creep, and
(ix) energy expenditure methods.

Some of these models have been examined in depth in previous reviews (e.g., Leven, 1973; Tomkins and Wareing, 1977; Priest and Ellison, 1982; Batte 1983; Miller *et al.*, 1984; Priest, 1989). Models are considered reasonably accurate if, on a logarithmic plot of predicted life against actual life, the data lie within a factor of 2 scatterband. Conservatism is generally favored if the models are used to predict remaining life. Only the basic reasoning behind some models is given here as this review intends to concentrate on ductility exhaustion and energy methods.

5.02.4.1.1 Frequency-modified strain range

The basic equation for LCF endurance (Coffin, 1954):

$$N_f^{\gamma} \Delta\varepsilon_p = \text{constant} \tag{11}$$

where γ is a constant, has been modified (Coffin, 1973) by relating hold time effects to an effective frequency, later models discriminating between damage accrued in the tension-going and compression-going directions according to strain rate. There is a danger of double accounting if the creep strain produced during a dwell is included in the overall plastic strain range of the above Equation (11) (Batte, 1983).

5.02.4.1.2 Strain range partitioning

The strain range partitioning method (Manson *et al.*, 1971; Halford and Saltsman, 1983) is a damage summation method which takes into account the reversed deformation associated with each mechanistic process. Thus any completely reversed inelastic strain range in a hysteresis loop can be partitioned in the following ways:

(i) Tensile plastic flow reversed by compressive plastic flow, $\Delta\varepsilon_{pp}$.
(ii) Tensile creep reversed by compressive plastic flow, $\Delta\varepsilon_{cp}$.
(iii) Tensile plastic flow reversed by compressive creep, $\Delta\varepsilon_{pc}$.
(iv) Tensile creep reversed by compressive creep, $\Delta\varepsilon_{cc}$.

Damage is conceived as the reciprocal of the respective cycles to failure, so the basic equation becomes:

$$\frac{1}{N_f} = \frac{1}{N_{pp}} + \frac{1}{N_{cp}} + \frac{1}{N_{pc}} + \frac{1}{N_{cc}} \tag{12}$$

though it should be noted that only three components can exist for a given hysteresis loop. It is immediately seen that the LCF testing requirement is increased in order to obtain the basic curves for an assessment. A large investigation into the applicability of the rule in the late 1970s showed reasonable correlation for austenitic steels but not for ferritic steels (Brinkman *et al.*, 1978; Priest, 1989).

The model has been refined to "weight" each of the functions in the right-hand side of Equation (12) and further to "normalize" the strain range components where there is an imbalance in each half of the cycle (Batte, 1983). A version in terms of total strain range has been produced (Halford and Saltsman, 1983). The method has found application in the aeronautical industry, and in an example on 316 steel (Miyahara and Tokimasa, 1996), small crack growth curves predicted by the strain-range-partitioned crack growth rate equations were used as damage accumulation curves. The method required "two-step" tests of the "cp" and "pp" type on smooth specimens accompanied by crack growth tests on compact-tension specimens.

5.02.4.1.3 Double linear damage rule

Halford (1997) has undertaken a critical review of the linear damage rule which has been shown wanting in cases where load combinations differ by more than an order of magnitude, e.g., flight missions. He regards damage as the progress of a dimensionless crack a/a_f where a_f is a final depth defined by the user. A double linear damage rule is introduced where it is applied firstly to micro-crack (phase I) initiation and then to macro-crack (phase II) initiation. This is a useful concept, which does not appear to have been applied to the creep–fatigue interaction, see also Section 5.02.9.

5.02.4.1.4 Damage parameter techniques

In this model, material degradation occurs through a mechanically defined damage variable, introduced originally by Kachanov (1958) and Rabotnov (1969) for creep failure. Loss of mechanical properties is defined on a continuum basis, damage accruing as a function of effective stress, e.g.,

$$\frac{d\omega}{dt} = \frac{\sigma}{1-\omega} \tag{13}$$

Damage equations can be written in terms of strain, and modifications made for transgranular and intergranular damage (LeMaitre and Plumtree, 1979). There are broad similarities with the integrated crack growth methods (Cailletaud and Levaillant, 1984). The approach will not be considered further here since a similar damage term is incorporated into the ductility exhaustion method, see Section 5.02.4.4, and internal damage variables have been reviewed previously (e.g., Chaboche, 1982). Similarities between the equations of continuum damage mechanics and those derived from micromechanisms have been explored by Ashby and Dyson (1984).

5.02.4.1.5 Damage rate approach

Majumdar and Maiya (1980a, 1980b) obtained evidence that a growing crack in damaged material decelerated as cavities sintered under a compressive dwell. They proposed a set of equations describing the effect of current cavity size on the crack growth rate da/dN. A steady increase or decrease in crack growth rate occurred according to cavity shrinkage or growth. The model in its final interactive form required the determination of several constants. Priest and Ellison (1982) applied the model to their own dwell-time endurances on 1CrMoV steel at 565°C and obtained a very good correlation, commenting that the method seemed promising as a universal approach. Pineau (1983) drew strong analogies with his own approach, which involved the use of a creep damage per cycle, D_c, a parameter which will be used later. However, the method does not seem to have been widely applied, possibly because of the time-consuming determination of the experimental constants (Priest, 1989).

5.02.4.1.6 Integrated crack growth methods

In the original Tomkins model (Tomkins, 1968), the crack growth rate in LCF is given by

$$\frac{da}{dN} = \Delta\varepsilon_p D \tag{14}$$

where D is the size of the plastic zone at the crack tip in turn given by

$$D = \left[\sec\left(\frac{\pi\sigma}{2\sigma_{UTS}}\right) - 1\right]a \tag{15}$$

where a is current crack depth, σ is the maximum tensile stress in the cycle, and σ_{UTS} is the UTS of well-cycled material. Equation (15) can thus be integrated to give a lower bound endurance (cycles to propagate from a depth a_o).

Subsequent work involving tensile dwells in a range of austenitic steels at elevated temperatures (e.g., Tomkins and Wareing, 1977; Wareing, 1981; Livesey and Wareing, 1983; Wareing *et al.*, 1985, 1986) have indicated striations on near-surface cracks and cavitation in the bulk. Net cavity growth occurs as the test progresses and the growing crack interacts with damaged boundaries. Unstable growth takes place when half the CTOD equals the inter-cavity spacing

$$\tfrac{1}{2}\delta = \lambda - p \qquad (16)$$

where λ is the distance between cavity centers and p is the cavity diameter and

$$\frac{\delta}{2} = \ln \sec\left(\frac{\pi\sigma}{2\sigma_{UTS}}\right)\left[\frac{2\sigma_{UTS}}{E} + \left(\frac{2\Delta\varepsilon_p}{1+\beta}\right)a\right] \qquad (17)$$

These last two equations thus define a critical depth for rapid crack growth.

The above relations have formed the basis of the Tomkins/Wareing approach, where assessment is performed in terms of plastic strain range. An example is given in Figure 18. A family of lines is indicated for $\lambda - p = 0.5\,\mu m$, $1\,\mu m$, and $10\,\mu m$ for a starting crack depth of $10\,\mu m$. Experimental values of the parameter were of the order of $1\,\mu m$.

5.02.4.1.7 Time fraction

The concept of time fraction was discussed in Section 5.02.2.3 and it is open to various interpretations. Clearly, if peak stress is used as a representative stress under a strain-controlled relaxation condition, then the time fraction for a given cycle will be overestimated as deduced from the stress rupture plot. In order to obtain good agreement in the life prediction of 0.5CrMoV turbine casing steels at 550°C with tensile dwells of 16 h Batte *et al.* (1978) regarded the first 0.5 h as constituting the "fatigue" component, and the representative stress during the dwell itself as the average of:

(i) the relaxed stress after 0.5 h and
(ii) the further relaxed stress at 16 h.

A similar scheme for separating the cycle and time-dependent components of damage had been adopted earlier by Marshall and Cook (1971), who identified a time on the relaxation curve for several materials where fast relaxation was virtually complete, independent of the starting stress. In the case of Batte *et al.* (1978) more sophisticated averaging procedures in the "remaining" dwell time did not alter the value of the effective stress.

Mixed success has thus been achieved with the life fraction rule (Priest, 1989), a major factor no doubt being due to the assumption of "virgin" creep rupture data whereas it is known that creep life can be affected by periods of prior LCF.

5.02.4.1.8 Ductility exhaustion

Edmunds and Whites' (1966) idea of summing ductilities was taken up by Priest and Ellison (1981) who further took into account the strain-rate dependence of accumulated strain. (An example based on the current interpretation is given in Section 5.02.4.5.) Hysteresis loops were partitioned into "plastic" and "creep" components of strain. If the creep component was dominant, the endurance was simply given by

$$N_c \Delta\varepsilon_c = \varepsilon_f \qquad (18)$$

where $\Delta\varepsilon_c$ is the creep strain per cycle accumulated below some critical value and ε_f is the material ductility ("elongation to failure"). A very similar relation applied for the "fatigue" component where the "fatigue

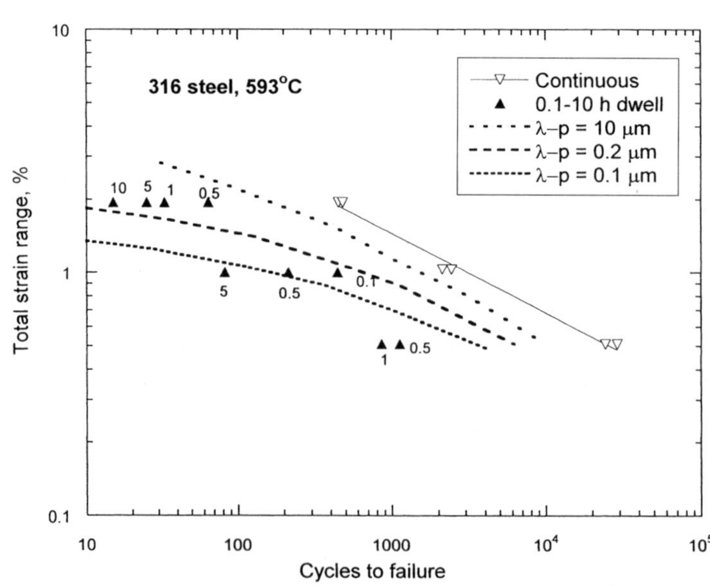

Figure 18 Predictions of creep–fatigue endurance from Equations (14) to (17) (after Wareing *et al.*, 1985).

ductility" could be identified from Equation (11), e.g., as the plastic strain range causing failure in one cycle. It was argued that, because creep and fatigue damage accumulated independently in their material (1CrMoV steel at 565°C), the mechanisms competed and failure was defined by the equation that predicted the least number of cycles. Agreement between predicted and actual endurance was within a factor of 2. In some later work on 316 steel at 625°C, however, an interaction between creep and fatigue damage mechanisms was identified in some tests (Priest *et al.*, 1983) and the "additive" form of the equations was found to give very good predictions:

$$\frac{\Delta\varepsilon_c}{\varepsilon_f}+\frac{\Delta\varepsilon_p}{\varepsilon_{fat}}=\frac{1}{N_f} \qquad (19)$$

Support for the ductility exhaustion approach was provided by the micromechanistic models of Miller *et al.* (1982). They showed that the ever-decreasing strain rate during stress relaxation passes through three regions which control the growth of grain boundary cavities:

(i) plastic deformation growth,
(ii) unconstrained diffusion growth, and
(iii) geometrically constrained growth.

The latter is the most damaging region because the reference ductility is limited to that given by Dyson (1979):

$$\varepsilon_f=\frac{\pi\lambda}{6d} \qquad (20)$$

where λ is the void spacing and d is the grain size. Thus taking a typical cavity spacing of 3 μm and a grain size of 100 μm, the ductility is limited to 1.6%. On a very simple model, if the relevant creep strain produced in one cycle of relaxation is 0.04%, then the permitted number of cycles is $1.6/0.04=40$.

There is no doubt that there is a direct correlation between the total amount of damaged boundary and the extent of creep strain accumulated during tensile hold periods, as demonstrated by Hales (1980). Successful application of cavity growth models in prediction, however, depends directly on an accurate knowledge of cavity spacing which could well change during the lifetime.

5.02.4.1.9 Energy expenditure

Interest in energy methods for creep–fatigue application were rekindled in a paper by Ostergren (1976). This author proposed that only the "tensile" hysteresis damage should be used on the grounds that the crack is open and is thus a measure of the energy put in to the material during crack propagation. Taking the data of Leven (1973) for 1CrMoV steel at 538°C, he showed that on a plot of tensile energy/cycle against cycles to failure (complete specimen rupture) the data fell into two groups according to cycle period (1 min continuous or 1 h dwell). A variation of the frequency-modified approach (Coffin, 1973) was then used to unify the data. Data for cast IN738 and René 80 both at 870°C could be, respectively, correlated without using a frequency modification, but the method did not distinguish between tension and compression hold times. Leis (1977) claimed that energy-partitioning methods have advantages over strain range partitioning, e.g., in that they can take account of mean stress and multiaxial effects. A further development was made in terms of a total energy partitioned into elastic, plastic, and creep components and good correlation obtained for various forms of cycling in 2(1/4)Cr1Mo steel at 538°C, Waspaloy and Astroloy at 650°C and MAR M002 between 850°C and 1,040°C, e.g., though the data did not always fall onto a single line (Radhakrishnan, 1982). A more complex model that incorporates strain range partitioning with the Ostergren (1976) model has been proposed (Duan *et al.*, 1988). (The ideas have been extended for thermomechanical fatigue of a range of superalloys (Radhakrishnan, 1996).) In other work, partitioning the energy according to the "strain range partitioning principles" did not change the merit order of endurances, these being in the order pp, cc, cp, and pc, i.e., pc cycling was the most damaging for a stainless steel at 600°C and a wrought Ni-base superalloy at 700°C (He *et al.*, 1983).

The repeated energy generation in successive hysteresis loops is mostly dissipated as heat and therefore cannot be linked directly with a damage mechanism. It is merely a device which is an indication of the much smaller damage that remains in the specimen with each passing cycle. There is some evidence, however, that the total energy expenditure to a given failure criterion is a constant, independent of strain range (Skelton, 1991). This is examined further in Section 5.02.4.6.

5.02.4.1.10 Comparison of models

Almost without exception each model, whether phenomenological or based on a physical mechanism, is an *ad hoc* approach to meet particular materials, temperatures, environment, and service cycle requirements and it is not always easy to make a direct comparison.

For René 95 at 650°C, e.g., Bernstein (1982) found that the strain range partitioning, frequency-modified, Ostergren, and damage-rate models all failed to correlate the fatigue life within a factor of 2.

Beech and Batte (1984) refined the models used in the earlier assessment (Batte *et al.*, 1978), using the Priest and Ellison (1982) argument that matrix damage processes cannot be reversed by grain boundary damage processes and vice versa. The strain range partitioning method was further subdivided into domains (defined by transition strain rates) of matrix-, wedge cracking-, and cavity-dominated creep deformation and fracture. (As a general comment, it appears to the writer that in order to maintain a factor of 2 between prediction and experimental values at low total strain ranges (<0.3%) all methods need extra sophistication which detracts from their original simplicity.) At the time, the authors concluded that all methods became unreliable, ductility exhaustion in particular producing conservative results just when endurance with 16 h dwell was showing signs of increasing, Figure 19.

Danzer and Bressers (1986) have pointed out that when partitioning parameters, even in continuous cycling, attention should be paid to the inelastic strain rate which should correspond to the minimum strain rate in the corresponding creep test. Referring to data on Waspalloy at 600–650°C and PM Astroloy at 730°C, the following mechanisms were found to be strain-rate dependent: dislocation motion, $M_{23}C_6$ precipitation on grain boundaries, dynamic strain aging, and oxidation—the latter referring to Reuchet and Rémy's (1983a, 1983b) work in MAR M509 where a 100-fold increase in the oxidation of interdendritic carbides was found compared to the unstrained condition. These authors quoted the constant crack growth rate regime (<600 μm, see Section 5.02.6) as an instance where damage is independent of previous damage.

In a series of strain-controlled tests with 1 h tensile dwell at 600°C on 9 Cr alloys, Taylor *et al.* (1994) showed that the endurance lay within the scatterband of continuous cycling data. The experimental findings were supported by application of the time fraction rule (fully integrated version). The creep damage fraction was found to be 0.01 in the longest tests, demonstrating the lack of an interactive effect. In contrast, the authors quoted other data on the same material where a reduction in endurance was demonstrated, but since the durations were very long (up to 1.5×10^4 h) they proposed that oxidation played a part in these latter tests.

Wareing (Wareing, 1999) considered AISI 316 and Modified 9Cr steel used in thermal nuclear reactors, comparing the alternative

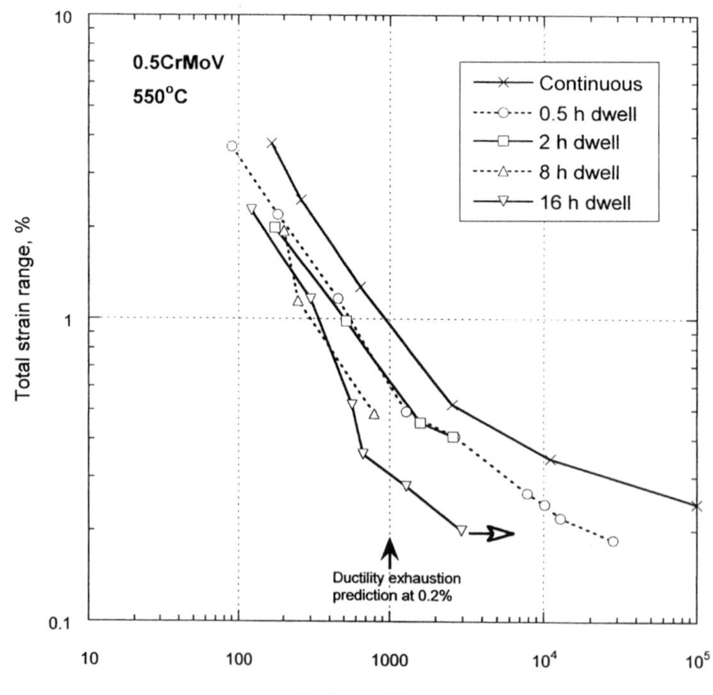

Figure 19 Dwell effect in a ferritic steel (from data in Batte *et al.*, 1978 and Beech and Batte, 1984).

strain-based approach to the time-based approach used in Code Case N47 and RCC-MR. His findings on austenitic steels are summarized below.

In cylindrical specimens, initiation of cracks was transgranular, becoming intergranular for low strain rates or tensile dwells. Whereas a single crack initiated at plastic strain ranges <0.3%, multiple initiation sites occurred in the same plane at higher strains, coalescing to a single crack. For higher strains still, circumferential cracking occurred. At a plastic strain range of 0.17% tensile or compressive dwells gave endurances around 10^4 cycles compared with fatigue "threshold" values out to 2×10^5 cycles in continuous cycling (316 steel at 550°C). Regarding the number of secondary cracks, initiation up to depths of 0.7 mm was shown to be particularly easy for both dwell types, the compressive mode giving transgranular cracking. A mean stress of 10 MPa occurred in this cycle. The author concluded that endurance with short dwells is dominated by crack growth since initiation to 5 μm depth is again easy.

At a total strain level of 0.6%, there was a factor of 32 reduction in endurance for a 48 h dwell time (cast BQ at 570°C). The author compared the R5 integrated strain-based approach during relaxation with a similarly integrated time-based calculation:

(i) At high (>0.8% total) strain range (strong interaction), small creep damage induces low fatigue failure damage.

(ii) For low strains creep damage at failure is unity, again with low fatigue damage.

This was supported by metallography and an L-shaped diagram was necessary to bring all the failure points outside the envelope, Figure 20(a) ("outside" because the specimens had failed, see Section 5.02.4.7). For the time-based diagram, Figure 20(b), there was no consistent relation between damage at failure and failure mechanism. Creep damage values exceeded unity and at 0.6% total strain range there was a factor 17 scatter compared with two for the strain-based approach. Neither linear nor the ASME (0.3, 0.3) lines were considered appropriate, see Figure 20(b). Hence strain together with an integrated crack growth law was favored for fatigue damage, taking 0.1 mm (the grain size) as a starting depth and 0.5 mm (detectable by NDT) as the final depth. For each material, including welds, upper or lower bound ductilities were used according to the propensity to cavitate.

Holdsworth (1998) noted that published design and assessment codes are necessarily conservative, but in today's climate of increased performance and reliability, in-house

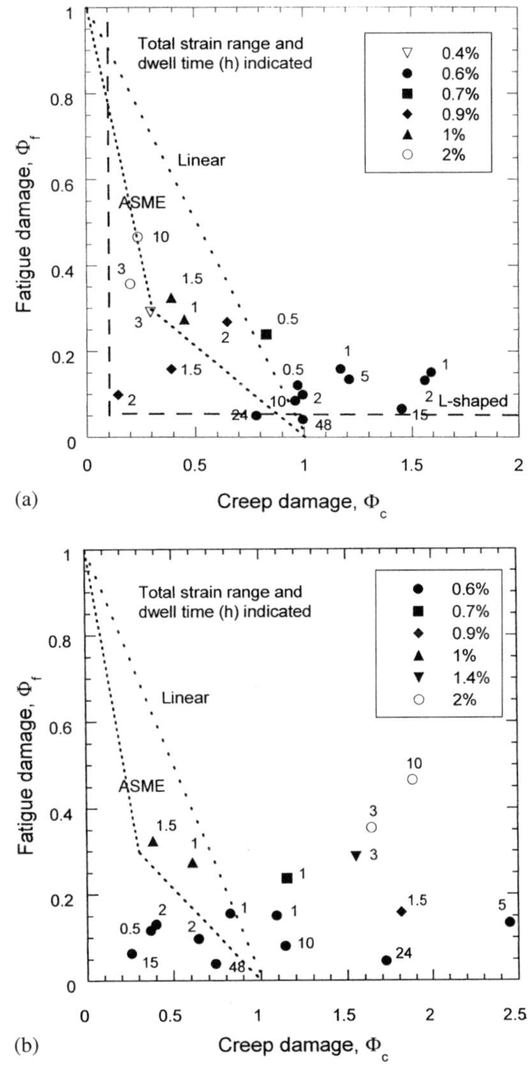

Figure 20 Damage assessment diagram based on: (a) strain and, (b) time (after Wareing, 1999).

procedures have been developed to predict creep–fatigue lifetimes with more precision. Anomalies arise because in time-fraction methods, strong materials resisting relaxation are attributed a high creep damage fraction (long time/low creep strain) but so too do weak materials on a strain fraction basis (short time/high relaxation strain). A variation of the time fraction rule is to partition the stress-rupture curves due to primary and secondary loadings. Tests on large features such as a large SENB specimen or a welded header geometry gave the opportunity to "tune" each model by selecting mean or lower bound properties of the property input parameters.

Comparing the properties of the new 9–11% Cr steels at 600°C with traditional 1CrMoV steels at 550°C, Holdsworth (2001) noted that nowadays (in Europe at least) the maximum dwell that is imposed in an experimental cycle is 0.5 h and creep–fatigue damage rules are used to predict endurance. For a bi-linear rule intersecting at the (x, y) co-ordinates (c, f) on a damage diagram:

$$\Phi_f = 1 - \frac{\Phi_c(1-f)}{c}, \quad \text{for } \Phi_c < c \qquad (21a)$$

$$\Phi_f = \frac{(1-\Phi_c)f}{1-c}, \quad \text{for } \Phi_c > c \qquad (21b)$$

As c and f both decrease, more of the data clearly lie outside the envelope. For ease in calculation, it may be noted that the above discontinuous function can be replaced by

$$\Phi_c^n + \Phi_f^n = 1 \qquad (21c)$$

to a reasonable approximation, see Figure 21, so that when $\Phi_c = \Phi_f = 0.3$ as in the case of the ASME N-47 intersection at (0.3, 0.3), then Equation (21c) is obeyed when $n = 0.58$.

Studies of lifetime assessment methods in the UK, mainland Europe, and the USA are sustained by an impressive body of work undertaken in Japan. For example, Yagi and Kubo (1992) tested 1CrMoV rotor steel, 2(1/4)Cr1Mo plate, 304 plate, 316 plate and alloy 800 bar in the range 500–800°C. For the 316 steel on a damage diagram:

(i) at 550°C ductility exhaustion gave less overall damage than the time fraction rule, all data lying within the envelope (owing to wedge cracking and low ductility);

(ii) the converse applied at 750°C, the time fraction data just straddling the linear line (owing to cavity growth and intermediate ductility); while

(iii) at the intermediate temperature of 650°C most data lay outside the envelope, the time fraction rule predicting far more damage (owing to transgranular fracture and intermediate ductility).

The authors found it necessary to introduce yet a third parameter to the rule ("interaction term") and empirical constants were provided. In an evaluation of the time fraction rule and ductility exhaustion methods (Takahashi, 1998) for a carbon-restricted 316 steel in the range 550–650°C, differences in ductility for two separate casts as a function of strain rate became apparent (not determinable by cursory examination of the stress–rupture plots). The higher ductility cast had consistently longer endurance under dwell conditions in the range 0.1–50 h, some tests lasting 3.5×10^4 h. The life reduction ratio was larger at 0.35% total strain range than at 1% range. Assessment of the ductility exhaustion method for the respective casts was made using reference values of 22% or 37% regardless of strain rate. The strain developed during relaxation was partitioned as "viscous" and "creep" strain, using a forward creep model. The

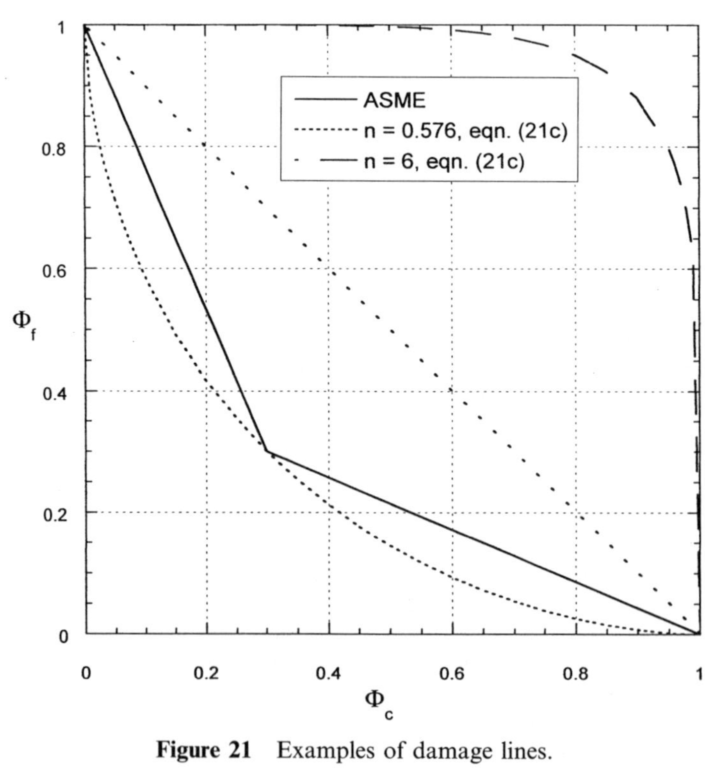

Figure 21 Examples of damage lines.

ductility exhaustion rule could predict lives to within a factor 1.5 whereas for low strain ranges and long hold times, the time fraction rule overestimated creep–fatigue endurance by a factor of 30. If all the strain during relaxation was included, the ductility exhaustion method became too conservative. In a related study, essentially the same conclusions were drawn for two further casts of 316, one a forging, the other a pipe (Takahashi, 1999). It was shown that if the Monkman–Grant (1956) expression for ductility was used, then the method effectively became equivalent to the time fraction rule based on secondary creep data.

A paper by Shimakawa *et al.* (2001) presents a comparison between the predicted creep–fatigue lives using the design Codes DDS (Japan), ASME N-47 (now referred to as subsection NH), RCC-MR, and R5. The benchmark systems were:

(i) a cylinder subjected to an axial temperature distribution,

(ii) an upper ring plate in a reactor vessel,

(iii) an upper tube sheet in an intermediate heat exchanger, and

(iv) a Z-shaped junction in an intermediate heat exchanger.

It may be noted that cyclic stress–strain properties are used in all Codes except for ASME which uses monotonic data and that the time fraction rule is used in all Codes except R5 which uses the ductility exhaustion method. Also the R5 Code:

(i) separates the initiation and propagation phases (whereas the others use rupture of a smooth specimen) and

(ii) uses a total damage factor, Φ, of unity, employing fatigue data directly (whereas the others use $\Phi < 1$ and also apply a further safety margin to fatigue data).

Not surprisingly, the R5 approach was the least conservative, nevertheless erring on the "safe side" by a factor of 2–5. In a similar approach, the accuracy of the strain range partitioning method was checked (Miyahara *et al.*, 2001). Actual ligament stub hole features were alternately heated and cooled between 260°C and 550°C with a 5 min dwell, typically for 2,500 cycles. Crack depths were typically 1 mm after 1,000 cycles (30 mm hole spacing, $\Delta\varepsilon_t = 1.14\%$) and 1.8 mm after 2,800 cycles (30 mm hole spacing with $\Delta\varepsilon_t = 0.49\%$). Supporting "pp," "pc," "cp," and "cc" data on 2(1/4)Cr1Mo steel at 550°C were provided. Two cases of:

(i) compressive creep damage equal to tensile creep damage and

(ii) compressive creep damage being negligible were examined.

It was found that the prediction accuracy of strain range partitioning was within a factor of 1.5. Linear damage methods (time based) on the other hand gave unconservative predictions, especially when compressive creep damage was ignored.

5.02.4.2 Scope of an Assessment

Many of the examples to be considered on creep–fatigue crack initiation and short-crack growth are based on concepts incorporated into the R5 Procedure (Ainsworth, 1996). This procedure is intended for ferritic and austenitic steels (and their weldments) in the nuclear industry but the initiation procedures could in principle be extended to other fields, e.g., superalloys operating at temperatures approaching 1,000°C. The component or structure is assumed to be defect-free but it may contain grooves, notches, or other changes in section. The loading may arise from thermal and/or mechanical sources and produce a residual stress state. Each operational cycle is assumed to induce cyclic plasticity and any creep component during steady running may involve a degree of elastic follow-up.

For an assessment, a procedure would generally follow the steps below:

(i) Determine the loading history of the component or feature.

(ii) Determine cyclic stress–strain deformation loops.

(iii) Obtain creep–fatigue endurances.

(iv) Calculate the total creep–fatigue damage, Φ.

(v) Assess whether crack initiation will occur.

(vi) Perform a sensitivity analysis.

(vii) Report the results.

These sequences are demonstrated in Section 5.02.8, where an actual service failure is retrospectively analyzed. Such an assessment may involve considerable laboratory testing on service-exposed material if similar data are unavailable in the literature.

5.02.4.3 Definitions

In this section, stress relaxation, fatigue, creep, and total damage parameters are defined for an appreciation of the remaining review. As already discussed, the parameters associated with the cyclic stress–strain loop are $\Delta\sigma$, $\Delta\varepsilon_t$, $\Delta\varepsilon_p$ as shown in Figure 22(a). In addition, if a dwell occurs somewhere in the loop, it is necessary to know the starting stress, σ_o and the follow-up factor, Z, given by:

$$Z = \frac{\varepsilon'}{\Delta\varepsilon'} \tag{22}$$

see Figure 22(b). The rate of stress relaxation is given by

$$\dot{\varepsilon} = -\frac{Z}{E}\frac{\mathrm{d}\sigma}{\mathrm{d}t} \tag{23}$$

A well-known equation of stress relaxation is due to Feltham (1960) (though of course other relations may be used, see Conway *et al.*, (1981)):

$$\sigma_o - \sigma = b\ln(at + 1) \tag{24}$$

where t is elapsed time, σ_o is the starting stress, σ is the stress at any time t, and a, b are constants. It can be shown that in the presence of elastic follow-up, Equation (24) is modified to

$$\sigma_o - \sigma = \sigma_o B''\ln\left(\frac{bt}{Z} + 1\right) \tag{25}$$

By differentiation of Equation (25) and substituting in Equation (23) the corresponding strain rate is given by

$$\dot{\varepsilon} = \frac{B''\sigma_o b}{E(bt/Z + 1)} \tag{26}$$

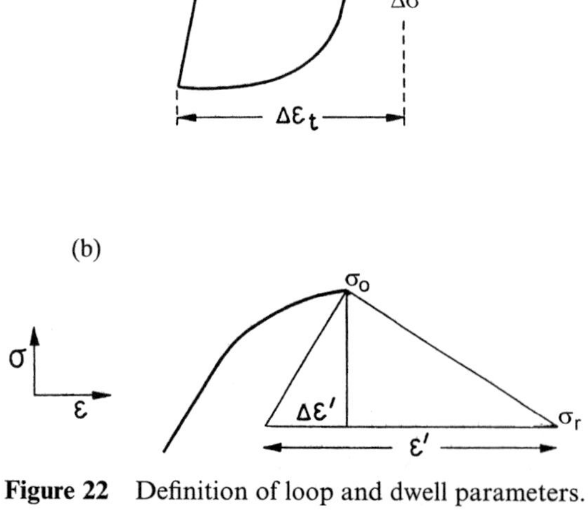

Figure 22 Definition of loop and dwell parameters.

5.02.4.3.1 Fatigue damage

The fatigue damage per cycle, D_f, is formally defined as the inverse of the continuous cycles to failure at some specified criterion. In the R5 procedures, e.g., Ainsworth (1996), at each respective strain range, laboratory failure data, N_f, are converted to an earlier failure criterion, N_o, linked to a specific embryo crack size, a_o, as defined by the user. Details are given in Section 5.02.5. Thus

$$D_f = \frac{1}{N_o} \tag{27}$$

5.02.4.3.2 Creep damage

In the period spanning the mid-1960s to the late 1980s, long-term creep–fatigue endurances were generated, e.g., Batte *et al.* (1978), Batte (1983) and assessments could be made directly from these data, see also Section 5.02.9. For example a 16 h dwell represents "two-shift" operation of power plant which runs steadily all day but shuts down at night, a 60 h dwell would represent a weekend shut down and so forth. Full analyses require data which, e.g., take account of:

(i) the position and duration of the dwell in the cycle,
(ii) elastic follow-up effects,
(iii) the creep strain rate during the dwell,
(iv) frequency effects, and
(v) the effect of prior service exposure.

If only short term dwell or other limited data are available, creep damage may be calculated, as has been discussed, by using existing long term creep failure and ductility data (where the average strain rate is the ductility divided by the time to rupture) and performing stress relaxation tests under cyclic conditions. The traditional damage method uses Robinson (1952) time summation of the form $\sum t/t_f$, but for creep-fatigue problems ductility exhaustion ($\sum \varepsilon/\varepsilon_f$) is generally preferred. Early models assumed a constant reference ductility (e.g. Priest and Ellison, 1981) but it is now appreciated that the ductility of many alloys decreases at low strain rates of testing. An example for a 17Cr8Ni2Mo weld metal at 625°C in forward creep is given in Figure 23, adapted from Wood *et al.* (1988). The form of ductility chosen for inclusion in an analysis (elongation, reduction of area, etc.) depends on the degree of conservatism required. For example, a ductility–strain rate diagram at 550°C shows (Kaneko *et al.*, 1992) that 316MN (low carbon) steel remains on the upper shelf compared with the conventional steel. In both cases, prior exposure to 'cp' type cycling to

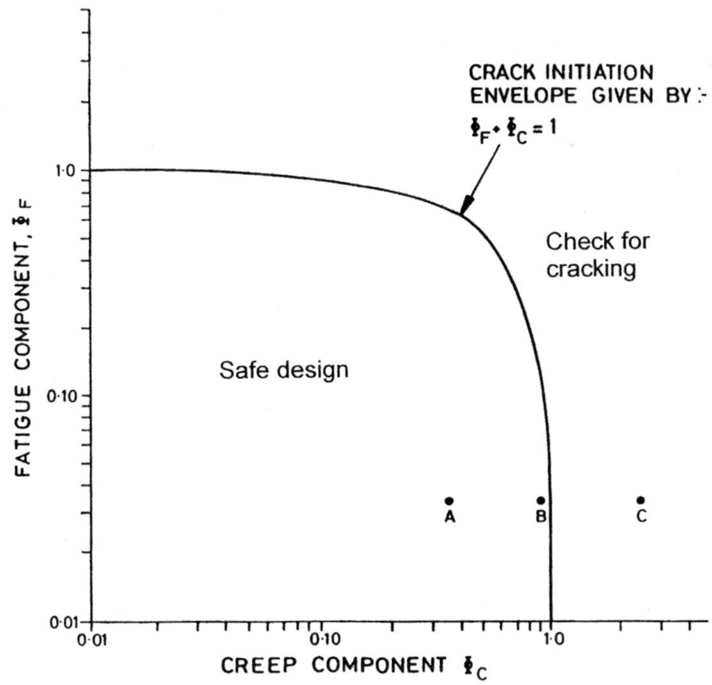

Figure 23 Linear damage diagram plotted in logarithmic co-ordinates, compare Figure 21.

increased fractions of life brought about a corresponding reduction in ductility (as measured by R of A).

During stress relaxation at constant strain, the associated strain rate decreases according to Equation (23) above. Typical examples for the weld metal just considered are $4 \times 10^{-2}\,h^{-1}$ at the start and $4 \times 10^{-6}\,h^{-1}$ at the end (Wood *et al.*, 1988). Consequently, a more sophisticated expression than Equation (18) is required, which calculates the total creep damage during the relaxation period by determining the ductility fraction at each time interval. The creep damage per cycle, D_c, is thus obtained from the expression (Hales, 1983b)

$$D_c = \int_0^{t_h} \frac{\dot{\varepsilon}}{\varepsilon_f \ln(\dot{\varepsilon})} dt \tag{28}$$

If compressive dwells are believed to be less damaging, Equation (28) can be simplified by retaining the upper bound (constant) value of ductility, in which case it reduces to Equation (18), but see Section 5.02.5.5.

5.02.4.3.3 *Total damage*

The total damage per cycle is the sum of Equations (27) and (28):

$$D_t = \frac{1}{N_o} + D_c \tag{29}$$

so that the creep–fatigue endurance is given by the reciprocal value:

$$N_o^* = \left(\frac{1}{N_o} + D_c\right)^{-1} \tag{30}$$

Equation (30) is also the basis for damage factors employed in the growth of short cracks, see Section 5.02.6.

5.02.4.3.4 *Elastic follow-up*

Elastic follow-up in service (defined by the term Z in the laboratory, see Equation (22) and Figure 22(b)), causes an additional forward creep component which has been examined experimentally. In some tests lasting up to 1 yr, Endo and Sakon (1984) tested two CrMoV steel forgings at 500°C and 304 steel at 600°C in a simple machine having an elastic follow-up factor, Z, of 5. The ferritic steel having the highest ductility (80%) did not show a significant effect of dwell times lasting up to 24 h whereas the low ductility version (60% to 20% according to the time of the stress rupture test) showed a reduction of up to a factor of 4. There was limited evidence that the austenitic steel showed a saturation effect for 6 h dwell tests. An attempt was made to improve the life fraction rule by taking cyclic hardening or softening into account (initial value of σ_o) but it made little difference to the outcome. Similarly, tests were conducted on 316 steel at

550°C for dwell times up to 10 h with Z factors of 1 and 3.2–6.2 (Takahashi *et al.*, 1999). For a total strain range of 1%, endurance progressively increased with enforced follow-up, being ~30% for a 10 h dwell. The stress relaxation curve with follow up lay above the "true" relaxation curve by the same amount at this time. The ductility exhaustion method gave good predictions of life and, as may be expected, the creep damage with follow up was always a factor of 3 or so greater than when $Z=1$. However, a partitioning version of ductility exhaustion, which separated a "viscous" component of creep strain, unified the damage independent of follow-up. Life predictions were well within a factor of 2. The authors noted that in plant application (say 200 MPa and 0.35% strain range) elastic follow-up effects only appear after 100 h and the additional creep damage is small. It may be pondered that an effect present in early tests (e.g., Walker, 1963) before the advent of closed-loop control is now deliberately introduced in order to simulate the service situation.

5.02.4.4 Damage Diagram

From Equation (29) the total damage, Φ, resulting from all service cycles of type j is given by

$$\Phi = \sum_j \frac{1}{N_{o,j}} \qquad (31)$$

If the total damage $\Phi > 1$ then initiation of a crack of depth a_o is predicted. Alternatively, the creep and fatigue components may be plotted separately on the axes of a damage diagram. This takes the form shown in Figure 23 when plotted in terms of logarithmic coordinates. Conventionally:

(i) The total fatigue damage (vertical axis) Φ_f for all service cycles is $1/N_{o,j}$.

(ii) The total creep damage (horizontal axis) Φ_c for all service cycles is $D_{c,j}$.

Results of typical calculations are shown schematically in Figure 23 as points A, B, and C. If they lie within the envelope (for the chosen crack size criterion), crack initiation is predicted not to occur.

5.02.4.5 Application of Ductility Exhaustion Method

In the procedure for calculating creep damage, a typical ductility–strain rate plot, as shown in Figure 24, is usually bounded by "lower shelf" and "upper shelf" ductilities, ε_L and ε_U, respectively, and joined with a transitional region. Calculations are then performed in three stages according to the prevailing strain rate. Hales (1983b) pointed out that a neater (continuous) description of the ductility dependence is afforded by a "Fermi" function:

$$\frac{\varepsilon_U - \varepsilon}{\varepsilon - \varepsilon_L} = \left(\frac{\dot{\varepsilon}}{\dot{\varepsilon}_c}\right)^{-\alpha} \qquad (32)$$

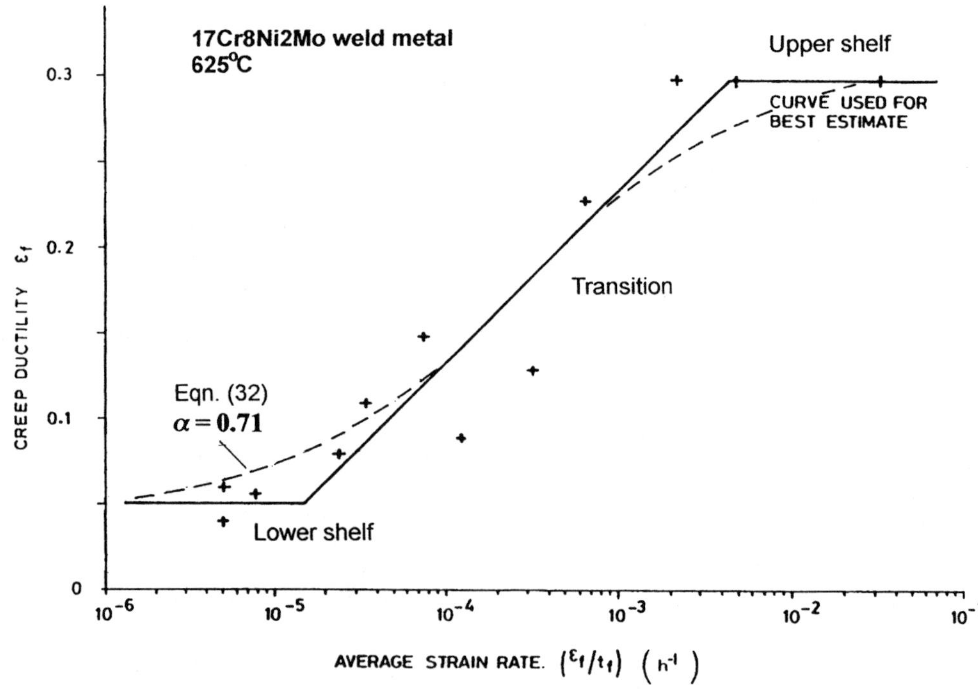

Figure 24 Example of ductility–strain-rate effect (after Wood *et al.*, 1988).

where ε_c is a critical strain rate at which the ductility is equal to the value $(\varepsilon_U + \varepsilon_L)/2$. The term α is a constant that is determined by matching slopes in the vicinity of this mid-range ductility. The higher the value of α, the more step-like the function becomes. The form of Equation (32) for the weld metal data of Wood *et al.* (1988) is shown in Figure 24. In this figure (and as a demonstration of a sensitivity analysis below) we have plotted the actual data in terms of *average* strain rate. This is obtained by dividing the ductility at the end of the test by the total *time* of test, and values are about a factor of 5 greater than the *minimum* strain rate adopted by Wood *et al.* (1988). For many test data this will in practice be the only method of obtaining ductility–strain-rate data since minimum (secondary) creep rates are not often quoted.

Using Equation (32) together with Equations (26) and (28) the required integral is

$$D_c = \int_0^{t_h} \frac{1 + [A\dot{\varepsilon}_c(Qt+1)]^\alpha}{A(Qt+1)\{\varepsilon_U + \varepsilon_L[A\dot{\varepsilon}_c(Qt+1)]^\alpha\}} \mathrm{d}t \quad (33)$$

where $Q = b/Z$, $A = E/B''\sigma_o b$. Hale's (1983b) solution for this relation may be conveniently expressed as

$$D_c = \frac{1}{Q\varepsilon_L A\alpha}\left\{N\alpha \ln(1+Qt) + (1-N) \times \ln\left[\frac{1 + N\{A\dot{\varepsilon}_c(1+Qt)\}^\alpha}{1 + N(A\dot{\varepsilon}_c)^\alpha}\right]\right\} \quad (34)$$

where $N = \varepsilon_L/\varepsilon_U$.

Brinkman (2000) took results from 22 creep–fatigue tests in the range 500–600°C and to test times of 5×10^4 h and they were analyzed using the "Fermi" slope method of Hales (1983b) above. Most of the data fell within a factor of 2 of the predicted cycles though no details were given of the working. The method was very unconservative at 500°C, but metallographic back-up indicated that the cracking was transgranular.

In a creep–fatigue test on 17Cr8Ni2Mo weld metal at 625°C and 0.6% total strain range incorporating a 24 h peak tension dwell, the specimen failed after 377 cycles using the 25% load drop criterion (Wood *et al.*, 1988). Using the three-step (discontinuous) method for ductility exhaustion, these authors' values for total creep damage over all cycles were as follows:

(i) $\Phi_c = 0.46$ when based on reduction of area,

(ii) $\Phi_c = 1.19$ when based on elongation to failure, and

(iii) $\Phi_c = 10.02$ when based on minimum creep rate × time to failure

The latter value is thus based on the Monkman–Grant (1956) interpretation of ductility which is much smaller than the reduction of area, for example. This gives a correspondingly greater value for the total creep damage. Wood *et al.* (1988) reported a typical relaxation stress drop of 87 MPa from a starting value of 167 MPa. From these and similar data it can be shown that in Equation (25), $b = 25$ and $B'' = 0.055$. Substituting in Equation (34) we find $D_c = 3.67 \times 10^{-3}$ so that over 377 cycles, $\Phi_c = 1.39$, compare the values above. This calculation is based on total elongation coupled with an average strain rate.

Wood *et al.* (1988) have listed some of the uncertainties associated with the ductility exhaustion method. These include:

(i) definition of cycles to failure (but see this review),

(ii) scatter in ductility,

(iii) lack of precision and reproducibility of stress relaxation curve,

(iii) categorizing initial rapid relaxation rate as "creep" damage,

(iv) definition of ductility (as shown above), and

(v) constraint (no necking) in creep–fatigue tests compared with forward creep.

Regarding points (iv)–(vi) the mechanistic justification for identifying forward creep with relaxation creep needs further examination. As an example, Beech and Batte (1984) found for a turbine casing steel at 550°C that the exponent n in the equation:

$$\dot{\varepsilon} = A\sigma^n \quad (35)$$

was 15–20 for stress relaxation during the 16 h dwell period in creep–fatigue tests whereas it was 2–4 for static load behavior, see Figure 25. A similar effect was found in a type 304 steel at 550°C (Meurer and Breitling, 1986), see Figure 25. The strain rates are seen to converge only at low stresses. The same trend has also been demonstrated after 5 h tension dwell in a 1CrMoV steel at 550°C (Jeong *et al.*, 1999).

Equation (34) can also be used for dwells in compression, and so the net damage for a hysteresis loop involving peak tension and compression dwells is thus $D_{ct} - D_{cc}$ (Hales, 1984). Strain control tests up to 3,000 cycles on 316 steel at 600°C were carried out with a combination of 30 min/2 min, 30 min/5 min, and 15 min/5 min dwell, the numbers referring to tension and compressive dwells, respectively (Hales, 1984). By use of a ductility–strain-rate plot in the earlier work (Hales, 1983b), cyclic damage factors were evaluated and are given in Table 1. The net damage was then calculated as indicated above and each value multiplied by

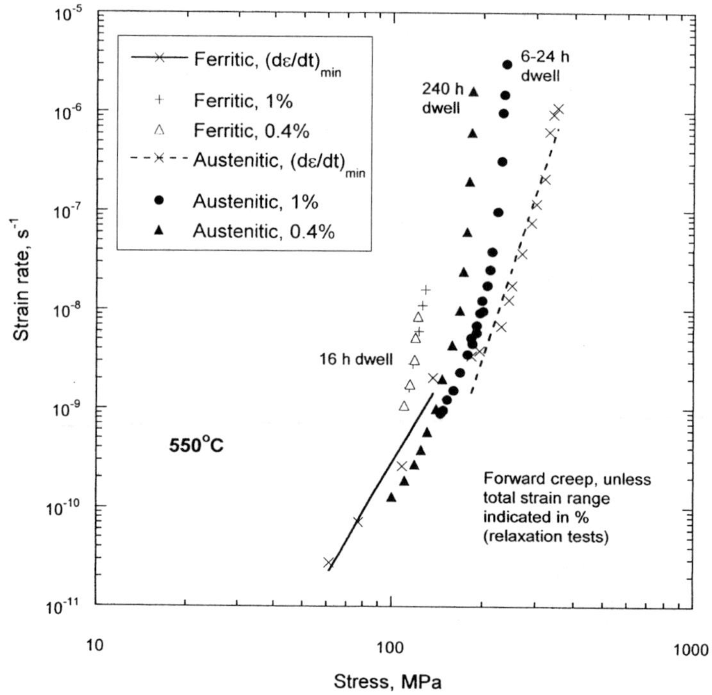

Figure 25 Comparison of forward creep rates with relaxation creep rates in 0.5CrMoV steel (after Beech and Batte, 1984) and 304 steel (after Meurer and Breitling, 1986).

Table 1 Calculated ductility exhaustion damage for respective tensile and compressive dwell periods.

Tension		*Compression*		*Predicted cycles*	
Dwell time (min)	D_{ct}	*Dwell time* (min)	D_{cc}	$(D_{ct}+D_{cc})^{-1}$	$(D_{ct}-D_{cc})^{-1}$
15	9.80×10^{-4}	5	8.2×10^{-4}	555	6,250
30	9.10×10^{-4}	2	6.0×10^{-4}	6,678	3,333
30	1.11×10^{-3}	5	8.80×10^{-4}	502	4,347

Source: Hales (1984).

the number of cycles in the (interrupted) tests. As before (Hales, 1980), the extent of damaged boundary (measured in μm mm^{-2}) was determined metallographically. A plot of this factor against net damage showed good agreement with previous work which involved very long tension dwells. Thus, it was argued, compressive creep strains may be regarded as not contributing to creep damage. Cyclic predictions of life assuming damage to be additive are also provided in Table 1 which shows this assumption to be conservative.

It should be pointed out, however, that to include a degree of conservatism the R5 Code (Ainsworth, 1996) at present provides that compressive damage should be added, but the reference ductility is always referred to the upper shelf, see Figure 24, for example. Hence the damage is reduced according to the simplified Equation (18). This may still be too conservative and an alternative explanation of the compressive dwell effect is given in Section 5.02.5.5.

5.02.4.6 Application of Energy Criterion

5.02.4.6.1 Continuous cycling

We return to Miner's (1945) idea that a material will fail when the net work expended (area of hysteresis loop) reaches a critical value, but now expressed in energy terms:

$$\sum_{i=1} \frac{w}{W_f} = 1 \quad (36)$$

over each cycle of type *i*. Under some circumstances at high temperature, the term W_f has been shown to be a material property, reasonably independent of strain range (Skelton, 1991). In LCF this result may be anticipated since energy dissipation arises as the product of:

(i) the stress range,
(ii) the plastic strain range, and
(iii) the number of cycles.

When the first two parameters are high the endurance is low, the converse applying for high values of endurance. Once the energy value has been established, it may be taken out of the summation sign as follows:

$$\frac{1}{W_f}\sum_{i=1} w = 1 \tag{37}$$

This leads to considerable reductions in experimental testing since W_f does not require to be established at each strain range. There is however some variation with temperature (Skelton, 1991). Typical values of W_f for austenitic and ferritic steels are given in Table 2.

These data were determined as follows. It may be shown that the energy expenditure in a single cycle is given by $\Delta\sigma\Delta\varepsilon_p(1-\beta)/(1+\beta)$. Since plastic strain range and stress range are connected by the cyclic form of the Ramberg–Osgood (1943) relation:

$$\Delta\sigma = A\Delta\varepsilon_p^{\beta} \tag{38}$$

where A and β are constants, this may be expressed as:

$$w = \frac{(1-\beta)\Delta\sigma^{(1+\beta)/\beta}}{(1+\beta)A^{1/\beta}} \tag{39}$$

and so the total energy to failure is simply $N_f w$.

Unfortunately, there is sometimes a variation of W_f with strain range. In Nazmy's (1983) work on IN738LC at 850°C a range of strain rates was used as follows:

"Fast–fast"	a–a	$10^{-2}\,s^{-1}$
"Slow–slow"	b–b	$10^{-5}\,s^{-1}$
"Fast–slow"	a–b	10^{-2}–$10^{-5}\,s^{-1}$
"Slow–fast"	b–a	10^{-5}–$10^{-2}\,s^{-1}$

From the published data it is thus possible from Equation (39) to determine failure energies at specimen separation and these are shown in Figure 26(a) as a function of strain range. It is noted that for the fastest tests, the parameter increases as the strain range decreases. At low strain rates, whether in the tension-going or compression-going direction, an approximately constant lower energy consumption of $\sim 0.2\,J\,mm^{-3}$ is entailed (corresponding to fewer cycles to failure). In this work, Nazmy (1983) also reported cycles to an earlier criterion of a 5% difference in strain measuring signals. Crack depths at this juncture were generally less than 1 mm in specimens of 8 mm diameter. The corresponding energy plot for this earlier "initiation" criterion is given in Figure 26(b) showing little difference for this alloy. This indicates that the crack propagation period beyond 1 mm was relatively rapid.

Bartolini and Ragazzoni (1984) tested a 1CrMoV rotor steel at 540°C from 0.5% to 5% total strain range and found the energy to (complete) failure approximately constant. Between $10^{-2}\,s^{-1}$ and $10^{-6}\,s^{-1}$ the failure energy decreased with strain rate according to

$$W_f = 12.4\dot{\varepsilon}^{0.143} \tag{40}$$

where W_f is expressed in $J\,mm^{-3}$. For tensile dwells up to 180 min the failure energy (with no partitioning) decreased with increasing dwell time thus,

$$W_f = 9.60t_h^{-0.245} \tag{41}$$

where t_h is expressed in min. In an extension of the work (Bicego *et al.*, 1988) at 480°C, as the strain rate in separate tests decreased from $3\times10^{-3}\,s^{-1}$ to $3\times10^{-6}\,s^{-1}$ the accumulated failure energy decreased from $6.3\,J\,mm^{-3}$ to $3.1\,J\,mm^{-3}$. At 540°C these values became $3.2\,J\,mm^{-3}$ and $1.4\,J\,mm^{-3}$,

Table 2 Average energy values in continuous cycling.

Material	*Temperature* (°C)	W_f ($J\,mm^{-3}$)
1CrMo0.25V, N&T	482	4.0
1CrMo0.25V rotor	482	2.1
1CrMo0.25V rotor	538	1.8
2.25Cr1Mo, annealed	538	2.1
2.25Cr1Mo, annealed	566	1.2
2.25Cr1Mo, N&T	538	2.0
2.25Cr1Mo, Q&T	482	3.1
9Cr1Mo, N&T	550	3.0
304 stainless steel, ST	566	3.0
304 stainless steel, ST	649	1.6
316 stainless steel, ST	550	0.7
Nimonic 101	850	0.5

Source: Skelton (1991).

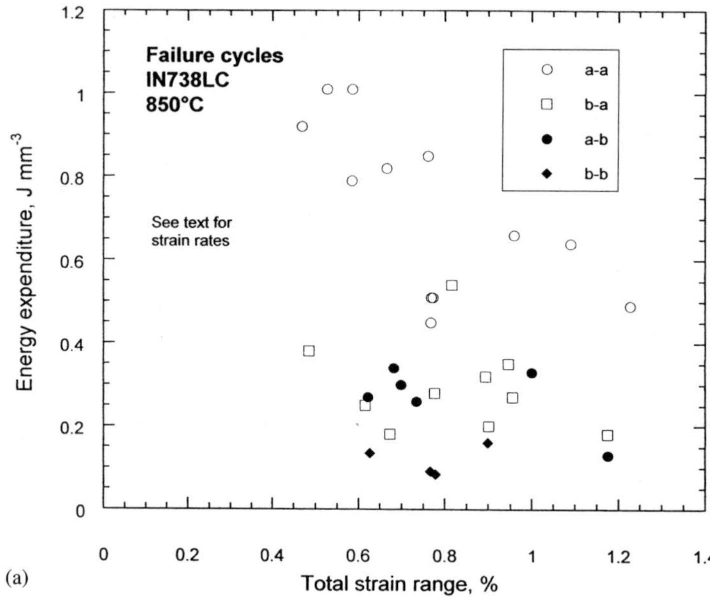

(a)

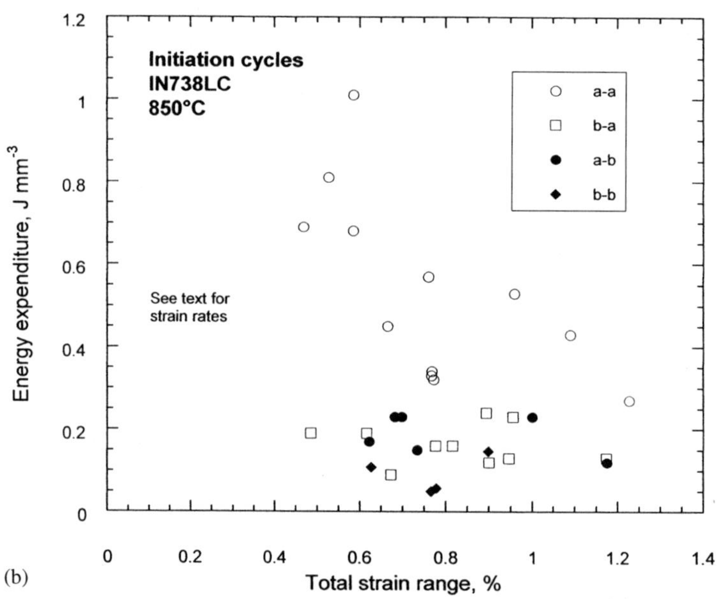

(b)

Figure 26 Data of Nazmy (1983) replotted in energy terms: (a) to complete fracture and (b) to crack initiation.

respectively. Typical scatter at 540°C from material taken from three different regions of a rotor was 3.2–4.3 J mm^{-3} at the intermediate strain rate. The authors confirmed (for given conditions) the strain-range independence of failure energies within a $\pm 50\%$ scatterband around the mean. Significantly, for creep rates at 550°C in the range 2×10^{-8}–10^{-9} s^{-1} the associated failure energies reduced from 0.07 J mm^{-3} to 0.03 J mm^{-3}. The trend of some of their data is shown in Figure 27, which has clear analogies with a ductility–strain-rate plot.

5.02.4.6.2 Example using energy criterion

The advantage of the energy criterion is its extreme simplicity; it has been applied in the life prediction of cast iron exhaust manifolds (Charkaluk and Constaninescu, 2000). We now give an example in continuous cycling where several strain ranges are given. Suppose a component is required to operate under mixed cycling according to the scheme in the first three columns of Table 3. Assume the material obeys the Ramberg–Osgood law (Equation (38)) having the constants $E = 1.5 \times 10^5$ MPa,

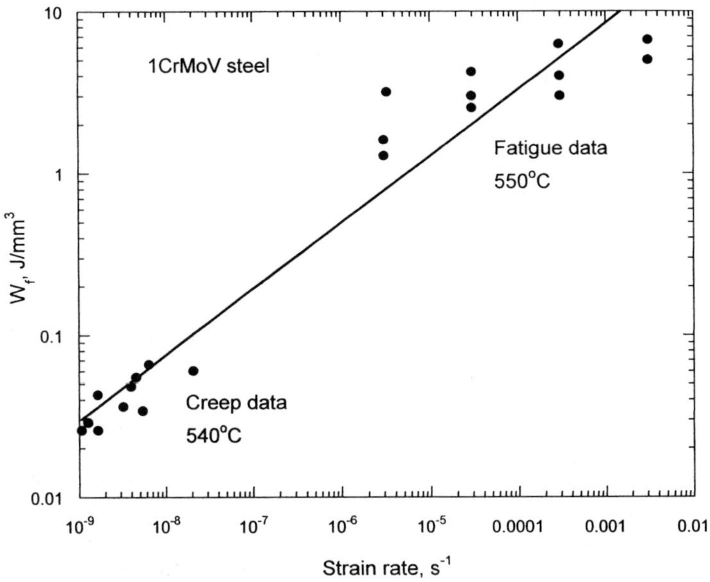

Figure 27 Energy failure data for 1CrMoV rotor steel (after Bicego *et al.*, 1988).

Table 3 Data for energy calculation.

	Given parameters		*Calculated parameters*	
Cycle type	*No. of cycles*	$\Delta\varepsilon_t$ (%)	$\Delta\sigma$ (MPa)	W (J mm^{-3})
Major	50	0.6	593	0.053
Intermediate	200	0.5	554	0.157
Minor	300	0.4	500	0.115
Total energy consumption				0.325

$A = 1{,}500$ MPa, and $\beta = 0.15$. The characteristic energy for crack initiation is 1.0 J mm^{-3}. Is crack initiation likely to occur in the component?

The "total strain range" version of Equation (38) is equal to the sum of elastic and plastic strain ranges and is thus given by

$$\Delta\varepsilon_t = \frac{\Delta\sigma}{E} + \left(\frac{\Delta\sigma}{A}\right)^{1/\beta} \qquad (42)$$

Substituting the values of E, A, and β in Equation (42) and solving numerically we have the corresponding stress ranges in column 4 of Table 3. The energy in each cycle is given by Equation (39). Multiplying respective energies by the appropriate number of major, intermediate, and minor cycles we then have the accumulated energies for each cycle type in the final column of Table 3.

Summing these energies, the total is 0.325 J mm^{-3}, i.e., <1.0 J mm^{-3}. Crack initiation is thus not likely to occur. This problem is returned to in Section 5.02.6.4.3 where short crack growth from an assumed defect size is assessed under the same regime.

5.02.4.6.3 Cycles with dwell

Attempts to partition the area of hysteresis loops along the lines of strain rate partitioning would detract from its ease of use but nevertheless the creep component of "energy," δ, should be assessed separately This has been done, e.g., on the data of Leven (1973) for a 1CrMoV steel at 538°C and the results are given in Table 4 (Skelton, 1991). Thus the extra energy expenditure during the dwell can account for a reduced number of cycles to failure (i.e., the creep–fatigue interaction).

However, this partitioning needs examining more carefully. Referring to the loop in Figure 28, area "A" is of fatigue type at relatively fast rate and so the cyclic energy is given by Equation (39) above. Area "B" is typically of creep type during stress relaxation, accompanied by an ever decreasing strain rate. It can be shown (Skelton, 1991) that the hysteresis

Table 4 Energy dwell values for 1CrMoV steel at 538°C.

$\Delta\varepsilon_t$	$\Delta\sigma$ (MPa)	σ_o (MPa)	W_f (J mm^{-3})	w (J mm^{-3})	δ (J mm^{-3})
0.051	947	363	1.17	4.49×10^{-2}	1.11×10^{-3}
0.023	825	391	1.22	1.45×10^{-2}	0.86×10^{-3}
0.012	692	307	1.87	0.42×10^{-2}	0.41×10^{-3}

Source: Skelton (1991).

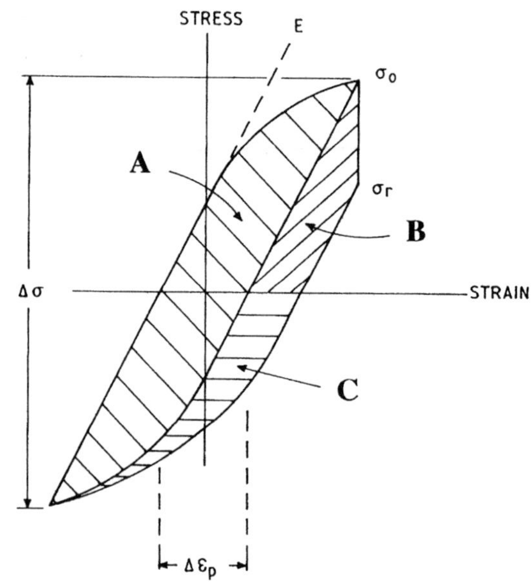

Figure 28 Energy terms introduced by dwell (after Skelton, 1991).

energy expended in this part is

$$w_{\text{creep}} = \frac{\sigma_o^2 - \sigma_r^2}{2E} \tag{43}$$

Area "C" is again generally again of a fatigue type at a fast strain rate. It can be shown (Skelton, 1991) that this area required to close the loop is equal to area "B" and can be given by

$$w_{\text{fatigue}} = \frac{\Delta\sigma(\sigma_o - \sigma_r)}{2E(1+\beta)} \tag{44}$$

Thus if areas A, B, and C are taken to represent actual energies, Equation (10) becomes for N cycles of this loop:

$$\frac{N \times (\text{A} + \text{C})}{W_{\text{f,fatigue}}} + \frac{N \times \text{B}}{W_{\text{f,creep}}} = 1 \tag{45}$$

Other combinations may be written for different loops, e.g., for a compressive dwell, or for a dwell in the middle of the strain range as indicated in Figures 4 and 22. Equation (45) will be used in a case study, see Section 5.02.8.4.

By means of Equation (45) (un-normalized form) it is possible to construct a detailed table of associated creep failure energies from the long-term creep–fatigue tests of Batte *et al.* (1978). Details are provided in Table 5. It is noted that:

(i) for 0.5 h dwell the failure energy increases for decreasing total strain range,

(ii) for 16 h dwell the energy is approximately constant, and

(iii) if the 16 h creep energy is measured from the end of the rapid relaxation period (i.e., 0.5 h) the failure energy is again approximately constant but at a much reduced value of $\sim$0.01 J mm^{-3}.

The latter is therefore comparable to typical failure energies encountered in conventional forward creep (e.g., a material under a creep stress of 100 MPa failing at a ductility of 10% has expended this same value of energy, see also Figure 27).

If elastic follow-up is present, then Equation (43) must be multiplied by the term Z to increase the creep contribution and so in turn should Equation (44) in order to close the loop. If it is further required to partition the first part of the relaxation curve into "fatigue" damage (as was done for $Z = 1$ above) then the critical strain rate at which this occurs can be identified from Equation (26). Clearly in the above case this would be greater than 0.5 h because strain rates are higher with elastic follow-up present, see Section 5.02.4.3.

It must again be noted that the energy values are a measure of the heat that has been dissipated in the specimen or volume element of the component. The amount causing actual damage with each cycle is a very small fraction of this. Further, the model cannot account for the underlying microstructural changes occurring, which as our review has shown, is generally transgranular cracking for continuous cycling and intergranular cracking for creep. Jeong and Nam (1999) claim, from their tests with dwell at 550°C on a 1CrMoV steel, that the ratio of "damaging energy" to "total energy" is of the order of 0.04 but they do this by taking the energy of hot tensile tests as a reference. The latter also produce dissipated heat, so their value is simply equivalent to that

Table 5 Creep–fatigue energy values.

$\Delta\varepsilon_t$ (%)	σ_o (MPa)	σ_r (MPa)	W (J mm^{-3})	N_f	W_f (J mm^{-3})
			0.5 h dwell		
0.8	224	127	5.37×10^{-5}	689	0.037
0.4	185	118	6.76×10^{-5}	2,637	0.178
0.3	162	108	4.86×10^{-5}	5,611	0.272
0.2	116	86	5.02×10^{-5}	18,233	0.368
			16 h dwell		
1.0	217	76	1.38×10^{-4}	319	0.044
0.8	228	99	1.40×10^{-4}	385	0.054
0.8	(127[a])	(98[a])	(2.18×10^{-5})	385	(0.008)
0.6	179	85	8.27×10^{-5}	511	0.042
0.4	199	76	1.13×10^{-4}	617	0.070
0.4	(118[a])	(94.5[a])	(1.66×10^{-5})	617	(0.01)
0.3	134	75	4.11×10^{-5}	1,000	0.041
0.3	(108[a])	(91[a])	(1.13×10^{-5})	1,000	(0.011)

Source: Batte *et al.* (1978).
Values in brackets refer to σ_o taken at end of 0.5 h dwell period.
[a] Single-specimen rather than average data.

obtained by the ratio of creep to fatigue energies presented in Figure 27, for example.

5.02.4.7 Design versus Assessment

Thus far a number of linear damage theories have been considered and all calculations can in principle be plotted on a damage diagram according to Equation (10), Section 5.02.4.1. If linear summation is indeed true then the damage factor is unity and interaction is taken into account by allowing the processes to become additive rather than competitive. It could thus be argued that empirical allowance for the damage factor $\Phi < 1$ is some measure of non-linear damage summation. In what follows, however, we shall always assume $\Phi = 1$ as adopted by the R5 procedure (Ainsworth, 1996).

In design, safety against crack initiation can be assured by using "best estimate" methodology and pessimistic materials data. In retrospective or diagnostic analysis it is appropriate to use "best estimate" methodology and "best estimate" materials data since it is required to explain as far as possible how a particular failure came about. Equation (10) is the failure locus on a damage diagram, as shown in Figure 29. Thus if laboratory creep–fatigue endurances are plotted and lie wholly outside the envelope, then $\Phi_{total} > 1$ and Equation (10) is a safe bound. Similarly if a retrospective analysis of a failed component also gives a point outside the envelope by applying laboratory data in Equation (10), then the model has been validated. Conversely, when assessing life at the design stage, all future operating cycles/steady running periods should combine to produce a point within the envelope for failure avoidance. This is the role of a sensitivity analysis. For example, if N_f from laboratory data were replaced by some earlier criterion N_i, the fatigue component Φ_f would move all data points towards unity, so that a point initially within the envelope might end up outside it. Depending on the component, the actions to be taken could include:

(i) repairing the damaged section,
(ii) input data reappraisal,
(iii) changing operating procedures, or
(iv) undertaking a crack growth assessment.

It is emphasized that two major assumptions of the review have been:

(i) that isothermal data may be applied to situations where the temperature is changing and

(ii) that cyclic history effects may be neglected although the latter can to some extent be allowed for by employing service-exposed data.

5.02.5 PREDICTION OF INITIATION: SIZE, ENVIRONMENTAL, AND OTHER EFFECTS

In this section we review the interplay between cycles to initiation and short crack growth, i.e., initiation criteria to different depths are considered without specific attention to the form of crack growth law (in contrast to Section 5.02.6). It is shown that a reference endurance, N_l, is required (usually in continuous cycling) for analysis, but inevitable

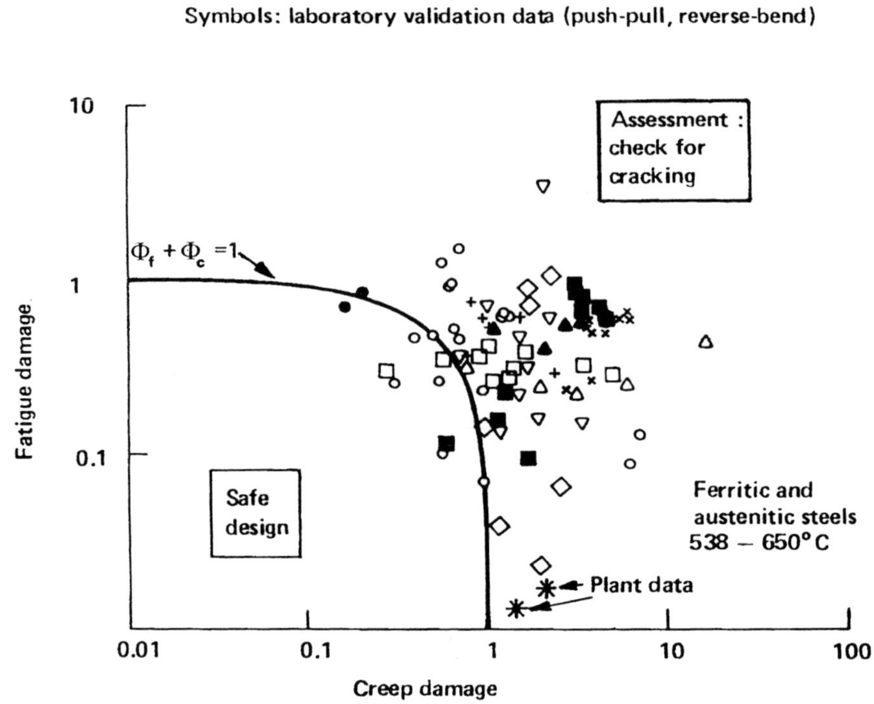

Figure 29 Damage diagram (after Ainsworth, 1989; Rémy and Skelton, 1992).

variation in such reference endurances can occur as follows.

5.02.5.1 Characteristics Affecting Endurance

5.02.5.1.1 Material scatter

In Figure 30 we plot a typical endurance curve for a service material (Batte *et al.*, 1978). There is typically half an order of magnitude variation in endurance at a given strain level. Material scatter may either be intrinsic (i.e., invested in the response of the material itself) or extrinsic (i.e., arising due to uncertainties of measurement in the testing system, especially strain measurement, (Kandil and Dyson, 1995)). Whichever is the case, the assumption of a mean line would give best estimate prediction of behavior while a lower 95% confidence limit line would be used for a pessimistic or safe prediction.

5.02.5.1.2 Frequency effect

It is generally observed that a decrease in frequency of testing gives rise to a reduction in endurance. An example at a single strain range has already been given in Figure 8. A further example across a range of strain is given in Figure 31 for an austenitic steel. For most steels an upper bound may be set at 0.1 Hz at temperatures less than 600°C, but there is in principle no lower bound as the frequency is reduced in continuous cycling. It has already been discussed how reduction in endurance is due to creep and oxidation effects. As will be shown, the effect of creep is assessed separately. There is thus a danger of "double accounting" if low frequency fatigue data are assumed as a reference but clearly this would be acceptable in a pessimistic prediction.

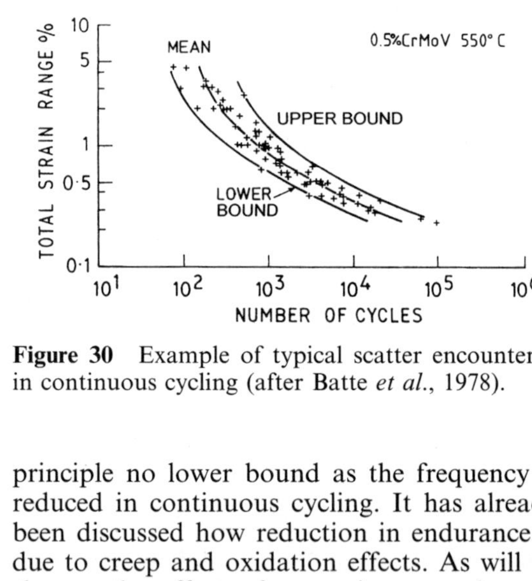

Figure 30 Example of typical scatter encountered in continuous cycling (after Batte *et al.*, 1978).

5.02.5.1.3 Effect of temperature

It is generally observed that an increase in temperature gives rise to a reduction in endurance. In ferritic steels, e.g., the reduction factor is ~4 over the temperature range from

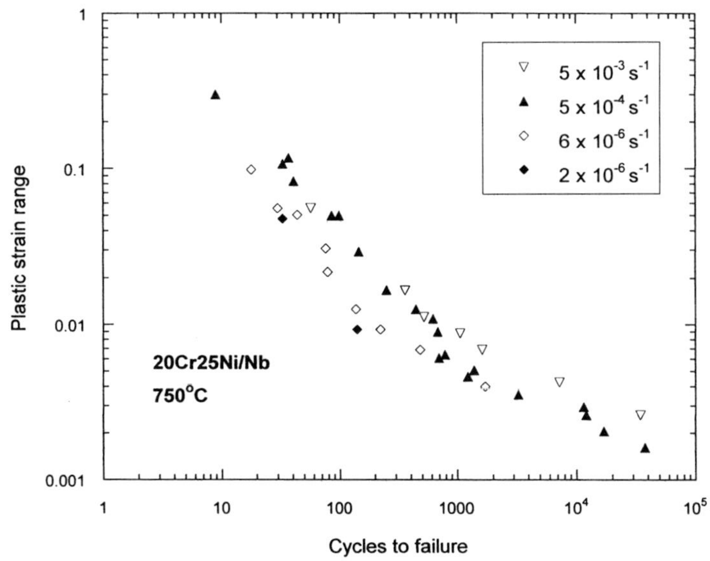

Figure 31 Strain rate effect on total endurances (after Wareing *et al.*, 1985).

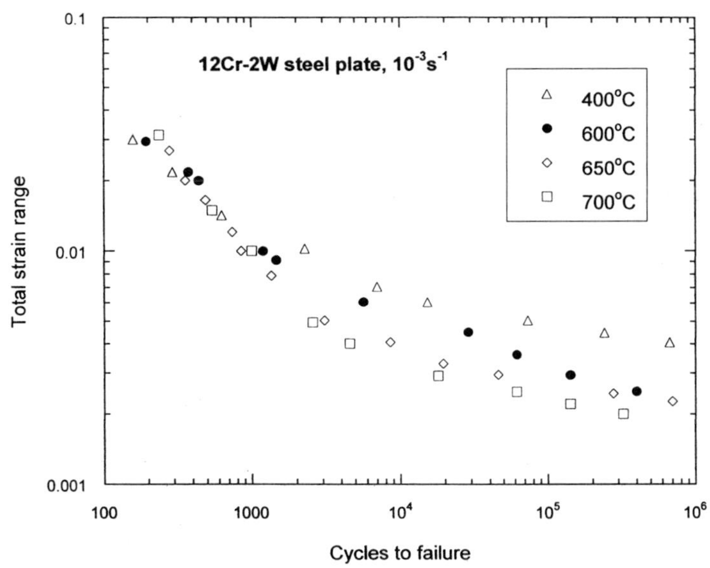

Figure 32 Effect of temperature on total endurance (after Kobayashi and Yamaguchi, 2001).

ambient to 550°C. The same factor of 4 applies to austenitic steels up to 850°C (Wareing *et al.*, 1985). An example of the effect in a 12Cr steel in the range 400–700°C is provided in Figure 32 (Kobayashi and Yamaguchi, 2001). As a rule of thumb, the effect is small in most steels when the change is limited to ~100°C.

5.02.5.1.4 Service exposure

Surprisingly, when plotted on a total strain range basis, the endurance of service-exposed materials sometimes appears to lie within the scatterband of virgin material. This has been demonstrated for material taken from a 316 steel turbine valve which had been in operation for 1.3×10^5 h at ~600°C. It was subsequently tested at 650°C (Argo *et al.*, 1984), see Figure 33. Corresponding creep tests at the same temperature showed that there was a pronounced decrease in life on a stress-rupture plot (Masuyama *et al.*, 1985) though re-solution treating restored the lifetime towards reference data on similar materials. In another example, a 1CrMoV rotor steel which was service-exposed to 536°C for 1.34×10^5 h and subsequently fatigue tested at 566°C again showed no

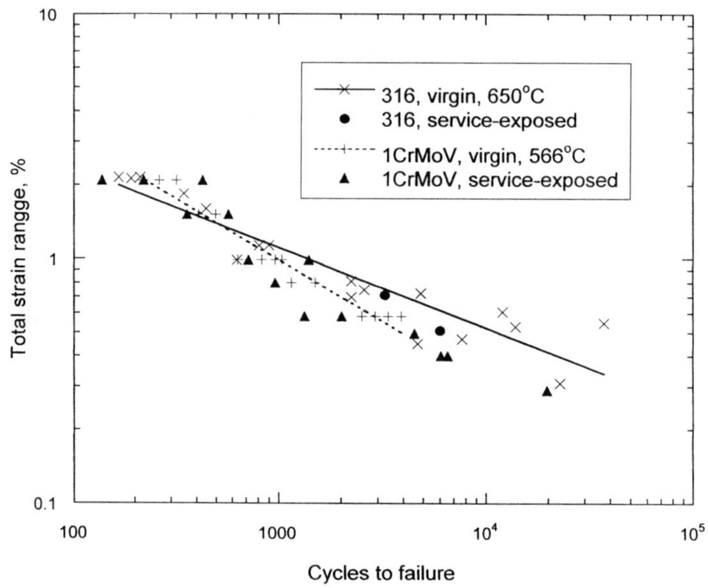

Figure 33 Effect of service exposure in an austenitic piping steel (after Argo *et al.*, 1984) and a ferritic rotor steel (after Kimura *et al.*, 1988).

significant difference in endurance compared with virgin material, see Figure 33 (Kimura *et al.*, 1988). This was despite the fact that:

(i) Creep rupture life of exposed material was significantly below that of virgin material in the range 70–300 MPa.

(ii) Secondary creep rates at 566°C were over an order of magnitude greater than for virgin material.

(iii) As judged from a comparison of cyclic stress–strain curves at 566°C, and other hardness data, service-exposed material had considerably softened.

The above fatigue tests were undertaken in continuous cycling and it would thus appear that cracking was transgranular (though no details were given), ignoring possible sites of intergranular damage which occurred in service. Since service-exposed material is not easily obtainable (especially where components are to continue in use) it is suggested that to explore the full scope of creep–fatigue interactions, laboratory fatigue tests on such material be conducted at low strain rates or should include a dwell period.

In an investigation (Vogt *et al.*, 2001) tests were conducted at 565°C on a 2(1/4)Cr1Mo header steel which had been service exposed for 1.1×10^5 h and 1.6×10^5 h, respectively at the same temperature, and a regenerative heat treatment was applied to some specimens. The service-exposed material with a higher ferrite content was weaker and cyclically hardened whereas the other material was initially stronger and so softened. Regenerated material first hardened and then softened. Tests in vacuum (10^{-4} Pa) gave about a factor of 5 increase in endurance at the lowest strain range (0.4%) and the surface density of cracks was very much reduced.

5.02.5.1.5 Environment

For fast reactor lifetime assessments, experiments in vacuum, sodium, helium, etc. eliminate the oxidation component of fatigue damage. As expected, testing in such environments prolongs the endurance of materials compared with standard tests in air. An order of magnitude increase is typical, see Figure 8, the same applying for 2(1/4)Cr1Mo steel at 593°C (Viswanathan, 1989). Fatigue and creep–fatigue tests were carried out on 316 and 9Cr1Mo steels in flowing sodium containing 1 ppm oxygen (Asayama *et al.*, 1999). Strain ranges varied from 0.4% to 1% at a strain rate of $10^{-3}\,s^{-1}$. For the austenitic steel, fatigue life in sodium was an order of magnitude greater at high strain ranges, converging at low strain range. A 1 h tension dwell reduced endurance whether in air or sodium, all data lying a factor of 2 below the continuous cycling air data. For the ferritic steel, continuous cycling data in air were a factor of 3 lower than in sodium whatever the strain range (0.5–1.7%). Dwells of 1 h in tension reduced endurance irrespective of environment while a compressive dwell in sodium restored the endurance completely to the continuous cycling curve in sodium.

Tests have been carried out (Ioka *et al.*, 2001) on 304 steel at 550°C with tension hold times of 10 min or 1 h in a vacuum of 10^{-1} Pa and at total strain ranges of 0.8–1.6%. Some specimens had previously been irradiated at temperature to a level of 2 dpa. Irradiation reduced endurance by a factor of 2–5 for continuous cycling with further reductions in the case of hold times. Fractography of irradiated specimens showed striations for continuous cycling, with increasing amounts of intergranular damage with increase in hold time. The trend was thus the same as for nonirradiated specimens. Both time fraction and ductility exhaustion methods underestimated the creep damage (i.e., the total damage lay within the ASME N-47 envelope) though no supporting data was given to check the effect of irradiation on ductility.

In some LCF tests at 600°C on a 12Cr steel Earthman *et al.* (1989) showed that vacuum (1 mPa) endurances were over twice those in air. At a total strain range of 2% the macrocrack density at the surface in air (500 mm^{-2}) greatly exceeded the population in vacuum (20 mm^{-2}). Investigating a little further, these authors found that at 0.4, 0.5, and 0.6 of the life in air, most of the cracks lay below 50 μm length in the surface while under vacuum at life ratios of 0.4, 0.6, and 0.8 the relative distribution was more spread out into the ranges 50–100 μm and 100–200 μm. The major crack to failure (600 μm in size) occurred at 60% of the life in air whereas it occurred at 80% of the life in vacuum.

Using a modified slow–fast cycle, Renner *et al.* (1989) carried out tests on 12 mm diameter specimens of 304 and 347 steels in the range 680–740°C, both in air and under vacuum. Though the absolute cycles were a factor of 5 above air data—the dwell effect persisted in vacuum. The authors concentrated on surface crack behavior but established an aspect ratio, a/c, of 1.04, i.e., the cracks were semicircular. By monitoring several surface cracks on separate specimens, the growth rates were similar, but as may be expected, for growth in the range 0.2–1 mm they did not occur at identical cycles, pointing out some scatter in the initiation cycles. For dwells of 7 min, however, this scatter was much reduced in the range 20–200 μm.

5.02.5.1.6 *Weldments*

Weld metal and weldments often show significantly lower endurance than base metal. Further, in 316 material at 550°C, e.g., circumferential weldments in cylindrical specimens gave endurances lower than for longitudinal weldments (O'Connor and Corum, 1986). It is believed that this reduction is due to a metallurgical notch effect produced by yield strength variations across the weld. Microstructural gradients in welds produce additional effects at high temperature to which it would be impossible to do justice in this review.

5.02.5.2 Size Effect—Smooth Specimen Crack Initiation

There are several empirical expressions for the number of cycles, N_o, for crack initiation to a specified depth, a_o, in smooth specimens. Most results have been determined by careful measurement of striation spacings, assuming one cycle per spacing. From LCF data on several types of 304 and 316 austenitic steels in the range 550–700°C, Pineau (1983) deduced that up to a depth of 20 μm:

$$N_o = N_l + 0.226N_l^{0.9} - 12N_l^{0.62} + 74 \quad (46)$$

where N_l is the number of cycles taken to fail a laboratory specimen (typically at 5–10% load drop rather than complete severance). The relation is shown plotted in Figure 34.

For alloy 800 at 550°C, initiation cycles up to a crack depth of 60 μm are given (El Gharad, 1986) by

$$N_o = N_l - 2.63N_l^{0.9} + 0.69N_l^{1.01} \quad (47)$$

In type 304 steel at 593°C Maiya (1975) found a relation (which must be solved numerically) valid for a crack depth of 100 μm:

$$N_l = N_o + 31.52N_o^{0.44} \quad (48)$$

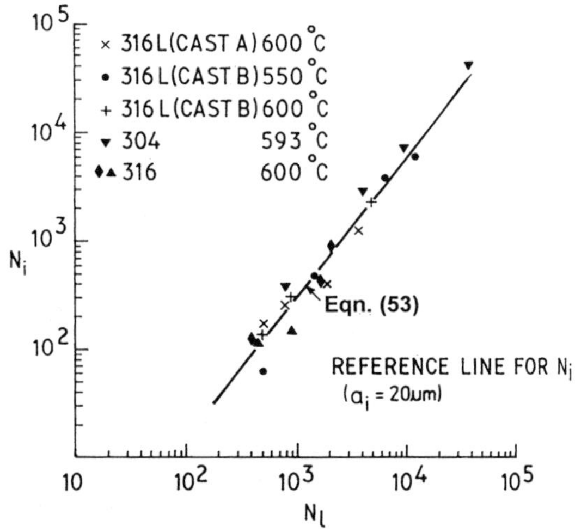

Figure 34 Data forming the basis of Equation (46) (after Pineau, 1983).

For 321 and 316 steels in the range 450–700°C the following relation was quoted (Yamaguchi and Kanazawa, 1979):

$$N_o = N_f - 1.44N_f^{0.91} \quad (49)$$

It appears to be valid "up to one grain size depth," which for the materials considered varied from 45 μm to 250 μm depending on heat treatment.

Finally, the often quoted relation of Manson and Hirschberg (1966) which was proposed for room temperature materials:

$$N_o = N_l - 4N_l^{0.6} \quad (50)$$

gives the number of cycles to observe a surface crack some 0.75 mm in length. Assuming a thumbnail crack, this is equivalent to a crack depth of 0.375 μm. Equation (50) was found to be applicable to 0.5CrMoV steels at 550°C even when a dwell period was included (Batte *et al.*, 1978) see Figure 35.

Equations (46)–(50) are shown plotted in Figure 36 for an assumed reference endurance, N_l, shown as the thick line. They are thus useful for estimating cycles to an earlier condition when only the total endurance is known.

It is emphasized that Equations (46)–(50) strictly apply to continuous cycling only, and there appear to be few quantitative relations catering for the effect of dwell. However, Levaillant *et al.* (1980) have shown directly that the term N_i itself is likely to be lower for dwell times in air than in pure fatigue. A specimen of 316 steel tested for 50 cycles with a 10 min tensile dwell at 0.9% plastic strain range indicated oxidized microcracks in the surface, showing that stage I cracking had already begun, i.e., $N_i/N_f < 0.1$. For companion specimens tested in continuous cycling this ratio was 0.3. Similarly, Reger and Rémy (1986) in their experiments on IN100 at 1,000°C demonstrated a reduction in N_i as the testing frequency was decreased. Their results may be expressed as

$$N_i = \psi \Delta\varepsilon_t^{-5/2} \nu^{1/4} \quad (51)$$

for frequencies ν between 10^{-1} Hz and 10^{-3} Hz. For a crack depth, a_o, of 200 μm, $\psi = 1.8 \times 10^{-4}$ whereas for $a_o = 300$ μm then $\psi = 4.0 \times 10^{-4}$. The authors were at pains to point out, however, that the frequency effect was due to strain-enhanced oxidation (see Section 5.02.3.1) and not to a creep–fatigue interaction. Thus in Equation (51) when $\Delta\varepsilon_t = 0.3\%$ (0.003) and $\nu = 10^{-3}$ Hz some 70 initiation cycles are predicted for a crack depth of 200 μm.

5.02.5.3 Size Effect—Choice of Specimen

In Figure 37, ABC is a strain-concentrating feature in a component which produces a plastic zone of depth X as a result of temperature or mechanical strain cycling. The effects of a strain gradient in the zone are

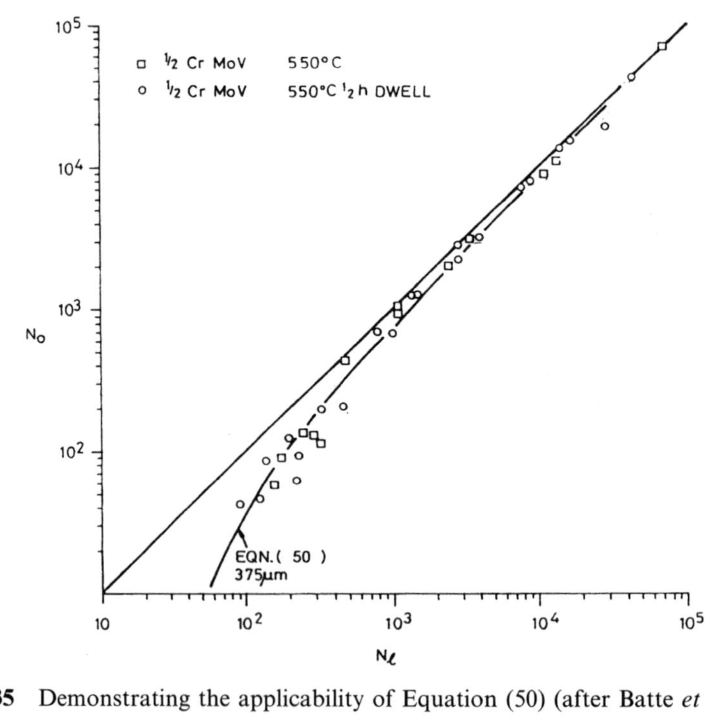

Figure 35 Demonstrating the applicability of Equation (50) (after Batte *et al.*, 1978).

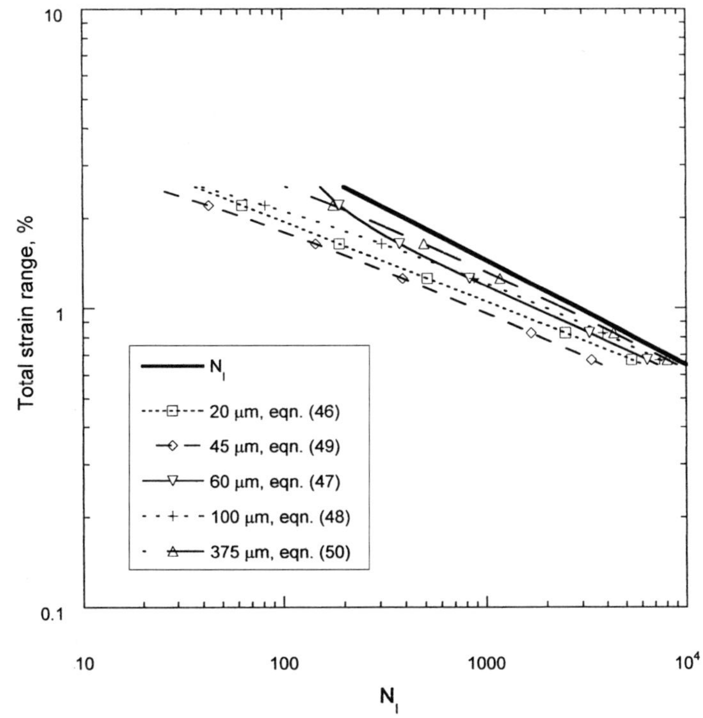

Figure 36 Initiation criteria to different crack depths.

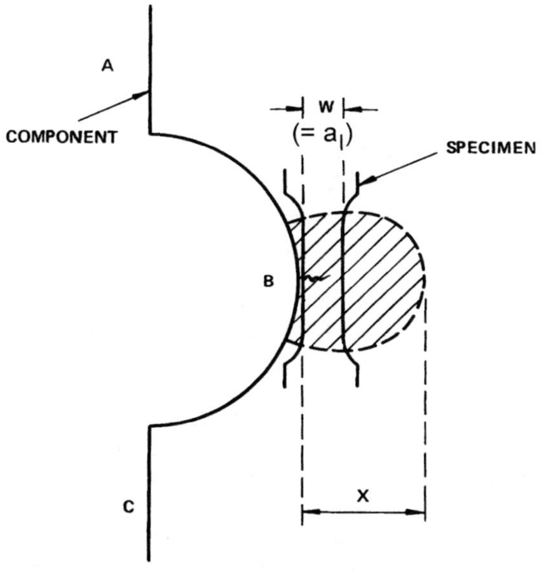

Figure 37 Laboratory specimen modeling fatigue–crack growth in a component (after Skelton, 1988a).

minimized by making the specimen width or diameter $W \ll X$ (i.e., uniform strain conditions across the specimen). If $X < W$ then either the specimen diameter may be further reduced or the test may be terminated at a specified load drop, as discussed previously. Thus initiation in both "thin" components (e.g., nuclear fuel element cladding) and "thick" components (e.g., the region ahead of a heat relief groove in a rotor) may be modeled (Skelton, 1985). From these arguments it appears that early growth rates in the component may also be reproduced by monitoring those occurring in the specimen. However, it is clear from Figure 37 that when the crack is at a depth of, say, $0.75W$, the specimen will indicate a notable increase in compliance (inverse stiffness), whereas a very much smaller increase would be observed in a thick component for the same crack depth. Beyond a crack depth of about $0.25W$, therefore, corresponding propagation rates are not necessarily equivalent.

An earlier failure criterion may also be obtained by testing thin components. In Figure 38(a) is shown the earlier failure of tubular specimens of 0.65 mm and 1.27 mm wall thickness compared with failure of standard 6.35 mm diameter specimens (van den Avyle, 1983). It may also be noted that the various stages of short crack growth are manifest in the stress response from a single test, as shown schematically in Figure 38(b).

5.02.5.4 Incorporation of Early Crack Growth

This section outlines the rationale behind the R5 high temperature assessment code dealing with crack initiation (Ainsworth, 1996). A comparison with other codes has been made

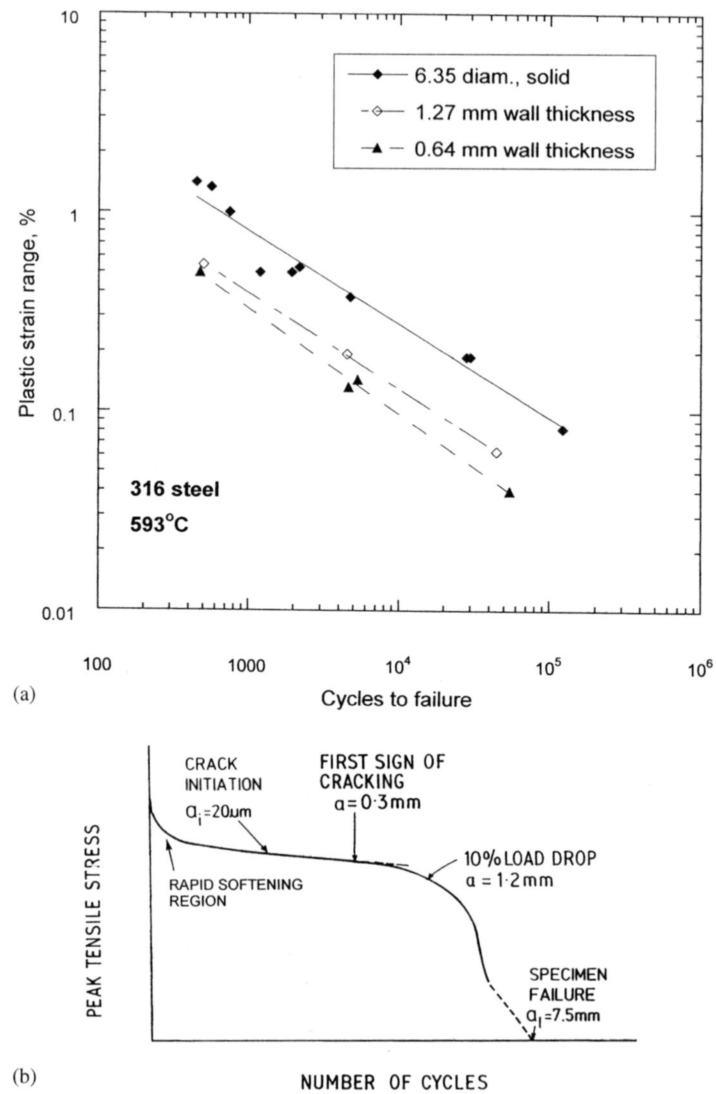

Figure 38 (a) Size effect alternatively demonstrated by tests on solid and tubular specimens (after van den Avyle, 1983) and (b) Likely crack depths during a typical fatigue test on cyclically softening material.

before (Skelton, 1996); the R5 approach is different in that:

(i) the value of the damage factor in Equation (10) is always set to unity and not <1 as in some codes;

(ii) no safety factors are applied to fatigue endurance data, assessments being made either on a best-fit curve or a lower bound to the data;

(iii) fatigue damage is not applied to the original database (that is generally based on laboratory specimen "failure" to some specified load drop criterion, thus containing an element of crack growth) but to an earlier initiation point; and

(iv) the creep damage summation in Equation (10) is effected with respect to strain (ductility exhaustion), not time.

The "fatigue" element thus contains a size effect whereas the "creep" element occurs throughout the bulk. The relation for estimating initiation cycles, N_o, relies on the propagation characteristics of cracks in the "short crack growth" regime immediately following initiation (discussed in detail in Section 5.02.6). However the crack growth constants do not appear explicitly in the formulas for crack initiation.

Partition of the fatigue cycles is indicated in Figure 39. A term N_l is used which represents the number of continuous cycles to failure at crack depth, a_l, in a laboratory specimen (this could represent 5% or 10% load drop rather than complete severance). It is necessary to establish the quantity, N_o, at a corresponding crack depth, a_o. The value of a_o is chosen by

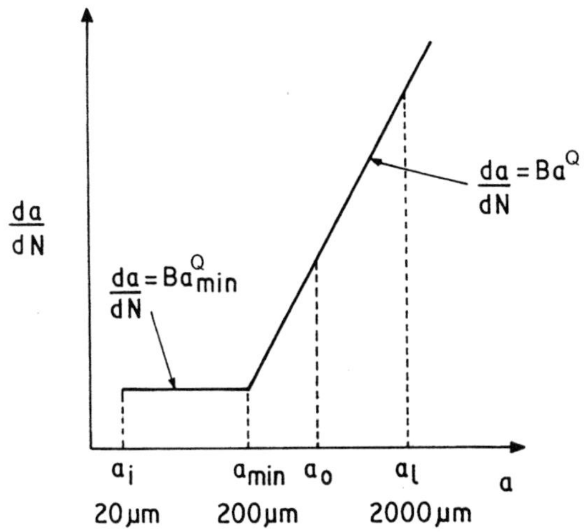

Figure 39 Definition of stages used in crack initiation/short crack growth calculations.

the user, having regard to the component. The term N_i corresponding to a_i in Figure 39 represents the number of cycles to a very early initiation depth a_i ($=20\,\mu m$) for use as a fiducial reference, by adapting the empirical Equation (46).

Along the sloping line in Figure 39 crack growth rate in LCF is given (Skelton, 1982, 1988a) by

$$\frac{da}{dN} = Ba^Q \tag{52}$$

where B and Q are empirical constants. It is known that up to a certain crack depth, a_{min}, experimental cyclic growth rates are constant (Smith, 1996), as indicated in Figure 40. In the R5 procedure, a_{min} is taken to be $200\,\mu m$. By identifying, at a depth of $20\,\mu m$, the term N_o with the reference value N_i, Equation (46) is more conveniently expressed (Ainsworth, 1989):

$$N_i = 0.0366 N_l^{1.306} \tag{53}$$

other slight variations occurring in successive editions of R5 (Hales and Ainsworth, 1995).

Upon integrating Equation (52) over the constant- and increasing-crack-growth-rate regions, eliminating the term B and assuming $Q=1$ (valid for many alloys) then (Ainsworth, 1989):

$$N_o = N_i + (N_l - N_i) \times \left[\frac{(a_{min} - a_i) + a_{min}\ln(a_o/a_{min})}{(a_{min} - a_i) + a_{min}\ln(a_l/a_{min})}\right] \tag{54}$$

provided $a_o > a_{min}$. A very similar relation applies for $a_o < a_{min}$ as follows:

$$N_o = N_i + (N_l - N_i) \times \left[\frac{a_o - a_i}{(a_{min} - a_i) + a_{min}\ln(a_l/a_{min})}\right] \tag{55}$$

A very similar result may be derived when $Q > 1$ and is given by

$$N_o = N_i + (N_l - N_i) \times \left\{\frac{a_{min}^Q(a_{min}^{1-Q} - a_o^{1-Q}) + (Q-1)(a_{min} - a_i)}{a_{min}^Q(a_{min}^{1-Q} - a_l^{1-Q}) + (Q-1)(a_{min} - a_i)}\right\} \tag{56}$$

and when $a_o < a_{min}$:

$$N_o = N_i + (N_l - N_i) \times \left\{\frac{(Q-1)(a_o - a_i)}{a_{min}^Q(a_{min}^{1-Q} - a_o^{1-Q}) + (Q-1)(a_{min} - a_i)}\right\} \tag{57}$$

In summary, the predicted endurance, N_o, in Equations (54) and (55) (or Equations (56) and (57), respectively) required for a damage calculation depends on the crack size criterion, a_o, adopted for initiation.

Taking $Q=1$ and using Equation (53) as the reference depth, Equations (54) and (55) as appropriate are validated against the respective Equations (47), (48), and (50) in Table 6. Similar comparisons may also be undertaken for Q values >1.

5.02.5.5 Sensitivity Analysis

Experimental variations in the value of Q in Equation (52) at high temperature will be surveyed in Section 5.02.6 but as a demonstration of the use of Equations (54)–(57) we take $Q=1$ and $Q=1.45$ as limiting values. Assuming the smooth specimen failure criterion (a_l) to be 2 mm after 1,000 cycles (N_l), the extremes in initiation cycles, N_o, for chosen values a_o of $60\,\mu m$ and $375\,\mu m$, respectively can be deduced as follows. First, the term N_i is calculated by means of Equation (53). Substituting in the relevant Equations (54)–(57) and taking $a_i = 20\,\mu m$, $a_{min} = 200\,\mu m$ as discussed above, we find the following:

To initiate a $60\,\mu m$ crack, 346 cycles are predicted when $Q=1$, while if $Q=1.45$, then 363 cycles are predicted. For initiation of a $375\,\mu m$ deep crack, the predicted cycles are 635 and 735 when $Q=1$ and 1.45, respectively. It thus appears that the calculations are reasonably insensitive to the value of the parameter Q. Equations (53)–(55) may similarly be manipulated (Goodall, private communication) to provide an "endurance reduction factor" for compressive dwell, if it is believed

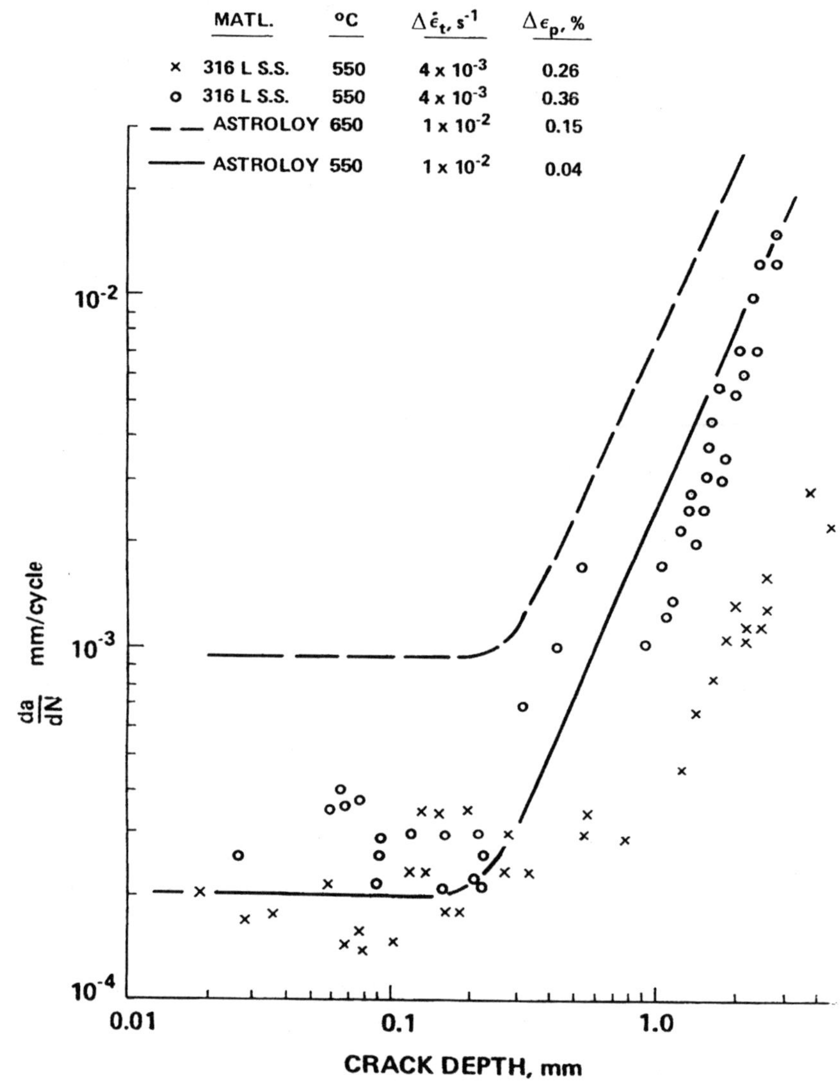

Figure 40 Up to a certain crack depth, crack growth rates in LCF tests are constant (after Skelton, 1988a).

Table 6 Comparison of crack initiation criteria for $Q = 1$.

	N_o					
	$a_o = 60\ \mu m < a_{min}$		$a_o = 100\ \mu m > a_{min}$		$a_o = 375\ \mu m > a_{min}$	
N_1	*Eqn. (55)*	*Eqn. (47)*	*Eqn. (54)*	*Eqn. (48)*	*Eqn. (54)*	*Eqn. (50)*
200	47	36	57	40	114	104
300	76	73	93	82	176	177
400	110	115	130	131	239	254
500	146	161	170	186	302	333
600	183	209	211	245	367	414
700	221	260	254	308	433	496
800	261	312	298	373	500	579
900	303	366	343	441	567	663
1,000	346	421	390	510	635	748
1,500	575	715	638	878	984	1,178
2,000	826	1,029	905	1,269	1,345	1,618
3,000	1,377	1,700	1,488	2,089	2,095	2,512
4,000	1,982	2,409	2,120	2,941	2,875	3,420
5,000	2,631	3,146	2,794	3,813	3,679	4,337
10,000	6,372	7,096	6,612	8,327	7,977	8,995

that stage I crack growth (up to a depth of a_{min}) is eliminated by the dwell effect.

5.02.6 PREDICTION OF SHORT-CRACK GROWTH (SEE ALSO CHAPTER 4.05)

From this point forwards in the review it is accepted that a small flaw is present and thus the concept of creep–fatigue crack "initiation" is extended to include the cyclic growth rate of a short crack. It will be shown that it is possible to calculate the penctration for a given number of cycles or, for a specified short depth, the allowable number of cycles. It is emphasized that growth is governed by the surface strain range (i.e., not stress intensity as in deep crack growth) and indeed compliance changes do not enter into the consideration since a "short" crack is effectively propagating in a semi-infinite body. This contrasts with the approach of Ohtani and Kitamura (1994), e.g., who use a ΔJ term partitioned into creep and fatigue components, very often based on strain information measured directly at the crack tip.

5.02.6.1 Definition of Validity Range

Crack growth experiments are thus undertaken to support the contention that structures can perform their design function even though they may contain defects, provided only that cyclic propagation behavior is well characterized. From the size effect argument in Section 5.02.5.3 (Figure 37) it appears that early growth rates in the component may also be reproduced by monitoring those occurring in the specimen. However, it is re-iterated from Figure 37 that crack growth in the specimen must be restricted to depths of $\sim 0.25W$, beyond which a notable increase in specimen compliance (inverse stiffness) would be observed. In a thick component, no such increase in compliance would be observed for the same crack depth.

In the following, as in most experimental work supporting lifetime assessments, it is assumed that simulation may be effected by testing at the maximum temperature of exposure, thus giving a conservative (pessimistic) prediction of crack growth rates.

5.02.6.2 Characteristics of Crack Growth Law

5.02.6.2.1 Empirical relation

In a typical LCF growth test, a smooth specimen of circular or rectangular cross-section is lightly notched and crack progress monitored by one or several of the following techniques:

(i) potential drop,
(ii) striation counting,
(iii) "beach marking"/heat tinting,
(iv) load drop changes, and
(v) surface observation/replica measurements.

Testing is normally done between limits of total strain range, $\Delta\varepsilon_t$. From such studies an empirical growth law is found (Skelton, 1982), given by Equation (52) where in more detail da/dN is the cyclic growth rate, B is a constant which depends on the value of $\Delta\varepsilon_t$, a is crack depth, and Q is a constant. Values of Q for several materials are summarized in Table 7 (Skelton, 1988a). A typical laboratory test starts with a chordal notch of depth 0.1 mm in a cylindrical specimen of 10 mm diameter and finishes when a significant tensile load drop of 5% or thereabouts has occurred (Raynor and Skelton, 1985). At this point a/W approximately equals 0.25 depending on crack shape. It may be noted that fundamental theory (e.g., Tomkins, 1968) always predicts

Table 7 Typical crack growth parameters.

Material	*Temperature* (°C)	*Q*	*Notes*
20Cr/25Ni/Nb	750	1.0	
316 steel	625	1.0	
High N 316 steel	600	1.0	
Aged 316 steel	625	1.9	0.5 h dwell, vacuum
Hastelloy X	760	1.0	
Alloy 800	500–760	1.0	
A286	593	1.0	
9Cr1Mo, annealed	550	1.9	Vacuum
9Cr1Mo, N&T	550	1.3	Vacuum
2(1/4)Cr1Mo, annealed	525	0.74	
2(1/4)Cr1Mo, N&T	525	0.86	
(1/2)CrMoV	550	1.0	
(1/2)CrMoV	550	1.5	0.5 h dwell

Source: Skelton (1988a).

$Q=1$ (exponential growth law) in Equation (52) but many workers have reported otherwise for the empirical case, as discussed in the previous reviews (Skelton, 1982, 1988a).

For assessing the behavior of short cracks in a component, therefore, we need to know the total strain range at the surface, $\Delta\varepsilon_t$. Examples of the variation of the constant B with total strain in Equation (52) when growth is expressed in mm cycle^{-1} are given in Figure 41 (Skelton, 1988a) for continuous cycling data or for dwells not exceeding 0.5 h. It may be noted that the value of B is the growth rate at unit crack depth (i.e., 1 mm), independent of Q. Diagrams such as that given in Figure 41 are a succinct way of summarizing crack growth data and, as will be seen, form an extremely useful way to assess crack growth in components.

Since the last reviews of short crack growth (Skelton, 1982, 1983, 1988a), few other data have appeared in the literature. Nonaka and Kitagawa (1992) examined crack growth in

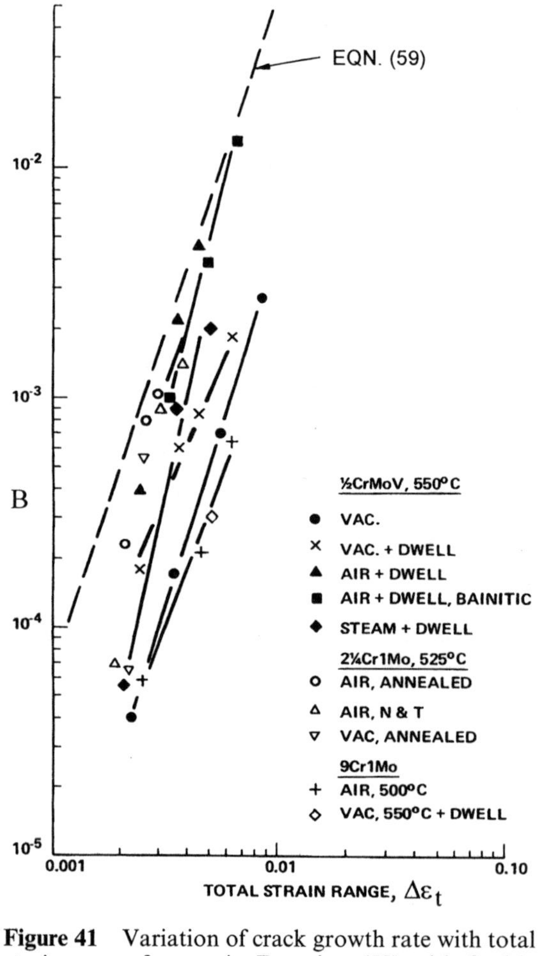

Figure 41 Variation of crack growth rate with total strain range for use in Equation (52) with ferritic steels—corresponding values of Q are given in Table 7 (after Skelton, 1988a).

2(1/4)Cr1Mo steel, and also in the heat-affected zone (HAZ) form. Strain-controlled tests were performed at 570°C between 0.4% and 2% total strain range, cooling to ambient temperature periodically and checking for surface cracking by replication techniques. Multiple cracking was observed at high strains, the initial direction being perpendicular to the stress for the base metal but at 45° to the stress direction for the HAZ at high strains, reverting to 90° for low strains. Propagation rates were the same for base metal as for the HAZ. These authors found,

$$\frac{da}{dN} = Ba = 1.18 \times 10^3\, \Delta\varepsilon_t^{2.69} a \qquad (58)$$

for growth in mm cycle^{-1}. The data are summarized in Figure 42.

Tests on crack growth in superalloys are equally sparse, no doubt because significant cracking, if any, cannot be permitted in aeroengines, for example. Data are shown in Figure 43 for IN100 in the range 700–900°C (Skelton, 1996), Hastelloy X at 760°C (Huang and Pelloux, 1980) and MAR M509 (Reuchet and Rémy, 1979). This figure also includes data on a single crystal Ni-base superalloy CMSX-2 (with the ⟨100⟩ axis in the stress direction) and the directionally solidified alloy CM247LC-DS, both at 500°C (Okazaki *et al.*, 1992). There was much scatter below a crack depth of 140 μm, and this aspect is discussed further in Section 5.02.6.2.3. Very limited dwell data on an Ni-base single crystal alloy at 900°C are also shown (Yamamoto and Nitta, 2001).

Okazaki *et al.* (1990) also tested René 80 + Hf and CM247LC directionally solidified 7 mm diameter materials at 600°C with the ⟨100⟩ axis in the stress direction. The respective average grain sizes were 0.5 mm and 2.0 mm. Crack initiation for stage I growth took place at micropores, sites of grain boundary carbide precipitation or in the slip plane itself. There were periods of cessation as a grain boundary was approached, or as the propagation switched from one {111} plane to another. The strain range was always positive (minimum strain/maximum strain, $R_\varepsilon = 0$) and no plastic hysteresis was observed.

5.02.6.2.2 *Effect of low strain rates and dwell*

Many references were cited in a previous review (Skelton, 1982) where lowering the strain rate of testing increased the cyclic crack growth rate. For example, Yamaguchi and Kanazawa (1979) found that upon reducing the strain rate from $10^{-3}\,s^{-1}$ to $10^{-5}\,s^{-1}$ in type 321 steel at 700°C there was a factor of 8

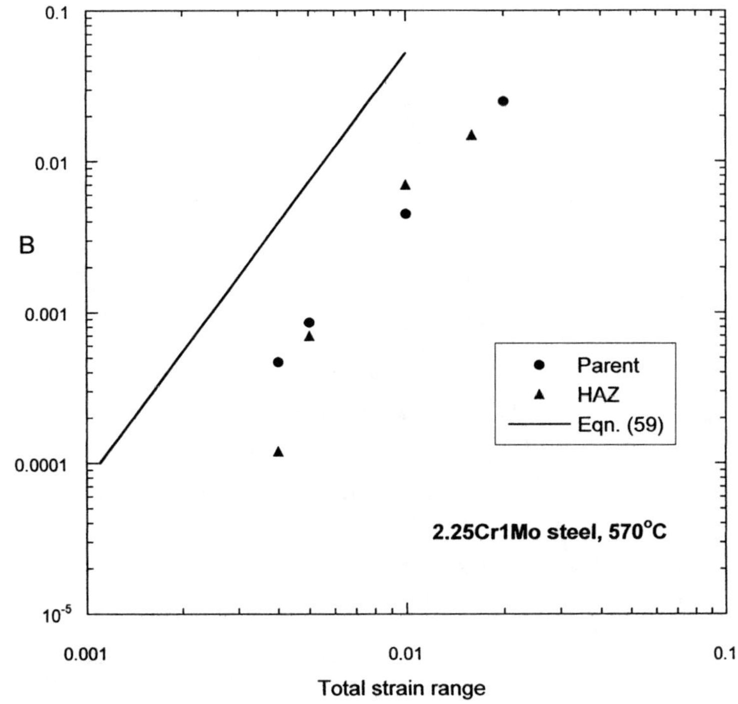

Figure 42 Short crack growth data (after Nonaka and Kitagawa, 1992).

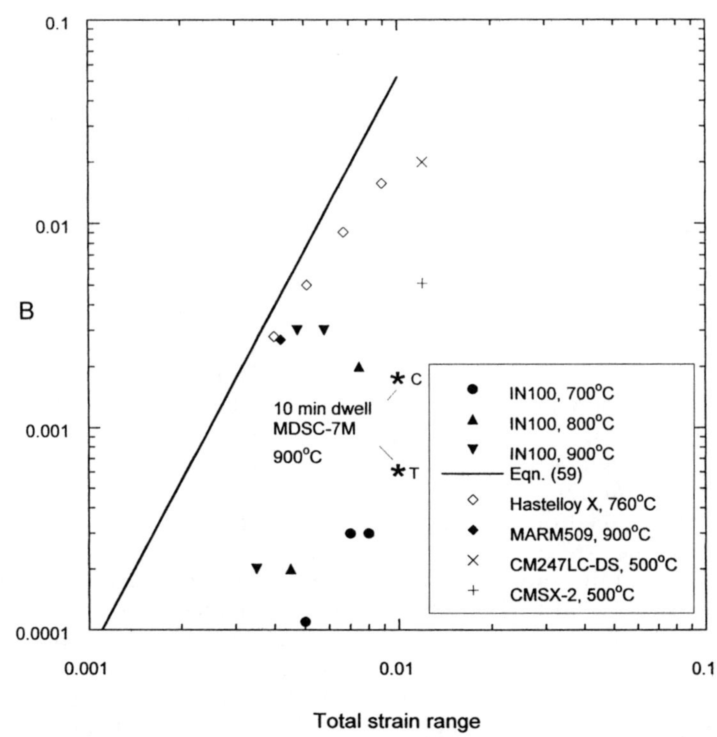

Figure 43 Limited crack growth data on superalloys.

increase in cyclic growth rate. Similarly in A286 at 593°C a factor of 4 orders of magnitude in frequency increased the growth rate by two orders of magnitude, cracking type changing from transgranular to mixed to intergranular (Solomon and Coffin, 1973). The same effects occurred in vacuum, see also Section 5.02.3.1 on crack initiation.

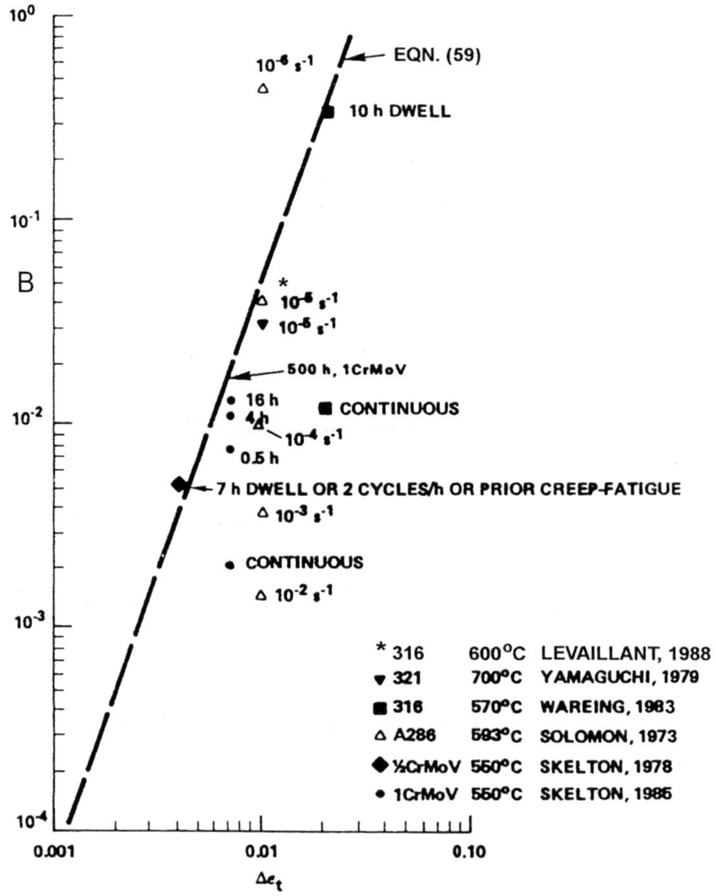

Figure 44 Crack growth data in austenitic alloys at high temperatures/low strain rates.

An upper bound value for B for growth in mm cycle^{-1} is given by the relation (Skelton, 1988a):

$$B = 2.61 \times 10^4 \Delta\varepsilon_t^{2.85} \tag{59}$$

This relation has been adopted in the R5 Assessment Code (Ainsworth, 1996). Although intended mainly for continuous cycling data in ferritic steels at temperatures up to 550°C, Figure 44 shows that the relation may safely be applied to austenitic steels at higher temperatures and low strain rates and for 316 steel with a 24 h dwell at 600°C (Levaillant *et al.*, 1988). In Figure 45 we have an example (Skelton, 1978, 1983) of crack growth rates measured during thermomechanical fatigue cycling, lying well below the upper bound relation of Equation (59), demonstrating that isothermal data at the maximum temperature are indeed pessimistic. It may be noted that for data lying well below the upper bound line, no implication is made as to whether crack growth is transgranular or intergranular.

In some fast–fast tests ($\sim 10^{-3}\,s^{-1}$) on type 304 steel at 600°C it has been shown that temporary excursions to slow–fast ($\sim 10^{-5}/10^{-4}\,s^{-1}$) cycling increased crack growth rates by a factor of 3, the original rate being restored on reverting to fast–fast cycles (Okazaki *et al.*, 1985). Thus there was little history effect after the 25 excursion cycles had taken place. Appropriate values of B at 0.8% total strain range for comparison with Figure 44 are 5×10^{-3} and 2×10^{-2}, respectively, at tensile strain rates of $2 \times 10^{-4}\,s^{-1}$ and $2 \times 10^{-5}\,s^{-1}$, respectively (Okazaki *et al.*, 1983a) although it should be noted that these hollow specimens of 13 mm diameter and 1.5 mm wall thickness had a 0.5 mm diameter through notch, crack "lengths" being measured on the surface. Other tests on this geometry employed an artificial surface notch of initial depth 0.15 mm and surface length, $2b$, of 0.3 mm (Okazaki and Koizumi 1983b). From various markings in subsequent crack growth tests these authors usefully demonstrated that the aspect ratio remained constant such that $a/2b = 0.36$.

In some tests on 304 steel at 593°C at 1% total strain range, Majumdar and Maiya

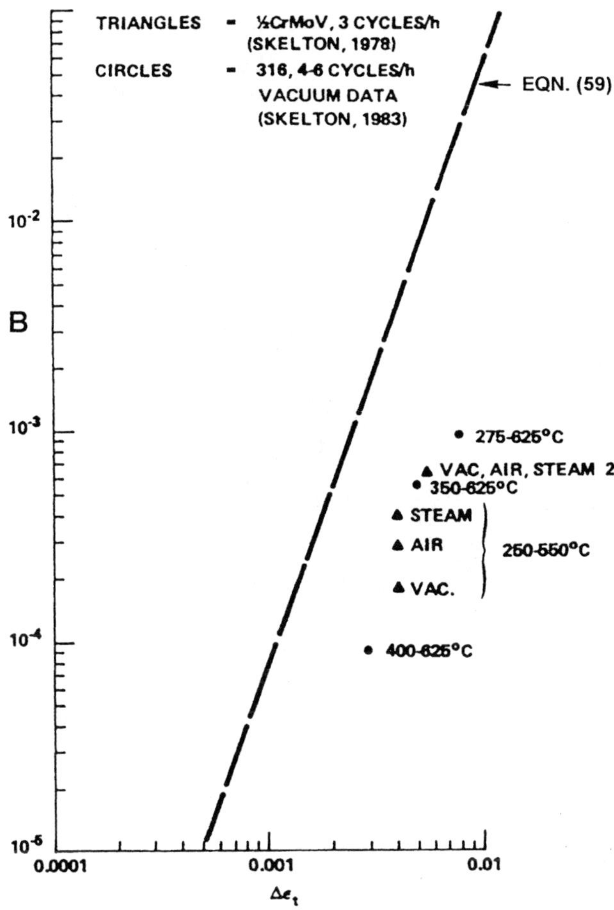

Figure 45 Crack growth constants in thermomechanical fatigue for comparison with upper bound line.

(1980a) propagated a crack to a depth of 1 mm under a 10 min tensile dwell condition ($B = 5.1 \times 10^{-3}$ in Equation (52)). At this point cycling continued with a 4 min compression only dwell. The crack growth rate was observed to slow down ($B = 3.5 \times 10^{-3}$ in Equation (52)) towards the continuous cycling growth rate condition ($B = 3.2 \times 10^{-3}$ in Equation (52)) observed in a separate test. The effect was attributed to a shrinkage of cavities under the compressive hold. This and similar experiments has formed the basis of the authors' interactive model of crack growth, see Section 5.02.4.1.5.

Since these observations apply for quite small cracks, it is seen that the now familiar creep–fatigue interaction whereby total endurances are decreased with decrease in strain rate (or increase in dwell) can equally be explained in terms of crack growth.

5.02.6.2.3 Very low crack depths

At very small specimen (and hence component) crack depths ($a/W < 0.01$) Equation (52) breaks down and growth rates appear to be independent of crack depth for a given strain level. The critical depth, $a_{\min}$, as discussed in Section 5.02.5.2, appears to be in the region of 200 μm as shown in Figure 40 (Pineau, 1983) for an austenitic steel and a superalloy (Bressers *et al.*, 1984). Abundant data have been provided on austenitic steels (Ohtani and Kitamura, 1994). In Figure 46 are shown the results of "cp" type tests on type 304 steel at 650°C, at a total strain range of 1% (Ohtani and Kitamura, 1994). They arise from painstaking surface observations on cracks in specimens of 10 mm diameter. In terms of surface crack density (number of cracks/mm^2/cycle) it was observed that slow–fast tests gave the greatest number and fast–fast tests the least. Slow–slow testing was intermediate. The half crack length ("c" as used by these authors) can be taken as the crack depth for thumbnail cracks. There is much scatter at early crack depths which is due to periods of propagation and arrest at triple points/grain boundaries, some cracks coalescing. Other information has been provided at different strain ranges and under vacuum steels (Ohtani and Kitamura,

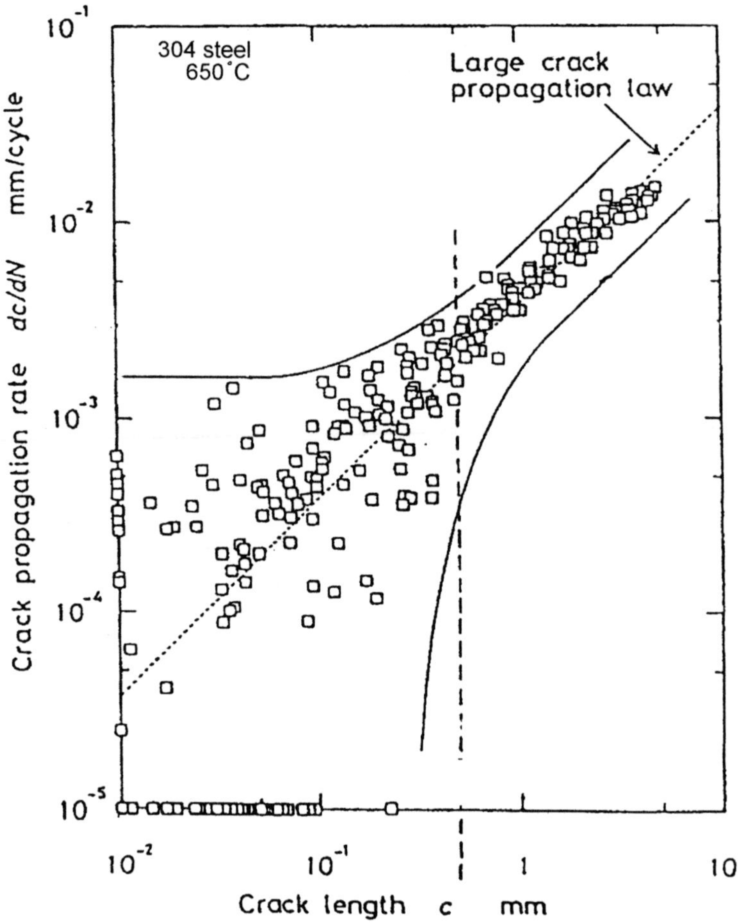

Figure 46 Behavior of cracks at very low depths (after Ohtani and Kitamura, 1994).

1994), where, e.g., the threshold at which the macrocrack propagation law, Equation (52), takes over is 100 μm for fast–fast cycling but is increased to ~200 μm for slow–fast cycling. For assessment purposes therefore it is conservative to assume an upper bound (constant) growth rate equal to that given by Equation (52) at the threshold depth, a_{min}. For general application, there may well be a case for reducing the value of a_{min} to 140 μm.

5.02.6.2.4 Integration of growth law

Equation (52) may be expressed in its integrated form:

$$N_g = \frac{1}{(Q-1)B}\left(a_o^{1-Q} - a_f^{1-Q}\right) \tag{60}$$

if $Q = \pm 1$ or,

$$N_g = \frac{1}{B}\ln\left(\frac{a_f}{a_o}\right) \tag{61}$$

if $Q = 1$, where N_g is the number of cycles involved in propagation, a_o is the initial depth, and a_f is an accepted final depth. These equations may of course be applied directly via the information in Figures 41–45 if a single value of surface strain range is known for the component. In practice a mixture of cycles of varying magnitude will occur and these will be considered several times henceforward.

5.02.6.2.5 Worked example

Suppose a component is 10 mm thick and it is required to operate under severe conditions with a surface total strain range of 1%. Assuming a surface defect of 20 μm and that the material obeys an exponential growth law ($Q = 1$) we calculate how many cycles would be required to propagate the crack to 2.5 mm depth using a conservative (upper bound) growth rate.

Putting $\Delta\varepsilon_t = 0.01$ in Equation (59), the required value of B is 5.21×10^{-2}. The calculation is next performed in two parts. Between 20 μm and 200 μm ($a_{min} = 0.2$ mm) the growth rate is constant. Substituting the value of a_{min}

in Equation (52), this growth rate is 1.04×10^{-2} mm cycle^{-1}. Over the range 20–200 μm this represents some 17 cycles. The number of cycles over the range 0.2–2.5 mm is given by Equation (61) and is some 49 cycles The total number of cycles is thus $17+49=66$.

It should be noted that if the starting depth (0.02 mm) and final depth (2.5 mm) are instead substituted directly into Equation (61), the cycle number is overestimated at 93 cycles. In the early crack growth region, therefore, propagation rates are faster than would be determined by backwards extrapolation from greater depths.

5.02.6.3 Industrial Examples on Short Crack Growth

5.02.6.3.1 Crack growth in a rotor

This example has been described previously (Rémy and Skelton, 1992). Intergranular cracks some 8 mm deep were discovered in a 1CrMoV rotor after some 6.5×10^4 h service (Priest *et al.*, 1989) with steady running periods of 325 h each. Finite element analysis showed that the peak surface strain range was ~0.7%. Simulative LCF crack growth tests with dwells up to 16 h at 550°C showed that the damage mechanism was intergranular (Skelton, 1988b), similar to that in service. The value of B was given by

$$B = 8.86 \times 10^{-3}(1 + 0.39 \log t) \qquad (62)$$

where t is the dwell time expressed in hours, see Figure 47. Extrapolated values of B out to 300 h lie below Equation (59). Using Equations (62) and (60), Table 8 summarizes likely service cycles to propagate a crack up to a final depth of 8 mm assuming the experimentally observed value of $Q=1.75$.

In the associated laboratory tests, p.d. records indicated quite convincingly that crack advance took place during the reversal and not during the dwell period itself (Skelton, 1988b). Thus the increasing values of B with time in Figure 47 are a reflection of the damage accrued during the previous dwell. Metallography confirmed that the degree of intergranular cracking was greater after a 16 h dwell than was the case with a dwell of 0.5 h (Skelton, 1988b).

Table 8 Calculated cycles to propagate crack to 8 mm depth.

Tension dwell (h)	*Propagation cycles*		
	$a_o = 0.2$ mm	$a_o = 0.5$ mm	$a_o = 1.0$ mm
16	321	151	81
100	265	124	67
500	229	108	58

Source: Rémy and Skelton (1992).

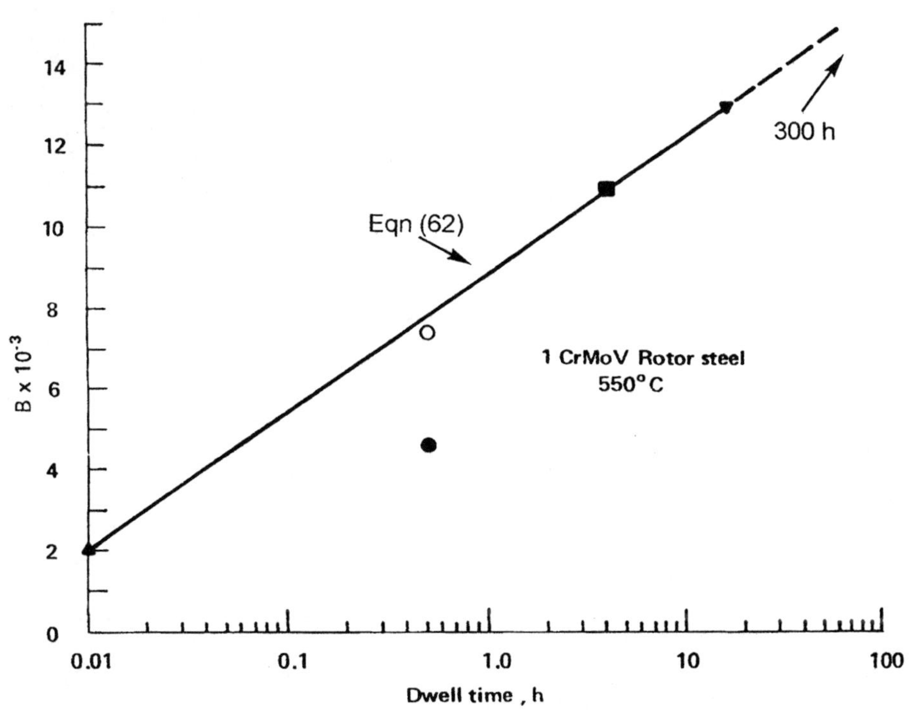

Figure 47 Extrapolation of B values to long dwell. Symbols refer to separate crack growth tests (after Rémy and Skelton, 1992).

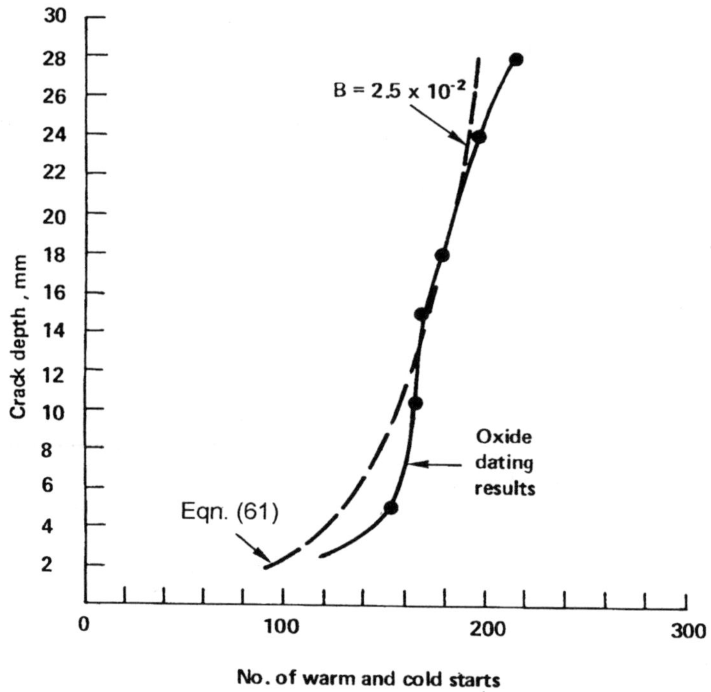

Figure 48 Progress of crack growth in a ligament (after Rémy and Skelton, 1992).

5.02.6.3.2 Crack growth in a ligament

The causes of interligament cracking in superheater headers in power stations has been discussed previously (Viswanathan, 1989; Viswanathan and Stringer, 2000). Transgranular cracking was observed in a ligament area between adjoining 2(1/4)CrMo steam tubes after 8.5×10^4 h of service at around 570°C (Rémy and Skelton, 1992). The unit had experienced some 360 start-ups ("cold," "warm," and "hot") during service and the severest thermal loading was estimated to occur in the range 350–450°C. It was possible to perform a retrospective analysis of crack progress using an oxide dating technique (Pinder, 1981). Estimated progress with the number of "cold" and "warm" starts (which were considered chiefly responsible) is shown in Figure 48.

The ligament may be regarded as a large laboratory specimen undergoing reversed loading, see Figure 49. A test on service-exposed material in continuous cycling at 400°C and 570°C at a total strain range of 1.3% indicated a transgranular cyclic growth rate, independent of temperature and that $B = 6.6 \times 10^{-3}$, $Q = 1$ in Equation (52), lying below the upper bound Equation (59). It may be noted that the best fit of Equation (61) to the oxide dating line in Figure 48 is by assuming $B = 2.5 \times 10^{-2}$. Separate crack growth tests were undertaken and these indicated that the total strain range in service must sometimes have exceeded 2%.

Figure 49 Typical ligament cracking. Inset shows detail of 3 mm deep transgranular crack.

5.02.6.4 Enhancement of Growth Law—Damage Factors

Our considerations so far have suggested that the basic LCF crack growth law in continuous cycling must be modified to take account of creep–fatigue damage. Reuchet and Rémy (1979) found that for MAR M509 alloy at 900°C integration of Equation (52) with $Q = 1.8$ and $a_o = 60$ μm gave a predicted life of 3,900 cycles compared with an observed plain specimen endurance of 1,700 cycles. They suggested that an exponential growth law applies for undamaged material, but as cracking proceeds into cumulatively damaged regions the relation becomes:

$$\frac{da}{dN} = Ba(1 + qf) \tag{63}$$

where q is a constant and f is the volume fraction of cavities, which in their experiments developed as $(\text{time})^{4/3}$, i.e., as $N^{4/3}$. Thus, the value of B in Equation (52) effectively increases with time. Upon integrating Equation (63) instead, a propagation life of only 1,350 cycles was predicted (Reuchet and Rémy, 1979).

5.02.6.4.1 Crack growth rate

If N denotes the number of cycles involved in a continuous cycling step and N^* the reduced number of cycles in that same step due to creep–fatigue interaction, then it is possible to show by differentiation of Equation (30) (Skelton, 1993b) or otherwise (Levaillant and Pineau, 1982) that the following equivalent relations apply:

$$\left(\frac{da}{dN}\right)^* = \left(\frac{da}{dN}\right)(1 + ND_c)^2 = \left(\frac{da}{dN}\right)\frac{1}{(1 - N^*D_c)^2} \tag{64}$$

The second relation shows a classic damage function ω increasing towards unity as the number of creep–fatigue cycles accumulates, i.e., in this linear-damage model the term D_c now explicitly accounts for the creep damage per cycle during the crack growth phase. Under these conditions therefore the basic LCF continuous cycling law given by Equation (52) becomes enhanced owing to the creep damage accumulated during the dwell. The result when $Q = 1$ is

$$\left(\frac{da}{dN}\right)^* = Ba\left\{1 + \frac{D_c}{B}\ln\left(\frac{a}{a_o}\right)\right\}^2 \tag{65}$$

the asterisk again indicating the creep–fatigue modification. A very similar relation applies when $Q > 1$. Values of D_c have been determined empirically for creep–fatigue growth in a variety of alloys (Skelton, 1993b) and the results may be summarized as a plot of damage factor versus effective frequency (Skelton and Byrne, 1994) where the latter combines the hold time and reversal period, see Figure 50. It may be noted that Equation (64) has similarly been adapted to cater for deep creep–fatigue crack growth with dwell in the LEFM range where the region of interest lies in the plastic zone ahead of the crack tip (Skelton, 1993b). Such considerations however lie outside our present scope.

5.02.6.4.2 Propagation depth

Equation (65) is most useful in its integrated form for block loading over N_i cycles. In

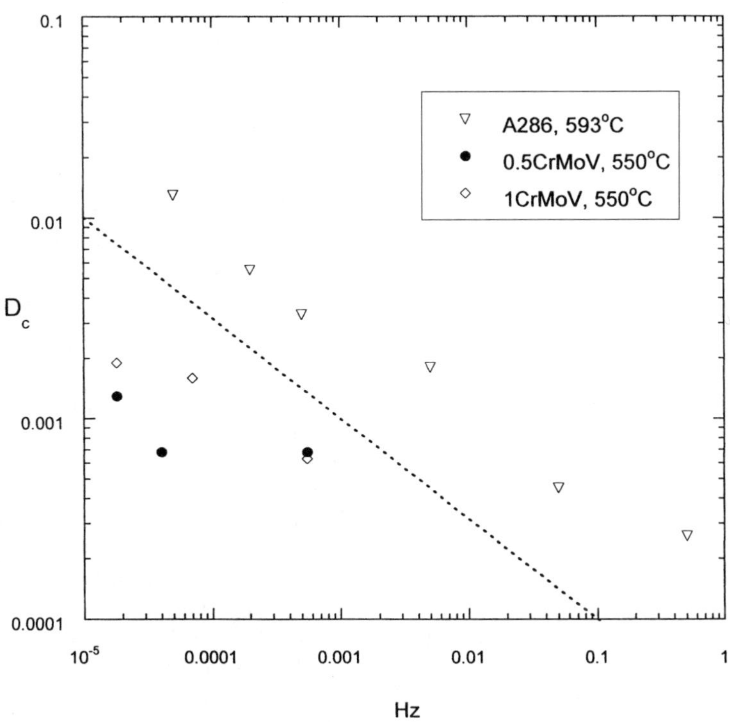

Figure 50 Trend of damage factor with decreasing testing frequency (after Skelton and Byrne, 1994).

Table 9 Postulated loading regime for component.

Application	*Cycles/block*	$\Delta\varepsilon_t$ (%)	D_c	*B in Eqn. (59)*
1st application	50	0.6	3×10^{-4}	1.21×10^{-2}
2nd application	200	0.5	2×10^{-4}	7.22×10^{-3}
3rd application	300	0.4	1×10^{-4}	3.82×10^{-3}

Table 10 Predicted crack growth from separate damage models.

Type of damage	*Penetration depth* (mm)
None (continuous cycling)	4.88
Noncumulative	5.41
Cumulative	6.56

continuous cycling this takes the form:

$$a_f = a_o \exp\left(\sum_{i=1} B_i N_i\right) \qquad (66)$$

where the damage function has been set to zero.

The relation may also be used in a "noncumulative damage" mode:

$$a_f = a_o \exp\left\{\sum_{i=1} \frac{B_i}{D_{ci}}\left(\frac{1}{1-N_i D_{ci}} - 1\right)\right\} \qquad (67)$$

This form appears in the R5 Code (Ainsworth, 1996). It allows creep damage to accumulate over each block of identical cycles only, resetting the value of D_c for each block. Damage in previous cycles is thus ignored, and this could occur physically when a period of intergranular cracking is followed by transgranular cracking.

For true cumulative damage, all damage that has occurred previously in the specimen or component must be accounted for. It can be shown (Skelton, 1993b) that

$$a_f = a_o \exp\left\{\sum_{i=1} \frac{B_i N_i}{(1-\omega_o)(1-\omega_t)}\right\} \qquad (68)$$

where ω_o is all previous cumulative damage up to the current ith block and ω_t is this damage plus that accumulated in the ith block itself.

5.02.6.4.3 *Worked example*

This case is an extension of the energy worked example in Section 5.02.4.6.2, where an initial defect size is now assumed. A component containing a surface crack of initial depth 0.2 mm receives the loading regime summarized in Table 9. It is required to calculate the final depth of crack penetration for the cases of Equations (66)–(68), respectively.

This case has been discussed in some detail previously (Rémy and Skelton, 1992; Skelton, 1993b). The results are given in Table 10 where the deeper penetration is demonstrated for the two instances of damage. These equations offer a very practical method of engineering assessment and the term D_c can be deduced from metallography (Levaillant *et al.*, 1988) or by means of Figure 50, for example.

5.02.6.5 Discussion

This section (Section 5.02.6) has demonstrated that penetration of surface defects may be calculated quite simply from a knowledge of the surface (total) strain range undergone by the component. The advantages of the techniques are:

(i) Generally, no complex finite element analyses or compliance considerations are required for the component.

(ii) Appropriate values of the parameter *B* may be read off from the "B–$\Delta\varepsilon_t$" diagram such as provided in Figures 41–45, or alternatively an upper bound relation (Equation (59)) may be used for a conservative assessment.

The relations may only be used to a limited depth ($a/W<0.25$) in the component, or, if beneath a stress concentrator, to the depth of a calculated plastic zone.

5.02.7 CASE STUDY OUTLINE—THICK SECTION 316 AUSTENITIC STEEL COMPONENTS

The discovery of a major steam leak in the main steam piping of Eddystone Unit 1 supercritical power station in March 1983 after 1.3052×10^5 h service provided a great impetus for metallurgical investigation and remanent life techniques on these pipes and other components. The findings were reported at some length as regards the failure investigation itself (DeLong *et al.*, 1985), metallurgical observations on failed components (Ellis and Bynum, 1983; Masuyama *et al.*, 1985), other components at risk (DeLong *et al.*, 1984a; Argo *et al.*, 1984), operational experience and lessons learned (DeLong *et al.*, 1984c; DeLong *et al.*, 1984b) and finally the choice of replacement material (McGough *et al.*, 1985). Table 11 provides further information on components where cracking in components led to remedial action. As will be discussed, cracking was either due to thermal fatigue, or a creep–fatigue interaction in the residual stress state. A relatively small incident of pipe cracking after 1.25×10^5 h, see Table 11, led to samples being removed for creep tests, and as a result, a predicted remaining life of some 2.3×10^5 h (DeLong *et al.*, 1984c). In fact only ~7,000 h elapsed before the major steam leak. In the following, two distinct forms of failure are compared and contrasted.

5.02.7.1 Layout and Operating Schedule

After leaving the outlet header, steam passes through the main steam pipes through boiler stop and by-pass valves into the junction header. From here, it then goes via the turbine stop and control valves into the HP turbine itself. A full description is provided by DeLong *et al.* (1984a). The 325 MW unit was commissioned in February 1960 with a steam temperature of 650°C and pressure of 34.5 MPa, though these were lowered to 610°C and 32.4 MPa, respectively following the onset of thermal fatigue cracking in the control valves and junction header. In Table 11 it is seen from the relatively small number of shutdowns that the unit was operating mostly on base load. During shutdown periods there was rapid cooling of the main steam pipes between the outlet header and the boiler stop valves, but not downstream of this point. Conversely, during start-up when the boiler stop valves opened, supercritical steam caused compressive yielding at the inner surface of the turbine stop and control valves.

5.02.7.2 Cracking in Boiler Stop and Control Valves

The two halves of the boiler stop and control valves are connected by a transition pipe of outside diameter 249 mm and wall thickness 64 mm, i.e., very similar to that of the main steam pipe to be considered below. The service history was also identical to that of the main steam pipe (ii) given in Table 11. Surface cracking was detected on the insides of the whole assembly to a depth of ~10 mm (Argo *et al.*, 1984; DeLong *et al.*, 1984c). Cracking in the bowl regions was more severe and in the range 16–35 mm.All cracking was intergranular and at the surface had a "crazed" appearance (DeLong *et al.*, 1984a) typical of thermal shock. Voids near the crack tips suggested that crack growth occurred at high temperature. There was no evidence of service-induced sigma-phase formation in contrast to the pipes, see below.

In supporting mechanical testing, the creep life of specimens taken from the transition pipe was improved after re-solution treating (DeLong *et al.*, 1984c). Similarly, the ductility of service-exposed material at 650°C (~30%) was markedly increased to ~55% after a solution treatment. Surprisingly, this heat treatment did not affect the fatigue life in strain-controlled tests (Argo *et al.*, 1984), see also Section 5.02.5.1.4 and Figure 33.

Two-dimensional thermal elastic–plastic-creep analysis was performed, assuming the valve body to be an axi-symmetric structure (DeLong *et al.*, 1984a). The maximum temperature difference between the inner and outer surface occurred in 0.075 h for the stop valve and 0.1 h for the control valve. These gave rise to inner–outer temperature differences of 130°C and 325°C, respectively. This was because the stop valve was to some extent pre-warmed, whereas the control valve experienced a sudden entrance of steam flow. Hysteresis loops were computed hysteresis loops for both valve parts. The initial stress rapidly became tensile, reaching a maximum in excess of 150 MPa after 0.2–0.3 h. After a long period this stress reduced under creep relaxation. Similar loops were created under repeated service cycles, giving rise to a creep–fatigue interaction. There were no major engineering

Table 11 Summary of austenitic components experiencing cracking.[a]

Component	*Hours service*	*Cycles*	*Notes*
Boiler stop/by-pass valve	0.512×10^5	121	20 mm crack depth. Ground out. Later, redesigned with thinner wall thickness.
Junction header	0.25×10^5	77	20 mm intergranular cracks. Redesigned, wall thickness reduced from 184 mm to 130 mm.
Turbine stop/control valve	1.304×10^5	326	See text
Main steam pipe (i)	1.25×10^5	305	1st incident
Main steam pipe (ii)	1.3052×10^5	315	Exterior cracking, see text.

[a] Average dwell time = 395 h.

implications except to note that the effects of thermal fatigue may be offset by pre-warming the control valve before admitting steam or increasing the temperature of this steam more slowly (DeLong *et al.*, 1984a).

5.02.7.3 Main Steam Pipe Failure

The steam pipe of austenitic steel had an outside diameter of 230 mm and a wall thickness of 64 mm.Intergranular crack failure initiated at the outside of the pipe normal to the hoop stress as shown in Figure 51. This contrasts with the transition pipe of the turbine stop/control valve, which was of similar dimensions but had cracked from the inside. Sectioning and metallographic polishing of the main steam pipe revealed extensive grain boundary cavitation adjacent to the main crack and to within 12.5 mm of the inside diameter. Closer examination revealed cavitation to occur in the ligament region between large sigma particles (1.4% precipitation) (DeLong *et al.*, 1985). There was no evidence of ratcheting (gross creep swelling).

5.02.7.3.1 Supporting mechanical testing

Accelerated creep tests at 750°C on samples taken from the failed pipe indicated extensive creep damage (Bynum *et al.*, 1984; DeLong *et al.*, 1985). Tests at 650°C indicated a much reduced time to failure compared with control data (Ellis and Bynum, 1983). In addition, ductilities were severely reduced at all testing temperatures. Accompanying room temperature Charpy tests, however, showed no propensity for rapid crack propagation.

5.02.7.3.2 Failure scenario

It was proposed (Bynum *et al.*, 1984; DeLong *et al.*, 1984c) that:

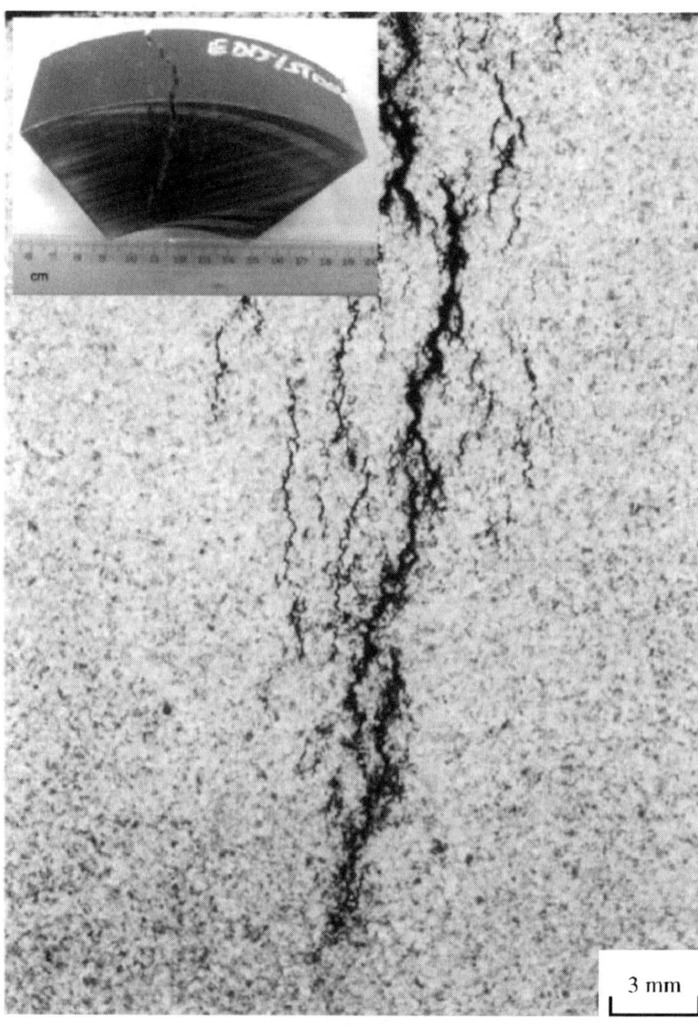

Figure 51 Intergranular cracking on outside of main steam pipe (see inset) (after Masuyama *et al.*, 1985. Reproduced with permission of ASME International).

(i) Residual plus primary stresses caused failure.

(ii) In the steady-state (base load) condition, the outer two-thirds of wall thickness must have been in tension and the inner one-third in compression for self-equilibrium.

(iii) Severe downshock occurred at the inner surface during closure of the boiler stop valve owing to relatively cold fluid impingement during shutdown.

To augment this hypothesis, residual stress measurements were conducted at selected steam pipe surfaces on site. The results are summarized in Table 12.

5.02.7.3.3 *Stress analysis*

An elastic–plastic analysis was undertaken (DeLong *et al.*, 1985) for the following conditions:

(i) 111°C h^{-1} start up,
(ii) 2,000 h dwell period at 616°C,
(iii) Slow cool to 538°C, and
(iv) Thermal downshock of 278°C in 5 min.

Table 12 Residual stresses, main pipe outside diameter (on site).

Steam line	*Tensile stress* (MPa)
1B	117
3B	79
2A	146
2B	209
1A	86

The maximum thermal gradient between the outer and inner diameter was calculated to be 217°C. The elastic–plastic hoop stress distribution resulting from this gradient is shown in Figure 52. After cooling to room temperature and removing the pressure the residual stress state is also shown in Figure 52, where it is noted that the stress is tensile at the outer surface. On re-applying pressure and heating to 616°C minor changes occurred to this distribution which also was demonstrated to relax a little after 2000 h creep (DeLong *et al.*, 1985).

5.02.7.4 *Comparison of Cases and Damage Diagram*

Detailed calculated hysteresis loops were provided (DeLong *et al.*, 1984c) for the transition pipe in the turbine control valve (successive upshocks) which contrasted with those for the steam pipe (successive downshocks). They indicated that the steady state of tension in the residual state of base load running of the power plant was at the bore for the control valve and at the surface for the steam pipe. Both these states induced intergranular cracking. Using Code ASME Case N-47, the creep–fatigue damage endured by several components was determined (DeLong *et al.*, 1984c) and these are summarized in Figure 53. It is seen that the steam pipe experienced mostly creep damage while the transition pipe also received an element of fatigue damage. The junction header (not considered here), e.g., received about equal components of creep and fatigue damage.

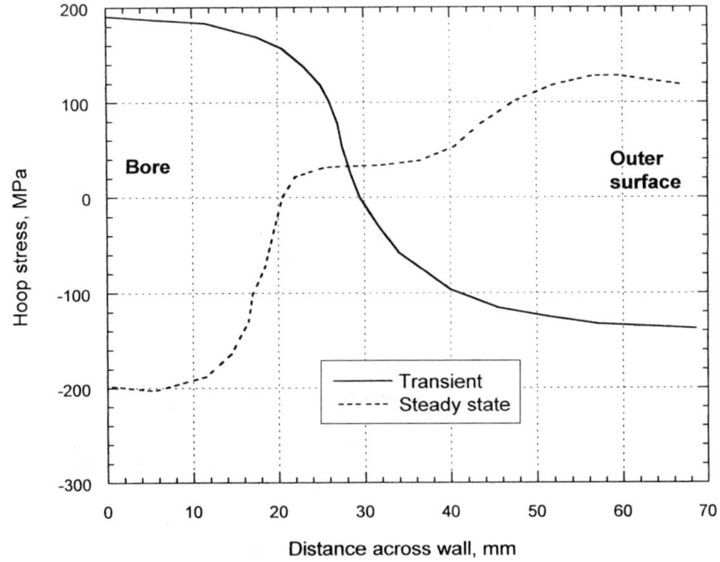

Figure 52 Stress profiles in main steam pipe during transient down shock and in the steady running period (after DeLong *et al.*, 1985).

The engineering decision following this large investigation was as follows:

(i) It was decided to replace all steam piping between the outlet headers to the boiler stop valves (McGough *et al.*, 1985).

(ii) Steam piping downstream of the boiler stop valves was left intact because there were no signs of cracking, creep damage or residual tensile stresses.

The key factor in this case study is that, if a thermal transient causes yielding at a location, the sign of the residual stress at that location is generally opposite to that induced there during the thermal transient. The same concerns continue to be debated in the design of modern supercritical plant (Bertilsson and Scarlin, 1990; Nakabayashi *et al.*, 1990; Blum, 1994).

5.02.8 Validation—Creep–Fatigue Crack Initiation in a Thick Component

In this detailed example we draw together many of the aspects covered in this review. The case has been given previously (Skelton, 1998) but further validation is supplied here in the form of an extra energy calculation and a revisited experiment on service-exposed material taken from the failed component. The methods closely follow the procedures on creep–fatigue initiation recommended in the R5 Assessment Procedure (Ainsworth, 1996).

5.02.8.1 Component History

This retrospective study concerns the failure of an actual steam chest that had been in service at 570°C peak temperature in power plant. The component was a control valve in a steam line. Its shape can be approximated by a cylinder of outside diameter 800 mm having a wall thickness of 155 mm and internal height 300 mm. Cracks initiated at a change of section at the corners of the vessel under the influence of a small stress concentration. The component appeared to have failed across the whole section by intergranular creep–fatigue crack growth accompanied by grain boundary cavitation.

Before entering service, the 1CrMoV component had been heat treated thus: normalized at 1,036°C followed by an oil quench. It was subsequently tempered for 10 h at 704°C giving a bainitic microstructure of grain size 200–400 μm. The operating history at the time of failure was as follows: hours run 7×10^4; number of hot starts 106; number of cold starts 127; therefore giving a total number of starts of 233 and average "dwell" time of 300 h.

It is required to predict the number of service cycles to initiate a crack 9.5 mm deep. As will be shown, this corresponds to the depth of the plastic zone generated by temperature changes at the inner surface of the wall, which is 155 mm thick.

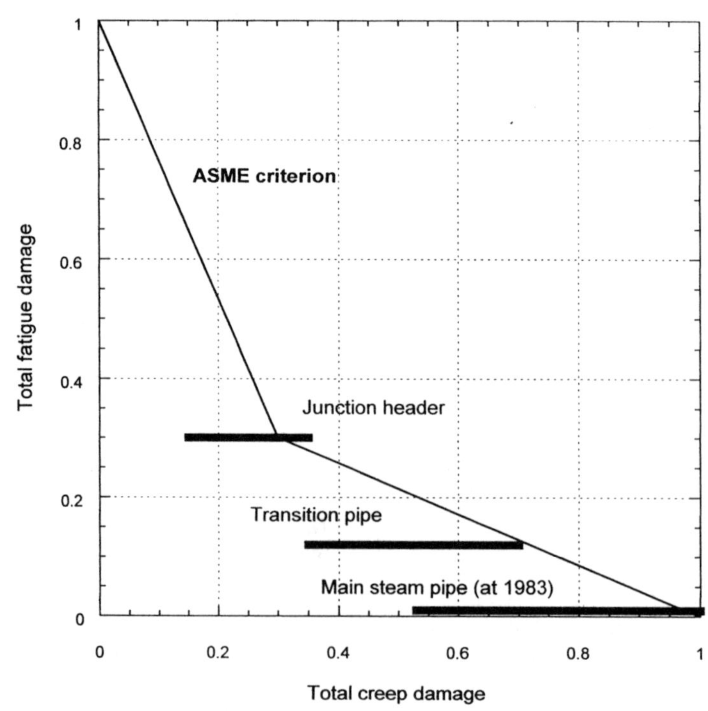

Figure 53 Creep–fatigue analyses of plant components ranked on the ASME diagram (after Masuyama *et al.*, 1985).

5.02.8.2 Preliminary Calculation

There are several requirements to be fulfilled before the calculations can be carried out. The procedure requires service data, elastic stress analyses for peak transients, materials properties, and elastic–plastic stress analyses as a prelude to the damage calculations. Many of the calculations were performed on material-specific properties. These were obtained on specimens machined from a wedge of material taken from the failed chest and included creep, fatigue (continuous cycling) endurance, cyclic stress–strain, and cyclic stress relaxation tests.

An outline of the procedure runs as follows:

(i) Obtain the service load history and perform stress analysis of significant cycle types.

(ii) Obtain cyclic stress–strain loops including creep during dwell.

(iii) Obtain cyclic endurances relevant to initiation depth, a_o, cyclic stress–strain data, relaxation properties, etc.

(iv) Calculate service cycles to initiate the crack (to depth of surface plastic zone only) and express result in terms of total damage, Φ.

(v) Plot result on damage diagram (optional).

(vi) Perform sensitivity analysis with respect to (i) service cycle type and (ii) materials properties.

(vii) Report results of the assessment.

For this example, it is appropriate to revisit the key definitions as certain qualifications are necessary:

a_o Initiation crack depth (variable, incorporates size effect). Changed to a_c (suffix "component") in this analysis since $a_c >$ specimen width.

N_o Continuous cycling endurance corresponding to a_o. Changed to N_c to correspond with a_c defined above.

D_c Creep damage per cycle (using ductility exhaustion). A bulk failure mechanism.

N_o^* Required parameter (creep–fatigue endurance).

From the above, the procedure may be roughly broken down into seven steps. These are now considered in turn, though some spilling over into other conventional section headings will occur. The last step (reporting) is self-explanatory and this case study is a precis of a much larger investigation that was undertaken.

Step (1). In Figure 54 is shown, from service records, the inner surface metal temperature rise during a typical start up together with the time of the peak transient. In this case the peak through-wall temperature differential was calculated to be 200°C. In other cases, the severity of the starts according to time off load was defined as "major," "intermediate," and "minor" cycles as shown in Figure 55. Typical peak elastic transients calculated from finite element analyses are shown in Figure 56. A smaller profile of opposite sign also occurred during cooling transients when coming off load. This is shown in Figure 57. To this profile is added the hoop stress due to steam pressure, giving a resultant stress profile, Figure 57.

Step (2). In order to obtain typical hysteresis loops experienced at the component surface during service, cyclic stress relaxation and cyclic stress–strain data are required. A typical relaxation result for use with Equation (25) obtained from steady-state cycling of the bainitic alloy at 565°C is shown in Figure 58. Similarly, typical cyclic stress–strain results for two service strain rates at 500°C for use with Equation (42) are given in Figure 59. Using the Neuber (1961) relation for a stress-concentrating feature, the compression-going and

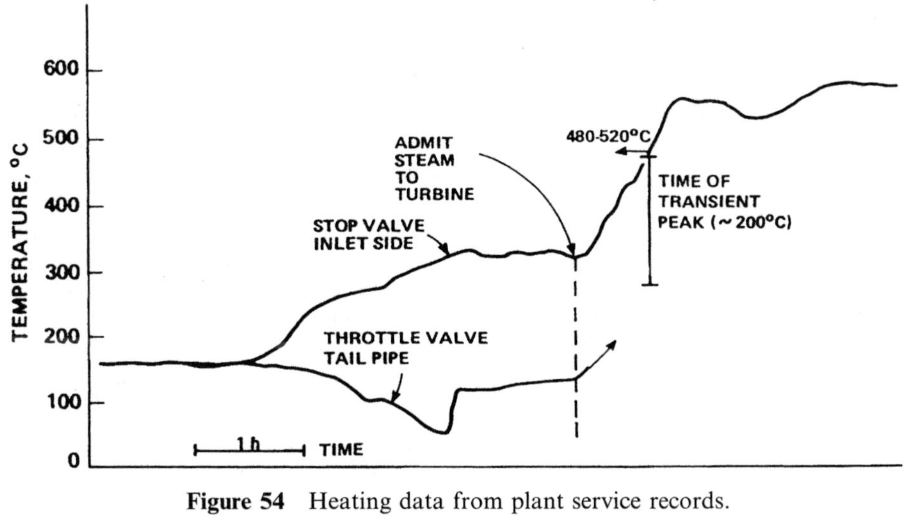

Figure 54 Heating data from plant service records.

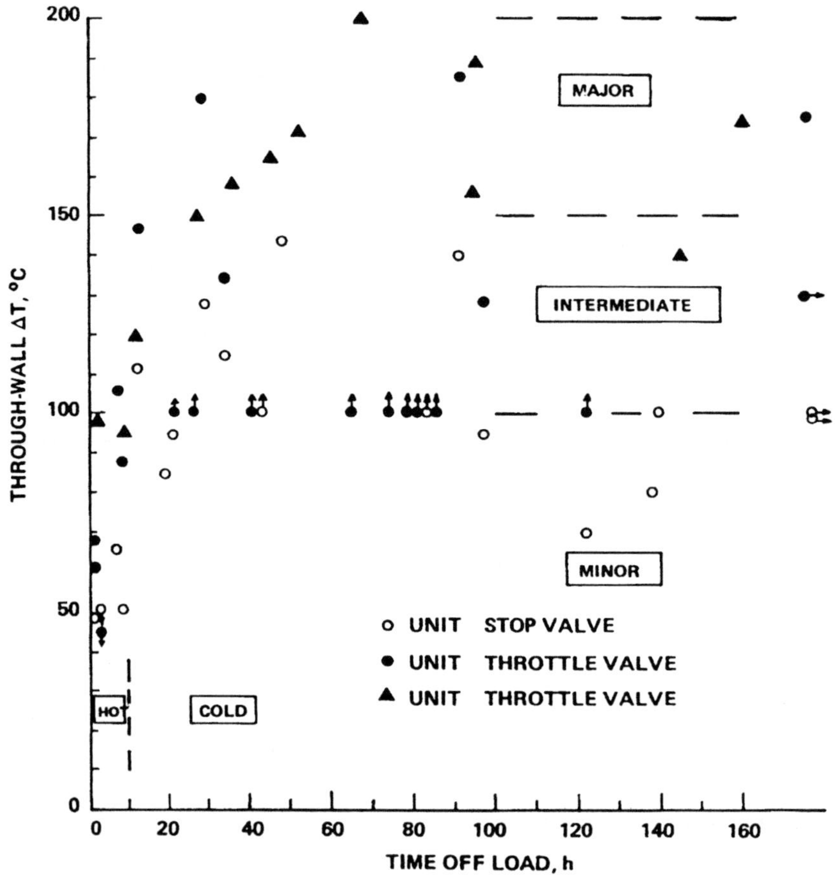

Figure 55 Temperature differentials at start-up as a function of time "off-load."

tension-going attributes of the loop were calculated using the data in Figures 58 and 59, for example. These calculations were pursued cycle by cycle until loop closure (shakedown) had been demonstrated. An example for a major cycle at a plastic strain range of 0.3% is shown in Figure 60. It is emphasized that the "ratcheting" effect into compressive strain is not real but is a feature of the computer system. Summarizing data for other cycle types are given in Table 13.

In this example, an extra step is required, namely the estimation of the surface plastic zone size after the stress profiles have shaken down to a steady state. The case for major cycling is shown in Figure 61. A summary of results for other cycle types is given in Table 14. It will be appreciated that, according to the cycle type chosen, there will be the possibility of overlapping "zones within zones" depending on the cycle mix.

5.02.8.3 Application of Linear Damage Rule

Step (3). In order to calculate creep–fatigue damage, a continuous-cycling endurance plot is required. Clearly, to have determined the complete curve over all strain ranges would have required prolonged experimental effort. Instead, an established curve for a similar material was used and a single endurance obtained for the valve material at the calculated strain range of the major cycle. The result is shown in Figure 62. Similarly, a ductility–strain rate curve is required for summing the creep damage. An example for two extreme heat treatments of the chest material (Neate, 1985) is given in Figure 63. Referring to Section 5.02.4.5 it is noted that in its re-heat treated form the alloy demonstrates a transition between upper and lower shelf behavior whereas the service-exposed from has a very low ductility on a separate lower shelf.

Step (4). As discussed in Section 5.02.4.3.3, the number of creep–fatigue cycles to fail the zone is given by

$$N_o^* = \left(\frac{1}{N_o} + D_c\right)^{-1} \qquad (69)$$

In this case, however, we note that the maximum plastic zone size is somewhat

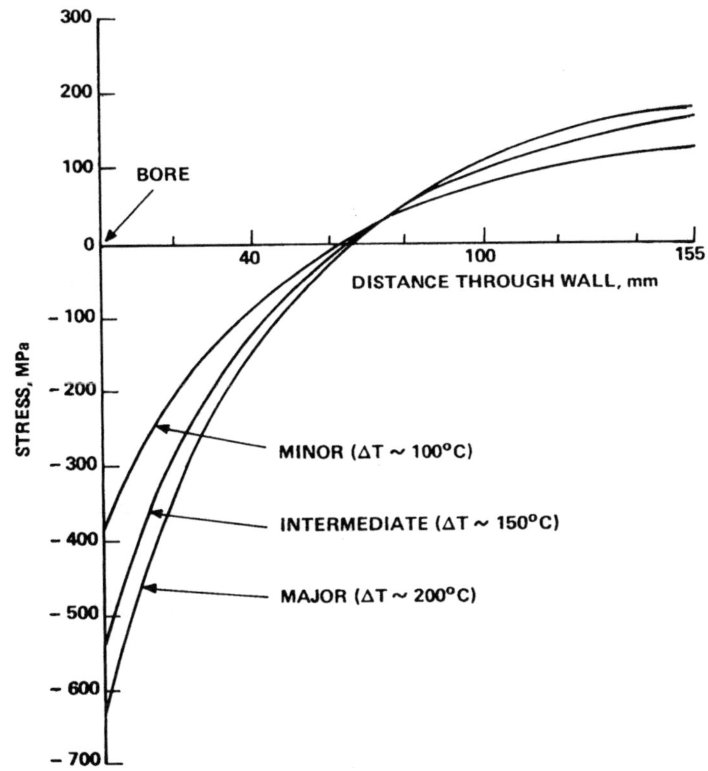

Figure 56 Peak (elastic) thermal stresses in chest at start-up.

Figure 57 Stresses arising in chest after cooling.

greater than the diameter of a typical LCF specimen (6–8 mm). Since $Q = 1$ for the present material, referring to Equation (54) it may be shown that the original terms N_o, a_o now become N_l, a_l and replacing the original terms N_l, a_l by the terms N_c, a_c (suffix "c" for component) then,

$$N_c = \frac{N_l - (1 - M)N_i}{M} \tag{70}$$

where

$$M = \frac{(a_{min} - a_i) + a_{min}\ln(a_l/a_{min})}{(a_{min} - a_i) + a_{min}\ln(a_c/a_{min})} \tag{71}$$

The maximum value of plastic zone size (a_c) in this example is 9.5 mm. The results of the damage calculation (i.e., cycles to fail the maximum zone size) are given in Table 15.

One method of mixing cycles for a damage calculation is based on the time that the unit is off load. In this case, only major and minor cycles are involved. Using the data in Table 15 therefore and assuming an 80%

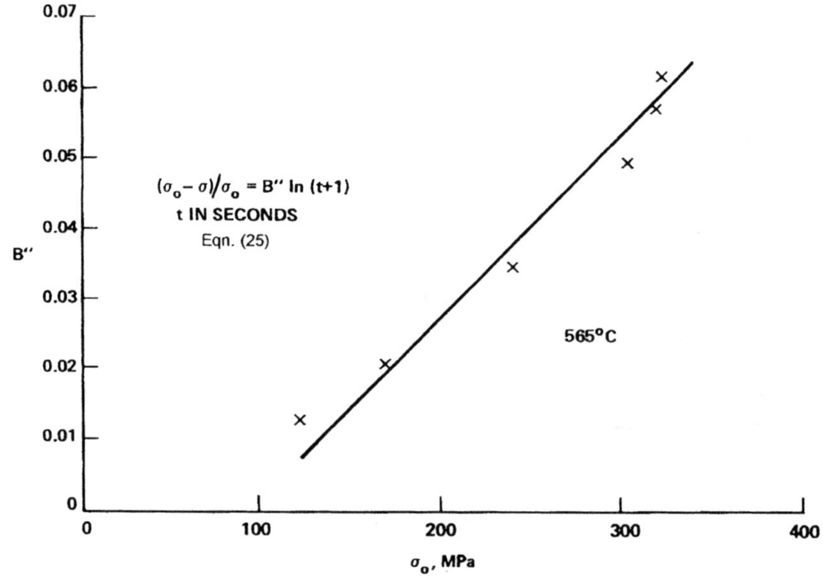

Figure 58 Constants required for stress relaxation under cyclic conditions.

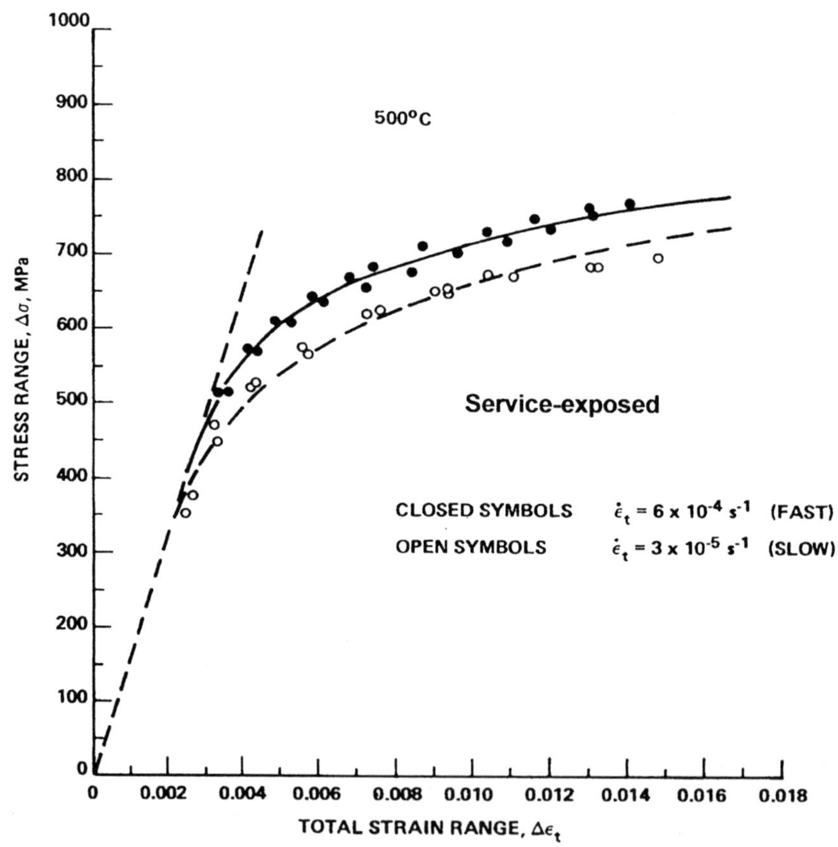

Figure 59 Cyclic stress–strain data for use in analysis of transient.

major ("cold start") and 20% minor ("hot start") distribution we have to solve the linear damage equation:

$$\frac{0.8N}{26} + \frac{0.2N}{483} = 1 \qquad (72)$$

giving $N = 32$ cycles. Thus a 9.5 mm crack initiates in the plastic zone by 32 service start-ups.

Step (5) (Optional). This step is presented last of all in order to compare alternative calculations.

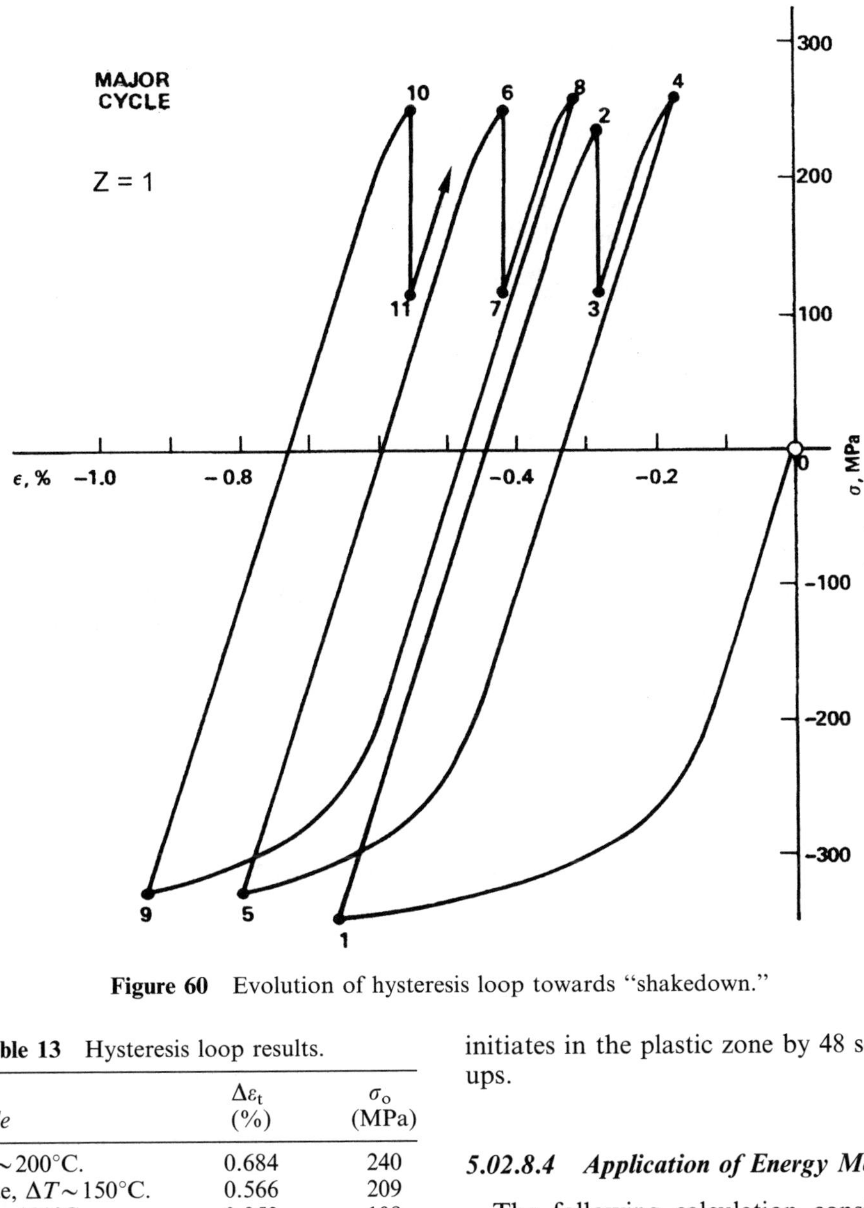

Figure 60 Evolution of hysteresis loop towards "shakedown."

Table 13 Hysteresis loop results.

Service cycle	$\Delta\varepsilon_t$ (%)	σ_o (MPa)
Major, $\Delta T \sim 200$°C.	0.684	240
Intermediate, $\Delta T \sim 150$°C.	0.566	209
Minor, $\Delta T \sim 100$°C.	0.353	109

Step (6). This step deals with a sensitivity analysis. Another method of mixing cycles for a damage calculation is based on the "temperature differential" distribution of data, see Figure 55. In this case, major, intermediate, and minor cycles are involved. Using the data in Table 15 therefore and assuming a 19% major ("cold start"), 48% intermediate ("warm start"), and 33% minor ("hot start") distribution we have to solve the linear damage equation:

$$\frac{0.19N}{26}+\frac{0.48N}{37}+\frac{0.33N}{483}=1 \qquad (73)$$

giving $N=48$ cycles. Thus a 9.5 mm crack initiates in the plastic zone by 48 service start-ups.

5.02.8.4 Application of Energy Method

The following calculation considers major cycles only and belongs to step (6) as part of a sensitivity analysis. In Figure 60 it is noted that the initial stress is 250 MPa and that from Equation (25) and Figure 58 the calculated stress after 300 h of relaxation is 110 MPa. From Equation (43) the creep energy dissipated per cycle is thus found to be 1.68×10^{-4} J mm^{-3}. There are no long-term creep–fatigue endurance data for this service-exposed steel but from the information in Table 5 we may assume a pessimistic value of $W_{f,\,creep}$ of 0.01 J mm^{-3} and, as discussed in Section 5.02.4.6.3, accord all the area "B" in Figure 2 to creep damage.

The fatigue component of damage is calculated as follows. From the "slow" cyclic stress–strain curve in Figure 59 it may be shown that $A=1{,}499$ MPa, $\beta=0.161$. From Equation (39)

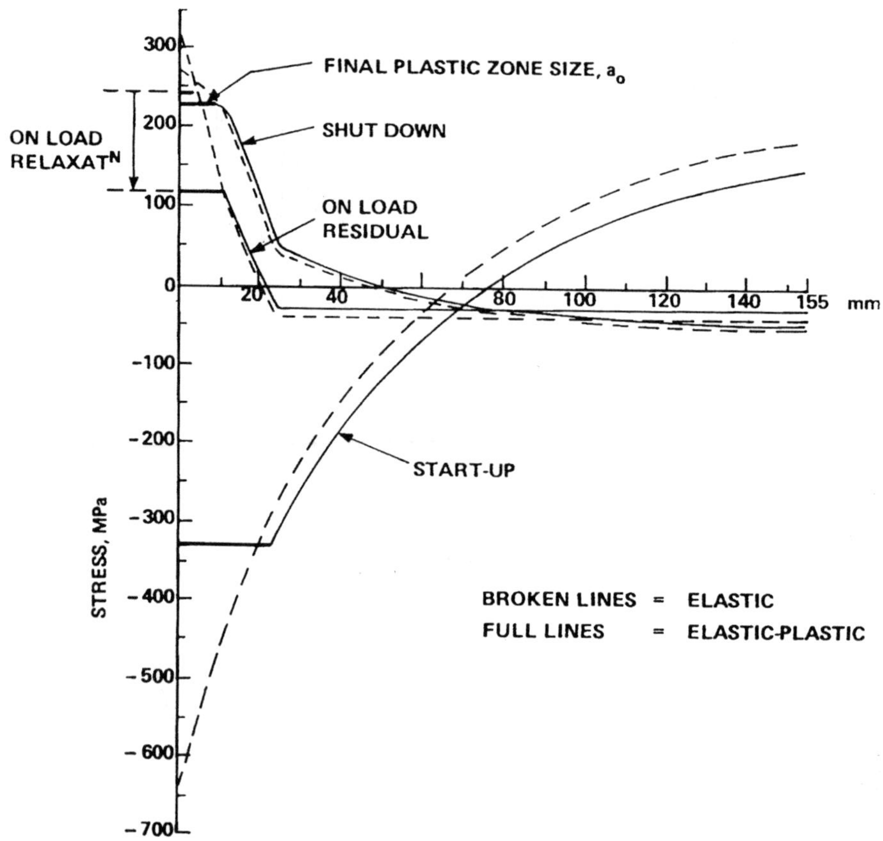

Figure 61 Calculation of plastic zone size at inner surface of chest after shakedown.

Table 14 Depth of plastic zone.

Service cycle	$\Delta\varepsilon_t$ (%)	*Zone size* (mm) ($=a_o$)
Major	0.684	9.5
Intermediate	0.566	7.5
Minor	0.353	5.0

the fatigue energy dissipated per cycle is found to be 1.29×10^{-3} J mm^{-3} and this constitutes the area "A" in Figure 28. The remaining area "C" is also considered to be a fatigue component, the associated value being equal to area "B" as discussed in Section 5.02.4.6.3. The fatigue failure energy, W_f calculated from the continuous cycling test in Figure 62 and from similar data (Skelton, 1991) may be taken as 1 J mm^{-3}. Substituting the above values in Equation (45) and solving it is found that $N=55$ cycles, in very fair comparison with values obtained by the ductility exhaustion method in Table 15.

As a bound to the calculation and as part of a sensitivity analysis we assume, with Batte *et al.* (1978), that the first 0.5 h of the dwell is accorded to "fatigue" damage on account of the relatively high strain rates. From Equation (25) and Figure 58 the calculated stress after 0.5 h of relaxation is 175 MPa. From Equation (43) the fatigue energy dissipated in this part of the cycle is now found to be 1.06×10^{-4} J mm^{-3}, calculated as if it were creep energy. By differences, the partitioning of area "B" in Figure 28 gives a creep damage of $(1.68–1.06)\times10^{-4}=6.17\times10^{-5}$ J mm^{-3}. Appropriate substitution of the new parameters into Equation (45) gives $N=129$ cycles, this calculation therefore indicating an insufficiency in assumed creep damage during the dwell.

5.02.8.5 Experimental Verification

5.02.8.5.1 Oxide dating

Oxide dating is a technique whereby the age of a crack in service can be deduced from the variation of oxide depth along its length (Pinder, 1981). The method assumes that spalling of the oxide (as discussed in Section 5.02.3.1) does not take place in the close confines between the crack surfaces. From the technique, it was established that out of the total of 233 service cycles that actually propagated the crack across the whole wall section, ~60 cycles were required to attain the

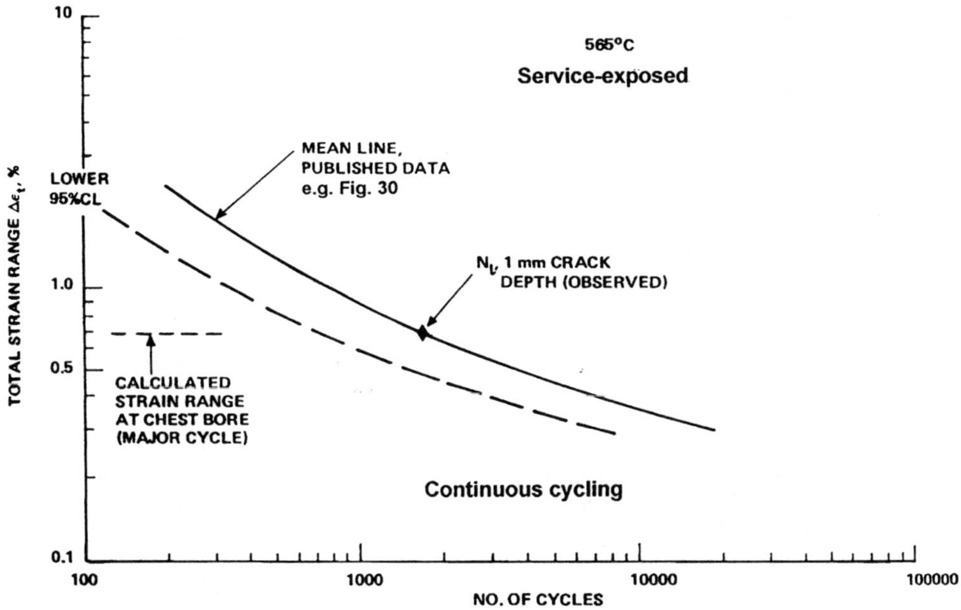

Figure 62 Check of service-exposed endurance against mean data, see also Figure 30.

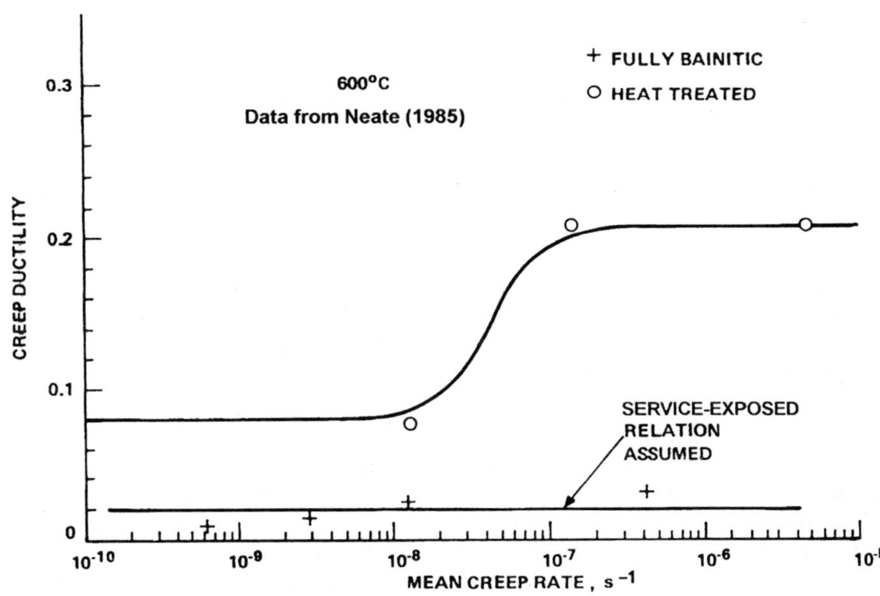

Figure 63 Ductility–strain rate data. A constant value of 2% was used in ductility exhaustion calculations (after Neate, 1985).

edge of the calculated plastic zone, see Figure 64. Both predictions (32 or 48) cycles are thus considered safe since they are less than the observed cycle number.

5.02.8.5.2 Crack growth test

At the time of the assessment, the only crack growth data available on service-exposed material at 565°C was for 0.5 h dwells at 0.7% total strain, i.e., approximating to the major cycle. This gave a value of B in Equation (52) of 8.5×10^{-3} and a Q value of 1. Assuming an initial defect size of 0.5 mm the number of cycles to propagate over a plastic zone depth of 9.5 mm is given from Equation (61) as 346. This exceeds observations. Similarly, use of the "upper bound" Equation (59) for the B value

Table 15 Results of damage calculation (cycles to fail 9.5 mm plastic zone size).

Service cycle	$\Delta\varepsilon_t$ (%)	N_o	$1/N_o$	*Creep damage* (D_c)	N_o^*(*predicted*)
Major	0.684	2,712	0.00037	0.039	26
Intermediate	0.566	4,339	0.00023	0.027	37
Minor	0.353	14,290	0.000069	<0.002	483

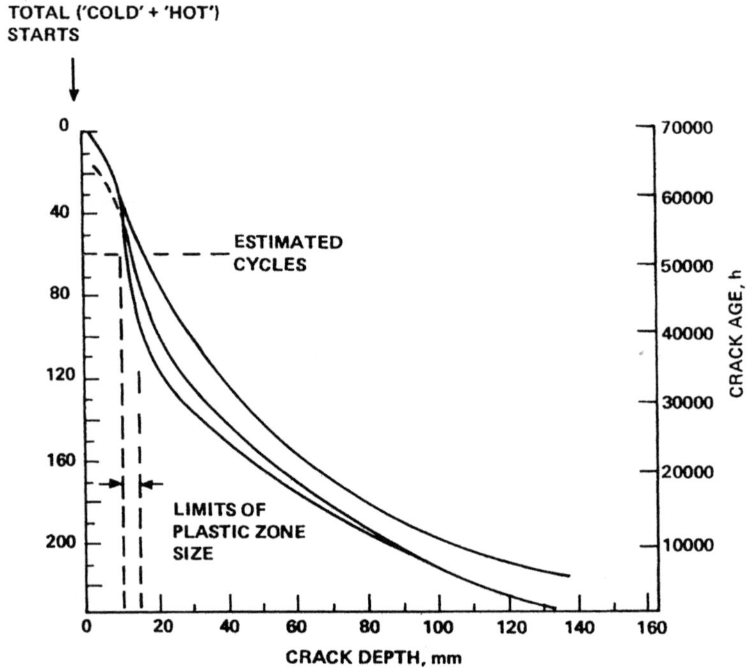

Figure 64 Assessment of cycles to traverse plastic zone using oxide-dating techniques.

predicts 156 cycles which again is unconservative for this example.

A 12.7 mm diameter specimen, which was used previously to generate the service-exposed cyclic stress–strain data of Figure 59, was retested with a 16 h tension dwell. The temperature was 565°C and the total strain range (including some elastic follow-up with $Z=9$) was 0.68%. Because of the previous testing, there was a circumferential starter crack estimated at some 0.2 mm deep. The specimen failed completely after 23 cycles, during which the crack front had traveled $12.7/2=6.35$ mm. In Figure 65 is shown the sudden fall off in the tension stress and accompanying increase in the crack p.d. monitoring signal which occurred in the dwell during the 19th cycle. Catastrophic crack growth thus appeared to occur suddenly in this material over the whole gauge section. (A formal value of B calculated from Equation (52) is 5.5×10^{-1}). This contrasts with earlier work on this and other materials (e.g., Skelton 1978, 1988a) and the investigations covered in Section 5.02.6 where at lower total strain ranges crack growth increased in a steady manner with increasing crack depth. Such studies have indeed formed the basis of Figures 41–45.

5.02.8.6 Damage Diagram

We end with plotting the present crack initiation results (from the ductility exhaustion method) on the damage diagram, thus constituting step (5). Referring to Equations (72) and (73) and Table 15 and assuming a calculated (best estimate) 60 service cycles for the crack to traverse the plastic zone, in Figure 66 points are plotted on the diagram for:

(i) major cycles only,

(ii) a mix of 80% major and 20% minor cycles, and

(iii) a mix of 19% major, 48% intermediate, and 33% minor cycles.

These are listed in decreasing order of severity but in each case they lie in the "check for cracking" part of the diagram (compare Figures 23 and 29), i.e., the calculations on a linear damage theory have been validated.

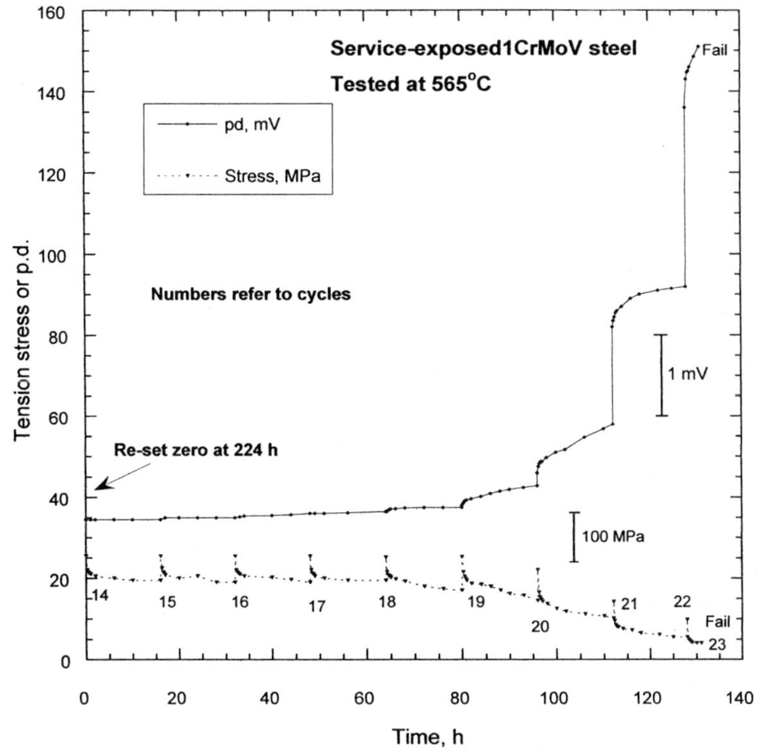

Figure 65 Unstable crack growth in service-exposed ferritic material. Compare also with Figure 15.

Owing to the nature of the failure, the results are heavily biased towards creep rather than fatigue damage. It would be profitable to seek examples of the diagram where the fatigue element is significant, as has been done using the time-fraction rule in Figure 53.

5.02.8.7 General Observations

A sensitivity analysis with respect to materials data was not carried out in the present instance, which has mostly used best-estimate data and actual data on the service-exposed material, see below. Changes in one parameter may promote or demote the effects of others. For example, a lower bound yield stress could cause an increase in strain and maximize fatigue damage but since the stress at the beginning of dwell is reduced, this may reduce the creep damage. Conversely, an upper bound yield assumption would tend to decrease the fatigue damage and increase that of the creep component.

It is of course essential to report the results of an analysis so that an independent assessment is possible. The present problem has not been sensitive to the choice of continuous cycling fatigue data because the problem is creep-dominated. The example has demonstrated the "failure avoidance" capability of the R5 procedure (Ainsworth, 1996). It may also be noted that the example has used service-exposed material properties and that this supporting materials data (continuous cycling endurance, stress relaxation and cyclic stress–strain) can be obtained relatively quickly. In the present example, the greatest sensitivity lies in the choice of service cycle mix, underlying the importance of temperature and strain monitoring in operation.

5.02.9 Concluding Remarks on Damage Methods

In the 1960s, when extensive creep–fatigue data were being gathered in tests lasting up to 10 yr, it was possible to calculate damage components directly, see, e.g., the data in Figure 67(a) for a rotor steel at 550°C incorporating peak tension dwells up to 16 h (Thomas and Dawson, 1980). These authors assumed:

$$N_{16}(d_0 + d_{0.5} + d_{16}) = 1 \qquad (74)$$

where d is a partitioned damage factor in a given time interval such that $d_0 = 1/N_0$, $d_{0.5} = (1 - N_{0.5}d_0)/N_{0.5}$, and $d_{16} = [1 - N_{16}(d_0 + d_{0.5})]/N_{16}$. The suffices for N indicate the endurances for a given dwell as supplied in Figure 67(a). The individual damage factors

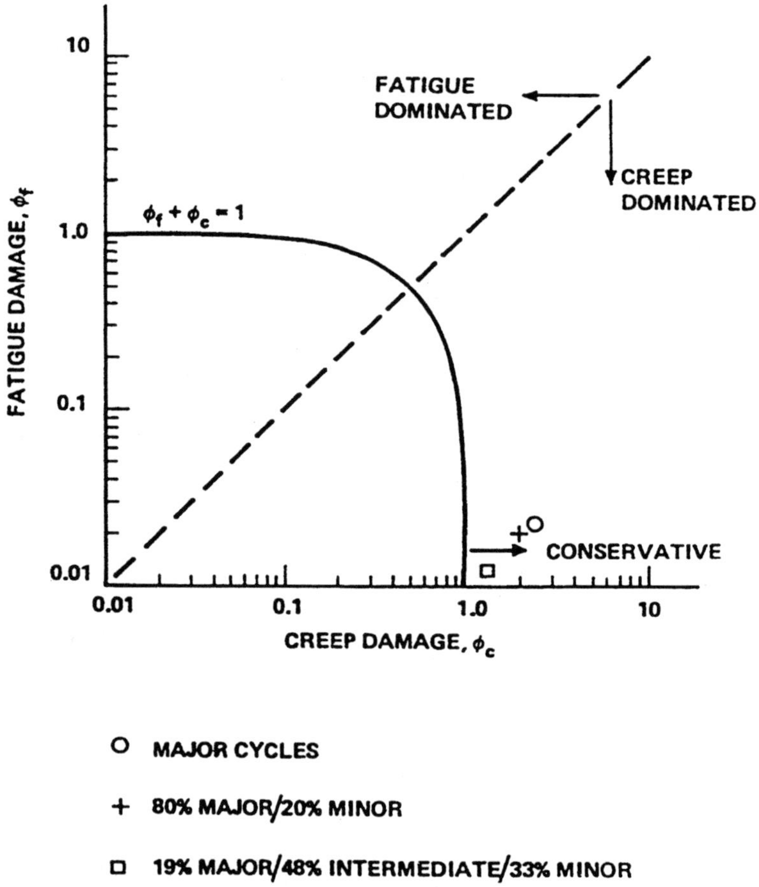

Figure 66 Damage diagram indicating sensitivity to cycle mix.

given in Figure 67(b) show quite clearly how creep damage for dwells >0.5 h dominate as the strain range is increased. This type of calculation is not possible nowadays since reliance is placed on extrapolation of short-term dwell data.

Our review has implied that, whichever damage model is chosen, investigators have usually demonstrated an agreement with experimental data to within a factor of 2. This has generally been managed by increasing use of "partitioning" rules or other degrees of sophistication so that the original models begin to lose their attractiveness—the most robust have proved to be those which are the easiest to use. It has not always been easy to trace the precise history of the damage laws but it appears that the time fraction rule originally took peak stress as the reference and was therefore conservative. Attempts to integrate down a relaxation curve and refine the appropriate time led to nonconservative predictions of life. Thus the interaction diagram was made bi-linear which restored conservatism. There have been similar difficulties with analyzing ductility data for strain fraction—which is the most physically correct ductility to take? The assumption of an average strain rate by taking ductility divided by time to failure is not necessarily an accurate reflection of secondary creep rate if both primary and tertiary stages of creep are large.

We return to the damage diagram and its likely shape as a function of cycle type or sequence of loading. The argument simplifies those of Chaboche (1982) and Cailletaud and Levaillant (1984). The "competitive" model of Priest and Ellison (1981, 1982) detaches any interaction between creep and fatigue and can be modeled, e.g., by putting $n=6$ in Equation (21c). The curve is plotted in Figure 21. Linear accumulation is obtained when the evolution of damage versus life is independent of loading, and the line thus connects the creep damage/fatigue damage (1/1) points directly as shown in Figure 21. Any other sequence effect attributed to nonlinear accumulation is shown, e.g., as the concave curve in Figure 21, the effect being larger when the damage "evolution" curves are farther apart. This is made more clear by referring to Figure 68 which shows schematically this evolution

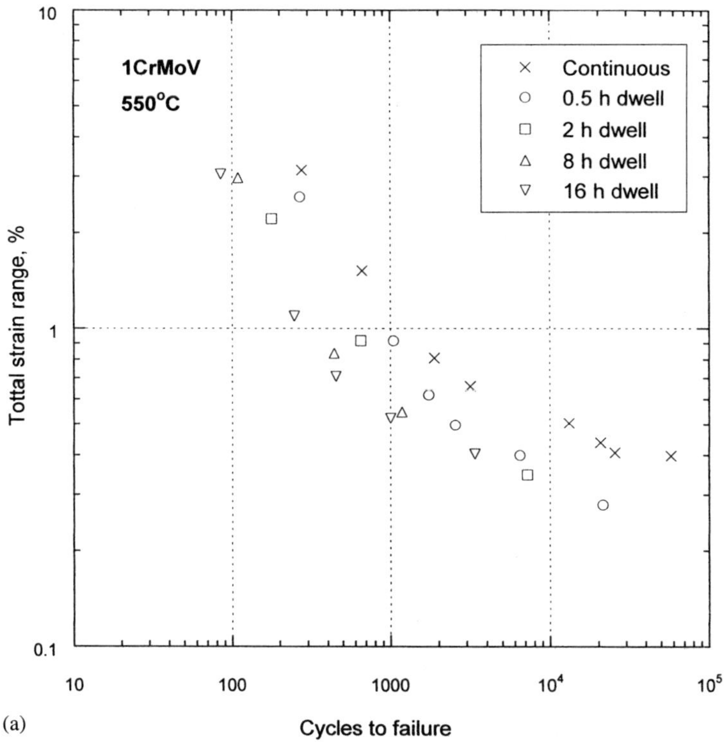

(a)

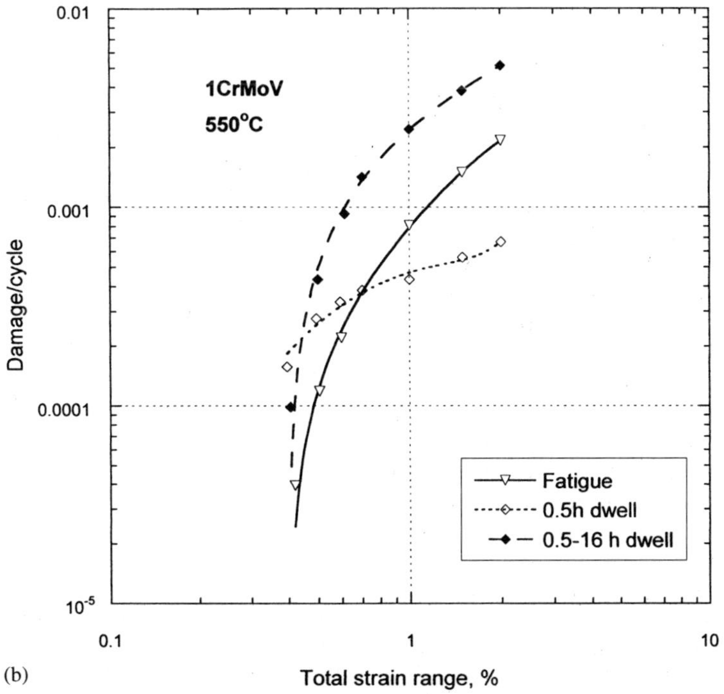

(b)

Figure 67 (a) Dwell effect in a rotor steel, compare with Figure 19. (b) Damage factors deduced from data in (a) showing preponderance of creep damage (after Thomas and Dawson, 1980).

of creep and fatigue damage. The fatigue element has been put below the creep component on the assumption that the fatigue crack initiation phase can accommodate fairly high levels of damage whereas the propagation phase is rapid. Suppose in a sequential test a creep fraction f_c is consumed. If the subsequent fatigue phase is one of fast continuous cycling then cracking would generally be transgranular and the damage allowed is $(1 - f_c)$. But if the LCF phase is at low strain rate the fatigue crack initiation phase is likely

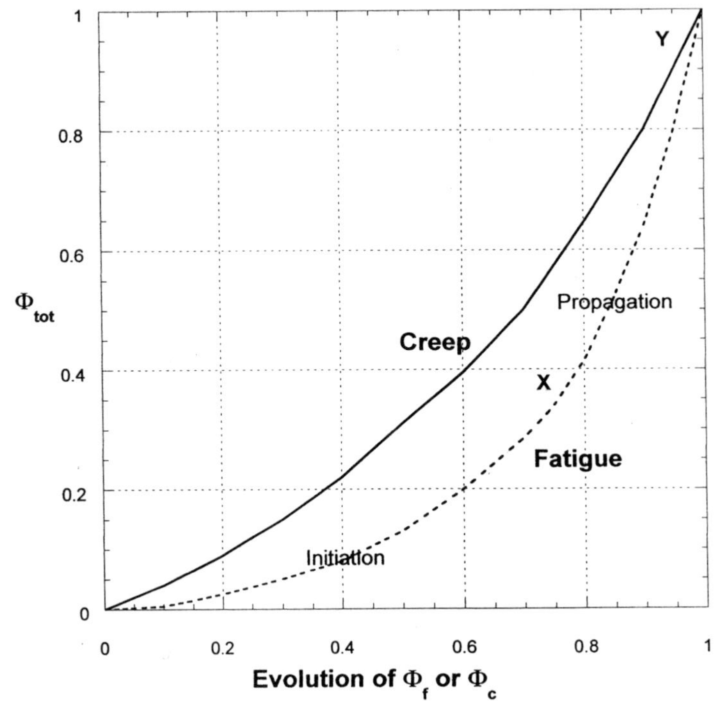

Figure 68 Schematic diagram of damage evolution.

to be by-passed or at most accelerated, as will be the grain boundary crack propagation rate. The original fatigue damage curve along OXY in Figure 68 has been perturbed by the interaction and the allowable damage is $<(1-f_c)$. This provides some insight behind the power law of Equation (21c) and Figure 21 and in this review the argument has been be applied to smaller and smaller sequences, culminating in the "standard" creep–fatigue cycle.

In a fatigue–creep single sequence the interaction is less clear. It is known that prior fatigue produces permanent microstructural changes such that, e.g., austenitic steels harden and bainitic steels soften: the resulting secondary creep rates are slower and faster than their respective rates in virgin steels (Plumbridge, 1987a). Such effects cannot be interpreted on the evolutionary plot of Figure 68. It would be expected, however, that the pre-exposed austenitic steels have a lower ductility (and conversely for the bainitic steels) and this would be reflected by the use of the ductility exhaustion approach.

Even when loading does not contain an element of creep, difficulties can be experienced with linear damage summation in two-level tests. In sequential LCF tests of low strain amplitude followed by high strain amplitude and vice versa, Miller and Ibrahim (1981) showed that,

$$\frac{n_1}{N_{f1}} + \frac{n_1}{N_{f2}} = \Phi_f \quad (75a)$$

such that when the "high" amplitudes were applied first, the damage factor, Φ_f, was less than unity whereas it became greater than unity when the "low" amplitude was applied first. Similar effects have been observed in repeated blocks of "major" and "minor" cycles (Skelton *et al.*, 1996). Miller and Ibrahim accounted for the discrepancy by reformulating the damage expression in terms of crack initiation and growth. It is seen immediately that when Equation (75a) is rewritten as

$$\frac{n_1}{N_{i1} + N_{g1}} + \frac{n_1}{N_{i2} + N_{g2}} = \Phi_f \quad (75b)$$

then if crack initiation has taken place during the "high" stage 1 then $N_{i2} = 0$ in the following "low" stage 2. This indicates that the denominator N_{f2} in Equation (75a) thereby becomes too large and artificially reduces the damage component. Similarly, when the "low" stage occurs first and there is no apparent initiation, it is nevertheless reasonable to suppose that the term N_{i2} in Equation (75b) requires factoring by the term $(1-n_1/N_{i1})$. Complex expressions may be developed and pursued for each individual case (Miller and Ibrahim, 1981) and also for creep–fatigue (Cailletaud and Levaillant, 1984) but this review has shown the undoubted benefits of applying crack initiation and growth criteria directly.

Our survey has attempted to demonstrate that assessment of damage in the early stages may either be performed in terms of cycles or increments of crack growth according to the preferences of the user. It may finally be noted that crack growth increments, n_g, can equally be expressed in damage form. Thus assuming exponential growth ($Q=1$ in Equation (52)) and that under given conditions it takes N_g cycles to propagate a crack from depth a_o to a_f then it may be shown for cycles of type j (Skelton *et al.*, 1996) that,

$$\sum \frac{n_{gi}}{N_{gi}} = \frac{\ln(a_k/a_o)}{\ln(a_f/a_o)} \qquad (76)$$

where a_k is an intermediate crack depth. Since the respective terms in B_i have cancelled out, see Equation (61), e.g., the right-hand side of Equation (75) is independent of cycle type. Damage is thus envisaged as the progress of a dimensionless crack of magnitude a_k/a_f according, e.g., to the amplitude of the "major" and "minor" block loadings, see also Cailletaud and Levaillant (1984). It is further noted that Equation (30) may also be expressed in the form $(N^*/N+N^*D_c)=1$ which describes the creep–fatigue state for any incremental stage of crack growth, since this must occur when the local damage sums to unity.

ACKNOWLEDGMENTS

This extended survey is based in part on a lecture series presented at CISM (International Center for Mechanical Sciences), Udine (Skelton, 1998). It is a pleasure to thank Dr. I. W. Goodall, Mr. M. W. Spindler, Dr. S. R. Holdsworth, and Dr. R. Hales for useful discussions during the preparation of the review.

5.02.10 REFERENCES

P. Agatonvic and N. Taylor, 1990, Establishing a reliable life prediction method for creep–fatigue interaction at high temperatures. In: "Creep and Fracture of Engineering Materials and Structures," eds. B. Wilshire and R. W. Evans, Institute of Metals, London, pp. 803–816.

R. A. Ainsworth, 1989, Defect assessment procedures at high temperature. In: "Proceedings of SMIRT 10 Conference, vol. L," Anaheim, CA, pp. 79–90.

R. A. Ainsworth (ed.), 1996, "R5: Assessment procedure for the high temperature response of structures, Issue 2," British Energy Generation Ltd., Barnwood, UK.

American Society of Mechanical Engineers, 1991, "Boiler and Pressure Vessel Code, Case N-47 (29), Class I Components In Elevated Temperature Service," Section III, Division I, ASME, New York (subsection NH in 2001 version).

C. Amzallag B. N. Leis and P. Rabbe (eds.), 1982, "Low Cycle Fatigue and Life Prediction, ASTM STP 770," ASTM, Philadelphia, PA.

Anon, 1985, "RCC-MR: Technical Appendix A3, Section 1, Subsection Z, Materials Design and Construction Rules for Mechanical Components of FBR Nuclear Test Islands," AFCEN, Paris.

M. Antill and D. J. Smith, 1996, The interaction of creep and fatigue in IN738LC at 850°C. In: "6th International Conference on Creep and Fatigue," Inst. Mech. Eng., London, pp. 231–240.

M. Arai, T. Ogata and A. Nitta, 1996, Continuous observation cavity growth and coalescence by creep–fatigue test in SEM. In: "6th International Conference on Creep and Fatigue," Inst. Mech. Eng., London, pp. 241–250.

H. C. Argo, J. F. DeLong, Y. Kadoya, M. Nakamura and K. Ando, 1984, Eddystone experience on long-term exposed 316SS steam turbine valve components. ASME Paper 84-JPGC-Pwr-15, ASME, New York, pp. 1–11.

Y. Asada and S. Mitsuhashi, 1980, Creep fatigue interaction of 304 and 316 stainless steels in air and vacuum. In: "4th International Conference on Pressure Vessel Technology," Inst. Mech. Eng., London, vol. 1, pp. 321–327.

T. Asayama, T. Furukawa and E. Yoshida, 1999, Evaluation procedure for sodium environmental effects on 316FR and modified 9Cr1Mo steel for the design of Japanese fast breeder reactors. In: "Advances in Life Prediction Methodology," ASME PVP-vol 391, ASME, New York, pp. 61–66.

M. F. Ashby and B. F. Dyson, 1984, "Creep Damage Mechanisms and Micromechanisms," Report DMA (A) 77, National Physical Laboratory, Teddington, UK.

J. A. van den Avyle, 1983, Low cycle fatigue of tubular specimens. *Scr. Metall.*, **17**, 737–740.

G. Bartolini and S. Ragazzoni, 1984, Low-cycle fatigue of 1CrMoV turbine rotor steel. In: "Proceedings of the Second International Conference on Creep and Fracture of Engineering Materials and Structures, Part II," eds. B. Wilshire and D. R. Owen, Pineridge Press, Swansea, pp. 1029–1042.

A. D. Batte, 1983, Creep–fatigue life predictions. In: "Fatigue at High Temperature," ed. R. P. Skelton, Applied Science, London, pp. 365–401.

A. D. Batte, M. C. Murphy and M. B. Stringer, 1978, High-strain fatigue properties of a 0.5CrMoV turbine casing steel. *Met. Technol.*, **5**, 405–5413.

S. M. Beech and A. D. Batte, 1984, An evaluation of the long term creep–fatigue behavior of a 0.5%CrMoV steam turbine casing steel. In: "Proceedings of the 2nd International Conference on Creep and Fracture of Engineering Materials and Structures, Part II," eds. B. Wilshire and D. R. J. Owen, Pineridge Press, Swansea, pp. 1015–1027.

P. Bensussan and J. P. Mascarell (eds.), 1990, "High Temperature Fracture Mechanisms and Mechanics," EGF 6, Mechanical Engineering Publications, London.

J. T. Berling and J. B. Conway, 1970, Effect of hold time on the low-cycle fatigue resistance of 304 stainless steel at 1200°F. In: "1st International Conference on Pressure Vessel Technology," ASME, New York, pp. 1233–1246.

J. T. Berling, and T. Slot, 1969, Effect of temperature and strain rate on low-cycle fatigue resistance of AISI 304, 316 and 348 stainless steels. In: "Fatigue at High Temperature, STP 459," ASTM, Philadelphia, PA, pp. 3–30.

H. L. Bernstein, 1982, An evaluation of four creep–fatigue models for a nickel-base superalloy. In: "Low-cycle Fatigue and Life Prediction, ASTM STP 770," eds. C. Amzallag, B. N. Leis and P. Rabbe, ASTM, Philadelphia, PA, pp. 105–134.

J. E. Bertilsson and B. Scarlin, 1990, Design requirements and materials for advanced steam turbines. In: "High

Temperature Materials for Power Engineering, Part 1," eds. E. Bachelet, R. Brunetand, D. Coutsouradis *et al.*, Kluwer Academic, Dordrecht, The Netherlands, pp. 281–312.

V. Bicego, A. Del Puglia, M. Marchionni and E. Vitale, 1992, Creep–fatigue–oxidation interactions in materials for high temperature components. In: "High Temperature Structural Design, ESIS 12," ed. L. H. Larsson, Mechanical Engineering Publications, London, pp. 259–282.

V. Bicego, C. Fossati and S. Ragazzoni, 1988, Low cycle fatigue characterization of a HP-IP steam turbine rotor. In: "Low Cycle Fatigue, ASTM STP 942," eds. H. D. Solomon, G. R. Halford, L. R. Kaisand and B. N. Leis, ASTM, Philadelphia, PA, pp. 1237–1260.

R. Blum, 1994, Materials development for power plants with advanced steam parameters. In: "Materials for Advanced Power Engineering, Part 1," eds. D. Coutsouradis, J. H. Davidson, J. Ewald *et al.*, Kluwer Academic, Dordrecht, The Netherlands, pp. 15–30.

J. Bressers, R. De Cat and E. Fenske, 1984, "Crack Initiation and Growth in High Temperature Low Cycle Fatigue," CEC report EUR 8808.

C. R. Brinkman, 1983, Creep–fatigue effects in structural materials used in advanced nuclear power generating systems. In: "The 27th Sagamore Army Materials Research Conference, Fatigue, Environment and Temperature Effect," ed. V. Weiss, Plenum, New York, pp. 241–263.

C. R. Brinkman, 1999, Elevated temperature mechanical properties of an advanced type 316 stainless steel. In: "Advances in Life Prediction Methodology," ASME PVP, ASME, New York, vol. 391, pp. 87–94.

C. R. Brinkman, 2000, Long-term elevated-temperature mechanical properties of an advanced type 316 stainless steel. In: "Understanding and Predicting Material Degradation," ASME PVP, ASME, New York, vol. 213, pp. 97–103.

C. R. Brinkman, J. P. Strizak and M. K. Booker, 1978, Experiences in the use of strain range partitioning for predicting time-dependent strain-controlled cyclic lifetimes of uniaxial specimens of 2(1/4)Cr-1Mo steel, type 316 steel and Hastelloy X. In: "AGARD Conference. No. 243: Characterization of Low Cycle High Temperature Fatigue by the Strain Range Partitioning Method," NATO, pp. 15.1–15.8.

T. Bui-Quoc, R. Gomuc, A. Biron, H. L. Nguyen and J. Masounave, 1988, Elevated temperature fatigue–creep behaviour of nickel-base superalloy IN 625. In: "Low Cycle Fatigue," ed. H. D. Solomon, STP 942, ASTM, Philadelphia, pp. 470–486.

J. E. Bynum, F. V. Ellis, M. H. Rafiee, W. F. Siddall, T. Daikoku, H. Haneda and J. F. DeLong, 1984, Failure investigation of Eddystone main steam piping. In: "American Power Conference," Chicago, IL, pp. 300–308.

G. Cailletaud and C. Levaillant, 1984, Creep–fatigue life prediction: what about initiation? *Nucl. Eng. Des.*, **83**, 279–292.

A. E. Carden (ed.), 1973, "Fatigue at Elevated Temperature, ASTM STP 520," ASTM, Philadelphia, PA.

J. L. Chaboche, 1982, Lifetime predictions and cumulative damage under high-temperature conditions. In: "Low-Cycle Fatigue and Life Prediction, ASTM STP 770," eds. C. Amzallag, B. N. Leis and P. Rabbe, ASTM, Philadelphia, PA, pp. 81–104.

K. D. Challenger, A. K. Miller and C. R. Brinkman, 1981, An explanation for the effects of hold periods on the elevated temperature fatigue behavior of 2(1/4)Cr-1Mo steel. *J. Eng. Mater. Technol.*, **103**, 7–14.

E. Charkaluk and A. Constaninescu, 2000, An energetic approach in thermomechanical fatigue for silicon molybdenum cast iron. *Mater. High Temp.*, **17**, 373–380.

T. E. Chung, J. J. Smith and R. G. Faulkner, 1990, Fatigue design limits based on microstructural damage criteria. *La Metallurgia Italiana*, **82**, 143–146.

L. F. Coffin, 1954, A study of the effects of cyclic thermal stresses on a ductile metal. *Trans. ASME Ser.*, **A76**, 931–950.

L. F. Coffin, 1963, Cyclic-strain-induced oxidation of high-temperature alloys. *Trans. Quart. ASM*, **56**, 339–344.

L. F. Coffin, 1973, Fatigue at high temperatures. In: "Fatigue at Elevated Temperatures, ASTM STP 520," ASTM, Philadelphia, PA, pp. 5–36.

A. Coles, G. J. Hill, R. A. T. Dawson and S. J. Watson, 1967, The high-strain fatigue properties of low-alloy creep-resisting steels. In: "Thermal and High Strain Fatigue," Monograph and report series no. 32, The Metals and Metallurgy Trust, London, pp. 270–294.

L. W. Collins, D. Sidey and D. M. R. Taplin, 1979, Fatigue maps and wave-shape effects on fatigue. *Can. Metall. Quart.*, **18**, 231–238.

J. B. Conway, R. H. Stentz and J. T. Berling, 1981, "Fatigue, Tensile, and Relaxation Behavior of Stainless Steels," US Atomic Energy Commission, Ohio.

R. M. Curran (ed.), 1976, "ASME MPC-S Symposium on Creep Fatigue Interaction," ASME, New York.

P. Dabo, C. Levaillant, A. Pineau, J. Grattier and M. Mottot, 1986, Endommagement intergranulaire en fatigue relaxation d'un acier inoxydable austenitique. In: "Fatigue à Haute Temperature," Societé Française de Metallurgie, Paris, pp. 40–52.

R. Danzer and J. Bressers, 1986, A new method to predict the life under high temperature low cycle fatigue conditions. *Fatigue Fract. Eng. Mater. Struct.*, **9**, 151–168.

R. A. T. Dawson, W. J. Elder, G. J. Hill and A. T. Price, 1967, High-strain fatigue of austenitic steels. In: "Thermal and High Strain Fatigue," Monograph and report series no. 32, The Metals and Metallurgy Trust, London, pp. 239–269.

J. F. DeLong, J. E. Bynum, F. V. Ellis, M. H. Rafiee, W. F. Siddall, T. Daikoku and H. Haneda, 1985, Failure investigation of Eddystone main steam piping. *Weld. Res. Suppl.*, **64**, 271s–280s.

J. F. DeLong, R. Ishimoto, K. Tokimasa, A. Ohtomo and T. Honda, 1984a, Failure of 316 stainless turbine valves in long term elevated temperature service. In: "Fatigue and Creep Characteristics of Materials for Transportation and Power Industries, New Orleans, 1984," ASME, New York, pp. 173–191.

J. F. DeLong, S. Kihara, K. Yoshikawa, O. Watanabe and T. T. Honda, 1984b, Lessons learned from failure of type 316 stainless steel components after long service at high temperature in Eddystone Unit No. 1. In: "Fatigue and Creep Characteristics of Materials for Transportation and Power Industries, New Orleans, 1984," ASME, New York, pp. 193–213.

J. F. DeLong, W. F. Siddall, F. V. Ellis, H. Haneda, T. Tsuchiya, T. Daikoku, F. Masuyama and K. Setoguchi, 1984c, "Operation Experiences and Reliability Evaluation on Main Steam Line Pressure Parts of Philadelphia Electric. Co., Eddystone No. 1," Mitsubishi Heavy Industries Ltd., Tokyo, Japan, Boiler Bulletin MBB-84112E.

Z. Duan, J. He, Y. Ning and Z. Dong, 1988, Strain energy partitioning approach and its application to low cycle fatigue life prediction for some heat-resistant steels. In: "Low Cycle Fatigue, ASTM STP 942," eds. H. D. Solomon, G. R. Halford, L. R. Kaisand and B. N. Leis, ASTM, Philadelphia, PA, pp. 1133–1143.

B. F. Dyson, 1979, Constrained cavity growth, its use in quantifying recent creep fracture results. *Can. Metall. Quart.*, **18**, 31–38.

B. F. Dyson, 2000, Use of CDM in materials modeling and component creep life prediction. *ASME J. Pres. Ves. Technol.*, **315**, 281–296.

J. C. Earthman, G. Eggeler and B. Ilschner, 1989, Deformation and damage processes in a 12%Cr-Mo-V steel under high temperature low cycle fatigue conditions in air and vacuum. *Mater. Sci. Eng.*, **A110**, 103–114.

H. G. Edmunds and D. J. White, 1966, Observations on the effect of creep relaxation on high strain fatigue. *J. Mech. Eng. Sci.*, **8**, 310–321.

A. El Gharad, 1986, Comportement en fatigue-relaxation de l'alliage 800 (Z 5 NCT A 33-21) à 550°C—influence du vieillissement et prévision a long term. Ph.D. thesis, Université de Metz.

F. V. Ellis and J. E. Bynum, 1983, Evaluation of TP 316 stainless steel main steam line after 125,000 hours of service. In: "Advances in Life Prediction Methods," ASME, Albany, NY, pp. 337–345.

E. G. Ellison, 1969, A review of the interaction of creep and fatigue. *J. Mech. Eng. Sci.*, **11**, 318–339.

E. G. Ellison and A. J. F. Paterson, 1976, Creep fatigue interactions in a 1CrMoV steel. *Proc. Inst. Mech. Eng.*, **190 12/76**, 321–350.

T. Endo and T. Sakon, 1984, Creep–fatigue life prediction using simple high-temperature low-cycle fatigue testing machines. *Met. Technol.*, **11**, 489–496.

E. P. Esztergar and J. R. Ellis, 1971, Cumulative damage concepts in creep–fatigue life prediction. In: "Thermal Stresses and Thermal Fatigue," ed. D. J. Littler, Butterworths,, London, pp. 128–155.

I. Ewald, W. Jakobeit, K. H. Mayer and W. Wiemann, 1977, A study on prevention of creep and fatigue fracture of steels for turbine application. In: "Fracture 1977," ICF4, Waterloo, Canada, vol. 2, pp. 777–784.

P. Feltham, 1960, Stress relaxation in copper and alpha brasses at low temperatures. *J. Inst. Met.*, **89**, 210–214.

J. H. Gittus, D. A. Howl and H. Hughes, 1970, Theoretical analysis of cladding stresses and strains produced by expansion of cracked fuel elements. *Nucl. Appl. Technol.*, **9**, 40–46.

D. Gladwin and D. A. Miller, 1982, The effect of cycle waveshape on the low cycle fatigue behavior of 20%Cr-25%Ni-Nb stainless steel at 650°C. *Fatigue Eng. Mater. Struct.*, **5**, 275–286.

J. Granacher and A. Scholz, 1992, Creep–fatigue behavior of heat resistant steels under service-type long-term conditions. In: "Low Cycle Fatigue and Elasto-plastic Behavior of Materials—3," eds. K. T. Rie, H. W. Grünling, G. König *et al.*, Elsevier, London, pp. 235–241.

J. Granacher, A. Scholz, H. Moehlig and C. Berger, 2000, Heat resistant power plant steels under variable long term conditions. In: "Advanced Materials for Twenty First-century Turbine and Power Plants," eds. A. Strang, W. M. Banks, R. D. Conroy *et al.*, IOM Communications, London, pp. 516–531.

R. Hales, 1980, A quantitative metallographic assessment of structural degradation of type 316 stainless steel during creep–fatigue. *Fatigue Eng. Mater. Struct.*, **3**, 339–356.

R. Hales, 1983a, Fatigue testing methods at elevated temperatures. In: "Fatigue at High Temperature," ed. R. P. Skelton, Applied Science, London, pp. 63–96.

R. Hales, 1983b, A method of creep damage summation based on accumulated strain for the assessment of creep–fatigue endurance. *Fatigue Eng. Mater. Struct.*, **6**, 121–135.

R. Hales, 1984, The application of creep strain as a parameter for estimating creep–fatigue endurances. In: "Proceedings of the 2nd International Conference on Creep and Fracture of Engineering Materials and Structures, Part II," eds. B. D. Wilshire and R. Owen, Pineridge Press, Swansea, pp. 1015–1027.

R. Hales and R. A. Ainsworth, 1995, Multiaxial creep–fatigue rules. *Nucl. Eng. Des.*, **153**, 257–264.

G. R. Halford, 1997, Cumulative fatigue damage modeling—crack nucleation and early growth. *Int. J. Fatigue*, **19**(1), S253–S259.

G. R. Halford and J. F. Saltsman, 1983, Strain range partitioning—a total strain range version. In: "International Conference on Advances in Life Prediction methods," ASME, New York, pp. 17–26.

J. He, Z. Duan, Y. Ning and D. Zhao, 1983, Strain energy partitioning and its application to GH33A nickel-base superalloy and 1Cr18NiTi stainless steel. In: "International Conference on Advances in Life Prediction methods," ASME, New York, pp. 17–26.

R. L. Hecht and J. R. Weertman, 1998, The effect of environment on the high temperature hold time fatigue behavior of annealed 2.25pctCr1pctMo steel. *Metall. Trans.*, **29A**, 2137–2145.

G. J. Hill, 1967, The failure of wrought 1%CrMoV steels in reverse-bending high strain fatigue at 550°. In: "Thermal and High Strain Fatigue," Monograph and report series no. 32, The Metals and Metallurgy Trust, London, pp. 312–327.

S. R. Holdsworth, 1998, Validation of creep–fatigue assessment procedures using the results of component-targeted feature-specimen tests. In: "Proceedings of the 4th International Conference on Low Cycle Fatigue," eds. K. T. Rie and P. D. Portella, Elsevier, Oxford, pp. 787–800.

S. R. Holdsworth, 2001, Creep–fatigue properties of high temperature turbine steels. *Mater. High Temp.*, **18**, 261–265.

J. S. Huang and R. M. Pelloux, 1980, Low cycle fatigue crack propagation in Hastelloy-X at 25 and 760°. *Metall. Trans.*, **11A**, 899–904.

D. Hull and D. E. Rimmer, 1959, The growth of grain boundary voids under stress. *Phil. Mag.*, **4**, 673–687.

I. Ioka, Y. Miwa, H. Tsuji, M. Yonekawa, F. Takada and T. Hoshiya, 2001, Study on creep–fatigue life of irradiated austenitic stainless steel. In: "Proceedings of Creep 7," Japan Soc. Mech. Eng., Tokyo, pp. 59–62.

C. E. Jaske, H. Mindlin and J. S. Perrin, 1972, Influence of hold-time and temperature on the low-cycle fatigue of Incoloy 800. *Trans. ASME Ser. B*, **94**, 930–934.

C. Y. Jeong and S. W. Nam, 1999, Estimation of the damaging energy under creep–fatigue interaction conditions in 1CrMoV steel. *Scr. Mater.*, **40**, 623–629.

C. Y. Jeong, S. W. Nam and J. Ginsztler, 1999, Stress dependence on stress relaxation creep rate during tensile holding under creep–fatigue interaction in 1CrMoV steel. *J. Mater. Sci.*, **34**, 2513–2517.

L. M. Kachanov, 1958, Time of the rupture process under creep conditions. *Izv. Akad. Nauk. SSR Otd. Tekh. Aauk.*, **8**, 26–31.

K. Kanazawa, K. Yamaguchi and S. Nishijima, 1988, Mapping of low cycle fatigue mechanisms at elevated temperatures for an austenitic stainless steel. In: "Low Cycle Fatigue, ASTM STP 942," eds. H. D. Solomon, G. R. Halford, L. R. Kaisand and B. N. Leis, Philadelphia, PA, pp. 519–530.

F. A. Kandil and B. F. Dyson, 1995, Uncertainties in uniaxial low cycle fatigue measurements due to load misalignment. In: "Materials Metrology and Standards for Structural Performance," eds. B. F. Dyson, M. S. Loveday and M. G. Gee, Chapman and Hall, London, pp. 134–149.

H. Kaneko, T. Sakon, H. Kaguchi, T. Nakazawa, N. Fujita and H. Ueda, 1992, Study on fracture mechanism and a life estimation method for low cycle creep–fatigue fracture of type 316 stainless steels. In: "Low Cycle Fatigue and Elasto-Plastic Behavior of Materials-3," ed. K. T. Rie, Elsevier, London, pp. 229–234.

S. Kim and J. R. Weertman, 1988, Investigation of microstructural changes in a ferritic steel caused by high temperature fatigue. *Metall. Trans.*, **19A**, 999–1007.

K. Kimura, K. Fujiyama and M. Muramatsu, 1988, Creep and fatigue life prediction based on the nondestructive assessment of material degradation for steam turbine rotors. In: "High Temperature Creep–Fatigue," ed. R. Ohtani, Elsevier, London, pp. 247–270.

K. Kobayashi and K. Yamaguchi, 2001, Long-term fatigue properties at high temperatures for ferritic heat-resistant 12Cr-2W steel. In: "Proceedings of Creep 7," Japan Soc. Mech. Eng., Tokyo, pp. 485–488.

E. Krempl and B. M. Wundt, 1971, Hold-time effects in high temperature low-cycle fatigue. ASTM STP 489, ASTM, Philadelphia, PA, pp. 1–29.

B. Kschinka and J. F. Stubbins, 1989, Creep–fatigue-environment interaction in a bainitic 2.25wt.%Cr–1wt.%Mo steel forging. *Mater. Sci. Eng.*, **A110**, 89–102.

B. F. Langer, 1937, Fatigue failure from stress cycles of varying amplitude. *Trans. ASME J. Appl. Mech.*, **59**, A160–A162.

B. N. Leis, 1977, An energy-based fatigue and creep–fatigue damage parameter. *Trans. ASME, J. Pres. Ves. Technol.*, Paper 77-PVP-3, 524–533.

J. LeMaitre and A. Plumtree, 1979, Application of damage concepts to predict creep–fatigue failures. *Trans. ASME J. Eng. Mater. Technol.*, **101**, 285–292.

C. Levaillant, J. Grattier, M. Mottot and Pineau, 1988, Creep and creep–fatigue intergranular damage in austenitic stainless steels: discussion of the creep-dominated regime. In: "Low Cycle Fatigue, ASTM STP 942," eds. H. D. Solomon *et al.*, Philadelphia, PA, pp. 414–437.

C. Levaillant and A. Pineau, 1982, Assessment of high temperature low cycle fatigue life of austenitic stainless steels by using intergranular damage as a correlating parameter. In: "Low Cycle Fatigue and Life Prediction, ASTM STP 770," eds. C. Amzallag, B. N. Leis and P. Rabbe, Philadelphia, pp. 169–193.

C. Levaillant, B. Rezgui and A. Pineau, 1980, Effects of environment and hold times on high temperature low cycle fatigue behaviour of 316L stainless steel. In: "Mechanical Behaviour of Materials (ICM3)," eds. K. J. Miller and R. F. Smith, Pergamon, Oxford, UK, pp. 163–172.

M. M. Leven, 1973, The interaction of creep and fatigue for a rotor steel. *Exper. Mech.*, **13**, 353–372.

S. X. Li and D. J. Smith, 1996, High-temperature fatigue-creep initiation and propagation of a single crystal nickel base superalloy. In: "The 6th International. Conference on Creep and Fatigue," Inst. Mech. Eng., London, pp. 279–288.

V. B. Livesey and J. Wareing, 1983, Influence of slow strain rate tensile deformation on creep–fatigue endurance of 20Cr-25Ni-Nb stainless steel at 593°C. *Met. Sci.*, **17**, 297–303.

K. Maile, 1982, Einfluss des Verformungsverhaltens der Oxydation und der Temperatur auf das Langzeitdehnungswechselverhalten der warmfesten Stähle. Ph.D. thesis, University of Stuttgart.

P. S. Maiya, 1975, Considerations of crack initiation and crack propagation in low-cycle fatigue. *Scr. Metall.*, **9**, 1141–1146.

P. S. Maiya, 1981, Effects of wave shape and ultra high vacuum on elevated temperature low cycle fatigue in type 304 stainless steel. *Mater. Sci. Eng.*, **47**, 13–21.

S. Majumdar and P. S. Maiya, 1980a, A mechanistic model for time-dependent fatigue. *J. Eng. Mater. Technol.*, **102**, 159–167.

S. Majumdar and P. S. Maiya, 1980b, An interactive damage equation for creep–fatigue interaction. In: "Mechanical Behaviour of Materials, Proceedings of the 3rd International Conference," Pergamon, Oxford, UK, vol. 2, pp. 101–109.

S. S. Manson, G. R. Halford and M. H. Hirshberg, 1971, "Creep–Fatigue Analysis by Strain-range Partitioning," NASA Technical Memorandum TMX-67838, Washington, DC.

S. S. Manson and M. H. Hirschberg, 1966, Crack initiation and propagation in notched fatigue specimens. In: "Proceedings of the 1st Conference on Fracture," eds. T. Yokobori, *et al.*, Japanese Society for Strength and Fracture of Materials, Sendai, Japan, vol. 1, pp. 479–499.

P. Marshall and T. R. Cook, 1971, Prediction of Failure of Materials under cyclic loading. In: "Thermal Stresses and Thermal Fatigue," ed. D. J. Littler, Butterworths, London, pp. 81–88.

F. Masuyama, K. Setoguchi, H. Haneda and F. Nanjo, 1985, "Findings on Creep–fatigue Damage in Pressure Parts of Long-term Service-exposed Thermal Power Plants," Paper No. PVP-MF-84-015, ASME, New York.

T. C. McGough, J. V. Pigford, P. A. Lafferty, S. Tomasevich, H. E. Zielke, C. T. Ward and J. F. DeLong, 1985, Selection and fabrication of replacement main steam piping for the Eddystone no. 1 supercritical pressure unit. *Weld. J.*, **65**, 29–36.

H. P. Meurer and H. Breitling, 1986, Evaluation of the creep–fatigue behavior of X6CrNi1811. In: "Fatigue à Haute Températures," Societé Française de Metallurgie, Paris, pp. 158–175.

D. A. Miller, 1982, A creep–fatigue failure mechanism map for 1Cr-Mo-V at 565°C. *Mater. Sci. Eng.*, **54**, 273–278.

D. A. Miller, 1983, Fatigue–creep failure mechanism maps for 20%Cr/25%Ni/Nb stainless steel at 593°C and 650°C. In: "International Conference on Advances in Life Prediction Methods," ASME, New York, pp. 157–163.

D. A. Miller, C. D. Hamm and J. L. Phillips, 1982, A mechanistic approach to the prediction of creep-dominated failure during simultaneous creep–fatigue. *Mater. Sci. Eng.*, **53**, 233–244.

D. A. Miller, R. H. Priest and E. G. Ellison, 1984, A review of material response and life prediction techniques under creep–fatigue loading conditions. *High Temp. Mater. Process.*, **6**, 155–194.

K. J. Miller and M. F. E. Ibrahim, 1981, Damage accumulation during initiation and short crack growth regimes. *Fatigue Eng. Mater. Struct.*, **4**, 263–277.

B. K. Min, 1980, Deformation and intergranular fracture during high temperature creep and creep fatigue. In: "Mechanical Behaviour of Materials 3," eds. K. J. Miller and R. F. Smith, Pergamon, Oxford UK, vol. 2, pp. 151–162.

M. A. Miner, 1945, Cumulative damage in fatigue. *J. Appl. Mech.*, **12**, A159–A164.

M. Miyahara, K. Okamura, K. Shinohara, H. Hayakawa, M. Hirota and K. Tanaka, 2001, Life prediction of ligament cracking in thermal fatigue test simulating boiler header. In: "Proceedings of Creep 7," Japan Soc. Mech. Eng., Tokyo, pp. 33–38.

M. Miyahara and K. Tokimasa, 1996, A new creep–fatigue damage rule and its application to 316LC steel under variable strainings at elevated temperature. In: "The 6th International Conference on Creep and Fatigue," Inst. Mech. Eng., London, pp. 211–220.

F. C. Monkman and N. J. Grant, 1956, An empirical relationship between rupture life and minimum creep rate in creep-rupture tests. *Proc. ASTM Quart.*, **56**, 593–605.

Z. Mu, K. Bothe and V. Gerold, 1994, Damage mechanisms in alloy 800 H under creep–fatigue conditions. *Fatigue Fract. Eng. Mater. Struct.*, **17**, 523–537.

Y. Nakabayashi, S. Ikeda, T. Yoshioka, A. Hizume, K. Shimomura, N. Yamada and T. Fujikama, 1990, Japanese developments in high temperature steam cycles. In: "High Temperature Materials for Power Engineering Part 1," eds. E. Bachelet, R. Brunetand, D. Coutsouradis *et al.*, Kluwer Academic, Dordrecht, The Netherlands, pp. 313–343.

T. Nakazawa, N. Fujita, H. Kaguchi, H. Kaneko and H. Ueda, 1992, Study on metallography of low cycle creep fatigue fracture of type 316 stainless steels. In: "Low Cycle Fatigue and Elasto-plastic Behaviour of Materials-3," eds. K. T. Rie, H. W. Grünling, G. König *et al.*, Elsevier, London, pp. 88–93.

S. W. Nam, J. W. Hong and K. T. Rie, 1988, Creep–fatigue life prediction in terms of nucleation and growth of fatigue crack and creep cavities. *Metall. Trans.*, **19A**, 121–127.

S. W. Nam, Y. J. Oh and B. O. Kong, 1992, Crack growth in low cycle fatigue with compressive hold for 1CrMoV and 12CrMoV steels. In: "Low Cycle Fatigue and Elasto-plastic Behaviour of Materials-3," eds. K. T. Rie, H. W. Grünling, G. König *et al.*, Elsevier, London, pp. 216–221.

S. W. Nam, Y. C. Yoon, B. G. Choi, J. M. Lee and J. W. Hong, 1996, The normalized Coffin–Manson plot in terms of a new damage function based on grain boundary creep condition. *Metall. Trans.*, **27A**, 1273–1281.

M. Y. Nazmy, 1982, Sequential effect of creep and fatigue in a cast Ni-base alloy. *Scr. Metall.*, **16**, 823–826.

M. Y. Nazmy, 1983, High temperature low cycle fatigue of IN738 and application of strain range partitioning. *Metall. Trans.*, **14A**, 449–461.

G. J. Neate, 1985, Crack growth in a CrMoV steel under cyclic loading at 873 K. *High Temp. Technol.*, **3**, 195–202.

A. Needleman and J. R. Rice, 1980, Plastic creep flow effects in the diffusive cavitation of grain boundaries. *Acta Metall.*, **28**, 1315–1332.

H. Neuber, 1961, Theory of stress concentration for shear-strained prismatical bodies with arbitrary non-linear stress–strain law. *Trans. ASME Ser. E*, **28**, 544–550.

I. Nonaka and M. Kitagawa, 1992, Small crack behavior and assessment of high temperature fatigue damage for 2.25Cr-1Mo steel. In: "Low Cycle Fatigue and Elasto-plastic Behavior of Materials-3," eds. K. T. Rie, H. W. Grünling, G. König *et al.*, Elsevier, London, pp. 527–532.

D. G. O'Connor and J. M. Corum, 1986, Design rule for fatigue of welded joints in elevated-temperature nuclear components. In: "Symposium on ASME Codes and Recent Advances in PVP and Valve Technology Including a Survey of Operations Research Methods in Engineering," PVP, vol. 109, ASME, New York, pp. 69–75.

R. Ohtani and T. Kitamura, 1994, Creep–fatigue interaction under high temperature conditions. In: "Crack Propagation in Metallic Structures," ed. A. Carpinteri, Elsevier, Amsterdam, vol. 2, pp. 1347–1383.

R. Ohtani, T. Kitamura, N. Tada and W. Zhou, 1996, Characterization of creep fatigue cracking in type 304 stainless steel. In: "Fatigue under Thermal and Mechanical Loading," eds. J. Bressers and L. Rémy, Kluwer Academic, Dordrecht The Netherlands, pp. 199–208.

M. Okazaki, I. Hattori, T. Ikeda and T. Koizumi, 1985, Low cycle fatigue crack growth behaviour at elevated temperature under the combined strain-wave-shape cycling in type 304 stainless steel. *Trans. ASME J. Eng. Mater. Technol.*, **107**, 346–355.

M. Okazaki, I. Hattori, F. Shiraiwa and T. Kozumi, 1983a, Effect of strain wave shape on low-cycle fatigue crack propagation of SU 304 stainless steel at elevated temperatures. *Metall. Trans.*, **14A**, 1649–1659.

M. Okazaki, T. Imai and S.-I. Nohmi, 1992, Naturally initiated fatigue small crack growth in a single crystal Ni-base superalloy at elevated temperature. In: "Low Cycle Fatigue and Elasto-Plastic Behavior of Materials-3," eds. K. T. Rie, H. W. Grünling, G. König *et al.*, Elsevier, London, pp. 539–544.

M. Okazaki and T. Koizumi, 1983b, Crack propagation during low cycle thermal-mechanical and isothermal fatigue at elevated temperatures. *Metall. Trans*, **14A**, 1641–1648.

M. Okazaki, T. Tabata and S. Nohmi, 1990, Intrinsic stage I crack growth of directionally solidified Ni-base superalloys during low-cycle fatigue at elevated temperature. *Metall. Trans.*, **21A**, 2201–2208.

W. J. Ostergren, 1976, A damage function and associated failure equations for predicting hold time and frequency effects in elevated temperature low cycle fatigue. *J. Testing Eval.*, **4**, 327–339.

A. Palmgren, 1924, Die Lebensdaur von Kugellagern. *Z. Vereines Deutscher Ing.*, **68**, 339–341.

R. M. Pelloux and N. Stoloff (eds.), 1980, "Creep–Fatigue Environment Interaction, Proceedings of the TMS-ASME Fall Meeting," Milwaukee, ASME, New York.

L. Pinder, 1981, Oxide characterization for service failure investigations. *Corr. Sci.*, **21**, 749–763.

A. Pineau, 1983, High temperature fatigue behavior of engineering materials in relation to microstructure. In: "Fatigue at High Temperature," ed. R. P. Skelton, Applied Science, London, pp. 305–364.

W. J. Plumbridge, 1987a, Metallography of high temperature fatigue. In: "High Temperature Fatigue: Properties and Prediction," ed. R. P. Skelton, Elsevier, London, pp. 177–228.

W. J. Plumbridge, 1987b, Damage production during high temperature low cycle fatigue of a titanium alloy (IMI 829). *Fatigue Fract. Eng. Mater. Struct.*, **10**, 385–398.

W. J. Plumbridge, 1991, Cyclic damage and fracture in a cast nickel-base superalloy at high temperature, endurance, damage, and life prediction. *Fatigue Fract. Eng. Mater. Struct.*, **14**, 721–739.

W. J. Plumbridge, M. S. Dean and D. A. Miller, 1982, The importance of failure mode in fatigue–creep interaction. *Fatigue Eng. Mater. Struct.*, **5**, 101–114.

W. J. Plumbridge, R. H. Priest and E. G. Ellison, 1980, Damage formation during fatigue–creep interactions. In: "Mechanical Behaviour of Materials 3," eds. K. J. Miller and R. F. Smith, Pergamon, Oxford, UK, vol. 2, pp. 129–139.

D. J. Powell, R. Pilkington and D. A. Miller, 1984, The effect of prestrain on the creep and creep fatigue properties of a 20%Cr-25%Ni-Nb-stabilised stainless steel. In: "Proceedings of the 2nd International Conference on Creep and Fracture of Engineering Materials and Structures, Part II," eds. B. D Wilshire and R. Owen, Pineridge Press, Swansea, pp. 989–1001.

R. H. Priest, 1989, Crack initiation and growth during creep–fatigue. In: "Proceedings of Engineering Applications of Modern Plasticity," Nagoya, Japan, (workshop session).

R. H. Priest, D. J. Beauchamp and E. G. Ellison, 1983, Damage during creep–fatigue. In: "International Conference on Advances in Life Prediction Methods," ASME, New York, pp. 115–122.

R. H. Priest and E. G. Ellison, 1981, A combined deformation map—ductility exhaustion approach to creep–fatigue analysis. *Mater. Sci. Eng.*, **49**, 7–17.

R. H. Priest and E. G. Ellison, 1982, An assessment of life analysis techniques for fatigue–creep situations. *Res. Mechanica*, **4**, 127–150.

R. H. Priest, D. A. Miller, D. H. Gladwin and J. Maguire, 1989, The creep–fatigue crack growth behavior of a 1CrMoV rotor steel. In: "Fossil Power Plant Rehabilitation," ASM, Ohio, pp. 31–37.

Y. N. Rabotnov, 1969, "Creep Problems in Structural Members," North-Holland, Amsterdam.

V. M. Radhakrishnan, 1982, Damage accumulation and fracture life in high-temperature low-cycle fatigue. In: "Low-Cycle Fatigue and Life Prediction, ASTM STP 770," eds. C. Amzallag, B. N. Leis and P. Rabbe, ASTM, Philadelphia, PA, pp. 131–151.

V. M. Radhakrishnan, 1996, An energy based analysis of thermomechanical fatigue. *Trans. Indian Inst. Met.*, **49**, 357–369.

R. Raj (ed.), 1985, "Flow and Fracture at Elevated Temperature," ASM, Ohio.

W. Ramberg and W. R. Osgood, 1943, Description of Stress–Strain Curves by Three Parameters," Tech. Note No. 902, NACA.

D. Raynor and R. P. Skelton, 1985, The onset of cracking and failure criteria in high strain fatigue. In: "Techniques for High Temperature Fatigue Testing," eds. G. Sumner and V. B. Livesey, Elsevier, London, pp. 143–166.

M. Reger and L. Rémy, 1986, Influence de l'oxydation sur le comportement en fatigue à haute température d'un alliage à base de nickel, IN100. In: "Fatigue à Haute Température," ed. C. Amzallag, Societé Française de Metallurgie, Paris, pp. 176–184.

M. Reger and L. Rémy, 1988, Fatigue oxidation interaction in IN100 superalloy. *Metall. Trans.*, **19A**, 2259–2268.

L. Rémy, 1996, Damage modeling in thermal fatigue. In: "Mechanical Behavior of Materials at High Temperature," eds. C. M. Branco, R. Ritchie and V. Sklenička, Kluwer Academic, Dordrecht, The Netherlands, pp. 399–417.

L. Rémy, F. Rézai-Aria, R. Danzer and W. Hoffelner, 1988, Evaluation of life prediction methods in high temperature fatigue. In: "Low Cycle Fatigue, ASTM STP 942," eds. H. Solomon, G. R. Halford, L. R. Kaisand and B. N. Leis, ASTM, Philadelphia, PA, pp. 1115–1132.

L. Rémy and R. P. Skelton, 1992, Damage assessment of components experiencing thermal transients. In: "High Temperature Structural Design, ESIS 12," ed. L. H. Larsson, Mechanical Engineering Publications, London, pp. 283–315.

E. Renner, H. Verhoff and P. Neumann, 1989, Life prediction for creep–fatigue based on the growth of short cracks. *Fatigue Fract. Eng. Mater. Struct.*, **12**, 569–584.

J. Reuchet and L. Rémy, 1979, High temperature fatigue behavior of a cast cobalt-base alloy. *Fatigue Eng. Mater. Struct.*, **2**, 51–62.

J. Reuchet and L. Rémy, 1983a, Fatigue oxidation interaction in a superalloy—application to life prediction in high temperature low cycle fatigue. *Metall. Trans.*, **14A**, 141–149.

J. Reuchet and L. Rémy, 1983b, High temperature low cycle fatigue of MAR-M509 superalloy: I. The influence of temperature on the low cycle fatigue behavior from 20 to 1100°C. *Mater. Sci. Eng.*, **58**, 19–32.

J. Reuchet, M. Réger, F. Rezai-Aria and L. Rémy, 1982, Cumulation of high-temperature low-cycle fatigue damage in two-temperature tests. In: "Low Cycle Fatigue and life Prediction, ASTM STP 770," eds. C. Amzallag, B. N. Leis and P. Rabbe, ASTM, Philadelphia, PA, pp. 194–211.

B. Rezgui, 1982, Interaction fatigue-fluage-environnement dans un acier inoxydable austenitique Z2 CND 17-13 (Type 316L) à 600 et 650°C: Evolution microstructurale et endommagement. Ph.D. thesis, Université de Paris-Sud, CentreD'Orsay.

B. Rezgui, P. Petrequin and M. Mottot, 1981, Hold time effects on low cycle fatigue properties of 316L stainless steel at 600°C and 650°C. In: "Advances in Fracture Research," ed. D. François, Pergamon, London, vol. 5, pp. 2393–2402.

K. T. Rie, R. M. Schmidt, B. Ilschner and S. W. Nam, 1988, A model for predicting low cycle fatigue life under creep–fatigue interaction. In: "Low Cycle Fatigue, ASTM STP 942," eds. H. Solomon, G. R. Halford, L. R. Kaisand and B. N. Leis, Philadelphia, PA, pp. 313–328.

E. L. Robinson, 1952, Effect of temperature variation on the long-time rupture strength of steels. *Trans. ASME*, **74**, 777–781.

R. Sandhya, K. Bhanu Sankara Rao and S. L. Mannan, 1989, Low cycle fatigue and creep–fatigue environment interaction of bainitic 2(1/4)Cr-1Mo steel. *Trans. Indian Inst. Metals*, **42**(Suppl.), S217–S227.

A. Scholz, J. Granacher and C. Berger, 1998, Simulation of long term creep fatigue behaviour by multi-stage service type strain cycling. In: "Proceedings of the 4th Conference on Low Cycle and Elasto-Plastic Behavior of Materials," eds. K. T. Rie and P. D. Portella, Elsevier, London, pp. 741–746.

T. Shimakawa, Y. Tanaka, H. Kaguchi, K. Imou and S. Kishi, 2001, Benchmark study of creep–fatigue life predicted by elevated temperature design standards of every country. In: "Proceedings of Creep 7," Japan Soc. Mech. Eng., Tokyo, pp. 173–177.

D. Sidey and L. F. Coffin, 1979, Damage mechanisms at high temperature. In: "Fatigue Mechanisms, ASTM STP 675," ASTM, Philadelphia, PA, pp. 528–568.

R. P. Skelton, 1978, Environmental crack growth in a 0.5CrMoV steel during isothermal high strain fatigue and temperature cycling. *Mater. Sci. Eng.*, **35**, 287–298.

R. P. Skelton, 1982, Growth of short cracks during high strain fatigue and thermal cycling. In: "Low Cycle Fatigue and Life Prediction, ASTM STP 770," eds. C. Amzallag, G. R. Halford, L. R. Kaisand and B. N. Leis, ASTM, Philadelphia, PA, pp. 337–381.

R. P. Skelton, 1983, Crack initiation and growth in simple metal components during thermal cycling. In: "Fatigue at High Temperature," ed. R. P. Skelton, Applied Science, London, pp. 1–62.

R. P. Skelton, 1985, High strain fatigue testing at elevated temperature: a review. *High Temp. Technol.*, **3**, 179–194.

R. P. Skelton, 1988a, Application of small specimen crack growth data to engineering components at high temperatures: a review. In: "Low Cycle Fatigue, ASTM STP 942," eds. H. Solomon, G. R. Halford, L. R. Kaisand and B. N. Leis, ASTM, Philadelphia, PA, pp. 209–235.

R. P. Skelton, 1988b, Fatigue crack growth. In: "Characterisation of High Temperature Materials: Mechanical Testing," ed. I. Curbishley, Institute of Metals, London, pp. 108–172.

R. P. Skelton, 1990, Introduction to thermal shock. *High Temp. Technol.*, **8**, 75–97.

R. P. Skelton, 1991, Energy criterion for high temperature low cycle fatigue failure. *Mater. Sci. Technol.*, **7**, 427–439.

R. P. Skelton, 1993a, Cyclic hardening, softening, and crack growth during high temperature fatigue. *Mater. Sci. Technol.*, **9**, 1001–1008.

R. P. Skelton, 1993b, Damage factors during high temperature fatigue crack growth. In: "Behaviour of Defects at High Temperatures, ESIS 15," eds. R. A. Ainsworth and R. P. Skelton, Mechanical Engineering Publications, London, pp. 191–218.

R. P. Skelton, 1996, Developments in creep–fatigue crack initiation and growth procedures in high temperature codes. In: "Mechanical Behavior of Materials at High Temperatures," ed. C. M. Branco, Kluwer Academic, Dordrecht, The Netherlands, pp. 281–297.

R. P. Skelton, 1998, Crack initiation and growth during thermal transients. In: "Component Reliability under Creep–fatigue Conditions, CISM Courses and Lectures No. 389," eds. J. Ginsztler and R. P. Skelton, Springer, Wien, pp. 17–86.

R. P. Skelton and J. I. Bucklow, 1978, Cyclic oxidation and crack growth during high strain fatigue of low alloy steel. *Met. Sci.*, **12**, 64–70.

R. P. Skelton and J. Byrne, 1994, Prediction of frequency effect in high temperature fatigue crack growth using damage factors. *Mater. High Temp.*, **12**, 67–74.

R. P. Skelton and K. D. Challenger, 1984, Fatigue crack growth in 2(1/4)Cr1Mo steel at 525°C I: effects of environment and dwell. *Mater. Sci. Eng.*, **65**, 271–281.

R. P. Skelton and C. A. P. Horton, 1999, The effect of thermal aging and mechanical exposure on low cycle creep–fatigue strength of 316 steel at 625°C. *Mater. High Temp.*, **16**, 87–97.

R. P. Skelton, C. J. Rees and G. A. Webster, 1996, Energy damage summation methods for crack initiation and growth during block loading in high temperature low cycle fatigue. *Fatigue Fract. Eng. Mater. Struct.*, **19**, 287–297.

D. J. Smith, 1996, The behavior of short cracks at elevated temperatures. In: "Mechanical Behavior of Materials at High Temperatures," ed. C. M. Branco, Kluwer Academic, Dordrecht, The Netherlands, pp. 195–215.

H. D. Solomon, G. R. Halford, L. R. Kaisand and B. N. Leis (eds.), 1988, "Low Cycle Fatigue, ASTM STP 942," ASTM, Philadelphia, PA.

H. D. Solomon and L. F. Coffin, 1973, Effects of frequency and environment on fatigue crack growth in A286 at 1100 F. In: "Fatigue at Elevated Temperatures, ASTM STP 520," ASTM, Philadelphia, PA, pp. 112–122.

R. C. Spencer, and D. P. Timo, 1974, Starting and loading of large steam turbines. In: "Proceedings of the American Power Conference," Chicago, vol. 36, pp. 511–521.

R. W. Swindeman and D. A. Douglas, 1959, The failure of structural metals subjected to strain-cycling conditions. *Trans. ASME Ser. D, J. Basic Eng.*, **81**, 203–208.

Y. Takahashi, 1998, Evaluation of creep–fatigue life prediction methods for low-carbon nitrogen-added 316 stainless steel. *Trans. ASME J. Eng. Mater. Technol.*, **120**, 119–125.

Y. Takahashi, 1999, Further evaluation of creep–fatigue life prediction methods for low-carbon nitrogen-added 316FR stainless steel. *Trans. ASME J. Eng. Mater. Technol.*, **121**, 142–148.

Y. Takahashi, S. Date and H. Kaguchi, 1999, Effect of elastic follow-up on creep–fatigue life of 316FR stainless steel. In: "Advances in Life Prediction Methodology," ASME PVP, vol. 391, pp. 151–157.

M. Tanaka and H. Iizuka, 1988, A micromechanics model of the initiation of grain boundary crack in high-temperature fatigue. In: "Low Cycle Fatigue, ASTM STP 942," eds. H. Solomon, G. R. Halford, L. R. Kaisand and B. N. Leis, ASTM, Philadelphia, PA, pp. 611–621.

K. Tanaka and T. Mura, 1981, A dislocation model for fatigue crack initiation. *J. Appl. Mech.*, **48**, 97–103.

D. M. R. Taplin, N. Y. Tang and H. H. E. Leipholz, 1984, On fatigue–creep–environment interaction and the feasibility of fatigue maps. In: "The 6th International Conference on Fracture," New Delhi, pp. 127–142.

N. Taylor, P. Bontempi, C. Maciga and M. Livraghi, 1994, Creep and fatigue characterization of advanced 9Cr steel base and welds. In: "Materials for Advanced Power Engineering, Part I," eds. D. Coutsouradis, J. H. Davidson, J. Ewald *et al.*, Kluwer Academic, Dordrecht, The Netherlands, pp. 341–350.

H. Teranishi and A. J. McEvily, 1979, The effect of oxidation on hold time fatigue behavior of 2.25Cr1Mo steel. *Metall. Trans.*, **10A**, 1806–1808.

G. Thomas and R. A. T. Dawson, 1980, The effect of dwell period and cycle type on high strain fatigue properties of a 1CrMoV rotor forging at 500–550°C. In: "Engineering Aspects of Creep," Inst. Mech. Eng., London, vol. 1, Paper C335/80.

B. Tomkins, 1968, Fatigue crack propagation in metals: an analysis. *Phil. Mag.*, **18**, 1041–1066.

B. Tomkins and J. Wareing, 1977, Elevated-temperature–fatigue interactions in engineering materials. *Metal Sci.*, **11**, 414–424.

R. Viswanathan, 1989, "Damage Mechanisms and Life Assessments of High Temperature Components," ASM International, Ohio.

R. Viswanathan and J. Stringer, 2000, Failure mechanisms of high temperature components in power plants. *Trans. ASME J. Eng. Mater. Technol.*, **122**, 246–255.

J. B. Vogt, J. L. Sosa and S. Argillier, 2001, High temperature low cycle fatigue of 2(1/4)Cr1Mo steels: role of microstructure and environment. In: "Proceedings of Creep 7," Japan Soc. Mech. Eng., Tokyo, pp. 327–330.

C. D. Walker, 1963, Strain-fatigue properties of some steels at 950°F (510°C)—with a hold in the tension part of the cycle. In: "Joint Conference on Creep," Inst. Mech. Eng., London, Paper 24, pp. 3.49–3.56.

J. Wareing, 1981, Creep–fatigue behavior of four casts of type 316 stainless steel. *Fatigue Eng. Mater. Struct*, **4**, 131–145.

J. Wareing, 1983, Mechanisms of high temperature fatigue and creep–fatigue failure in engineering materials. In: "Fatigue at High Temperature," ed. R. P. Skelton, Elsevier, pp. 135–185.

J. Wareing, 1999, Design against elevated temperature creep–fatigue: the role of failure modeling. In: "Engineering against Fatigue," eds. J. H. Beynon, M. W. Brown, R. A. Smith *et al.*, Balkema, Rotterdam, The Netherlands, pp. 547–556.

J. Wareing, I. Bretherton and V. B. Livesey, 1986, Life prediction for elevated temperature components subjected to cyclic deformation. In: "Conference on Fatigue of Engineering Materials and Structures," Inst. Mech. Eng., London, Paper C269/86, pp. 147–163.

J. Wareing, B. Tomkins and I. Bretherton, 1985, Life prediction in austenitic stainless steel. In: "Flow and Fracture at Elevated Temperatures," ASM, Ohio, pp. 251–278.

D. J. White, 1969, Effect of environment and hold time on the high strain fatigue endurance of 1/2 per cent molybdenum steel. *Proc. Inst. Mech. Eng.*, **184**, 223–240.

D. S. Wood and P. Banks, 1990, The creep/fatigue behavior of type 316 steel under reverse bending conditions. *Fatigue Fract. Eng. Mater. Struct.*, **13**, 551–561.

D. S. Wood, J. Wynn, C. Austin and J. G. Green, 1988, A ductility exhaustion evaluation of some long term creep/fatigue tests on austenitic steel. *Fatigue Fract. Eng. Mater. Struct.*, **11**, 371–381.

K. Yagi and K. Kubo, 1992, Evaluation of creep–fatigue interaction based on creep damage mode. In: "Low Cycle Fatigue and Elastoplastic Behavior of Materials 3," eds. K. T. Rie, H. W. Grünling, G. König *et al.*, Elsevier, London, pp. 242–247.

M. Yaguchi and Y. Asada, 1996, Creep–fatigue life prediction of G91 steel based on mechanistic model. In: "The 6th International Conference on. Creep and Fatigue," Inst. Mech. Eng., London, pp. 193–201.

M. Yaguchi, Y. Okamoto, T. Nakamura and Y. Asada, 1992, Creep–fatigue damage evaluation of modified 9Cr1Mo steel based on the overstress concept. In: "Low Cycle Fatigue and Elastoplastic Behavior of Materials 3," eds. K. T. Rie, H. W. Grünling, G. König *et al.*, Elsevier, London, pp. 210–215.

K. Yamaguchi and K. Kanazawa, 1979, Crack propagation rates of austenitic steels under high temperature low-cycle fatigue conditions. *Metall. Trans.*, **10A**, 1445–1451.

K. Yamaguchi and K. Kanazawa, 1980, Effect of strain wave shape on high temperature fatigue life of a type 316 steel and application of the strain range partitioning method. *Metall. Trans.*, **11A**, 2019–2027.

K. Yamaguchi and S. Nishijima, 1986, Prediction and evaluation of long-term creep–fatigue life. *Fatigue Fract. Eng. Mater. Struct.*, **9**, 95–107.

M. Yamamoto and A. Nitta, 2001, Effects of strain holding on creep–fatigue microcrack initiation life in Ni-based single crystal superalloy. In: "Proceedings of Creep 7," Japan Soc. Mech. Eng., Tokyo, pp. 647–652.

M. Yamamoto and T. Ogata, 2000, Microscopic damage mechanism of nickel based superalloy Inconel 738LC under creep–fatigue conditions. *Trans. ASME J. Eng. Mater. Technol.*, **122**, 315–320.

Comprehensive Structural Integrity
ISBN (set): 0-08-043749-4

Volume 5; (ISBN: 0-08-044154-8); pp. 25–112

5.03
Thermal–Mechanical Fatigue (Including Thermal Shock)

L. RÉMY
Centre des Matériaux, UMR CNRS 7633, Ecole des Mines de Paris, France

NOMENCLATURE

a	crack length (a_0, initial; a_f, final)
a	thermal diffusivity ($a = k/\rho C_p$)
A_{II}	octahedral shear stress amplitude
C_p	specific heat at constant pressure
da/dN	crack growth rate (per cycle)
$(da/dN)_c$	crack growth rate due to creep effects
$(da/dN)_f$	crack growth rate due to pure fatigue
$(da/dN)_{ox}$	crack growth rate due to oxidation effects
da/dt	crack growth rate (per unit time)
dD/dN	damage rate per cycle
D	damage (D_f—fatigue, D_c—creep, D_{ox}—oxidation)
E	Young's modulus
G	shear modulus
I_{ox}	depth of interdendritic or intergranular localized oxidation
J	Rice integral
J^*	modified J-integral for creep loading
J_{cycl}	J-integral value for cyclic loading
$J_{cycl,c}$	range of modified J-integral for creep loading over a fatigue cycle
J_0, J_1, J_2	maximum principal stress, first and second invariants of the stress tensor
k	thermal conductivity
K_0	drag stress
K_{ox}	parabolic oxidation constant ($kg^2\,s^{-1}$)
K_{IC}	fracture toughness
M	strain triaxiality factor, defined by Manson
N	number of cycles
N_f	number of cycles to failure in a pure fatigue test
$N_{f,ox}$	number of cycles to fracture of oxide scale
N_i	number of cycles to crack initiation
N_r	number of cycles to failure
p	cumulative viscoplastic strain
q	heat flux (vector)
r^s	size of the elastic shear stress domain for slip system s (isotropic hardening variable)
R	stress ratio ($R = \sigma_{min}/\sigma_{max}$) in a uniaxial fatigue test
R_p	radius of the elastic domain in deviatoric space (isotropic hardening variable)
R_ε	strain ratio ($R_\varepsilon = \varepsilon_{min}/\varepsilon_{max}$) in a uniaxial fatigue test
s_{ij}	components of the deviatoric stress tensor
S	normalized stress (ratio σ_{eq}/σ_u, equivalent stress/ultimate tensile strength)
t	time
t^*	time to fracture of oxide scale
t_h	hold time
t_r	rupture time in a uniaxial creep test
W_f	total tensile strain energy density
x^s	internal variable describing the position of the center of the elastic domain for slip system s (nonlinear kinematic component of shear stress)
X	internal variable describing the position of the center of the elastic domain (kinematic hardening component of deviatoric stress)
X_{ij}	deviatoric back stress components
α	linear coefficient of thermal expansion
α_M	matrix oxidation constant
α_{ox}	interdendritic or intergranular oxidation constant
γ^s	resolved viscoplastic shear strain on a slip system s
$\Delta D_c/\Delta N$	creep damage per loading cycle
ΔK	range of stress intensity factor
ΔK_{eff}	effective value of the range of stress intensity factor
ΔK_{eq}	range of equivalent stress intensity factor
ΔK_ε	range of strain intensity factor
Δt	period of a loading cycle
ΔW_{in}	inelastic strain energy density dissipated per cycle
$\Delta W_{in,t}$	range of density of tensile inelastic strain energy dissipated per cycle
ΔW_t	range of tensile strain energy density dissipated per cycle

$\Delta\gamma$	shear strain range	ν	frequency of loading cycle ($\nu = 1/\Delta t$)
$\Delta\varepsilon$	mechanical strain range	ν	Poisson's ratio (should not be confused with frequency)
$\Delta\varepsilon_{eq}$	equivalent strain range	ρ	density (mass per unit volume)
$\Delta\varepsilon_{in}$	inelastic strain range	σ	stress
$\Delta\varepsilon_{ij,in}$	component of inelastic strain range in strain-range partitioning (i, j = c for creep i, j = p for plasticity)	$\bar{\sigma}$	mean stress in uniaxial fatigue test
$\Delta\varepsilon_{ij,in}$	components of the inelastic strain tensor	σ_a	Stress amplitude in a fatigue test
$\Delta\varepsilon_{\alpha}$	thermal strain expansion mismatch between two different materials (substrate and coating)	$\tilde{\sigma}$	effective stress in the Rabotnov–Kachanov sense
ε	strain	σ_c^*	critical value of maximum principal stress to fracture
ε_a	alternate mechanical strain	σ_{eff}	effective stress variable
ε_p	plastic strain	σ_{eq}	equivalent stress
$\varepsilon_{p,a}$	alternate plastic strain	σ_{ij}	components of the stress tensor
ε_r	ductility in a uniaxial creep test	σ_l	fatigue limit
ε_{th}	thermal strain	σ_{max}	maximum stress in a uniaxial fatigue test
ε_u	ductility in a uniaxial tensile test	σ_{op}	stress at crack opening
$\varepsilon_{u,ox}$	strain to fracture of oxide scale	σ_u	ultimate tensile strength
ε_v	viscous strain	σ_{yy}	normal tensile stress ahead of a mode I crack
$\dot{\varepsilon}$	mechanical strain rate	τ^s	resolved shear stress on a slip system s
$\dot{\varepsilon}_{in}$	inelastic strain rate	Φ_{ox}	phasing factor for oxidation–fatigue interaction
$\dot{\varepsilon}_s$	secondary creep strain rate in uniaxial creep test stress		

5.03.1 INTRODUCTION

Components operating at high temperature are subjected to heating and cooling transients during startup and shutdown and to regime changes during operation. Thermal stresses arise when thermal strain, which can be either expansive or contractive, is totally or partly constrained. Repetition of these thermal transients can result in thermal fatigue (TF) of the component. An example of TF is thermal shock (TS), where sudden temperature changes occur at the internal or external surface of a component leading to cracking. It is important to keep in mind that the rate of change in temperature influences the damage inflicted on the component. In many engineering applications, thermal stresses are superimposed on primary stresses, which arise from pressures, mechanical end loads (including centrifugal loading in rotating components), and constraint. Severe temperature changes in the presence or absence of mechanical loading may induce plastic deformation in surface layers, which in turn can result in initiation of cracks in 5×10^4 or less cycles. Thermal and thermal–mechanical fatigue (TMF) is therefore mostly an anisothermal low-cycle fatigue (LCF) process.

TF and creep fatigue are life limiting for turbine blades and vanes in gas turbines used in jet, aircraft engines (cobalt and nickel alloys, Glenny and Taylor, 1959–1960; Spera 1969) and land-based gas turbines used in power plants, combustion chamber (Castelli *et al.*, 1993), railways wagon wheels (Neu and Sehitoglu, 1989a, 1989b), inner casings, high-pressure and intermediate-pressure rotors in steam turbines used in fossil-fuel-fired plants (Cr–Mo–V forged steel), components of diesel engines, pistons (Al-based alloys), cylinder heads (Cailletaud *et al.*, 1990), exhaust systems (cast iron) used in automotive industry, braking systems in automotive and railways (Dufrénoy *et al.*, 2002), rolling mill rolls, hot working tools (steels, Northcott and Baron, 1956; Rousseau *et al.*, 1975), heat exchanger tubes in power and chemical plants.

Due to the cost and safety considerations in the above components which are made from ferritic and austenitic steels, forged and cast aluminum alloys, cast iron, titanium alloys, forged or cast cobalt, and nickel-based superalloys, our knowledge of TF is mostly limited to these industrial alloys. By comparison, only limited work has been conducted on pure metals such as copper, nickel, iron, and titanium that are of academic interest.

Spera (1976) has defined TF as a case of LCF, because TF cracks usually start in less than 5×10^4 cycles and involve inelastic strains. This definition is consistent with most studies published. However, smaller temperature fluctuations in components may result in much larger number of cycles (Stephan *et al.*, 2002). TF always involves the constraining of free thermal expansion and contraction. TF with externally imposed constraints is known as TMF and is encountered more frequently in laboratory than in actual components.

Thermal (stress) fatigue, alternatively, corresponds to the case where constraint is internal, as encountered mostly in components. These internal constraints are typically found in components of high-temperature machinery, which are subjected to rapid heating and cooling, frequently due to interaction with some fluid.

In the design of high-temperature systems, one must consider cycle- and time-dependent fatigue damage. Material testing is a very important aspect of accurately predicting high-temperature fatigue in components. When thermal stress dominates over mechanical stress, or near stress concentrations, tests are conducted under strain-controlled conditions, with a reference curve under axial tension–compression loading. When steady operation is much longer than transients, LCF tests under isothermal conditions and creep fatigue including a strain hold at maximum strain (or elsewhere) are used. When mechanical stress dominates, stress-controlled tests and "cyclic creep" are used.

Isothermal tests are usually LCF tests conducted under uniaxial strain control using closed-loop electromechanical or servohydraulic systems. Typical strain vs. time waveforms used to characterize materials subjected to fully reversed strain-controlled loading at temperatures within the creep range are shown in Figure 1. Strain holds are usually imposed at peak strains, which results in stress relaxation. This stress relaxation converts elastic strain to creep strain, which accumulates at decreasing strain rates during the hold period.

When the strain rate is varied, or when a strain hold is added to an LCF test, the number of cycles to fracture can be plotted as a function of mechanical strain or inelastic strain range ($\Delta\varepsilon_{in}$). Using the Manson–Coffin plot vs. $\Delta\varepsilon_{in}$,

$$\Delta\varepsilon_{in}/2 = \varepsilon_f'(2N_f)^{-c} \qquad (1)$$

the number of cycles to failure, N_f, usually decreases with decreasing frequency, i.e., the constant c tends to increase. Constants in the Manson–Coffin equation are frequency dependent or strain rate dependent, at variance with the low-temperature behavior ($T/T_m \leq 0.4$, where T_m and T are melting and test temperature, respectively) (Coffin, 1973). Conversely, the reverse is true on a time-to-rupture (t_f) basis. If $t_f = N_f\,\Delta t$, where Δt is the cycle period, the time to failure increases with cycle period and an upper limit often corresponds to pure creep.

In ductile materials, when temperature increases, the number of cycles to failure tends to decrease at increasing temperature, except in high-strength materials, where ductility may increase at high temperatures and anomalous

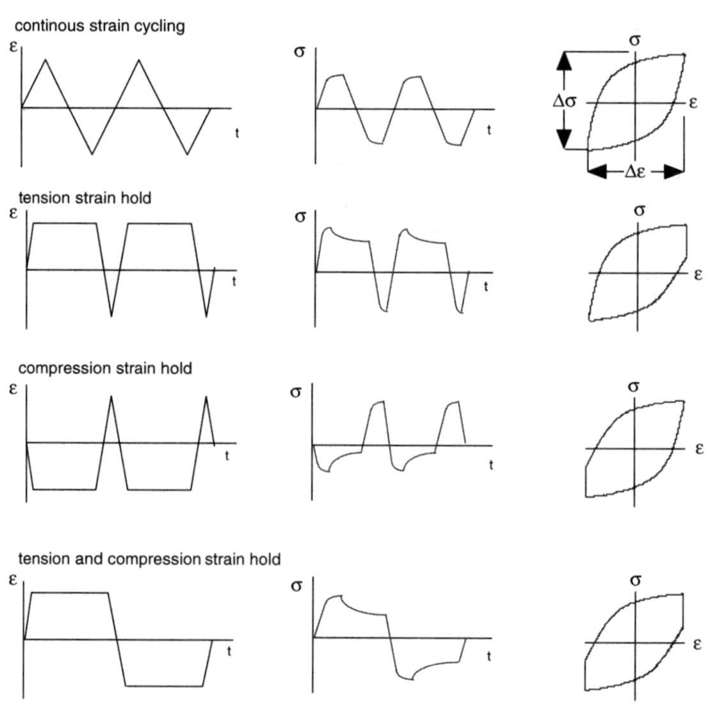

Figure 1 Typical waveform of mechanical strain cycles used in high-temperature strain-controlled fatigue.

temperature variations may occur with some materials causing a reduction in life. An increase in temperature and a decrease in strain rate or frequency usually give rise to a decrease in flow stress and cyclic flow stress. Therefore, the temperature and frequency dependence are in most cases amplified using mechanical strain range instead of inelastic strain range.

When temperature variations are too large to be neglected, tests have to be conducted under variable temperature. For a long time, these transient situations were simulated by TS experiments using thermal quenching or/and rapid heating of cylindrical billets, wedges, or tapered disks. These tests simulate component-like situations in a fairly realistic manner, but they require a complete thermal and mechanical analysis. When thermal stress is essential, TMF tests are used to check the behavior of a volume element, with or without strain holds.

This chapter specifically addresses the case of significant temperature variations, but even in that case when long-time operation is involved, long-term creep damage may dominate as in isothermal creep fatigue. The reader will find more extensive information about creep fatigue in Chapter 5.02 for crack initiation and in Chapter 5.06 for crack growth, and many concepts are common to all these areas.

Most design rules against LCF use strains and strain ranges. Since TMF involves temperature variation, much effort has been devoted to the comparison of TMF results or endurance of components with isothermal fatigue results. In numerous cases, it has been shown that TMF life, defined by failure or by some other criterion, is not shorter than the LCF life at some equivalent or maximum temperature.

Considerable data have been accumulated for isothermal LCF at high temperature, and creep fatigue on stainless steels and Cr–Mo steels used in power plants and experience is embodied in design codes (Ainsworth and Budden, 1993). The reader should consult Chapter 5.02 in this volume, where design is treated in the context of crack initiation.

Due to the difficulty of testing, many studies only focus on the thermal transient aspect of TF. Creep-fatigue loading is mostly investigated under isothermal conditions and the reader should refer to Chapter 5.02 for initiation and Chapter 5.06 for crack propagation. Models have been developed that allow us to account for creep fatigue under TMF loading in a number of cases.

One of the important aspects of TF is due to the fact that most materials are heterogeneous and comprised of multiple phases with different coefficients of thermal expansion. This is particularly the case for composite materials with fibers, and, to keep things brief, will not be considered here. Some relatively minor considerations under isothermal conditions can become essential under thermal transient conditions. For example, high-temperature components often oxidize under service conditions, or are even corroded, and during thermal transients large stresses can occur in the oxide or at the interface between oxide and substrate and promote damage mechanisms, which would not be observed under isothermal testing. Components when exposed to aggressive environments are often protected by specific coatings. Internal stresses, which develop on thermal transients, can significantly alter the life of coated components under thermal–mechanical loading. Oxidation and surface effects are therefore specifically addressed, as they form the basis of some of the damage models used for life prediction.

Thermal stresses in components as in TS experiments cannot be measured and can only be estimated through finite element (FE) analysis using appropriate constitutive models that are of tremendous importance in estimating inelastic strains and stress redistributions. A brief outline of unified viscoplastic models with internal variables used under TF is given in Section 5.03.5.

Most components are designed to resist crack initiation in the engineering practice, which is often taken as the life to failure of a laboratory specimen. Crack initiation models are then reviewed, though a number of them are presented already in Chapter 5.02. Application of models to nonisothermal conditions often involves specific difficulties. A number of models, however, try to predict a large number of situations in oxidation–creep-fatigue interactions under nonisothermal situations. Testing difficulties in TF push toward the extrapolation from short-term tests to long-time service operations, which implies that predictions heavily rely on soundness of damage models and on their relevance to service conditions. Some examples of applications of these models are shown using instances from the literature, including a few which were already quoted in a previous review (Rèmy and Skelton, 1992).

Crack growth is only briefly mentioned to show some of the specific aspects, which occur under TF. This topic is treated in much greater depth in other chapters.

5.03.2 LABORATORY TESTS

5.03.2.1 Thermal–Mechanical Fatigue

The TMF test is basically a strain-controlled LCF test where the temperature varies continuously or it includes strain hold or temperature hold times. Thus, strain and temperature profiles with time are imposed while stress is a response. This is a volume element test, like the LCF test, which implies that there is no temperature gradient across the specimen cross-section or along the gauge length.

Temperature ($T(t)$) and mechanical strain $\varepsilon_m(t)$, which simulates the thermal loading of a part of a component, are imposed simultaneously on the specimen. Total strain is imposed in practice and not mechanical strain:

$$\varepsilon_t(t) = \varepsilon_m + \varepsilon_{th} \quad (2)$$

where ε_{th} is thermal strain

$$d\varepsilon_{th} = \alpha(T)\, dT \quad (3)$$

and α is thermal expansion coefficient at temperature T.

Most facilities use servohydraulic machines and less frequently electromechanical machines that can be more reliable for long test durations. An induction heater is often used but lamp furnaces are also used. Natural cooling or forced cooling is used to achieve the needed temperature variation.

In all modern facilities, a microcomputer is an essential feature. It generates the two synchronous temperature and mechanical strain signals. The microcomputer imposes a temperature cycle through the temperature controller after digital-to-analog conversion and simultaneously imposes a strain cycle on the specimen, adding a mechanical strain to the previously memorized thermal strain.

A few authors use a two-bar structure to impose some phasing between thermal strain and mechanical strain, where a bar with a large cross-section is used to constrain the thermal displacement of the test specimen (Sehitoglu, 1985). Under total constraint, out-of-phase cycle of thermal and mechanical strain is obtained when the constraining bar is held at minimum temperature. An in-phase cycle results when the temperature of both bars varies identically. Some authors actually use two servohydraulic machines to experimentally load the two bars (Angarita *et al.*, 2000).

A large variety of specimen geometries have been used. Several authors have used solid specimens similar to these for isothermal LCF, which are only convenient for materials that have a good thermal diffusivity: carbon steels and aluminum alloys behave much better than austenitic stainless steels or superalloys. Hollow specimens were introduced by Hopkins (1976) and are often preferred in order to reduce thermal gradients. Care must be taken to avoid surface effects due to machining marks, as well as in ensuring the right ratio of grain size to thickness, especially for cast materials.

The testing procedure is rather complex, as opposed to conventional LCF at constant temperature, but now a consensus exists among researchers. An ISO standard for TMF testing is being developed. The very first TMF test facilities used data tracking systems to control the thermal cycle, but with poor accuracy. The introduction and development of microcomputers have considerably increased possibilities and accuracy (Köster *et al.*, 1994): almost any TMF cycle can be imposed which is compatible with the thermal performance of the testing system.

Early leaders in the field (Taira, 1973; Hopkins, 1976) used two basic cycles (Figure 2): the in-phase cycle, where mechanical strain is maximum at maximum temperature; and the out-of-phase cycle, where mechanical strain is maximum at minimum temperature. Fully reversed mechanical strain cycling is generally used for these tests. They are very popular and are incorporated in the ISO TMF standard being developed. Most authors follow this practice to investigate crack initiation or crack growth and often use it to determine if in-phase cycle is more detrimental than out-of-phase cycles.

Authors working in closer connection with design engineers prefer to more closely simulate the strain–temperature history of critical parts (Hopkins, 1976; Malpertu and Rémy, 1990; Jordan and Meyers, 1986). A so-called diamond type was proposed early by Hopkins (1976): a counterclockwise diamond cycle is displayed in Figure 2, which has been used to simulate thermal transients occurring in blades and vanes in gas turbines (Malpertu and Rémy, 1990; Guedou and Honnorat, 1993; Esslinger

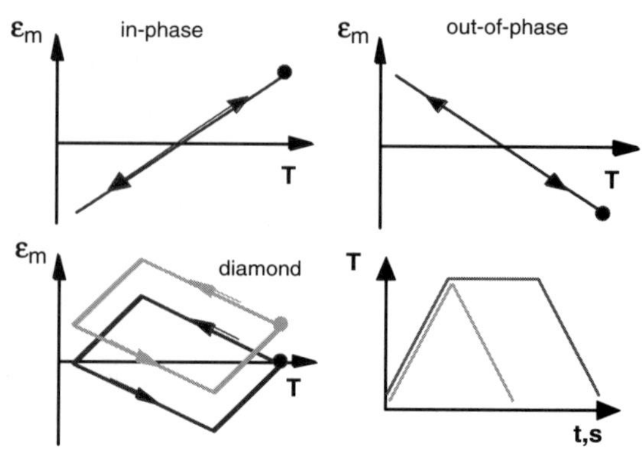

Figure 2 Examples of TMF cycles; the solid circle indicates the usual position of strain hold time.

and Affeldt, 1998). It was pointed out that the TMF cycle has a large influence on mean stress present in high-strength alloys. In superalloys a diamond cycle was shown to give insignificant mean stress, while an in-phase cycle yields a large compressive mean stress and an out-of-phase cycle yields a large tensile mean stress (Malpertu and Rémy, 1990; Rémy 1990). The mean stress is important to consider when conflicting observations from tests conducted under different phase conditions are encountered.

The concept of a volume element test nevertheless imposes a constraint of no stress gradient in the gauge length. The longitudinal and transverse temperature variations along the gauge length and across specimen section should be kept as small as possible. This places, in turn, some limitations on the temperature rate that can be achieved. Therefore, it is often impossible to simulate the actual loading of critical parts because of the temperature rate and the duration of hold time under load during steady operation, as for creep fatigue.

TMF tests can be used to compare different candidate materials. Some authors believe that there is a basic cycle which can be used like isothermal LCF to characterize the material behavior and give master curves directly useful for design. Others consider that this test is often too complex to be used directly in design. It must be used to validate constitutive models or damage models under standardized conditions that are more representative of service loading than isothermal LCF at maximum temperature or at an average temperature (Rèmy, 1990). The validation of each model can then be conducted independently by using results from these tests, since they are volume element tests. Such a validation cannot be conducted by using TS tests, since the predicted lifetime is the convolution of predictions of each type of model, and can be biased by inaccuracy in thermal analysis as in actual components. After comparison with test results under complex conditions, these models can be used to predict component behavior.

TMF testing involves the same difficulties as LCF at high temperature, but since temperature also varies with time, the control of test parameters is more critical and additional problems also arise (see, e.g., Rémy, 1996).

Different practices are followed regarding when to stop the test for damage assessment. The most common practice is to continue the TMF test up to failure. As is the case for LCF tests, there are questions about the effect of specimen size on the number of cycles to failure and its relationship to crack initiation in engineering components. The size of the crack at failure, and more significantly the number of cycles to failure, N_f, may also depend on specimen geometry. This may or may not correspond to the size of cracks in components at the instance of crack initiation, and may be rather inappropriate for the FE model used to design components. This difficulty can be circumvented by conducting the test up to a crack size relevant to the FE analysis, continuously monitoring the evolution of microcracks by an *in situ* visual system or a plastic replication technique using test interruptions.

All these techniques, though useful, are fairly expensive. Some authors (Cailletaud *et al.*, 1983) used a constraining frame to generate thermal (mechanical) fatigue data on a uniaxial specimen, in which stress induced by thermal cycling simulates pure thermal stress in structures, i.e., compressive at maximum temperature and tensile at minimum temperature.

The majority of TMF tests are carried out on uniaxial specimens, but in practice the multiaxial loading cannot always be ignored, as suggested by the results of inelastic stress analysis. Biaxial TMF tests are only scarcely reported (Ziebs *et al.*, 1992, 1994) and are used for the case of tension–torsion. Three control loops are used for axial strain, torsional strain, and temperature, respectively. Simple in-phase, out-of-phase, or diamond cycles of diamond axial mechanical strain vs. temperature cycles can be combined with proportional or non-proportional loading when the shear strain is superimposed on axial strain.

A multiaxial test rig has been designed (Chieragatti and Paun, 2000) using a rotating bending machine. A tubular specimen is used with a cylindrical hole (6 mm diameter) and a conical outer geometry. The central part of the specimen is exposed to two thermal sources fixed at 180° along specimen axis: a heating source which is a high-frequency induction coil (400 kHz generator) and a cooling source which is a cold air jet. During rotation, a given volume of material is exposed alternately to heating and cooling sources, which allows one to achieve fairly high frequencies for TMF testing, e.g., a thermal transient from 600°C to 1,050°C in 5 s for a superalloy. Loading is multiaxial because the specimen behaves as a structure due to the high thermal gradient in the specimen.

5.03.2.2 Thermal Shock

Since tests on actual components in service-like conditions are very expensive, tests using simplified structures have been used to simulate thermal stress cycling of components.

5.03.2.2.1 Facilities developed for gas turbine applications

The aircraft and jet engine industry has long been concerned with TF problems, especially with blades and nozzle guide vanes in gas turbines. Several facilities have been developed to simulate the specific conditions that occur in such components.

Glenny in the late 1950s introduced a fluidized bed technique using tapered disks (Glenny and Taylor, 1959–1960). This technique has subsequently been used in other laboratories (Woodford and Mowbray, 1974; Bizon and Spera, 1976). Others have used the specimen with disk geometry proposed by Glenny by increasing the dimensions by a factor of 1.5 (Woodford and Mowbray, 1974, Figure 3). Some authors tend to use single-wedge (Bizon and Spera 1976) or double-wedge geometries. These disks and wedges are generally smooth but some authors have also used notched specimens.

Specimens are loosely supported by end grooves or through hole supports, to allow free thermal expansion. Two beds are used: one for heating that is kept at maximum temperature, and the other for cooling that is kept at the minimum temperature. Both heating and cooling beds consist of retorts filled with a few μm diameter alumina particles through which air is pumped. The airflow is adjusted to allow the particles to circulate and behave as a fluid. The large number of particles in the beds and their fluid action promote uniform, high heat transfer rates. The specimens (or group of specimens) are transferred alternately from one bed to the other by means of a pneumatic actuator, which is automatically controlled. Typical test conditions used heating for 2–8 min and cooling for 2–8 min with transfer times of just a few seconds. The usual practice is to run tests and to interrupt them at regular number of cycles. The thin peripheral edge of disk or the edge of single- or double-wedge specimens are examined prior to testing and periodically during the tests at low magnification (×30 or 100).

For a long time these tests were used to rank alloys and compare the TS resistance of candidate materials for turbine components. A large number of test parameters can be changed. The maximum temperature is a primary variable. The hold time in both the hot and cold beds can be varied. Since the work of Spera (1969), these tests are used to assess the validity of a complete life prediction chain, including constitutive and damage models.

The radius of the thin edge is often varied at a constant T_{max} to investigate different thermal stresses (or mechanical strains) at the thin edge, or peripheral area (Glenny, 1967; Woodford and Mowbray, 1974). The dimensions of the bulk part can be changed, or the wedge opening angle, but this is very seldom made (Glenny, 1967).

Engine manufacturers also use burner rigs. Such a rig was described by Rézaï-Aria *et al.* (1988a, 1988b) (Figure 4). Wedge-type TF specimen is simply placed on three vertical silica rods to impose no constraint on the specimen. The specimen is uniformly heated along its thin edge by the flame of a burner using a mixture of air and propane gas for ~60 s and then it is cooled by compressed air via a cooling nozzle for 20 s. The vertical displacement of the cooling and heating nozzles is made via pneumatic actuator.

To avoid the limitations of the above simple tests, laboratory test facilities have been developed that use conventional heating techniques. Meyer-Obersleben *et al.* (1990, 1996) and Blümm *et al.* (1994) used a radiofrequency

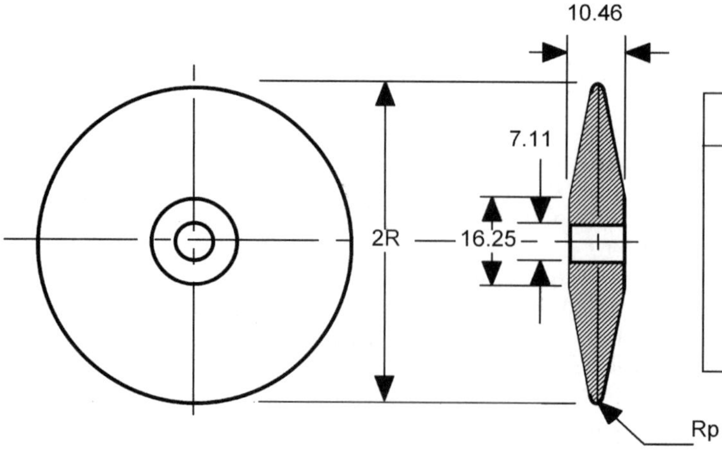

PART	Rp, mm	2R, mm
1	0.254	60.45
2	0.508	60.96
3	0.762	61.47
4	1.016	61.98

Figure 3 Geometry of the disk specimens used for TF with fluidized beds (Woodford and Mowbray, 1974).

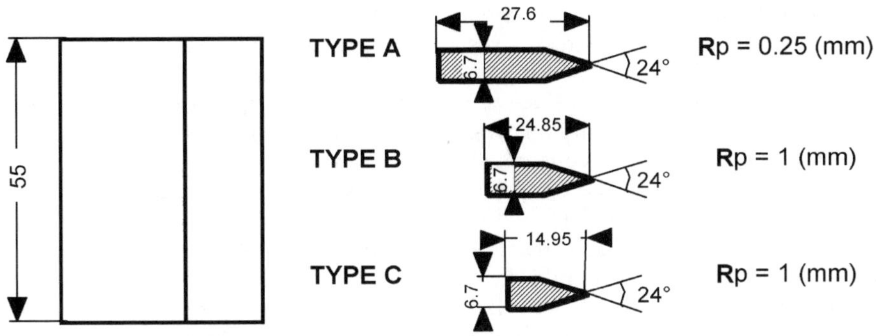

Figure 4 Geometry of the wedge specimens used for TF with a burner rig (sources Rezai-Aria *et al.*, 1988a, 1988b).

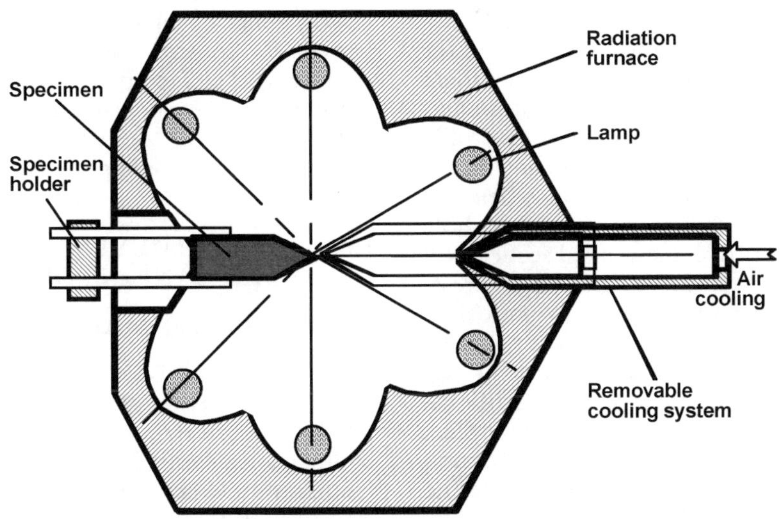

Figure 5 Schematic drawing of the TF rig used to test the wedge-shaped specimen.

(RF) induction heater and a cooling nozzle: induction generators with frequencies 200 kHz and 3,000 kHz, which can generate the same temperature–time cycle but achieve different temperature gradients during transients, and thus different mechanical strain–temperature cycles. An extensometer was mounted to measure the total strain at the thin edge and deduced an experimental measurement of mechanical strain to rationalize the influence of maximum temperature at the thin edge on the lifetime to crack initiation in single-crystal superalloys (Meyer-Obersleben *et al.*, 1996).

Another type of test facility uses wedge specimens that are heated by a lamp furnace utilizing 1,500 W light bulbs (Figure 5, Köster *et al.*, 1996a, 1996b). The specimen leading edge is placed at the center on an adjustable holder that kept the tested structure in position without mechanical loading. The furnace is designed to focus lamp radiation on a fairly small area. Cooling is ensured by a removable cooling nozzle (via a pneumatic actuator) with air compressed at 8 bar.

5.03.2.2.2 Facilities motivated by other applications

In numerous applications the overall surface of a component is exposed to a fluid that is quickly heated or cooled, wherein craze cracking or the so-called heat-checking crack networks may form. A two-dimensional (2D) network of surface cracks appears in these conditions while in gas turbines c, a 1D network is more often found to exist on blades. The latter case has been widely studied, and reported above.

Skelton (1980, 1983, 1990) and Skelton and Nix (1987) used billet of materials using typical 92 mm outside diameter and 36 mm wall thickness. A so-called up-shock configuration is obtained when a thick structure is at a uniformly low temperature and when hot fluid impinges at the surface. The surface experiences compressive stresses as it attempts to expand against the remaining material. Heat flows slowly in the bulk, and as the temperature gradient decreases the whole system expands, putting the surface into tensile stress.

If steady-state conditions are realized at high temperature, the residual tensile stress may relax but will favor intergranular cracking in many polycrystalline alloys.

This situation is simulated by rapidly heating the outside surface of the cylinder, by means of RF (of 35 kW). For ferritic steels, the temperature of the surface starts from a uniform value of 200°C (as indicated by spot-welded thermocouple) and reaches rapidly to 565°C in 8–12 s. The temperature becomes uniform in the billet after 3 min at this control temperature. Then the billet is cooled with compressor as quickly as possible, but avoiding a TS on cooling. The airstream is divided at the bore and outer surface and cooling is allowed for 15–20 min before the next heating shock from 200°C.

The same geometry was used also for simulating the so-called down-shock, i.e., thick structure at a uniformly high temperature and when the fluid suddenly cools (or is shut off). The surface layer goes in tension as contraction is attempted. Peak tension occurs when the thermal gradient is maximum. When the whole component cools down, the yielded section will be submitted to residual compressive stress. When induction heating is used to uniformly heat the billet, water flow or air stream can be used to rapidly cool down the bore of the structure. This results, under severe conditions, in a 2D network of cracks on the inner surface.

Other facilities were described by Marsh (1981) and Fissolo *et al.* (1996), which use the same basic concept. A TF facility uses a thick specimen with a rectangular cross-section. The specimen is heated by the passage of an electrical current. Two opposite forces are submitted to a down-shock by spraying water over a central area 9 mm wide and 30 mm high. A typical cycle consists of a period of 4 s with a quenching time of 0.25 s. Temperature always remains constant in the core. Thus, quenching leads to a temperature decrease at the surface by an amount equal to the temperature gradient between surface and core. Typical maximum temperatures investigated are 320–550°C and temperature variations range from 100°C to 300°C.

Single crack propagation from a circular groove machined in the heated length can be studied as described by Skelton (1990), by using a thick tube with (a) permanent cooling by water flow inside and (b) periodic heating by induction (1 MHz) at the outside (Fissolo *et al.*, 1996).

A similar facility has been used to investigate the TS resistance of tool steels (Miquel *et al.*, 2002). The specimen geometry is that of a hollow billet of varying thickness that can be varied to control the thermal stress level (Figure 6). The facility shown in Figure 7 used permanent internal water cooling, and alternates heating for ~6 s via an induction coil and external air cooling via compressed air.

5.03.3 EXPERIMENTAL RESULTS AND MECHANISMS OF DAMAGE

5.03.3.1 Thermal–Mechanical Fatigue

TMF in steels. Taira (1973) has investigated the TMF behavior of various steels including 304 stainless steel and has shown that TMF life compares well with LCF at some intermediate temperature when maximum temperature is low, or with LCF at maximum temperature on a strain basis. Out-of-phase tests with a small temperature range, 90–165°C, on 304 steel were found to give a TMF life almost identical to LCF life at maximum temperature (Figure 8) (Haddar *et al.*, 2002). A similar conclusion was obtained using a somewhat larger temperature range on a rotating TMF facility by Hayashi *et al.* (1994).

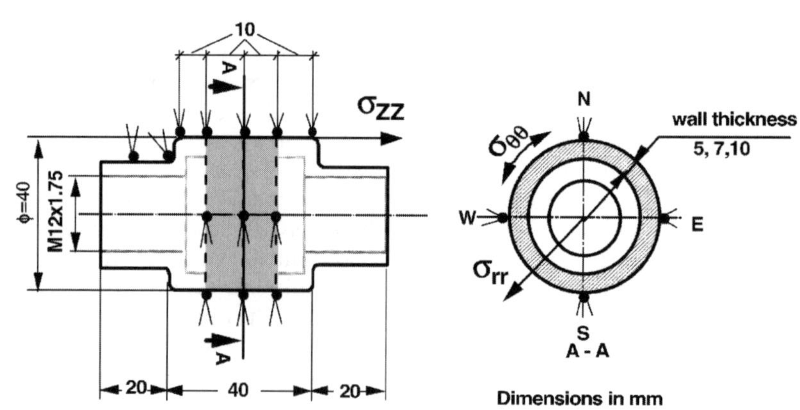

Figure 6 TS specimens with different thicknesses and the locations of the thermocouples, used to investigate heat checking in tool steels (source Miquel *et al.*, 2002).

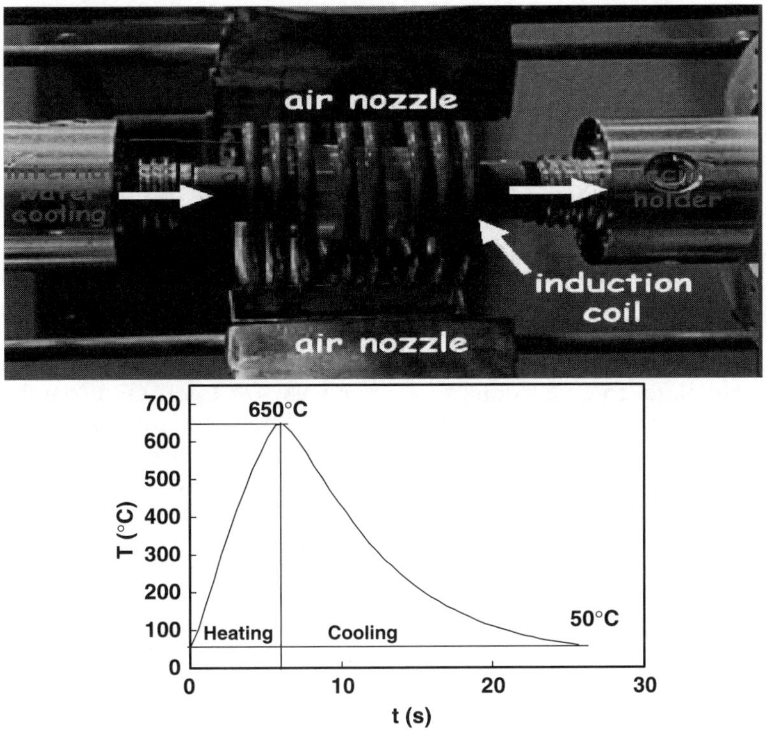

Figure 7 TS rig (induction coil, pressure air cooling, and specimen) and a typical temperature–time cycle at the center of the specimen used to investigate heat checking in tool steels (source Miquel *et al.*, 2002).

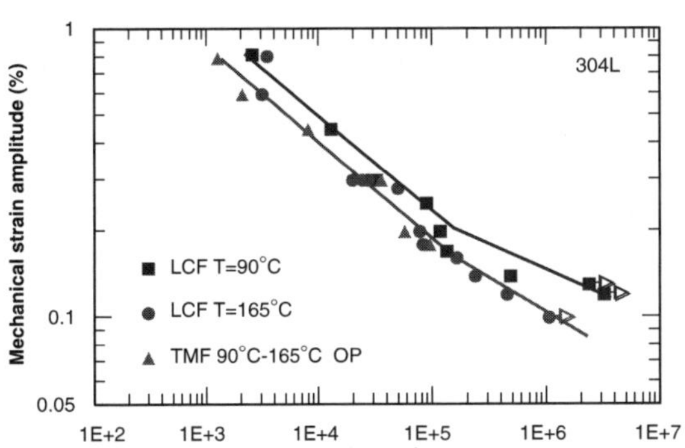

Figure 8 Number of cycles to failure as a function of mechanical strain range in 304 stainless steel: comparison between out-of-phase 90–165°C TMF tests and tests at minimum and maximum temperature (after Haddar *et al.*, 2002).

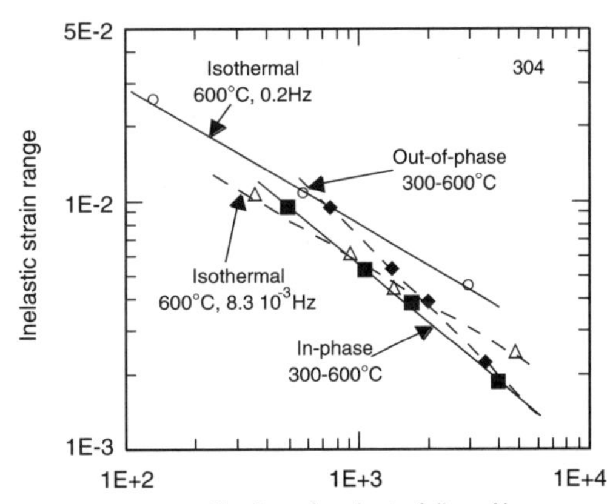

Figure 9 Number of cycles to failure as a function of inelastic strain range in 304 stainless steel: comparison between in-phase, out-of-phase 300–600°C TMF tests, and tests at maximum temperature (after Kuwabara and Nitta, 1980).

Kuwabara and Nitta (1980) have reported that the influence of frequency on LCF life, as well as the difference between out-of-phase, in-phase TMF (300–600°C), and LCF (600°C) depends on the composition of the specific heat of 304 stainless steel (most tests being carried out at a frequency of 8.3×10^{-3} Hz). Out-of-phase cycling gives a longer life than in-phase cycling as a function of inelastic strain range (Figure 9). Both TMF cycles can yield shorter life than isothermal fatigue at maximum temperature, at the relevant frequency. The slope of the Manson–Coffin curves is higher than that for LCF, whatever the frequency. These authors investigated the proportion of intergranular cracks and observed variations with cycle period: the proportion of intergranular cracks was negligible for the out-of-phase cycle for a 10–20 min period and increases from isothermal, to isothermal with 18 min tensile strain hold and continuous in-phase cycle, to in-phase with 18 min tensile strain hold (Figure 10). A physical argument for the strain range partitioning method proposed by Manson (1973) was found to correlate very well their experimental results.

Fujino and Taira (1980) tried to elucidate phasing effects in TMF of 304 stainless steel using tests under vacuum to make metallographic observations of grain boundary (GB) sliding, using a cycle between 400°C and 700°C (with a cycle period of 300 s, due to the difficulty of cooling in vacuum). They show that GB sliding accumulates with the number of cycles to give a tensile residual strain for in-phase TMF; it accumulates to give a compressive residual strain for out-of-phase TMF;

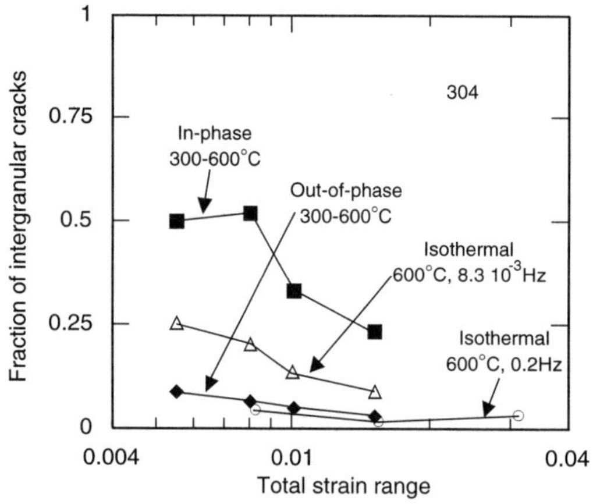

Figure 10 Fraction of intergranular cracks vs. total strain range in 304 stainless steel: comparison between in-phase, out-of-phase 300–600°C TMF tests, and tests at maximum temperature (after Kuwabara and Nitta, 1980).

high-temperature isothermal results in less GB sliding but with tensile and compressive components.

This mechanism is linked with the occurrence of GB cracking and is considered to be the reason for the shorter TMF life in in-phase cycle than with out-of-phase and LCF at T_{max}, for a given inelastic strain amplitude.

Both out-of-phase and in-phase cycling give a TMF life shorter than LCF life at T_{max} for a low carbon steel (Fujino and Taira, 1980). Jaske (1976) has conducted TMF tests on 1010 low carbon steel. For temperature ranges 93–316°C and 93–427°C, TMF life was found to be in good agreement with that under LCF at T_{max} using the empirical parameter $\sigma_{max} \cdot \Delta\varepsilon_m/2$, proposed by Smith *et al.* (1970). However, in the range 93–538°C it yields a life much lower than isothermal LCF at 538°C. This failure was associated with the occurrence of time-dependent phenomena at higher peak temperatures. Carbon steels have been shown to be sensitive to aging phenomena and to exhibit strong interactions between fatigue oxidation and creep, depending on the phasing between strain and temperature (Neu and Sehitoglu, 1989a, 1989b).

Tool steel 0.38%C–5Cr–1.25Mo–0.5V has been investigated using out-of-phase TMF for a minimum temperature and various temperature ranges (Oudin *et al.*, 2002). For a given mechanical strain range, at lower-temperature ranges, out-of-phase TMF results in a life longer than under isothermal fatigue at maximum temperature and at the same frequency, while it yields a shorter life than LCF at high maximum temperature (Figure 11). Oxide-scale

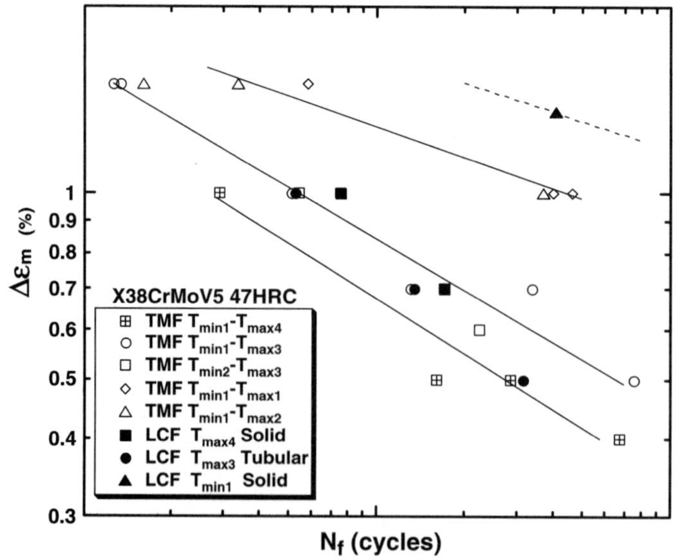

Figure 11 TMF life in a tool steel as a function of the mechanical strain amplitude and compared with LCF life at maximum temperature. The minimum temperature is constant in all TMF cycles (source Oudin *et al.*, 2002).

cracking perpendicular to the loading axis is the dominant damage mechanism for this type of loading, while secondary cracks are observed, as well as oxide spalling. On a section, this results in numerous oxide intrusion growing into the alloy and internally cracked (Figure 12).

Chromium steels (9–12%) are used in power plants for their resistance to LCF, high-cycle fatigue (HCF) as against oxidation and corrosion up to 550°C. In order to improve the efficiency of steam turbines, there is a worldwide effort to raise their operating temperature up to 650°C. Swindeman (1988) investigated 9Cr–1Mo–V–Nb steel in two-bar ratcheting tests with thermal cycling between 150°C and long dwell times at 550°C. Thorough investigation was carried out on the same steel by Säuberlich (1998). This steel has a tempered lath martensite microstructure with carbides (mostly $M_{23}C_6$) precipitated at subgrain boundaries and former austenite grain boundaries. T_{min} was kept at 150°C and T_{max} was varied in out-of-phase TMF tests between 450°C and 650°C, using a temperature rate of $10°C\,s^{-1}$. Tests were performed with dwell times between 3 min and 20 min at $T_{max} = 600°C$. With T_{max} up to 500°C, compressive mean stresses develop and vanish

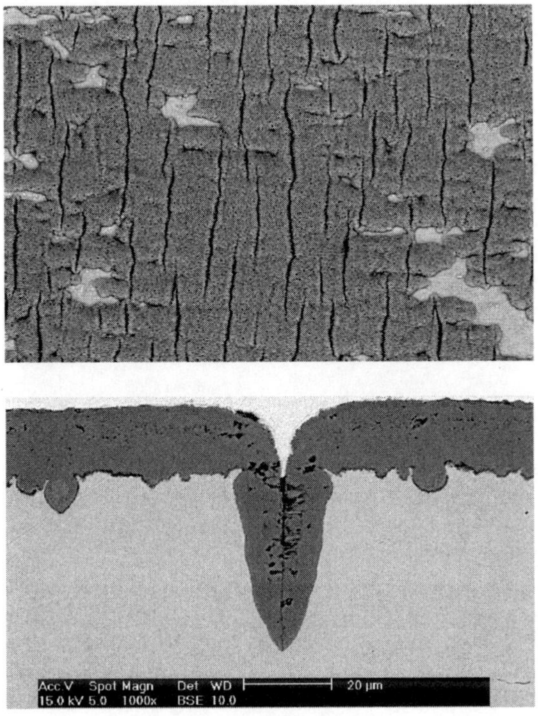

Figure 12 General features of the external crack initiation, oxide-scale spalling (top), and cracking in depth (bottom) in a tool steel subjected to out-of-phase TMF; stress axis is horizontal (source Oudin *et al.*, 2002).

progressively with softening of material. For T_{max} above 500°C, tensile mean stresses are observed (larger than 100 MPa), which increase with increasing maximum temperature.

Subgrain size varies and the mean carbide size increases with increasing maximum cycle temperature, which results in softening (peak stress and hardness reduction).

A comparison with 12Cr steel shows that different mean stresses occur in the two steels for the same TMF conditions. The 12Cr steel exhibits a shorter TMF life when using a plastic strain amplitude criterion. The Smith–Watson–Topper parameter

$$P_{SWT} = \sqrt{(\sigma_a + \bar{\sigma})\varepsilon_a E} \tag{4}$$

where σ_a is stress amplitude, $\bar{\sigma}$ mean stress, ε_a alternate mechanical strain, and E Young's modulus, fails to rationalize the differences in steels due to mean stresses, while the difference was much smaller using plastic strain amplitude. Ostergren parameter, which uses

$$P_{OST} = (\sigma_a + \bar{\sigma})\varepsilon_{p,a} \tag{5}$$

where $\varepsilon_{p,a}$ is plastic strain amplitude, alternatively, is able to give a unique master curve. This difference in life between 12Cr and 9Cr steels is mostly due to the larger mean stresses occurring in 12Cr steels. This is favored by their higher strength at low temperatures, where tensile stresses develop during out-of-phase TMF, and their stronger temperature dependence of yield strength.

TMF in aluminum alloys. Sehitoglu and coworkers have investigated the TMF behavior of cast Al–Si–Mg alloy used for automotive applications (Engler-Pinto *et al.*, 2002). Their results on one of the alloys investigated are summarized in Figure 13. LCF results at room temperature give a maximum fatigue life. When TMF results for the temperature range 100–300°C are compared with isothermal LCF at various temperatures and frequencies, there is at most a decrease by a factor of 5 in life in LCF at 300°C and at low frequencies for a given mechanical strain range; out-of phase TMF is comparable with LCF at maximum temperature but in-phase TMF gives shorter lives by an order of magnitude than LCF at maximum temperature.

The same comparison has been made using stress range taken at midlife, since there was no stabilized cycle (Figure 14). There is much larger influence of temperature and strain rate in such a plot for isothermal tests. For a given mechanical strain range, the stress range at midlife is now higher for in-phase than for out-of-phase, so that only minor difference is

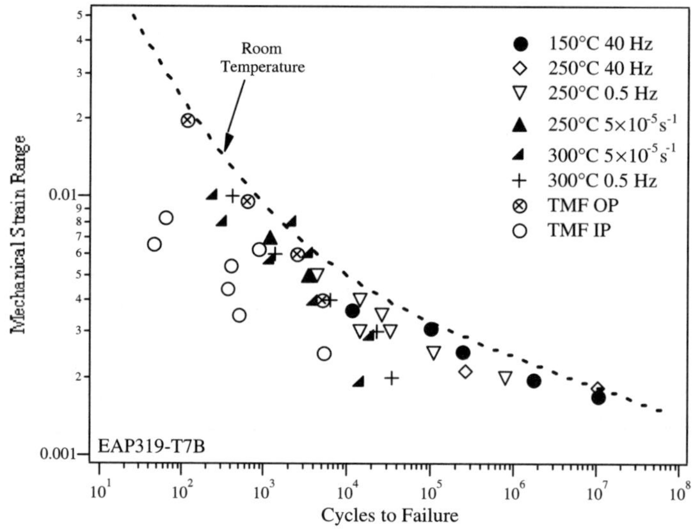

Figure 13 High-temperature fatigue life results for all tests with the EAP319-T7B alloy (after Engler-Pinto *et al.*, 2002).

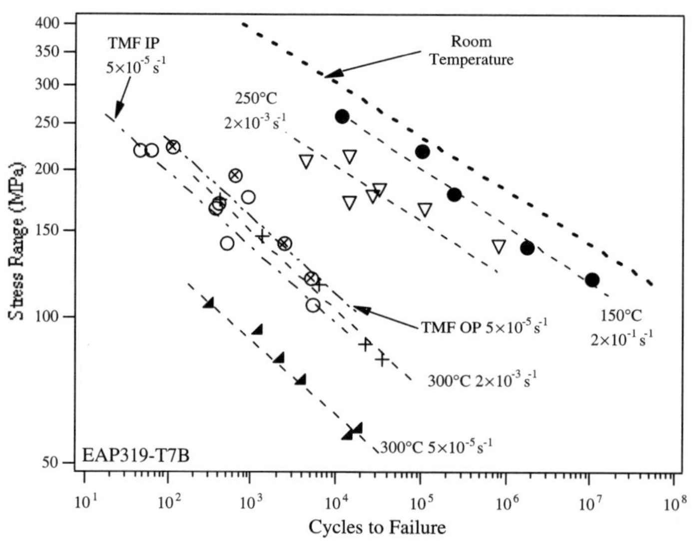

Figure 14 Stress range vs. fatigue life showing the effect of temperature and strain rate for the EAP319-T7B alloy (after Engler-Pinto *et al.*, 2002).

observed between these tests. It must be emphasized that this is fortuitous but more important on a stress range basis is that both TMF tests give a life longer than LCF life at maximum temperature and at low frequency.

A better correlation for all fatigue lives from different test conditions is obtained by using inelastic strain range vs. fatigue life (Figure 15). All fatigue lives except in-phase TMF are on a single Manson–Coffin curve, which implies little difference between pc and pp components in strain-range partitioning framework (Manson, 1973; Halford and Manson, 1976). The in-phase TMFs are much shorter but the slope is smaller, which is very much different from the case of austenitic stainless steels where Manson *et al.* reported for the cp component and the in-phase cycling a similar slope but with lower ductility coefficient. The fracture surface analysis reveals transgranular fracture for all test conditions with intergranular fracture for in-phase TMF cycling.

TMF in superalloys. Crack initiation tests were carried out on superalloy IN718 which has been investigated under stress-controlled TMF crack growth by Cook *et al.* (1996). For TMF cycles between 300°C and 649°C, the trends are basically the same for in-phase and out-of-phase cycles. In-phase cycles yield lower lives, almost as short as the lives observed at maximum temperature. Out-of-phase cycles result in longer lives, approximately half

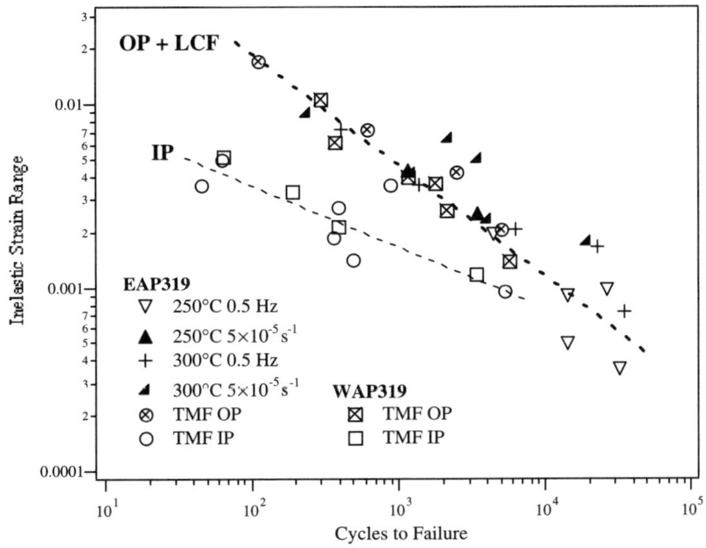

Figure 15 Inelastic strain range vs. fatigue life for the WAP319 and EAP319 alloys. Note that TMF–IP lives are significantly shorter than TMF–OP lives (after Engler-Pinto *et al.*, 2002).

Table 1 Influence of TMF phase angle on the number of cycles to failure of IN718 for a mechanical strain range $\Delta\varepsilon_m = 1.3\%$ in the temperature range 300–649°C (strain ratio $R_\varepsilon = -1$).

Cycle type	N_f (cycles)	*Peak strain temperature* (°C)	*Fraction of period above* $\bar{T}$ *and* $\bar{\varepsilon}$	*Mean stress*	*Fracture mode*
Isothermal at T_{min}	3,460	300		0	Transgranular
270° TMF counterclockwise diamond	1,940/2,800	474	1/4	<0	Predominantly transgranular
180° TMF out-of-phase	1,750	300	0	>0	Transgranular
90° TMF clockwise diamond	750	474	0	0	Predominantly transgranular
0° TMF in phase	500	649	1/4	≪0	Intergranular
Isothermal at T_{max}	350	649			Intergranular

After Cook *et al.* (1996).
$\bar{T}, \bar{\varepsilon}$ refer to mean temperature and mean mechanical strains, respectively.

of those observed at the minimum temperature.

Isothermal results at maximum temperature would predict the in-phase cycle life correctly but would be very conservative for the out-of-phase cycle. Lives during out-of-phase, clockwise diamond, and in-phase cycles rank in the order of the peak strain temperature. However, there is no rationale for the counterclockwise diamond, which yields longer lives, near the minimum temperature. Mean stress effects are small between clockwise and counterclockwise diamond. In all cases, the out-of-phase and clockwise diamond cycles had the highest stress and the longest life.

The only clear difference comes from the fracture mode and the effect of cycle shape is actually the ranking according to the proportion of intergranular and transgranular areas on the fracture surfaces (see Table 1).

A counterclockwise diamond-type cycle has been used for Waspaloy in the temperature range 100–750°C (Köster *et al.*, 1996a, 1996b). At high temperatures (100–750°C), this alloy is prone to intergranular cracking near 750°C and the LCF lives decrease by an order of magnitude between 1 Hz and 6×10^{-3} Hz (Figure 16). TMF life is longer than LCF life at maximum temperature at the same frequency for a given mechanical strain range. In spite of this, tests under vacuum suggest that both creep and oxidation could interact with fatigue damage (Köster, 1997).

The Ostergren parameter was used by Cook *et al.* (1996) to correlate isothermal results of directionally solidified Rene 80 H and conventionally cast Rene 80, in the original frequency modified form:

$$\sigma_{max}\Delta\varepsilon_{in} = A(N_i\nu^k)^c \qquad (6)$$

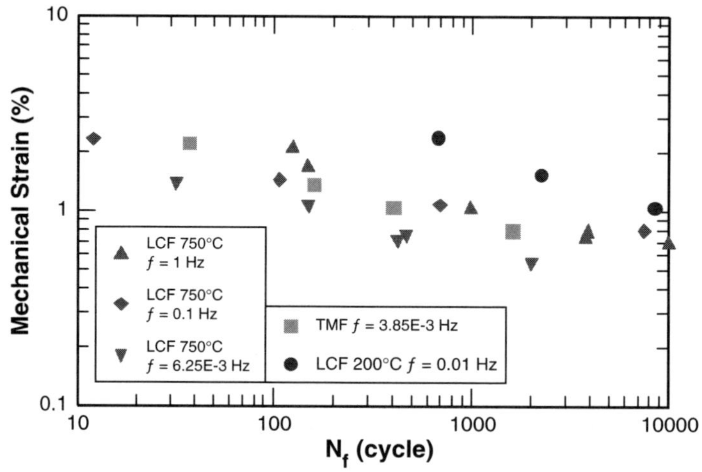

Figure 16 Lifetime comparison between LCF (200°C and 750°C) and TMF (100–750°C) in Waspaloy (after Koster, 1997).

where σ_{max} is maximum tensile stress, $\Delta\varepsilon_{in}$ the inelastic strain range, v the frequency, N_i the number of cycles to crack initiation, and A, k, c are three constants. The results of more than 120 tests at various temperatures (538°C, 649°C, 760°C, 871°C, 982°C) with a wide range of hold times and strain rates were correlated within a factor of 3.

A limitation to this approach is that beyond ~1,000 cycles, the inelastic strain range becomes very small and the equation tends to overpredict LCF life. A second concern is that the use of such a method requires accurate predictions of maximum tensile stress and inelastic strains, which, in turn, requires accurate constitutive models.

Application to in-phase, out-of-phase, and clockwise diamond TMF cycles over the temperature range of the isothermal tests captures most data within a factor of 2. Data with larger temperature excursions and higher strain rates are overpredicted by a factor of 4–10. These data were related to the occurrence of annular coating cracks, whereas the database was established for uncoated material. Combination of low-temperature brittle coating and high-temperature stress relaxation can lead to specimen life lower than that encountered in isothermal test conditions.

Out-of-phase TMF tests on IN738 superalloy in the temperature range 482–816°C give shorter lives than in-phase tests by at least a factor higher than 2 (Bernstein *et al.*, 1993). Increasing the temperature range to 482–871°C reduces fatigue life by a factor of 5 for out-of-phase tests.

Ratchev *et al.* (2002) have carried out work on 706 and IN718 wrought superalloys using out-of-phase TMF tests between 150°C and T_{max} (in the range 650°C and 800°C). These authors concluded that the Ostergren parameter yields a better correlation of their results than the Smith–Watson and Topper parameter.

Zamrik and Renauld (2000) have compared all the data generated on this alloy under TMF, since it is widely used in stationary gas turbines (Kuwabara *et al.*, 1983; Engler-Pinto *et al.*, 1994; Bernstein *et al.*, 1993). The life prediction is based on the frequency-modified energy failure criterion proposed by Ostergren (1976):

$$N_f = A(\Delta W_{in,t})^{\beta} v^{c} \tag{7}$$

where $\Delta W_{in,t}$ is the tensile inelastic strain energy density that should be known accurately, but large errors arise when attempting to measure it or predict it from the inelastic strain range. These authors proposed to use tensile strain range instead of the inelastic strain range as

$$N_f = A(\Delta W_t / W_r)^B [h(t)]^C [r(T)]^D \tag{8}$$

The ratio of cyclic fatigue energy $\Delta W_{in,t}$ to static energy W_r as measured under monotonic tensile loading at the temperature of maximum tensile stress is defined as

$$\frac{W_t}{W_r} = \frac{\sigma_{max}\varepsilon_{ten}}{\sigma_u \varepsilon_u} \tag{9}$$

where σ_{max} is maximum tensile stress (defined at midlife), ε_{ten} the tensile strain range for which the stress is tensile, σ_u the ultimate strength, and ε_u the elongation to failure measured under monotonic tensile loading:

$$h(t) = 1 + t_h / \Delta t \tag{10}$$

where t_h is compressive hold time, Δt the total cycle time, and $r(T)$ is an inverse Arrhenius function of temperature. Data from different authors (Kuwabara *et al.*, 1983; Engler-Pinto *et al.*, 1994; Bernstein *et al.*, 1993) lie within a factor of 5 using the exponential temperature function.

There has been considerable debate on whether an equivalent LCF temperature could be defined using mechanical or inelastic strain range in steels when there is dynamic strain aging or when there is intergranular creep-fatigue damage (see Jaske, 1976; Fujino and Taira, 1980; Spera and Mowbray, 1976). In some superalloys the same debate occurs where creep-fatigue-type damage dominates such that in-phase cycling can be more detrimental than isothermal cycling and also to a degree more than out-of-phase cycling. The comparison actually depends on different damage types interacting with fatigue as well as on mean stress relaxation.

To circumvent the above difficulty, energy concepts as illustrated above are often used to rationalize the temperature and strain rate dependence of lifetimes: either the hysteretic energy dissipated in the loop or the sum of elastic and inelastic energy (assuming life is spent mostly in crack growth) (Ohtani *et al.*, 1988; Ohtani and Kitamura, 1994). This improves life correlations when different damage forms are weakly coupled, but this may not hold for complex situations when there are strong interactions: the use of damage accumulation models becomes necessary. One fundamental reason is that the life to fracture of a laboratory specimen, which is usually assumed to represent the life to crack initiation in a component in an engineering sense, encompasses both physical crack initiation and early crack growth.

Examples of tests interrupted at regular intervals to monitor crack growth are used to separate initiation from early crack growth. Changing the parameters of a TMF cycle within the same temperature range gives TMF lives that can be shorter or longer than LCF at maximum temperatures (depending upon strain level and cycle shape and period) (Malpertu and Rèmy, 1990). When one uses a life criterion to a short crack depth (0.1 mm or less), the influence of TMF cycle upon life to crack initiation vanishes and TMF life compares fairly well with isothermal fatigue life at maximum temperature (Figure 17). The same effect had been previously reported for damage accumulation in sequential experiments at two temperature levels in Co-based superalloy (Reuchet *et al.*, 1982). Miner's rule of fatigue damage accumulation was found to be very conservative using conventional fatigue life, whereas it was applicable or yielded conservative predictions when life was defined to a short crack depth.

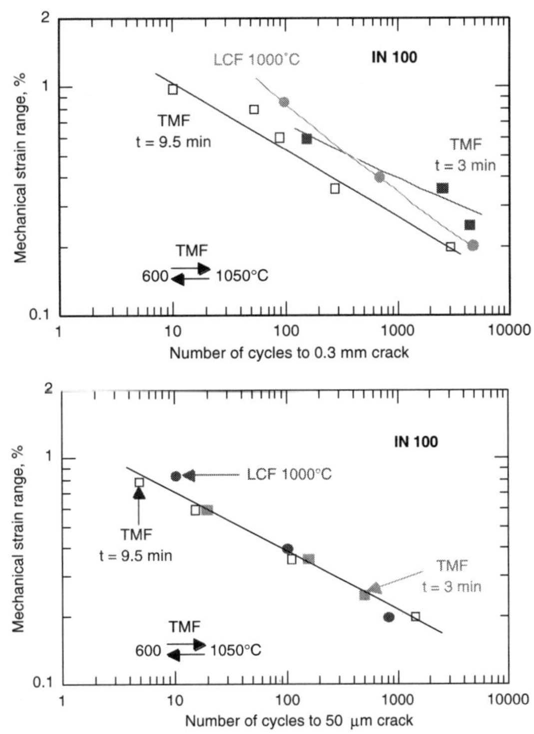

Figure 17 Comparison between lives under isothermal LCF and TMF in IN100 superalloy: (top) to 0.3 mm crack depth and (bottom) to 50 μm crack depth (after Malpertu and Rémy, 1990).

Fleury and Rémy (1993, 1994) have investigated microcrack growth on AM1 superalloy single crystals in the uncoated condition in LCF and in counterclockwise TMF. In high-temperature LCF, as in TMF in the range 600–1,100°C, life is mostly spent in the growth of microcracks. This has been investigated for different crystallographic orientations of the loading axis with respect to cubic axes of the crystals. Figure 18 shows the behavior observed in a TMF life range between 100 cycles and 2×10^4 cycles. The shape of the major crack growth curve depends on the mechanical strain range, and the early growth rate is constant up to crack sizes of 0.1 mm or more. A similar investigation was carried out on coated specimens but using destructive observations of tests interrupted before failure instead of plastic replicas of a single-specimen interrupted tests at regular intervals (Chataigner and Rémy, 1996) and reached the same conclusions.

Ariana *et al.* (1996) have reported rather similar observations on the TMF damage mechanisms of uncoated and coated SRR 99 nickel-based superalloy single crystals. They employed a lag of −135°C between temperature and mechanical strain, with a

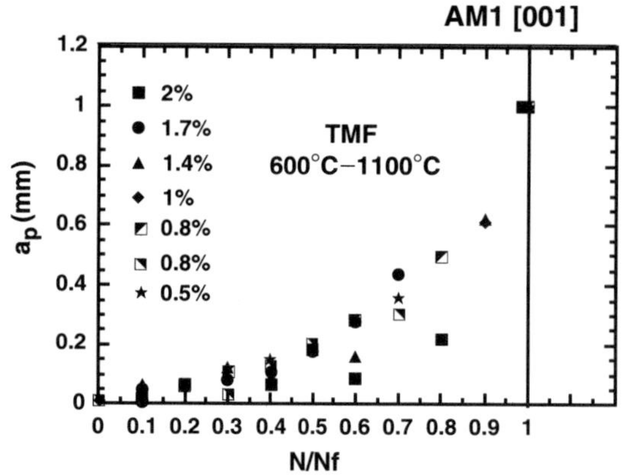

Figure 18 Depth of the major crack as a function of expected life fraction for a ⟨0 0 1⟩ superalloy single crystal tested under TMF (600–1,100°C), for different mechanical strain ranges (after Fleury and Rémy, 1994).

temperature range 300–1,050°C and a period of 90 s; the mechanical strain range varied between 0.5% and 1% and two different strain ratios were used, $R=0$ and $R=-\infty$.

Failure of uncoated specimens is due to multiple initiation of cracks at oxidation spikes on the surface, followed by growth and crack coalescence. The analysis of microcrack growth revealed that crack growth rate was constant up to an average crack length of 0.3 mm, and then crack growth rate increased rapidly with crack length.

Thermal–mechanical creep fatigue. High-temperature components of steam power plants normally operate under complex and partly variable loading conditions. At the surfaces of heated components, the temperature transients cause strain cycling with variable thermal (secondary) stresses. In addition, pressure loading or, for rotors, centrifugal loading leads to quasistatic (primary) stresses. Strain cycling can be the dominant loading condition at the surface of large hot section components. In addition, creep can occur and cause stress relaxation during the hold period during strain cycles.

Granacher and Scholz (1992) and Scholz *et al.* (1998) have developed a service strain cycle with a compressive strain hold phase simulating start-up conditions, a zero strain hold period during steady-state operating conditions, and a tensile strain hold phase simulating shutdown condition. Tests were conducted mostly on steels, typical of heat-resistant turbine rotor steels: a bainitic 1Cr–1Mo–0.7Ni–0.3V steel at 525°C and a martensitic 12Cr–1Mo–1V steel at 550°C.

For comparison purposes, thermal–mechanical creep-fatigue tests with compensated thermal strain, as well as isothermal tests at the maximum temperature of the nonisothermal tests, were used. TMF tests were carried out up to 8,000 h, with typical 10 h hold times and temperature range 150°C, and gave only insignificantly higher number of cycles to failure than comparable isothermal tests (Granacher and Scholz, 1996).

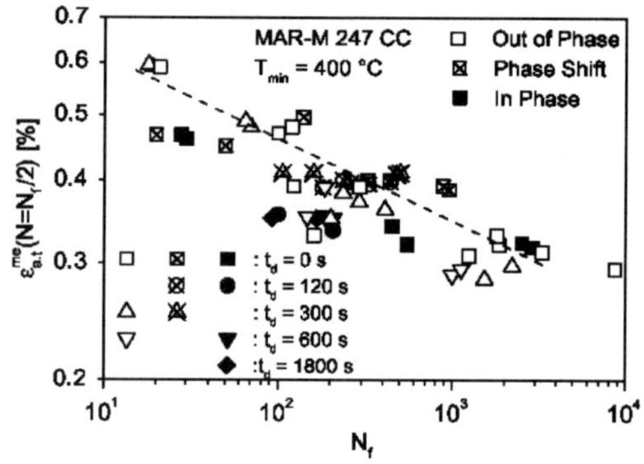

Figure 19 Lifetime behavior at in-phase, out-of-phase, and phase-shift TMF in the representation $\varepsilon^{me}_{a,t}$ vs. N_f (source Beck *et al.*, 2002).

Cycle mixing using different strain cycles was analyzed using a generalized linear damage accumulation rule that has been modified to account for the transition transgranular and intergranular fracture on the creep rupture curve, a mean stress effect, and an effect of prior creep on fatigue life as stated in the following equation:

$$\sum N/N_f + \sum t/t_r = L \qquad (11)$$

It has been concluded that for long period of cycles and long total exposure times, the nonisothermal character of transient strains can be neglected and complex thermal–mechanical creep fatigue reduces to isothermal creep-fatigue interactions, as demonstrated by tests carried out up to 9,000 h (for moderate temperature variations up to 150°C to be compared with T_{max} above 500°C).

The behavior of nickel-based conventionally cast MAR-M247 superalloy has been investigated using in-phase (strain ratio $R=0$), out-of-phase, and out-of-phase with a phase shift (strain ratio $R=-\infty$) (Beck *et al.*, 2002). The temperature investigated was 400–950°C and mechanical strain range varied between 0.3% and 0.6%. Hold time at maximum temperature up to 30 min was also used. Figure 19 depicts the results in terms of mechanical strain amplitude. Dwell times reduce the lifetime of out-of-phase cycle by a factor of 2–3. Even under continuous cycles, in-phase tests result in life shorter than out-of-phase and phase-shift tests. The influence of hold time on the life of in-phase tests is less important. The fracture surfaces display intergranular fracture for continuous in-phase tests, which increases with increasing hold times; the cracks remain transgranular for out-of-phase cycles but become partly intergranular for phase-shift tests with increasing dwell time.

Multiaxial TMF. Few investigations have been conducted to investigate damage under multiaxial conditions. Ziebs *et al.* (2000) have investigated the TMF behavior of superalloys used in stationary gas turbines and they have run two kinds of experiments: (1) simple uniaxial and tension–torsion TMF tests with linear, diamond, and simusoidal cycling and (2) complex uniaxial and tension–torsion TMF tests simulating the cycle endured by actual "buckets," with a complex variation of strain with time and temperature (axial and tensional strain).

Comparison with isothermal fatigue on a strain amplitude basis shows that TMF life is always shorter than isothermal fatigue at maximum temperature, and a von Mises hypothesis adequately describes the deformation behavior (Ziebs *et al.*, 1994, 2000).

Very few studies have investigated fatigue life prediction under complex bucket-type TMF path. As cumulative fatigue and creep damage mechanisms are intimately coupled under TMF loading and as environment may play a role in the temperature range investigated (T_{max} was varied between 750°C and 950°C), an engineering approach is needed to handle the problems. The driving force for the propagation of nucleated cracks in metals is a function of the shear and normal stresses on the crack plane (Socie, 1987), which depend on the strain paths in multiaxial loading.

A global measure can be chosen to account for all these parameters as the inelastic strain range is a strain energy criterion in the following form:

$$\sum \sigma_{ij}\Delta\varepsilon_{ij,\mathrm{in}} = AN_{\mathrm{f}}^{\mathrm{m}} \tag{12}$$

where A and m are two constants which were identified by linear regression analysis for five different TMF histories (N_f has been defined at a load drop of 1% from the steady-state value). Using this criterion, biaxial TMF tests are in close agreement with uniaxial TMF tests, with a factor of ± 2 between predicted and actual life. The factor introduced by Manson and Halford (1977) to account for the change in ductility with stress triaxiality does not improve the accuracy of predictions.

Chieragatti and Paun (2000) analyzed tests on IN100 cast superalloy subjected to superimposed heating and cooling between 600°C and 1,050°C and rotating bending. They used an equivalent stress based on the Hayhurst (1972) criterion:

$$\sigma_{\mathrm{eq}} = \alpha\ \mathrm{sign}(\sigma_{\mathrm{i}})J_0 + \beta J_1 + (1-\alpha-\beta)J_2 \tag{13}$$

where J_0 is the maximum principal stress $\mathrm{Max}(\sigma_i)$ (σ_i = principal stress tensor components), J_1 is the first invariant of the stress tensor, J_2 is the second invariant of the stress tensor, and α and β are two coefficients.

They used a normalized stress $S = \sigma_{eq}/\sigma_u(T)$ where σ_{eq} is the variable equivalent stress, and σ_u is the ultimate tensile strength, as proposed by Lemaître and Chaboche (1985), and assumed a simple power law between the normalized stress amplitude and fatigue life:

$$\left(\frac{\Delta S}{2}\right)^{\beta(T)} N_{\mathrm{f}}(\Delta S, \bar{S}, T) = A(\bar{S}, T) \tag{14}$$

where $\Delta S/2$ and $\bar{S}$ are the amplitude and mean value of normalized stress in a cycle. By integration over the TMF cycle, a good agreement was obtained with LCF life at 1,000°C and the location of cracks on the inner surface as opposed to the outer surface of hollow specimens, depending upon deflection.

Bonacuse and Kalluri (1996) have investigated the axial-torsional TMF behavior of Co-based superalloy Haynes 188. A constant-rate thermal cycle was used between 316°C and 760°C with a 600 s cycle period. All tests were carried out under strain control. Four basic types of tests including mechanically in-phase (proportional loading), thermally in-phase, mechanically in-phase but thermally out-of-phase, mechanically out-of-phase (90°, non-proportional), and thermally out-of-phase.

A von Mises equivalent strain range $\Delta\varepsilon_{eq}$ was used as

$$\Delta\varepsilon_{\mathrm{eq}} = \left(\Delta\varepsilon^2 + \frac{3}{4(1+\nu_{\mathrm{eff}})^2}\Delta\gamma^2\right)^{1/2} \tag{15}$$

where $\Delta\varepsilon$ and $\Delta\gamma$ are the maximum of the differences between a peak strain (axial and shear) and the recorded strains around a cycle. For a TMF test

$$\nu_{\mathrm{eff}} = \frac{\Delta\varepsilon_{\mathrm{e}}\nu_{\mathrm{e}} + \Delta\varepsilon_{\mathrm{p}}\nu_{\mathrm{p}}}{\Delta\varepsilon} \tag{16}$$

with

$$\Delta\varepsilon_{\mathrm{e}} = \left|\frac{\sigma}{E}(T_{\max}) - \frac{\sigma}{E}(T_{\min})\right| \tag{17}$$

and ν_e and ν_p are elastic and plastic Poisson's ratios. The von Mises equivalent stress was also modified by a multiaxiality factor deduced from Manson's triaxiality factor (Manson and

Halford, 1977) and TMF data correlate better with isothermal life at maximum temperature. More complex models were shown to perform worse (Socie, 1987; Fatemi and Socie, 1988).

5.03.3.2 TMF of Coated Materials

Coatings are used to protect against oxidation and corrosion of vanes and blades in gas turbines. The most frequently used intermetallic coatings are overlay plasma-sprayed Ni or Co–Cr Al-Y coatings or chemical-vapor-deposited Ni–Al aluminides. All these alloys are rich in aluminum to produce an Al_2O_3 oxide layer, which protects the substrate from the aggressive environment.

In bare materials, it is often assumed that the life to initiation under TMF cycling is at least equal to the life to crack initiation under LCF conditions at maximum temperature. However, this is frequently not the case for these intermetallic coatings that are brittle or have a low ductility up to medium temperatures, and their behavior depends strongly on the TMF cycle.

Leverant *et al.* (1976) have investigated Ni–Co–Cr–Al–Y overlay coating deposited on DS Mar-M200 + Hf and conventionally cast B-1900 + Hf and NX-188 for various TMF cycles, out-of-phase, in-phase or diamond, and they have shown that the more detrimental cycle was the out-of-phase cycle. Further, the influence of cycle shape on TMF life could be rationalized when the mismatch strain between coating and substrate, *s*, due to the differences in coefficients of thermal expansion, was taken into account.

The effect of the degradation of the number of cycles to engineering crack initiation induced by out-of-phase cycling between 593°C and 1,038°C is illustrated in Figure 20 (after Bain, 1985). For a given mechanical strain range, the TMF lifetime is much lower than the isothermal LCF life, especially for low strain amplitudes.

The thermal strain expansion mismatch $\Delta\varepsilon_\alpha$ is given by

$$\Delta\varepsilon_\alpha = \int_{T_0}^{T} (\alpha_s - \alpha_c)\, dT \qquad (18)$$

where α_s and α_c are coefficient of thermal expansion (CTE) of substrate and coating, respectively, and T_0 the reference temperature where coating is stress free. As CTE of coating is often smaller than substrate, this results in a strain range in the coating $\Delta\varepsilon^c_m = \Delta\varepsilon^s_{m,s} + \Delta\varepsilon_\alpha$ higher than for the substrate. This favors much higher strains in the coating than in the substrate at low temperature.

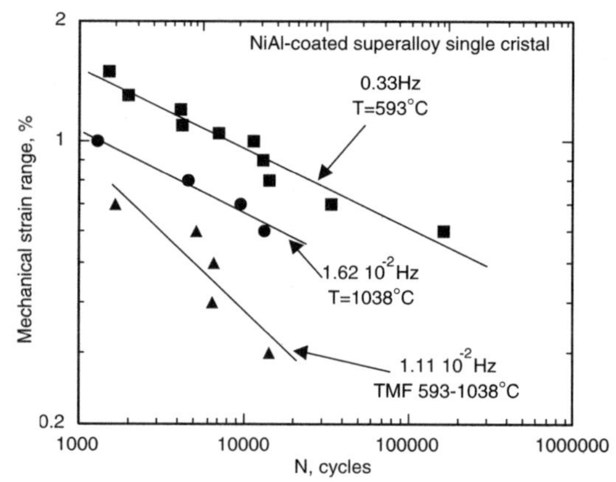

Figure 20 Variation of lifetime vs. strain range in NiAl-coated superalloy single crystal under out-of-phase cycling between 593°C and 1,038°C; comparison with isothermal test data at maximum and minimum temperature (after Bain, 1985).

If the mismatch strain is a primary variable, coating properties become very important. The low ductility at lower temperatures will control the propensity to form brittle cracks at higher strain ranges, and tensile strains at lower temperatures. The creep resistance or the creep fatigue at high temperature is very important that causes relaxation of stresses at high temperatures. The lower creep resistance will favor rumpling of the coating surface in compression, which will, in turn, lead to the formation of cracks during the tension phase of the cycle. The relaxation of stresses at high temperatures will lead to the development of tensile stresses in the coating upon cooling.

Martinez-Esnaola *et al.* (1996) and Ariana *et al.* (1996) have investigated the TMF of smooth NiAl-coated specimens of SRR 99 nickel-based superalloy single crystals, using flat specimens to monitor surface crack nucleation and early growth. A lag of –135°C between temperature and mechanical strain was used. The temperature range was 300–1,050°C with a period of 90 s, mechanical strain range varied between 0.5% and 1%, and two different strain ratios were used ($R = 0$ and $R = -\infty$). A comparison was made on the basis of a stress range. For compressive strain ($R = -\infty$), coating life is reduced by a factor of 2. For fully tensile strain ($R = 0$), the reduction in coating life increases with increasing strain or stress range, and may reach as much as a factor of 7. An in-phase cycle with a fully tensile strain results in a life reduction by less than a factor of 2, or is insignificant at the highest stress range. On the uncoated alloy, the effect of TMF cycle shape is almost negligible.

We now turn to the influence of TMF cycle. An NiAl coating was previously shown to give a significant increase in the TMF life to a few tenths of mm in crack depth for conventionally cast IN100 with respect to the uncoated material: a counterclockwise diamond cycle was used between 600°C and 1,050°C (Bernard and Rémy, 1990). A 600–1,100°C diamond cycle was used for a modified NiAl coating on AM1 single crystals and no difference could be observed between bare and coated alloy (Chataigner and Rémy, 1996) (see Figure 21). Cracking initiated mostly at subsurface pores due to the casting process and was fairly insensitive to microcracks in coating.

A different result was obtained with a TS on a wedge specimen submitted to a temperature range 30–1100°C (Figure 21) (Köster *et al.*, 1996a, 1996b). Early crack initiation was observed in coated structure and attributed to the occurrence of brittle cracking in coating.

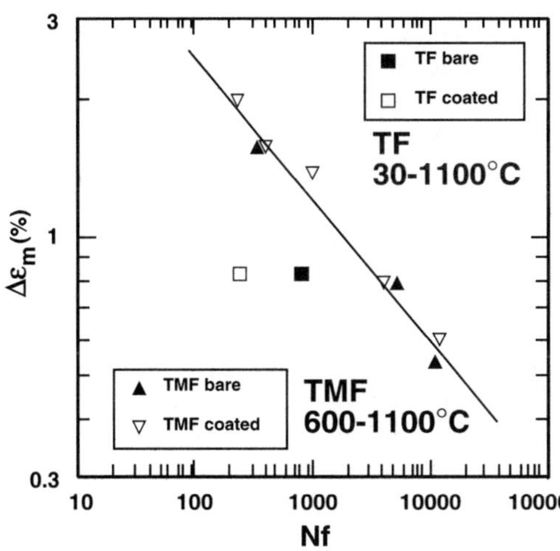

Figure 21 Variation of lifetime of AM1 superalloy single crystals as a function of mechanical strain range (left) and stress range (right) for bare and coated conditions, using the TMF cycle of Figure 6. Points labeled TF refer to TF of standard wedge specimens cycled between 30°C and 1,100°C (source Koster *et al.*, 1996a, 1996b).

Kowalewski and Mughrabi (2000) used an unsymmetrical out-of-phase TMF cycle from 400°C to 1,000°C for an Ni–Cr–Al–Y coating deposited by plasma spray on directionally solidified MAR-M247. A decrease in life by a factor of 2 was observed, corresponding to the development of multiple and fairly long cracks in the coating.

More recent investigations suggest that more realistic thermal–mechanical creep-fatigue cycles (Figure 22) should be used to reflect the variety of situations occurring in actual components and capture finer details of interactions between loading segments and more detrimental damage mechanisms that limit the life of a component (Rémy and Guédou, 2002; Rémy *et al.*, 2002).

It is worth noting that very little information is available on TMF tests of thermal barrier coatings (TBCs) despite their technical importance (Wright, 1998; Baufeld *et al.*, 2001; Rémy *et al.*, 2002).

5.03.3.3 Thermal Shock

Tests on tapered disks and wedges have been used to study the TF resistance of superalloys used in gas turbines, for aeroengines or stationary gas turbines. These TS tests were first used to rank candidate materials and eventually coated materials for a given component. Therefore, several material variables were tested using this technique that include the influence of thermal treatments and microstructural changes, the effect of grain size, and the effects of surface treatment including protective coatings against oxidation or corrosion.

Woodford and Mowbray (1974) have reported systematic investigations of such tests for cast cobalt- and nickel-based superalloys.

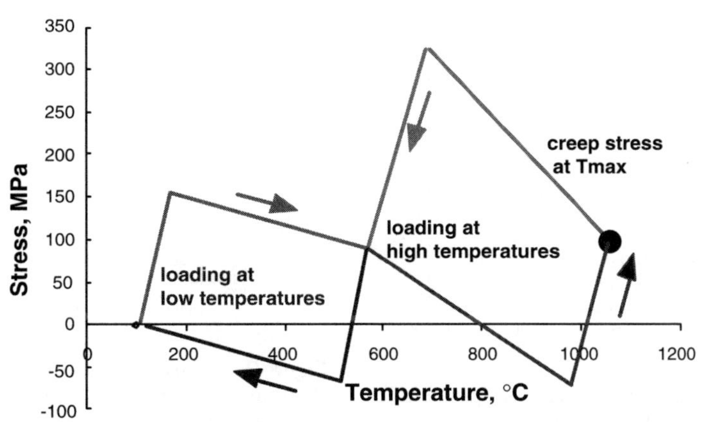

Figure 22 Example of a creep-fatigue cycle used to investigate the degradation of CVD-C1A coating on AM1 under TMF conditions (after Rémy *et al.*, 2002).

They investigated the effect of thermal cycle parameters such as maximum temperature and hold time at maximum temperature.

The effect of maximum temperature on the number of cycles to crack initiation life is reported in Figure 23 for cast superalloy IN100 with two grain sizes, a fine-grained and a large grained condition (Meyer-Obersleben *et al.*, 1990). The coarse-grain material displays a shorter life than the fine-grain one and the effect increases with decreasing maximum temperature. Similar observations were also reported for other nickel-based superalloys by Woodford and Mowbray (1974), and for IN738, Mar M247, and other alloys by Blümm *et al.* (1994).

In most cases, intergranular crack growth has been reported in such alloys during thermal cracking as well as oxidation along the crack surfaces, with eventual alloy depletion beneath the oxide layers (Woodford and Mowbray, 1974). An increasing life to crack initiation with decreasing maximum temperature was observed for superalloy single crystals (Meyer-Olbersleben *et al.*, 1996), but with much longer lives, by more than an order of magnitude, for the same experimental conditions.

The influence of maximum temperature is shown on the evolution of the depth of the dominant crack in TS experiments in Figure 24 for cast IN100 (Meyer-Olbersleben *et al.*, 1990). During an initial stage, the crack growth rate increases up to ~1 mm in depth, then it remains nearly constant and finally starts to decrease. This behavior is discussed in detail through a fracture mechanics analysis (Mowbray and Woodford, 1973). A reduction in maximum temperature tends to reduce crack growth rate in the linear stage. The temperature dependence was found to be reduced in cobalt-based alloys (Woodford and Mowbray, 1974). Crack propagation seems to be more transgranular in these alloys but tends to occur along interdendritic regions that are decorated with carbides and are often oxidized.

Reducing the maximum temperature decreases the importance of time-dependent damage mechanisms and simultaneously the amplitude of thermal strain and stresses. Glenny (1967) was the first to maintain the thermal cycle at the thin edge constant and investigate different temperature gradients across the specimen by changing the specimen geometry. Woodford and Mowbray (1974) mainly used a variation in peripheral radius between 0.25 mm and 1 mm, which results in a moderate change in mechanical strain range at the thin edge.

Rézaï-Aria *et al.* (1988a, 1988b) increased the thin edge radius and reduced the specimen width to achieve less severe mechanical strain, whereas others (Köster *et al.*, 1996a, 1996b) increased the severity of loading using wider specimens. The effect of a variation in specimen geometry is illustrated in Figure 25 for a cobalt-based superalloy tested in the temperature range 200–1,100°C.

For design purpose, the main use of such experiments is to check the validity of the whole life prediction chain using constitutive models and damage equations. The computation of stress–strain loading at the thin edge has been carried out in several works (Spera, 1969; Mowbray and Mc Connelee, 1976; Rézaï-Aria *et al.*, 1988b; Köster *et al.*, 1996a, 1996b). Meyer-Olbersleben *et al.* (1990) have measured the displacement along the thin edge of the wedge to have some check on total strain

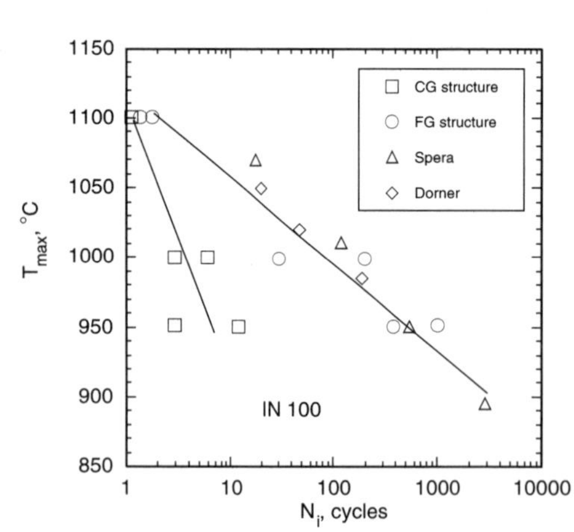

Figure 23 TF crack initiation life to 0.4 mm crack depth (N_i) as a function of maximum temperature in coarse grain (CG) and fine grain (FG) cast IN100 superalloy (after Meyer-Obersleben *et al.*, 1990).

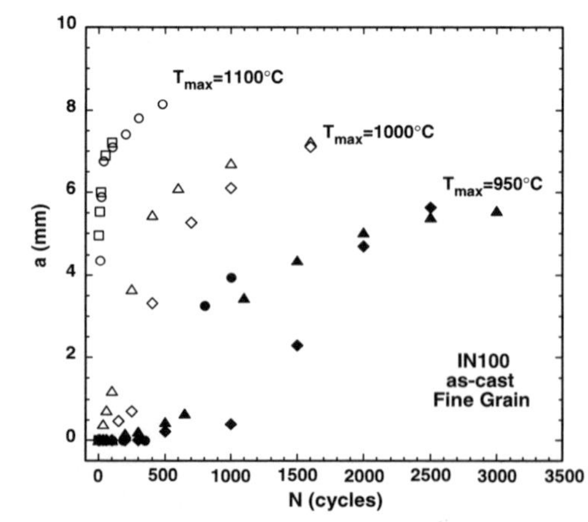

Figure 24 Variation of crack length with number of thermal cycle in fine grain (FG) as-cast IN100 superalloy for different maximum temperatures and minimum temperature at 200°C (Meyer-Obersleben *et al.*, 1990).

and thus on mechanical strains experienced in this area. A worked example is depicted later in this chapter.

In wedge or disk specimens submitted to TS, as in blades and vanes in gas turbines, TF gives rise to a single family of cracks and often a dominant crack or a few major cracks. In most other applications when a metal surface is exposed more uniformly to high temperature, a more complex network of cracks develops.

This is, for instance, the case for hot work tool steels (Northcott and Baron, 1956; Rousseau *et al.*, 1975). During hot metal forming, the surface of the tool in contact with the hot workpiece (in forging) or with molten alloys (aluminum or brass in pressure die casting) is heated in a very short time. When the part is formed and during its ejection from a die, the surface of the tool is rapidly cooled down, in particular, when a cooling or lubricating liquid is used. Biaxial thermomechanical loading of the surface gives rise to a biaxial cracking network (the so-called heat checking or crazing). This can be simulated in the laboratory by using hollow cylindrical specimens which are internally cooled continuously by water, with heating and cooling shocks by alternating between high-frequency induction heating and cooling by compressed air flow (Miquel *et al.*, 2002).

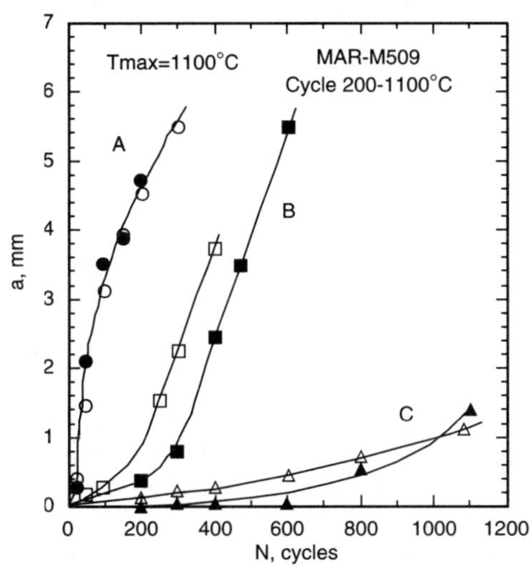

Figure 25 Variation of crack length with number of thermal cycles in MAR-M 509 submitted to a TS between 200°C and 1,100°C: influence of changes in specimen geometry (after Rezai-Aria *et al.*, 1988a, 1988b).

Observations of the heat-checking pattern can be made on the external surface of such a simulation test cylinder. In the central part of the specimen, a classical heat-checking pattern can be observed, as in actual tools with a cell-type (almost square type) pattern on 0.38%C–5Cr–1Mo–0.56V steel. Near the extremities of the specimens, the isotropic cell pattern turns into a rectangular type pattern (Figure 26). These changes are in agreement with X-ray residual stress measurements and FE analysis, which have shown that the ratio of longitudinal stress to hoop stress varies slowly from unity (equibiaxial stress state) in the central part to very low values with mostly an array of long cracks perpendicular to hoop stress (or the radial direction) (almost uniaxial stress state) in the outer region of the specimen. Despite intense oxidation at the surface, cracks are mainly transgranular.

Crack initiation under TS is defined from specimen observation of interrupted tests.

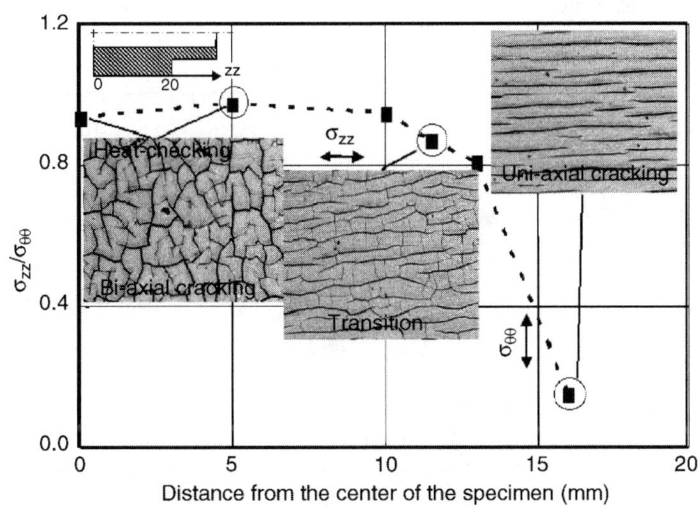

Figure 26 Variation of the stress ratio (longitudinal/hoop) and corresponding evolution of the heat-checking pattern plotted along the z-direction from the center (biaxial loading) to the extremity (uniaxial loading) for a 7 mm wall thickness specimen (Miquel *et al.*, 2002).

Cracks are detected using optical microscopes and crack initiation is defined as the occurrence of at least one crack of length 50–150 μm. This often corresponds to a few grain sizes and thus encompasses some early crack growth.

For a given T_{max}, 550°C for steel 316 (Fissolo *et al.*, 1996), a decrease in temperature range from 300°C to 100°C increases the number of cycles to crack initiation from $\sim 5 \times 10^3$ up to 10^6 cycles. For such a downshock, a hold at T_{max} of 20 s for a transient of ~4 s increases the number of cycles to initiation, as does a decrease of T_{max} for the same temperature range. These results have been explained on the basis of an approximate estimation of strain range that was deduced from a thermoelastic computation, assuming an equibiaxial stress state and using the cyclic stress–strain curve at maximum temperature (Fissolo *et al.*, 1996). The experimental results for TS initiation were close to isothermal LCF near maximum temperature.

Crack networks or crazing were characterized by a number of authors, using appropriate facilities, or component tests. The typical procedure is to interrupt and to make a map of the specimen surface when it can be easily observed. Image analysis can be made and used to investigate the evolution of crack length, density (total length per unit area), the number of triple junctions between cracks, etc. Figure 27 shows the skeleton of a crack network observed in 304 stainless after 1.5×10^5 TSs (maximum temperature 320°C, minimum temperature 120°C) observed at the surface and at different depths below the surface 0.36 mm and 1 mm, while the maximum crack depth is ~2.4 mm (Maillot *et al.*, 2002). It is clearly observed that cracking occurs approximately along two directions that reflect the approximately equibiaxial thermal stress field. The mean distance between cracks increases when crack depth increases.

These observations suggest that there is a strong interaction between the various cracks and, of course, the deeper cracks are the more dangerous for practical applications. Observations after the same number of cycles using a higher T_{max} show a much higher density of cracks, but the cracks are of shallower depth. It is fairly difficult to observe the variation in depth of cracks since destructive observations must be made on sections to get the distribution of depth. Such curves are shown in Figure 28 for thermal downshocks on steel 316 between 550°C and 300°C (Fissolo *et al.*, 1996) (Figure 29). The major cracks tend to grow rapidly and then slow down progressively, whereas the numerous secondary cracks appear to stop more rapidly, as reflected by the average crack depth variation.

Malm and Norström (1979) reported similar results for 0.4C–5Cr–1Mo–1V tool steel using TSs between 650°C and 750°C and room temperature. The rate of leading cracks and secondary cracks decreases with the number of TSs and can be rationalized using inelastic strain range estimates at the surface. Howes (1976) investigated the same steel as well as a maraging 18Ni–9Co–5Mo–Ti steel using mild cycle, 43–649°C, and severe cycle, 43–1,030°C. The severe cycle especially brought severe microstructure changes on TS cycling and gives rise to a large softening of tested steels.

Similar observations can be made more easily using TS on wedges, and the same trend is observed for a population of cracks (Köster, 1997).

Multiaxiality effects on crack initiation. It is often very difficult to compare directly the

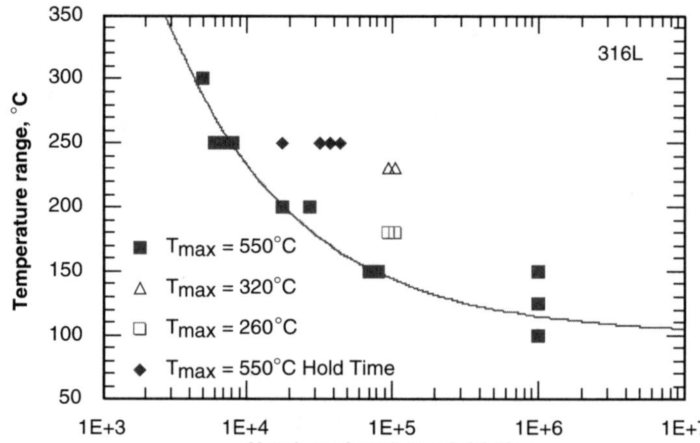

Figure 27 Temperature range vs. number of cycles to initiation in downshock experiments on 316L stainless steel (after Fissolo *et al.*, 1996).

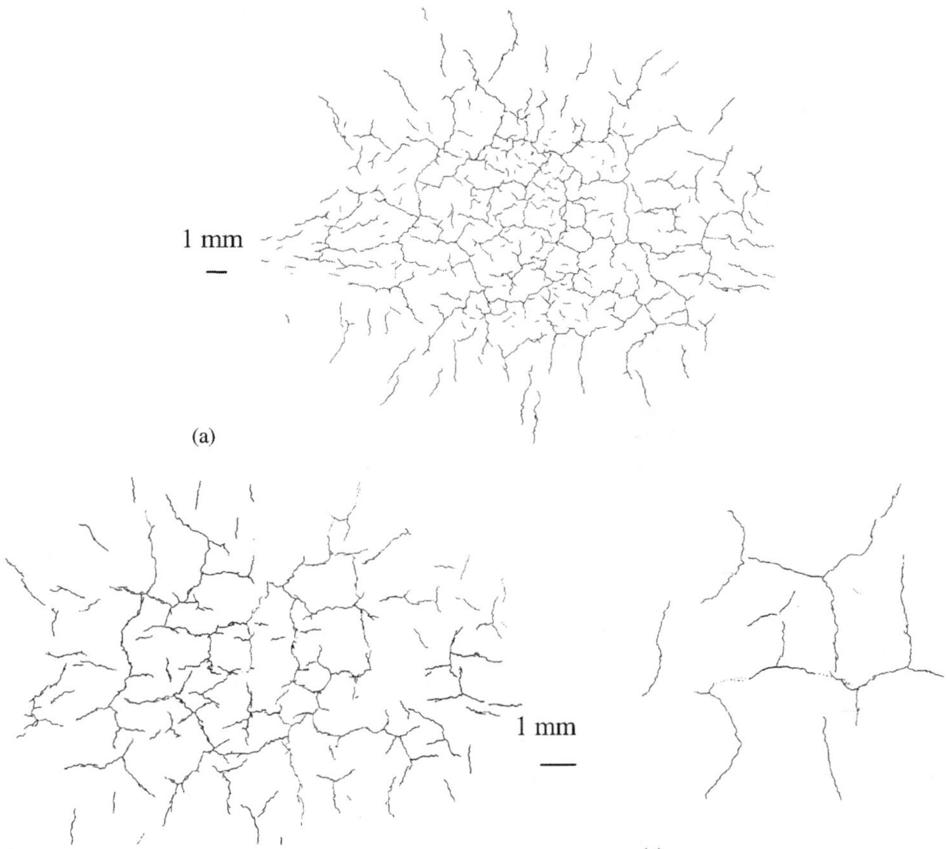

Figure 28 Crack network after 1.5×10^5 downshock cycles, 120–320°C, in 304 stainless steel: (a) at the surface, (b) 360 μm below the surface, and (c) 1 mm below the surface (source Maillot *et al.*, 2002).

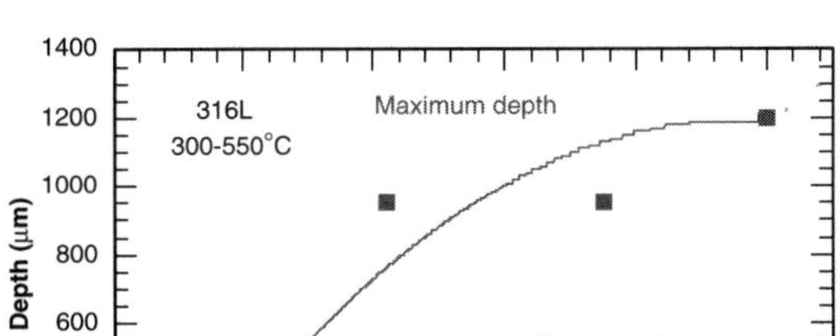

Figure 29 Variation of the dominant crack depth and of the mean crack depth as a function of the number of TS 300–550°C for 316 stainless steel (after Fissolo *et al.*, 1996).

results of TMF and TS tests, since frequently they are conducted under very different conditions. Such a comparison was made by Yamauchi *et al.* (1996) using tests that simulate the thermal striping problem in fast breeder reactors, on cylindrical specimen (155 mm outer diameter, 5 mm wall thickness, and 300 mm length) machined in a 304 steel pipe. In the plant, the wall below the sodium level in the main vessel is heated by hot liquid sodium, but the upper part is exposed to low-temperature nitrogen cover gas, which is cooler; there is a temperature peak with a sharp gradient in the longitudinal direction, which gives a peak stress in the wall. The vessel wall is subjected to cyclic thermal stress due to movement of the temperature distribution during startup and shutdown operations, which cause ratchetting and creep-fatigue damage. The facility combines external induction heating and internal water cooling (with a water level near the midheight of the tube) to produce a longitudinal temperature distribution with a sharp gradient and temperature range 20–600°C, and specimen axial to-and-fro motion, which gives a temperature gradient of 20°C mm^{-1}.

The conditions for the points that experience most severe loading are reported at the outer and inner surface, using FE analysis (Table 2). Transgranular cracking occurs in the circumferential direction at the outer surface, which is consistent with almost uniaxial stress state and out-of-phase TMF cycle. A 2D crack network is observed at the inner surface, which is

Table 2 Analysis of longitudinal cyclic movement of axial temperature on 304 pipe specimen tested between 20°C and 600°C.

Location	*von Mises equivalent strain range*	*Thermal-strain cycle*	*Stress state*	*Triaxiality factor at peak strain, M*	*Surface crack distribution*
Outer surface	0.57%(#1) 0.58%(#1)	In-phase	Uniaxial	1	1D circumferential
Inner surface	0.54%(#1) 0.55%(#1)	Out-of-phase	Biaxial	2	2D network

		Structural model (20–600°C)		*Uniaxial TMF (300–600°C)*	
Thermal-strain cycle	*Strain range*	*Cycles to crack initiation*		*Strain range*	*TMF life, cycles*
In-phase	0.57%(#1) 0.58%(#1)	4,500 2,500	(outer surface)	0.7%	1,930
Out-of-phase	0.54%(#1) 0.55%(#1)	3,000 1,500	(inner surface)	0.7%	3,850

After Yamuchi *et al.* (1996).

consistent with an equibiaxial stress state and in-phase TMF cycle.

Uniaxial TMF tests were conducted to simulate the two critical locations in component tests with same heating/cooling rate: in-phase uniaxial TMF life is consistent with the number of cycles to initiation at the outer surface. However, out-of-phase uniaxial TMF life is longer than in-phase uniaxial test, while the reverse is true in the component test. This is simply a multiaxial effect and component result is quantitatively accounted for using the multiaxiality factor of Manson and Halford (1977).

5.03.4 OXIDATION AND SURFACE EFFECTS

5.03.4.1 Oxidation Effects

Damage models often refer to the interaction between fatigue and creep or oxidation. Though oxidation can be very important under TMF testing and TS, most of our knowledge comes from isothermal tests. Testing at high temperatures in air with some level of humidity may strongly influence the life of smooth creep specimens; fatigue damage involving surface cracks is especially sensitive to oxidation effects. Oxidation effects have been reviewed in the past (Cook and Skelton, 1974; King and Cotterill, 1991; Rémy, 1993) emphasizing creep or fatigue behavior.

The influence of oxidation on lifetime is shown mostly by comparing tests in air and in vacuum even though the residual atmosphere in vacuum is seldom "inert." The understanding of oxidation–deformation interactions is highlighted by considering the behavior of oxide scales and by emphasizing the role of preferential oxidation paths and grain boundaries in wrought alloys and interdendritic areas in cast alloys. Modeling the interaction between oxidation and thermal–mechanical loading to predict lifetime is still in its infancy even as of early 2000s.

5.03.4.1.1 Oxidation effects on creep

Extrapolation of actual creep lives to low service loads can be biased by oxidation effects, especially using uniaxial specimens of small dimensions, except for the case of thin wall components. Oxidation can waste metal on the overall surface by GB penetration or by internal oxidation, and reduce specimen diameter, increase the applied stress, and reduce creep life. Cane and Manning (1981) have made a thorough study of oxidation effects in ferritic steels, like 0.5Cr–0.5Mo–0.25V steel, used in power plants. A 2.5 times increase in rupture life is thus observed for this steel (at 70 MPa and 675°C) when the specimen diameter increases from 2.5 mm to 50 mm due to the reduction of the surface area/volume ratio of the specimen (see Figure 38). Actually there is almost no difference between the lifetime of a specimen in vacuum and a 50 mm diameter specimen in air. This effect is attributed to a progressive decrease in cross-section of the specimen together with alloy degradation in the vicinity of the surface caused by decarburization. At intermediate diameter sizes, the oxide can fracture and the test load is retransferred to the metal core that undergoes an enhanced strain rate. Lifetime reductions are thus due to oxide fracture events.

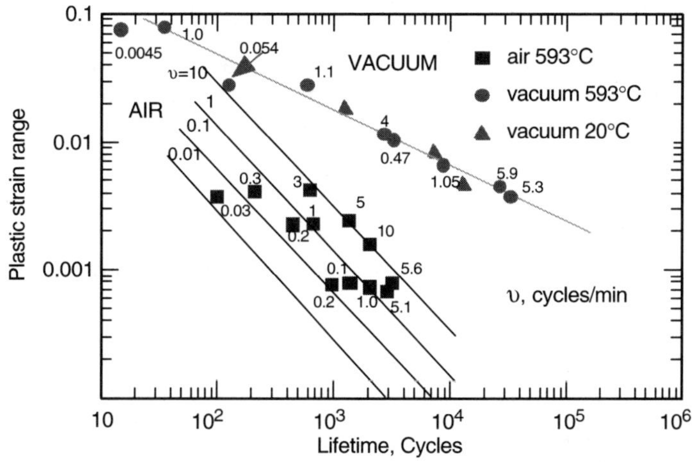

Figure 30 Plastic strain range vs. fatigue life for alloy A 286 in air and in vacuum at 593°C and at 20°C. Numbers adjacent to test points refer to frequency in cycles per minute (after Coffin, 1972).

The decrease in cross-section by corrosion in various environments has been emphasized and analyzed by Grünling *et al.* (1980). The influence of environmental effects due to high-temperature exposure before testing has been investigated in various superalloys, including IN 738 (Woodford, 1981). A discussion of detailed mechanisms can be found elsewhere (Dyson and Osgerby, 1987; Rémy, 1993): the resultant creep damage due to this effect is enhanced cavitation near the specimen surface instead of uniform cavitation throughout usual creep test pieces (Dyson and Osgerby, 1987).

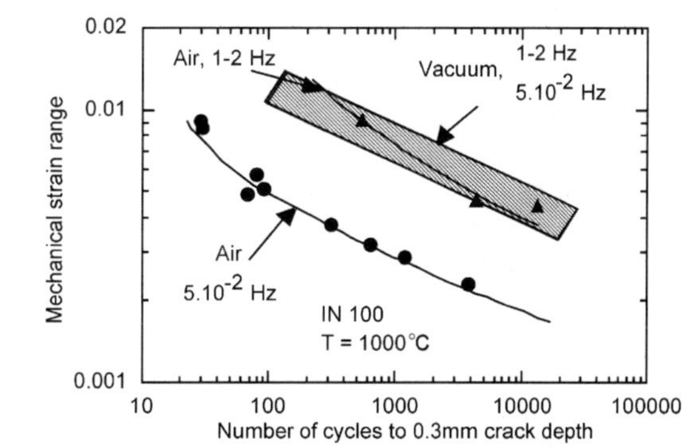

Figure 31 Fatigue life to 0.3 mm crack depth vs. total strain range for cast IN100 in air and in vacuum at 1,000°C, for two frequencies 1–2 Hz and 5×10^{-2} Hz (after Reger and Remy, 1988a, 1988b, 1988c).

5.03.4.1.2 Oxidation effects on high-temperature fatigue

The influence of oxygen pressure on fatigue was studied in the pioneering work of Achter *et al.* (1963) on nickel and stainless steels, where the existence of two lateaus in fatigue life at low and high pressure was shown. They interpreted this result with an "adsorption" model considering the coverage of fresh surface at the crack tip by oxygen atoms.

Coffin (1972, 1973) carried out a thorough study of the influence of frequency in LCF for a wrought Fe–Ni-based superalloy A 286 in air and in high vacuum (1.3 μPa). Fatigue life in air strongly decreases with decreasing test frequency when plotted as a function of plastic strain range (Figure 30). The influence of frequency is almost negligible under vacuum, and life compares fairly well with that obtained at room temperature for the same plastic strain range. In air, crack initiation takes place at grain boundaries due to stress-assisted GB oxidation, but in vacuum A 286 displays usual crystallographic crack initiation, as at room temperature.

This was confirmed in different wrought or cast superalloys at high temperature (Sidey and Coffin, 1979; Antolovich *et al.*, 1981; Reuchet and Rémy, 1983a, 1983b, 1983c; Reger and Rémy, 1988a, 1988b, 1988c). The life to initiate a major crack 0.3 mm deep (as deduced from potential drop measurements) is plotted vs. total strain range for bare IN100 superalloy at 1,000°C in Figure 31. Tests were run at various frequencies, 2.5×10^{-2} Hz, 5×10^{-3} Hz, and with tensile strain holds of 3 min and 10 min. The larger frequency dependence was actually observed in the frequency range 5×10^{-2} Hz to 2 Hz, where a 20 times increase in frequency enhances the lifetime by a factor of 20–50. This strong frequency dependence of the lifetime to 0.3 mm crack depth almost vanishes in vacuum (1 mPa) and life at high frequency in air is almost identical to that in vacuum.

Observations of LCF or TMF specimens tested in air in cast superalloys have shown strong evidence of the effects of oxidation. In air, surface cracks nucleate preferentially at oxidized MC carbides in interdendritic areas, due to the high carbon content of these conventionally cast alloys. Microcracks once nucleated grow nominally in mode I (stage II) but tend to follow predominantly an interdendritic path in air, in low-frequency LCF, TMF (Malpertu and Rémy, 1990; Vasseur, 1993), and TS tests (Rézaï-Aria *et al.*, 1988a, 1988b; Rémy *et al.*, 1993). Therefore, in these coarse-grained alloys the crack path tends to stay transgranular but interdendritic in air. By contrast, under vacuum, IN100 at 1,000°C displays an intergranular crack path but with a much longer life than in air (Reger and Rémy, 1988a, 1988b, 1988c). In addition, outward diffusion of elements like Al and Ti in nickel-based alloys gives rise to a zone depleted of γ' strengthening precipitates in front of the oxide at high temperature, as shown in Figure 32 (Vasseur, 1993). It is worth emphasizing that oxidation forms a scale on the surface of these alloys, but in addition oxide spikes form preferentially at interdendritic areas rich in MC carbides and segregated in alloy elements.

Fatigue and creep-fatigue crack growth laws were extensively studied using fracture mechanics specimens in alloys for turbine disk applications, such as wrought IN718 (Pédron and Pineau, 1982) or powder metallurgy superalloys (Soniak and Rémy, 1990; Bernède and Rémy, 1992). The frequency dependence of fatigue crack growth rate (FCGR) was investigated in a Cr–Mo–V steel at 550°C at low R-values (Haigh *et al.*, 1976). At high ΔK values, the crack growth rate increases with decreasing frequency in air but this influence of frequency is much smaller in vacuum. However, there is a crossover of the curves at low ΔK and slow crack growth rates, with an abrupt decrease at the threshold value, while there is none in vacuum. Such behavior was observed in other steels and superalloys (Soniak and Rémy, 1990; King and Cotterill, 1991): in particular, the threshold increases with decreasing frequency, due to the occurrence of crack branching in air and to crack closure induced by the oxide filling the crack at low ΔK values.

The influence of frequency above the threshold has been studied in various superalloys. The case of short cracks was studied, for instance, in fine-grained Astroloy at 650°C (Bernède and Rémy, 1992): FCGR increases with cycle period above a few seconds almost in a linear manner in air, i.e., $da/dN = (da/dt)\Delta t$ with a constant da/dt, which supports the view that the creep-fatigue crack growth rate is controlled by pure time-dependent creep- or oxidation-assisted behavior. In this time-dependent regime the crack path becomes intergranular, but in vacuum the frequency dependence of FCGR almost vanishes. Changing the grain size in wrought alloys such as IN718 at 650°C (Pédron and Pineau, 1982) gives rise to a strong variation in FCGR: a fine-grained structure gives a much larger growth rate than a coarse-grained structure. A necklace microstructure with coarse grains surrounded by small grains gives still a lower growth rate. This effect observed for a 300 s dwell almost vanishes under vacuum.

5.03.4.1.3 *Influence of deformation on oxidation*

Steady-state oxidation is associated with a parabolic kinetics through weight gain or thickness measurements. In engineering alloys, this parabolic kinetics is even observed when the scale has a layered structure. Protective behavior is often associated with such kinetics, but real protection is only obtained with the formation of a continuous layer of protective oxide such as Cr_2O_3, Al_2O_3, or SiO_2.

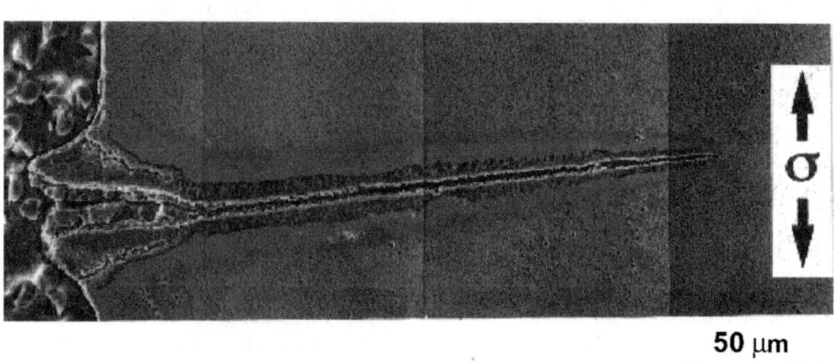

Figure 32 Oxidized crack in a directionally solidified superalloy MAR-M247 in LCF at 950°C. Note the layered structure of the oxide and the formation of a zone depleted of γ' precipitates in front of the oxide (Vasseur, 1993).

Fracture of the oxide scales can thus impair the creep ductility and time to rupture, leading at least temporarily to nonprotective behavior, but except for the case of severe internal oxidation or corrosion, explained later, only a limited effect is expected. This becomes important for thermal cycling where severe stresses can arise from thermal expansion mismatch between alloy and oxide.

Fatigue life seems to be mainly spent in the growth of microcracks in an oxidizing atmosphere. Therefore, with the creation of fresh surface in fatigue cracking, steady-state oxidation is not likely to be established and transient oxidation conditions will prevail. In the transient oxidation regime, kinetic effects are essential as selective oxidation of alloying element will predominate, as illustrated by Andrieu *et al.* (1992) in IN718, where a few min exposure at 650°C gives selective oxidation of Cr at oxygen pressures less than 10^{-4} torr while Ni and Fe were oxidized above 10^{-2} torr. In addition, all the diffusion short circuits should play a major role in the transient oxidation regime such as grain boundaries of the substrate, and of the oxides.

The behavior of scales is discussed under monotonic and cyclic loading; intergranular oxidation and interdendritic oxidation are then considered.

(i) Behavior of oxide scales under monotonic loading

When the oxide scale at the surface of a testpiece is intact, oxidation proceeds with parabolic kinetics as in a stress-free specimen, at least in the steady-state regime. If failure occurs for a given strain level and repeats for the same strain increment in creep, the oxidation rate is increased and can follow pseudo-linear kinetics. Provided that time is a multiple of $t_{r,ox}$, the time to fracture the oxide scale, the thickness h varies linearly with time as

$$h = (kt_{r,ox})^{1/2}\, t/t_{r,ox} \tag{19}$$

where k is the parabolic oxidation constant.

The onset of through-thickness cracking was determined by Manning (1981) from fracture surface energy criteria, who considered that spallation occurred when the strain energy per unit area is greater than the interfacial energy between oxide and substrate. Hancock and Nicholls (1988) pointed out the importance of pores, since the defect size consumed ~80% of the oxide layer thickness in Nimonic 75. Assuming the existence of residual stress σ_r and strain ($\varepsilon_r = \sigma_r/E_{ox}$ with E_{ox} Young's modulus of the oxide), a simple fracture mechanics analysis yields

$$\varepsilon_{u,ox} = \frac{K_{IC}}{E_{ox} Y\sqrt{0.8\pi}}\frac{1}{h^{1/2}} + \varepsilon_r \tag{20}$$

This value compares fairly well with experimental strain to fracture as a function of scale thickness on Nimonic 75 at 950°C with $K_{Ic} = 4.5\,\mathrm{MPa\,m^{1/2}}$ or the chromia scale, in spite of incorrect trends (Figure 33).

Manning (1981) and Robertson and Manning (1990) analyzed tensile cracking and spalling of oxides along such approximations distinguishing oxide and interfacial toughness between oxide and substrate, and using an empirical equation for the fracture surface energy of oxides $\gamma_{ox}/E_{ox} \cong 0.04$ nm.

Viscoplasticity in oxide and substrate can alter these conclusions (Riedel, 1982) and crack healing effects can occur at temperatures where oxide scales are growing, when they are "living": while the ductility of the scale is very low at high strain rates, at low strain rates healing can occur and increase ductility as shown by constant strain rate extension tests on alloy 800 and Cr ferritic steels (Schütze, 1985, 1990, 1997).

(ii) Behavior of scales under cyclic loading and early cracking

Thermal cycling alone can have a very large influence on the adherence of oxide scale due to the difference in thermal expansion coefficients ($\Delta\alpha$) between the substrate and the oxide. Thermal strains vary as $\Delta\varepsilon_m = \Delta\alpha\Delta T$, where ΔT is the temperature difference and $\Delta\alpha$ the difference in thermal expansion coefficients. Generally, the oxide is under compression due to its lower expansion coefficient and can spall

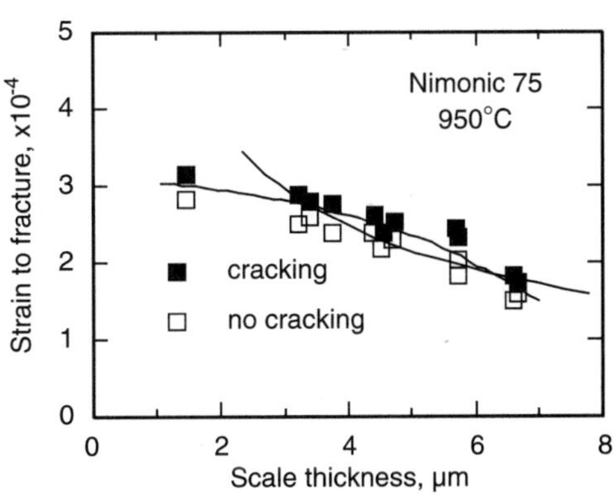

Figure 33 Fracture strain of the oxide formed on Nimonic 75 at 950°C (and measured at 950°C) as a function of scale thickness. The solid curve refers to Equation (6) deduced from a fracture mechanics analysis (after Hancock and Nicholls, 1988).

more easily under thermal cycling. Complex behavior can be observed in layered structure of oxide on engineering alloys as shown in scales of a 2.25Cr–1Mo steel thermally cycled in oxygen, air, and water vapor (Christl *et al.*, 1989).

The influence of high-frequency cyclic stressing on the growth of oxide scale on fatigue specimens was studied by Ward *et al.* (1974) on a low carbon steel EN2 under push–pull stress control at 500°C and 570°C. At sufficiently low stresses, the scale thickness displays a usual parabolic kinetics as for an unstressed alloy, whereas at higher stresses the scale thickness shows an apparent linear variation with cycling time in agreement with Equation (26), Figure 34. A banded structure was observed at 570°C, which indicates that continual loss of adhesion at the metal scale interface has occurred throughout the test, and scale cracking occurs for a threshold strain between 2.5×10^{-4} and 5.1×10^{-4} at 500°C.

Skelton and Bucklow (1978) made similar observations on oxidation kinetics in a ferritic (1/2)Cr–Mo–V steel at 550°C under high-strain low-frequency fatigue. At a plastic strain range of 2×10^{-4}, the thickness of the oxide layer is the same as in a stress-free specimen exposed for the same time. At plastic strain in excess of 10^{-3}, the oxide becomes layered due to periodic spalling and the oxide growth per cycle increases linearly with the plastic strain: the time to break the oxide scale decreased with increasing strain amplitude.

Reuchet and Rèmy (1983a, 1983b, 1983c) in cast Co-based alloy MAR-M509 observed a strong enhancement of oxidation rate after LCF at 900°C. Fracture of the oxide scale can

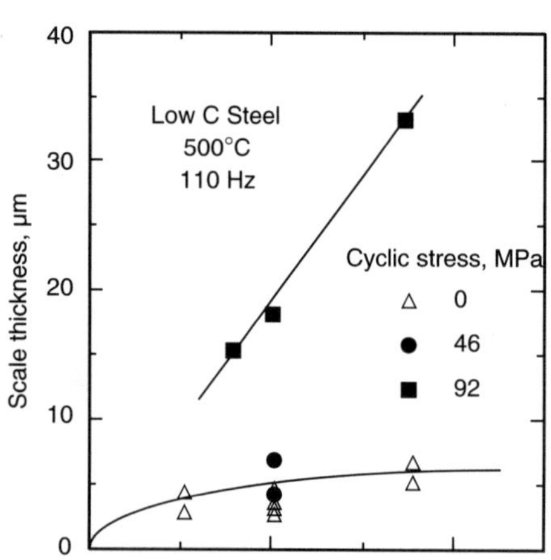

Figure 34 Oxide-scale thickness on fatigue stressed low carbon EN2 steel at 500°C above the fracture stress of the oxide (92 MPa) and below this stress (after Ward *et al.*, 1974).

occur after $n_{r,ox}$ fatigue cycles, for a given strain range, by microcracks perpendicular to oxide scale and/or local loss of adhesion between oxide and substrate (Reger and Rèmy, 1988a, 1988b, 1988c). After fracture of the oxide scale, direct access of oxygen to the metal is achieved, so that oxidation kinetics will resume from this new time origin. In such a situation, the local matrix oxide thickness, $h_{f,m}$, after N fatigue cycles will be

$$h_{f,m} = (N/n_{r,ox})(kn_{r,ox}\Delta t)^{1/2} \quad (21)$$

where Δt is cycle period, and $n_{r,ox}$ the number of cycles to fracture of the oxide scale ($n_{r,ox} < N$). Using a statistical analysis of oxide thickness measurements, the minimum number of cycles to fracture the oxide scale $n_{r,ox}$ is found consistent with a power-law relationship:

$$\Delta\varepsilon(n_{r,ox})^r = \varepsilon_f \quad (22)$$

where r is $\sim$0.25. This number of cycles is $\sim$10% of the lifetime of LCF specimens, and is weakly dependent upon frequency and hold time. The number of cycles to initiate a crack 50 μm in depth, for low-frequency fatigue tests in IN100 at 1,000°C, is good agreement with the minimum number of cycles to fracture of the matrix oxide scale (Reger and Rèmy, 1988a, 1988b, 1988c), i.e., the nucleating event in LCF is the fracture of the oxide scale in most critical areas.

As emphasized previously (Rèmy, 1993), this power-law relationship is partly in agreement with the equation $\Delta\varepsilon_m^2 h =$ const., since the exponent is correctly predicted, which corresponds to Equation (27). However, $n_{r,ox}$ should vary as frequency, which is clearly not observed in cast superalloys. Data from Skelton and Bucklow (1978) are consistent with a variation as ε_t^3 and show no time dependence from a 200 s period to 16 h hold times. In MAR-M509, no strain effect on oxidation kinetics is observed below a threshold strain of 0.45% and healing effects on oxide ductility need to be taken into account (Schütze, 1997).

(iii) Behavior of scales under thermomechanical cyclic loading and early cracking

Neu and Sehitoglu (1989a) investigated the evolution of oxide layers in 1070 carbon steel used in railroad wheels under isothermal fatigue at various temperatures between 150°C and 600°C. They distinguish the type I oxidation-induced growth as characterized by the formation of an oxide intrusion by a process of oxide rupture (see, e.g., Figure 12).

When this oxide layer reaches a critical thickness h_f, the oxide breaks and crack nucleation has occurred. Type II growth corresponds to a case where for a critical thickness fracture of the oxide occurs but part of the layer detaches from the surface. This gives rise to a multilayer or stratified oxide growth, as seen by Skelton and Bucklow (1978), and this results in exposing a layer fresh surface area to environment, which accounts for the wider intrusions observed in that case.

Type I growth is typical of out-of-phase TMF, while multilayered oxide growth type II is observed under isothermal loading. Measurements of the average critical oxide thickness at rupture are plotted vs. mechanical strain range in Figure 35. Much smaller oxide thickness at fracture is observed for out-of-phase TMF, than for isothermal while fracture cannot be achieved under in-phase TMF. The relationship between mechanical strain range and critical oxide thickness h_f is given by

$$h_f \Delta\varepsilon^2 \Phi_{ox} \dot{\varepsilon}^a = \delta_0 \qquad (23)$$

This equation shows the same relationship as previous equations. However, a phasing factor Φ_{ox} is introduced which is a Gaussian-type function of $\dot{\varepsilon}_{th}/\dot{\varepsilon}_m$, where peak damage corresponds to fully constrained out-of-phase loading ($\dot{\varepsilon}_{th}/\dot{\varepsilon}_m = -1$). It accounts for an increase of oxide ductility with decreasing strain rate.

A large effect of phasing between temperature (or thermal strain) and mechanical strain has been reported. Measurement of thickness of oxide layers has been used to estimate the critical thickness of oxide at fracture. For a mechanical strain range of 0.7%, this results in an average initial thickness of 0.08 μm for out-of-phase TMF, 1.2 μm for isothermal fatigue at 600°C, and 18 μm for in-phase cycling. This means that the effective oxide ductility is decreased by a factor of 15 from in-phase to out-of-phase cycling.

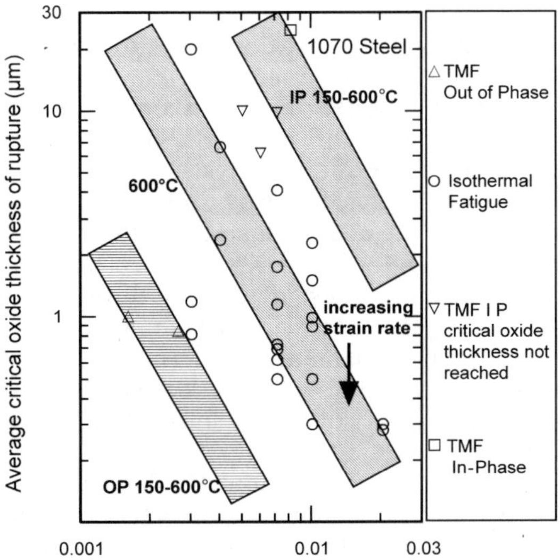

Figure 35 Plot of average critical oxide thickness at rupture from in-phase, out-of-phase 150–600°C TMF tests and LCF tests at 600°C on 1070 steel (after Neu and Sehitoglu, 1989).

This repeated microrupture process forms the basis of the oxide growth contribution of a TMF damage model that assumes an apparent kinetic law with an exponent of 1.5, while unstressed oxide kinetics was parabolic; this is different from previous authors, who assumed pseudolinear kinetics and a constant contribution for oxide cracking at each cycle (Skelton and Bucklow, 1978; Reuchet and Rèmy, 1983a, 1983b, 1983c).

Ariana *et al.* (1996) invoked such repeated oxide cracking mechanism in their TMF investigation of uncoated and coated (NiAl coating) SRR 99 nickel-based superalloy single crystals in the temperature range 300–1,050°C. Failure of uncoated specimens was shown to result from the multiple initiation of cracks at oxidation spikes on the surface, followed by growth and crack coalescence, as monitored by a video camera system. Microcrack growth rate was constant up to an average crack length of 0.3 mm: the rate in this regime of repeated oxide cracking was related to maximum tensile stress via a power law, consistent with the fatigue-oxide cracking superposition model (Reuchet and Rémy, 1983a, 1983b, 1983c; Chataigner *et al.*, 1996). For deeper crack the behavior was explained by the Paris law using an effective stress range of stress intensity factor and an exponent of 2. TMF life for uncoated and coated specimens was described using a larger closure term for coated specimens.

(iv) Intergranular and interdendritic oxidation

Intergranular oxidation often occurs with internal oxidation in austenitic alloys with high chromium content, like alloy 800 (Shida and Moroishi, 1992); the depth of intergranular oxides l_{ox} seems to correspond to the extent of the zone depleted of chromium and obeys parabolic kinetics.

The kinetics of intergranular oxidation was studied in a nickel-based superalloy TAZ-8A in the temperature range 600–1,000°C by Oshida and Liu (1988), which occurs below an external scale composed primarily of Cr_2O_3 or $(Cr–Al)_2O_3$ at high temperature with lower amounts of NiO or spinels. A statistical analysis of oxide depth measurements has shown that this depth l_{ox} varies as

$$dl_{ox}^4 = \alpha_{ox}^4(T)\,dt \qquad (24)$$

where α_{ox} is a constant which varies with temperature according to Arrhenius's law. Such kinetics is in reasonably good agreement with models for intergranular diffusion.

In cast superalloys, oxidation occurs preferentially in interdendritic areas, which are heavily segregated and rich in carbides, especially MC carbides. These carbides are quite susceptible to oxidation and may form CO_2/CO gas on oxidation, which then makes oxygen ingress and carbon–oxygen gas outward transport easier (Courtright *et al.*, 1991). Figure 36 (Reger *et al.*, 1984a) shows the depth of interdendritic oxide spikes l_{ox}, compared with *oxide-scale* thickness, as a function of exposure time in IN100 at 1,000°C: interdendritic oxidation proceeds according to Equation (24), as in MAR-M509 (Reuchet and Rémy, 1983c).

Interdendritic oxidation in C-bearing cast alloys seems to be affected by strain cycling in cobalt-based superalloy MARM-M509, whereas it does not in nickel-based IN100.

Almost no data are available concerning the influence of deformation on intergranular oxidation, in spite of its importance for creep-fatigue and TMF cracking. Investigations in nickel-based alloys, PM Astroloy and Waspaloy (Bernède, 1994; Rémy, 1995; Köster and Rémy, 1999), have shown that intergranular oxidation, as interdendritic oxidation, obeys a $t^{1/4}$ kinetics, and that the oxidation constant α_{ox} is strongly dependent upon plastic strain ε_p under continuous fatigue and creep-fatigue loading. Intergranular cracking under creep loading was studied in wrought superalloy

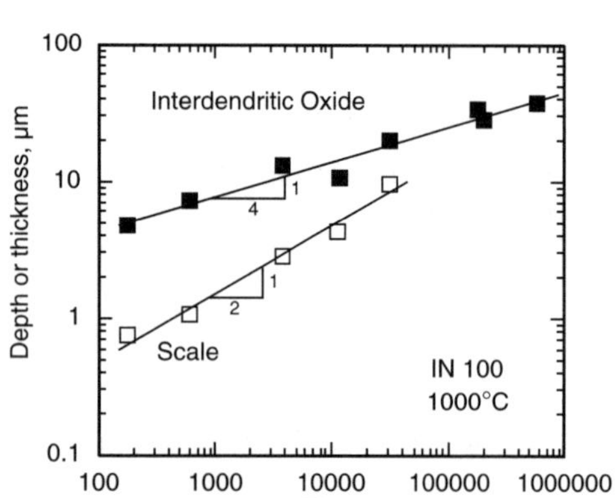

Figure 36 Depth of interdendritic oxide and oxide-scale thickness as a function of time in air at 1,000°C in IN100 superalloy (after Reger and Rémy, 1988a, 1988b, 1988c).

Udimet 720 LI (Chateau and Rémy, 2001). Oxidation of this alloy under creep at 850°C gives rise to a parabolic kinetics for the external scale without load and under load, without any preferential oxidation at grain boundaries. Under creep, grain boundaries appear to be oxidized from the specimen surface at much greater depth than internal oxidation, which results in intergranular crack initiation and multiple cracking under creep, until a major crack grows and leads to sample failure.

The depth of oxide spikes exhibits a time dependence similar to creep-elongation curves, and oxidation constant under load can be written as

$$\alpha_{ox}^4(T) = \alpha_{ox,o}^4(T)[1 + (\varepsilon_p/\varepsilon_{po})^2] \qquad (25)$$

where $\alpha_{ox,o}(T)$ and ε_{po} are two constants at a given temperature. Combining Equations (24) and (25) and integrating as a function of time, which is done using a creep constitutive model, results in predicted curve shown in Figure 37.

5.03.4.1.4 Modeling oxidation effects on lifetime

Continuum damage mechanics provides a framework especially convenient to incorporate different damage rate equations to account for different damage mechanisms. This method has largely been used for modeling creep damage by Ashby and Dyson (1984), and in subsequent work (Osgerby and Dyson, 1990; Dyson and McLean, 1998). They extended the continuum damage mechanics initially proposed by Rabotnov and Kachanov, but they identified damage equations for different micromechanisms. The interaction with oxidation has been considered in creep damage models considering internal oxidation from surface and failure of protective oxide. The oxidized zones were assumed to carry no load, so that they can be treated as damage in the original sense of Rabotnov and Kachanov. The effect of damage is thus described through an effective stress concept, which implies a load transfer on undamaged material, which is treated in a global manner. Damage factors form the basis of most physically based damage models.

Application of continuum damage mechanics to life prediction for creep-fatigue, creep-oxidation-fatigue, and TMF is treated in some detail in Section 5.03.6, dealing with crack initiation models. Here, only a brief account of oxidation–creep interaction is given.

Early work explained the influence of oxidation on creep life and ductility, but

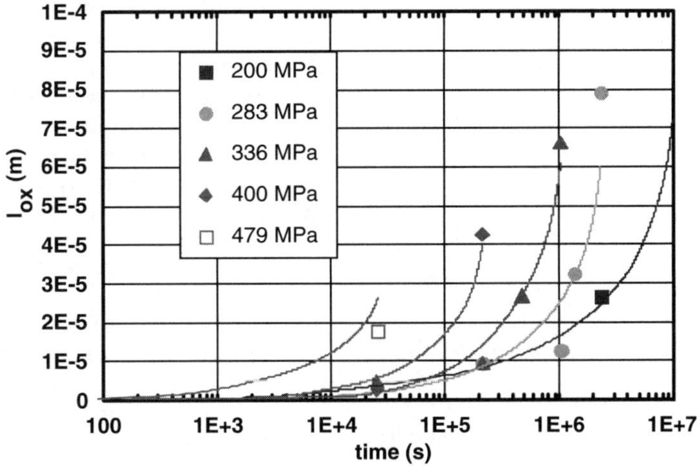

Figure 37 Evolution of the mean depth of grain boundary oxide spikes as a function of creep time at 850°C for Udimet 720LI (after Chateau and Rémy, 2001).

quantitative agreement was only achieved using a better constitutive creep model than simple Norton's law. Osgerby and Dyson (1990) proposed a different damage equation for the case where the oxide scale undergoes spallation and considered different criteria for the onset of through-thickness crack (strain energy as in Manning's original model, a constant defect size, and a defect size proportional to scale thickness). Much better description of experimental data in air and vacuum was achieved, and the influence of testpiece size on creep behavior was quite well represented (Figure 38).

The effect of intergranular oxidation in Udimet 720 LI on creep behavior reported earlier (Chateau and Rémy, 2001) has been accounted for along these lines. Drawing from the constitutive model of Dyson and Mc Lean (1998), creep strain rate was described by the following equation (in uniaxial form), assuming that precipitate coarsening increases inelastic strain rate, through the increase of interparticle distance λ:

$$\dot{\varepsilon}_{\text{in}} = \dot{\varepsilon}_0(\lambda/\lambda_0)\sinh[\sigma(1 - H)/\sigma_0] \qquad (26)$$

where $\dot{\varepsilon}_0$ and σ_0 are two constants that account for the behavior under relaxation as well as the early stages of the creep-elongation curves. H accounts for back stress and is constant at the end of primary creep stage. Tertiary creep can be described by Equation (26), replacing applied stress σ by an effective stress $\tilde{\sigma}$, which takes into account surface crack growth. Failure is assumed to be the consequence of ductility exhaustion, using an average ductility, which decreases with the amount of crack growth from the surface, according to a mean Weibull function. The ductility of a micro-

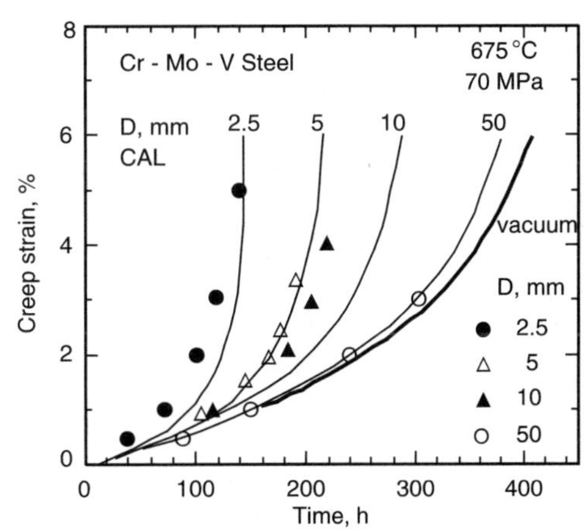

Figure 38 Influence of testpiece size on the creep curves of a Cr–Mo–V steel at 675°C and 70 MPa: comparison between experimental data of Cane and Manning (1981) (points) and predicted curves from Osgerby and Dyson (1990).

structural element in air is reduced due to intergranular oxidation.

The oxidation creep damage model predicts a size and specimen geometry effect due to oxidation-assisted crack growth. Validation tests were carried out using thin flat specimens with a rectangular cross-section to maximize surface effects (0.3 mm thick and 12 mm wide, which yields approximately two grains in thickness). Stress-rupture data of thin rectangular specimens and bulk cylindrical specimens are compared in Figure 39. A reduction in rupture time is observed by a factor of 3 or more as a significant decrease in ductility. The model is able to predict the rupture lifetime fairly well, though the model was identified only on bulk specimens and was extrapolated to thin rectangular specimens with no extra parameter fitting. The model is effective in predicting creep elongation vs. time, as well the drastic reduction of strain of failure.

5.03.4.2 Damage Mechanisms in Coatings

The intrinsic resistance of alloys to aggressive environment (involving oxidation as well as corrosion effects) is insufficient especially with advanced high-strength directionally solidified superalloy poly- and single crystals. Overlay coatings deposited by plasma technique and aluminide coatings (chemical vapor deposition CVD) are widely used owing to their good oxidation resistance. They prevent loss of section, which can be especially important for thin-walled component operating for long service conditions. A common practice is to use a coating a few tens of μm in thickness and to design a component ignoring the presence of coating. Such procedure can be conservative or nonconservative according to the substrate–coating system and to the thermal–mechanical loading path.

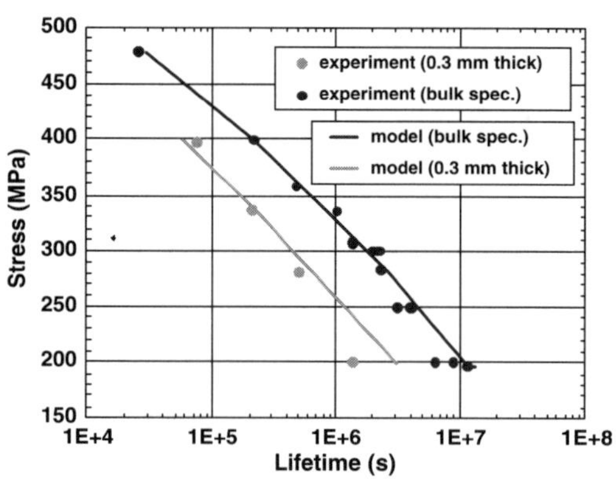

Figure 39 Creep lifetime for Udimet 720LI at 850°C as a function of stress: experiment and model prediction for bulk and thin specimens (after Chateau and Rémy 2001).

In these multilayered systems, thermal strains $\Delta\varepsilon_{th}$ (and stresses) arise from the differences in thermal expansion coefficients between substrate and coating ($\Delta\alpha$) as

$$\Delta\varepsilon_{th} = \Delta\alpha\Delta T \tag{27}$$

where ΔT is the difference between deposition temperature and operating temperature. These thermal strains are a major parameter in the integrity of coating systems under service loading (Grünling *et al.*, 1987; Strang and Lang, 1982; Schneider and Grünling, 1983; Stringer, 1987).

The situation is like that of the natural oxides that form when a material is exposed to an oxidizing environment. The mismatch is, however, much smaller in artificial coating–substrate systems than in natural oxide–alloy systems.

Damage mechanisms are dependent upon thermal–mechanical loading path as well as upon mechanical strain range (e.g., Bernard and Rémy, 1990). High strain ranges and medium maximum temperature favor the formation of multiple transgranular cracks in the coatings, which are mostly arrested at substrate–coating interface (Figure 40); lower strain ranges and/or medium-to-high maximum temperature can give rise to oxidation-assisted intergranular cracking of the coating and surface rumpling. Low strain ranges and large number of cycles on high maximum temperature favor oxide-induced delamination and spallation of coating. Holmes *et al.* (1988)

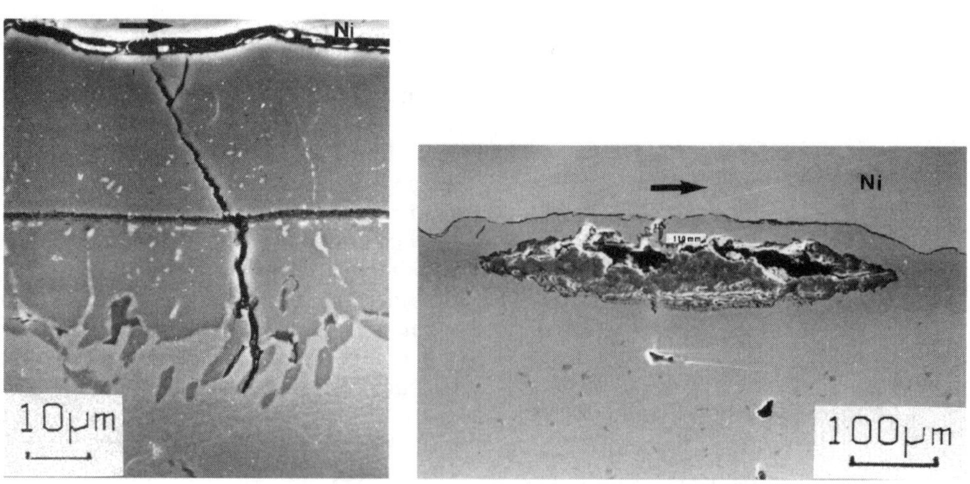

Figure 40 Coating damage under TMF cycling: transverse cracks at high strain ranges and delamination at low strain ranges in NiAl-CVD on IN100 superalloy (counterclockwise diamond 600–1,050°C) (after Bernard and Rémy, 1990).

and Holmes and McClintock (1990) have investigated coating degradation by surface rumpling and oxidation in an NiCoCrAlY coating deposited on Réne N4 single-crystal superalloy using disks submitted to TS between 520°C and 1,080°C. Periodic surface oxidation and roughening result in "scalloping" of coating which correlates with peak compressive substrate strain. This mechanism of degradation was attributed to cyclic breakdown of surface oxide and irreversible cyclic creep of the coating, as supported by a cyclic creep inelastic analysis (Busso and McClintock, 1988).

Due to their low ductility at low temperature, thin layers of intermetallic compounds can sometimes give early initiation of cleavage or intergranular cracks for severe TMF loading in the low ductility range.

Totemeier and King (1996) have carried out a fairly comprehensive investigation of high activity CVD NiAl coating on a single-crystal superalloy under isothermal fatigue conditions. At 600°C in the brittle domain (ductility ~0.6%), the coating reduces fatigue life with respect to uncoated specimens. Initiation occurs in a wide zone along the coated surface and involves intergranular and transgranular cracking in the coating. The reduction in life is more important at low strain ranges and can reach a factor of 10. The Basquin exponent varies from 0.21 to 0.17. At 800°C and 1,000°C, due to the strong reduction in flow stress of coating, cracks nucleate very early in the coating and no significant difference is observed between coated and uncoated specimens. Coating has a detrimental effect on fatigue life at 800°C for strain ranges exceeding 1.2%. Coating cracks nucleate early and most of them are arrested at the interface between coating and substrate, as noted by Chataigner and Rémy (1996) in other alloy systems with a low activity coating. They concluded the occurrence of crack closure due to oxidation, but viscoplasticity should also play a major role, especially for lower-frequency tests used in some investigations.

To get some insight into the consequences of brittle cracking in a TMF situation, Totemeier *et al.* (1996) used different precracks. Narrow precracks at room temperature have no detrimental effect due to an oxidation-induced crack closure mechanism at 600°C and 800°C. Wide precracks at room temperature or elevated-temperature precracks reduced the life at 800°C in particular, while no effect of precracking was observed at 1,000°C. The importance of crack closure is also supported by the observation on a similar system that, under vacuum at 950°C, the life of an aluminide-coated specimen is shorter than in air (Perruchaut *et al.*, 1996).

Martinez-Esnaola *et al.* (1996) have investigated in some detail the occurrence of cracking in coated SR99 superalloy single crystals for −135° phase lag between strain and temperature, in the range 300–1,050°C. For strain ranges smaller than 0.8% they found crack initiation at some sites giving rise to usual semi-elliptical cracks initiating from the surface. For strain ranges equal or larger than 0.8%, circumferential cracks were observed in the coating with a crack density, which tends to increase with strain range, within some range of scatter.

These authors used the 2D analysis of multiple transverse cracking by Nairn and Kim (1992) for brittle coatings on substrates, assuming linear elastic fracture mechanics. It relates the strain to first crack and the distribution of coating cracks, as a function of applied strain, to the fracture toughness of the coating. The energy release rate G is given as

$$G = [E_c\varepsilon_0 + C_1(\alpha_s - \alpha_c)(T - T_0)]C_2 t_c f(D) \quad (28)$$

where ε_0 is applied strain, α_s and α_c are linear coefficients of thermal expansion for the substrate and coating, respectively, T_0 is the equilibrium temperature where there are no residual thermal stresses, C_1 and C_2 are functions of Young's modulus, Poisson's ratio, and thickness of the substrate (E_s, ν_s, and t_s) and of the coating (E_s, ν_s, and t_c). $f(D)$ is a function of crack density, which is unity for the first crack. Analytical expressions are given in Nairn and Kim (1992) for C_1, C_2, and $f(D)$. The analysis yields a fracture toughness of 6 MPa m$^{1/2}$ and a crack density at the temperature of peak strain, consistent with the distribution of surface cracks monitored by the optical equipment on TMF specimens. 2D computer simulations suggest that crack shielding effects play a secondary role on the evolution of crack distribution.

Brittle cracking is a primary risk for premature fatigue failure of coated components, which is analyzed by engineers in terms of tensile ductility, defined as the strain to a macroscopic crack, as a function of temperature. Figure 41 shows the ductility of C1A coating on AM1 single crystals for [0 0 1] orientation: the tensile ductility varies from 0.65% to 1% when temperature increases from 20°C to 750°C (Chataigner and Rémy, 1995). Brittle cracking under tensile loading gives rise to large transgranular cracks typical of cleavage in aluminide coating at the surface of coated specimens. There is a very large variability in fracture strains measured as a function of temperature: the temperature below which the strain to cracking does not exceed 1% varies between 280°C and 820°C, for high-activity coatings, depending upon conditions and superalloys. Duret-Thual *et al.* (1986) have pointed out the extreme sensitivity of NiAl-type intermetallic compounds to composition, which is reflected in the details of the deposition process, coating thickness variation, as well as substrate composition.

TMF is, in principle, especially convenient to demonstrate synergy effects, which can be expected with such a transition in ductility of the coating with temperature. In the case of [0 0 1] C1A-coated single crystals of AM1 superalloy, counterclockwise diamond TMF tests in the range 600–1,100°C bring no detrimental effect of coating even for large strain range (2%) where the peak tensile strain at 700°C slightly exceeds the ductility, on cooling (see Figure 21). No brittle crack is actually observed on the surface of the highly strained TMF specimens using this diamond-type cycle, and crack initiation in this material occurs mostly at subsurface casting pores whatever the condition, bare or aluminized (Chataigner and Rémy, 1996).

This tensile strain ductility criterion does not seem to be very reliable to account for brittle fracture in the coating. In order to demonstrate it, a TMF cycle was used with a purely compressive strain cycle with identical temperature range 600–1,100°C and period, depicted in Figure 42. When the mechanical strain varies from zero to a minimum value of −1%, brittle cracking does not occur, while long brittle cracks are observed for a minimum strain of −1.2%. Cleavage seems to be the dominant cracking mechanism in the aluminide, which is well described using a critical principal stress criterion (Knott, 1973). The assessment of a critical stress criterion requires first to estimate the stress state in the coating for the different thermal–mechanical loading paths. This was done using literature data on viscoplastic behavior and assuming same strains in coating and substrate, but accounting for residual stresses (Chataigner and Rémy, 1996).

The stress–temperature loops for C1A and the substrate AM1 are shown in Figure 43 when a TMF cycle with a compressive strain is used, together with the critical stress corresponding to the tensile ductility curve of the coating. TMF gives rise to a large relaxation of stresses at high temperature in the coating, but at temperatures lower than ~800°C, a tensile stress develops (even though mechanical strain is negative). A cycle with a minimum compressive strain of −1% gives a tensile stress which does not exceed the tensile stress for cleavage in the coating. On the opposite, for a minimum strain of −1.2%, the peak tensile stress at 600°C exceeds the tensile stress for cleavage. These computations are consistent with the observation of brittle cleavage cracks for a

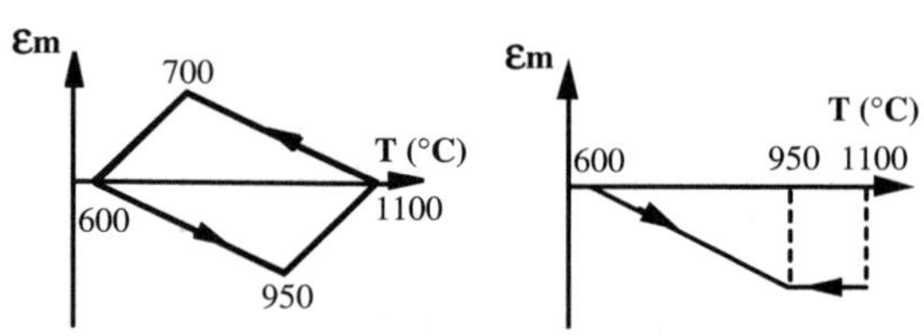

Figure 42 Standard TMF diamond cycle used for C1A-coated single-crystal AM1 superalloy (left) and compressive strain cycle used to check fracture criterion (right) (Chataigner and Rémy, 1995).

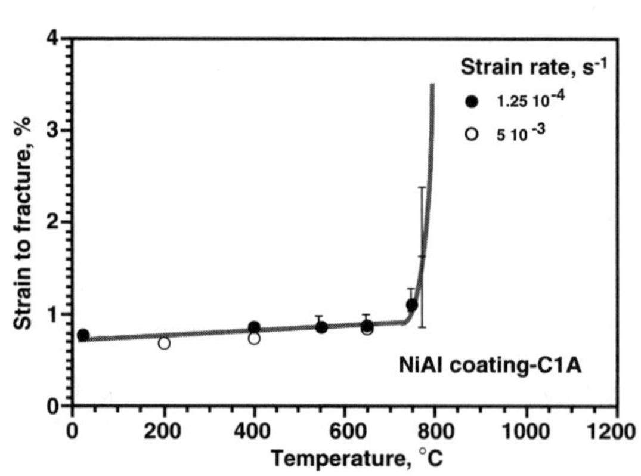

Figure 41 Variation of the ductility of C1A NiAl coating applied on single-crystal AM1 substrate as a function of temperature ([0 0 1] orientation) (after Chataigner and Rémy, 1995).

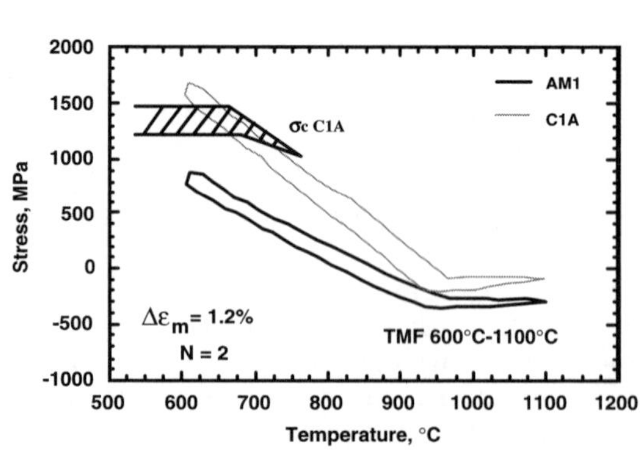

Figure 43 Stress vs. temperature hysteresis loop for AM1 (experimental) and C1A (computed) using the TMF cycle shown in Figure 8 for a minimum strain −1.2%. Note the position of critical stress computed for C1A coating (after Chataigner and Rémy, 1995).

peak strain of –1.2%, and of no crack for a peak strain of –1%.

Application to 600–1,100°C diamond tests has shown that even for the largest strain range, these TMF cycles generate smaller tensile stresses, and the tensile stress for cleavage is not reached even for a strain range as large as 2%. Therefore, a critical tensile stress can be used to rationalize the onset of brittle cracking in aluminide coatings.

A similar methodology has been used to quantify the effect of fatigue cracking below the critical stress level for brittle cracking, and a satisfying agreement has been reported for TS tests (Rémy *et al.*, 2000).

5.03.5 CONSTITUTIVE BEHAVIOR

Only a short description of constitutive models that have been used under nonisothermal conditions is given here. Only thermoelastic computations have been used for a long time and are still used in many cases to describe nonisothermal loading conditions. The analysis of TS experiments and of real components in gas turbines has stimulated a lot of work. Spera (1969) used crude approximations in his attempt to describe Glenny's experiments. Later Mowbray and McConnellee (1976) have made a detailed analysis of the TS test based on their own fluidized bed facility. The heat transfer and linear stress analysis were made using the same FE model. At that time a nonlinear FE analysis of the specimen was only carried out at a demonstration stage for preliminary comparisons with a simpler model, to reduce cost analysis and computer time. The FE computer program was based on incremental plasticity theory and a von Mises yield criterion and associated flow rule. The analysis was restricted to isotropic hardening.

Mechanical strain was partitioned into elastic (ε_e), plastic (ε_p), and viscoplastic (ε_v) strain:

$$\varepsilon_m = \varepsilon_e + \varepsilon_p + \varepsilon_v \qquad (29)$$

The instantaneous plastic strain was deduced from simple tensile test at six temperatures. Linear interpolation was used at intermediate temperature. Viscoplastic strain was specified in the form of a hyperbolic sine law:

$$\varepsilon_v = 2At^{1/2} \sinh G\sigma \qquad (30)$$

where A and G are functions of temperature, evaluated by nonlinear regression analysis of available creep data. A strain-hardening assumption was used to determine creep rate after a change of stress level occurs.

A simplified analysis was made using the elastic strain-invariance principle (Manson, 1966), which assumes that strains computed by the elastic analysis are correct mechanical strains, even though the stresses are affected by plastic flow. This principle yields accurate predictions of mechanical strain for purely thermal axisymmetric loading such as in the disk test. Mechanical strains were thus obtained by the thermoelastic FE analysis. The peripheral fiber was treated as a uniaxial bar subject to prescribed strain–time–temperature history, in the nonlinear part of the analysis. Comparison with the full FE model assuming isotropic hardening was considered as satisfying for the cobalt alloy used in the analysis, FSX414.

The cyclic stress–strain behavior is often described at temperatures below the creep range, by a single power law between stress and plastic strain amplitude. At high temperature, thermal activation promotes dislocation viscous glide and all the deformation mechanisms usually associated with creep. Early work used an empirical equation, to describe the usual softening with decreasing strain rate (Coffin, 1973; Lemaître and Chaboche, 1985):

$$\Delta\sigma/2 = K(\Delta\varepsilon_p/2)^{n'}(\dot{\varepsilon}_p)^m \qquad (31)$$

or use a frequency dependence instead of strain rate.

A number of difficulties are associated with the partition of inelastic strain between plastic and viscous components. From the late 1970s, unified constitutive models were largely developed which use only viscoplastic strain and assume a partition of stress into components described by internal variables: a kinematic-hardening component, an isotropic-hardening component, and a viscous stress component (Lemaître and Chaboche, 1985). In uniaxial form, the viscoplastic strain rate is expressed as a modified Norton's law:

$$\dot{\varepsilon}_p = \left\langle \frac{|\sigma - X| - R}{K} \right\rangle^n \mathrm{sign}(\sigma - X) \qquad (32)$$

where K and n are constants at a given temperature, $\langle\,\rangle$ are Mc Caulay brackets ($\langle U \rangle = U$ if $U>0$ else 0), R_v is the radius of the elastic domain, X_v is an internal variable that describes the position of the center of the elastic domain in stress space. R_v and X_v are given by differential rate equations; X_v describes nonlinear kinematic hardening. A large temperature dependence of n accounts for plasticity at low temperature and creep at high temperature.

In the structure computations, the 3D formulation is used assuming a power-law equation for the dissipation potential. The strain rate tensor is then given by

$$\dot{\varepsilon}_p = \frac{3}{2}\dot{p}\,\frac{\sigma' - X}{J_2(\sigma - X)} \tag{33}$$

$$\dot{p} = \left\langle \frac{J_2(\sigma - X) - R}{k} \right\rangle^N \tag{34}$$

where

$$\dot{p} = \left(\frac{2}{3}\dot{\varepsilon}_p : \varepsilon_p\right)^{1/2} = \frac{2}{3}J_2(\dot{\varepsilon}_p) \tag{35}$$

and the "prime" refers to the deviatoric part of the tensor (X is a deviatoric tensor); p is the cumulative viscoplastic strain.

The description of kinematic hardening is a key issue in cyclic loading due to the importance of Bauschinger effect, especially for nonsymmetrical straining or stress loading, and is essential for thermal–mechanical loading which often involves stress relaxation in high-temperature part of a cycle and stress redistribution during a thermal–mechanical cycle, especially in components. Nonlinear kinematic is usually written as

$$X = \frac{2}{3}C\alpha \tag{36}$$

with

$$\dot{\alpha} = \dot{\varepsilon}_p - D\alpha\,\dot{p} \tag{37}$$

for a simple nonlinear kinematic variable.

It has to be noted that the introduction of the α variable is compatible only with variable temperature loading. A rapid temperature without viscoplastic flow results in a constant α, but the back stress X can change immediately since C is usually temperature dependent (see Lemaître and Chaboche, 1985).

Such equations account fairly well for the stress–strain behavior under isothermal and thermal mechanical fatigue, and they are identified from isothermal LCF tests including continuous sawtooth cycles or stress or strain hold time. Such models were previously shown to predict fairly well the stress–strain behavior of various alloys, steels, aluminum alloys, and superalloys under isothermal fatigue (Lemaître and Chaboche, 1985), and under varying temperature conditions in conventional superalloys, see Figure 44 (Dambrine and Mascarell, 1990; Rémy, 1990; Köster *et al.*, 1996a, 1996b), and in aluminum alloys (Verger *et al.*, 2002).

Ratcheting which occurs under asymmetrical stress loading can be described to some extent by using these equations. These equations are especially useful for creep-fatigue situation involving stress relaxation at peak strain in strain-controlled loading. Unified models including the description of creep and fatigue at high temperature need further development. It is worth emphasizing that due to the usual decrease of yield strength with increasing temperature, as well as due to strain rate effects at higher temperature, the in-phase results in a compressive mean stress. This mean stress is partly responsible for the difficulties in comparing isothermal and TMF lifetimes. Diamond-type cycles on the contrary give rise to much smaller mean stresses.

For anisotropic single-crystal superalloys, Cailletaud and co-workers (Méric *et al.*, 1991; Hanriot *et al.*, 1991) have proposed a crystallographic model which uses a viscoplastic phenomenological equation like Equation (25) but at the level of the slip system. On each slip system s, Norton's law is written as

$$\dot{\gamma}^s = \left\langle \frac{|\tau^s - x^s| - r^s}{K} \right\rangle^n \operatorname{sign}(\tau^s - x^s) \tag{38}$$

where τ^s is the shear stress on the slip system, x^s is a kinematic variable, and r^s describes isotropic hardening. K and n are viscosity constant and exponent, respectively, and two sets of values are used for octahedral cube slip.

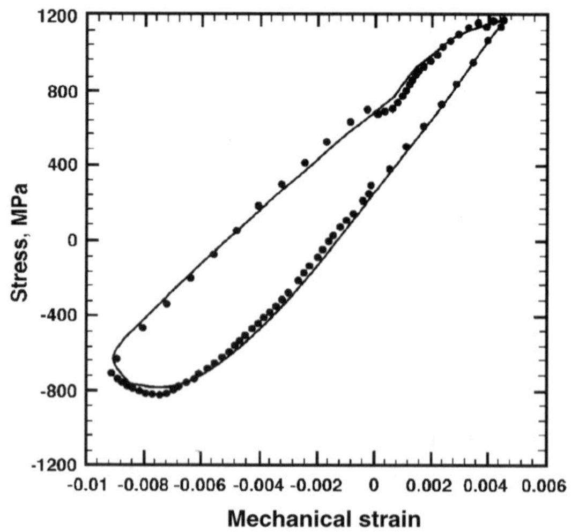

Figure 44 Comparison between constitutive model solid line and experiment (symbols) for hysteresis stress–strain loops of Superwaspaloy under thermal mechanical fatigue from 100°C to 750°C (test frequency 5×10^{-3} Hz) (source Koster, 1996a, 1996b).

No direct equation is postulated at the macroscopic level: the macroscopic stress is projected on to the slip systems using a generalized Schmid law, and the macrocopic strain is the sum of all the contributions of the different slip systems. Cubic elastic anisotropy is assumed. This model is implemented in an FE code.

This model is identified from isothermal tests for $\langle 0\,0\,1 \rangle$ and $\langle 1\,1\,1 \rangle$ crystals. It can predict the uniaxial tests in other orientations or multiaxial stress situations. One of the major advantages of a crystallographic model is that it can predict the active slip systems; predictions were successfully compared with experimental observations of slip traces under various conditions on AM1 superalloy (Hanriot *et al.*, 1991). Predictions are in good agreement with experimental stress–strain hysteresis loops under TMF cycling with a diamond-type cycle between 600°C and 1,100°C (Figure 45) (Hanriot, 1993; Rémy, 1998).

For long-term operation, thermal–mechanical loading can give rise to rafting of γ' precipitates (Arrel *et al.*, 1996). Such an evolution of microstructure can give rise to some discrepancy between experiment and predictions and may necessitate a description of the microstructure evolution. The influence of γ' rafting on the mechanical behavior has been reported (Espié, 1996). Composite models can be made using an FE analysis of a periodic composite structure, but they will give rise to rather long structure computations for components (Espié, 1996).

These models were identified in the 1990s for single crystals and could be used in isolating a

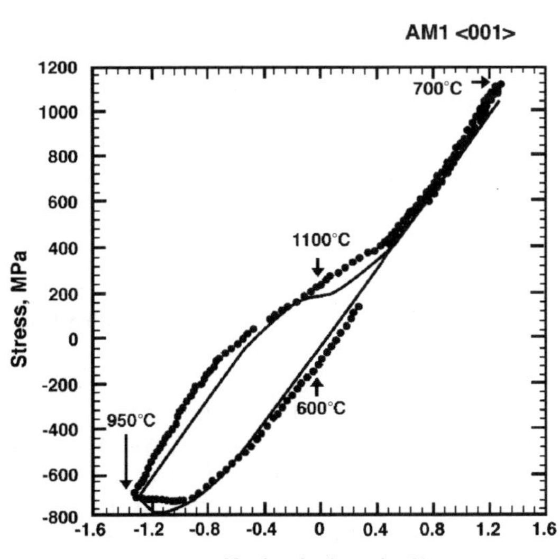

Figure 45 Crystallographic viscoplastic constitutive model. Comparison between model (solid line) and experiment (symbols) for stress vs. mechanical strain loops of [0 0 1] single-crystal AM1 superalloy under TMF.

plane section of the blade, which remains plane during the deformation process. Macroscopic viscoplastic Chaboche model was used in such a simplified assumption for aeroengines for 20 years (Dambrine and Mascarell, 1990). Crack initiation is then computed using a damage model. Full 3D analyses were also carried out in the 1990s starting with a 3D section; transverse stress and geometrical factors can be used to account for local stress concentrations.

Due to the increased complexity of blade shape and cooling technologies, critical areas are more sensitive to temperature field and external loading (Shepherd *et al.*, 2001). Large 3D computations are now necessary and are carried out using different codes or appropriate constitutive model material libraries (Foerch *et al.*, 2000; Cailletaud *et al.*, 2001, 2003), with eventually parallel computation to describe finer geometrical details in the blades. An example of such computations is depicted in Figure 46, with a complete computation of the temperature of an internally cooled hollow blade, and a section of the airfoil showing the equivalent stress computed from the viscoplastic crystallographic model. Integration schemes and constitutive models are described in more detail elsewhere (Chaboche and Cailletaud, 1996; Besson *et al.*, 2001).

Sehitoglu and co-workers (Smith *et al.*, 1999; Sehitoglu *et al.*, 2000a, 2000b; Sehitoglu, 1996) have made extensive use of unified models that have the potential to predict creep–plasticity interaction and strain rate effects more accurately than more conventional models in which a time-dependent creep strain is adding to the plastic strain. The model used is composed of two state variables, a back stress and a drag stress.

As largely emphasized by Chaboche, the back stress is essential to predict the Bauschinger stress as well as transient and steady-state creep response at high temperature. The flow rule is described as

$$\dot{\varepsilon}_{ij,\mathrm{in}} = \frac{3}{2} A f\left(\frac{s_{\mathrm{eff}}}{K_0}\right) \frac{s_{ij} - X_{ij}}{s_{\mathrm{eff}}} \qquad (39)$$

where $\dot{\varepsilon}_{ij,\mathrm{in}}$ is the inelastic strain rate, A is a material constant, K_0 is the drag stress, $\bar{\sigma}$ is the effective stress, s_{ij} is the deviatoric stress, and X_{ij} is the deviatoric back stress.

The effective stress is given as

$$s_{\mathrm{eff}} = [3/2(s_{ij} - X_{ij})(s_{ij} - X_{ij})]^{1/2} \qquad (40)$$

The flow function depends upon the active deformation mechanism (power-law creep, plasticity, or diffusional flow) and is motivated by Ashby's deformation mechanisms map (Frost and Ashby, 1982).

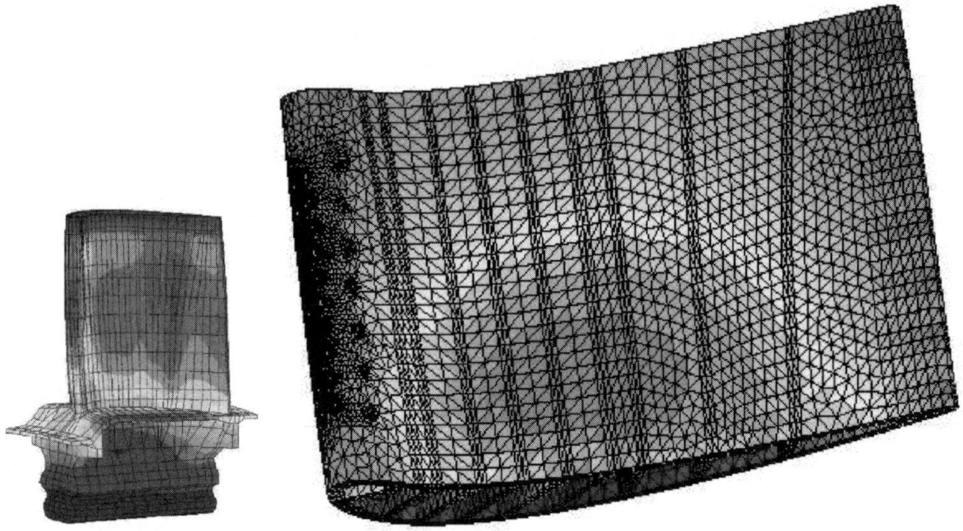

Figure 46 Crystallographic viscoplastic constitutive model. Application in a 3D FE analysis of a cooled turbine blade under thermomechanical loading: (left) temperature field, and (right) equivalent stress map in the airfoil (courtesy of Cailletaud).

The flow rule is determined directly from isothermal experiments conducted at constant strain rates (tensile tests, isothermal fatigue, and creep tests). The yield strength under various conditions is then normalized using the initial drag stress at the intersection point of the transition from power-law creep to plasticity (K_i) or when there is no transition in mechanisms, K_i/E is assumed to be independent of temperature.

The normalizing technique results in curves such as those shown in Figure 47 for an aluminum alloy: plasticity dominates for combinations of large stresses and high strain rates, whereas power-law creep dominates for smaller stress and slower strain rates. The flow function is chosen as

$$f\left(\frac{s_{\text{eff}}}{K_0}\right) = \left(\frac{s_{\text{eff}}}{K_0}\right)^{n_1} \quad \text{when} \quad \left(\frac{s_{\text{eff}}}{K_0}\right) < 1 \qquad (41)$$

$$f\left(\frac{s_{\text{eff}}}{K_0}\right) = \exp\left[\left(\frac{s_{\text{eff}}}{K_0}\right)^{n_2} - 1\right] \quad \text{when} \quad \left(\frac{s_{\text{eff}}}{K_0}\right) > 1 \qquad (42)$$

n_1 and n_2 are power-law creep and plasticity exponents, respectively. The first function is of generalized Norton's type as used by Lemaître and Chaboche (1985).

The evolution of back-stress-state variable has the form of a nonlinear kinematic-hardening equation, as used by Lemaître and Chaboche (1985), among others:

$$\dot{X}_{ij} = \frac{2}{3} h_\alpha \dot{\varepsilon}_{ij,\text{in}} - r_\alpha X_{ij} \qquad (43)$$

where the hardening function h_α describes the evolution of back stress during high strain rate

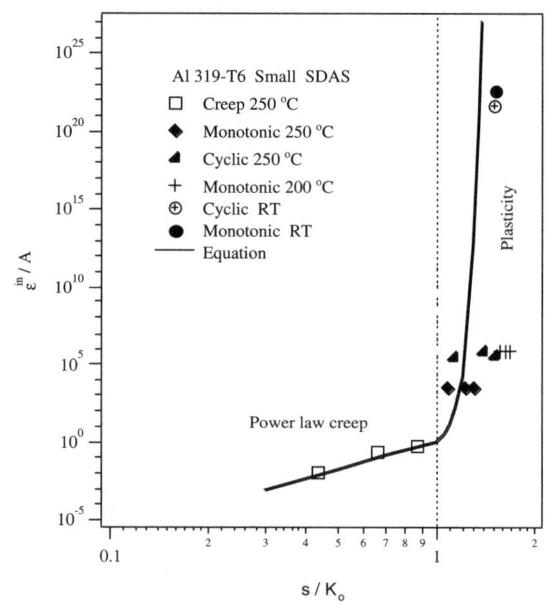

Figure 47 Normalized inelastic strain rate vs. normalized stress for Al 319-T6 small SDAS (after Sehitoglu *et al.*, 2000a, 2000b).

tests and increases the back stress. r_α is a recovery function which can incorporate static as well as dynamic recovery (hence a function of the von Mises inelastic strain rate).

A similar equation is used for the evolution of drag stress:

$$\dot{K}_0 = h_K - r_K + \theta \dot{T} \qquad (44)$$

where $\dot{K}_0$ is the drag stress rate. The drag-stress-hardening term (h_K) describes hardening or softening due to plastic deformation (proportional to the von Mises inelastic strain rate). The recovery term (r_K) corresponds to change in K_0 caused by microstructural changes resulting from high-temperature exposure. The θ term gives the temperature dependence of the drag stress and $\dot{T}$ is temperature rate.

Detailed expressions for various quantities are given, e.g., in Smith *et al.* (1999). The method results in a fairly high number of coefficients to be identified for a complete model useful in a large temperature and loading range. A number of exponential functions are used which may yield some numérical difficulties in FE computations.

This model has been modified to account for the variation of microstructure parameters in cast aluminum alloys, in particular in an Al–7%Si–3Cu–Mg alloy. Specifically, the stress–strain behavior was found to be rather strongly dependent upon the size of casting parameters which can be captured by the variation of the constant A in the flow rule (Equation (39), Sehitoglu *et al.*, 2000a, 2000b). The constant A is thus written as

$$A = A_0(L/L_0)^m \exp(-\Delta H/RT) \qquad (45)$$

where L is the size of secondary dendrite arm spacing (SDAS), and L_0, A_0, m_0, ΔH are four constants.

Application to the description of stress–strain behavior under TMF loading is shown for out-of-phase tests in the range 100–300°C, for small out-of-phase tests in the range 100–300°C, and for small and large secondary dendrite sizes (Figure 48).

At high temperatures, the microstructure of materials can become unstable. Aging may

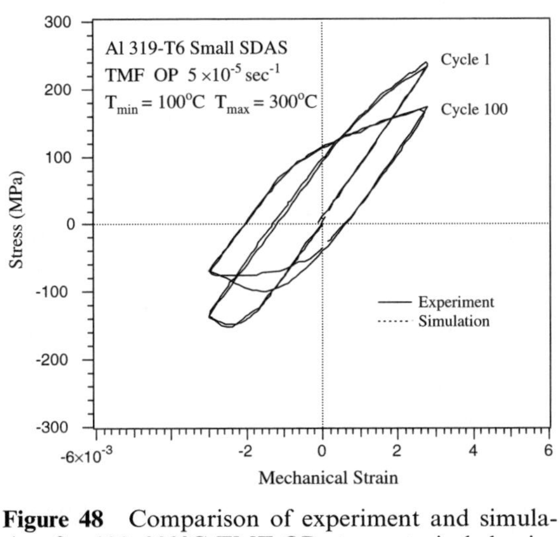

Figure 48 Comparison of experiment and simulation for 100–300°C TMF OP stress–strain behavior of Al 319-T6 small SDAS (after Sehitoglu *et al.*, 2000a, 2000b).

result in precipitation in grains and/or at grain boundaries in near-solid-solution alloys, solute–dislocation interaction giving rise to dynamic strain aging. Precipitate-strengthened alloys may soften due to coalescence or partial dissolution.

When temperature is cyclically varied, thermal strains may arise from mismatch in thermal expansion coefficients of different phases. These microscopic strains can favor the evolution of microstructure. This area still needs much effort especially for the long-term operation of components.

Dynamic strain aging is known to occur in austenitic stainless steel and gives rise to anomalous variation of yield strength or cyclic stress amplitude in some temperature range, as reported by Zauter *et al.* (1993). Dynamic strain aging occurs in numerous nickel-based alloys used as sheet material for combustion chambers in aeroengines, or in stationary gas turbines, like Hastelloy X (Castelli *et al.*, 1993) and IN617, an Ni–22Cr–Co12–Mo9 alloy (Kleinpass *et al.*, 1996; Moalla *et al.*, 2002). Both groups of these authors investigated in some details the in-phase and out-of-phase TMF deformation behavior of these alloys, as well as under isothermal conditions. Typical irregularities on the TMF stress–strain loops were observed (Moalla *et al.*, 2002).

A systematic comparison can be made between isothermal cyclic stress–strain behavior at minimum and maximum temperatures and minimum and maximum stresses in in-phase and out-of-phase cycles. This has been done, in particular, for Hastelloy X (Castelli *et al.*, 1993) by exploring different temperature ranges. Hastelloy, which is prone to the precipitation of $M_{23}C_6$ carbides, exhibits anomalous thermal–mechanical hardening with respect to isothermal conditions, in particular for TMF test in the range 200–400°C and 400–600°C, where the maximum temperature did not exceed that at which microstructure overaging began (i.e., 650°C).

A very strong hardening was actually observed in cast cobalt-based superalloy MAR-M509 under isothermal conditions, which peaks at 700°C, due to precipitation of carbides onto dislocations (Reuchet and Rémy, 1983a). The kinetics of precipitation and overaging was investigated by metallography in the temperature range of service, up to 1,100°C temperature excursions (Rézaï-Aria *et al.*, 1988a). Heat-treatment simulation of actual precipitation and aging effects occurring in TS specimens and in actual components have shown that overaging was such that, under TMF and TS, after a few number of cycles the effect on the stress–strain hardening was not very significant. For practical applications, the constitutive model could, therefore, be used as identified on virgin materials, with sufficient confidence.

In all alloys strengthened by precipitates, TMF often results in overaging of the microstructure. In nickel-based superalloys, even a small temperature variation close enough to precipitate solvus temperature can lead to overaging at high temperature and re-precipitation at a lower temperature: this was investigated in cast IN100 by Cailletaud and Chaboche (1980). ODS superalloys with a high volume fraction of strengthening γ'–precipitates and further hardening by Y_2O_3 particles like MA760 have very good creep properties but can be very unstable when large temperature variations are combined with high maximum temperatures as for heavy-duty stationary gas turbine blades (Vasseur and Rémy, 1994). When T_{max} exceeds 1,000°C, complete dissolution of precipitates can occur and a very fine microstructure is obtained on temperature cycling due to mismatch strains and dislocation punching at Y_2O_3 particles during TMF cycling. Inelastic straining enhances this effect but microstructure changes can be already observed after simple thermal cycling without load.

An efficient way of handling overaging effects was introduced by Cailletaud and Chaboche (1980). This method has largely been used to investigate the progressive reduction in strength due to high-temperature exposure of Al–Cu–Mg alloys used for cylinder heads (Cailletaud *et al.*, 2000; Nicouleau *et al.*, 2002).

A thermodynamic model was established which introduces a thermal contribution R^* in the isotropic-hardening variable R:

$$R = R_0 + R^* \tag{46}$$

with

$$\dot{R}^* = a\langle T_f - T\rangle - \frac{R^*}{\beta}, \quad R^* \geq 0 \tag{47}$$

where

$$\dot{T}_f = \frac{T - T_f}{\alpha^*} \tag{48}$$

T_f is a fictitious temperature or delayed temperature, which corresponds to the fact that, when a temperature change occurs, some time is necessary so that aging or overaging is completed; α^*, β, and a are strongly temperature-dependent coefficients.

Such a model is able to describe overheating ($T_f - T < 0$) and overaging effects, and aging on cooling (if $T_f - T$ is great enough).

For aluminum, overaging effects are very important and this model has been used to describe the evolution of peak stress and stress range in aluminum alloys (Nicouleau *et al.*, 2002). It has been used in a parallel computation of the whole cylinder block and shown to account for location of actual failure under thermal–mechanical loading.

Different strategies are used in other groups to run computations of complete components in reasonable computation time. Verger *et al.* (2002) chose to make a simulation of hardness variations induced by thermal exposure from a thermal analysis of the cylinder head. Boundary can be established between overaged material and almost unaffected areas. A much simpler constitutive model can then be used with only two sets of parameters, one for the virgin alloy, and another for the overaged stabilized alloy.

5.03.6 CRACK INITIATION MODELS

Damage modeling is necessary under TMF loading as under pure TF or TS. Component behavior can be simulated using simple structure tests or TMF tests on a volume element to model material degradation in the laboratory (Rémy, 1990; Rémy and Skelton, 1992).

It is, however, often impossible to simulate accurately the actual loading conditions using laboratory specimens and testing conditions (the temperature transient has often an unrealistic period, hold times under load are generally too short, environmental conditions are inadequate, etc.).

It is, therefore, necessary to extrapolate results from laboratory tests to actual service conditions, and this can only be achieved using damage models. This section is focused on models of crack initiation, which are used to design components, and relies partly on previous reviews (Rémy and Skelton, 1992; Rémy, 1996). A fairly extensive but inexhaustive review of these models will be given here.

Most models used for TMF/TF were initially developed for LCF at high temperature. They can be classified as continuum based and physically based. Continuum-based models are mostly phenomenological and can use either parametric equations or accumulation of creep and fatigue damage. They are general and easy to use; there is no physical meaning of damage and their extrapolation is questionable, outside from the range of conditions for which they have been identified.

Physically based models attempt to describe damage mechanisms. They are specific to a material and to a range of testing conditions. Their identification is often expensive in terms of manpower, since they often involve metallographic observations. They have a better capability of extrapolation than continuum-based models, at least in principle.

In the following, we shall see first the continuum-based models considering those based on parametric equations (using mostly inelastic strain range) and those based on damage accumulation (using mostly stress and stress range). Physically based models will then be described with different attempts to describe mechanisms of interactions between fatigue and oxidation and/or creep.

Background of models. Most physically based models dealing with fatigue consider that fatigue damage is essentially due to the initiation and growth of cracks and often focus on the second phase. A single-crack situation is often used to describe fatigue life occurring in laboratory specimens. Only in some instances, the effect of a population of cracks is considered. Ratchetting in some thermal loading situations has to be considered (see Skelton, 1983; Manson, 1973), but prediction of ratchetting behavior for viscoplastic materials at high temperature and under thermal–mechanical loading is quite difficult to achieve (see, e.g., Chaboche, 1991).

The description of fatigue microcraks has received much attention under isothermal LCF conditions at moderate and at high temperatures. In a volume element test, such as an LCF or a TMF test, initiation of a small crack from a defect (metallurgical imperfection, second phase particle, etc.) usually takes only a small fraction of lifetime. Therefore, several life prediction models used a fatigue life estimate based on the integration of small crack growth law. This law can be established from surface replicas of smooth specimens interrupted from time to time, or direct observations of specimen surface. This can also be derived from striation measurements, on the fracture surfaces, as done in ductile alloys such as austenitic stainless steels.

In numerous cases, there is a plateau where the crack growth rate is constant for crack depth smaller than a critical value a_t; above this critical value, crack growth rate increases with crack length:

$$\mathrm{d}a/\mathrm{d}N = (\mathrm{d}a/\mathrm{d}N)_o \quad \text{for } a < a_t \tag{49}$$

$$\mathrm{d}a/\mathrm{d}N = f(a) \quad \text{for } a < a_t \tag{50}$$

Different approaches are used to describe the variation of crack growth rate with crack length.

Levaillant *et al.* (1988) observed this plateau from striation spacing measurements on 316 stainless steel only at 550°C, while there was a smooth increase of da/dN with crack length (assuming a one-to-one correspondence with crack growth rate) at higher temperatures. Danzer and Bressers (1986) reported a plateau up to 0.1–0.2 mm in small grain sized superalloys, Astroloy and Waspaloy. Similar observations were made in cast cobalt- and nickel-based alloys in a wide range of LCF conditions (Reuchet and Rémy, 1983a, 1983b, 1983c; Reger and Rémy, 1988a, 1988b, 1988c) and in single crystals (Fleury and Rémy, 1993; Li *et al.*, 1993) or in wrought alloys (Köster, 1997).

There are not many studies of crack growth under TMF conditions, but the transition crack length for TMF loading was nearly equal to that under LCF on superalloys (Fleury and Rèmy, 1994).

Under LCF conditions, it is observed that, for most of the life, no detectable change of compliance can be detected with a thumbnail surface crack. Compliance changes are detected only for a significant decrease in tensile load where crack size is often ~1 mm in depth for a specimen 6–10 mm in diameter. For the case where compliance changes occur, many authors have applied cyclic J concepts and used experimental measurement of displacement in the middle of the crack along the loading direction, as suggested by Dowling (1977). It is worth noting that, in this regime, experimental increase of plastic strain amplitude under applied constant stress is used as a measure of damage, in identifying coefficients of fatigue damage equations in continuous damage mechanics (Lemaître and Chaboche, 1985; Chaboche, 1974).

Manson *et al.* (1967) introduced a relationship between the number of cycles to initiate a crack defined for an arbitrary crack length and the number of cycles to failure of a smooth specimen, at ambient temperature:

$$N_i = N_f - AN_f^p \tag{51}$$

with $p \sim 0.6$. The underlying reasons for this approach must be recalled: first it was necessary to account for the numerous observations of fatigue mechanisms that microcrack growth could represent a very large fraction of fatigue life especially in the LCF regime. Second, it was necessary to account for nonlinear damage accumulation in fatigue sequences at different strain ranges. This definition enables one to justify the so-called double linear cumulative damage rule, using crack initiation (N_i) and crack propagation (N_p) periods.

Various empirical equations have been proposed for steels to obtain an expression for $N_i(a_i)$, the number of cycles to initiate a crack of specified depth a_i. These have been deduced mostly from striation spacing measurements on fracture surfaces of laboratory specimens. A summary of equations proposed can be found in Chapter 5.02.

Measurements of striation spacing on fracture surfaces and of crack size distribution using plastic replicas of interrupted tests were made to estimate crack growth law, for a major crack or for several cracks in smooth specimens. The major purpose was to achieve a computation of specimen fatigue or creep-fatigue life and to predict at which rate damage progresses in the specimen. This gives a much richer prediction capability than total life in conventional models (Manson, 1973; Coffin 1973, and others) but this is also the aim of continuum damage mechanics.

The traditional fatigue strategy assumes that significant crack growth cannot be allowed in a component. The engineering life to initiation for a critical area in a component is modeled by the life of a smooth specimen that is conducted to failure (N_f).

The life to initiate a crack in a component is assumed to be the N_f for the computed loading conditions at a given location. It is implicitly assumed that laboratory specimen size is relevant to the critical size in component. To account for initiation in thin-walled or in thick components, it may be necessary to use an earlier failure criterion defined from the specimen itself. Potential drop measurements can help to define initiation to 0.25 mm crack depth as used, e.g., in superalloys (Rémy *et al.*, 1982). Such a definition was shown to be useful to handle with damage accumulation in variable temperature loading (Reuchet *et al.*, 1982) and life prediction in components but necessitates adequate information monitoring during the tests used to establish a database for life prediction.

All engineering models that employ a simple number of cycles N_f to describe damage have to use a cumulative damage like that of Miner or the double linear cumulative damage rule (Manson *et al.*, 1967; Manson and Halford, 1981) to deal with cycle mixing.

All models that are based on the integration of a microcrack growth law (using strain range, cyclic J or else) or continuum damage mechanics have all the necessary ingredients to deal with cycle mixing and damage accumulation under complex loading.

Models presented below vary from very simple to fairly complex ones with a large number of coefficients to identify. In the 1980s, identification of nonlinear models was quite a problem and still requires some experience in coefficient "tuning." Though all these models have very different predictive capabilities, attention must always be paid to the relevance of laboratory test conditions used in the database to component service life. Extrapolation from fairly short-time test results using continuum damage mechanics or physically based models will be more reliable than using engineering models, provided only that coefficients are identified from a database which encompasses the mechanisms operating in service.

5.03.6.1 Continuum-based Models: Using Parametric Equations

Most models based on parametric equations use inelastic strain range as a primary variable such as proposed by Coffin (1973) and Manson (1973). Some attempts to use total strain range were nevertheless reported, since total strain could be more readily estimated from a computation in a component than inelastic strains. Energy expenditure models also used inelastic strain and all these models usually require a nonlinear stress analysis in components. The engineering practice often used mechanical strains and thermoelastic analyses.

5.03.6.1.1 Models using inelastic strain range

Coffin and Manson proposed two models concurrently for describing isothermal LCF at high temperatures. Each model proposed by the authors of the famous damage law for LCF stems from different standpoints: Coffin, being involved in research on iron- and nickel-based superalloys, has assumed that fatigue damage at high temperature is mostly due to oxidation effects (McMahon and Coffin, 1970; Coffin, 1972) and fatigue life is mostly frequency dependent (there is no major influence of strain hold time *per se* except the reduction of cycle period). Manson, in his study of austenitic stainless steels, assumed that damage results mostly from creep fatigue, since fatigue life is strongly dependent upon cycle shape.

Coffin (1973) has thus proposed that a frequency-modified fatigue life should substitute for usual life (N_f) in the Manson–Coffin relationship, which holds at lower temperatures:

$$\Delta\varepsilon_{in} = C_2(N_f \nu^{k-1})^{-\beta} \tag{52}$$

where $\Delta\varepsilon_p$ is the plastic strain range (often called inelastic by other authors) and ν is frequency. This equation involves three constants: C_2 and β, which are the usual coefficients of ductility and exponent of Manson–Coffin relationship, and k, which accounts for time dependence ($0<k<1$). When k tends towards zero, there is no frequency dependence anymore, and when k reaches unity, fatigue lifetime is purely time dependent. Arguments for such an equation were provided by the vanishing influence of frequency upon fatigue life under vacuum of some superalloys, as shown for alloy A286 in Figure 30 (Coffin, 1973, 1972).

The identification of the coefficients for isothermal LCF requires running LCF tests at high frequency (to reach a "pure" fatigue damage law) and LCF tests at various frequencies. Application of such an equation to TMF/TF requires using an equivalent temperature concept and is mostly made at the maximum temperature for a given TMF cycle.

Manson (1973) has proposed to use a strain-range partitioning method. As in the previous equation the inelastic strain range is the relevant parameter, but time-independent plasticity is distinguished from time-dependent viscoplasticity (creep strain). Damage, in addition, is assumed to be dependent upon the direction of straining (tension or compression) as is pure creep damage. Four inelastic strain components are thus considered: $\Delta\varepsilon_{\mathrm{pp}}$ (plastic strain in tension reversed by plastic strain in compression), $\Delta\varepsilon_{\mathrm{pc}}$ (plastic strain in tension reversed by creep strain in compression), $\Delta\varepsilon_{\mathrm{cp}}$ (plastic strain in creep reversed by plastic strain in compression), and $\Delta\varepsilon_{\mathrm{cc}}$ (creep strain in tension reversed by creep strain in compression). Each pure basic of strain range $\Delta\varepsilon_{\mathrm{in},ij}$, where i and j are either p or c, is related to a fatigue life N_{ij} according to the classical Manson–Coffin equation:

$$N_{ij} = A_{ij}(\Delta\varepsilon_{\mathrm{in},ij})^{-\alpha_{ij}}, \quad \text{where } i, j = \mathrm{p} \text{ or } \mathrm{c} \tag{53}$$

For a complex cycle with an inelastic strain range $\varepsilon_{\mathrm{in}}$ partitioned in $\Delta\varepsilon_{ij}$ components, the life is given as

$$1/N = \sum B_{ij}/N_{ij}(\Delta\varepsilon_{\mathrm{in}}) \tag{54}$$

with $B_{ij}=1$ for the so-called "linear" or $B_{ij}=\Delta\varepsilon_{\mathrm{in},ij}/\Delta\varepsilon_{\mathrm{in}}$ for the so-called "linear" and "interaction" damage rule, respectively.

This model involves eight fitting parameters at one temperature, and its identification requires a lot of LCF tests. Pure pp cycles are identified as high-frequency tests. Low-frequency sawtooth cycles are partitioned into pp and cc components. The cp or pc component in hold-time test corresponds to the inelastic strain variation during the strain hold (i.e., the amount of stress relaxation divided by Young's modulus).

Application of SRP to thermal transients requires that the A_{ij} and α_{ij} coefficients in Equation (3) are temperature independent (Halford and Manson, 1976). However, a critical issue is the partitioning of inelastic strains in a TMF cycle into plasticity and creep components. The step-stress method was used in TMF tests in the laboratory: the stress is held constant at different points around the TMF cycle and steady-state creep rates are determined experimentally. Creep strain in tension and in compression is estimated by integrating these rates over a cycle period. Some success was reported for stainless steels under TMF as well as under biTF (Halford and Manson, 1976; Halford *et al.*, 1988). However, in a component, partitioning relies on constitutive equations and its accuracy is questionable. An inelastic strain analysis has to be used.

Problems arise in high-strength materials (superalloys) where for a given mechanical strain range, the inelastic strain range in TMF is significantly lower than in LCF at the maximum temperature in the cycle (Fischmeister *et al.*, 1986; Rémy *et al.*, 1988).

5.03.6.1.2 *Models using energy expenditure*

A main deficiency of models based on inelastic strain range is that no effect of mean stress is accounted for. For isothermal LCF at high temperature, often this is not a big problem, but under TMF condition phasing effects give rise to large mean stress effects. This has been emphasized by a number of authors under TMF loading where different phase lags between strain and temperature may result in very different mean stress levels, especially in high-strength alloys: in-phase loading often yielding compressive mean stress and out-of-phase loading yielding tensile mean stress; this effect is more important under thermal–mechanical creep fatigue with hold time at maximum temperature (Malpertu and Rémy, 1990; Rémy, 1990).

Ostergren has proposed to combine a hysteresis energy concept and frequency-modified Coffin approach. As a matter of fact, hysteresis energy concepts are especially attractive when one wants to use uniaxial tests in the laboratory and to extend these results to multiaxial situations in components. Previously Smith *et al.* (1970) have proposed a simple parameter for the same reason:

$$P_{\mathrm{SWT}} = \sqrt{(\sigma_{\mathrm{a}} + \bar{\sigma})\varepsilon_{\mathrm{a}} E} \tag{55}$$

where σ_{a} is stress amplitude, $\bar{\sigma}$ mean stress, ε_{a} alternate mechanical strain, and E Young's modulus. This parameter using mechanical strain range actually includes hysteresis energy as elastic energy in a basic form.

The variant proposed by Ostergren (1976) to account for mean stresses has been written in uniaxial form as

$$\sigma_{\max}\Delta\varepsilon_{\text{in}} = C_2(N_f\nu^{k-1})^{-\beta} \quad (56)$$

using only the tensile energy. Basically this form implicitly assumes that LCF damage is a microcrack growth and that microcracks are open only under tensile loading, which is a very crude approximation. This combination of Coffin frequency-modified law and energy concept is very simple and thus rather popular. Of course, it is only applicable, as any method based on inelastic strain, to the LCF range and it requires a nonlinear stress analysis for components using a constitutive model. This method is very simple, but it has only a limited physical basis since most hysteresis energy is dissipated as heat, and in most cases it appears merely as a correlating parameter.

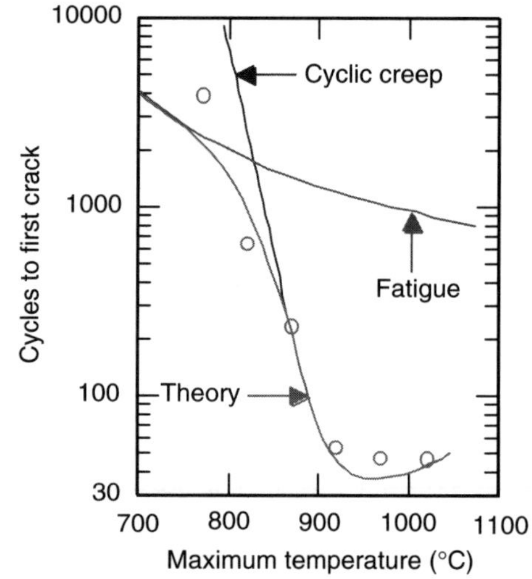

Figure 49 Number of cycles to first crack as a function of maximum temperature in TS experiment on Nimonic alloy: experiment and prediction from linear cumulation from creep and fatigue (after Spera, 1969).

5.03.6.2 Continuum-based Models: Accumulation of Creep and Fatigue Damage

5.03.6.2.1 Linear accumulation of creep and fatigue damage

Linear accumulation of fatigue and creep damage is the simplest way of handling creep-fatigue situations under isothermal conditions. Fatigue damage is described through the Miner (1945) rule, and creep damage is estimated by the equation of Robinson (1952) and Taira (1962). N_r is estimated as

$$\frac{1}{N_r} = \frac{1}{N_f} + \int_0^{\Delta t} \frac{dt}{t_r(\sigma, T)} \quad (57)$$

where N_f is the number of cycles in pure fatigue (high-frequency tests which are described by a Manson–Coffin equation using strain range or Basquin's equation using stress range), t_r is the time to rupture in creep experiments, and Δt is the period of the fatigue cycle. This equation can be changed into

$$\frac{1}{N_r} = \frac{1}{N_f} + \int_0^{\Delta t} \frac{\dot{\varepsilon}_s\, dt}{\varepsilon_r(\dot{\varepsilon}_s)} \quad (58)$$

where ε_r is the monotonic ductility for a secondary creep strain rate $\dot{\varepsilon}_s$ (Neate, 1985). This equation can be considered as a very simple case of summation of creep and fatigue damage increments used in the framework for continuum damage mechanics (see below).

Linear accumulation of creep and fatigue damage requires only pure fatigue and creep tests conducted to failure for Equation (47) and creep strain tests for Equation (48). Aging has been shown to give problems in isothermal LCF, but its application to TMF is simple and straightforward: fatigue damage is defined at an equivalent temperature and creep damage is integrated over the thermal cycle. Compression creep can be considered as damaging as tensile creep (Spera, 1969, 1973; Spera and Cox, 1976) or undamaging (Fischmeister *et al.*, 1986). Spera obtained good predictions for TF tests using Glenny's disks where creep damage dominates (Figure 49). However, in other instances this method can seriously overestimate fatigue life under thermal transients (Fischmeister *et al.*, 1986; Rémy *et al.*, 1988), sometimes by more than an order of magnitude.

5.03.6.2.2 Nonlinear accumulation of creep and fatigue damages

(i) Franklin's model

A fairly simple model has been proposed by Franklin based on the concept of forced creep (1978). The instantaneous inelastic strain rate in fatigue $\dot{\varepsilon}_p$, which often exceeds the secondary creep rate $\dot{\varepsilon}_s$ is supposed to enhance creep damage. The predicted life in a thermal strain cycle is given as

$$\frac{1}{N_r} = \int_0^{\Delta t} \frac{1}{t_r(\sigma, T)} \left[\dot{\varepsilon}_p(\sigma, T)/\dot{\varepsilon}_s(\sigma, T)\right]^{\nu} dt \quad (59)$$

where ν is a constant. This cyclic creep damage equation introduces a simple correction factor to the Robinson–Taira rule.

This model needs creep rupture tests, creep strain tests, and high-frequency fatigue tests to identify the single parameter v. When the secondary creep rate is not known from creep strain curves, the Monkman–Grant rule may be used.

Though much simpler than continuum damage theory, this model can bring significant improvement over the linear accumulation of creep and fatigue damage (Fischmeister *et al.*, 1986). Danzer and co-workers (Fischmeister *et al.*, 1986; Danzer and Bressers, 1986) used this model for many austenitic alloys and superalloys.

(ii) Majumdar and Maiya model

Extensive work has been reported for creep fatigue of austenitic stainless steels. Specimens tested under LCF conditions when subjected to continuous and compressive strain hold cycles fail by transgranular crack growth, with striations appearing on the fracture surface (Cheng and Diercks, 1973). Specimens tested under tensile strain hold cycles fail by intergranular cracking. This is attributed to the nucleation and growth of GB cavities.

Striation spacing measurements on fracture surfaces of specimens submitted to different loading sequences suggest that transgranular crack growth rate is increased by the growth of GB cavities. A compressive hold after a tensile hold suggests that shrinkage of cavities does occur (Majumdar and Maiya, 1980). These authors described damage under creep fatigue by the growth of crack length influenced by the average cavity size c:

$$\frac{da}{dt} = aB_0[1 + \alpha \ln(c/c_0)]|\varepsilon_p|^m|\dot{\varepsilon}_p|^k \qquad (60)$$

where B_0 has a different value for tensile and compressive stress, α is a constant, which is zero for $c \leq c_0$, the initial cavity size:

$$d(\ln c)/dt = G_0|\varepsilon_p|^m|\dot{\varepsilon}_p|^{k_c} \qquad (61)$$

where G_0 may have a different value for tensile and compressive stress, and m, k, and k_c are three constants.

These equations provided a phenomenological way to describe the effect of plastic strain rate on damage development. The model has been compared with crack growth measurements under creep-fatigue loading, as well as LCF life.

(iii) Continuum damage mechanics

An elegant description of nonlinear accumulation of creep and fatigue damage is provided by continuum damage mechanics. This theory was developed first by Rabotnov (1969), who introduced the concept of effective stress $\tilde{\sigma}$ which would produce the same strain in an undamaged element:

$$\tilde{\sigma} = \frac{\sigma}{1 - D} \qquad (62)$$

where damage D is such that $0 \leq D \leq 1$. This theory was used first for creep; the creep damage equation is

$$dD = (\sigma/A)^r(1 - D)^{-p}\, dt \qquad (63)$$

where A, r, p are three constants at a single temperature. In the original papers p was taken as identical to r, which results in linear accumulation of creep damage in creep tests with different stress levels. Integration of the equation results in a simple power relationship between stress and time to failure, and assuming that creep behavior can be described as being mostly tertiary creep. It is often necessary to describe the rate of creep damage accumulation with a "p" value different from r (Chaboche, 1982). Identification of parameters at a given temperature is achieved using creep rupture tests and some tertiary creep strain curves.

Damage is basically undefined and is considered to be spread over the volume of a stressed element. For creep, nevertheless, Leckie and Hayhurst (1977) have shown that similar forms of tertiary creep curves can be derived from a continuum damage framework and from physical metallurgy models which use a continuous nucleation of cavities and a growth process.

Most authors assume linear summation of creep (dD_c) and fatigue (dD_f) damage increments:

$$dD = dD_f(\Delta\sigma/2,\ \bar{\sigma},\ D) + dD_c(\sigma, D) \qquad (64)$$

$\Delta\sigma$, $\bar{\sigma}$, σ, and D refer to the stress range, mean stress, current stress, and damage, respectively. Fatigue and creep damage obey two different rate equations, but they are considered as additive, with no distinction between them. Nonlinear accumulation of creep and fatigue damage follows from Equation (54), since the same damage D is involved in creep and fatigue damage.

Lemaître and Plumtree (1979) proposed a fatigue damage rate, similar to creep damage one as

$$dD_f/dN = (\Delta\tilde{\sigma}/H)^h(1 - D)^{-q} \qquad (65)$$

where H, h, q are temperature-dependent material constants.

A more sophisticated fatigue damage rate was proposed by Chaboche (1974, 1982) and Chaboche and Lesne (1988) to account for nonlinear accumulation of damage in block loading sequences as

$$dD = \left[1 - (1 - D)^{\beta+1}\right]^{\alpha(\alpha_{\max} - \bar{\sigma})} \left[\frac{\sigma_{\max} - \bar{\sigma}}{M(\bar{\sigma})(1 - D)}\right]^{\beta} \quad (66)$$

where β is a coefficient and $M(\bar{\sigma})$ and $\alpha(\sigma_{\max}, \bar{\sigma})$ are two functions of mean and maximum stress:

$$\alpha(\sigma_{\max}, \bar{\sigma}) = 1 - a\left\langle \frac{\sigma_{\max} - \sigma_{\ell}(\bar{\sigma})}{\sigma_u - \sigma_{\max}} \right\rangle \quad (67)$$

where σ_{ℓ} is the fatigue limit for a mean stress $\bar{\sigma}$:

$$\sigma_{\ell}(\bar{\sigma}) = \sigma_{\ell 0} + (1 - b\sigma_{\ell 0})\bar{\sigma} \quad (68)$$

$\sigma_{\ell 0}$ being the fatigue limit at zero mean stress, σ_u static fracture strength.

$$M(\bar{\sigma}) = M_0(1 - b\bar{\sigma}) \quad (69)$$

M_0 and b are two constants.

α must not to be constant in order to account for nonlinear accumulation of damage in fatigue. It yields correct predictions of changes of stress level and amplitude in block sequential loading, but the effects are not so different from those predicted by Manson *et al.* (1967) for instance.

High-frequency $\Delta\sigma$–N_f curves are needed to represent the variation of plastic strain range with the number of cycles in stress-controlled tests. Nonlinear accumulation of creep and fatigue damage follows from Equation (9), since the same damage D is involved in both creep and fatigue damage rate equations.

The identification of this model is not so easy, since it involves a number of parameters in strongly nonlinear equations. However, it is especially appropriate for complex thermal–mechanical loading.

The identification of the fatigue damage rate equations requires tests to be carried out at fairly high frequency, under push–pull loading as well as under some mean stress. In addition, a few curves giving the variation of plastic strain amplitude under stress control or of stress amplitude under strain control are necessary.

In the case of thermal transients, application of these equations is more difficult. One can introduce a normalized stress $S = \sigma/\sigma_u(T)$ where $\sigma_u(T)$ is tensile fracture strength, which depends upon temperature. Such normalized stress can account for most of the effect of temperature on $\Delta\sigma$–N_f curves. An equivalent mean temperature can also be used as proposed by Taira (1973), cumulating linearly the temperature effects over the whole thermal–mechanical cycles.

Chaboche and co-workers (Chaboche, 1982; Chaboche and Lesne, 1988; Chaboche and Gallerneau, 2000) among others used this approach extensively to predict the life to engineering crack initiation in gas turbine materials. Good predictions were achieved for specimens or components. This model has proved to be very effective when coupled with inelastic strain analyses using constitutive equations based on viscoplasticity unified theory.

For multiaxial loading, the mean hydrostatic stress can be substituted for mean stress in uniaxial loading. The stress amplitude $\sigma_{\max} - \bar{\sigma}$ should be replaced by an equivalent stress amplitude A_{II}:

$$A_{II} = \frac{1}{2} \underset{t_o}{\mathrm{Max}}\, \underset{t}{\mathrm{Max}}\, J_2(\sigma(t) - \sigma(t_0)) \quad (70)$$

In the damage rate, the effective stress amplitude $(\sigma_{\max} - \bar{\sigma})/(1-D)$ should thus be replaced by $\tilde{A}_{II}$, deduced from the equation where $\sigma(t)$ and $\sigma(t_0)$ are replaced by the effective tensors. Under most circumstances an isotropic damage theory is used and D is a scalar.

Savalle and Cailletaud (1982) introduced a refinement with a partition of lifetime in initiation and propagation phases with two damage forms, one for initiation, D_i, and one for propagation, D_p. This was done to handle creep-fatigue situations and damage accumulation in sequential tests (Manson *et al.*, 1967). From tests on 304 steels (Maiya, 1981) it was inferred that N_i was mostly controlled by fatigue and frequency, say environment. For propagation they used the classical creep–fatigue interaction of Chaboche (1982). Within such framework continuum damage mechanics can be closely linked with strain-based crack growth models in plain specimens as shown by Cailletaud and Levaillant (1984). The continuum damage law for the propagation phase was found to be equivalent to the strain-based microcrack growth law in fatigue using a fairly complex expression for damage as a function of crack length. This refinement is especially useful for complex situations as for coated components.

5.03.6.3 Physically Based models

5.03.6.3.1 Creep–fatigue models using microcrack growth law

Okazaki and co-workers have attempted to describe creep–fatigue interaction in various steels using a microcrack growth model

(Okazaki and Koizumi, 1983, 1987; Okazaki *et al.*, 1983):

$$\frac{da}{dN} = \left(\frac{da}{dN}\right)_f + \left(\frac{da}{dN}\right)_c \tag{71}$$

Their analysis is based on the experimental determination of crack growth laws under fatigue and creep, and uses the concept of cyclic J to rationalize the influence of loading parameters on FCGR or total J, or creep modified J-integral to rationalize the rate under creep or creep fatigue. Details are reported in Section 5.03.8 on crack growth, where further information can be found on fracture mechanics concepts applicable on the growth of long cracks under creep-fatigue and thermal–mechanical loading. It is worth emphasizing that these authors were able to predict total endurances of various ferritic, martensitic, and austenitic steels under TMF and creep fatigue with this approach.

Renner *et al.* (1989) used a similar summation for crack-tip opening displacement components instead of crack growth rates. Each component, cycle dependent and time dependent, was related either to cyclic J or C^* expressions for short cracks. Crack growth rate was assumed to be a power law of crack-tip opening displacement. By integration of such an equation, the authors derived an enriched energy expenditure parameter, which takes into account creep-fatigue interaction when life is controlled by microcrack growth. This does not apply when life is controlled by creep damage coalescence. This model was applied successfully to TMF of austenitic stainless steel (Schmitt *et al.*, 2002).

5.03.6.3.2 Creep–fatigue interaction damage model

A good example of a creep fatigue physically based model is that provided by the work of Levaillant and co-workers (Levaillant and Pineau, 1982; Levaillant *et al.*, 1988) on stainless steel 316, using two batches of materials at 550°C and 600°C. Striation spacing measurements were used to estimate the crack growth rate increment per cycle in continuous fatigue as

$$da/dN = i_0 \quad \text{for } a_0 < a < a_c \tag{72}$$

$$da/dN = A(\Delta\varepsilon_p/2)^{\alpha} a^{\beta} \quad \text{for } a_c < a \tag{73}$$

with $\alpha = 1.5$, $\beta = 1.45$, $a_0 = 20\,\mu m$, i.e., grain size, a_c the transition length between the plateau regime and the crack-size-dependent regime, which is ~ 0.2 mm at 550°C, while this plateau is no longer observed at higher temperatures.

Levaillant has investigated the occurrence of wedge-type cracking in low-cycle creep fatigue. From an extensive investigation of hold-time conditions, he was able to identify conditions of creep–fatigue tests when creep–fatigue interaction occurs, i.e., when intergranular damage was enhanced at the tip of fatigue cracks. This creep–fatigue interaction occurs for medium hold times. For a very long hold time (1 yr test), however, no enhancement of intergranular damage was detected at the tip of cracks and intergranular damage spreads over the whole specimen gauge length. In that case damage was clearly creep dominated.

Based on extensive metallography work, intergranular damage is defined, like in the bulk of specimens, as the ratio of cracked grain boundaries to total GB length. It was actually found that the rate of damage accumulation $\Delta D_c/\Delta N$ is constant in this regime, as found previously by Hales (1980). The crack growth was found to agree with a coupled equation with creep damage as

$$\left(\frac{da}{dN}\right)_{cf} = \left(\frac{da}{dN}\right)_f \frac{1}{(1 - N_{cf}\Delta D_c/\Delta N)^2} \tag{74}$$

where subscripts "cf" and "f" refer to creep fatigue and pure fatigue; N_{cf} is the number of accumulated creep-fatigue cycles.

The intergranular damage rate $\Delta D_c/\Delta N$ was further found to be simply related through a power-law equation to the amount of stress relaxation occurring during hold time.

Alternatively, a damage rate equation identified under pure creep conditions overestimates the actual intergranular damage in creep fatigue but yields satisfactory results for hold times longer than 1 h.

5.03.6.3.3 Oxidation–fatigue model

Numerous materials are prone to strong oxidation–fatigue interactions like superalloys (McMahon and Coffin, 1970; Coffin, 1972; Rémy, 1993) under some operating conditions.

Environmental attacks on the surface by oxidation or corrosion trigger surface crack initiation at high temperature, as shown by metallographic observations and by the reduction of life to crack initiation with respect to that in vacuum (Reuchet and Rémy, 1983a; Réger and Rémy, 1988a, 1988b). Rémy and co-workers (Reuchet and Rémy, 1983c; Rémy *et al.*, 1988) assumed that the engineering life to crack initiation in cast superalloys tested at high temperature was mainly spent in oxidation-assisted microcrack growth. This damage process results from an advance due to crack opening under cyclic loading and from a supplementary crack advance due to oxidation at the crack tip. This results in a crack growth rate equation:

$$\frac{da}{dN} = \left(\frac{da}{dN}\right)_{f} + \left(\frac{da}{dN}\right)_{ox} \tag{75}$$

The form of this equation is similar to that of Equation (9) with a cycle-dependent and time-dependent damage increments. The fatigue damage rate was deduced from the Tomkins (1968) model:

$$\left(\frac{da}{dN}\right)_{f} = Ba \tag{76}$$

with

$$B = \Delta\varepsilon_{p}\left[\frac{1}{\cos[(\pi/2)\sigma_{max}/2\sigma_{u}]} - 1\right] \tag{77}$$

where a is the crack depth, σ_{max} the peak tensile stress, and σ_{u} the tensile fracture stress. The contribution of crack advance due to oxidation was supposed to result from the fracture oxide spike at every tensile stroke. The equations proposed by Reuchet and Rémy (1983a, 1983b, 1983c) for the fatigue–oxide cracking superposition model were identified from metallography of fatigued specimens and written as a function of inelastic strain for matrix oxide and stress for interdendritic oxidation kinetics. These equations were reworked for nickel-based superalloys single crystals in the bare and coated conditions (Chataigner *et al.*, 1996; Chataigner and Rémy, 1996) as a function of peak tensile stress. The oxidation contribution is written as

$$(da/dN)_{ox} = \Delta l_{ox} = \alpha(\sigma_{max})\Delta t^{1/4} \tag{78}$$

with $\alpha(\sigma_{max}) = \alpha_{0}(\sigma_{max}/\sigma_{0})^{n}$ when $\sigma_{max} > \sigma_{0}$ and with $n = 4.5$ for 950°C and 1,100°C.

The application to thermal transients needs to define an equivalent temperature like that of peak maximum stress. An average oxidation constant α_{av} has to be substituted for isothermal α by integration of oxidation kinetic equation over any TMF cycle.

No TMF information is used to identify the coefficients but, as noted previously, this model is restricted to fairly large strain ranges and fairly low frequencies, due to the use of inelastic strain range in the fatigue contribution. Under these restrictions, such a model is actually predictive and can describe life to any crack depth (Chataigner and Rémy, 1996). High-frequency conditions and small strain ranges, which yield no significant inelastic strains, cannot be described.

At a given temperature, this model only requires the tensile strength for the fatigue term (which can be checked using high-frequency LCF tests), localized oxidation kinetics measurements using metallographic observations on specimens oxidized in a furnace (Reuchet and Rémy, 1983c) and LCF specimens tested at low fequency. However, this model can be identified simply from some curves of crack depth vs. the number of cycles for low-frequency LCF tests. This method was used, for instance, in anisotropic superalloys single crystals with a single [0 0 1] orientation (Chataigner and Rémy, 1996).

5.03.6.3.4 Oxidation–embrittlement fatigue damage model

The oxidation–fatigue model as described by Equations (65)–(68) can handle a variety of situations including TF or TMF, but it cannot predict oxidation embrittlement ahead of the oxide spike at the crack tip. Such a case can occur, however, as demonstrated by fracture mechanics experiments.

CT-type specimens can be precracked at a given temperature, oxidized at high temperature, and then cracked at the test temperature. A crack that grows under TF usually experiences oxidation at the crack tip during exposure to high temperature and then tensile loading at medium temperatures. This situation can be simulated in typical bithermal fracture mechanics tests, using, for instance, a CT specimen, which has been precracked by fatigue at low temperature, oxidizing it at high temperature for some time, and then growing the crack again at the test temperature. This procedure has been applied in a large number of superalloys, conventionally cast cobalt-based MAR-M509 and nickel-based IN100 with a moderate carbon content, wrought, or single-crystal nickel alloys with a low carbon content (Rézaï-Aria *et al.*, 1988a, 1988b; Rémy *et al.*, 1993; Chataigner *et al.*, 1996; Köster *et al.*, 2002). Figure 50 shows a typical FCGR curve at 700°C in precracked and oxidized (for 300 h at 950°C) CT specimen of single-crystal superalloy CMSX4 with that of a virgin specimen. A large increase in FCGR is observed, with a sawtooth variation, ahead of the crack tip, until the crack growth rate recovers the behavior of the virgin specimen. This behavior has been attributed to a local decrease of toughness of the alloy over some distance ahead of the oxidized crack tip, ~0.3 mm in the case illustrated. Whenever the depth of interdendritic or intergranular oxide spikes has been measured, the distance where the material has been embrittled is larger than the oxidized depth (up to a factor of 10 times or

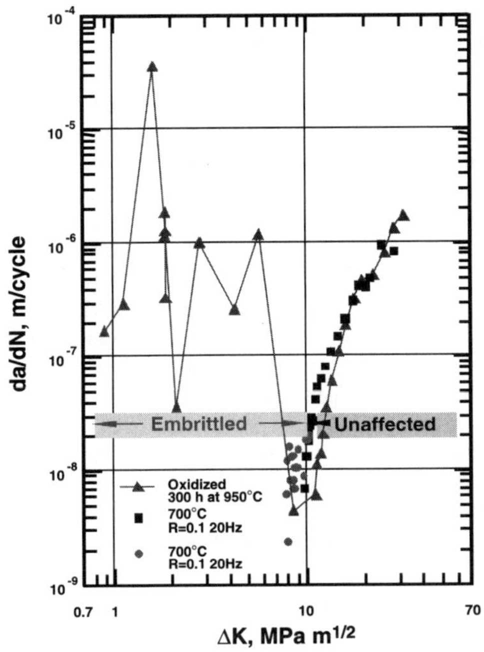

Figure 50 Comparison of crack growth rate data $da/dN = f(\Delta K)$ between nonoxidized and preoxidized CT specimens of CMSX4 superalloy (after Koster *et al.*, 2002).

more in cast alloys with MC carbides). In cast alloys, oxygen diffusion ahead of the crack was shown to be responsible for this behavior (Rémy, 1993; Rézaï-Aria and Rémy, 1989; Rémy *et al.*, 1993) and can be accounted for by a diffusion model. It is worth emphasizing that crack growth can be observed below the threshold of virgin material.

These experiments have been analyzed by assuming a discontinuous crack growth process with the existence of a process zone of a microstructure size λ ahead of the crack, as introduced by Mc Clintock (1963). Also, the crack growth is treated as repeated nucleation events, with $N(\lambda)$ the number of cycles to break a microstructure element. In this element, the crack growth rate is constant and is given by

$$da/dN = \lambda / N(\lambda) \tag{79}$$

where λ is the microstructure size and $N(\lambda)$ is the number of cycles to break a microstructure element. Using the stress–strain singularity at the crack tip, fracture of the microstructure element results from damage accumulation by fatigue. For cast alloys a coupled equation was used with such local approach to fracture as

$$dD/dN = (\Delta\sigma_{eq}/2S_0)^M / [(1-R)(\sigma_c - \sigma_{yy})/S_0]^{\alpha} \tag{80}$$

with $\Delta\sigma_{eq}$ the equivalent von Mises stress range, σ_{yy} the peak normal stress, R the load ratio $\langle \sigma_{yy,\min}/\sigma_{yy,\max} \rangle$ (with Mc Caulay brackets), S_0, M, α the constants, and σ_c the critical stress:

$$\sigma_c = \mathrm{Min}(\sigma_{c,v};\ \sigma_{c,ox}(t, T)) \tag{81a}$$

with

$$\sigma_{c,ox} = \sigma_{co}(T) f(\lambda / l_{ox}) \tag{81b}$$

where σ_{cv} is the critical stress to fracture for virgin specimens, $\sigma_{c,ox}$ the critical stress in the region embrittled by oxidation, σ_{co} a temperature-dependent constant, and f a function which describes the variation of critical stress to fracture with the distance ahead of the tip. l_{ox} is the depth of localized oxide (interdendritic) which obeys a kinetic equation as Equation (31). Integration of Equation (25) up to a damage unity gives the value $N(\lambda)$ and crack growth rate.

The identification of the model needs FCGR experiments on virgin CT (or other type) fracture mechanics specimens for two load ratios and push–pull LCF tests at high frequency. Coefficients of Equation (25) are deduced from FE solutions of the stress singularity. FCGR experiments using oxidized specimens without load are used to fit Equation (27).

The identification of this model is fairly difficult, since it involves fracture mechanics tests. Its applications (LCF, TMF, and TS experiments) have been quite successful for cast materials (Rézaï-Aria and Rémy, 1989; Rémy, 1993; Rémy *et al.*, 1993).

One limitation of this model is that it is fairly sensitive to the stress–strain singularity in fracture mechanics specimen. Further when applied to specimens, it describes only the regime where the growth rate of microcracks is fairly insensitive to crack length. In addition, if localized oxidation is very detrimental in alloys with a fairly high carbon content, alloys with a low carbon are less sensitive and creep damage often can no longer be neglected. For these alloys, creep–oxidation–embrittlement models have been derived.

5.03.6.3.5 Creep–oxidation–fatigue models

Models using total endurances. Neu and Sehitoglu (1989b) investigated a pearlitic steel and have used a linear summation of damage to account for the damage in isothermal LCF and in TMF, which has been used later for other materials, as

$$\frac{1}{N_r} = \frac{1}{N_f} + \frac{1}{N_{ox}} + \frac{1}{N_c} \tag{82}$$

These authors also investigated the evolution of crack under various loading cycles: in-phase, out-of-phase, and isothermal at maximum temperature (Figures 51 and 52). However, they use a total endurance formulation close to the usual engineering practice accumulated with Miner's rule.

N_f is the usual pure fatigue life, which is deduced from tests at room temperature and given as a function of mechanical strain range.

The summation of a pure fatigue term and an oxidation term is substantiated by metallographic observations (Neu and Sehitoglu, 1989a). As evidenced earlier in Cr–Mo steels (Skelton and Bucklow, 1978), they observed a layered structure in the oxide under some fatigue conditions and oxide intrusions (see Section 5.03.4) and proposed an oxide-cracking crack growth mechanism.

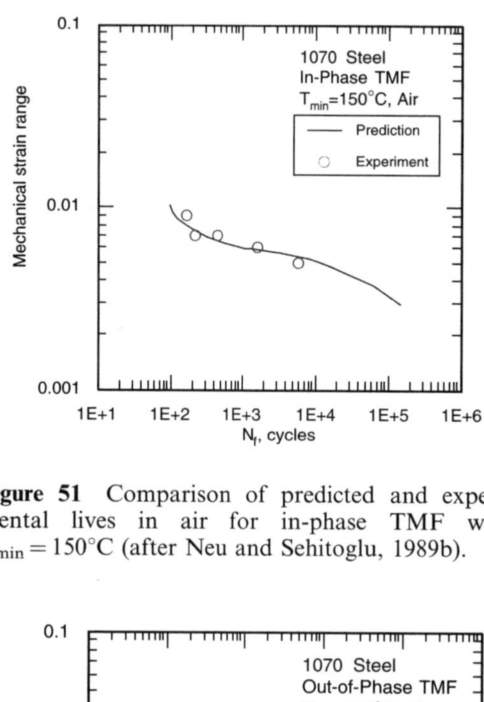

Figure 51 Comparison of predicted and experimental lives in air for in-phase TMF with $T_{min} = 150°C$ (after Neu and Sehitoglu, 1989b).

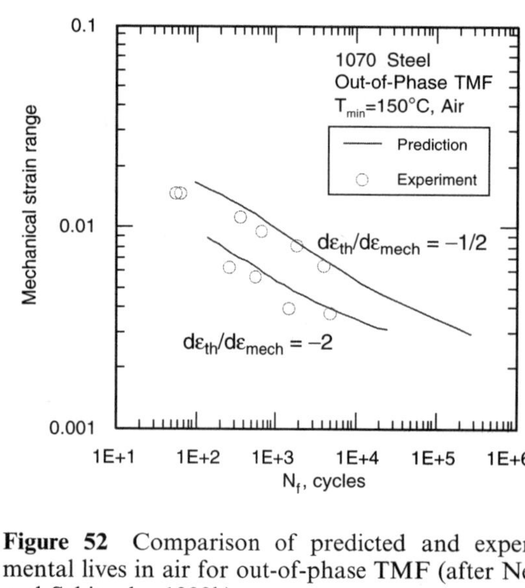

Figure 52 Comparison of predicted and experimental lives in air for out-of-phase TMF (after Neu and Sehitoglu, 1989b).

N_{ox}, the life to oxidation of surface layers, is described by a complex equation, which is based on a critical thickness to fracture, which decreases as the square of mechanical strain range (see Section 5.03.4). Parameters are identified from measurements of oxide layers at the crack tip, and connected directly with surface mechanical strain range. This contribution incorporates a phasing factor to account for the fact that strain–oxidation coupling is increased when going from in-phase TMF to LCF at T_{max}, and further on to out-of-phase TMF.

N_c, the life due to creep, is described by another equation. Coefficients are not identified from creep experiments but they are deduced from LCF data in inert environment at different strain rates. As for oxidation however, a phasing factor has to be introduced when integrating the creep damage over the TMF cycle. This is to be consistent with the experimental observation that intergranular creep-type damage occurs in the bulk especially under in-phase TMF cycling.

This model combines a fairly large number of parameters. They are identified not only from various LCF tests in air, under inert environment and metallography (for the oxidation kinetics equation) but also from TMF tests, in-phase and out-of-phase. A good description of TMF tests is achieved as shown in Figures 49 and 51. This model was used for steels and also for superalloys (Neu and Sehitoglu, 1989b; Sehitoglu and Boismier, 1990a, 1990b). Further improvement was made to describe the initation stage at oxide spikes, including application to coated components (Kadioglu and Sehitoglu, 1993): this was done by Eshelby using inclusion theory to account for strain concentration in the alloy in the vicinity of the oxide spikes.

Models based on integrated microcrack growth. Okazaki and Koizumi (1983, 1987) have attempted to describe creep–fatigue interaction in steels using a microcrack growth model. Miller *et al.* (1992) generalized this summation to three components for fatigue, oxidation, and creep, respectively:

$$\frac{da}{dN} = \left(\frac{da}{dN}\right)_f + \left(\frac{da}{dN}\right)_{ox} + \left(\frac{da}{dN}\right)_c \quad (83)$$

The pure fatigue contribution is deduced from a cyclic J argument:

$$\left(\frac{da}{dN}\right)_f = C_f(\Delta W a)^{m_f} \quad (84)$$

with

$$\Delta W = \frac{1}{2}\Delta\sigma\Delta\varepsilon_e + \frac{1}{2\pi}f(n')\Delta\sigma\Delta\varepsilon_p$$

C_f, m_f are two constants, and n' is the exponent of the cyclic stress–strain law.

The creep-crack growth rate is deduced from a C^* analogy (Landes and Begley, 1976; Riedel and Rice, 1980) and is written as

$$\left(\frac{da}{dN}\right)_c = C_c\left\langle\frac{1}{t_1}\int_0^{t_1}\sigma\,\dot{\varepsilon}_c\,dt - \frac{1}{t_2}\int_0^{t_2}\sigma\,\dot{\varepsilon}_c\,dt\right\rangle a \quad (85)$$

where t_1 and t_2 are hold times in tension and in compression, respectively, $\dot{\varepsilon}_c$ is the creep rate, and C_c is a constant; this is made to distinguish the effect of tension-going and compression-going loading paths.

The oxidation-crack growth rate is assumed to depend on crack length and strain energy density. It depends on temperature through the Arrhenius equation, and is a power law of cycle period Δt. It is written, for small cracks, as

$$\left(\frac{da}{dN}\right)_{ox} = C_{ox}\exp\left(-\frac{Q_{ox} - B\sigma_{eff}}{RT_{eff}}\right)(\Delta Wa)^{m_{ox}}\Delta t^{p} \quad (86)$$

where C_{ox}, m_{ox}, Q_{ox}, B, k, p are constants ($R = 8{,}315$ J), and σ_{eff} is an effective stress. Though $k = 2$ was first suggested from an energy argument, it turns out that $k = 1$ gives a best fit for oxidation contribution for MAR-M247 superalloy, which corresponds to a classical activation energy in creep, and might be related to some oxidation–creep strain interaction as suggested by metallographic investigations on other alloys and $p = 0.25$ was used (Chateau and Rémy, 2001).

This model uses LCF data at medium temperatures to get fatigue parameters. In-phase TMF tests are used to estimate the coefficients of creep equation. LCF tests at two strain rates at high temperature give part of the oxidation coefficients. The remaining coefficients B and k are deduced from out-of-phase TMF. As in the Neu and Sehitoglu (1989b) model, TMF tests are actually used to identify coefficients, which necessitates a fairly expensive database to have all the coefficients of the model. Real prediction was shown only for counterclockwise diamond tests, which is more of an interpolation than true extrapolation of the model.

This approach, which has been used by Maier *et al.* (2000, 2002) on titanium alloys, actually gives satisfactory results.

5.03.6.3.6 *Creep–oxidation–embrittlement fatigue damage model*

Damage is identified as the growth of a major microcrack, which is described as failure of microstructure elements of size λ, using a local stress approach. Computations of the stress–strain singularity in single-crystal superalloy have shown that in the zone embrittled by oxidation, the critical stress is now essentially constant. A constant critical stress was assumed to hold in the area embrittled by oxidation. To account for the influence of crack length, a continuum damage mechanics type framework was used where damage was assumed to be the proportion of cracked surface to specimen section.

A first version was written for polycrystalline material susceptible to intergranular oxidation embrittlement (Köster and Rémy, 1999) and for single-crystal material susceptible to interdendritic oxidation (Köster *et al.*, 2002). For the sake of brevity, only the case of the single crystal is outlined.

In the model, the interaction between oxidation and creep-fatigue damage is introduced by assigning a lower critical strength in the area embrittled by localized oxidation. This procedure was previously described for polycrystalline superalloys, in cast or wrought forms. In other words, it is considered that an oxide spike of length l_{ox} induces an embrittled zone of depth l_f which exhibits lower mechanical properties than a nonoxidized material. Embrittled zone is considered to be greater in size than the oxidized zone.

A crack growth rate increment by fatigue is considered to be the maximum value of discontinuous crack growth regime governed by Basquin's law and a continuous crack growth regime governed by Tomkins' model:

$$da_{fatigue} = \max\langle da_{Basquin} : da_{Tomkins}\rangle \quad (87)$$

Basquin's relation is considered to cater to stress-controlled crack propagation regime. However, Basquin's law is written in terms of resolved shear stress on cubic and octahedral slip systems:

$$\frac{\Delta\tilde{\tau}}{2}N^{b} = \tau_c \quad (88)$$

where $\Delta\tilde{\tau}$ is the effective shear stress range defined as follows:

$$\Delta\tilde{\tau} = \frac{\Delta\tau}{1 - D} \quad (89)$$

where D is the damage resulting from crack growth $D = f(a/w)$. Tomkins' model is adapted for single-crystal case, for crack growth under extended plasticity conditions, and is written in a compact form close to polycrystals (in fact this is a summation along all slip systems) as

$$\frac{da}{dN} = \Delta\varepsilon_p\left(\frac{1}{\cos(\pi/2)(\Delta\tilde{\tau}/2\tau_c)} - 1\right)a \quad (90)$$

where τ_c is the critical strength of the microstructural element λ, $\Delta\varepsilon_p$ is the plastic deformation range, and $\Delta\tilde{\tau}$ is the effective shear stress range defined above.

Interaction with creep is introduced in the effective shear stress as

$$\Delta\tilde{\tau} = \frac{\Delta\tau}{(1 - D_c)(1 - f(a/w))} \quad (91)$$

The creep damage is estimated using the classical Rabotnov (1969) law. This formulation is defined for isothermal loading conditions and is easily adapted for thermomechanical loading and thermal transients.

The depth of intergranular oxide l_{ox} is given by the rate equation, as shown previously for interdendritic oxide in cast alloys (Reuchet and Rémy, 1983b):

$$d(l_{ox}^4) = \alpha^4(T)\,dt \quad (92)$$

with

$$\alpha_{ox,o}(T) = \alpha_{ox,o} \exp(-Q/RT) f(\Delta\varepsilon_{in}) \quad (93)$$

where α_0 is a constant, Q is an activation energy, $R = 8.315$ J, T is temperature (in K), and f is a function which describes the interaction between oxidation kinetics and straining (see Equation (40)). α_0 and Q are deduced from metallography of specimen sections, which are oxidized in a furnace under no load, and f is deduced from the observation of strained specimens (if necessary, low-frequency isothermal LCF tests can be used to check parameters). The application to thermal transients requires the integration of Equation (93) over the whole thermal–mechanical cycle.

This model accounts for LCF lifetime under a large variety of testing conditions (Köster, 1997). It has been observed to give a very good prediction of TMF life for a temperature cycle between 100°C and 750°C.

Competition between damage mechanisms from the surface or from internal defects. In some cases it becomes necessary to consider the competition between surface cracks and internal cracking, in creep-fatigue situations, and even in fatigue-dominated situations. Casting pores are known to be important under isothermal fatigue in cast irons (Clément *et al.*, 1984) and in cast superalloys (Gell *et al.*, 1970); while fatigue cracking is usually associated with surface cracks, they can favor crack initiation at internal defects, especially under low strain ranges. The importance of casting pores has been emphasized in investigations of creep fatigue and TMF of advanced single-crystal superalloys used for stationery gas turbines (Bullough *et al.*, 1998) or aeroengines (Guedou and Honnorat, 1993). The importance of internal vs. surface defects has been shown to vary according to creep-fatigue and TMF conditions. This problem has been considered in depth for ceramic inclusions: probabilistic models have been proposed using a description of the growth of microcracks nucleated from inclusions based on linear elastic fracture mechanics (Lautridou *et al.*, 1993; Brückner-Foit *et al.*, 1993), eventually including crack closure, short crack, and environmental effects (Grison and Rémy, 1997).

High-strength alloys are rather sensitive to such defects in the HCF and LCF regimes, and in TMF.

A model has been proposed to handle creep–oxidation–fatigue situations for single-crystal superalloys accounting for the presence of a distribution of micropores (Köster *et al.*, 2002). This model considers damage as the growth of microcracks originating from casting pores. The size distribution of these pores is introduced in the model in a semi-probabilistic way and it is based on tedious microscopic observations (Probst-Hein and Eggeler, 2001). The whole distribution of micropores in a test specimen (or in a volume element of a structure) is described approximately by only three different pore sizes in the same section of the test specimen (Figure 53): medium-sized pores of initial radius a_{mean0} present uniformly everywhere in the material as one such pore per characteristic volume L^3; the largest-size pore of initial radius a_{max0} in a given volume; and an intermediate-size pore of radius a_{surf0} in the subsurface zone of the test specimen. This pore is slightly larger in volume than the medium-sized pore and is responsible for the initiation of only one surface crack.

A major surface crack, a, results from the failure of a microstructural element λ (Figure 53). The initiation time for this surface crack depends upon the probability of the presence of a pore near the surface. Its propagation is assisted by oxidation, which is introduced as the weakening of material due to embrittlement. While calculating damage of a microstructural element λ, the volume effect on its critical strength τ_c is introduced with the help of a Weibull-type defect distribution. Besides surface crack, other internal cracks propagate under creep-fatigue condition without any interaction with ambient environment. The global damage results from the propagation of all internal as well surface cracks and can be decomposed into fatigue and creep damages. The global damage variable D is calculated on the basis of crack areas nucleating at the different pores, using crack propagation laws under fatigue and creep loading, which have been described above.

This model has been applied to two single-crystal superalloys, AM1 and CMSX4. This model is able to predict not only life but also the phasing effects, with out-of-phase loading dominated by the propagation of oxidized surface cracks and in-phase loading giving more importance to internal damage. Comparison between model and TMF experiments is within a factor of 2 in most cases, as shown in Figure 54, despite the variety of

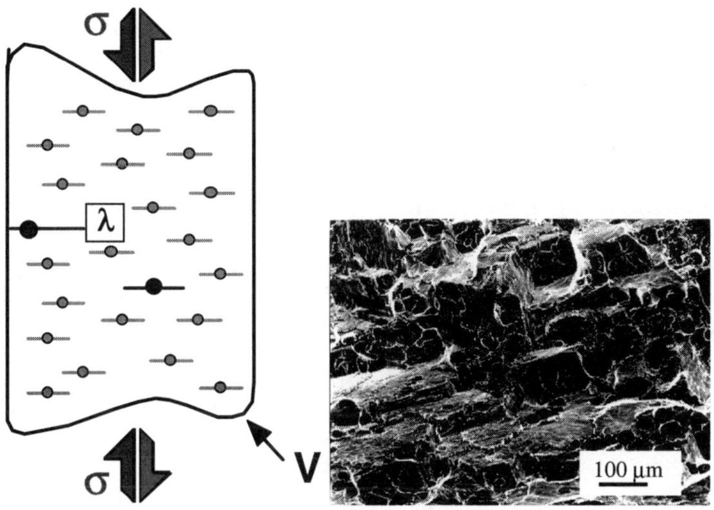

Figure 53 Schematic diagram of the physical model showing a volume element with internal and surface pores (left). Micrograph of the fracture surface of a specimen, tested in LCF with tension dwell at a constant 950°C temperature (right) (after Koster *et al.*, 2002).

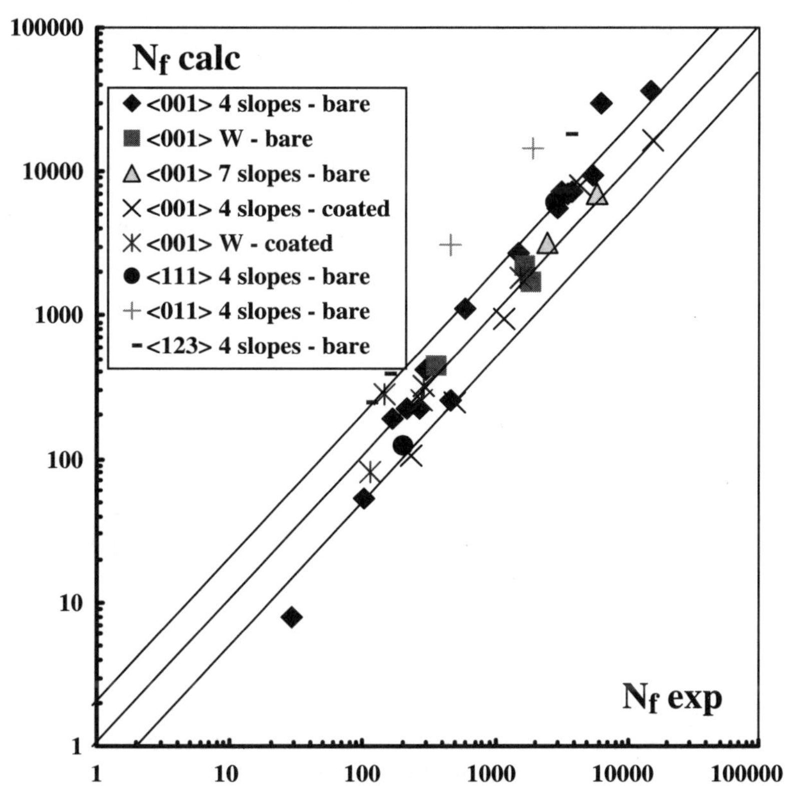

Figure 54 Comparison between model and experiment: TMF lifetime for various cycles and orientations of AM1 superalloy single crystals (courtesy of Alam).

crystallographic orientations of the loading axis with respect to crystal axes, and of TMF cycle shapes (Rémy and Guedou, 2002). Prediction of surface crack growth compares fairly well with measurements under TMF (Figure 55).

5.03.7 APPLICATION TO CRACK INITIATION

5.03.7.1 Energy Expenditure and Nonlinear Viscoplastic Stress Analysis of Automotive Components

Analyses of automotive components using energy expenditure were shown by Bignonnet and Charkaluk (2002) and Verger *et al.* (2002).

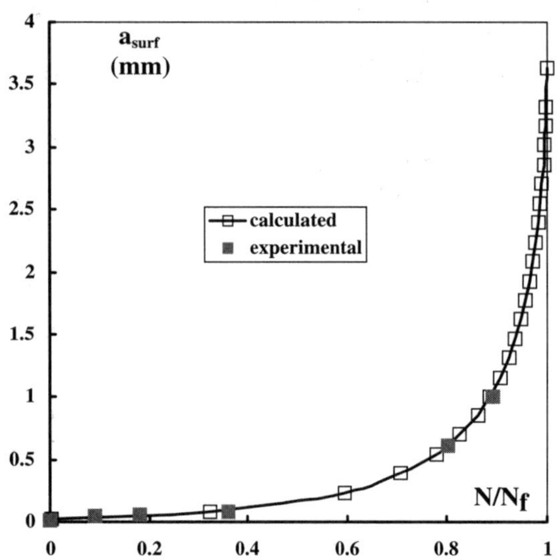

Figure 55 Comparison between model and experiment: surface crack growth curves of a diamond type 600–1,100°C TMF test on ⟨0 0 1⟩-orientated AM1 single-crystal superalloy (courtesy of Alam).

A first example is an exhaust manifold of a 2.1 l turbocharged diesel engine, that is made of 4Si–0.8Mo cast iron. A standard durability cycle is used at full load and light load at 4,000 rpm with 30 s transients and total period 700 s.

The temperature field distribution has been computed on the manifold from engine data, exhaust gas temperature, and mass flow rate in two steps to get the transient response. A unified Chaboche model with five temperature-dependent parameters was used, which was identified from isothermal cyclic tension–relaxation–recovery tests in the temperature range 200–700°C.

The manifold was modeled using 1.5×10^4 3D brick elements, and running the thermal–mechanical analysis after completion of the transient thermal response, which is completed within 6 h of CPU time.

Charkaluk and Constantinescu (2000) have shown that classical models using plastic strain range or including the maximal stress in the cycle (Ostergren, 1976; Smith *et al.*, 1970) are rather inadequate for LCF prediction under nonisothermal multiaxial conditions. The dissipated energy per cycle was chosen as the damage indicator to be used for crack initiation in this context. The fatigue criterion is

$$\Delta W_{\rm in}\, N_{\rm f}^{\beta} = C \qquad (94)$$

where β and C are two constants and the dissipated energy $\Delta W_{\rm in}$ is calculated over the stabilized cycle.

A typical nonisothermal stress-mechanical strain-loading path is shown in Figure 56 at critical areas of the manifold, as depicted in the inset. Figure 57 compares results obtained on specimens tested under isothermal LCF, and TMF, as well as those obtained on exhaust manifolds. Reasonably good predictions of

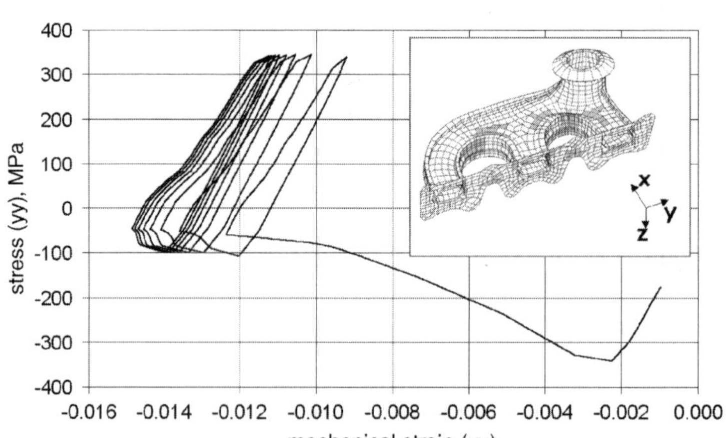

Figure 56 Thermomechanical calculation and map of dissipated energy in a cast iron manifold (source Bignonnet and Charkaluk, 2002).

crack initiation life in components can be deduced from the whole chain of thermal–mechanical analysis and postprocessor for TMF life computation.

Aluminum alloy cylinder heads were analyzed by Nicouleau *et al.* (2002) and Verger *et al.* (2002). The latter authors used the same methodology as defined above. Alloy A356 (7%Si–0.3%Mg–Al) was used and subjected, in critical areas, to temperature cycles ranging from 20°C to 260°C. A transient thermal computation was made in the same way. The problem raises two main extra difficulties. The first one is that cylinder heads have a very complex geometry: a numérical model of 8×10^5 degrees of freedom was actually used. A second difficulty is that, above 150°C, considerable overaging occurred, which can result in a 70% loss of yield strength and hardness at the higher temperatures. A maximum temperature field was computed, from

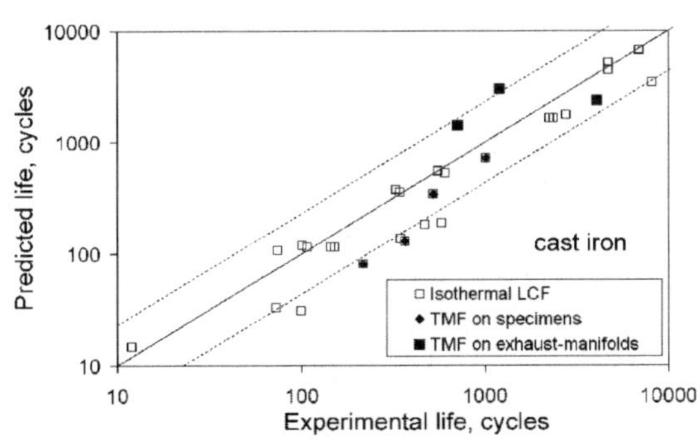

Figure 57 Predicted vs. experimental fatigue life for cast iron exhaust manifold (source Bignonnet and Charkaluk, 2002).

which a map of hardness distribution after 10^4 cycles can be computed (Figure 58).

Due to the size of the FE model, a submodel, where viscoplasticity was significant, was extracted from the total model after a global elastic computation. The area where aging was important was then delineated and a viscoplastic model identified in the fully aged conditions was used to run the thermal–mechanical analysis.

The energy fatigue criterion was identified on isothermal LCF tests, which were aged before testing at different temperatures between 20°C and 300°C. The lifetime can be predicted within a factor of 2.5 on specimens and slightly better on cylinder head tests (Figure 59). The distribution of critical energy yields the location of crack initiation at the intervalve bridge, in good agreement with the observed location (Figure 60).

A fairly similar but embellished approach was shown by Schmitt *et al.* (2002) for an austenitic steel thin-walled exhaust manifold, where severe thermal cycles, between ambient temperature and ~880°C, require new designs of the exhaust system with the catalyst being located as close as possible to the engine for better pollution control. The growing number of model variants makes the developments of suitable simulation models very attractive to reduce the number of necessary component tests, and thus development times and costs.

Schmitt *et al.* (2002) used a unified viscoplastic model, which was implemented into the FE code Abaqus, as a user-defined material routine. They assumed that the damage under TMF is controlled by the formation and growth of microcracks and used equations

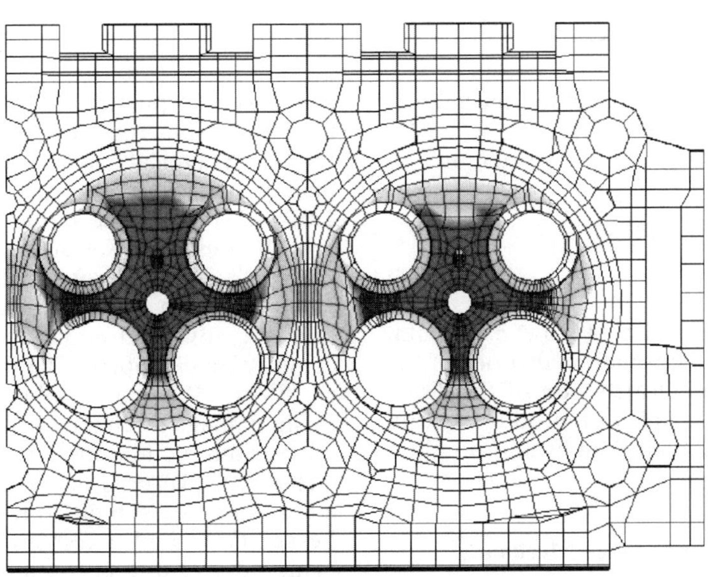

Figure 58 Hardness distribution after 10^4 cycles computed on a cylinder head in cast A356 (AlSiMg) alloy submitted to thermomechanical cycles (source Verger *et al.*, 2002).

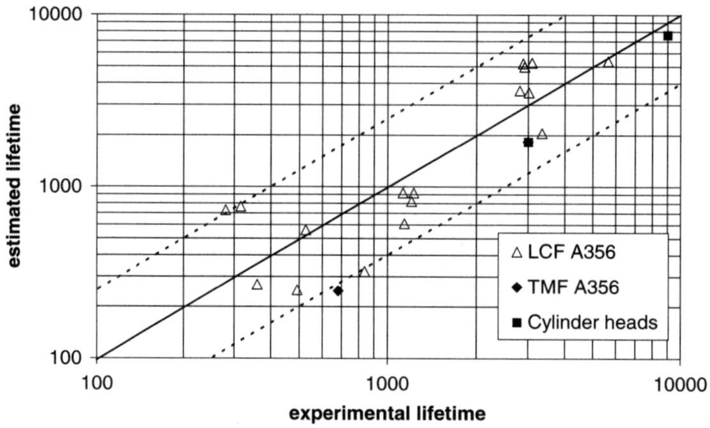

Figure 59 Comparison of the estimated life and experimental life for A356 alloy: LCF and TMF tests and several structures (source Verger *et al.*, 2002).

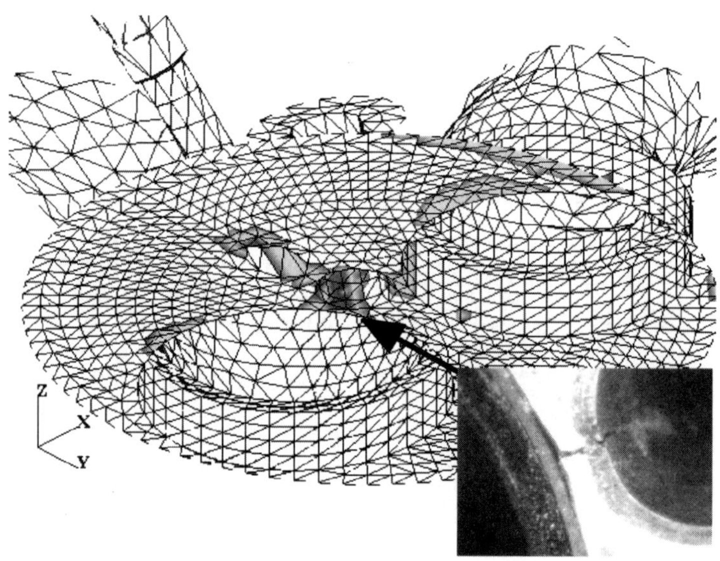

Figure 60 Crack initiation on a cylinder head prototype and distribution of the fatigue criterion on the FE model (source Verger *et al.*, 2002).

based on elastic–plastic fracture mechanics. The damage parameter D_{TMF} is a creep-fatigue-type damage parameter written as

$$D_{\mathrm{TMF}} = \left(1.45\,\frac{\Delta\sigma_{\mathrm{Ieff}}^2}{E\sigma_{\mathrm{cy}}} + \frac{2.4}{\sqrt{1+3N}}\,\frac{\Delta\sigma_{\mathrm{I}}^2}{\sigma_{\mathrm{cy}}\Delta\sigma_{\mathrm{eq}}}\,\Delta\varepsilon_{\mathrm{in,eq}}\right)F \qquad (95)$$

A good fit is achieved between experimental TMF stress–strain loops and predictions of the constitutive models. Distribution of the TMF damage parameter is able to reproduce, within good accuracy, the number of cycles to failure of engine tests.

5.03.7.2 Continuum Damage Mechanics

The continuum damage mechanics approach has widely been used by Chaboche and co-workers to predict life to engineering crack initiation. Numerous results were published on IN100, a typical alloy for turbine blades, mainly on laboratory specimens (Chaboche, 1982; Chaboch and Lesne, 1988). This approach is used by Société nationale d'étude et de construction de moteurs d'aviation (SNECMA) in a postprocessor of a computation code used for advanced cooled blades, vanes, and combustion chambers (Dambrine and Mascarell, 1990). Predictions obtained using continuum damage mechanics are shown in Table 3 for various conditions on superalloy IN100 including TMF tests on specimens, TF experiments on plates submitted to TS (Policella and Lesne, 1986) and a test on an actual cooled blade submitted to both centrifugal load and a thermal cycle (see Rémy and Skelton, 1992). Crack location in components seems to be

Table 3 Examples of life prediction for specimens and components in IN100 using continuous damage mechanics.

Test conditions	*Experimental life*	*Predicted life*	*Reference*
TMF 600–1,050°C (S) diamond shape cycle	300*	520	1
TMF 600–1,050°C (S) In phase cycle	200*	370	1
TF 550–1,030°C (C)	4,000	1,860	2
TF 550–1,050°C (C)	700	1,180	2
Cooled blade (C) $T_{max} \simeq 1{,}050$°C	210 + At trailing edge	490 At trailing edge	1

Ref. 1: Dambrine and Mascarell, 1990, Ref. 2: Policella and Lesne, 1986.
(S): specimen: stresses are known from experiment
(C): component: stresses are computed from a viscoplastic stress analysis.
*Life to 0.5 mm crack depth + life to a few mm crack at trailing edge.

fairly well predicted. It is worth emphasizing that simple models based on linear accumulation of creep and fatigue damage fail to predict actual life of components. Only oxidation–fatigue models (see below) seem to provide better life predictions. Application of continuum damage mechanics to the case of coated materials is presented later.

5.03.7.3 TS of Superalloys: A Worked Example with Creep–Oxidation–Fatigue Damage Models

The prediction of TF life in a real component is much more difficult than for isothermal LCF on laboratory specimens. The difficulty in real components, as in TS experiments, is that stress and strain are unknown and have therefore to be computed. The prediction of life depends as much upon the stress and strain analysis as upon the life prediction model that is used. But the major issue before entering mechanics is the computation of temperature, which is the key factor in thermal stresses.

The scheme of analysis is illustrated for TS experiments on Waspaloy (Köster *et al.*, 1996a, 1996b) where results of thermal and stress analysis are briefly presented, and early crack growth is predicted. This example is completed by previous results shown on cobalt-based superalloy MAR-M509 (Rémy and Skelton, 1992).

The thermal analysis is made to compute the temperature in the solid, which is of interest when considering components submitted to thermal transients. The unsteady heat conduction equation has thus to be solved:

$$\rho C_p \frac{\partial T}{\partial t} = -\nabla q \tag{96}$$

where T is temperature, ρ is density (mass per unit volume), C_p is specific heat at constant pressure, q is heat flux (a vector), and ∇ is the gradient operator. The heat flux is given by Fourier's equation, and Equation (96) then reads

$$\rho C_p \frac{\partial T}{\partial t} = \nabla(k \nabla T) \tag{97}$$

with k the thermal conductivity. If k is independent of temperature and position, which is applicable to a homogeneous material, this equation simplifies into the well-known form

$$\frac{\partial T}{\partial t} = a \Delta T \tag{98}$$

with thermal diffusivity $a = k/\rho C_p$. In most cases, however, ρ, C_p, and k are temperature dependent and their dependence must be accounted for.

A major problem in engineering structures is to evaluate initial and boundary conditions using fluid thermodynamic computations. Substantial efforts are often necessary to achieve reasonable accuracy under transient conditions. Such an analysis must be carried out for TS experiments as for actual components. Temperatures are usually measured at the surfaces and heat transfer coefficients h in the expression of heat flux at the surface:

$$Q = h(T - T_{air}) \tag{99}$$

considering convection, where T_{air} refers to the temperature of the external environment.

Köster *et al.* (1996a, 1996b) made such heat transfer analysis for TS tests using wedge-type specimens and a lamp-furnace facility. Examples of surface temperature variation as determined from 14 thermocouples along the midsection of a dummy specimen are shown in Figure 56, for a stabilized thermal cycle. The transient temperature map of the structures was obtained using a 2D analysis with the FE code Zebulon (Burlet and Cailletaud, 1986) developed at Ecole nationale supérieure des mines de Paris (ENSMP), in combination with a commercial software for identifying parameters. Only one-half of the midsection was modeled for symmetry reasons. The thermal

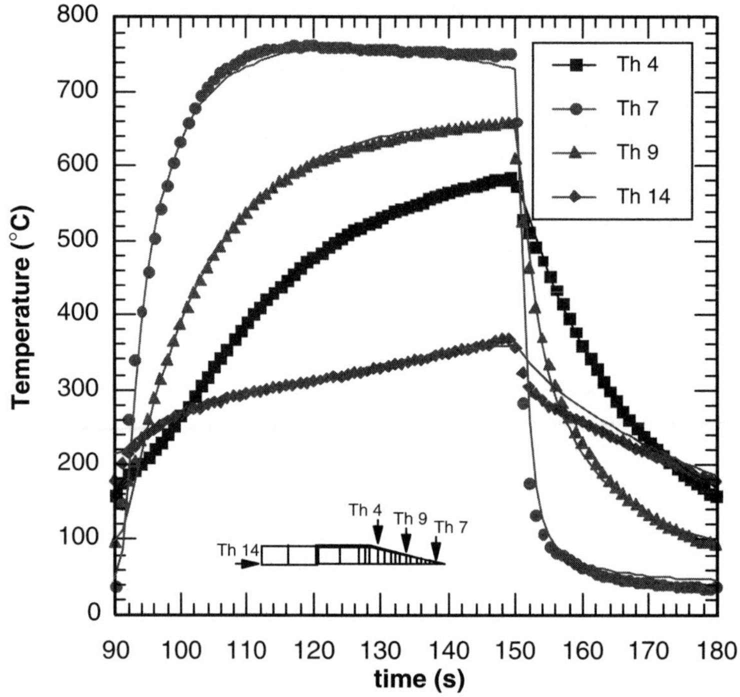

Figure 61 Experimental and computed temperature profiles for the FT-L wedge geometry of Waspaloy cycled in the temperature range 30–750°C (after Koster *et al.*, 1996a, 1996b).

loading was described with convective film cooling (h_i) and radiation heat flow (ϕ_I), which vary with the location on the specimen surface.

Temperatures were calculated with a set of coefficients (h_i, ϕ_I) and compared to experimental values using the SiDoLo software that computes an error function. An optimal set of coefficients is obtained by minimizing the error function. The best results are shown in Figure 61 for one specimen geometry and a good agreement is achieved along the whole specimen section.

A Chaboche-type viscoplastic constitutive model was used for wrought Waspaloy: while a single kinematic-hardening variable is often used in the form of Equation (25) in the usual Chaboche equations, two kinematic-hardening variables were necessary in this alloy for the temperature and stress range investigated (Köster *et al.*, 1996a, 1996b). The material parameters were identified from LCF tests with various strain rates at different temperatures, and 100–750°C TMF stress–strain loops were correctly predicted.

The results of the heat transfer analysis were used as input for the mechanical stress–strain analysis, using the identified constitutive equations. The computation is illustrated for a mechanically stabilized cycle (20th) for the two wedge geometries: Figure 62 shows the mechanical strain–temperature at various distances from the thin edge of the specimen.

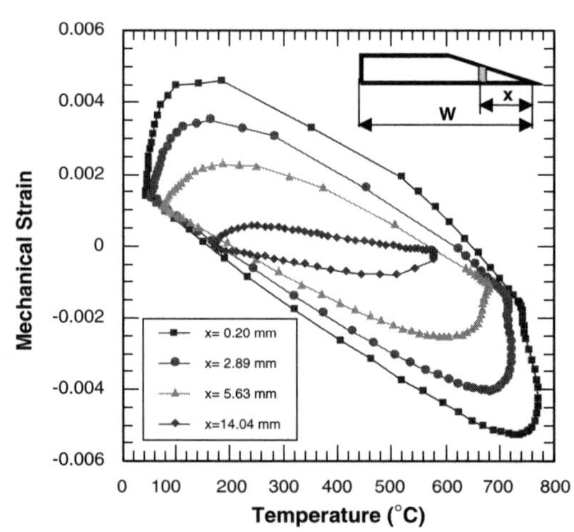

Figure 62 Computed mechanical strain–temperature cycles for the FT-L wedge geometry of Waspaloy cycled in the temperature range 30–750°C (after Koster *et al.*, 1996a, 1996b).

Stress–temperature loops as well as the stress–mechanical strain loops are shown, for the same locations, in Figures 63 and 64. The stresses are higher for the large geometry than for the standard geometry, and the same conclusions hold for the mechanical strain and inelastic strain as well. The loops exhibit a mean tensile stress and mean compressive strain, which is rather sensitive to the amount of viscoplasticity and kinematic hardening (Figure 64).

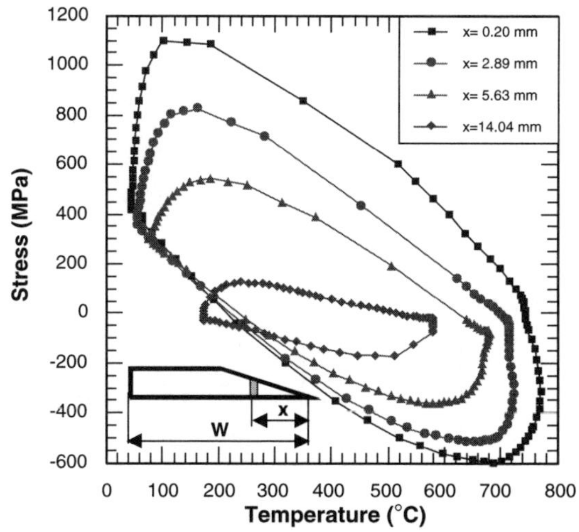

Figure 63 Computed mechanical stress–temperature cycles for the FT-L wedge geometry of Waspaloy cycled in the temperature range 30–750°C (after Koster *et al.*, 1996a, 1996b).

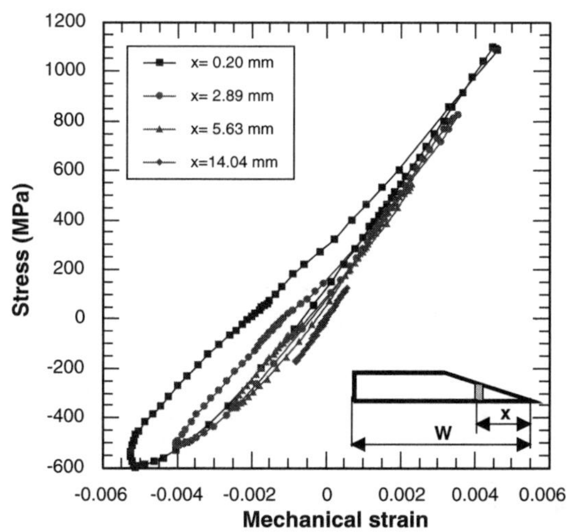

Figure 64 Calculated stress–mechanical strain loops for FT-L geometry (after Koster *et al.*, 1996a, 1996b).

These TS experiments were used to check the predictions of the creep–oxidation–embrittlement fatigue damage model in a component-like situation (Köster *et al.*, 1996a, 1996b). Figure 65 depicts the comparison between the computed and experimental growth curve of the major crack for the two different geometries of Waspaloy tested between 30°C and 750°C. A good description of early crack growth is achieved up to 3 mm and 1 mm for the standard and large geometry, respectively (Köster and Rèmy, 1999). One has to remember that this estimation is made in a post-processor of a stress analysis of an uncracked structure and it is therefore not expected to describe a long crack situation, since stress

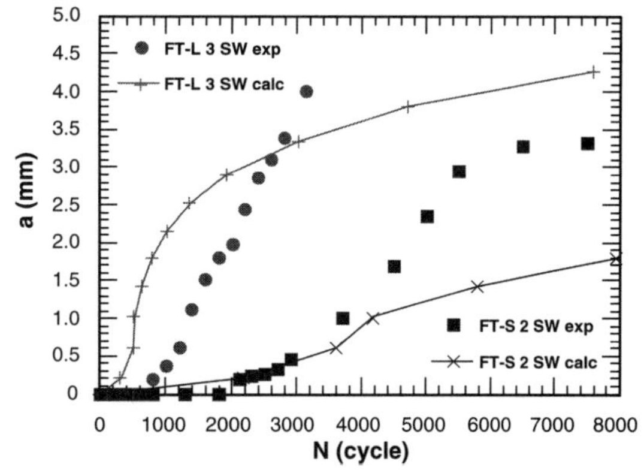

Figure 65 Depth of the major crack as a function of cycle number for the two wedge geometries in Superwaspaloy submitted to a TS between 30°C and 750°C: comparison between experiment and model predictions (source Koster and Rémy, 1999).

redistribution due to crack growth is not taken into account.

Accounting for the local stress redistributions due to the existence of a long crack, the application of these equations to a long crack in a real component would require to use local stresses ahead of the crack tip and thus an FE computation of the cracked structure. However, if the prediction is limited to short crack lengths (and thus to engineering crack initiation), the stresses computed for the uncracked structure could be used in a first approximation.

TS experiments on MAR-M509, a cobalt-based superalloy, were run using a burner rig and wedge-type specimens of various geometries (Rézaï-Aria *et al.*, 1988a, 1988b). The standard thermal transient at the thin edge is between 200°C and 1,100°C for a period of 80 s (60 s on heating, 20 s on cooling). Specimen geometries are shown in Figure 4, which enable one to decrease the maximum temperature difference between the thin edge and the bulk (from a standard specimen with 0.25 mm thin edge radius, a standard specimen with 1 mm thin edge radius and a narrow specimen with 1 mm thin edge radius). Tests were also run on the standard geometry with a maximum temperature of 950°C.

The experimental life to 1 mm crack depth is reported in Figure 66 as a function of mechanical strain range computed at the thin edge of wedge specimens and compared with isothermal LCF at various temperatures (Rézaï-Aria *et al.*, 1988b). TF life is always shorter than LCF life even at maximum temperature, which renders a strain-based comparison nonconservative.

The same comparison is made as a function of stress range in Figure 67. TF life compares now with LCF life at 900°C except at the lower

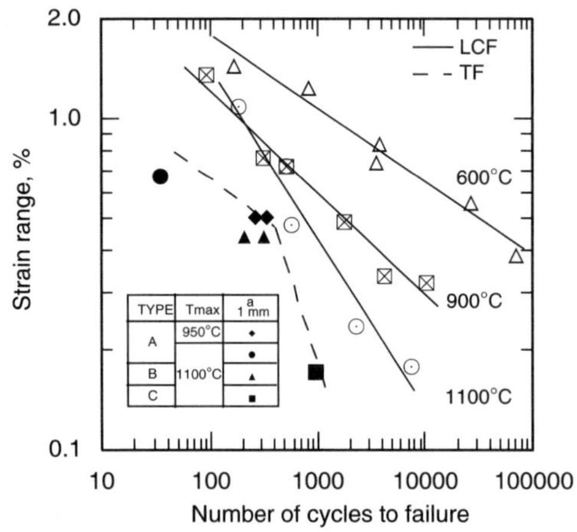

Figure 66 Strain range vs. number of cycles to 1 mm crack depth in TS experiments on wedge specimens tested between 200°C and 1,100°C, or 950°C for different geometries compared with life to failure in LCF on superalloy MAR-M509 (after Rezai-Aria *et al.*, 1988a, 1988b).

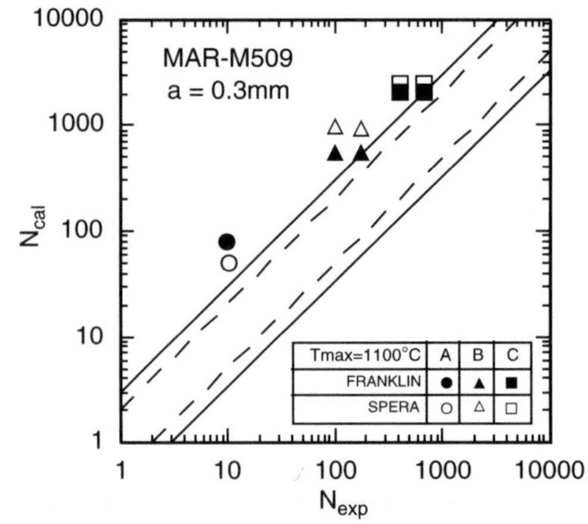

Figure 68 TF life to 0.3 mm and 1 mm crack depth predicted from the accumulated creep damage models, as proposed by Spera and Franklin (N_{cal}), as a function of experimental lifetime (N_{exp}) for superalloy MAR-M509 (after Remy *et al.*, 1988).

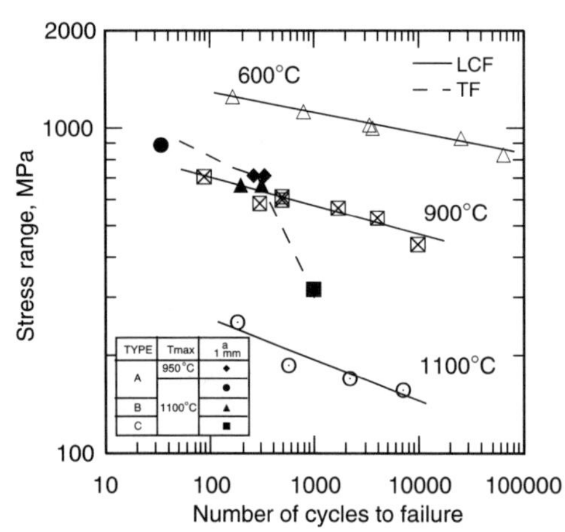

Figure 67 Stress range vs. number of cycles to 1 mm crack depth in TS experiments on wedge specimens tested between 200°C and 1,100°C, or 950°C for different geometries compared with life to failure in LCF on superalloy MAR-M509 (after Rezai-Aria *et al.*, 1988a, 1988b).

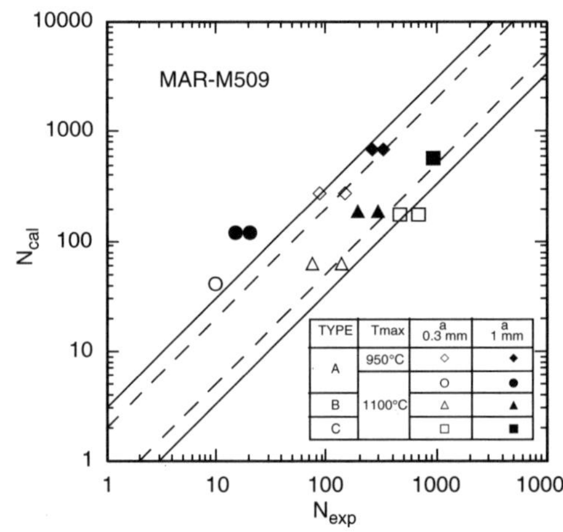

Figure 69 TF life to 0.3 mm and 1 mm crack depth predicted from the oxidation–fatigue model (N_{cal}) as a function of experimental lifetime (N_{exp}) for superalloy MAR-M509 (after Rémy *et al.*, 1988).

strain ranges and a conservative comparison can be made on a stress range basis.

The stress and inelastic strain rate variation at the thin edge have been used for creep damage computation. Negative stresses were assumed to be nondamaging as in isothermal fatigue, since this procedure gives good life predictions in isothermal fatigue. Spera's linear accumulation damage model and Franklin's (1978) model overestimated the life to 0.3 mm crack depth, by a factor of 3–6 for a maximum temperature of 1,100°C (Figure 68). For a maximum temperature of 950°C, creep damage is insignificant and these models failed to predict lifetime.

The simple oxidation fatigue model described above (Reuchet and Rémy, 1983c) has been used to analyze the same experiments. The contribution of oxidation to crack growth has been computed by integration of the static oxidation equation over the temperature–time cycle of each element of the structural analysis. Tomkins' model was applied at the temperature of the peak tensile stress. The number of cycles to form a 0.3 mm or a 1 mm deep crack so computed compares with experiment within a factor of 3, except for the most severe test conditions (Figure 69). The oxidation fatigue model, which has a physical basis for the alloy, yields much better prediction than creep–damage models.

The oxidation embrittlement–fatigue damage model has been used to describe the early growth behavior of TF cracks in MAR-M509 in wedge specimens with a thin edge radius of 0.25 mm (Rézaï-Aria and Rèmy, 1989). The influence of maximum temperature on the growth of cracks can be described up to a few mm in depth (Figure 70). It must be pointed out that this good agreement up to 3–5 mm is probably due to the high sensitivity of crack growth to oxidation, since crack growth is not accounted for in the stress analysis.

Actual components are generally cast to size. As they are cooled with more or less sophisticated techniques, they are often cast as thin walls so that their microstructure can be rather fine. Laboratory specimens, alternatively, are most often machined from bulk cast cylinders. Most data used in design thus come from specimens which can have a much coarser microstructure than cast-to-size components

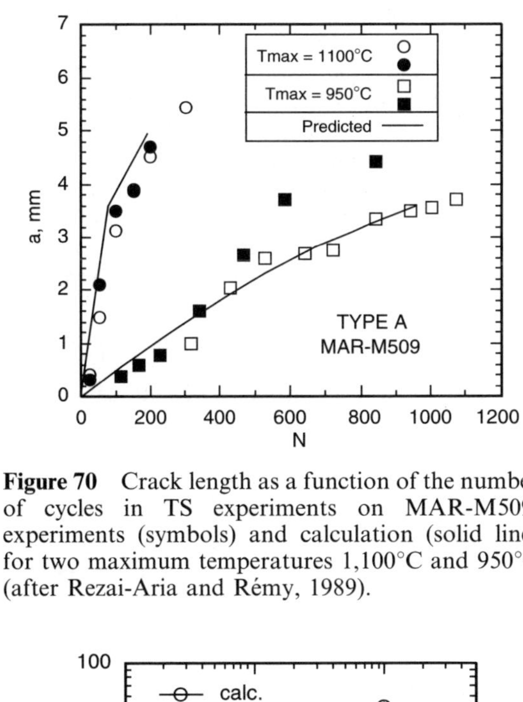

Figure 70 Crack length as a function of the number of cycles in TS experiments on MAR-M509: experiments (symbols) and calculation (solid line) for two maximum temperatures 1,100°C and 950°C (after Rezai-Aria and Rémy, 1989).

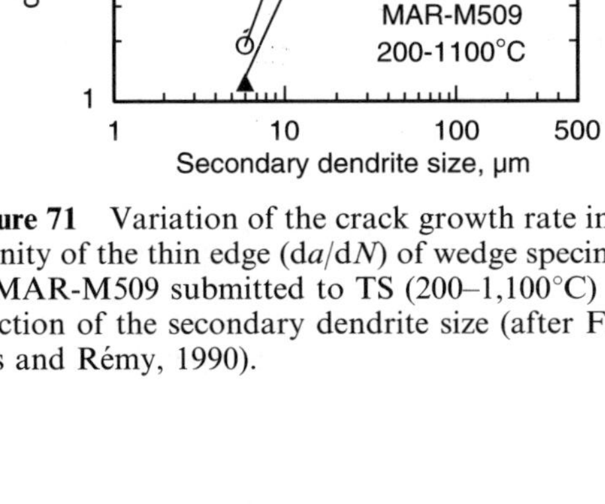

Figure 71 Variation of the crack growth rate in the vicinity of the thin edge (da/dN) of wedge specimens of MAR-M509 submitted to TS (200–1,100°C) as a function of the secondary dendrite size (after François and Rémy, 1990).

due to large differences in solidification conditions. The influence of dendrite size due to different casting conditions has been investigated in MAR-M509 (François and Rémy, 1990). A large range of dendrite sizes from 16 µm to 100 µm was used in TS experiments between 200°C and 1,100°C. This large variation in size results in a large increase of crack growth rate in the vicinity of the thin edge (Figure 71). It was assumed that this behavior could only be described by taking into account the decrease of the interdendritic oxidation constant with decreasing dendrite size, as deduced from metallography. The computed crack growth rates under this hypothesis are actually within a factor of 2 of experiment.

5.03.7.4 Coated Materials

5.03.7.4.1 An empirical model for coated components

Bernstein *et al.* (1993) reported a strain-based empirical model for the so-called buckets in stationery gas turbines, as a part of a life management system for industrial gas turbine, in the power range 60–80 MW of electricity.

An experimental program was made on IN738 LC, used for hot-section components in industrial gas turbines, which are coated against high-temperature environmental attack, by a plasma spray deposited CoCrAlY-type coating. Hollow specimens were submitted to out-of-phase TMF tests between a minimum temperature of 427°C and a maximum temperature of 871°C, 916°C, or 982°C. Compression hold times of 0 min, 2 min, and 15 min were applied for both coated and uncoated specimens at maximum temperatures and most tests were carried out between zero and a minimum value. A complex "bucket" TMF cycle was also used with a 2 min compression hold at maximum temperature and a strain ratio $A = \Delta\varepsilon_{min}/\varepsilon_{max}$ of −2.3, for different strain ranges. As often reported, coating gives a reduction in out-of-phase TMF life with no strain hold, but no difference is found for hold-time tests, between bare and coated conditions. The service-like "bucket" cycles with a hold time yield a significantly shorter life (often by a factor of 2). Out-of-phase TMF test on coated specimens gives always a shorter life than for isothermal tests at maximum temperature, as often reported in the literature.

The general description of TMF for in-phase, out-of-phase, and isothermal cycle was not considered. TMF life model using maximum tensile stress correlates life well, but is considered to be difficult to apply because of its strong dependence on the materials constitutive behavior. A strain-based approach was estimated to be more robust.

The goal was to use independent variables determined from the elastic FE analysis and to have a life prediction in order to model and manage components operating in plants. The equation used strain range, the strain ratio $A = \Delta\varepsilon/2\varepsilon_{\max}$, and the hold time t_h can be written as

$$N_f = C_0(\Delta\varepsilon)^{C_1}(t_h)^{C_2}\exp\left(C_3\frac{2\varepsilon_{\max}}{\Delta\varepsilon_m}\right) \quad (100)$$

where C_i are four empirical constants. A set of C_i was identified from out-of-phase tests, one for coated and one for uncoated specimens. This empirical model was found to correlate within a factor of 2, in most cases, the data published in the literature.

To predict field data, a major part of the effort was the calculation of heat transfer film coefficient and gas temperature distributions for starting, steady-state, and shutdown operating conditions. This requires numerous calculations of turbine performance, aerodynamic flow field, heat transfer boundary layer, and cooling airflow network to achieve time-dependent boundary conditions for each element of the model that is exposed to hot-gas or cooling-air flow streams.

The predictions are made for a leading edge of buckets, which corresponds to the location of actual cracks after service operation. Using a linear damage rule, most components corresponding to 100% life and more show cracks. Alternatively, components interrupted before 100% life consumption do not show cracks in most cases.

5.03.7.4.2 Strain-based methods for coatings

Strain-based models that incorporate a description of crack initiation and early growth were proposed for coated advanced thin wall hollow airfoil in aeroengines, where complex internal air cooling schemes are used to maintain the structural superalloy at acceptable temperature levels. Sections can experience widely different TMF cycles. Cracks in diffusion aluminide and (Ni, Co) CrAlY overlay coatings are frequency initiated in turbine blades at locations where significant out-of-phase thermal–mechanical strains are present, since tensile strain peaks at temperature where coating ductility is often very low.

The life to initiate a crack in the coating was described in a complete life model by Strangman (1992) as a Manson–Coffin equation:

$$(\Delta\varepsilon_\alpha + \Delta\varepsilon_m)N_i^c = \varepsilon_{pf} + \frac{\sigma_f}{E} \quad (101)$$

For coatings with significant ductility (1–10% elongation, i.e., (Ni, Co)CrAlY coatings with aluminum content below 12.5 wt.%), c is ~ 0.36 at high strain ranges.

For low TMF strains in ductile coatings or for brittle coatings in the low ductility temperature range, all strains are assumed to be elastic, the plastic contribution ε_{pf} is dropped, and the exponent c (~ 0.36 in the ductile regime) is reduced to Basquin-type value 0.12, as proposed in the universal-slope method of Manson.

Strangman (1992) has pointed out that at the tip of a TMF crack, the oxide scale is ruptured every cycle that the crack propagates. Therefore, a linear oxidation kinetics is used and deduced from short-time thermogravimetric data, which show a linear regime for short exposure time (less than 20 min in modern superalloys) before protective parabolic equation does establish. A similar approach is proposed for hot corrosion by molten salt deposits and TMF life is dependent upon engine design gas and metal temperature, turbine pressure fuel/air ratio, and mission usage parameters (fuel quality, salt in the air, and time at duty cycle points).

The relaxation of compressive strains in an out-of-phase TMF cycle is especially important when peak strains occur at fairly high temperatures, which will give rise to large mean stresses, and larger tensile stresses on cooling. Strangman (1992) proposed an empirical equation to describe the effective strain range, which should be substituted for applied mechanical strain range, and to incorporate the amount of stress relaxation from creep rupture data.

Early crack growth description uses the strain intensity factor, based on extensive TMF crack growth investigations by Rau *et al.* (1973): for crack growth rates below 2.5×10^{-5} m cycle^{-1} the rate of cracks in superalloys depends only on the strain intensity factor range ΔK_ε and is independent of strain range, mean strain, and mean stress. The magnitude of the thermal expansion mismatch strain between the coating and substrate was found to provide an additional localized driving force for base metal fatigue crack propagation (Strangman and Hopkins, 1976). For a single edge crack in a semi-infinite sheet, the strain intensity factor range ΔK_ε for a substrate crack is written as

$$\Delta K_\varepsilon = \frac{4}{\pi}\frac{E_c}{E_s}\Delta\varepsilon_\alpha(t+a)^{1/2}\sin^{-1}\left(\frac{t}{t+a}\right) + 1.12\Delta\varepsilon\left[\pi\left(t\frac{E_c}{E_s}+a\right)\right]^{1/2} \quad (102)$$

E_c, E_s are Young's modulus of coating and substrate at peak strain temperature, $\Delta\varepsilon_\alpha$ is thermal expansion mismatch at peak strain, t the coating thickness, a the crack depth into the substrate ($a+t$ being the total depth from surface), and $\Delta\varepsilon_m$ the mechanical strain range in the substrate. This solution is adequate for small cracks in large bodies but geometry correction factors can be added for accounting crack and part geometry.

For small surface connected substrate crack ($0<a<t$), the strain intensity factor is predominantly a function of coating thickness, mismatch strain range, and mechanical strain range. For longer crack, the mechanical strain term due to substrate becomes dominant. The life spent in crack propagation in the substrate once a crack is initiated in the coating is obtained by integration of the Paris law between da/dN and ΔK_ε to a given crack length. Such a model can be used to describe the improvement in fatigue life (typically for out-of-phase cycle), which is obtained through decreasing the elastic moduli of coating with respect to that of the substrate, decreasing coating thickness, or decreasing mismatch strains.

A strain-based approach has several drawbacks in spite of the fact that $\Delta K/E$ is a very effective parameter to rationalize crack growth rate between materials when time-dependent effects and closure effects are discarded.

Models have addressed the spallation of thermal barrier coatings developed for air-cooled turbine blades and vanes in aircraft gas turbines engines and in advanced stationary gas turbines but they seem still to be at an early stage due to the complexity of the process. Coating spalling is induced by the thermal expansion mismatch between the ceramic and metallic layers: these stresses are greatly influenced by processes controlled by time spent at high temperature such as oxidation and possibly sintering. Miller (1987, 1988) has reviewed models for plasma-sprayed zirconia. The same framework was basically used for other deposition processes. Meier *et al.* (1991) have proposed an engineering model to describe the life to spalling of TBC, for EB-PVD TBC (electron-beam physical vapor deposition) which is particularly used for blades since these TBCs are very tolerant to in-phase strains: ceramic spallation occurs at the thermally grown oxide (TGO) layer, at the junction between bondcoat deposited on the substrate alloy and EB-PVD zirconia. This model was identified using burner rig test over a wide range of conditions, with emphasis on strain, using many rapid thermal cycles, with emphasis on oxide growth, using hold time at maximum temperature, and mixed cycles. It combines oxide kinetic equations and Coffin–Manson-type equations and assumes that there is a critical oxide thickness at spalling, which may be questioned according to observations on oxidation under TMF loading reported in Section 5.03.4. This model was given by Singheiser *et al.* (2001), who pointed out the role of critical strain in Coffin–Manson-type equations in damage accumulation and sintering may occur and degrade this strain. They used a correction to ε_f' to account for sintering. These models offer only an approximate description of crack growth and spalling in layered materials according to the various mechanisms summarized by Evans *et al.* (1983). Linear elastic fracture mechanics studies have been dedicated to various aspects of cracking in dissimilar materials, along or near interfaces (Rice and Sih, 1965; Rice, 1988) accounting for mixed mode, delamination, and buckling (Hutchinson and Suo, 1992). The complexity of fracture mechanics tools in multi-material situations has, however, delayed their introduction into engineering models.

5.03.7.4.3 *Continuum damage mechanics for coated materials*

Gallerneau *et al.* (1996a, 1996b) proposed a different model for coated superalloys in the framework of continuous damage mechanics, based on a previous refinement by Savalle and Cailletaud (1982) inspired by accumulated experimental observations, which distinguishes a crack initiation phase ("microinitiation"), corresponding to a number or cycles N_i and a crack propagation phase ("micropropagation"), corresponding to a number of cycles N_p. This distinction is based on numerous observations made by materials science experimentalists.

For coatings, this approach is especially convenient, since the microinitiation zone can easily be identified as coating thickness. Gallerneau *et al.* have distinguished a microinitiation phase in coating mostly influenced by oxidation (oxidation damage D_{ox}), and a micropropagation phase controlled by creep and fatigue.

The damage rates are written as

$$dD_i = f_i(A_{II},\ \bar{\sigma}_H,\ \sigma_{eq,max},\ D_{ox})dN \quad (103)$$

$$dD_p = H(D_i - 1)f_p(A_{11},\ \bar{\sigma}_H,\ \sigma_{eq,max},\ D_p)dN \quad (104)$$

$$dD_{ox} = f_{ox}(X_{ox},\ D_{ox})\ dt \quad (105)$$

$$dD_c = f_c(X_c,\ D_c)\ dt \quad (106)$$

where $H(D_i-1)=0$ if $D_i<1$ and $H(D_i-1)=1$ if $D_i\geq 1$.

When the initiation damage variable equals 1, creep-fatigue interaction can start, otherwise only creep damage develops. The usual continuum damage mechanics equations are then applicable (cf. Equations (53), (54), (56)).

A_{II} is octahedral shear stress amplitude, and σ_{eq} is equivalent stress which for anisotropy used two fourth-order structure tensors, $\bar{\sigma}_H$ is mean hydrostatic pressure. A_{II} and σ_{eq} reduce to $\Delta\sigma/2$ and $|\sigma|$, respectively, in the uniaxial isotropic case.

In the creep and oxidation laws, a delayed stress X_i is introduced to saturate the frequency dependence of the fatigue damage process in high-frequency regime:

$$\frac{dX^i}{dt} = \frac{\sigma^i_{eq} - X_i}{\tau_i} \tag{107}$$

in the general case, or

$$\frac{dX^i}{dt} = \frac{|\sigma| - X_i}{\tau_i} \tag{108}$$

in the uniaxial case. The parameters τ_i defines the frequency above which creep or oxidation effects are ineffective. The use of these delayed stress tensors solved difficulties found in classical creep–fatigue interaction model (Chaboche and Gallerneau, 2000).

The fatigue damage for crack initiation is expressed here in the uniaxial isotropic case, for the sake of simplicity, as

$$dD_i = \frac{1}{C}\left\langle \frac{\Delta\sigma/2(1 - D_{ox}) - \sigma_{li}(\bar{\sigma})}{\sigma_u - \sigma_{max}/(1 - D_{ox})} \right\rangle^b dN \tag{109}$$

where oxidation damage is introduced through the effective stress concept, σ_u is the ultimate tensile strength, σ_{max} is maximum stress, and $\Delta\sigma$ is the stress amplitude. σ_{li} is the fatigue limit in initiation. The oxidation law is given by a second differential equation, which reflects the influence of loading on oxidation kinetics:

$$dD_{ox} = \frac{1}{2} D_{ox}^{-1} \left(\frac{k_{ox}}{l_0}\right)^2 \left[1 + \left\langle \frac{\sigma - \sigma_{lox}(\bar{\sigma})}{B} \right\rangle\right]^{2m} dt \tag{110}$$

where σ_{lox} is the stress above which the oxidation kinetics is accelerated and K_{ox} the oxidation constant. B and m are two constants. This equation is inspired from those identified for strain–oxidation interaction on superalloys (Reuchet and Rémy, 1983a, 1983b, 1983c; Bernède, 1994).

Application of this model for AM1 superalloy $\langle 0\,0\,1 \rangle$ single crystals with a diffusion aluminide coating is shown in Figure 72, under isothermal conditions.

For thermal–mechanical loading, the fatigue law for initiation (as for propagation) is normalized by the ultimate tensile strength, so that parameters become independent of temperature. Of course, creep parameters and oxidation parameters vary with temperature.

This is attractive for application in a large domain. As a matter of fact, microinitiation and micropropagation laws evolve with temperature only through variation of ultimate strength for initiation $\sigma_{u,i}(T)$ and for propagation $\sigma_{u,p}(T)$. The coating is treated as the substrate and only damage parameters differ. The ultimate strength for micropropagation $\sigma_{u,p}$ is that of the substrate. To account for the brittleness of the coating at low temperature,

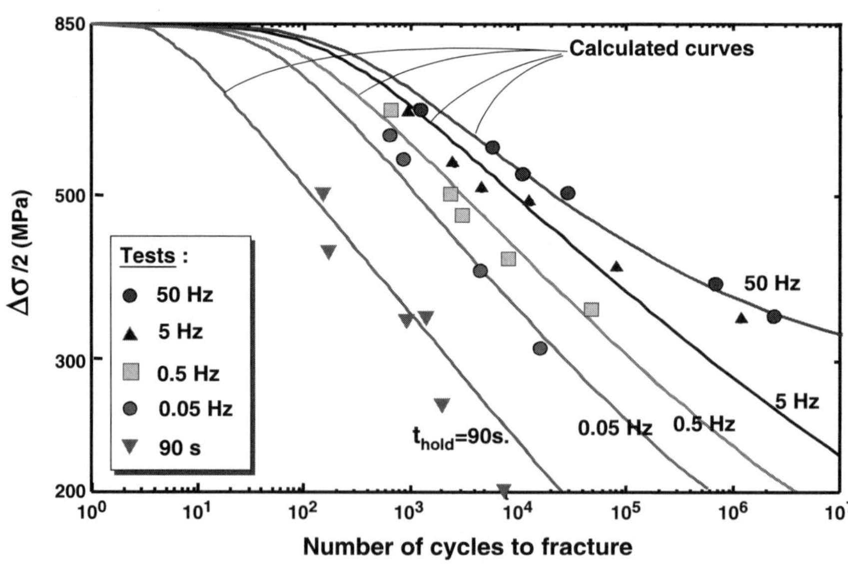

Figure 72 Comparison between continuum damage mechanics model and push–pull fatigue life of coated AM1 $\langle 0\,0\,1 \rangle$ crystals at 950°C (after Gallerneau *et al.*, 1996a, 1996b).

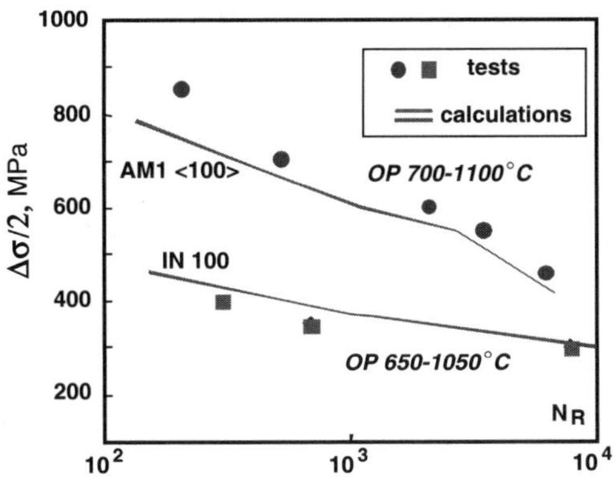

Figure 73 Comparison between continuum damage mechanics model and OP TMF life of coated superalloys: AM1 ⟨001⟩ crystals and IN100, for high thermal gradient in thin wall (above 200°C between inner and outer surfaces) (after Gallerneau, 1999).

the ultimate strength in microinitiation $\sigma_{u,i}(T)$ is taken to be lower than that of the superalloy. Pure fatigue tests at high frequency at a given temperature are then sufficient to identify all the parameters of microinitiation and micropropagation laws. Creep-fatigue tests at a low frequency are used to identify the oxidation damage law and pure creep tests at several temperatures are nevertheless required to identify the creep damage law.

Predictions of TMF life are shown for out-of-phase TMF tests on specimens with high thermal gradients, carried out on aluminized polycrystal IN100- and C1A-coated AM1 in Figure 73 (Gallerneau, 1999) and agree well with experiment.

5.03.8 CRACK PROPAGATION

5.03.8.1 Concepts Available to Analyze Crack Growth Under TF

Crack growth under TF is much more difficult to describe. In numerous practical situations multiple cracking may occur either when a principal stress component dominates over others or an equibiaxial stress field where a heat-checking crack network develops.

In components submitted to TF, a crack when initiated often grows in a yielded zone, where significant plastic or viscoplastic cyclic strains occur, and then also under small-scale yielding. Crack growth then involves stress gradients and/or temperature gradients. The long crack situation is actually an idealization of real situations.

5.03.8.1.1 Early crack growth

Several groups have addressed early growth under displacement control: Solomon and Coffin (1973), and especially Skelton (1982, 1988), have produced LCF crack growth tests using smooth specimen (cylindrical, rectangular cross-section) with a shallow edge notch. Typical tests start from a depth ~0.1 mm or more and are stopped when a significant load drop or cusp in the hysteresis loop has occurred (Skelton, 1982). Studies in steels and alloys with moderate strength can be described by the empirical law:

$$\frac{da}{dN} = Ba^q \qquad (111)$$

where q is often ~1 (Skelton, 1988) and B is given as

$$B = C(\Delta\varepsilon_p/2)^m \qquad (112)$$

A bit further from this standpoint, early crack growth in the component can be modeled from that occurring in a laboratory specimen, provided crack depth is such that no significant compliance change occurs in the specimen (which is in most cases smaller than the component). Application to thin wall components as in gas turbines requires still more care in transferring laboratory data to components. Skelton has found that an upper bound for B in ferritic steels can be expressed as a function of mechanical strain range as

$$B = 2.61 \times 10^4(\Delta\varepsilon/2)^{2.85} \qquad (113)$$

which is more useful for application in a component. At very small depth ($a/w<0.01$), equations often breakdown, and FCGRs appear to be fairly independent of crack depth (Levaillant and Pineau, 1982; Bressers *et al.*, 1986).

One has to note that most equations proposed to describe the growth of cracks in Section 5.03.6 can also be used to describe early crack growth in components.

5.03.8.1.2 Long crack growth

Therefore, to extend the description of crack growth beyond the range of applicability of Equation (99) (typically for $a/w\sim0.25$ in a specimen), other parameters need to be introduced.

Given its simplicity the range of stress intensity factor is the most widely used concept to analyze crack growth under TF. Its validity is limited if cracks are too small and if crack growth occurs in a yielded zone.

Strain intensity factor and cyclic J-integral are proposed when significant plasticity prevents the use of linear elastic stress intensity factor. The use of J despite a number of drawbacks (definition only for 2D situations, limitation to proportional loading, problems with unloading in fatigue) seems to be more popular.

Authors tried to use first a strain intensity factor ΔK_ε to address propagation in a generally yielded material. Rau *et al.* (1973) and Gemma *et al.* (1976) correlated their TMF crack growth data using ΔK_ε as

$$\Delta K_\varepsilon = \Delta\varepsilon(\pi a)^{1/2} f(a/w) \tag{114}$$

where $\Delta\varepsilon$ is the remote strain range (uniform over a TMF specimen) and $f(a/w)$ is a geometric correction factor. This factor is often assumed to be identical to that derived for the elastic stress intensity factor. However, it must be emphasized that this strain factor has no rigorous meaning.

Haigh and Skelton (1978) and Skelton (1988) used the strain extensity factor to correlate high-temperature LCF with TMF crack growth rates on various steels but suggested to use $\Delta\varepsilon_p + (1/2)\Delta\varepsilon_e$ instead of $\Delta\varepsilon$, using crack closure arguments.

An alternative fracture mechanics concept to handle situations where gross plasticity occurs is Rice's *J*-integral (Rice, 1968). As Paris did previously with *K*, Dowling and Begley (1976) and Dowling (1977) tried to use the *J*-integral to describe fatigue crack growth in yielded specimens. This approach was successfully applied originally for deep cracks, though no theoretical justification was available. Justifications were then set by Lamba (1975) and by Wütrich (1982). Lamba has shown that path independence of cyclic *J* is ensured provided the strain potential ψ is such that

$$\Delta\sigma_{ij} = \frac{\partial\psi}{\partial(\Delta\varepsilon_{ij})} \tag{115}$$

$$\psi(\Delta\varepsilon_{mn}) = \int_0^{\Delta\varepsilon_{mn}} \Delta\sigma_{kl}\, d(\Delta\varepsilon_{kl}) \tag{116}$$

This is obeyed using a Masing hardening rule and a power law between stress and strain components, for proportional loading situations.

Approximate values of cyclic *J* have been derived separating *J* into elastic and plastic components, following Shih and Hutchinson (1976) and Goldman and Hutchinson (1975), and using engineering computations from available solutions for simple crack geometries. Successful correlation has been observed between different geometries, or different strain ranges, but this approach failed to describe numerous short-crack situations (Reger *et al.*, 1984b).

This approximate derivation of the cyclic *J*-integral has been used by a number of authors to correlate TMF crack growth data. The usual definition of *J*-integral should apply to these situations when there is no temperature gradient across the specimen section. A number of authors have thus obtained good correlation of TMF crack growth data (Okazaki and Koizumi, 1983, 1987; Koizumi and Okazaki, 1979; Jordan and Meyers, 1986).

The application of *J* in a thermally loaded structure is more complex, since *J* becomes path dependent in a thermal field. As noted by Bui (1978), an extra term should be added to maintain path independence as

$$J = \int_C \left(W\, dx_2 - T_i \frac{\partial u_i}{\partial x_1}\, ds \right) + \iint_A \sigma_{ij} \frac{\partial \varepsilon_{th,ij}}{\partial x_1}\, dA \tag{117}$$

where, as usual, *C* is an integration contour surrounding the crack tip, *W* is the strain energy density, T_i are traction components corresponding to displacements u_i, and x_1, x_2 are local axes at the crack tip; $\varepsilon_{th,ij}$ is the thermal strain tensor and *A* is the area inside the contour *C*.

Modified *J* line integrals have thus been proposed to handle thermal fields, as well as direct computation of potential energy release rate *G* (examples of applications can be found in Kaguchi *et al.* (1998) and Kim and Van Stone (1998)).

Haigh and Skelton (1978) and Skelton (1988) have proposed to use an equivalent stress intensity factor,

$$\Delta K_{eq} = (q_0 \Delta\sigma + E\Delta\varepsilon_p) Y(a/w)\sqrt{\pi a} \tag{118}$$

with $\Delta\sigma$ the stress range and $Y(a/w)$ a compliance function. Starkey and Skelton (1982) have shown that a good agreement was therefore obtained with long-crack fatigue data obtained under load control and small-scale yielding conditions. These authors have shown that this method is in good agreement with the formula of Shih and Hutchinson (1976), widely used for estimating cyclic *J* values for short cracks.

Skelton (1988) has run LCF crack growth tests up to very large fractions of specimen size, where the crack-size-to-ligament ratio is such that compliance is significantly increased (stiffness decreased). Load–displacement curves were used to define the crack opening load, which was introduced through q_0 as

$$q_0 = (\sigma_{max} - \sigma_{op})/\Delta\sigma \tag{119}$$

which is the analogue of Elber's *U* for closure for stress-controlled situations, under small-scale yielding (Elber, 1971). Starting from a load ratio $R=1$, σ_{max} decreases in displacement control and the R-ratio becomes increasingly negative. q_0 starts from a value ~ 0.78 and then decreases with a/w.

From this result one can conclude that for TF cracks growing into structures, the effective stress range is only part of the total stress range. However, it must be emphasized that the good agreement between equivalent stress intensity factor and LCF crack growth data in steels is largely based on taking into account an effective stress range and a crack closure phenomenon, with opening occurring at negative stresses for push–pull conditions (Skelton, 1988).

In the same way, load-controlled tests which give rise to ratchetting on an austenitic and two ferritic steels at 550°C were found to merge with displacement-controlled tests, due to the crack closure correction (Skelton, 1988); in that case q_0 tends to increase with increasing a/w.

These different tools equivalent to ΔK and cyclic J can be used to compute the lifetime of components by integration of the crack growth laws, when the crack leaves the plastic zone (LCF range) and enters the small-scale yielding regime.

Otherwise, in the plastic zone, an upper bound can be obtained for crack growth rate using a strain-based LCF crack growth equation where surface strain is used as an input. Contribution of the growth from outside the plastic zone must be assessed separately using linear elastic fracture mechanics.

Applications of empirical equations relating da/dN, strain range, and crack length were criticized. Huang and Pelloux (1980) have criticized these laws, as the Tomkins (1968) model, which do not explicitly describe crack geometry, while J and cyclic J do, as fracture mechanics tools. Using the partition into elastic and plastic components as proposed by Shih and Hutchinson (1976), the compliance function $h(a/w, n)$, with exponent n of the Ramberg–Osgood equation, becomes strongly sensitive to n and the situation is even worse for a strip in phase strain (Goldman and Hutchinson, 1975) than under plane stress (He and Hutchinson, 1981). Application of elastic compliance corrections as in equivalent stress intensity can give erroneous descriptions of deep-crack growth.

5.03.8.2 TMF Crack Growth

Most of the TMF activity on crack growth was devoted to the comparison between fatigue crack growth under isothermal and varying temperature conditions.

Rau *et al.* (1973) correlated their TMF crack growth data using the strain range intensity. It must be recalled that their tests were, as for a large number of investigations, carried out under strain-controlled conditions. They produced results mostly on superalloys.

Rau *et al.* (1973) suggested that crack growth rate increases from isothermal at maximum temperature, to in-phase TMF and further to out-of-phase TMF, from tests on cobalt-based alloy MAR-M509 in the temperature range 315–900°C and in the superalloy B-1900 cycled in the range 427–926°C. Their conclusions were drawn using a ΔK_ε criterion for various strain ranges and different TMF cycles. Gemma *et al.* (1976) extended that work to other nickel-based superalloys including anisotropic directionally solidified superalloys (MarM200-Hf).

Assuming transverse isotropy in the plane maximal to the columnar grain orientation [0 0 1], they rationalize results obtained for different angles of loading axis with respect to the [0 0 1] direction and out-of-phase loading in the temperature range 427–1,038°C. The basis of this was a longitudinal modulus correction λ and the crack growth rate for an off-axis direction (θ):

$$da/dN = C[\lambda \Delta K_\varepsilon]^m \tag{120}$$

where $\lambda = E_1/E_L$ is the ratio of the loading axis 1 to the longitudinal modulus E_L (along the anisotropy axis) with

$$\begin{aligned}\lambda^{-1} &= \frac{E_L}{E_1} \\ &= \cos^4\theta + \frac{E_L}{E_T}\sin^4\theta + \left(\frac{E_L}{G_{LT}} - 2\nu_{LT}\right) \\ &\quad \times \cos^2\theta \sin^2\theta \end{aligned} \tag{121}$$

where θ is the angle between the load and grain growth axes, E_T the transverse modulus ($\theta = 90°$), G_{LT} the shear modulus, ν_{LT} the Poisson's ratio, and C and m are constants determined for tests where the loading axis coincides with grain growth direction ($\theta = 0$, $\lambda = 1$).

This strain intensity factor concept was later used by Gemma and Philips (1977) to predict the lifetime of TMF specimens with arrays of holes to simulate film and transpiration cooled configurations used in advanced turbine airfoils. The method was applied successfully to various hole configurations (hole size, skew angle, spacing between holes). This parameter was found to be able to correlate mean strain effects and hold times under constant stress.

Jordan and Meyers (1986) used TMF tests on Hastelloy-X under strain-controlled conditions to compare the usefulness of stress intensity factor, strain intensity factor range, and cyclic *J*-integral. Tests involve mostly out-of-phase cycling between 427°C and T_{max} (760°C, 871°C, 927°C, and 982°C) and strain ranges (0.4%, 0.25%, and 0.15%). The stress intensity factor ΔK was found to be the best parameter to reduce data for different strain ranges for out-of-phase cycles in the range 427–927°C, cyclic *J*-integral the worst.

None of the parameters can reduce completely the influence of maximum temperature on TMF crack growth rate for out-of-phase cycles. Comparison between out-of-phase, in-phase, and a faithful cycle with a broad hysteresis in the strain–temperature loop gives a crossover in crack growth rate curves whatever the correlating parameter. The in-phase and faithful cycle give higher crack growth rate than the out-of-phase one at low values of the loading parameter, while the opposite is true at high values of the loading parameter.

The authors concluded that from an engineering point of view all these parameters could be used, since they can correlate crack growth rate within a factor of 2 in all situations investigated.

Marchand *et al.* (1988) have investigated the behavior of the same alloy in the temperature range 427–927°C in strain-controlled loading and showed that in-phase cycling and complex temperature strain cycling could be predicted from isothermal data on a ΔK_ε basis. However, out-of-phase crack growth rates were higher than predicted and most of the complex effects of maximum temperature could be accounted for.

TMF experiments by Marchand *et al.* (1988) were carried out on superalloy B 1900-Hf using single edge specimens in the temperature range 400–925°C. Tests were also conducted at decreasing effective strain, since displacement is controlled behind the crack tip and encompasses some crack opening displacement when the compliance of the specimen decreases. Crack arrest at the detection limit of testing conditions was therefore corresponding to d*a*/d*N* smaller than 1 μm cycle^{-1}.

Crack growth rate was much better correlated using the effective stress intensity factor range instead of the strain intensity factor range or cyclic *J*-integral (Figure 74). The effective ΔK was indeed better than the usual ΔK. There are several reasons for this. First in-phase and out-of-phase cycling give rise to mean stress effect (tensile for OP cycle, compressive for IP cycle). Marchand used potential drop measurements along the loading cycle to estimate an opening stress.

Such measurements are not so easy but within a limited scatter, out-of-phase cycling gives the same crack growth rate as isothermal fatigue at minimum temperature. This difference in behavior has been explained in terms of the fracture mode. At minimum temperature (427°C) the fracture mode is transgranular and it remains transgranular for out-of-phase TMF

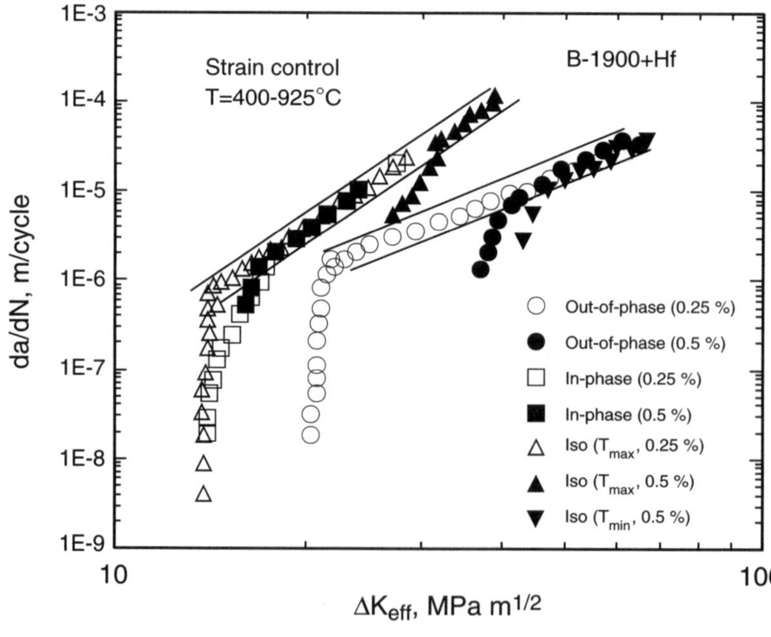

Figure 74 TMF crack growth rate in superalloy B-1900 (+Hf) tested under strain control between 400°C and 925°C for out-of-phase and in-phase cycling vs. effective range of stress intensity factor; comparison with isothermal tests (after Marchand *et al.*, 1988).

tests. In-phase TMF, alternatively, gives rise to intergranular fracture as isothermal fatigue at maximum temperature, the same rationale was found to apply to Hastelloy X (Marchand *et al.*, 1988).

Some bias in the analysis of crack growth can come from the solution used to correlate results. Marchand *et al.* (1988) took care to account for extra constraints imposed by the grips which can influence significantly the crack driving force (Marchand *et al.*, 1986; Yagawa *et al.*, 1985). Extra constraints seem to have been neglected in some work on tubular specimens (Jordan and Meyers, 1986; Okazaki and Koizumi, 1983, 1987).

Cook *et al.* (1996) have attempted to correlate TMF crack growth data on IN718 under inelastic loading conditions: using SEN geometry under strain control. A nonlinear FE analysis including node release to model crack advance was performed on each test, where the crack mouth opening displacement, load, and back face displacement were measured.

Different path-independent integrals were used to correlate the data and they yield satisfactory results, such as Kishimoto *et al.* (1980) cyclic J^*-integral. It was possible to correlate the isothermal base line results at 538°C, and in-phase and out-of-phase TMF data in the temperature range 427–649°C into a narrow scatter band. These methods were much less successful in the presence of a thermal gradient, which is nevertheless the condition encountered in components.

Load-controlled tests were conducted on superalloys IN718 by Nicholas *et al.* (1989) to evaluate crack growth rate as a function of the phase angle between load and temperature. This alloy has been crudely investigated under isothermal conditions by a number of authors (see, e.g., Clavel and Pineau, 1978; Pèdron and Pineau, 1982). At 427°C, the crack growth rate is only cycle dependent, and fracture is transgranular. At 649°C, the crack growth rate is highly time dependent, and for a frequency of 0.01 Hz, Nicholas *et al.* found a 100 times increase in FCGR for the same ΔK with respect to that observed at 427°C. This effect is known to be the consequence of an oxidation-assisted intergranular cracking, which is mainly due to the tension going part of the loading cycle.

Nicholas *et al.* varied the phase angle between temperature and strain in a systematic manner as a function of time. They found, using 0° (in-phase), 90° (clockwise diamond), 180° (out-of-phase), and 270° (counterclockwise diamond), that the in-phase cycle was the more detrimental. The counterclockwise diamond comes second but with an FCGR only five times that observed at T_{min} (see Table 4).

The most severe TMF cycles are ranked in the order of the temperature of the peak stress of the TMF cycle. However, the effect of cycling is slightly more complex since the counterclockwise diamond is more damaging than the clockwise diamond, although the peak stress temperature is the same. These cycles differ actually by the time spent above the mean temperature and mean stress.

The ranking of TMF crack growth rate closely corresponds to the average stress on the tension going part of the cycle above mean temperature. Nicholas *et al.* (1989) used a summation rule of the form

$$da/dN = (da/dN)_{\text{cycle-dep}} + \int \frac{da}{dt}(K, T(t))\, dt \quad (122)$$

da/dt was determined from sustained load experiments as a function of K and temperature.

The authors proposed to obtain the time-dependent damage term by integrating da/dt only along the load portion of the TMF cycle in use, while $da/dt(K, T)$ is a numerically increasing function. The integration was

Table 4 Influence of TMF phase angle on FCGR of IN718 at $\Delta K = 38\ \text{MPa m}^{1/2}$ in the temperature range 427–649°C.

Cycle type	(da/dN) (m/cycle)	*Peak stress temperature* (°C)	*Fraction of period above* $\bar{T}$ *and* $\bar{\sigma}$	*Effective tension ratio above* $\bar{T}$[a]
Isothermal at T_{min}	4.6×10^{-7}	427		
90° TMF (clockwise diamond)	7.5×10^{-7}	538	0	0
180° TMF (out-of-phase)	9.7×10^{-7}	427	0	1/4
270° TMF (counterclockwise diamond)	1.9×10^{-6}	538	1/4	1/2
0° TMF (in-phase)	1.6×10^{-5}	649	1/4	3/4
Isothermal at T_{max}	5.8×10^{-5}	649		

After Nicholas *et al.* (1989).
$\bar{T}$, average temperature of the cycle.
$\bar{\sigma}$, average stress of the cycle.
[a] Ratio of average stress on the tension going part of the cycle above mean temperature to peak stress.

terminated when da/dt starts to decrease, whether or not maximum load has been reached. A comparison between predicted FCGR and experimental results shows this to be a valid postulate.

Okazaki and co-workers (Okazaki and Koizumi, 1983, 1987; Okazaki *et al.*, 1983) have investigated the TMF behavior of various steels, a cast low-alloy ferritic steel (1Cr–1Mo steel), martensitic 12Cr–Mo–V–W, and austenitic 304 stainless steels.

For fairly high strain rate and out-of-phase 300–600°C strain cycling for the austenitic alloy, and 300–550°C for the others, a good correlation of crack growth rates was obtained using the cyclic J-integral:

$$da/dN = CJ_{\text{cycl}}^{m} \tag{123}$$

Cyclic J was estimated first using the partition of J into elastic and plastic components as proposed by Shih and Hutchinson (1976) and using the plane-stress approximation in the short crack limit where analytical expressions were proposed. Fatigue crack growth data from tubular specimens with small through-cracks emanating from a circular hole, or surface semi-elliptical cracks obtained for different strains, do agree with single power-law equation, with a narrow scatter band ($da/dN >$ a few $10^{-7}\,\text{m cycle}^{-1}$) and m in the range 1.2–1.5.

Using slow–fast strain cycling under isothermal loading or strain hold, creep can occur in the study of various steels. Crack growth rates do not correlate so well in that case and the crack growth rate increases with decreasing strain rate or decreasing frequency; this effect is increased at higher strain amplitudes (Okazaki *et al.*, 1983). This correlates with the increased proportion of intergranular facets at low frequency while at high strain rates at high temperature, fracture is transgranular and exhibits striations.

They used a creep-modified J-integral, J^*, defined in analogy to the J-integral, substituting strain rate for strains. This is the same concept as that of C^* introduced by Riedel and Rice (1980) and extensively used to describe static creep crack growth.

Application to creep in specimens with creep-fatigue loading and gross viscoplasticity is rather difficult. For a given stress level, J^* is given as

$$J^* = \sigma \dot{\delta}_c a f(1/n_c) \tag{124}$$

where n_c is the exponent of Norton's law ($\dot{\varepsilon} \neq \sigma^{n_c}$). There are, of course, severe limitations to the application of such a law. Experimentally, the authors used measurement of the creep rate on the very specimen used for crack growth testing to estimate the exponent.

Taira *et al.* (1979a, 1979b) used a total cyclic creep term $J_{\text{cycl,c}}$ by integration of J^* on the time interval where creep occurs:

$$J_{\text{cycl,c}} = \int J^* dt \tag{125}$$

Measurements of displacement rate in the center of the crack were used to check the predicted value of COD and were in good agreement with the model used (Okazaki and Koizumi, 1983).

Under creep-fatigue loading, a fairly good correlation was observed between crack growth rate (da/dN) and total cyclic J:

$$J_{\text{cycl,t}} = J_{\text{cycl,fat}} + J_{\text{cycl, creep}} \tag{126}$$

Observations on fracture surfaces show that slow–fast cycling gives rise to GB sliding, as already pointed out by Fujino and Taira (1980).

Okazaki *et al.* (1983) proposed that the crack growth rate could be estimated as

$$da/dN = da/dN[J_{\text{cycl,fat}}] + da/dN[J_{\text{cycl, creep}}] \tag{127}$$

and can be bracketed between a lower bound given by a pure FCGR described by da/dN $[J_{\text{cycl, fat}}]$, and an upper bound given by a pure creep-crack growth rate described by da/dN $[J_{\text{cycl, creep}}]$.

Application of this total J concept was investigated for in-phase TMF in these various steels. A better correlation was actually found using total J_{cycl} than the fatigue component alone. This is especially useful for in-phase TMF loading and thus improved correlation of crack growth rate is so achieved.

This approach, though valuable, does not take crack closure into account, which was considered in the Skelton (1983) scheme. Investigation of creep-fatigue crack growth under tension–compression by Koterazawa (1993) on stainless steel 304 (at 650°C), 2.25Cr–1Mo steel (at 565°C), low carbon steel (at 400°C), Hastelloy X (750°C), and Ti-6242 (500°C) using center cracked panels tends to support the use of total cyclic J-integral to correlate crack growth rate, irrespective of the loading type and of the intergranular or transgranular character of the crack growth. Different loading cycles were examined: pp (continuous fatigue loads), cp (creep in the tension reversed by plasticity in compression), and cc (creep in tension reversed by creep in compression). Tests under vacuum correlate well with those in air in terms of cyclic J. At high growth rates only, lower-creep-ductility materials tend to give higher crack growth rate for the same cyclic J value.

Crack growth under TMF was investigated in 1070 (eutectoid) steel, which is used in railroad wheel (Sehitoglu, 1987). A comparison was made between isothermal LCF at T_{max}, out-of-phase TMF (with T_{max} in the range 275–450°C and $T_{min} = 150$°C) and repeated strain aging conditions (alternating exposure at zero stress level and repeated strain cycling at room temperature) using double edge notch specimen (initial crack depth 0.6 mm for 15.5 mm width). The crack growth rate in out-of-phase TMF was found to increase drastically with the increase in maximum temperature. This alloy is very sensitive to strain aging and exhibits a strong increase in yield strength with increasing number of strain aging cycles. However, its crack growth rate increases drastically with respect to isothermal conditions, while aging should decrease the COD.

Sehitoglu used crack-tip opening displacement range, $\Delta\delta$, as estimated in the Dugdale model (plane stress):

$$\Delta\delta = \frac{8}{\pi}\frac{2s_0}{E}a \ln\left[\sec\frac{\pi}{2}\frac{\Delta s_{eff}}{2s_0}\right] \qquad (128)$$

where Δs_{eff} is the stress range over which the crack remains open, s_0 is the flow stress, and E the Young's modulus. The crack was assumed to be opened on the tensile portion of the hysteresis loop and, to account for strain hardening, an average value between yield and ultimate strength was used for s_0.

This COD parameter is rather effective for a single TMF history to correlate tests with different strain ranges. However, the very strong dependence of thermal–mechanical history cannot vanish. Dynamic strain aging as well as environmental effects strongly interact in such steels and should be responsible for coupling effects, which cannot be reduced by simple mechanical parameters.

5.03.8.3 Crack Growth under Thermal Transients

Conventional linear elastic fracture mechanics is widely used to investigate crack growth under TF. Most authors calculate the stress intensity factor range ΔK through a weighting function method and use an elasto- or viscoplastic analysis of the noncracked structure. Mowbray *et al.* (1973) used this approach to estimate ΔK in TS experiments on superalloy using double-edge notched Glenny-type specimens. They used the equation

$$\Delta K = (2/\pi)^{1/2}\int_0^a \Delta\sigma(x/a)x^{-1/2}m(x/a)\,dx \qquad (129)$$

where x is the distance along the crack, a the crack length, $\Delta\sigma(x/a)$ the range of normal stress as computed with no crack, and $m(x/a)$ a weighting function given by Bueckner (1972).

This equation was computed when the strain range was maximum at the thin edge. ΔK varies like the crack growth rate as a function of crack length, which is observed in Glenny-type specimens, Figure 75 (Mowbray and Woodford, 1973). It is questionable at small crack length, but it gives a rationale for the plateau in growth rate observed in such specimens. It was used, in particular, to correlate growth rate in the plateau regime for different specimen geometries with some success, Figure 76 (Mowbray *et al.*, 1973). A more difficult exercise is to predict the whole crack growth curve in a component under thermal transients.

Skelton (1980, 1990) has investigated the problem of crack arrest in TS experiments in stainless steel 316. Down-shock experiments were carried out on cylindrical specimens heated by induction at 625°C with a core cooled down for 2 s every 2 min. Longitudinal and circumferential cracks originate at the inner rim and grow to an identical depth. The growth rate of circumferential cracks was monitored using potential-drop measurements and was found to increase up to a maximum and then to decrease until arrest at midwall thickness. The crack growth rate curve at 625°C was used to describe the material behavior under thermal loading. The range of stress intensity factor values was computed using a weight function method. This conventional procedure yields the proper trend of a maximum in crack growth rate at intermediate depth and a decrease in it for larger depth; however, the crack growth rate

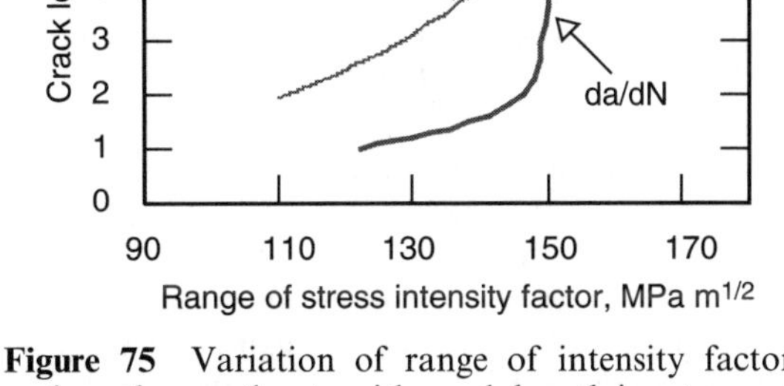

Figure 75 Variation of range of intensity factor and crack growth rate with crack length in a tapered disk specimen with a peripheral radius of 0.25 mm submitted to TS between room temperature and 920°C (after Mowbray *et al.*, 1973).

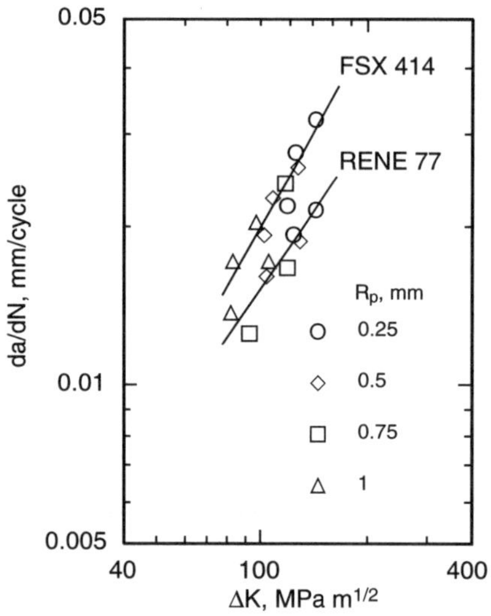

Figure 76 Variation of crack growth rate in the plateau regime with the range of stress intensity factor in tapered disk specimen of various peripheral radii R_p submitted to TS between room temperature and 920°C (after Mowbray *et al.*, 1973).

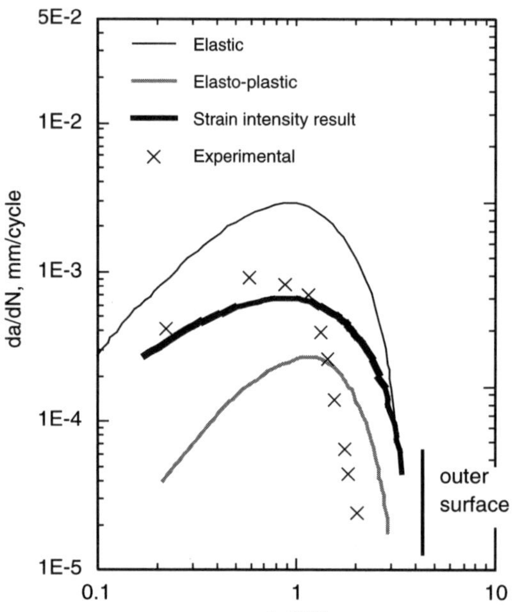

Figure 77 Comparison between predicted and experimental crack growth rate vs. crack depth in thermal downshock experiment on a 316 stainless steel hollow specimen (after Skelton, 1980).

is somewhat overestimated and the decrease of rate is achieved at much higher depth-to-thickness ratio than actually observed experimentally (Figure 77). Different improvements were attempted at that time using an elastic–plastic analysis of the cracked billet or using strain intensity but some discrepancy between experiment and calculation cannot be completely eliminated.

One of the reasons for this discrepancy was attributed to the occurrence of multiple cracking in the two principal growth directions, except when notches are introduced in specimens. The effect of an array of equally spaced transverse (circumferential) and radial (longitudinal) cracks is to reduce the stress intensity factor in a substantial way (Baratta, 1981). Various estimates of relief factors as a function of crack depth/crack spacing ratio were compared by Skelton (1983).

Both through a stress intensity factor analysis and by experimental observations, a multiple array of cracks tend to stop to some depth, for a pure TS loading. Interactions between cracks at that time consider simplified arrays with equally spaced cracks of the same length.

The analysis of crack growth under thermal transient conditions is often made using the model proposed by Haigh and Skelton (1978). This model considers that, when there is plasticity, a linear elastic fracture mechanics ΔK can be extended using an effective stress written as

$$\Delta K = \Delta\sigma_{\text{eff}}\sqrt{\pi a}\; Y\left(\frac{a}{w}\right) \qquad (130)$$

$$\Delta\sigma_{\text{eff}} = q\Delta\sigma + E\Delta\varepsilon_{\text{p}} \qquad (131)$$

The elastic term introduces a factor q to describe crack closure, which for tension–compression loading is often below 0.5 (i.e., opening of the crack tip occurs for compressive stresses). The second term is a pseudostress normal to the crack plane, which would produce an elastic strain equal to the plastic strain range.

This effective stress range can be estimated from FE computations, ignoring the presence of a crack. To estimate the effective stress intensity factor, the effective stress is fitted by a polynomial form as a function of depth x:

$$\Delta\sigma_{\text{eff}}(\text{x}) = \sum A_i x^i \qquad (132)$$

An effective ΔK is estimated using the superposition method of Buchalet and Bamford (1978) for cracks in cylinders:

$$\Delta K_{\text{eff}}(a) = \sqrt{\pi a}\left(A_0F_1 + \frac{2a}{\pi}A_1F_2 + \frac{a^2}{2}A_2F_3 + \frac{4a^3}{3\pi}A_3F_4\right) \qquad (133)$$

Fissolo *et al.* (1996) have shown that the growth of a single crack under thermal upshock between 300°C and 550°C in stainless steel 316 can properly be described using a constant opening coefficient q of 0.6 (Figure 78).

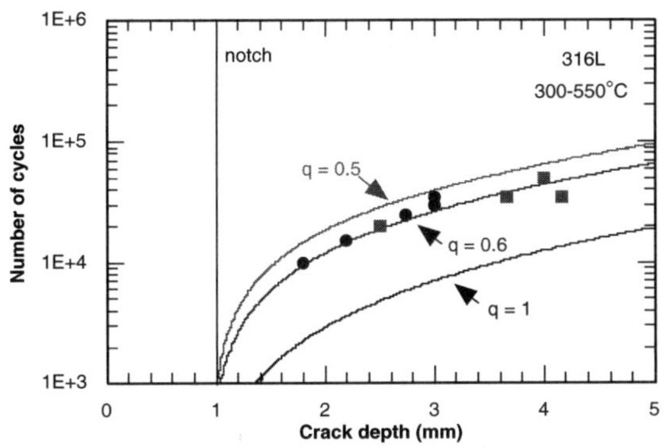

Figure 78 Number of cycles as a function of crack length for a single crack TS test on 316 steel with the Haigh and Skelton model: an opening coefficient q of 0.6 fits experimental results (courtesy of Fissolo).

The same approach has been used, together combining nonlinear analysis of stress–strain loading and this weight function method, to describe TF cracks in molds for centrifugal casting of large cast iron tubes (Burlet *et al.*, 1990). Life has been obtained by integration of a Paris-type law for crack propagation between an initial crack (0.5–2.5 mm, deduced from observation on in-service cracked molds) and final crack depth (17 mm observed in cracked components). In-service lives vary between 1,000 and 2,000 cycles: TF lives are over-estimated by a factor of 10 when the effect of plasticity is neglected in the computation of stress intensity factor range.

A much better agreement is observed when a Skelton's effective ΔK is used instead, using a q crack closure factor deduced from load–displacement loops measured near the crack tip in high strain fatigue crack growth experiments.

The development of a network of cracks under TS, and the observations of arrest of high density of secondary cracks suggest that it is necessary to account for multiple cracking. The relief factor due to the shielding of one crack by the presence of the others decreases with the ratio "crack depth/crack spacing" (Skelton, 1983). A denser array gives more reduction of the stress intensity factor as compared to a single crack situation. Recent work has considered more carefully the problem of nonuniform arrays and variation in crack growth rate from one crack to the other one.

More sophisticated calculations can be made using a Monte-Carlo method and remeshing in an FE method of crack growth. Such a 2D model was used by Fissolo and Marini (1998) and Maillot *et al.* (2002). First a Monte-Carlo randomization method is used to determine the size of grains according to measured distributions, and from the beginning an initial set of grains at the surface is assumed to be cracked.

The Skelton (1983) model with an equivalent stress factor is used along the lines described above. The growth of each crack follows a Paris law for a given number of cycles. An auto-adaptative meshing is used, refining only the crack-tip region to reduce computation time. The computation clearly shows that a smaller crack close to a deeper crack will slow the growth of the deeper one; the growth of the smaller one will be reduced even more, so that the smaller cracks eventually stop growing (Figure 79).

Mean and maximum crack depths computed are close to those observed experimentally. Such a model rationalizes the observed behavior of crack network in TS. 3D computations taking into account crack front effects (Dell'Erba and Aliabadi, 2000) should bring predictions closer to experimental results.

5.03.8.3.1 Effect of superimposed loading

In a number of cases, though crack growth can be tolerated, complete transverse crack growth must be avoided.

Marsh (1981) and Marsh *et al.* (1986) tested large tubes (152 mm inner diameter, 7 mm wall thickness) and carried experiments on tubes with a welded internal plug and crevice with down-shocks between 200°C and 400°C. A 900 s period was used for the larger-amplitude down-shock, using external cooling with a water spray. Fairly high crack growth rates were observed and a long inner circumferential crack could grow through wall thickness within 1,300 cycles (Marsh *et al.*, 1986). A combination of welded zones and defects gives rise to residual stress pattern, which superimposes on thermal stresses.

A number of investigations were carried out in austenitic stainless steels to assess the effect of end loading on the development of a cracking network. While a TF crack network tends to stabilize at some crack depth, external stresses tend to favor the growth of an individual crack.

Analyses were presented by Skelton (1990) which show that superposition of a static load tends to add a dangerous extra contribution to stress intensity factor which might increase when increasing static load values. A material with a low threshold for crack growth might yield a single crack breakthrough, as observed in some instances on small 316 specimens exposed to peak tension $\sim$60 MPa (Skelton, 1990) for a typical quench 625–300°C. Different investigators fail to achieve a single crack breaking through in hollow billets or prismatic bars (Hasan and Brown, 1990).

As reported by Skelton (1990), Marsh and Charlesworth have subjected 316 steel to TSs from 650°C under constant stress. Therefore, specimens give rise to ratchetting.

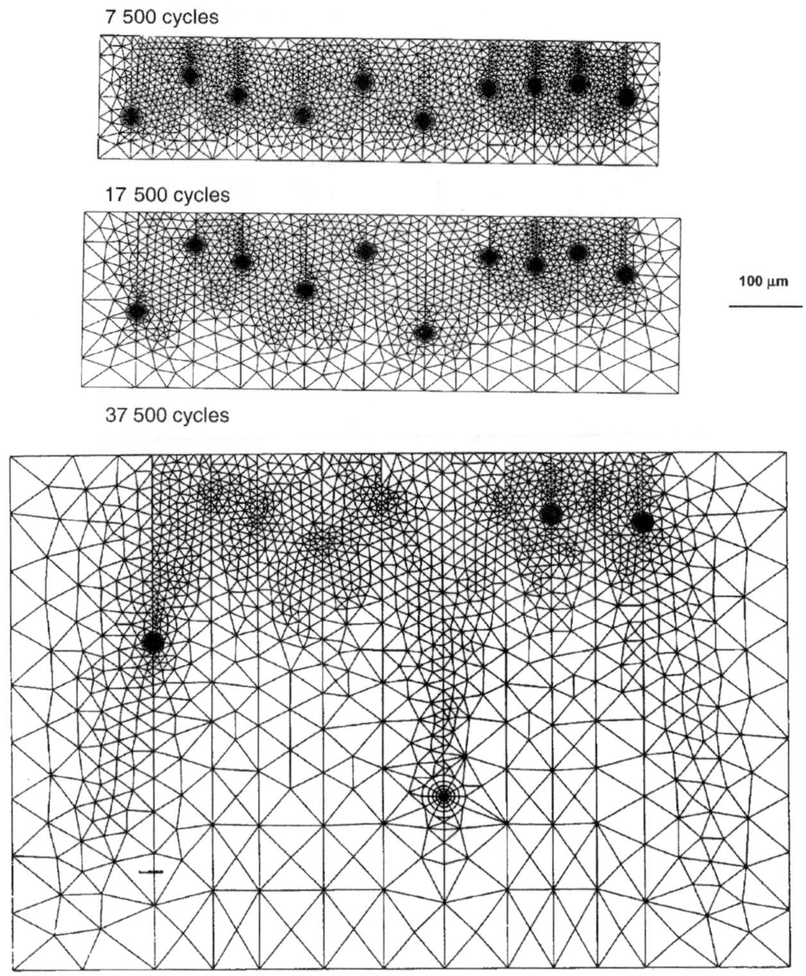

Figure 79 Numerical simulation of interactions between cracks in a network and their influence on the distribution of crack growth (courtesy of Fissolo).

Short crack growth rates were compared with the upper bound for B in Equation (99), $B = 2.61 \times 10^4 \Delta\varepsilon_m^{2.85}$, valid for continuous cycling in continuous fatigue. For end load tests under no ratchetting conditions, the crack growth rate is below the upper bound. When there is ratcheting, the crack growth rate constant tends to exceed the upper bound (Figure 80). Conditions of crack breakthrough are almost never reached.

This area needs further confirmation, since many situations in practice will involve superposition of a static load and temperature variations.

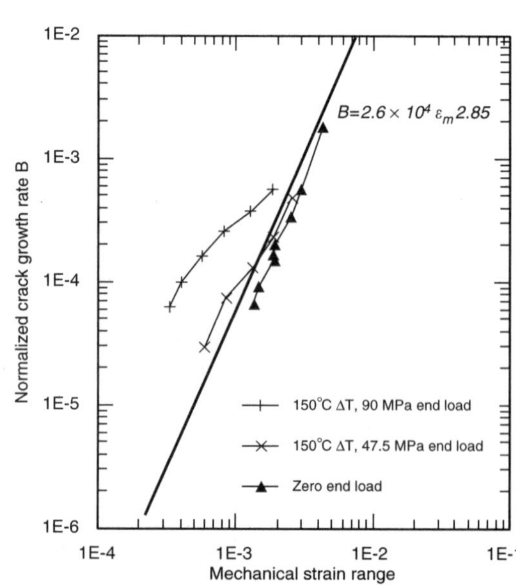

Figure 80 Influence on normalized crack growth rate vs. strain range in TS experiments: influence of end loading (after Skelton, 1990).

5.03.9 CONCLUSIONS

The field of TMFs (and TSs) has now achieved a large degree of maturity as compared to the situation in 1975, as can be seen from the reference STP edited by Spera and Mowbray (1976). At that time, only pioneering work had been carried out by several groups:

Taira (1973) and Jaske (1976) on steels and Hopkins (1976) and (Rau *et al.*, 1973) on superalloys. A major incentive for advances in TMF has been the development of closed-loop servohydraulic machines together with the introduction of microcomputers in the late 1970s. Only a combination of both technologies has made possible the necessary precision in test control as the monitoring and storage capabilities required by such complex tests.

The analysis of TS was also in infancy in 1975. The pioneering work of Spera (1969) and Mowbray and Mc Connelee (1976) introduced nonlinear stress analysis of the Glenny-type TS experiments. A fairly crude description of Bauschinger effect was used at that time.

Viscoplastic constitutive models, especially the unified models, appeared in the late 1970s, but in the 1980s the nonlinear stress analysis was still difficult to carry out and only 2D approximation could easily be used in nonlinear FE analysis.

Thermoelasticity was used in most instances. This explains why most design codes are still based on strain variables to describe fatigue in the LCF regime, even when high-temperature excursions incurred. In numerous austenitic and ferritic steels used in power plants, thermal–mechanical creep fatigue can be analyzed in such a way in ASME codes or equivalent European codes, R5 or RCC-MR (in nuclear power plants).

The definition of service loading at high temperatures is always a major issue. This is a key issue in every fatigue problem, but it is much more difficult to know thermal stresses in components under thermal transients, with secondary stresses as opposite to primary stresses in mechanical fatigue. Numerous sophisticated computations are often necessary in most TF problems, and complex 3D fluid dynamics and thermomechanics computations are necessary, given the use of gas in engines (automotive, aeroengines, or stationary gas turbines, hot and cold fluids in various circuits in power plants) and interaction between fluid and structures.

In other cases heating dissipation comes from friction like in brakes for automotive and railway industry. The determination of heat exchange coefficients is often fairly difficult but essential.

The continuous increase in computer capabilities make transient thermal analyses for many components possible.

The reduction of safety margins, due to the search for increased performance with increased temperatures, and increase in thermal stresses, or due to weight reductions, makes the use of nonlinear analyses more and more useful.

There are two major driving forces: standard codes are often conservative, though for high-strength materials and complex damage interactions they may yield nonconservative predictions; the reduction of development cost and the time to introduce new components into the market needs to use more computational steps and less component testing.

Nonlinear unified viscoplastic models developed in the 1980s or embellished more recently that employ internal variables are now used in various industries, at least for critical components. Nonlinear FE analyses are now made with a very large number of degrees of freedom (from 10,000 up to 1 million), using either work stations, parallel computers or PC clusters.

While stresses are now computed in a more reliable way for thermal mechanical loading, simple damage laws are used in numerous cases. One may anticipate that more performing analyses will be carried out at the design stage, making full use of continuum damage mechanics and physically based models.

Design engineers need robust and rapid computations. A lot of experience has accumulated in high-temperature fatigue, creep-fatigue, and TMF which shows that damage interactions often occur in real component loading which cannot be captured by simple models, mostly derived from knowledge under isothermal conditions.

Drawing from the progress made in stress analyses with the introduction, in industrial practice, of constitutive models developed at the laboratory stage in the late 1970s and 1980s, one may anticipate that performing damage models should spread over various industries in future.

The issue of crack propagation under thermal–mechanical loading still requires further work. Progress should follow that achieved for the description of crack growth under creep-fatigue conditions. Thermal transients involve, however, a number of specific difficulties such as contact between crack surfaces, which should be addressed. When crack growth can be tolerated in components, the transition from short cracks in yielded or viscoplastic zones to longer cracks under small-scale yielding is still an open problem.

ACKNOWLEDGMENTS

The author would like to thank Dr. A. Fissolo, Prof. F. Rezai-Aria, Prof. H. Sehitoglu, Dr. F. Gallerneau, Pr. G. Cailletaud, as well as Dr. S. Hertz-Clemens, Dr. A. Köster, N. Haddar, and L. Bucher, for useful discussions and help with illustrations during the preparation of this manuscript, and Mrs. V. Matos for help with the manuscript.

5.03.10 REFERENCES

M. R. Achter, G. J. Danek, Jr. and H. H. Smith, 1963, Effect on fatigue of gaseous environments under varying temperature and pressure. *Trans. Met. Soc. AIME*, **227**, 1296–1301.

R. A. Ainsworth and P. J. Budden, 1993, Assessment of defects at high temperatures: the r5 procedures. In: "Behaviour of Defects at High Temperatures, ESIS 15," eds. R. A. Ainsworth and R. P. Skelton, Mechanical Engineering Publications, London, pp. 453–465.

E. Andrieu, R. Molins, H. Ghonem and A. Pineau, 1992, Intergranular crack tip oxidation mechanism in a nickel-based superalloy. *Mater. Sci. Eng. A*, **154**, 21–28.

L. Angarita, G. Pitz, K.-H. Lang and D. Löhe, 2000, Realization of complex thermal–mechanical fatigue by a two-specimen testing system. In: "Thermo-mechanical Fatigue Behaviour of Materials: 3rd Volume, ASTM STP 1371," eds. H. Sehitoglu and H. J. Maier, American Society for Testing and Materials, West Conshohocken, PA, pp. 304–318.

S. D. Antolovich, S. Liu and R. Baur, 1981, Low cycle fatigue behavior of René 80 at elevated temperature. *Metall. Trans. A*, **12**, 473–481.

M. Ariana, J. M. Martinez-Esnaola and J. Bressers, 1996, Crack propagation and life prediction in a nickel-based superalloy under TMF conditions. In: "Fatigue under Thermal and Mechanical Loading," eds. J. Bressers, L. Rémy, M. Steen and J. L. Vallès, Kluwer, Amsterdam, pp. 393–402.

D. J. Arrel, K. M. Ostolaza, J. L. Vallès and J. Bressers, 1996, Solidification structures and post-fatigue precipitate morphology in a single crystal superalloy. In: "Fatigue under Thermal and Mechanical Loading," eds. J. Bressers, L. Rémy, M. Steen and J. L. Vallès, Kluwer, Amsterdam, pp. 295–304.

M. F. Ashby and B. F. Dyson, 1984, Creep damage mechanics and micromechanisms. In: "Advances in Fracture Research, Proceedings, ICF 6," eds. S. R. Valluri, D. M. R. Taplin, P. Rama Rao, J. F. Knott and R. Dubey, Pergamon, vol. 1, pp. 3–30.

K. R. Bain, 1985, The effect of coatings on the thermo-mechanical fatigue life of a single crystal turbine blade material. In: "American Institute of Aeronautics and Astronautics, AIAA/SAE/ASME/ASEE 21st Joint Propulsion Conference, Monterey, CA, July 8–10, 1985, AIAA-85-1366," pp. 85–1366.

F. I. Baratta, 1981, Stress intensity factors for internal multiple cracks in thick-walled cylinders stressed by internal pressure using load relief factors. *Eng. Fract. Mech.*, **14**, 237–241.

B. Baufeld, E. Tzimas, P. Hähner, H. Müllejans, S. Peteves and P. Moretto, 2001, Phase angle effects on damage mechanisms of thermal barrier coatings under thermo-mechanical fatigue. *Scr. Mater.*, **45**, 859–865.

T. Beck, R. Ratchev, M. Moalla, K.-H. Lang and D. Löhe, 2002, Lifetime, cyclic deformation and damage behaviour of Mar-M 247 CC under in-phase, out-of-phase and phase-shift TMF loadings. eds. L. Rémy and J. Petit, ESIS Publication 29, Elsevier, Amsterdam, pp. 115–124.

H. Bernard and L. Rémy, 1990, Thermal–mechanical fatigue of an aluminide coated nickel base superalloy. In: "Advanced Materials and Process EUROMAT 89," eds. H. E. Exner and V. Schumacher, DGM, Oberursel, vol. 1, pp. 529–534.

P. Bernède, 1994, Effet de l'oxydation sur la fissuration en fatigue-fluage d'un superalliage pour disques de turbines: l'Astroloy. Ph.D. thesis, Ecole des Mines, Paris.

P. Bernède and L. Rémy, 1992, Fatigue crack growth of small defects in a superalloy at high temperature in air and in vacuum. In: "Short Fatigue Cracks, ESIS 13," eds. K. J. Miller and E. R. de los Rios, Mechanical Engineering Publications, London, pp. 251–262.

H. Bernstein, T. S. Grant, R. C. McClung and J. P. Allen, 1993, Prediction of thermal–mechanical fatigue life for gas turbine blades in electric power generation. In: "Thermal Mechanical Fatigue Behavior of Materials, ASTM STP 1186," ed. H. Sehitoglu, American Society for testing and Materials, Philadelphia, pp. 212–238.

J. Besson, G. Cailletaud, J. L. Chaboche and S. Forest, 2001, Mécanique non linéaire des matériaux, Hermes.

A. Bignonnet and E. Charkaluk, 2002, Thermomechanical fatigue in the automotive industry. In: "Temperature-fatigue Interaction," eds. L. Rémy and J. Petit, ESIS Publication 29, Elsevier, Amsterdam, pp. 319–330.

P. J. Bizon and D. A. Spera, 1976, Thermal-stress fatigue behavior of twenty-six superalloys. In: "Thermal Fatigue of Materials and Components, ASTM STP 612," eds. D. A. Spera and D. F. Mowbray, ASTM, Philadelphia, pp. 106–122.

M. Blümm, C. C. Engler-Pinto, Jr., F. Meyer-Olbersleben and F. Rezai-Aria, 1994, Grain orientation and size effects on thermal fatigue behaviour of CM247 LC DS. In: "Materials for Power Engineering," eds. D. Coutsouradis, *et al.*, Kluwer, Dordrecht, Part II, pp. 1099–1108.

P. Bonacuse and S. Kalluri, 1996, Axial-torsional thermo-mechanical fatigue behavior of Haynes 188 superalloy. In: "Thermal Mechanical Fatigue of Aircraft Engine Materials, Agard Conference Proceedings 569, 81st Meeting of the AGARD Structures and Materials Panel, Banff, Canada, October 2–4, 1995," AGARD, Neuilly, France, pp. 15-1, 15-10.

J. Bressers, L. Rémy and W. Hoffelner, 1986, Fatigue dominated damage processes. In: "High Temperature Alloys 1986: Proceedings of a Conference held in Liege, Belgium, October 6–9, 1986," eds. W. Betz, R. Brunetaud, D. Coutsouradis, H. Fischmeister, T. B. Gibbons, I. Kvernes, Y. Lindblom, J. B. Marriott and D. B. Meadowcroft, Reidel, Dordrecht, vol. 1, pp. 441–468.

A. Brückner-Foit, H. Jäckels and U. Quadfasel, 1993, Prediction of the lifetime distribution of high-strength components subjected to fatigue loading. *Fatigue Fract. Mater. Struct.*, **16**, 891–908.

C. B. Buchalet and W. H. Bamford, 1976, Stress intensity factor solutions for continuous surface flaws in reactor pressure vessels. In: "Mechanics of Crack Growth, ASTM STP 590," eds. J. Rice and P. C. Paris, American Society for Testing and Materials, Philadelphia, pp. 385–402.

H. F. Bueckner, 1972, Field singularities and related integral representations. In: "Mechanics of Fracture," ed. G. C. Sih, Noordhoff, Leyden, vol. 1, pp. 239–314.

H. D. Bui, 1978, "Mecanique de la rupture fragile," Masson, Paris.

C. K. Bullough, M. Toulios, M. Oehl and P. Lukas, 1998, The characterisation of the single crystal superalloy CMSX-4 for industrial gas turbine applications. In: "Materials for Advanced Power Engineering 1998: Proceedings of the 6th Liege Conference," eds. J. Lecomte-Beckers, F. Schubert and P. J. Ennis, Forschungszentrum, Jülich, Jülich, Part II, pp. 861–878.

H. Burlet and G. Cailletaud, 1986, Numerical techniques for cyclic plasticity at variable temperature. *Eng. Comp.*, **3**, 143–153.

H. Burlet, S. Vasseur, G. Cailletaud and A. Pineau, 1990, Fatigue crack growth under thermomechanical loading application to life prediction of centrifugal casting equipment. In: "High Temperature Fracture Mechanisms and Mechanics, EGF 6," eds. P. Bensussan and J. P. Mascarell, Mechanical Engineering Publications, London, pp. 379–395.

E. S. Busso and F. A. McClintock, 1988, Stress–strain histories in coatings on single crystal specimens of a turbine blade alloy. *Int. J. Solids Struct.*, **11**, 1113–1130.

G. Cailletaud and J. L. Chaboche, 1979, Macroscopic description of the microstructural changes induced by varying temperature: example of IN100 cyclic behaviour of materials. In: "Proceedings of the 3rd International Conference," eds. K. J. Miller and R. F. Smith, Pergamon, Oxford, pp. 23–32.

G. Cailletaud, J. L. Chaboche, S. Forest and L. Rémy, 2003, On the design of single crystal turbine blades. Autumn Meeting SF2M, La Revue de Métallurgie—CIT/Science des Matériaux, 100th year, pp. 1–8.

G. Cailletaud, M. Chaudonneret, J. P. Culie, F. Darbois, P. Peruzetto and P. de Roo, 1990, Two- and three-dimensional calculations of anisothermal cyclic plasticity. In: "High Temperature Fracture Mechanisms and Mechanics, EGF 6," eds. P. Bensussan and J. P. Mascarell, Mechanical Engineering Publications, London, pp. 211–225.

G. Cailletaud, J. P. Culié and H. Kaczmareck, 1983, Thermal fatigue of a thermally unstable alloy. In: "Mechanical Behaviour of Materials—IV: Proceedings of the 4th International Conference, Stockholm, Sweden, August 15–19, 1983," eds. J. Carlsson and N. F. Ohlson, Pergamon, Oxford, vol. 1, pp. 225–261.

G. Cailletaud, C. Depoid, D. Massinon and E. Nicouleau-Bourles, 2000, Elastoviscoplasticity with aging in aluminum alloys. In: "Continuum Thermodynamics: The Art and Science of Modelling Material Behaviour," eds. G. A. Maugin, R. Drouot and F. Sidoroff, Kluwer, Dordrecht, pp. 75–86.

G. Cailletaud and C. Levaillant, 1984, Creep-fatigue life prediction what about initiation? *Nucl. Eng. Des.*, **83**, 279–292.

B. J. Cane and M. I. Manning, 1981, "Effect of Test Piece Oxidation on Rupture Data," Internal report lm/MATS/407, CEGB.

M. G. Castelli, R. V. Miner and D. N. Robinson, 1993, Thermomechanical deformation behavior of a dynamic strain ageing alloy, Hastelloy X. In: "Thermomechanical Fatigue Behavior of Materials, ASTM STP 1186," ed. H. Sehitoglu, American Society for Testing and Materials, Philadelphia, pp. 106–125.

J. L. Chaboche, 1974, Une loi différentielle d'endommagement de fatigue avec cumulation non linéaire. *Revue Française de Mécanique*, **50**, 71–82.

J. L. Chaboche, 1982, Lifetime predictions and cumulative damage under high-temperature conditions. In: "Low Cycle Fatigue and Life Prediction, ASTM STP 770," eds. C. Amzallag, B. N. Leis and P. Rabbe, American Society for Testing and Materials, Philadelphia, pp. 81–104.

J. L. Chaboche, 1991, On some modifications of kinematic hardening to improve the description of ratchetting effects. *Int. J. Plasticity*, **7**, 661–678.

J. L. Chaboche and G. Cailletaud, 1996, Integration methods for complex plastic constitutive equations. *Comp. Methods Appl. Mech. Eng.*, **133**, 125–155.

J. L. Chaboche and F. Gallerneau, 2000, Durability modelling at elevated temperature. *Fatigue Fract. Eng. Mater. Struct.*, **24**, 405–418.

J. L. Chaboche and P. M. Lesne, 1988, A non-linear continuous fatigue damage model. *Fatigue Fract. Eng. Mater. Struct.*, **11**, 1–17.

E. Charkaluk and A. Constantinescu, 2000, An energetic approach in thermomechanical fatigue for silicon molybdenum cast iron. *Mater. High Temp.*, **17**, 373–380.

E. Chataigner, E. Fleury and L. Rémy, 1996, Thermomechanical fatigue of coated and bare nickel base superalloy single crystals. In: "Fatigue under Thermal and Mechanical Loading," eds. J. Bressers and L. Rémy, Kluwer Academic, Dordrecht, The Netherlands, pp. 381–392.

E. Chataigner and L. Rémy, 1995, Influence d'un revêtement protecteur (C1A) sur la tenue en fatigue d'aubes de turbines monocristallines. In: "Fatigue et Traitements de surface, SF2M," Revue de Métallurgie, Paris, pp. 11–22.

E. Chataigner and L. Rémy, 1996, Thermomechanical fatigue of coated and bare nickel base superalloy single crystals. In: "Thermomechanical Fatigue Behaviour of Materials, ASTM STP 1263," eds. M. J. Verrilli and M. G. Castelli, American Society for Testing and Materials, Philadelphia, pp. 3–23.

E. Chateau and L. Rémy, 2001, Oxidation-assisted creep damage in Udimet 720 LI. In: "EUROMAT 2001: Proceedings of the 7th European Conference on Advanced Materials and Processes, Rimini, Italy, June 10–14, 2001," Associazione Italiana di Metallurgia, Milano, CD Rom, ISBN 88-85298.

C. Y. Cheng and D. R. Diercks, 1973, Effects of hold time on low-cycle fatigue behavior of AISI type 304 stainless steel. *Metall. Trans.*, **4**, 615–617.

R. Chieragatti and F. Paun, 2000, A new technique for high frequency multiaxial thermo-mechanical fatigue testing of materials. In: "Thermomechanical Fatigue Behavior of Materials: 3rd Volume, ASTM STP 1371," eds. H. Sehitoglu and M. J. Maier, American Society for Testing and Materials, West Conshohoken, PA, pp. 319–332.

W. Christl, A. Rahmel and M. Schütze, 1989, Behavior of oxide scales on 2.25Cr–1Mo steel during thermal cycling: I. Scales formed in oxygen and air, II. Scales grown in water vapor. *Oxidat. Metals*, **31**, 1–69.

M. Clavel and A. Pineau, 1978, Frequency and waveform effects on the fatigue crack growth behaviour of alloy 718 at 298 and 823 K. *Metall. Trans. A*, **9**, 471–480.

P. Clément, J. P. Angeli and A. Pineau, 1984, Short crack behaviour in nodular cast iron. *Fatigue Eng. Mater. Struct.*, **7**, 251–265.

L. F. Coffin, Jr., 1972, The effect of high vacuum on the low cycle fatigue law. *Metall. Trans.*, **3**, 1777–1788.

L. F. Coffin, Jr., 1973, Fatigue at high temperature. In: "Fatigue at Elevated Temperatures, ASTM STP 520," eds. A. E. Carden, A. J. Mc Evily and C. H. Wells, American Society for Testing and Materials, Philadelphia, pp. 5–34.

R. H. Cook and R. P. Skelton, 1974, Environment-dependence of the mechanical properties of metals at high temperature. *Int. Metall. Rev.*, **19**, 199–222.

T. S. Cook, J. H. Loflen, R. H. Van Stone and P. K. Wright, 1996, TMF experience with gas turbine engine materials. In: "Thermal Mechanical Fatigue of Aircraft Engine Materials, AGARD Conference Proceedings 569, 81st Meeting of the AGARD Structures and Materials Panel, Banff, Canada, October 2–4, 1995," pp. 2-1, 2-15.

E. L. Courtright, J. T. Prater, G. R. Holcomb, St. G. R. Pierre and R. A. Rapp, 1991, Oxidation of hafnium carbide and hafnium carbide with additions of tantalum and praseodymium. *Oxidat. Metals*, **36**, 423–437.

B. Dambrine and J. P. Mascarell, 1990, On the interest of using unified viscoplastic models in the life prediction of engine hot components. In: "High Temperature Fracture Mechanisms and Mechanics, EGF 6," eds. P. Bensussan and J. P. Mascarell, Mechanical Engineering Publications, London, pp. 195–210.

R. Danzer and J. Bressers, 1986, A new method to predict the life under high-temperature low cyle fatigue conditions. *Fatigue Fract. Eng. Mater. Struct.*, **9**, 151–168.

D. N. Dell'Erba and M. H. Aliabadi, 2000, Three-dimensional thermo-mechanical fatigue crack growth using BEM. *Int. J. Fatigue*, **22**, 261–273.

N. E. Dowling, 1977, Crack growth during low cycle fatigue of smooth axial specimens. In: "Cyclic Stress–

Strain and Plastic Deformation Aspects of Fatigue Crack Growth, ASTM STP 637," American Society for Testing and Materials, Philadelphia, pp. 97–121.

N. E. Dowling and J. A. Begley, 1976, Fatigue crack growth during gross plasticity and the *J*-integral. In: "Mechanics of Crack Growth, ASTM STP 590," eds. J. Rice and P. C. Paris, American Society for Testing and Materials, Philadelphia, pp. 82–103.

P. Dufrénoy, G. Bodovillé and G. Degallaix, 2002, Damage mechanisms and thermomechanical loading of brake discs. In: "Temperature-fatigue Interaction," eds. L. Rémy and J. Petit, ESIS Publication 29, Elsevier, Amsterdam, pp. 167–176.

C. Duret-Thual, R. Morbioli and P. Steinmetz, 1986, A guide to the control of high temperature corrosion and protection of gas turbine materials. In: "Commission of European Communities, Directorate General Science, Research and Development, EUR 10G82 EN CEC," ed. O. Morocutti, Office for Official Publications of European Communities, Luxembourg.

B. F. Dyson and M. McLean, 1998, Microstructural evolution and its effects on the creep performance of high temperature alloys. In: "Microstructural Stability of Creep Resistant Alloys for High Temperature Plant Applications," eds. A. Strang, J. Cawley and G. W. Greenwood, Institute of Materials, London, pp. 371–392.

B. F. Dyson and S. Osgerby, 1987, Modelling creep-corrosion interactions in nickel-base superalloys. *Mater. Sci. Technol.*, **3**, 545–553.

W. Elber, 1971, The significance of fatigue crack closure. In: "Damage Tolerance in Aircraft Structures, ASTM STP 486," American Society for Testing and Materials, Philadelphia, pp. 230–242.

C. C. Engler-Pinto, Jr., G. Harkegard, B. Ilschner, M. Y. Nazmy, C. Noseda and F. Rezai-Aria, 1994, Thermo-mechanical fatigue behavior of IN738LC. In: "Materials for Advanced Power Engineering," eds. D. Coutsouradis, *et al.*, Kluwer Academic, Dordrecht, The Netherlands, part I, pp. 853–862.

C. C. Engler-Pinto, Jr., H. Sehitoglu, H. J. Maier and T. J. Foglesong, 2002, Thermo-mechanical fatigue behaviour of cast 319 aluminum alloys. In: "Temperature-fatigue Interaction," eds. L. Rémy and J. Petit, ESIS Publication 29, Elsevier, Amsterdam, pp. 3–14.

L. Espié, 1996, Etude expérimentale et modélisation numérique du comportement mécanique de monocristaux de superalliage (in French), Thesis, Ecole des Mines de Paris.

J. Esslinger and E. E. Affeldt, 1998, Thermo-mechanical fatigue (TMF) in turbine airfoils: typical loading, 3D effects and adjusted testing. In: "Low Cycle Fatigue and Elasto-plastic Behaviour of Materials," eds. K. T. Rie and P. D. Portella, Elsevier, Amsterdam, pp. 131–136.

A. G. Evans, G. B. Crumley and R. E. Demaray, 1983, On the mechanical behaviour of brittle coatings and layers. *Oxidat. Metals*, **20**, 193–216.

A. Fatemi and D. F. Socie, 1988, A critical plane approach to multiaxial fatigue damage including out-of-phase loading. *Fatigue Fract. Eng. Mater. Struct.*, **11**(3), 149–165.

H. F. Fischmeister, R. Danzer and B. Buchmayr, 1986, Life time prediction models. In: "High Temperature Alloys 1986: Proceedings of a Conference held in Liege, Belgium, October 6–9, 1986," eds. W. Betz, R. Brunetaud, D. Coutsouradis, H. Fischmeister, T. B. Gibbons, I. Kvernes, Y. Lindblom, J. B. Marriott and D. B. Meadowcroft, Reidel, Dordrecht, vol. l, pp. 495–549.

A. Fissolo and B. Marini, 1998, Multiple cracking in thermal fatigue. In: "Fracture from Defects, ECF 12, Proc. 12th Conference on Fracture," eds. M. W. Brown, E. R. de los Rios and K. J. Miller, EMAS, Cradley Heath, Warley, vol. I, pp. 31–36.

A. Fissolo, B. Marini, A. Berrada, G. Nais and P. Wident, 1996, Initiation and growth of cracks under thermal fatigue loading for a 316L type steel. In: "Fatigue under Thermal and Mechanical Loading," eds. J. Bressers and L. Rémy, Kluwer, Dordrecht, pp. 67–77.

E. Fleury and L. Rémy, 1993, Low cycle fatigue damage in nickel-base superalloy single crystals at elevated temperature. *Mater. Sci. Eng. A*, **167**, 23–30.

E. Fleury and L. Rémy, 1994, Behaviour of nickel-base superalloy single crystals under thermal–mechanical fatigue. *Metall. Mater. Trans. A*, **25**, 99–109.

R. Foerch, V. Gros, V. Mounoury, S. Quilici and G. Cailletaud, 2000, Cyclic calculations and life prediction in thermomechanical fatigue using the Zmat library. Abaqus User's Meeting, USA.

M. François and L. Rémy, 1990, Influence of microstructure on the thermal fatigue behaviour of a cast cobalt based superalloy. *Metall. Trans. A*, **21**, 949–958.

C. J. Franklin, 1978, Cyclic creep and fatigue life prediction. In: "High Temperature Alloys for Gas Turbines: Proceedings of a Conference held in Liège, September 25–27, 1978," eds. E. Glenny, *et al.*, Applied Science Publishers, Barking, pp. 513–547.

H. J. Frost and M. F. Ashby, 1982, "Deformation-mechanism Map," Pergamon, Oxford, pp. 1–5.

M. Fujino and S. Taira, 1980, Effect of thermal cycle on low cycle fatigue of steels and grain boundary sliding characteristics. In: "Proceedings of the 3rd Int. Conf. on Mechanical Behaviour of Material," eds. K. J. Miller and R. F. Smith, Pergamon, Oxford, vol. 2, pp. 49–58.

F. Gallerneau, 1999, Modelling of anisotropy effects of a single crystal superalloy on its creep-fatigue resistance. In: "Proceedings ICAF'99, Seattle, July 12–16, 1999," TP ONERA no. 1999-95.

F. Gallerneau, D. Nouailhas and J. L. Chaboche, 1996a, Fatigue damage behaviour of a coated Ni-base single crystal superalloy. In: "Mechanisms and Mechanics of Damage and Failure, ECF 11, Proc. 11th Conference on Fracture," ed. J. Petit, EMAS, Cradley Heath, Warley, vol. II, pp. 1275–1280.

F. Gallerneau, D. Nouailhas and J. L. Chaboche, 1996b, A fatigue damage model including interaction effects with oxidation and creep damage. In: "Fatigue 96: Proceedings of the 6th International Fatigue Congress," eds. G. Lütjering and H. Nowack, Pergamon, Oxford, vol. II, pp. 861–866.

M. Gell, G. R. Leverant and C. H. Wells, 1970, The fatigue strength of nickel-base superalloys. In: "Achievement of High Fatigue Resistance in Metals and Alloys, ASTM STP 467," American Society for Testing and Materials, Philadelphia, pp. 113–153.

A. E. Gemma, B. S. Langer and G. R. Leverant, 1976, Thermomechanical fatigue crack propagation in an anisotropic (directionally solidified) nickel-base superalloy. In: "Thermal Fatigue of Materials and Components, ASTM STP 612," eds. D. A. Spera and D. F. Mowbray, American Society for Testing and Materials, Philadelphia, pp. 199–213.

A. E. Gemma and J. S. Phillips, 1977, The application of fracture mechanics to life prediction of cooling hole configuration in thermal mechanical fatigue. *Eng. Fract. Mech.*, **9**, 25–36.

E. Glenny, 1967, The influence of specimen geometry on thermal-fatigue behaviour. In: "Thermal and High Strain Fatigue," The Institute of Metals and the Iron and Steel Institute, London, pp. 346–363.

E. Glenny and T. A. Taylor, 1959–1960, A study of the thermal-fatigue behaviour of metals. *J. Inst. Metals*, **88**, 449–461.

N. L. Goldman and J. W. Hutchinson, 1975, Fully plastic crack problems: the centre-cracked strip under plane strain. *Int. J. Solids Struct.*, **11**, 575–591.

J. Granacher and A. Scholz, 1992, Creep-fatigue behaviour of heat resistant steels under service-type long-terms conditions. In: "Low Cycle Fatigue and Elasto Plastic Behavior of Materials-3," ed. K. R. Rie, Elsevier, London, pp. 235–241.

J. Granacher and A. Scholz, 1996, Creep-fatigue behaviour under service type strain cycling. In: "Fatigue under Thermal and Mechanical Loading," eds. J. Bressers and L. Rémy, Kluwer, The Netherlands, pp. 209–214.

J. Grison and L. Rémy, 1997, Fatigue failure probability in a powder metallurgy Ni-base superalloy. *Eng. Fract. Mech.*, **57**, 41–55.

H. W. Grünling, B. Ilschner, S. Leistikov, A. Rahmel and M. Schmidt, 1980, Interaction between creep deformation and hot gas corrosion. In: "Behaviour of High Temperature Alloys in Aggressive Environments: Proceed. Petten Inter. Conf," eds. I. Kirman, J. B. Marriott, M. Merz, P. R. Sahm and D. P. Whittle, The Metals Society, London, pp. 869–893.

H. W. Grünling, K. Schneider and L. Singheiser, 1987, Mechanical properties of coated systems. *Mater. Sci. Eng.*, **88**, 177–189.

J. Y. Guedou and Y. Honnorat, 1993, Thermomechanical fatigue of turbo-engine blade superalloys. In: "Thermomechanical Fatigue Behavior of Materials, ASTM STP 1186," ed. H. Sehitoglu, American Society for Testing and Materials, Philadelphia, pp. 157–175.

N. Haddar, L. Rémy, A. Koster and M. Akamatsu, 2002, Thermal–mechanical fatigue of 304 stainless steel. In: "Fatigue 2002: Proceedings of the 8th International Fatigue Congress, Stockholm, Sweden, June 3–7, 2002," ed. A. F. Blom, EMAS, Cradley Heath, UK, vol. 3, pp. 1665–1672.

J. R. Haigh and R. P. Skelton, 1978, A strain intensity approach to high temperature fatigue crack growth and failure. *Mater. Sci. Eng.*, **36**, 133–137.

J. R. Haigh, R. P. Skelton and C. E. Richards, 1976, Oxidation-assisted crack growth during high cycle fatigue of a 1%Cr–Mo–V steel at 550°C. *Mater. Sci. Eng.*, **26**, 167–174.

R. Hales, 1980, A quantitative metallographic assessment of structural degradation of type 316 stainless steel. *Fatigue Eng. Mater. Struct.*, **3**, 339–356.

G. R. Halford and S. S. Manson, 1976, Life prediction of thermal–mechanical fatigue using strain range partitioning. In: "Thermal Fatigue, ASTM STP 612," eds. D. A. Spera and D. F. Mowbray, American Society for Testing and Materials, Philadelphia, pp. 239–254.

G. R. Halford, M. A. McGaw, R. C. Bill and P. D. Fanti, 1988, Bithermal fatigue: a link between isothermal and thermomechanical fatigue. In: "Low Cycle Fatigue, ASTM STP 942," eds. H. D. Solomon, G. R. Halford, L. R. Kaisand and B. N. Leis, American Society for Testing and Materials, Philadelphia, pp. 1115–1132.

P. Hancock and J. R. Nicholls, 1988, Application of fracture mechanics to failure of surface oxide scales. *Mater. Sci. Technol.*, **4**, 398–406.

F. Hanriot, 1993, Etude du comportement du superalliage mono-cristallin AM1 sous sollicitations cycliques (in French)," Ph.D. thesis, Ecole des Mines, Paris.

F. Hanriot, G. Cailletaud and L. Rémy, 1991, Mechanical behaviour of a nickel base superalloy single crystal. In: "High Temperature Constitutive Modeling. Theory and Application," eds. A. D. Freed and K. P. Walker, American Society of Mechanical Engineers, New York, USA, vol. 121, pp. 139–150 MD-Vol. 26/AMD.

S. T. Hasan and M. W. Brown, 1990, Thermal downshock fatigue testing in AISI 316 stainless steel plate. *High Temp. Technol.*, **8**, 89–97.

M. Hayashi, K. Enomoto, T. Saito and T. Miyagawa, 1994, Development of thermal fatigue apparatus with BWR water environment and thermal fatigue strength of stainless steel. In: "Fracture Mechanics Application," ASME PVP-vol. 287/MD-vol. 47.

D. R. Hayhurst, 1972, Creep rupture inter multiaxial state of stress. *J. Mech. Solids*, **20**(6), 381–390.

M. Y. He and J. W. Hutchinson, 1981, The penny-shaped crack and the plane strain crack in an infinite body of power-law material. *J. Appl. Mech.*, **48**, 830–840.

J. W. Holmes and F. A. McClintock, 1990, The chemical and mechanical processes of thermal fatigue degradation of an aluminide coating. *Metall. Trans. A*, **21**, 1209–1222.

J. W. Holmes, F. A. McClintock, K. S. O'Hara and M. E. Conners, 1988, Thermal fatigue testing of coated monocrystalline superalloys. In: "Low Cycle Fatigue, ASTM STP 942," eds. H. D. Solomon, G. R. Halford, L. R. Kaisand and B. N. Leis, American Society for Testing and Materials, Philadelphia, pp. 672–691.

S. W. Hopkins, 1976, Low-cycle thermal mechanical fatigue testing. In: "Thermal Fatigue of Materials and Components, ASTM STP 612," eds. D. A. Spera and D. F. Mowbray, American Society for Testing and Materials, Philadelphia, pp. 157–169.

M. A. H. Howes, 1976, A study of thermal fatigue mechanisms. In: "Thermal Fatigue of Materials and Components, ASTM STP 612," eds. D. A. Spera and D. F. Mowbray, American Society for Testing and Materials, Philadelphia, pp. 86–105.

J. S. Huang and R. M. Pelloux, 1980, Low cycle fatigue crack propagation in Hastelloy X at 25 and 760°C. *Metall. Trans. A*, **11**, 899–904.

J. W. Hutchinson and Z. Suo, 1992, Mixed mode cracking in layered materials. In: "Advances in Applied Mechanics," eds. J. W. Hutchinson and T. Y. Wu, Academic Press, New York, vol. 29, pp. 63–191.

C. E. Jaske, 1976, Thermal–mechanical, low cycle fatigue of AISI 1010 steel. In: "Thermal Fatigue of Materials and Components, ASTM STP 612," eds. D. A. Spera and D. F. Mowbray, American Society for Testing and Materials, Philadelphia, pp. 170–198.

E. H. Jordan and G. J. Meyers, 1986, Fracture mechanics applied to nonisothermal fatigue crack growth. *Eng. Fract. Mech.*, **23**, 345–358.

Y. Kadioglu and H. Sehitoglu, 1993, Modeling of thermomechanical fatigue damage in coated alloys. In: "Thermomechanical Fatigue Behavior of Materials, ASTM STP 1186," ed. H. Sehitoglu, American Society for Testing and Materials, Philadelphia, pp. 17–34.

H. Kaguchi, H. Koto, S. Date and Y. Takahashi, 1998, Fatigue crack propagation under cyclic thermal stress. *Mater. High Temp.*, **15**, 285–290.

K. S. Kim and R. H. Van Stone, 1998, Characterization of elevated temperature crack growth in Hastelloy X using integral parameters. *ASME J. Eng. Mater. Technol.*, **117**, 299–304.

J. E. King and P. J. Cotterill, 1991, Role of oxides in fatigue crack propagation. *Mater. Sci. Technol.*, **6**, 19–31.

K. Kishimoto, S. Aoki and M. Sakada, 1980, On the path-independent integral. *J. Eng. Fract. Mech.*, **13**, 841–850.

B. Kleinpass, K. H. Lang, D. Löhe and E. Macherauch, 1996, Thermal–mechanical fatigue behaviour of NiCr 22–Ci–12Mo9. In: "Fatigue under Thermal and Mechanical Loading," eds. J. Bressers, L. Rémy, M. Steen and J. Valles, Kluwer, Dordrecht, pp. 327–337.

J. Knott, 1973, "Fundamentals of Fracture Mechanics," Butterworths, London, p. 227.

T. Koizumi and M. Okazaki, 1979, Crack growth and prediction of endurance in thermal–mechanical fatigue of 12Cr–Mo–V–W steel. *Fatigue Eng. Mater. Struct.*, **1**, 509–520.

A. Koster, 1997, Fatigue thermique d'un alliage pour aubes de turbopompage astronautique: le Superwaspaloy (in French), Ph.D. thesis, Ecole des Mines de Paris.

A. Koster, A. Alam and L. Rémy, 2002, A physical-base model for life prediction of single crystal turbine blades under creep-fatigue loading and thermal transient conditions. In: "Temperature-fatigue Interaction, 9th International Spring Meeting, Paris, May 29–31, 2001," eds. L. Rémy and J. Petit, ESIS Publication 29, Elsevier, Amsterdam, pp. 203–212.

A. Koster, E. Chataigner and L. Rémy, 1996a, Thermal fatigue, a useful tool to assess low cycle fatigue damage in superalloys for components experiencing thermal transient. In: "81st AGARD Structures and Materials Panel, Thermal–mechanical Fatigue of Aircraft Engine Materials, Banff, Canada, October 2–4, 1995, AGARD-CP-569," AGARD, Neuilly, France, pp. 8-1, 8-8.

A. Koster, E. Fleury, E. Vasseur and L. Rémy, 1994, "Thermal–mechanical Fatigue Testing, Automation in Fatigue and Fracture: Testing and Analysis, ASTM STP 1231," ed. C. Amzallag, American Society for Testing and Materials, West Conshohocken, pp. 559–576.

A. Koster, G. Laurent, G. Cailletaud and L. Rémy, 1996b, Development of a thermal fatigue facility to simulate the behaviour of superalloy components. In: "Fatigue under Thermal and Mechanical Loading," eds. J. Bressers and L. Rémy, Kluwer Academic, Dordrecht, The Netherlands, pp. 25–35.

A. Koster and L. Rémy, 1999, An oxidation-creep-fatigue damage model for fatigue at high temperature and under thermal transients. In: "Fatigue 99: Proceedings of the 7th International Fatigue Congress, Beijing, China, June 8–12, 1999," eds. X. R. Wu and Z. G. Wang, Higher Education Press and EMAS, Cradley Heath, vol. 4, pp. 2139–2144.

R. Koterazawa, 1993, Creep-fatigue crack growth and fractography of metallic materials in air and vacuum at elevated temperatures. *Fatigue Fract. Eng. Mater. Struct.*, **16**, 619–630.

R. Kowaleski and H. Mughrabi, 2000, Thermal–mechanical and isothermal fatigue of a coated columnar-grained directionally solidified nickel-base superalloy. In: "Thermo-mechanical Fatigue Behaviour of Materials: 3rd Volume, ASTM STP 1371," eds. H. Sehitoglu and H. J. Maier, American Society for Testing and Materials, West Conshohocken, PA, pp. 223–238.

K. Kuwabara and A. Nitta, 1980, Thermal–mechanical low cycle fatigue under creep fatigue interaction on type 304 stainless steels. In: "Proceedings of the 3rd Int. Conf. on Mechanical Behaviour of Material," eds. K. J. Miller and R. F. Smith, Pergamon, Oxford, vol. 2, pp. 69–78.

K. Kuwabara, A. Nitta and T. Kitamura, 1983, Thermal–mechanical fatigue life prediction in high-temperature component materials for power plant. In: "Advances in Life Prediction Methods," eds. D. A. Woodford and J. R. Whitehead, American Society of Mechanical Engineers, New York, pp. 131–141.

H. S. Lamba, 1975, The *J*-integral applied to cyclic loading. *Eng. Fract. Mech.*, **7**, 693–703.

J. D. Landes and J. A. Begley, 1976, A fracture mechanics approach to creep crack growth. In: "Mechanics of Crack Growth, ASTM STP 590," eds. J. R. Rice and P. C. Paris, American Society for Testing and Materials, Philadelphia, pp. 128–148.

J. C. Lautridou, A. Debussac and F. Soniak, 1993, A probabilistic model for fatigue life prediction of P. M. Ni base superalloy containing inclusions. In: "Fatigue 93," eds. J. P. Bailon and J. I. Dickson, EMAS, Cradley Heath, vol. 1, pp. 285–290.

F. A. Leckie and D. R. Hayhurst, 1977, Constitutive equations for creep rupture. *Acta Metall.*, **25**, 1059–1070.

J. Lemaitre and J. L. Chaboche, 1985, "Mécanique des matériaux solides," Dunod, Paris.

J. Lemaitre and A. Plumtree, 1979, Application of damage concepts to predict creep-fatigue failures. *ASME J. Eng. Mater. Technol.*, **101**, 284–292.

C. Levaillant, J. Grattier, M. Mottot and A. Pineau, 1988, Creep and creep-fatigue intergranular damage in austenitic stainless steels: discussion of the creep-dominated regime. In: "Low Cycle Fatigue ASTM STP 942," eds. H. D. Solomon, G. R. Halford, L. R. Kaisand and B. N. Leis, American Society for Testing and Materials, Philadelphia, pp. 414–437.

C. Levaillant and A. Pineau, 1982, Assessment of high-temperature low-cycle fatigue life of austenitic stainless steel using intergranular damage as a correlating parameter. In: "Low-cycle Fatigue and Life Prediction, ASTM STP 770," eds. C. Amzallag, B. N. Leis and P. Rabbe, American Society for Testing and Materials, Philadelphia, pp. 169–193.

G. R. Leverant, T. E. Strangman and B. S. Langer, 1976, Parameters controlling the thermal fatigue properties of conventionally-cast and directionally-solidified turbine alloys. In: "Proceedings of the 3rd International Symposium on Superalloys, Metallurgy and Manufacture, September 12–15, 1976," eds. B. H. Kear, D. R. Muzyka, J. K. Tien and S. T. Wlodek, Claitor, Seven Springs, pp. 285–295.

S. X. Li, D. J. Smith and E. G. Ellison, 1993, Creep-fatigue crack growth in a single crystal nickel base superalloy. In: "Behaviour of Defects at High Temperatures, ESIS 15," eds. R. A. Ainsworth and R. P. Skelton, Mechanical Engineering Publications, London, pp. 351–365.

H. J. Maier and H.-J. Christ, 2002, Validating the predictive capabilities: a key issue in modelling thermo-mechanical fatigue life. In: "Temperature–Fatigue Interaction," eds. L. Rémy and J. Petit, ESIS Publication 29, Elsevier, Amsterdam, pp. 25–35.

H. J. Maier, R. G. Teteruk and H. J. Christ, 2000, Modeling thermomechanical fatigue life of high-temperature titanium alloy IMI 834. *Metall. Mater. Trans. A*, **31**, 431–444.

V. Maillot, A. Fissolo, G. Degallaix, S. Degallaix, S. Degallaix, B. Marini and M. Akamatsu, 2002, Thermal fatigue of a 304 type steel. In: "Temperature-fatigue Interaction," eds. L. Rémy and J. Petit, ESIS Publication 29, Elsevier, Amsterdam, pp. 135–193.

P. S. Maiya, 1981, Effect of wave shape and ultrahigh vacuum on elevated temperature low cycle fatigue in type 304 stainless steel. *Mater. Sci. Eng.*, **47**, 13–21.

S. Majumdar and P. S. Maiya, 1980, A mechanistic model for time-dependent fatigue. *ASME J. Eng. Mater. Technol.*, **102**, 159–167.

S. Malm and L.-A. Norström, 1979, Material-related model for thermal fatigue applied to tool steels in hot-work applications. *Metal Sci.*, **13**, 544–550.

J. L. Malpertu and L. Rémy, 1990, Influence of test parameters on the thermal–mechanical fatigue behavior of a superalloy. *Metall. Trans. A*, **21**, 389–399.

M. I. Manning, 1981, Limits to the integrity of oxide scales growing on curved surfaces and distending metal substrates. In: "Corrosion and Mechanical Stress at High Temperatures," eds. V. Guttmann and M. Merz, Applied Science Publishers, London, pp. 323–338.

S. S. Manson, 1966, "Thermal Stress and Low-cycle Fatigue," McGraw-Hill, New York.

S. S. Manson, 1973, The challenge to unify treatment of high temperature fatigue a partisan proposal based on strain range partitioning. In: "Fatigue at Elevated Temperature, ASTM STP 520," eds. A. E. Carden, A. J. McEvily and C. H. Wells, American Society for Testing and Materials, Philadelphia, pp. 744–782.

S. S. Manson, J. C. Freche and C. R. Ensign, 1967, Application of a double linear damage rule to cumulative fatigue. In: "Fatigue Crack Propagation, ASTM STP

415," American Society for Testing and Materials, Philadelphia, pp. 384–412.
S. S. Manson and G. R. Halford, 1981, Practical implementation of the double linear damage rule and damage curve approach for treating cumulative damage. *Int. J. Fract.*, **17**, 169–192.
S. S. Manson and G. R. Halford, 1977, Discussion on multiaxial low cycle fatigue of type 304 stainless steel. *ASME J. Eng. Mater. Technol.*, **99**, 283–285.
N. Marchand, D. M. Parks and R. M. Pelloux, 1986, K_I-solutions for single edge notch specimens under fixed-end displacements. *Int. J. Fract.*, **31**, 53–65.
N. J. Marchand, R. M. Pelloux and B. Ilschner, 1988, A fracture mechanic criterion for thermal–mechanical fatigue crack growth of gas turbine materials. *Eng. Fract. Mech.*, **31**, 535–551.
D. J. Marsh, 1981, A thermal shock fatigue study of type 304 and 316 stainless steels. *Fatigue Eng. Mater. Struct.*, **4**, 179–195.
D. J. Marsh, D. Green and R. Parker, 1986, Comparison of theoretical estimates and experimental measurements of fatigue crack growth under severe thermal shock conditions: Part I. Experimental observations. *ASME J. Pres. Ves. Technol.*, **108**, 501–506.
J. M. Martinez-Esnaola, M. Ariana, J. Bressers, J. Timm, A. Martin-Meizoso, A. Bennet and E. E. Affeldt, 1996, Crack initiation in an aluminide coated single crystal during thermo-mechanical fatigue. In: "Thermomechanical Fatigue Behavior of Materials: 2nd Volume, ASTM STP 1263," eds. M. J. Verrilli and M. G. Castelli, American Society for Testing and Materials, Philadelphia, pp. 68–81.
F. A. Mc Clintock, 1963, On the plasticity of the growth of fatigue cracks. In: "Fracture of Solids," eds. D. C. Drucker and J. J. Gilman, Interscience, New York, pp. 65–102.
C. J. Mc Mahon and L. F. Coffin, Jr., 1970, Mechanisms of damage and fracture in high-temperature, low-cycle fatigue of a cast nickel-based superalloy. *Metall. Trans.*, **1**, 3443–3450.
S. M. Meier, D. M. Nissley, K. D. Sheffler and T. A. Cruse, 1991, Thermal barrier coating life prediction model development. In: "Int. Gas Turbine and Aeroengine Congress, Orlando, FL, June 3–6, 1991."
L. Méric, P. Poubanne and G. Cailletaud, 1991, Single crystal modeling for structural applications. *ASME J. Eng. Mater. Technol.*, **113**, 165–170.
F. Meyer-Obersleben, C. C. Engler-Pinto, Jr. and F. Rezai-Aria, 1996, On thermal fatigue of nickel based superalloys. In: "Thermomechanical Fatigue Behavior of Materials: 2nd Volume, ASTM STP 1263," eds. M. J. Verrilli and M. G. Castelli, American Society for Testing and Materials, Philadelphia, pp. 41–55.
F. Meyer-Obersleben, F. Rezai-Aria and B. Ilschner, 1990, The effect of grain size and upper cycle temperature in thermal fatigue behaviour of IN 100 superalloy. In: "High Temperature Alloys for Gas Turbines and other Applications," eds. W. Betz, *et al.*, Reidel, Dordrecht, vol. II, pp. 1121–1129.
M. P. Miller, D. L. Mc Dowell and R. L. T. Oehmke, 1992, A creep-fatigue-oxidation microcrack propagation model for thermomechanical fatigue. *J. Eng. Mater. Technol.*, **114**, 282–288.
R. A. Miller, 1987, Progress toward life modeling of thermal barrier coatings for aircraft gas turbine engines. *ASME J. Eng. Gas Turbines Power*, **109**, 448–451.
R. A. Miller, 1988, Life modeling of thermal barrier coatings for aircraft gas turbine engines. In: "32nd ASME International Gas Turbines and Aeroengine Congress, Amsterdam, June 5–9, 1988," pp. 109–115.
M. A. Miner, 1945, Cumulative damage in fatigue. *J. Appl. Mech.*, **12**, A159–A164.
B. Miquel, S. Jean, S. Le Roux, P. Lamesle and F. Rezai-Aria, 2002, Heat-checking of hot work tool steels. In: "Temperature-fatigue Interaction," eds. L. Rémy and J. Petit, ESIS Publication 29, Elsevier, Amsterdam, pp. 185–193.
M. Moalla, K. H. Lang and D. Löhe, 2002, Cycle deformation and life time behaviour of NiCr22–Co12Mo isothermal and thermal–mechanical fatigue. In: "Temperature-fatigue Interaction," eds. L. Rémy and J. Petit, ESIS Publication 29, Elsevier, Oxford, pp. 85–94.
D. F. Mowbray and J. E. Mc Connelee, 1976, Nonlinear analysis of a tapered disk thermal fatigue specimen. In: "Thermal Fatigue of Materials and Components, ASTM STP 612," eds. D. A. Spera and D. F. Mowbray, American Society for Testing and Materials, Philadelphia, pp. 10–29.
D. F. Mowbray and D. A. Woodford, 1973, Observations and interpretation of crack propagation under conditions of transient thermal strain. In: "International Conference on Creep and Fatigue," Institute of Mechanical Engineers, Conference Publication 13, pp. 179.1–179.11.
D. F. Mowbray, D. A. Woodford and D. E. Brandt, 1973, Thermal fatigue characterization of cast cobalt and nickel-base superalloys. In: "Fatigue at Elevated Temperatures, ASTM STP 520," eds. A. E. Larden, A. J. McEvily and C. H. Wells, American Society for Testing and Materials, Philadelphia, pp. 416–426.
J. A. Nairn and S. R. Kim, 1992, A fracture mechanics analysis of multiple cracking in coatings. *Eng. Fract. Mech.*, **42**, 195–208.
G. J. Neate, 1985, Crack growth in a CrMoV steel under cyclic loading at 873 K. *High Temp. Technol.*, **3**, 195–202.
R. W. Neu and H. Sehitoglu, 1989a, Thermomechanical fatigue, oxidation and creep: Part I. Damage mechanisms. *Metall. Trans. A*, **20**, 1755–1767.
R. W. Neu and H. Sehitoglu, 1989b, Thermomechanical fatigue, oxidation and creep: Part II. Life prediction. *Metall. Trans. A*, **20**, 1769–1783.
T. Nicholas, M. L. Heil and G. K. Haritos, 1989, Predicting crack growth under thermo-mechanical cycling. *Int. J. Fract.*, **41**, 157–176.
E. Nicouleau, F. Feyel, S. Quilici and G. Cailletaud, 2002, Structural calculation and lifetime-prediction in thermo-mechanical fatigue of engine components. In: "Temperature-fatigue Interaction," eds. L. Rémy and J. Petit, Elsevier, Amsterdam, pp. 331–340.
L. Northcott and M. G. Baron, 1956, The craze-cracking of metals. *J. Iron Steel Inst.*, **184**, 385–408.
R. Ohtani and T. Kitamura, 1994, Creep-fatigue interaction under high-temperature conditions. In: "Handbook of Fatigue Crack Propagation in Metallic Structures," ed. A. Carpinteri, Elsevier, Amsterdam, pp. 1347–1383.
R. Ohtani, T. Kitamura, A. Nitta and K. Kuwabara, 1988, High-temperature low cycle fatigue crack propagation and life laws of smooth specimens derived from the crack propagation laws. In: "Low Cycle Fatigue, ASTM STP 942," eds. H. D. Solomon, G. R. Halford, L. R. Kaisand and B. N. Leis, American Society for Testing and Materials, Philadelphia, pp. 1163–1180.
M. Okazaki, I. Hattori, F. Shiraiwa and T. Koizumi, 1983, Effect of strain wave shape on low-cycle fatigue crack propagation of SUS 304 stainless steel at elevated temperatures. *Metall. Trans. A*, **14**, 1649–1659.
M. Okazaki and T. Koizumi, 1983, Crack Propagation of steels during low cycle thermal–mechanical fatigue and isothermal fatigue at elevated temperatures. *Metall. Trans. A*, **14**, 1641–1648.
M. Okazaki and T. Koizumi, 1987, Relationship of crack growth between thermal–mechanical and low cycle

fatigue at elevated temperatures. *ASME J. Eng. Mater. Technol.*, **109**, 114–118.

S. Osgerby and B. F. Dyson, 1990, Effect of oxygen on creep performance: mechanisms and predictive modelling. *Mater. Sci. Technol.*, **6**, 2–8.

Y. Oshida and H. W. Liu, 1988, Grain boundary oxidation and an analysis of the effects of oxidation on fatigue crack nucleation life. In: "Low Cycle Fatigue, ASTM STP 942," eds. H. D. Solomon, G. R. Halford, L. R. Kaisand and B. N. Leis, American Society for Testing and Materials, Philadephia, pp. 1199–1217.

W. J. Ostergren, 1976, A damage function and associated failure equations for predicting hold time and frequency effects in elevated temperature, low-cycle fatigue. *J. Test. Evaluat.*, **4**, 327–339.

A. Oudin, P. Lamesle, L. Penazzi, S. Le Roux and F. Rezai-Aria, 2002, Thermomechanical fatigue behaviour and life assessment of hot work tool steels. In: "Temperature-fatigue Interaction," eds. L. Rémy and J. Petit, ESIS Publication 29, Elsevier, Amsterdam, pp. 195–201.

J. P. Pédron and A. Pineau, 1982, The effect of microstructure and environment on the crack growth behaviour of Inconel 718 alloy at 650°C under fatigue, creep and combined loading. *Mater. Sci. Eng.*, **56**, 143–156.

P. Perruchaut, P. Villechaise and J. Mendez, 1996, Influence of a coating on the fatigue damage of the ⟨001⟩ AM1 single crystal superalloy at 950°C. In: "Mechanisms and Mechanics of Damage and Failure, ECF 11, Proc. 11th Conference on Fracture," ed. J. Petit, EMAS, Cradley Heath, vol. II, pp. 1219–1224.

H. Policella and P. M. Lesne, 1986, Amorçage et propagation de fissures en fatigue thermique, calculs de prevision. In: "Fatigue at High Temperature: Proceed. Int. Spring Meeting SFM, Paris, June 9–11, 1986," pp. 275–289.

M. Probst-Hein and G. Eggeler, 2001, Predictive microstructural assessment and micromechanical modelling of deformation and damage accumulation in single crystal gas turbine blading, Micromod-SX, Brite-Euram III, final report.

Y. N. Rabotnov, 1969, "Creep Problems in Structural Members," North-Holland, Amsterdam.

R. Ratchev, T. Beck, K.-H. Lang and D. Löhe, 2002, Lifetime, cyclic deformation and damage behaviour of IN706 and IN718 under out-of-phase TMF loadings. In: "Fatigue 2002: Proceedings of the 8th International Fatigue Congress, Stockholm, June 3–7, 2002," ed. A. F. Blom, EMAS, Cradley Heath, vol. 2, pp. 1267–1274.

C. A. Rau, Jr., A. E. Gemma and G. R. Leverant, 1973, Thermal–mechanical fatigue crack propagation in nickel and cobalt base superalloys under various strain temperature cycles. In: "Fatigue at Elevated Temperatures, ASTM STP 520," eds. A. E. Carden, A. J. McEvily and C. H. Wells, American Society for Testing and Materials, Philadelphia, pp. 166–178.

M. Reger, M. François and L. Rémy, 1984, Interaction between oxidation and cyclic straining at high temperature. In: "Physical Chemistry of the Solid State: Applications to Metals and their Compounds," ed. P. Lacombe, Elsevier, Amsterdam, pp. 413–420.

M. Reger and L. Rémy, 1988a, Fatigue oxidation interaction in IN100 superalloy. *Metall. Trans. A*, **19**, 2259–2268.

M. Reger and L. Rémy, 1988a, High temperature low cycle fatigue behaviour of IN100 superalloy: I. Influence of temperature on the low cycle fatigue behaviour. *Mater. Sci. Eng. A*, **101**, 47–54.

M. Reger and L. Rémy, 1988b, High temperature low cycle fatigue behaviour of IN100 superalloy: II. Influence of frequency and environment at high temperature. *Mater. Sci. Eng. A*, **101**, 55–63.

M. Reger, F. Soniak and L. Rémy, 1984c, Propagation of short cracks in low cycle fatigue. In: "Fatigue 84: Proceedings of the 2nd International Conference on Fatigue and Fatigue Thresholds, Birmingham, September 3–7, 1984," ed. C. J. Beevers, EMAS, London, vol. II, pp. 797–806.

L. Rémy, 1990b, Recent developments in thermal fatigue. In: "High Temperature Fracture Mechanisms and Mechanics, EGF 6," eds. P. Bensussan and J. P. Mascarell, Mechanical Engineering Publications, London, pp. 353–377.

L. Rémy, 1993, Oxidation effects in high temperature creep and fatigue of engineering alloys. In: "Corrosion–Deformation Interactions CDI'92," eds. T. Magnin and J. M. Gras, Les Editions de Physique, Courtaboeuf, pp. 425–459.

L. Rémy, 1995, Oxidation and damage in turbojet superalloys under cyclic loading. In: "Actes du 16e colloque Matériaux pour l'Aéronautique et l'Espace: Proceedings of a Conference organized by AAAF, Paris, June 13–14, 1995," pp. 45–62.

L. Rémy, 1996, Thermal–mechanical fatigue and thermal fatigue experiments: damage modelling in thermal mechanical fatigue. In: "Mechanical Behaviour of Materials at High Temperature, NATO ASI Series," eds. C. Moura-Branco, T. Ritchie and V. Sklenicka, Kluwer, Dordrecht, pp. 381–418.

L. Rémy, 1998, Thermal and thermal–mechanical fatigue of superalloys, a challenging goal for mechanical tests and models. In: "Low Cycle Fatigue and Elasto-plastic Behaviour of Materials: Proceedings of the 4th International Conference, September 7–11, 1998," eds. K.-T. Rie and P. D. Portella, Elsevier, Garmisch-Partenkirchen, Germany, pp. 119–130.

L. Rémy, A. Alam and A. Bickard, 2002, Thermomechanical creep-fatigue of coated systems. In: "Thermomechanical Fatigue Behavior of Materials: 4th Volume, ASTM STP 1428," eds. M. A. McGaw, S. Kalluri, J. Bressers and S. D. Peteves, American Society for Testing and Materials, West Conshohocken, PA.

L. Rémy, H. Bernard, J. L. Malpertu and F. Rezai-Aria, 1993, Fatigue life prediction under thermal–mechanical loading in a nickel-base superalloy. In: "Thermomechanical Fatigue Behaviour of Materials, ASTM STP 1186," ed. H. Sehitoglu, American Society for Testing and Materials, Philadelphia, pp. 3–16.

L. Rémy and J.-Y. Guédou, 2002, Thermal–mechanical fatigue problems in aero-engine components. In: "Fatigue 2002: Proceedings of the 8th International Fatigue Congress, Stockholm, Sweden, June 3–7, 2002," ed. A. F. Blom, EMAS, Cradley Heath, UK, vol. 3, pp. 1045–1062.

L. Rémy, A. Koster, E. Chataigner and A. Bickard, 2000, Thermal–mechanical fatigue and the modelling of materials behaviour under thermal transients. In: "Thermo-mechanical Fatigue Behaviour of Materials: 3rd Volume, ASTM STP 1371," eds. H. Sehitoglu and H. J. Maier, American Society for Testing and Materials, West Conshohocken, PA, pp. 223–238.

L. Rémy, M. Reger, J. Reuchet and F. Rezai-Aria, 1982, An operational definition of life to crack initiation in high temperature fatigue. In: "High Temperature Alloys for Gas Turbines," eds. R. Brunetaud, D. Coutsouradis, T. B. Gibbons, Y. Lindblom, D. B. Meadowcroft and R. Stickler, Reidel, Dordrecht, pp. 619–632.

L. Rémy, F. Rézaï-Aria, R. Danzer and W. Hoffelner, 1988, Evaluation of life prediction methods in high temperature fatigue. In: "Low Cycle Fatigue, ASTM STP 942," eds. H. D. Solomon, G. R. Halford, L. R. Kaisand and B. N. Leis, American Society for Testing and Materials, Philadelphia, pp. 1115–1132.

L. Rémy and P. Skelton, 1992, Damage assessment of components experiencing thermal transients. In: "High Temperature Structural Design, ESIS 12," ed. L. H.

Larsson, Mechanical Engineering Publications, London, pp. 283–315.

E. Renner, H. Vehoff and P. Neumann, 1989, Life prediction for creep-fatigue based on the growth of short cracks. *Fatigue Fract. Eng. Mater. Struct.*, **12**, 569–584.

J. Reuchet, M. Reger, F. Rézaï-Aria and L. Rémy, 1982, The cumulation of damage in high temperature low cycle fatigue. In: "Low Cycle Fatigue and Life Prediction, ASTM STP 770," eds. C. Amzallag, P. Rabbe and B. N. Leis, American Society for Testing and Materials, Philadelphia, pp. 194–211.

J. Reuchet and L. Rémy, 1983, Fatigue oxidation interaction in a superalloy. Application to life prediction in high temperature low cycle fatigue. *Metall. Trans. A*, **14**, 141–149.

J. Reuchet and L. Rémy, 1983a, High temperature low cycle fatigue of superalloy Mar-M509: Part I. The influence of temperature on the low cycle fatigue behaviour. *Mater. Sci. Eng.*, **58**, 19–32.

J. Reuchet and L. Rémy, 1983b, High temperature low cycle fatigue of superalloy Mar-M509: Part II. The influence of oxidation at high temperatures. *Mater. Sci. Eng.*, **58**, 33–42.

F. Rézaï-Aria, B. Dambrine and L. Rémy, 1988c, Thermal fatigue of MAR-M509 superalloy: II. Evaluation of life prediction models. *Fatigue Fract. Eng. Mater. Struct.*, **11**, 291–302.

F. Rézaï-Aria, M. François and L. Rémy, 1988a, Thermal fatigue of MAR-M509 superalloy: I. The influence of specimen geometry on the thermal fatigue behaviour. *Fatigue Fract. Eng. Mater. Struct.*, **11**, 277–289.

F. Rézai-Aria and L. Rémy, 1989b, An oxidation fatigue interaction damage model for thermal fatigue crack growth. *Eng. Fract. Mech.*, **34**, 283–294.

J. R. Rice, 1968, A path independent integral and the approximate analysis of strain concentration by notches and cracks. *ASME J. Appl. Mech. E*, **35**(2), 379–386.

J. R. Rice, 1988, Elastic fracture mechanics concepts for interfacial cracks. *Trans. ASME, J. Appl. Mech.*, **55**, 98–103.

J. R. Rice and G. C. Sih, 1965, Plane problems of cracks in dissimilar media. *Trans. ASME, J. Appl. Mech.*, **32**, 418–423.

H. Riedel, 1982, Deformation and cracking of thin second-phase layers on deforming metals at elevated temperatures. *Metal Sci.*, **16**, 569–574.

H. Riedel and J. R. Rice, 1980, Tensile cracks in creeping solids. In: "Fracture Mechanics: 12th Conference, ASTM STP 700," American Society for Testing and Materials, Philadelphia, PA, pp. 121–130.

J. Robertson and M. I. Manning, 1990, Limits to adherence of oxide scales. *Mater. Sci. Technol.*, **6**, 81–91.

E. L. Robinson, 1952, Effect of temperature variation on the long-time rupture strength of steels. *Trans. ASME*, **74**, 777–781.

D. Rousseau, J. P. Riegert, L. Séraphin and R. Tricot, 1975, Fatigue thermique des aciers à outils pour travail à chaud. *Revue de Métallurgie*, 875–889.

T. Säuberlich, G. Pitz and D. Löhe, 1998, Thermal–mechanical fatigue behaviour of a 9% chromium steel. In: "Materials for Advanced Power Engineering 1998: Proceedings of the 6th Liege Conference," eds. J. Lecomte-Beckers, F. Schubert and P. J. Ennis, Forschungszentrum Jülich, Jülich, Part 1, pp. 341–350.

S. Savalle and G. Cailletaud, 1982, Microinitiation, micropropagation and damage. *La Recherche Aérospatiale (french and english editions)*, **6**, 395–411.

W. Schmitt, R. Mohrman, H. Riedel, A. Dietsche and A. Fischersworring-Bunk, 2002, Modelling of the fatigue life of automobile exhaust components. In: "Fatigue 2002: Proceeding of the 8th International Fatigue Congress, Stockholm, June 3–7, 2002," ed. A. F. Blom, EMAS, Cradley Heath, vol. 2, pp. 781–788.

K. Schneider and H. W. Grünling, 1983, Mechanical properties of high temperature coatings. *Thin Solid Films*, **107**, 395–416.

A. Scholz, J. Granacher and C. Berger, 1998, Simulation of long term creep-fatigue behaviour by multiatage service type strain cycling. In: "Low Cycle Fatigue and Elasto-plastic Behaviour of Materials," eds. K. T. Rie and P. D. Portella, Elsevier, Amsterdam, pp. 741–746.

M. Schütze, 1985, Deformation and cracking behavior of protective oxide scales on heat-resistant steels under tensile strain. *Oxidat. Metals*, **24**, 199–232.

M. Schütze, 1990, Plasticity of protective oxide scales. *Mater. Sci. Technol.*, **6**, 32–38.

M. Schütze, 1997, "Protective Oxide Scales and their Breakdown," Wiley, New York.

H. Sehitoglu, 1985, Constraint effect in thermo-mechanical fatigue. *ASME J. Eng. Mater. Technol.*, **107**, 221–226.

H. Sehitoglu, 1987, Crack growth studies under selected temperature–strain histories. *Eng. Fract. Mech.*, **26**, 475–489.

H. Sehitoglu, 1996, Thermo-mechanical deformation of engineering alloys and components—experiments and modelling. In: "Mechanical Behaviour of Materials at High Temperature," eds. C. Moura-Branco, T. Ritchie and V. Sklenicka, NATO ASI Series, Kluwer, Dordrecht, pp. 349–379.

H. Sehitoglu and D. A. Boismier, 1990, Thermo-mechanical fatigue of Mar-M247: Part 1. Experiments. *ASME J. Eng. Mater. Technol.*, **112**, 68–79.

H. Sehitoglu and D. A. Boismier, 1990a, Thermo-mechanical fatigue of Mar-M247: Part 2. Life prediction. *ASME J. Eng. Mater. Technol.*, **112**, 80–89.

H. Sehitoglu, X. Qing, T. Smith, H. J. Maier and J. A. Allison, 2000b, Stress–strain response of a cast 319-T6 aluminium under thermomechanical loading. *Metall. Mater. Trans. A*, **31**, 139–151.

H. Sehitoglu, T. J. Smith and H. J. Maier, 2000a, Thermo-mechanical deformation of Al 319-T7B with small secondary dendrite arm spacing. In: "Thermo-mechanical Fatigue Behavior of Materials: 3rd Volume, ASTM STP 1371," eds. H. Sehitoglu and H. J. Maier, American Society for Testing and Materials, West Conshohocken, PA, pp. 53–68.

D. P. Shepherd, A. Wisbey, G. F. Harrison, T. J. Ward and B. Vermeulen, 2001b, Cyclic operation in aero gas turbines—materials and component life implications. *Mater. High Temp.*, **18**, 231–239.

Y. Shida and T. Moroishi, 1992, Oxidation behaviour of alloy 800 in an impure helium atmosphere at high temperatures. *Corros. Sci.*, **33**, 211–224.

C. F. Shih and J. W. Hutchinson, 1976, Fully plastic solutions and large scale yielding estimates for plane stress crack problems. *J. Eng. Mater. Technol.*, **98**, 289–295.

D. Sidey and L. F. Coffin, Jr., 1979, Low-cycle fatigue damage mechanisms at high temperature. In: "Fatigue Mechanisms, ASTM STP 675," ed. J. T. Fong, American Society for Testing and Materials, Philadelphia, pp. 528–568.

L. Singheiser, R. Steinbrech, W. J. Quadakkers and R. Herzog, 2001, Failure aspects of thermal barrier coatings. *Mater. High Temp.*, **18**, 249–259.

R. P. Skelton, 1980, Crack propagation in metals during thermal shock. In: "Proceedings of the 3rd Int. Conf. on Mechanical Behaviour of Materials," eds. K. J. Miller and R. F. Smith, Pergamon, Oxford, vol. 2, pp. 13–22.

R. P. Skelton, 1982, Growth of short cracks during high strain fatigue and thermal fatigue and thermal cycling. In: "Low-cycle Fatigue and Life Prediction, ASTM STP 770," eds. C. Amzallag, B. N. Leis and P. Rabbe, American Society for Testing and Materials, Philadelphia, pp. 337–381.

R. P. Skelton, 1983, Crack initiation and growth in simple metal components during thermal cycling. In: "Fatigue at High Temperature," ed. R. P. Skelton, Applied Science Publishers, London, pp. 1–62.

R. P. Skelton, 1988, Application of small specimen crack growth data to engineering components at high temperature: a review. In: "Low-cycle Fatigue, ASTM STP 942," eds. H. D. Solomon, G. R. Halford, L. R. Kaisand and B. N. Leis, American Society for Testing and Materials, Philadelphia, pp. 209–235.

R. P. Skelton, 1990, Introduction to thermal shock. *High Temp. Technol.*, **8**, 75–88.

R. P. Skelton and J. I. Bucklow, 1978, Cyclic oxidation and crack growth during high strain fatigue of low alloy steel. *Metal Sci.*, **12**, 64–70.

R. P. Skelton and K. J. Nix, 1987, Crack growth behaviour in austenitic and ferritic steels during thermal quenching from 550°C. *High Temp. Technol.*, **5**, 3–12.

K. N. Smith, P. Watson and T. M. Topper, 1970, A stress strain function for the fatigue of metals. *J. Mater.*, **4**, 767–778.

T. J. Smith, H. J. Maier, H. Sehitoglu, E. Fleury and J. Allison, 1999, Modeling high-temperature stress–strain behavior of cast aluminium alloys. *Metall. Mater. Trans. A*, **30**, 133–146.

D. F. Socie, 1987, Multiaxial fatigue damage models. *J. Eng. Mater. Technol.*, **109**, 293–298.

H. D. Solomon and L. F. Coffin, 1973, Effects of frequency and environment on fatigue crack growth in A286 at 1100F. In: "Fatigue at Elevated Temperatures, ASTM STP 520," eds. A. E. Carden, A. J. McEvily and C. H. Wells, American Society of Testing and Materials, Philadelphia, pp. 112–121.

F. Soniak and L. Rémy, 1990, Fatigue crack growth of small defects in a powder metallurgy superalloy at high temperature. In: "Fatigue 90," eds. H. Kitagawa and T. Tanaka, Materials and Component Engineering Publications, Edgbaston, UK, vol. 3, pp. 1583–1588.

D. A. Spera, 1969, The calculation of thermal fatigue life based on accumulated creep damage, NASA TMX-52558, National Aeronautics and Space Administration.

D. A. Spera, 1973, Comparison of experimental and theoretical fatigue lives for five nickel-base alloys. In: "Fatigue at Elevated Temperatures, ASTM STP 520," eds. A. E. Carden, A. J. McEvily and C. H. Wells, American Society for Testing and Materials, Philadelphia, pp. 744–782.

D. A. Spera, 1976, What is thermal fatigue? In: "Thermal Fatigue of Materials and Components, ASTM STP 612," eds. D. A. Spera and D. F. Mowbray, American Society for Testing and Materials, Philadelphia, pp. 3–9.

D. A. Spera and E. C. Cox, 1976, Description of a computerized method for predicting thermal fatigue life of metals. In: "Thermal Fatigue of Materials and Components, ASTM STP 612," eds. D. A. Spera and D. F. Mowbray, American Society for Testing and Materials, Philadelphia, pp. 69–85.

D. A. Spera and D. F. Mowbray (eds.), 1976, "Thermal Fatigue of Materials and Components, ASTM STP 612" American Society for Testing and Materials, Philadelphia, p. 262

M. S. Starkey and R. P. Skelton, 1982, A comparison of the strain intensity and cyclic *J* approaches to crack growth. *Fatigue Eng. Mater. Struct.*, **5**, 329–341.

J. M. Stephan, F. Curtit, C. Vindeirinho, S. Taheri, M. Akamatsu and C. Peniguel, 2002, Evaluation of the risk of damages in mixing zones: EDF R&D programme. In: "Fatigue 2002: Proceedings of the 8th International Fatigue Congress, Stockholm, June 3–7, 2002," ed. A. F. Blom, EMAS, Cradley Heath, vol. 3, pp. 1707–1714.

A. Strang and E. Lang, 1982, Effect of coatings on the mechanical properties of superalloys. In: "High Temperature Alloys for Gas Turbines, Proc. Conf. Liege, Belgium," eds. Brunetaud, *et al.*, Reidel, Dordrecht, pp. 469–506.

T. E. Strangman, 1992, Thermal–mechanical fatigue life model for coated superalloy turbine components. In: "Superalloys 1992," eds. S. D. Antolovitch, R. W. Stusrud, R. A. Mackay, D. L. Anton, T. Khan, R. D. Kissinger and D. L. Klarstrom, The Minerals, Metals and Materials Society, Warrendale, pp. 795–804.

T. E. Strangman and S. W. Hopkins, 1976, Thermal fatigue of coated superalloys. *Ceram. Bull.*, **55**, 304–306.

J. Stringer, 1987, Role of coatings in energy-producing systems: an overview. *Mater. Sci. Eng.*, **87**, 1–10.

R. W. Swindeman, 1988, Cyclic stress–strain time response of a 9Cr–1Mo–V–Nb pressure vessel steel at high temperature. In: "Low Cycle Fatigue, ASTM STP 942," eds. H. D. Solomon, G. R. Halford, L. R. Kaisand and B. N. Leis, American Society for Testing and Materials, Philadelphia, pp. 107–122.

S. Taira, 1962, Lifetime of structures subjected to varying load and temperature. In: "Creep in Structures," ed. N. J. Hoff, Springer, Berlin, pp. 96–124.

S. Taira, 1973, Relationship between thermal fatigue and low-cycle fatigue at elevated temperature. In: "Fatigue at Elevated Temperature, ASTM STP 520," eds. A. E. Carden, A. J. Mc Evily and C. H. Wells, American Society for Testing and Materials, Philadelphia, pp. 744–782.

S. Taira, R. Ohtani and T. Kitamura, 1979, Application of *J*-integral to high temperature crack propagation: Part I. Creep crack propagation. *J. Eng. Mater. Technol., Trans. ASME*, **101**, 154–161.

S. Taira, R. Ohtani and T. Komatsu, 1979a, Application of *J*-integral to high temperature crack propagation: Part II. Fatigue crack propagation. *J. Eng. Mater. Technol., Trans. ASME*, **101**, 162–167.

B. Tomkins, 1968b, Fatigue crack propagation—an analysis. *Phil. Mag.*, **18**, 1041–1066.

T. C. Totemeier, W. F. Gale and J. E. King, 1996, Isothermal fatigue of an aluminide coated single crystal superalloy: Part II. Effects of brittle precracking. *Metall. Mater. Trans. A*, **27**, 363–369.

T. C. Totemeier and J. E. King, 1996, Isothermal fatigue of an aluminide coated single crystal superalloy: Part I. *Metall. Mater. Trans. A*, **27**, 353–361.

E. Vasseur, 1993, Fatigue anisotherme at microstructure de superalliages colonnaires (in French). Ph.D. thesis, Ecole des Mines, Paris.

E. Vasseur and L. Rémy, 1994, High temperature low cycle fatigue and thermal–mechanical fatigue behaviour of an oxide-dispersion-strengthened nickel-base superalloy. *Mater. Sci. Eng. A*, **184**, 1–15.

L. Verger, A. Constantinescu and E. Charkaluk, 2002, Thermomechanical fatigue design of aluminum components. In: "Temperature-fatigue Interaction," eds. L. Rémy and J. Petit, ESIS Publication 29, Elsevier, Amsterdam, pp. 309–318.

G. Ward, B. S. Hockenhull and P. Hancock, 1974, The effect of cyclic stressing on the oxidation of a low-carbon steel. *Metall. Trans.*, **5**, 1451–1455.

D. A. Woodford, 1981, Environmental damage of a cast nickel base superalloy. *Metall. Trans. A*, **12**, 299–308.

D. A. Woodford and D. F. Mowbray, 1974, Effect of material characteristics and test variables on thermal fatigue of cast superalloys. *Mater. Sci. Eng.*, **16**, 5–43.

P. K. Wright, 1998, Influence of cyclic strain on life of a PVD TBC. *Mater. Sci. Eng. A*, **245**, 191–200.

C. Wütrich, 1982, The extension of the *J*-integral concept to fatigue cracks. *Int. J. Fract.*, **20**, R35–R37.

G. Yagawa, Y. Takahashi and H. Ueda, 1985, Three-dimensional fully plastic solutions for plates and cylinders with through-wall cracks. *J. Appl. Mech.*, **52**, 319–325.

M. Yamauchi, T. Ohtani and Y. Takahashi, 1996, Thermal fatigue behavior of a SUS 304 pipe under longitudinal cyclic movement of axial temperature distribution. In:

"Thermomechanical Fatigue Behavior of Materials: 2nd Volume, ASTM STP 1263," eds. M. J. Verrilli and M. G. Castelli, American Society for Testing and Materials, Philadelphia, pp. 117–129.

S. Y. Zamrik and M. L. Renauld, 2000, Thermo-mechanical out-of-phase fatigue life of overlay coated IN-738LC gas turbine material. In: "Thermomechanical Fatigue Behavior of Materials: 3rd Volume, ASTM STP 1371," eds. M. Sehitoglu and H. J. Maier, American Society for Testing and Materials, West Conshohocken, PA, pp. 119–137.

R. Zauter, F. Petry, H.-J. Christ and H. Mughrabi, 1993, Thermomechanical fatigue of the austenitic stainless steel AISI 304L. In: "Thermomechanical Fatigue Behaviour of Materials, ASTM STP 1186," ed. H. Sehitoglu, American Society for Testing and Materials, Philadelphia, pp. 70–90.

J. Ziebs, J. Meersmann, H. J. Kühn, N. Hülsmann and J. Olchewski, 1994, Multiaxial thermomechanical behaviour of 738LC. In: "Proceedings of 4th International Conference on Biaxial/Multiaxial Fatigue, St. Germain en Laye, France, May 31–June 3, 1994," eds. A. Pineau and G. Cailletaud, SF2M, Paris, vol. 2, pp. 247–259.

J. Ziebs, J. Meersmann, H.-J. Kühn and H. Klingelhöffer, 2000, Multiaxial thermo-mechanical deformation behavior of 738LC and SC16. In: "Thermomechanical Fatigue Behavior of Materials: 3rd Volume, ASTM STP 1371," eds. H. Sehitoglu and H. J. Maier, American Society for Testing and Materials, West Conshohocken, PA, pp. 257–278.

J. Ziebs, J. Meersmann, H.-J. Kühn and S. Ledworoski, 1992, High temperature inelastic deformation of IN738 LC under uniaxial and multiaxial loading. In: "Low Cycle Fatigue and Elasto-plastic Behaviour of Materials, LCF3," ed. K. T. Rie, Elsevier, Amsterdam, pp. 248–255.

Comprehensive Structural Integrity
ISBN (set): 0-08-043749-4

Volume 5; (ISBN: 0-08-044154-8); pp. 113–199

5.04

Analysis of Cracks under Creep Conditions

A. SAXENA

Georgia Institute of Technology, Atlanta, USA

5.04.1 BACKGROUND

Several components of power plants, chemical reactors, and land-, air-, and sea-based gas turbines operate at temperatures where creep deformation and fracture are design concerns. Several of these high-temperature components, especially those containing thick sections, are subjected to stress and temperature gradients and do not fail by creep rupture. It is more likely that at the end of the predicted creep rupture life, a crack develops at a high-stress location, propagates, and ultimately causes fracture. Failures can also result from preexisting defects, in which case the entire life is consumed by crack propagation. Therefore, it is important to develop the capability to predict crack propagation life at elevated temperatures in the presence of creep deformation.

Linear-elastic and elastic–plastic fracture mechanics concepts are unable to predict crack growth in the presence of significant creep strains. Thus this chapter will focus on developing the concepts of time-dependent fracture mechanics (TDFM) to account for the role of creep in the crack growth behavior in elevated-temperature components. In looking to extend the fracture mechanics concepts to conditions where time-dependent creep deformation is no longer a limitation, we will be taking advantage of considerable analogies that exist between TDFM and elastic–plastic fracture mechanics (EPFM).

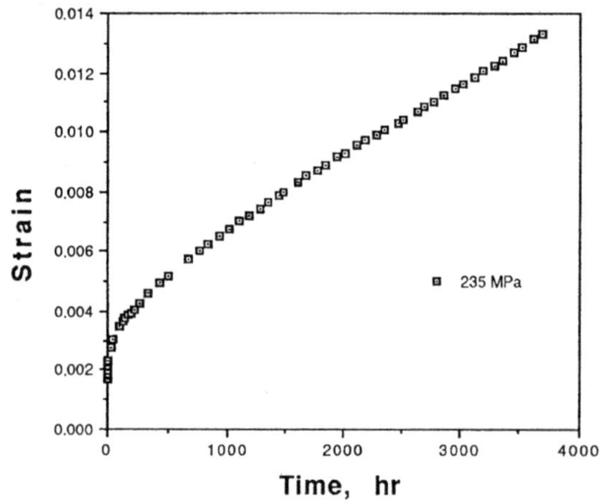

Figure 1 Creep deformation behavior of 1Cr–1Mo–0.25V steel at a stress level of 235 MPa (Saxena, 1998; reproduced with permission of CRC Press).

Figure 1 shows creep deformation as a function of time from a uniaxial specimen of a 1Cr–1Mo–0.25V steel subjected to constant stress. The creep behavior can be divided into three regions; the first region is called the primary creep region, in which the strain rate continuously decreases with time until it reaches a steady-state value from where it enters into the second region, called the steady-state creep region. In the tertiary creep region that follows the steady-state region, the creep strain rates begin to rise with time as necking develops and is followed by rupture. Tertiary creep regime is not significant in practical applications, because it must be avoided. Primary creep in several materials is short-lived, thus making the steady-state creep region the most important in design considerations. As mentioned earlier, the strain rates in this region remain constant with time, but they change significantly with stress and are described by a power-law (also known as Norton's Law), Equation (1):

$$\dot{\varepsilon}_{ss} = A\sigma^n \tag{1}$$

where $\dot{\varepsilon}_{ss}$ is the steady-state strain rate and σ = applied stress. Initially in this chapter we will be using Equation (1) to describe the creep deformation kinetics. Later, we will generalize the creep deformation equation to also include primary creep in our consideration of crack-tip parameters.

5.04.2 BEHAVIOR OF CRACKS UNDER STEADY-STATE CREEP AND THE C^*-INTEGRAL

To identify relevant field parameters for characterizing crack growth at elevated temperature, the following process should be considered (Saxena, 1980). Figure 2 shows a schematic of the deformation zones ahead of a stationary crack tip subjected to load in the creep regime. A load, P, is assumed to be applied instantaneously and is sustained indefinitely. Upon loading, a plastic zone is formed at the crack tip. The size of the plastic zone depends on the applied K level and the yield strength of the material. Also shown in the figure is a K-zone in which the K-controlled elastic stress and deformation fields hold, if small-scale yielding conditions are maintained during the initial loading. If extensive plastic

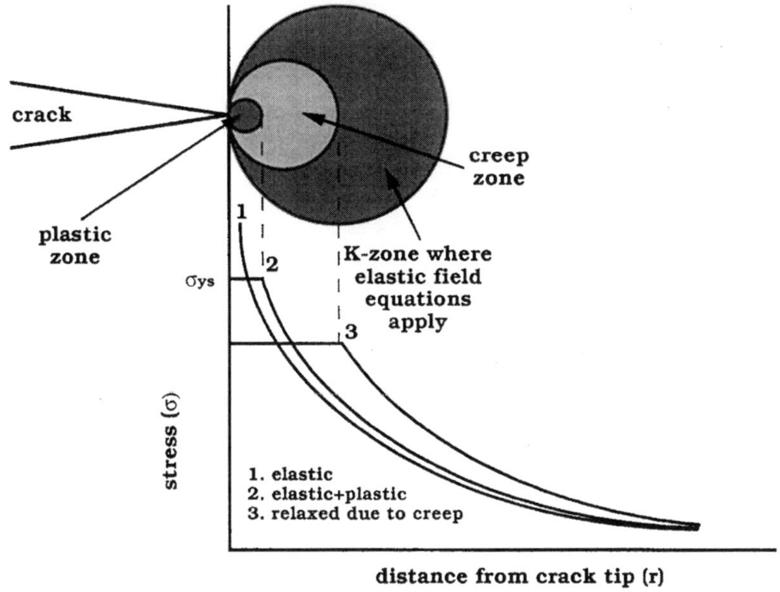

Figure 2 Schematic of the deformation zones ahead of the crack tip and the associated crack-tip stress fields (Saxena, 1980; reproduced with permission of ASTM).

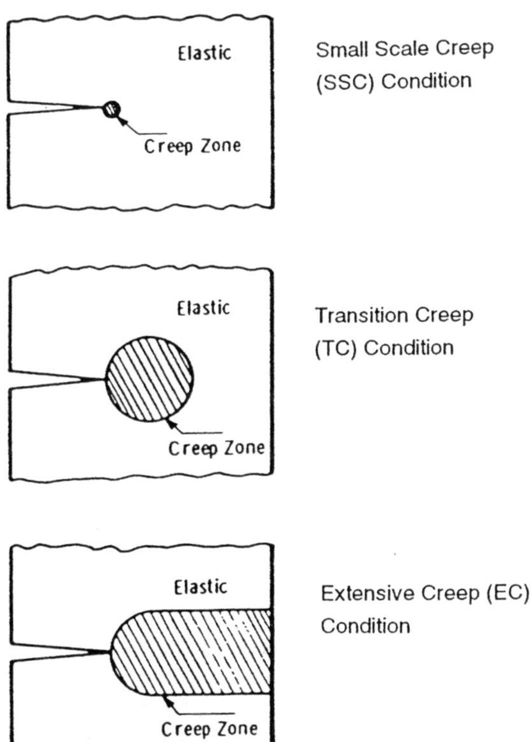

Figure 3 Schematic representation of the levels of creep deformation under which creep crack growth can occur.

deformation occurs during initial loading, K loses its significance as the dominating crack-tip parameter. Under such conditions, the crack-tip stresses and strains are characterized by the J-integral. With time, in either case, the stresses in the vicinity of the crack begin to relax due to creep deformation, and the size of the relaxation (or creep) zone increases with time if the crack is assumed to remain stationary. Neither K nor J is expected to uniquely characterize the crack-tip stress relaxation (or redistribution) behavior within the creep zone, because creep deformation is not admitted in their formulation. However, if the creep zone is small, K or J will continue to characterize the crack-tip stresses outside the creep zone. When the creep zone becomes comparable to the dimensions of the cracked body, K and J completely lose their significance as crack-tip parameters. Therefore, when creep is present we have to look for new crack-tip parameters.

Figure 3 schematically shows the regimes of small-scale creep (SSC), transition creep (TC), and extensive creep (EC) for a body with constant crack size. Under SSC, the creep zone size is small in comparison to the crack length and pertinent dimensions of the body. Alternatively, under EC conditions the creep zone completely engulfs the uncracked ligament. The TC condition represents the intermediate regime. The SSC and TC conditions are transient conditions, characterized by crack-tip stresses that vary with time as the stress redistributes within the uncracked ligament. Alternatively, the EC condition is a steady-state condition, because the crack-tip stresses remain constant with time for a constant crack size.

The steady-state, EC region is the simplest for analysis purposes; hence, we will begin the crack-tip stress analysis under those conditions. The SSC and TC conditions will be considered later in the chapter. In developing these analyses, it must be pointed out that the crack tip is assumed to be stationary. This assumption places fundamental limitations in the use of crack-tip parameters resulting from these analyses. These limitations will be discussed in detail when the crack-tip parameters are formulated.

We assume that a cracked body is subjected to a static load at elevated temperatures and that the load has been applied for a sufficiently long time for steady-state creep conditions to engulf the entire remaining ligament. The stress and strain rates everywhere in the body are related by Equation (1), which is analogous to the relationship between plastic strain and stress under monotonic loading given by Equation (2), for which the crack-tip stress and strains are described by the J-integral (Hutchinson, 1968; Rice and Rosengren, 1968):

$$\varepsilon = \varepsilon_e + \varepsilon_p = \frac{\sigma}{E} + \alpha\varepsilon_0\left(\frac{\sigma}{\sigma_0}\right)^m \tag{2}$$

where ε_e is the elastic strain and ε_p is the plastic strain, E is the elastic modulus, σ_0 is the yield strength, $\varepsilon_0 = \sigma_0/E$, and α and m are regression constants obtained from the stress–strain behavior of the material. In the analogy, strain is replaced by strain rate, the term $\alpha\varepsilon_0/\sigma_0^m$ is replaced by the constant A, and m is replaced by n. Recognizing the above analogy between the power-law descriptions of plasticity and creep, Landes and Begley (1976) and Nikbin *et al.* (1976) independently defined an integral analogous to Rice's J-integral (Rice, 1968). Landes and Begley called this new integral C^* and defined it in the following manner:

$$C^* = \int_\Gamma W^*\,dy - T_i\left(\frac{\partial \dot{u}_i}{\partial x}\right)ds \tag{3}$$

where

$$W^* = \int_0^{\dot{\varepsilon}_{ij}} \sigma_{ij}\,d\dot{\varepsilon}_{ij} \tag{4}$$

and Γ is a line contour shown in Figure 4, taken counterclockwise from the lower crack surface to the upper crack surface. W^* is the strain energy rate density associated with the point stress, σ_{ij}, and strain rate, $\dot{\varepsilon}_{ij}$. T_i is the traction vector defined by the outward normal, n_j, along Γ, Figure 4. Thus, $T_i = \sigma_{ij}n_j$. The displacement vector is denoted by u_i and s is the arc length along the contour. In the analogy between J and C^*, the strain and displacement quantities are replaced with their respective time rates, and stress remains as stress. Like J, the C^*-integral is also path

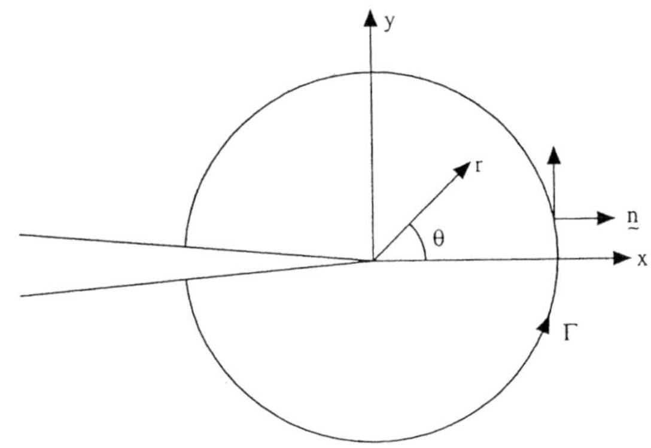

Figure 4 Crack-tip coordinate system and arbitrary contour along which C^*-integral is evaluated.

independent. Two other very important consequences of the analogy between J and C^* also follow: (i) the energy-rate interpretation of C^* and (ii) the relationship between the crack-tip stress fields and C^*. These are discussed below.

5.04.2.1 Energy Rate Interpretation of C^*

The energy rate definition of C^* was given as follows (Landes and Begley, 1976). Consider two cracked bodies that are identical in all respects with one exception. The crack length of one body is a and the crack length of the other is incrementally longer and is $a + \Delta a$. Both bodies are subjected to identical static loads, P_1; their load deflection rates, $\dot{V}_c$, are monitored after sufficient time has been allowed for steady-state conditions to develop in the specimens. The subscript c is added to V to designate that this deflection is entirely due to creep and the dot refers to rate with time. Recall that the crack sizes are assumed to be constant, so the entire deflection beyond the instant of load application ($t = 0^+$) is due to creep. Next, we consider several other identical pairs of cracked bodies and load them to different levels: P_2, P_3, ..., etc. We can then plot the relationship between the load, P, and the displacement rate, $\dot{V}_c$, separately for all specimens of crack size a and for those with crack size $a + \Delta a$. The areas under the load–displacement rate diagrams will be $U^*(a)$ and $U^*(a + \Delta a)$ for a fixed displacement rate, as shown in Figure 5. The difference between the two areas is ΔU^*. U^* is the energy rate or stress-power input into the cracked bodies and ΔU^* is the stress-power difference between the two bodies. The C^*-integral can be shown to be equal to

$$C^* = -\frac{1}{B}\frac{dU^*}{da} \tag{5}$$

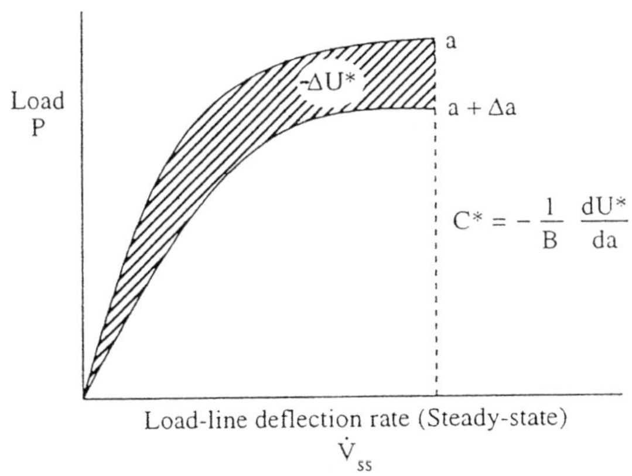

Figure 5 Schematic illustration of the energy rate interpretation of the C^*-integral.

where $B =$ the thickness of the body. Again, the proof of the above relationship is left as an exercise. Equation (5) provides the basis for measuring the value of C^* at the loading pins.

5.04.2.2 Relationship between C*-integral and the Crack-tip Stress Fields

Using the analogy with the HRR crack-tip stress fields for elastic–plastic and fully plastic conditions, the following relationships between crack-tip stress and strain rate and the C^*-integral for extensive steady-state creep can be derived (Goldman and Hutchinson, 1975):

$$\sigma_{ij} = \left(\frac{C^*}{I_n A r}\right)^{1/(1+n)} \hat{\sigma}_{ij}(\theta, n) \tag{6a}$$

$$\dot{\varepsilon}_{ij} = A\left(\frac{C^*}{I_n A r}\right)^{n/(n+1)} \hat{\varepsilon}_{ij}(\theta, n) \tag{6b}$$

where $r =$ distance from the crack tip, and I_n can be obtained from the following equations (Saxena, 1998):

For plane strain conditions:

$$I_n = 6.568 - 0.478n + 0.040n^2 - 0.001262n^3 \tag{7a}$$

For plane stress conditions:

$$I_n = 4.546 - 0.2827n + 0.0175n^2 - 0.45816 \times 10^{-4}n^3 \tag{7b}$$

$\hat{\sigma}_{ij}(n,\theta)$ and $\hat{\varepsilon}_{ij}(n,\theta)$ are angular functions defined elsewhere (Rice and Rosengren, 1968; Saxena, 1998). The above equations establish the unique relationship between crack-tip stress and C^*. C^*-integral has been shown to have the following properties: (i) It is a path-independent integral which can be computed along contours remote from the crack tip. (ii) It can be measured at the loading points. (iii) It can be uniquely related to the magnitude of crack-tip stress and strain rate. Thus, it is an attractive parameter for characterizing the creep crack growth rate under conditions of extensive steady-state creep. There is, therefore, considerable interest in methods for determining its value.

5.04.2.3 Methods of Determining C^*

The methods for determining C^* follow the methods for determining the J-integral. We will see that by making use of appropriate analogies, we can derive a C^* expression for any geometry and loading configuration for which we have an available expression for the J-integral. However, as a caution, the following must always be kept in mind while using analogies to develop expressions for C^*. The analogy between J and C^* is limited to the plastic part of the J-integral, J_p. Thus, we will need to separate the plastic part of J from the expression for the total J before proceeding with the analogy.

C^* can be obtained analytically for some very limited geometries. The methods that can be applied to complex geometries and loading conditions include: (i) experimental method, (ii) semi-empirical methods, and (iii) numerical solutions. These methods are discussed further.

5.04.2.3.1 Experimental methods for determining C^*

This method is based on the energy rate (or stress-power) interpretation of C^* given in Equation (5) (Landes and Begley, 1976). The method requires several sets of identical specimens. The crack size is varied between different sets of specimens. Let us arbitrarily choose five specimens per set, in each of five sets. The crack sizes of specimens in the various sets are a_1, a_2, a_3, a_4, a_5. Thus, a total of 25 specimens are prepared for testing. We select five load levels, P_1, P_2, ..., P_5, subject a specimen from each set to one of the five load levels, and measure the steady-state deflection rates. We will use the designation $\dot{V}_{ss}$ to represent the steady-state deflection rate that is determined as shown in Figure 6. Note that the steady-state conditions are realized in the specimen only after some time. Next, the P vs. $\dot{V}_{ss}$ relationship for each crack size can be plotted and the area under the curve for different values of $\dot{V}_{ss}$ can be obtained to determine U^*, as shown in Figure 7. We then plot U^* as a function of crack size for fixed values of $\dot{V}_{ss}$, as shown in Figure 8. The slope of the curve of U^* vs. crack size, a, can be related to C^*. This

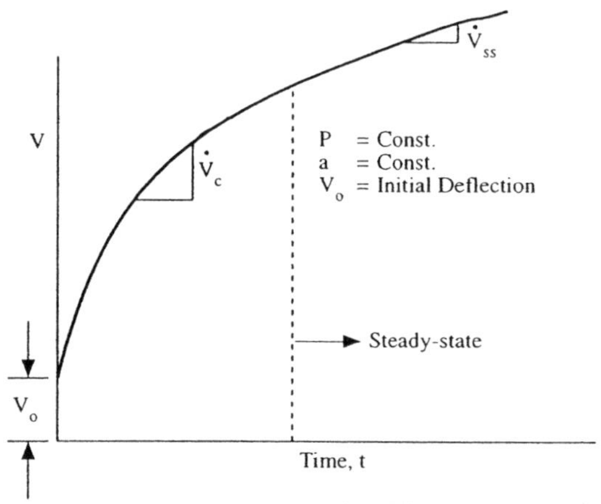

Figure 6 Load-line deflection as a function of time for a cracked body for a fixed applied load. The determination of $\dot{V}_{ss}$ is also shown in the figure.

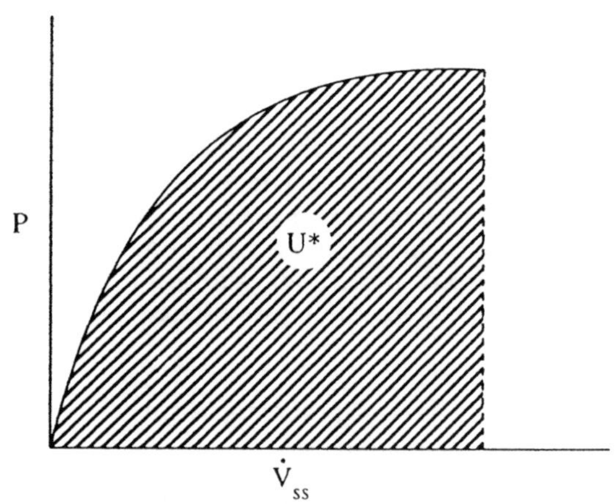

Figure 7 $P-\dot{V}_{ss}$ relationship for a fixed crack size and the definition of stress-power, U^*.

method of determining C^* is not practical due to the extensive amount of testing that is required. The C^* values obtained are not only for a specific geometry and loading configuration, they are also specific to the material being tested. Although this method was used for determining C^* in the early studies of creep crack growth (Saxena, 1980; Landes and Begley, 1976), it was quickly abandoned in favor of more direct methods. These are the semi-empirical methods described next.

5.04.2.3.2 *Semi-empirical methods of determining* C^*

By analogy to the method for estimating the J-integral, expressions for estimating C^* can be obtained quite readily. In the equations for estimating J, we must substitute $\dot{V}_{ss}$ in place of the displacement V. The following equations

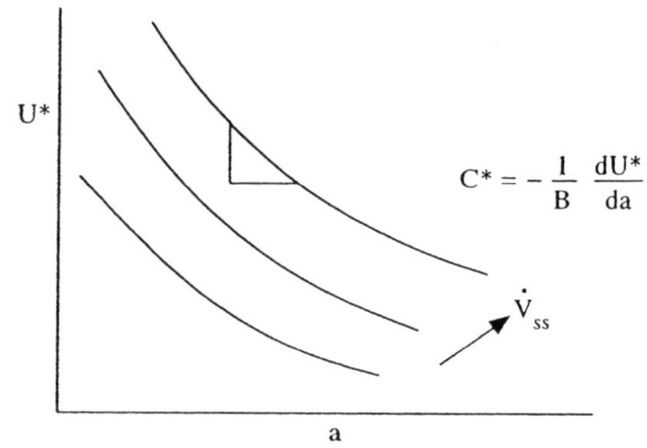

Figure 8 U^* as a function of crack size for various values of $\dot{V}_{ss}$.

are obtained for estimating C^* (Saxena, 1998):

$$C^* = -\frac{1}{B}\int_0^{\dot{V}_{ss}} \left(\frac{\partial P}{\partial a}\right)_{\dot{V}_{ss}} \mathrm{d}\dot{V}_{ss} \tag{8a}$$

$$C^* = \frac{1}{B}\int_0^{P} \left(\frac{\partial \dot{V}_{ss}}{\partial a}\right)_{P} \mathrm{d}P \tag{8b}$$

During the extensive steady-state creep conditions, the following equation holds between load and displacement rate:

$$\dot{V}_{ss} = \phi(a/W)P^n \tag{9}$$

Combining Equation (9) with Equation (8a) leads to an equation for estimating C^*. For a compact-type (CT) specimen for $a/W > 0.40$, the following equation is derived:

$$C^* = \frac{P\dot{V}_{ss}}{B(W-a)}(2 + 0.522(1 - a/W))\frac{n}{n+1} \tag{10}$$

where W = width of the specimen. The equivalent expression for center crack tension (CCT) specimens (Saxena, 1998) is

$$C^* = \frac{P\dot{V}_{ss}}{2BW(1-a/W)}\frac{n-1}{n+1} \tag{11}$$

where W = half width of the specimen. Similar expressions can also be derived for other specimen geometries such as the single-edge notch bend (SENB) and the double-edge notch (DEN) specimens.

5.04.2.3.3 C^* *based on numerical solutions*

Numerical solutions are available for estimating the fully plastic J-integral values (Kumar *et al.*, 1981). These equations can be modified to estimate C^* (Saxena, 1980; Riedel and Rice, 1980). Such equations are necessary for estimating C^* in components where no measurements of the load-line deflection rate,

$\dot{V}_{ss}$, are available. As with the J-integral, these solutions are restricted to either plane stress or plane strain conditions.

In this section, we will develop the equations for estimating C^* in the CT specimen. The derivations of equations for other geometries are similar. The equations for estimating J for fully plastic conditions, J_p, and the fully plastic portion of the load-line deflection, V_p, in CT specimens (Kumar *et al.*, 1981) are reproduced as follows:

$$J_p = \alpha\sigma_0\varepsilon_0(W-a)h_1(a/W,m)(P/P_0)^{m+1} \qquad (12)$$

$$V_p = \alpha\varepsilon_0 a h_3(a/W,m)(P/P_0)^m \qquad (13)$$

where, for plane strain conditions,

$$P_0 = 1.455\eta(W-a)\sigma_0 \qquad (14a)$$

and for plane stress conditions,

$$P_0 = 1.071\eta(W-a)\sigma_0 \qquad (14b)$$

and

$$\eta = \left[\left(\frac{2a}{W-a}\right)^2 + 2\left(\frac{2a}{W-a}\right) + 2\right]^{1/2} - \left[\left(\frac{2a}{W-a} + 1\right)\right] \qquad (14c)$$

Substituting the value of P_0 from Equations (14a) or (14b) into Equation (12) and replacing $\alpha\varepsilon_0/\sigma_0^m$ with A and m with n, we get the following equation:

$$C^* = A(W-a)h_1(a/W,n) \times \left(\frac{P}{1.455\eta B(W-a)}\right)^{n+1} \qquad (15)$$

Equation (15) is for the plane strain condition, but by substituting the P_0 and h_1 values for plane stress, the corresponding expression for plane stress can be derived. The expressions for load-line displacement rate and the CTOD rate due to creep, $\dot{\delta}_{ss}$, are given by the following equations:

$$\dot{V}_{ss} = Aah_3(a/W,n)\left(\frac{P}{1.455\eta B(W-a)}\right)^n \qquad (16)$$

$$\dot{\delta}_{ss} = Aah_2(a/W,n)\left(\frac{P}{1.455\eta B(W-a)}\right)^n \qquad (17)$$

The functions of h_1, h_2, and h_3 in the above equations are listed in tabular form for several geometries (Kumar *et al.*, 1981). If we substitute Equation (16) into (15) and compare the terms with Equation (10), we can show that,

$$\frac{h_1}{h_3}\frac{(1-a/W)}{1.455\eta(a/W)} \approx (2 + 0.522(1-a/W))\frac{n}{n+1} \qquad (18)$$

In fact, the term on the right-hand side of Equation (18) has been derived by fitting the chosen form of the equation to the results obtained by calculating the values of the left-hand side for different a/W values.

5.04.2.4 Correlation between Creep Crack Growth Rates and C^*

To demonstrate the validity of C^* for characterizing creep crack growth rates, it is necessary to show that the crack growth rates measured from specimens of the same geometry and size but loaded to different load levels, as well as those from specimens of different geometry and size, uniquely correlate with C^*. The early attempts (Landes and Begley, 1976) to correlate $\mathrm{d}a/\mathrm{d}t$ with C^* under creep conditions met with only limited success. These experiments were conducted on A286 creep-resistant alloy, in which it was difficult to obtain the conditions of extensive steady-state creep within the time period of the experiment. Similar experiments were then conducted later (Saxena, 1980) on 304 stainless steel at 594°C to promote extensive steady-state creep in the specimens. These experiments were conducted using CT and CCT specimen geometries. Figure 9 shows the results of this study, which demonstrated the validity of C^* for correlating creep crack growth rates. Similar correlations have been obtained by several other investigators since these early experiments and the validity of C^* for characterizing creep crack growth rates under extensive creep conditions is now widely accepted.

5.04.3 ANALYSIS OF CRACKS UNDER SSC AND TC

The validity of the C^*-integral presented in the previous section is limited to extensive steady-state creep conditions. In practice, this condition may not always be realized, because components contain stress and temperature gradients and are designed to resist widespread creep deformation. Therefore, in this section we will derive the crack-tip stress fields for the conditions of SSC and TC. In these analyses, we will continue to assume that creep deformation occurs by power-law creep and that the crack is stationary. However, the elastic stress redistributions due to creep deformation will be considered. Thus the material is modeled as elastic and power-law creep material for which

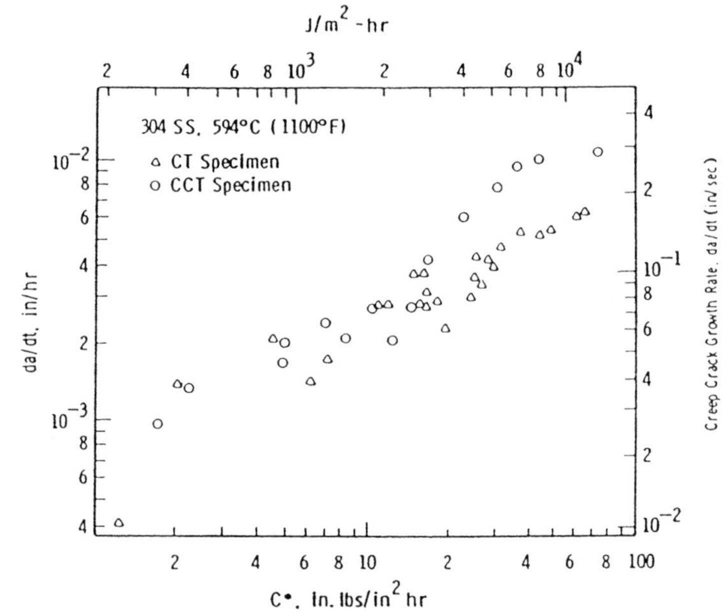

Figure 9 Creep crack growth rate as a function of the C^*-integral for 304 stainless steel (Saxena, 1980; reproduced with permission from ASTM).

the uniaxial version of the constitutive equation is given by

$$\dot{\varepsilon} = \frac{\dot{\sigma}}{E} + A\sigma^n \tag{19}$$

It is further assumed that the elastic and creep strains are additive, thus,

$$\dot{\varepsilon} = \dot{\varepsilon}_e + \dot{\varepsilon}_c \tag{20}$$

where ε_e = elastic strain and ε_c = accumulated time-dependent creep strain, and the dots indicate derivatives with time. Integrating Equation (20) gives

$$\varepsilon(t) = \frac{\sigma(t)}{E} + \int_0^t A[\sigma(\tau)]^n \,\mathrm{d}\tau \tag{21}$$

where τ is a dummy integration variable and $\sigma(t)$ is the time-dependent stress at any time t.

5.04.3.1 Small-scale Creep

5.04.3.1.1 Crack-tip stress fields

Riedel and Rice (1980) and Ohji *et al.* (1979) independently derived the nature of the crack-tip stress fields under SSC conditions and reached identical results. In the following discussion, we will follow the 1979 derivation by Ohji *et al.* (1979) because of its simplicity.

Equation (21) can be written as

$$\varepsilon = \frac{\sigma(t)}{E} + Af(t)(\sigma(t))^n \tag{22}$$

where

$$f(t) = \int_0^t [\sigma(\tau)/\sigma(t)]^n \,\mathrm{d}\tau \tag{23}$$

Since $\sigma(t)$ is a function of time only, it is valid to include it inside the integral and also to apply it to all the area in the crack-tip region. For constant time, the form of Equation (22) is identical to the Ramberg–Osgood Equation (2). We note also that the relationship between stress and strain is univalued for a constant time. This would imply that the J-integral is path independent. Hence, we can write the HRR stress fields as

$$\sigma_{ij} = \left[\frac{J}{I_n Af(t)r}\right]^{1/(1+n)} \hat{\sigma}_{ij}(\theta, n) \tag{24a}$$

$$\varepsilon_{ij} = Af(t)\left[\frac{J}{I_n Af(t)r}\right]^{n/(n+1)} \hat{\varepsilon}_{ij}(\theta, n) \tag{24b}$$

In Equation (24a), σ_{ij} is proportional to $(1/f(t))^{1/(n+1)}$. Thus, Equation (23) can be written as

$$f(t) = \int_0^t [f(t)/f(\tau)]^{n/(n+1)} \,\mathrm{d}\tau \tag{25}$$

By trial and error, we find that the following solution fits the integral Equation (25):

$$F(t) = (n+1)t \tag{26}$$

If we substitute Equation (26) into Equation (24a), and also recognize that for small-scale-yielding and elastic conditions $J = (K^2/E)(1 - \nu^2)$, the result can be written as follows:

$$\sigma_{ij} = \left[\frac{K^2(1-\nu^2)}{EI_nA(n+1)tr}\right]^{1/(1+n)} \hat{\sigma}_{ij}(\theta, n) \quad (27a)$$

$$\varepsilon_{ij} = \left[\frac{K^2(1-\nu^2)}{EI_nA(n+1)tr}\right]^{n/(n+1)} A(n+1)t\hat{\varepsilon}_{ij}(\theta, n) \quad (27b)$$

$$\dot{\varepsilon}_{ij} = A\left[\frac{K^2(1-\nu^2)}{EI_nA(n+1)tr}\right]^{n/(n+1)} \hat{\varepsilon}_{ij}(\theta, n) \quad (27c)$$

In the above equations, the time-dependence of the crack-tip stress, strain and strain rate fields is quite evident. The above analysis lends itself to the estimation of the creep zone size and transition time, t_T, which is the time needed for EC conditions to develop from SSC conditions. These are discussed in the subsequent sections.

5.04.3.1.2 Estimation of the creep zone size

The creep zone boundary was defined (Riedel and Rice, 1980) as the locus of points where time-dependent effective creep strains equal the instantaneous effective elastic strains in the cracked body. This leads to

$$r_c(\theta, t) = \frac{1}{2\pi}\left(\frac{K}{E}\right)^2 \left[\frac{(n+1)I_nE^nAt}{2\pi(1-\nu^2)}\right]^{2/(n-1)} F_{cr}(\theta, n) \quad (28a)$$

This equation can also written as

$$r_c(\theta, t) = \frac{1}{2\pi}\left(\frac{(n+1)^2}{2n\alpha_n^{n+1}}\right)^{2/(n-1)} K^2(EAt)^{2/(n-1)}F_{cr}(\theta, n) \quad (28b)$$

where α_n^{n+1} has an approximate value of 0.69 for $3 \leq n \leq 13$ and F_{cr} is a function dependent on θ, n and the state of stress, as shown in Figure 10. The creep zone size is dependent on time as $t^{2(n-1)}$ and on K as the second power.

An alternative definition was formulated for the creep zone boundary by choosing the boundary to be a locus of points where the time-dependent strain has a value of 0.2% (Adefris *et al.*, 1998). This leads to

$$r_c(\theta, t) = \left[\frac{(n+1)A}{0.002}\right]^{(n+1)/n} \frac{(1-\nu^2)K^2}{(n+1)AEI_n} t^{1/n}\hat{r}_c(\theta, n) \quad (29)$$

Equations (28a) and (28b) become unstable for $n = 1$. Note that for real materials n cannot be less than 1. For $n < 3$, Equations (28a) and

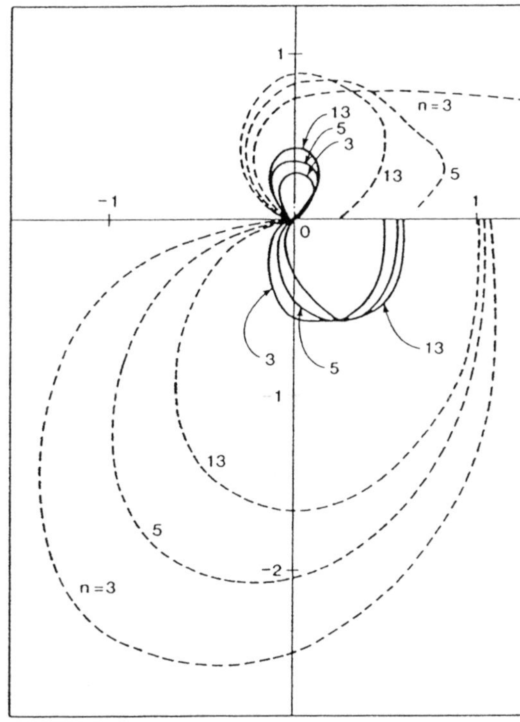

Figure 10 Function F_{cr} (θ, n) for plane stress and plane strain for various values of n (Riedel and Rice, 1980; reproduced with permission from ASTM).

(28b) predict very rapidly growing creep zone size. Both these results are an artifact of the Riedel–Rice definition of the creep zone. When the alternate creep zone definition is considered, such as in Equation (29), this problem no longer exists. For $n = 1$, a linear-viscous material, the creep zone size increases linearly with time which is a reasonable prediction. Thus, the Riedel–Rice definition of creep zone size is not recommended for n values approaching 3, or less than 3. For most metals n is significantly greater than 3, and the differences between the two definitions of creep zone size are not of major significance.

5.04.3.1.3 Transition time (t_T)

Riedel and Rice (1980), as well as Ohji *et al.* (1979), in their original analyses presented a concept of transition time, t_T. They defined the transition time as the time when the SSC stress fields equal the extensive steady-state creep fields characterized by C^*. Therefore, equating Equations (6a) and (29a) at time t_T, we get,

$$t_T = \frac{K^2(1-\nu^2)}{E(n+1)C^*} \quad (30)$$

We can then state that if $t \ll t_T$, Equation (27) describes the crack-tip conditions, and for $t \gg t_T$ Equation (6) determines the crack-tip conditions. An interpolation scheme for determining the crack-tip stress fields for $t/t_T \approx 1$ will be described later in this section.

5.04.3.2 $C(t)$-integral and the Stress Fields in the TC Region

Bassani and McClintock (1981) recognized that the crack-tip stress fields under SSC can also be characterized by a time-dependent C-integral, whose value is determined along a contour taken very close to the crack tip:

$$C(t) = \int_{\Gamma \to 0} W^* \, dy - T_i \left(\frac{\partial \dot{u}_i}{\partial x} \right) ds \tag{31}$$

Note that $C(t)$ is the same as C^* but its value is determined close to the crack tip within a region where the creep strains dominate over the elastic strains. Recall that the value of C^* could be determined along any contour which originated at the lower crack surface, ended on the upper crack surface, or enclosed the crack tip. Thus, determining the C-integral requires accurate solutions of stress and strain near the crack tip. Bassani and McClintock (1981) further related the value of $C(t)$ with the HRR-type stress fields as follows:

$$\sigma_{ij} = \left(\frac{C(t)}{AI_n r} \right)^{1/(n+1)} \hat{\sigma}_{ij}(\theta, n) \tag{32a}$$

$$\dot{\varepsilon}_{ij} = A \left(\frac{C(t)}{AI_n r} \right)^{n/(n+1)} \hat{\varepsilon}_{ij}(\theta, n) \tag{32b}$$

By comparing Equation (27a) with Equation (32a), we get the following result for SSC:

$$C(t) \approx \left(\frac{K^2(1 - \nu^2)}{E(n+1)t} \right) \tag{33}$$

The reason for the approximate equality in Equation (33) is that both sets of equations, i.e., Equations (27) and (32), are strictly valid only as $r \to 0$. For finite values of r, they are only approximately valid. Equation (33) has been verified by numerical analysis and has been shown (Bassani and McClintock, 1981; Ehlers and Riedel, 1981) to hold true. The validity of the $C(t)$-integral is not simply limited to the SSC conditions. It becomes equal to C^* by definition for extensive SSC, with the additional property that its value becomes path independent. Hence, $C(t)$ characterizes the amplitude of the HRR field for all conditions ranging from SSC to extensive secondary-state creep, and also including the TC conditions in between. Ehlers and Riedel (1981) and Bassani and McClintock (1981) independently proposed the same interpolation formula for approximating the HRR stress fields in the TC region as follows:

$$C(t) \approx \frac{K^2(1 - \nu^2)}{E(n+1)t} + C^* \tag{34}$$

Equation (34) can be combined with Equation (30) to develop an alternate expression for $C(t)$:

$$C(t) \approx (1 + t_T/t)C^* \tag{35}$$

The accuracy of Equation (35) has been proven by several finite element studies, as seen in Figure 11 (Bassani *et al.*, 1989; Leung *et al.*, 1989; Ehlers and Riedel, 1981). The $C(t)$-integral, by virtue of its ability to characterize the HRR fields from SSC to extensive steady-state creep, is an attractive parameter for correlating creep crack growth rate data. However, it has a major drawback in that it cannot be measured at the loading pins under SSC and TC conditions. Its value must always be calculated from Equation (35). The calculated value is very sensitive to the accuracy of the creep constants, A and n, and the assumptions regarding the state of stress. Figure 12 shows the creep crack growth rate plotted with $C(t)$ for several CT specimens of 1Cr–1Mo–V steel showing considerable variability in the data from different specimens. It cannot be determined with certainty if the poor correlation is due to inaccuracy of Equation (35) in estimating the value of $C(t)$, or whether the parameter itself is not suitable for characterizing creep crack growth. Nonetheless, it is a shortcoming of the approach which makes it less desirable in practical applications.

We also note that during SSC and TC, the crack-tip stresses are significantly higher than under extensive steady-state creep conditions. The crack growth rates during that period are expected to be higher as well. Therefore,

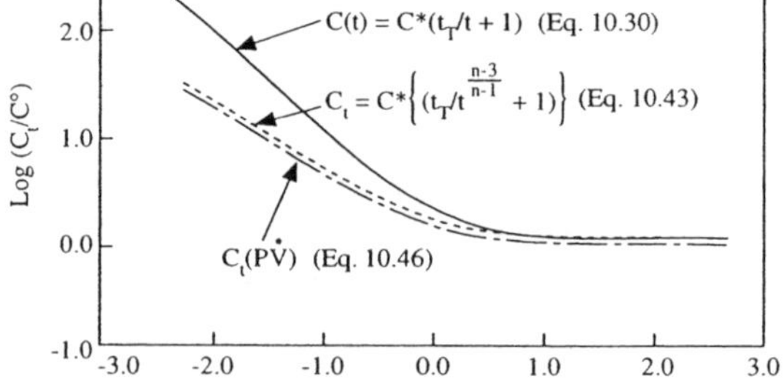

Figure 11 Comparison between the numerically calculated values of the C-integral and those predicted by the interpolation formula, Equation (34) (Bassani *et al.*, 1989; reproduced with permission from ASTM).

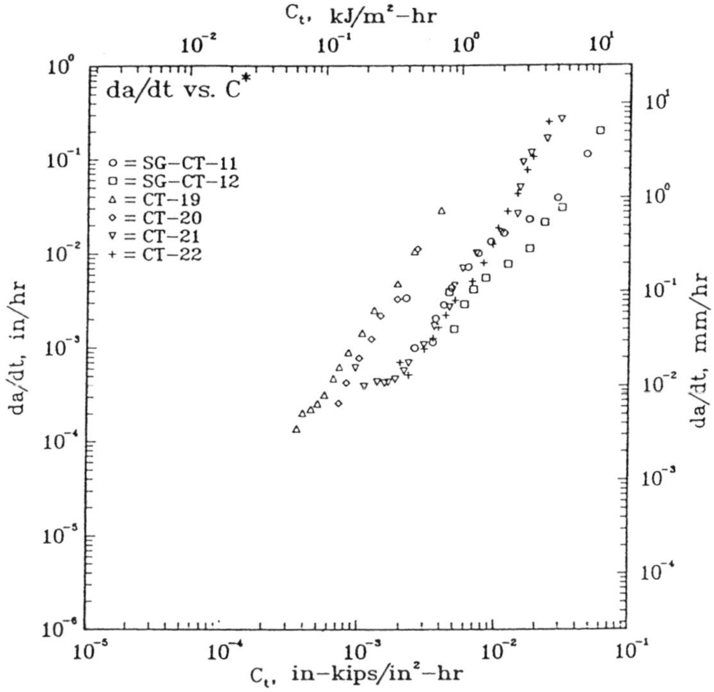

Figure 12 Creep crack growth rate as a function of the C_t parameter for 1Cr–1Mo–0.25V steel at 593°C (Bassani *et al.*, 1989; reproduced with permission from ASTM).

ignoring the SSC regime during analysis of components can be quite nonconservative.

5.04.3.3 The C_t-Parameter

It is evident that SSC and TC regions cannot be ignored in component analysis. Also, recall that the results from the applications of the $C(t)$ integral to characterize crack growth rates were not very successful. Therefore, alternate parameters are needed to characterize creep crack growth rates under a wide range of conditions, from SSC to extensive steady-state creep. The C_t parameter proposed by Saxena (1986) fulfills this need and is discussed in this section.

In searching for a new parameter, it is important to keep in mind that the candidate parameter must be identical to C^* in the EC regime because of its proven value. If we further impose the condition that the parameter be measurable at the load-line, it seems reasonable to generalize the stress-power definition of C^* into the SSC and TC region. Once such a parameter is defined, its relationship to crack-tip conditions can be explored.

Consider several identical pairs of cracked specimens. Within each pair, one specimen has a crack length a, and the other has an incrementally differing crack length $a+\Delta a$. The specimens of each pair are loaded to various load levels: P_1, P_2, $P_3,\ldots$, P_i, etc. at elevated temperatures, and the load-line deflection due to creep, $\dot{V}_c$, is recorded with time (Figure 13(a)). It is assumed that no crack extension occurs in any of the specimens and the instantaneous response is linear elastic. In this discussion, let us first limit our discussion to the SSC condition characterized by $t/t_T \ll 1$. At a fixed time t, the load vs. creep deflection rate, $\dot{V}_c$, behavior is plotted for all these specimens. A schematic of the expected behavior is shown in Figure 13(b). Several such plots for varying time can be generated from the above tests by varying time. When steady-state conditions are reached, the plots will become identical and will be similar to the ones shown earlier in Figure 5. The area between the P–$\dot{V}_c$ curves for specimens with crack length a and $a+\Delta a$ is called ΔU_t^*. Physically, ΔU_t^* (the subscript denotes that this value is at a fixed time t) represents the instantaneous difference in the stress-power supplied to two cracked bodies with identical creep deformation histories, as they are loaded to different deflection-rate levels. The C_t parameter is given by

$$C_t = -\frac{1}{B}\frac{\partial U_t^*(a,t,\dot{V}_c)}{\partial a} \tag{36}$$

As $t/t_T,\rightarrow\infty$, $C_t = C^*$ by definition. Thus, the C_t parameter satisfies one of the conditions that we specified earlier. However, it still remains to be seen what relationship exists between C_t and the crack-tip stress and

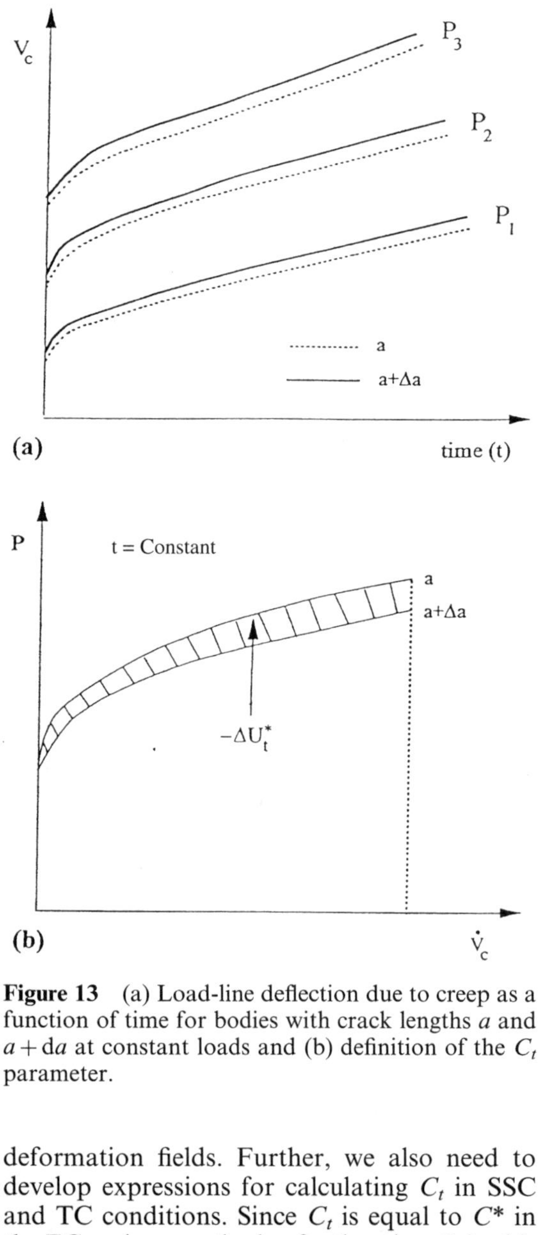

Figure 13 (a) Load-line deflection due to creep as a function of time for bodies with crack lengths a and $a + \mathrm{d}a$ at constant loads and (b) definition of the C_t parameter.

deformation fields. Further, we also need to develop expressions for calculating C_t in SSC and TC conditions. Since C_t is equal to C^* in the EC regime, methods of estimating C_t in this regime have already been covered.

5.04.3.3.1 Estimation of C_t for SSC

In order to derive an expression for estimating C_t, it is necessary to derive a relationship amongst the applied load, P, load-line deflection rate, $\dot{V}_c$, crack size, a, and the time t. This will allow us to determine U_t^* as a function of a and t for given values of P and subsequently U_t^* can be differentiated to determine the value of C_t (Saxena, 1986, 1989, 1991), as briefly described below.

We invoke the Irwin concept (Saxena, 1986) of effective crack length, a_{eff}, for calculating load-line deflection as a function of P, a, and t. We write

$$a_{\text{eff}} = a_0 + \beta r_c \tag{37}$$

where r_c is the creep zone size as defined in Equation (28) and β is a scaling factor. Since the creep zone can sustain stresses, the entire creep zone does not act as a crack. Therefore, the value of β is less than one. If we choose the value of r_c corresponding to $\theta = 90°$, it has been shown by finite element analysis (Bassani *et al.*, 1989) that $\beta = 1/3$. The additional load-line displacement due to creep, V_c, at any time t is given by

$$V_c = V - V_0 = P\frac{\mathrm{d}C}{\mathrm{d}a}\beta r_c \tag{38}$$

where P = applied load, V = total displacement, V_0 = instantaneous deflection, and C is the elastic compliance of the cracked body. The following relationship exists between compliance and K for plane strain:

$$(1 - \nu^2)\frac{K^2}{E} = \frac{P^2}{2B}\frac{\mathrm{d}C}{\mathrm{d}a} \tag{39}$$

Substituting Equations (28a) and (39) into Equation (37) and differentiating with time, we get,

$$\dot{V}_c = \frac{4\alpha(1 - \nu^2)}{E(n-1)}\beta F_{\text{cr}}\left(\frac{P}{B}\right)^3\frac{F^4}{W^2}(EA)^{2/(n-1)} \times t^{-(n-3)/(n-1)} \tag{40}$$

Equation (40) in the functional form can be written as

$$\dot{V}_c = \phi(a/W, t)\left(\frac{P}{B}\right)^3 \tag{41}$$

where

$$\phi = \frac{4\alpha(1 - \nu^2)}{E(n-1)}\frac{F^4}{W}(EA)^{2/(n-1)}t^{-(n-3)/(n-1)} \tag{42}$$

and

$$\alpha = \frac{1}{2\pi}\left(\frac{(n+1)^2}{2n\alpha_n^{n+1}}\right)^{2/(n-1)} \tag{43}$$

The C_t parameter can be determined by

$$(C_t)_{\text{ssc}} = -\frac{1}{B}\frac{\partial U_t^*}{\partial a} = \frac{B}{B\,\mathrm{d}a}\left[\int_0^{\dot{V}_c}\left(\frac{\dot{V}_c}{\phi}\right)^{1/3}\mathrm{d}\dot{V}_c - \int_0^{\dot{V}_c}\left(\frac{\dot{V}_c}{\phi + (\partial\phi/\partial a)\,\mathrm{d}a}\right)^{1/3}\mathrm{d}\dot{V}_c\right] \tag{44}$$

This finally reduces to

$$(C_t)_{\text{ssc}} = \frac{P\dot{V}_c}{BW}\frac{F'}{F} \tag{45}$$

where $F' = \mathrm{d}F/\mathrm{d}(a/W)$. Substituting for $\dot{V}_c$ in Equation (45) from Equation (40), we get Equation (46), which is an alternative expression for determining C_t:

$$(C_t)_{\text{ssc}} = \left(\frac{2K^2(1-\nu^2)}{EW}\right)\left(\frac{F'}{F}\right)\beta\dot{r}_c \tag{46}$$

$$(C_t)_{\text{ssc}} = \frac{4\alpha(1-\nu^2)}{E(n-1)}\beta F_{\text{cr}}(\theta,n)\frac{K^4}{W}(EA)^{2/(n-1)} \times \left(\frac{F'}{F}\right)t^{-(n-3)/(n-1)} \tag{47}$$

Equation (46) relates $(C_t)_{\text{ssc}}$ to the creep zone expansion rate, which is a crack-tip quantity. This relationship qualifies $(C_t)_{\text{ssc}}$ as a crack-tip parameter, so the next step in the evaluation of $(C_t)_{\text{ssc}}$ will be to conduct an experimental study. Equation (47) is a method by which $(C_t)_{\text{ssc}}$ can be computed for any geometry and loading configuration provided the *K*-calibration expression is available and also the creep constants *A* and *n* are known. If we choose $\theta = 90°$, as mentioned before, $\beta = 1/3$ (Bassani *et al.*, 1989) and $F_{\text{cr}}(\theta, n) = 0.4$ for *n* values of ~10. In general, $F_{\text{cr}}(\theta)$ will depend on *n*, but, the dependence is weak, as seen in Figure 10.

5.04.3.3.2 Estimation of C_t for wide range of creep conditions

It is convenient to have a single expression for estimating C_t for conditions ranging from SSC to EC. An approach similar to the one used for estimating the elastic and plastic part of the *J*-integral has also been successfully used for estimating C_t for a wide range of creep conditions:

$$C_t \approx (C_t)_{\text{ssc}} + C^* \tag{48a}$$

Another approximate expression for estimating C_t over a wide range was proposed by Bassani *et al.* (1989) as follows:

$$C_t \approx \left(1 + \left(\frac{t_T}{t}\right)^{(n-3)/(n-1)}\right)C^* \tag{48b}$$

Equation (47) can also be implemented for test specimens where the load-line displacement is available by recognizing that

$$\dot{V}_c \approx \dot{V}_{\text{ssc}} + \dot{V}_{\text{ss}} \tag{49}$$

Let us express C^* for any crack geometry in the following form:

$$C^* = \frac{P\dot{V}_{\text{ss}}}{BW}\eta_1(a/W, n) \tag{50}$$

The term η_1 can be derived for CT and CCT specimens easily by comparing Equations (10) and (11), respectively, with Equation (50). Then C_t can be written (Saxena, 1989) as

$$C_t \approx \frac{P\dot{V}_c}{BW}\frac{F'}{F} - C^*\left(\frac{F'}{F\eta_1} - 1\right) \tag{51}$$

The values of F'/F and η_1 for CT and CCT geometries are calculated and compared in Table 1. For CT geometries, the difference between F'/F and η_1 is no more than 20% for $0.4 \leq a/W \leq 1$ and for $3 \leq n \leq 20$. Thus, for this geometry, C_t can be estimated by either Equation (50) or (45) within an accuracy of ~20%. However, for the CCT geometry, Equation (51) must be used.

In SSC, C_t is directly related to the rate of expansion of the creep zone size as it evolves with time. We have also shown, in Equation (47), that

$$(C_t)_{\text{ssc}} \propto \frac{K^4}{(t)^{(n-3)/(n-1)}} \tag{52}$$

But also,

$$(C(t))_{\text{ssc}} \propto \frac{K^2}{t} \tag{53}$$

Table 1 Comparison of F'/F and η_1 $(a/W, n)$ for CT and CCT specimens.

		CT specimen η $(a/W, n)$					CCT specimen η_1 $(a/W, n)$					
a/W	F'/F	$n=3$	5	7	10	20	F'/F	$n=3$	5	7	10	20
0.2	2.976						2.755	0.312	0.417	0.468	0.511	0.565
0.3	2.590						2.067	0.357	0.476	0.537	0.584	0.646
0.4	2.644	2.725	3.211	3.373	3.504	3.67	1.821	0.417	0.555	0.625	0.682	0.753
0.5	3.078	3.274	3.767	3.957	4.111	4.307	1.785	0.500	0.667	0.750	0.818	0.905
0.6	3.920	4.063	4.60	4.83	5.020	5.259	1.914	0.625	0.833	0.937	1.022	1.131
0.7	5.330	5.340	5.99	6.29	6.536	6.847	2.255	0.833	1.111	1.250	1.363	1.508
0.8	7.949	7.856	8.763	9.205	9.563	10.019	3.042	1.25	1.667	1.875	2.045	2.262

Thus, $(C_t)_{\mathrm{ssc}}$ and $(C(t))_{\mathrm{ssc}}$ are not uniquely related to each other. This is somewhat of a negative for $(C_t)_{\mathrm{ssc}}$ as a candidate parameter for characterizing creep crack growth. But Figure 14 shows a plot of da/dt vs. C_t using the same data as shown in Figure 12, and the consolidation of all data into a single trend is evident, validating the use of C_t for creep crack growth.

Figure 15 shows creep crack growth data obtained on CT specimens of 1Cr–1Mo–0.25V that were 254 mm wide and 63.5 mm thick (Saxena *et al.*, 1994). The purpose of testing large specimens was to ensure significant crack

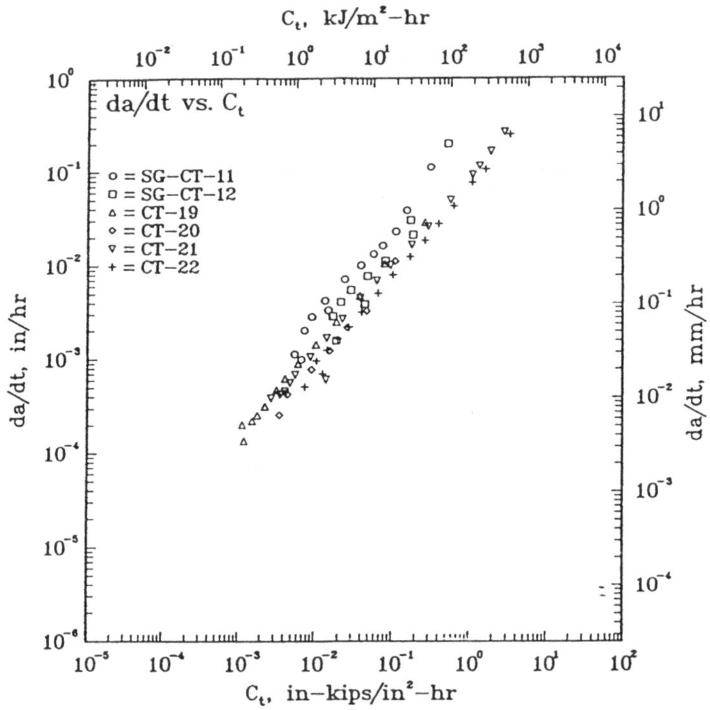

Figure 14 Creep crack growth rate as a function of C_t for the same data correlated with $C(t)$ in Figure 12 (Bassani *et al.*, 1989; reproduced with permission from ASTM).

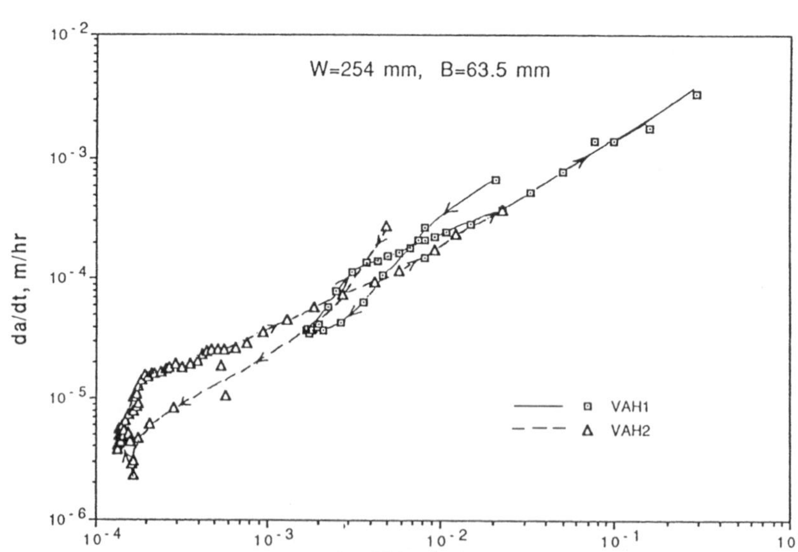

Figure 15 Creep crack growth rate as a function of C_t for 63.5 mm thick and 254 mm wide CT specimens. The arrows indicate the direction of change in C_t and da/dt from the beginning of the test to its end (Saxena *et al.*, 1994; reproduced with permission from ASTM).

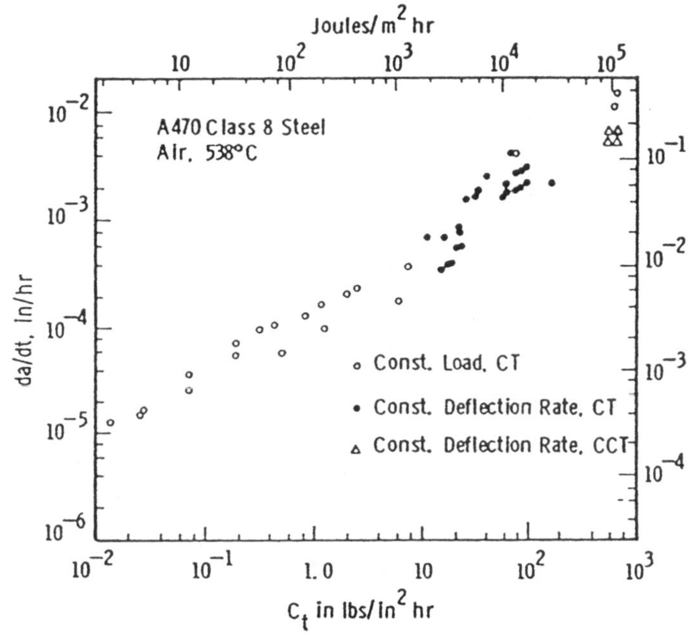

Figure 16 Creep crack growth rate as a function of C_t from several tests conducted on CT and CCT specimens under conditions of constant load and constant deflection rate (Saxena, 1986; reproduced with permission from ASTM).

extension that occurs under both SSC and EC conditions. The crack growth rates in these specimens begin at a higher rate, decrease progressively until they reach a minimum, and then increase. During both the increasing and decreasing portions of the test, the growth rates correlate with C_t. The decreasing deflection rate portion of the test is dominated by SSC and TC. Some decrease in C_t may also have been caused by primary creep deformation in the early part of the test, as will be discussed in the next section. The increasing crack growth rate portion of the test is dominated by EC conditions. These results nicely support the validity of C_t for uniquely characterizing the creep crack growth behavior under a wide range of creep conditions. Figure 16 shows similar data from CT and CCT geometries and obtained from tests conducted under constant load conditions and also from tests conducted under constant deflection rate conditions (Saxena, 1986). Again these results strongly support the viability of C_t as a crack-tip parameter.

5.04.4 CONSIDERATION OF PRIMARY CREEP

Several high-temperature materials exhibit significant primary creep at stress levels that are typical of service operation. Therefore, consideration of primary creep in estimating C_t can be significant. In the subsequent discussion, we will first consider primary creep in isolation in cracked bodies and then consider the combination of primary and secondary creep and elastic stress redistribution all together.

5.04.4.1 Creep Constitutive Law Including Primary Creep

A creep constitutive law that combines elastic deformation, primary creep, and steady-state creep can be written as follows:

$$\dot{\varepsilon} = \dot{\sigma}/E + A_1\varepsilon^{-p}\sigma^{n_1(1+p)} + A\sigma^n \qquad (54)$$

In the above equation, the first and third terms on the right-hand side are the elastic deformation and the secondary-creep deformation terms, respectively, described in the previous section. The middle term accounts for creep strains and contains three constants, p, A_1, and n_1, that are derived from regression of the creep data. Figure 17 shows a plot of the fitted curves from Equation (54) and the creep data for a 1.25Cr–1Mo–0.25V, showing the ability of Equation (54) to accurately represent the creep data at different stress levels. We will be exclusively using this form of equation in our discussion regarding primary creep.

5.04.4.2 Crack-tip Parameters for Extensive Primary Creep

If we consider extensive primary creep condition to dominate or, in other words,

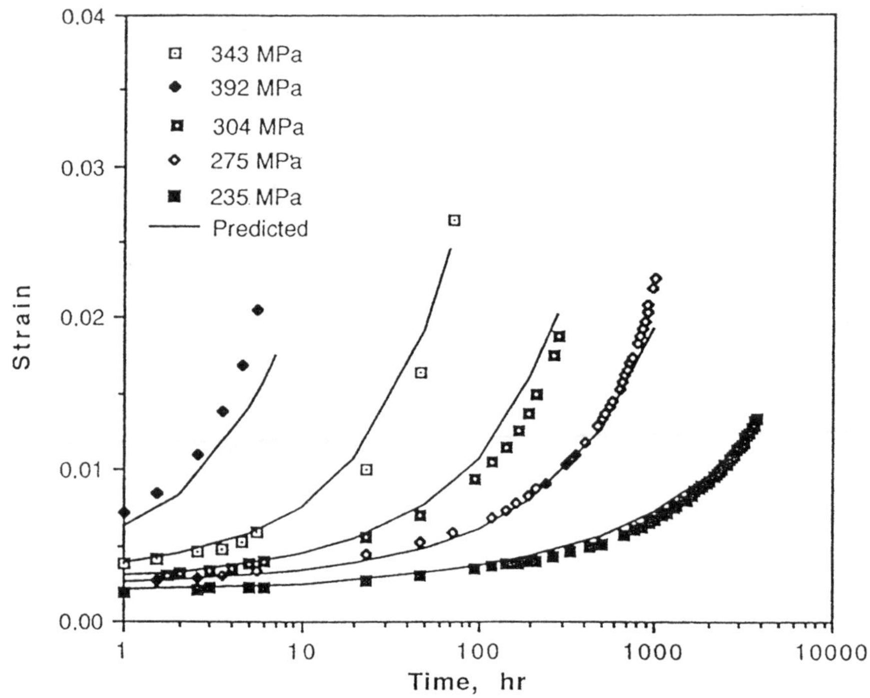

Figure 17 Creep strain as a function of time at various stress levels. The line represents predicted values from Equation (54) with $A_1 = 1.49 \times 10^{-31}$ (for σ in MPa and $\dot{\varepsilon}$ in h^{-1}), $n_1 = 3.54$, $p = 1.55$, $A = 9.51 \times 10^{-41}$ (for σ in MPa and $\dot{\varepsilon}$ in h^{-1}), and $n = 14.35$ (Saxena *et al.*, 1994; reproduced with permission from ASTM).

primary creep in isolation, we can simplify Equation (54) to

$$\dot{\varepsilon} = A_1 \varepsilon^{-p} \sigma^{n_1(1+p)} \tag{55}$$

which can also be expressed as

$$\varepsilon = [(1+p)A_1 t]^{1/(1+p)} \sigma^{n_1} \tag{55a}$$

and as

$$\dot{\varepsilon} = A_1[(1+p)A_1 t]^{-p/(1+p)} \sigma^{n_1} \tag{55b}$$

If we hold time as constant in Equation (55b), Riedel (1981) was first to recognize that the form of this equation is identical to Equation (1) for pure steady-state creep. That is, the strain rate is only a function of stress at a fixed time. Therefore, C^*-integral is path independent for a fixed time and its value can be calculated at a time t, by simply replacing A with $A_1\,[(1+p)A_1 t]^{-p/(1+p)}$ and n with n_1 in the expressions for calculating the C^*-integral. By examining Equation (55a), we can see that for a fixed time, the stress–strain relationship is also unique. That is, strain is only a function of stress. Therefore, the J-integral is also path independent and can be obtained by replacing $\alpha\varepsilon_0/\sigma_0^m$ in the equation for estimating J with $[(1+p)A_1 t]^{1/(1+p)}$ and m with n_1. Riedel (1981) defined yet another path-independent integral, C_h^*, which can be estimated by replacing $\alpha\varepsilon_0/\sigma_0^m$ in the equations for estimating J by $[(1+p)A_1]^{1/(1+p)}$ and m by n_1. The relationship between C_h^* and J is then given by

$$J = C_h^* t^{1/(1+p)} \tag{56}$$

The value of C_h^* is independent of time. Therefore, the HRR crack-tip field equation for stress is given by

$$\sigma_{ij} = \left[\frac{C_h^* t^{1/(1+p)}}{I_{n_1} r((1+p)A_1 t)^{1/(1+p)}}\right]^{1/(n_1+1)} \hat{\sigma}_{ij}(\theta, n_1) \tag{57a}$$

or

$$\sigma_{ij} = \left[\frac{C_h}{I_{n_1} r((1+p)A_1)^{1/(1+p)}}\right]^{1/(n_1+1)} \hat{\sigma}_{ij}(\theta, n_1) \tag{57b}$$

or

$$\sigma_{ij} = \left[\frac{C^*(t)}{I_{n_1} r A_1[(1+p)A_1 t]^{-p/(1+p)}}\right]^{1/(n_1+1)} \hat{\sigma}_{ij}(\theta, n_1) \tag{58}$$

where $C^*(t)$ is the time-dependent C^*-integral for primary creep. By comparing Equations (57b) and (58), one can show that,

$$C^*(t) = \frac{C_h^*}{(1+p)t^{p/(1+p)}} \tag{59}$$

The above analogies have worked well because of the separability of stress and time dependence in strain and strain rate in the formulation of primary creep. Therefore, by a judicious choice of the form of constitutive equations, the solutions available in the literature for estimating the J-integral and C^* can be directly used for estimating $C^*(t)$ for primary creep. Also note, in Equation (57b), that the crack-tip stress field for primary creep is independent of time and is only a function of spatial coordinates, and its amplitude is characterized by C_h^*.

Since the units of C_h^* are different from $C^*(t)$, it is preferable to characterize creep crack growth rates with $C^*(t)$, which has the same units and stress-power dissipation rate interpretation as does C^* for steady-state creep (Saxena, 1991). During extensive primary creep conditions, even though the crack-tip stress field is constant, the rate of creep strain accumulation varies with time and it is found that the creep crack growth rates also vary with time. Thus, no unique correlations between C_h^* and $\mathrm{d}a/\mathrm{d}t$ can be expected. However, good correlations between $C^*(t)$ and $\mathrm{d}a/\mathrm{d}t$ have been observed in the presence of primary creep (Saxena, 1991). The relevance of C_h^* in including primary creep is mostly convenience in the computation of $C^*(t)$. Another important observation is that $C^*(t)$, because of its relationship to the instantaneous stress-power dissipation rate, is identical to C_t. Thus, for extensive primary creep

$$C_t \equiv C^*(t) \tag{60}$$

Following the analogy further, the load-line deflection rate due to primary creep, $\dot{V}_{\mathrm{cp}}$, can be given by making the same substitutions for creep constants as for estimating $C^*(t)$ (Saxena, 1998).

5.04.4.3 Small-scale Primary Creep

In this discussion, we will consider the conditions of small-scale primary creep. Therefore, in Equation (54), we neglect the third term on the right-hand side and retain the other two, while also assuming the creep zone size is small in comparison to the other characteristic dimensions of the cracked body. We go back to Equations (44) and (45), which were derived for SSC conditions. In deriving either of those equations, no material constitutive laws were assumed. In other words, the validity of these equations holds for any general creep law and is not restricted to power-law creep, even though the equations were derived in the context of the considerations of power-law creep. In the case of Equation (44), the correct creep constitutive law, whatever it is, is considered rigorously in the measured value of $\dot{V}_{\mathrm{c}}$. Therefore, Equation (44) can be used to estimate $(C_t)_{\mathrm{ssc}}$ under primary creep also. Similarly, Equation (45) is valid for any creep law. However, an analytical expression for estimating $\dot{r}_{\mathrm{c}}$ for elastic–primary creep must be derived. This equation is given by (22)

$$\dot{r}_{\mathrm{c}}(\theta, t) = \frac{K^2}{2\pi}\left[\frac{I_{n_1} E}{2\pi(1-\nu^2)}\right]^{2/(n_1-1)} \times [(1+n_1)(1+p)A_1]^{2/(1+p)(n_1-1)} \times \frac{2F_{\mathrm{cr}}(\theta, n_1)}{(1+p)(n_1-1)} t^{[(2/(1+p)(n_1-1))-1]} \tag{61}$$

Substitution of Equation (61) into Equation (45) will give us an analytical expression for estimating $(C_t)_{\mathrm{ssc}}$ under elastic and primary creep.

5.04.4.4 Combined Primary and Secondary Creep

If we assume that EC conditions exist in the cracked body but it consists of a mixture of primary plus secondary creep, the constitutive equation will consist of the second and third terms on the right-hand side of Equation (54). In this equation the stress and time dependence of strain rate is not separable. Therefore, a unique relationship between strain rate and stress cannot be defined. Therefore, no path-independent integral exists which uniquely characterizes the crack-tip stress and strain fields. However, approximate path independence (within 2%) of $C^*(t)$ has been shown for one case (Leung and McDowell, 1990). This result, although derived for one material and geometry, is expected to be quite general. This extends the validity of $C^*(t)$, and therefore C_t, as crack-tip parameters for the case of any EC condition, regardless of whether it is steady-state creep, primary creep, or a combination of both.

5.04.4.4.1 Transition time between extensive primary creep and extensive secondary creep

A transition time may be defined for extensive secondary creep conditions to develop from extensive primary creep conditions. This transition time, t_2, can be estimated by equating the value of $C^*(t)$ for primary creep, Equation (59) to C_s^* (Riedel, 1981). The latter is the C^*-integral corresponding to steady-state condition. Since this value is fixed, it is prudent to use a special designation for its value that is distinct from the time-dependent value of C^*. Thus

$$t_2 = \left[\frac{C_h^*}{(1+p)C_s^*}\right]^{1+1/p} \tag{62}$$

In this regime, the values of C_t and $C^*(t)$ can be estimated by the following equations:

$$C^*(t) = C_t = [1 + (t_2/t)^{p/(1+p)}]C_s^* \tag{63}$$

In test specimens in which the load-line deflection rate can be measured, expressions can be derived for estimating $C^*(t)$ and C_t. For example, for a CT specimen Equation (15) can be modified in the following way (Saxena, 1991)

$$C^*(t) = \frac{P\dot{V}_c}{B(W-a)}(t)(2 + 0.522(1 - a/W)) \times f(n, n_1) \tag{64a}$$

or

$$C^*(t) = \frac{P\dot{V}_c}{BW}\eta_1(a/W, n_1, n) \tag{64b}$$

$$f(n, n_1) = \frac{n}{n+1}(t/t_2) + \frac{n_1}{1+n_1}(1 - t/t_2) \tag{65}$$

5.04.4.4.2 *Transition from small-scale to extensive primary creep*

The extensive primary creep and small-scale primary creep conditions were considered separately in earlier sections. The transition time, t_{TP}, from small-scale to extensive primary creep, can be obtained by following the analysis of Riedel (1981), which equates the amplitudes of the HRR fields predicted by the two limiting conditions:

$$t_{TP} = \left[\frac{K^2(1-v^2)}{EC_h^*}\right]^{1+p} \frac{1}{1+n_1} \tag{66}$$

5.04.4.4.3 *Elastic, primary and secondary creep combined*

The value of C_t can be estimated by adding the C_t for SSC, $(C_t)_{ssc}$ with the C_t for EC, $C^*(t)$. Obtaining the value of $C^*(t)$ for combined primary and secondary creep is quite straightforward from Equation (63). Analytical expressions for estimating $(C_t)_{ssc}$ for combined primary and secondary creep have not been derived. In the case of experiments, Equation (44) remains valid and combined contributions from secondary and primary creep will be represented in the measured value of $\dot{V}_c$. One approach to analytically estimate $(C_t)_{ssc}$ for combined primary and secondary creep is to add the values of the creep zone expansion rates, $\dot{r}_c$, for primary and secondary creep. For secondary creep, $\dot{r}_c$ is given by differentiating Equation (28a) and for primary creep it is given by Equation (61). These combined rates can be substituted into Equation (45) to obtain $(C_t)_{ssc}$.

5.04.5 EFFECTS OF CRACK GROWTH ON THE CRACK-TIP STRESS FIELDS

Crack growth is expected to perturb the stress fields for stationary cracks. Since the analysis in the previous sections was limited to stationary cracks, an obvious question which must be addressed is what restrictions must be placed on stationary crack parameters such as C_t, $C(t)$, and $C^*(t)$ for characterizing crack growth rates. The stress fields in the vicinity of a crack growing at a velocity $\dot{a}$, in an elastic, power-law creeping material, was addressed by Hui and Riedel (1981). They argued that stresses and elastic strains near the moving tip will be different from the HRR field due to the instantaneous elastic response of the material. They derived what is termed as the HR (Hui–Riedel) field, and found that two distinctly different singularities develop depending on the steady-state creep exponent, n. For $n<3$, the asymptotic stress field is dominated by elastic strain rates, so that $r^{-1/2}$ singularity develops. For $n>3$, which is typical of metals, the competing effects of stress elevation due to crack growth and stress relaxation due to creep give rise to a field with an amplitude related to the crack growth rate as described below

$$\sigma_{ij} = \alpha_n\left(\frac{\dot{a}}{AEr}\right)^{1/(n-1)}\bar{\sigma}_{ij}(\theta, n) \tag{67}$$

where, $\bar{\sigma}_{ij}(\theta, n)$ is an angular function and α_n depends on, and is approximately unity for, plane strain. For $n=4$ and 6, the values of $\alpha_n = 0.815$ and 1.064, respectively. The amplitude of the HR field is completely determined by the crack growth rate and is independent of the loading and crack growth history. The domain of dominance of the HR field is estimated by equating the effective stress predicted by the HRR field with that corresponding to Equation (67). Such nesting of the singular crack-tip solutions (HR, HRR, and K for SSC conditions) was found to be accurate by Hawk and Bassani (1986), who conducted numerical analysis to verify the concept. We will first derive the equations for extensive secondary creep conditions and then for SSC.

5.04.5.1 Crack Growth under Extensive Steady-state Creep

The HRR field for extensive steady-state creep conditions are given by Equation (6a). Equating Equations (6a) and (67) and solving for r gives the size of the domain in which the HR field dominates (Riedel and Wagner, 1981):

$$r_{\text{HR}} = \frac{1}{A}\left(\frac{\dot{a}}{E}\right)^{(n+1)/2}\left(\frac{I_n}{C^*}\right)^{(n-1)/2}\beta(\theta, n) \quad (68)$$

where β is an angular function.

In order for C^* to be a valid parameter for characterizing creep crack growth, $r_{\text{HR}} \ll r_{\text{HRR}}$. In EC, the extent of r_{HRR} is of the order of the crack size and the value of r_{HR} can be calculated. For example, for 304 stainless steel the $\dot{a}$ vs. C^* relationship is shown in Figure 9, and the E, A, and n values for this material are $E = 172 \times 10^3$ MPa, $A = 2.13 \times 10^{-18}$ and $n = 6$. If one estimates a crack growth rate of $2.5 \times 10^{-4}\,\text{m}\,\text{h}^{-1}$ at a C^* value of $15\,\text{kJ}\,\text{m}^{-2}\,\text{h}^{-1}$ we get,

$$r_{\text{HR}} \simeq \frac{1}{2.13 \times 10^{-18}}\left(\frac{2.5 \times 10^{-4}}{1.72 \times 10^5}\right)^{3.5}\left(\frac{I_n}{0.015}\right)^{2.5} \times \beta(\theta, n) \quad (69)$$

I_n for $n = 6$ from Equation (7) is 4.9. If we assume that the angular term is on the order of unity, we get $r_{\text{HR}} \sim 10^{-6}$ m. The crack size in the specimen is of the order of 10^{-2} m. Therefore, we can make the argument that of the HR field is not significant or the effects of the growing crack are not important in the case of these tests. In general, if crack growth rate occurs under EC conditions, the effects of crack growth are not likely to be significant. Therefore, C^* is a valid parameter for characterizing creep crack growth under the conditions of extensive secondary creep.

5.04.5.2 Crack Growth under SSC

In this section, we will examine the limits of $(C_t)_{\text{ssc}}$ derived for stationary crack conditions for characterizing creep crack growth if SSC conditions exist in the specimen (or component). We postulate that for $\text{d}a/\text{d}t$ to be characterized by $(C_t)_{\text{ssc}}$, we must meet the condition that $\dot{r}_c \gg \dot{a}$.

From differentiating Equation (37) we can write

$$\dot{V}_c = P\frac{\text{d}C}{\text{d}a}\beta\dot{r}_c$$

By combining the above equation with Equation (38), we get

$$\dot{V}_c = \frac{2BK^2}{PE}(1 - \nu^2)\beta\dot{r}_c \quad (70)$$

The above equation provides a relationship between the deflection rate due to creep deformation and the creep zone expansion rate. In the case of a growing crack in the absence of creep, the relationship between elastic compliance and deflection, V_e, can be expressed by the following relationship

$$V_e = PC$$

For constant load, P,

$$\dot{V}_e = P\frac{\text{d}C}{\text{d}t} = P\dot{a}\frac{\text{d}C}{\text{d}a}$$

Combining the above equation with Equation (38), we get

$$\dot{V}_e = P\dot{a}\left(\frac{2BK^2}{EP^2}(1 - \nu^2)\right)$$

or

$$\dot{V}_e = \frac{2B\dot{a}}{P}\frac{K^2}{E}(1 - \nu^2) \quad (71)$$

Comparing Equations (70) and (71), and noting that β for $\theta = 0°$ is approximately unity, we can meet the condition $\dot{r}_c \gg \dot{a}$ by requiring that $\dot{V}_c \gg \dot{V}_e$. The total measured deflection rate in a test is

$$\dot{V} = \dot{V}_c + \dot{V}_e$$

Thus, the condition, $\dot{V}_c \gg \dot{V}_e$ can also be written as

$$\dot{V} \gg \dot{V}_e = \frac{2B\dot{a}}{P}\frac{K^2}{E}(1 - \nu^2) \quad (72)$$

The above equation, known as deflection-rate partitioning, was first intuitively derived (Saxena *et al.*, 1983) and proposed as a condition for validating $\text{d}a/\text{d}t$ vs. C_t correlations of the type shown in Figures 15 and 16.

If the condition of Equation (72) is not met, the stationary crack-tip parameters such as C_t should not be used to correlate creep crack growth rate. The materials in which creep crack growth occurs while satisfying Equation (72) are known as creep-ductile materials, and the ones in which this condition is not satisfied are known as creep-brittle materials. In the latter, the crack growth effects are quite significant in determining the crack-tip stress fields. The next section discusses approaches for characterizing creep crack growth in such materials.

5.04.5.3 Crack Growth in Creep-brittle Materials

Creep-brittle behavior is said to occur when $\dot{a}$ becomes comparable in magnitude to $\dot{r}_c$ such that the growing crack effects on the crack-tip stress field are no longer negligible. As mentioned in the previous section, the stationary crack tip parameters, C^* and C_t are no longer valid for characterizing the creep crack growth behavior under these circumstances. In this section, we will focus our discussion on crack-tip parameters in the presence of crack growth effects. There are several high-temperature materials, such as high-temperature Al alloys, titanium alloys, nickel-based superalloys, intermetallics, and ceramics, whose behavior at their intended service temperatures and environments leads to creep-brittle crack growth conditions. Therefore, modeling of such behavior is of considerable technological interest.

Figure 18 shows the development of creep deformation of the crack tip in specimens of two hypothetical materials. The specimens are identical in all respects and are subjected

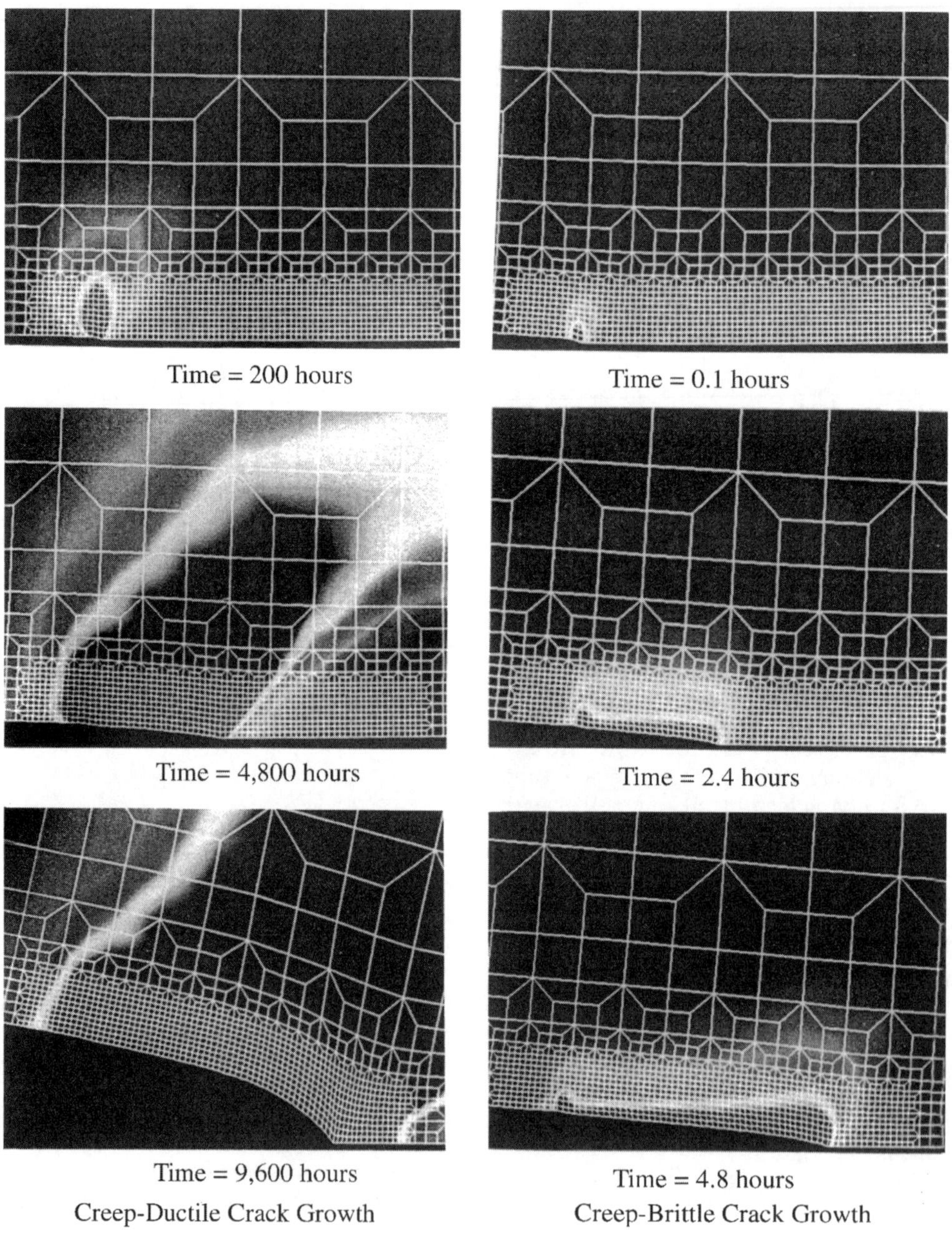

Figure 18 Contour plots of effective creep strain for creep-ductile and creep-brittle crack growth at three times during crack growth (Hall, 1995). Reproduced with permission from the author.

to the same loads. The creep deformation behavior of the two materials is also the same. In one material, $\dot{a}$ is ten times larger than in the other. Hall (1995) analyzed the specimens by conducting finite element analysis which simulated the stress fields at various times in the presence of growing cracks (Figure 18). In the specimen with the slower crack growth rate, the creep-ductile conditions prevailed and EC conditions developed progressively. In the specimen with the higher crack growth rate, creep-brittle conditions developed and the specimen always remained in the SSC condition. This simulation brings out the role of crack extension in determining the crack-tip conditions. One can imagine a steady-state condition for which $\dot{a}$ and $\dot{r}_c$ are equal along $\theta = 0°$. In this case, to an observer stationed at the moving crack tip, the creep zone size is constant along the x-axis. Under these circumstances, da/dt is likely to correlate with K. However, prior to reaching the steady-state conditions, considerable crack growth can occur under transient conditions. Also, even if $\dot{r}_c = \dot{a}$ at $\theta = 0°$, the self-similarity in the growth of the creep zone is not necessarily assured. In the subsequent sections, both the steady-state and transient conditions are explored further.

5.04.5.3.1 Steady-state creep crack growth under SSC

Steady-state creep crack growth under SSC in elastic, power-law creeping material has been analytically treated (Hui, 1983, 1986). Hui assumed that away from the immediate vicinity of the crack tip, the stress field is characterized uniquely by K. By dimensional analysis, he showed (Hui, 1983) that the crack-tip stress field in the immediate vicinity of the crack tip is described by

$$\sigma_{ij} = \left[\frac{\dot{a}}{EAK^2}\right]^{1/(n-3)} \Sigma_{ij}(R, \theta) \tag{73}$$

where the dimensionless radical coordinate, R, is given by

$$R = r\left(\frac{EAK^{n-1}}{\dot{a}}\right)^{-2/(n-3)} \tag{74}$$

A measure of the characteristic length over which the asymptotic field, Equation (73), dominates is obtained by choosing $R = 1$. For K to be a valid parameter, this distance must be small in comparison to the crack length and the uncracked remaining ligament. Figure 19 shows da/dt vs. K for an ASTM SA-106C

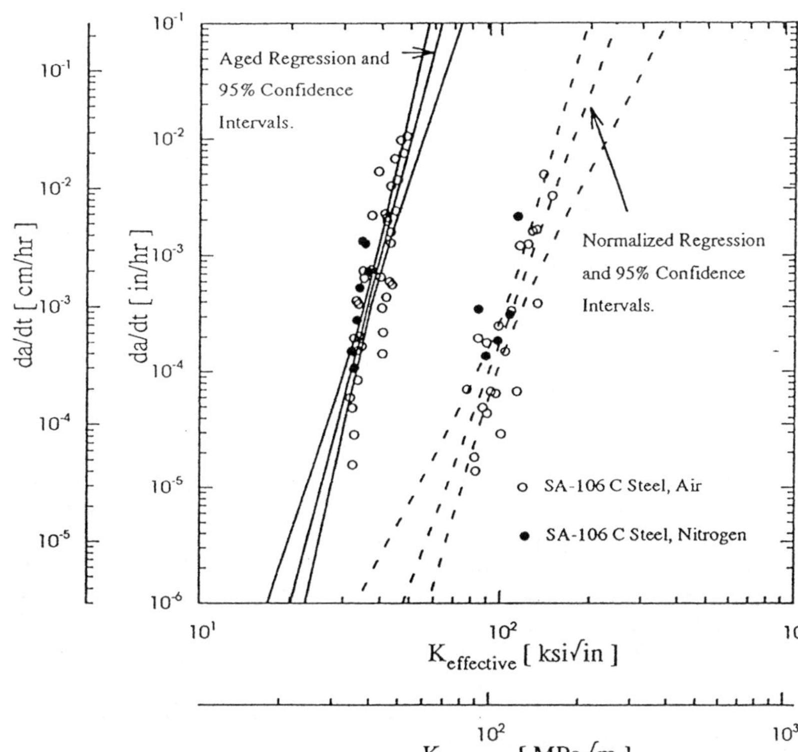

Figure 19 Creep crack growth rates as a function of K for SA-106C carbon steel (Saxena, 1998; reproduced with permission from CRC Press).

carbon steel at 360°C (Gill *et al.*, 1994). Thus, in principle, steady-state crack growth in small-scale creep is likely when these special conditions are met. However, a steady state is likely to be preceded by transient crack growth, which is discussed next.

5.04.5.3.2 Transient crack growth under SSC

Owing to the difficulty of obtaining analytical solutions for the transient crack growth under SSC accounting for crack growth effects, this problem has been simulated numerically (Hall *et al.*, 1997). Experimentally observed crack growth histories of four compact type specimens (BCH-1,4,6,7) of aluminum alloy 2519-T87 tested at 135°C were numerically enforced to study the evolution of crack-tip fields and far-field fracture parameters.

The time-dependent deformation characteristics for the aluminum alloy 2519-T87 at 135°C were modeled using primary and steady-state creep material constants. This model was incorporated in the finite element analyses conducted. A constant external load of 10.5 kN was applied to one of the specimens, BCH-6, whose width and thickness were 5.08 cm and 2.21 cm, respectively. Crack growth began at a crack length of 2.25 cm and ended 105 h later at a crack length of 2.52 cm when the specimen became unstable. A plot of the crack growth history, Figure 20, reveals a relatively long incubation period during which the crack grew either very slowly or not at all. The incubation period was followed by a period of crack growth. Figure 21 compares the experimentally observed deflection as a function of time with that predicted by the finite element analysis for this specimen. The good agreement between the experimental results and the finite element analysis results are clear indications of the accuracy of the analysis.

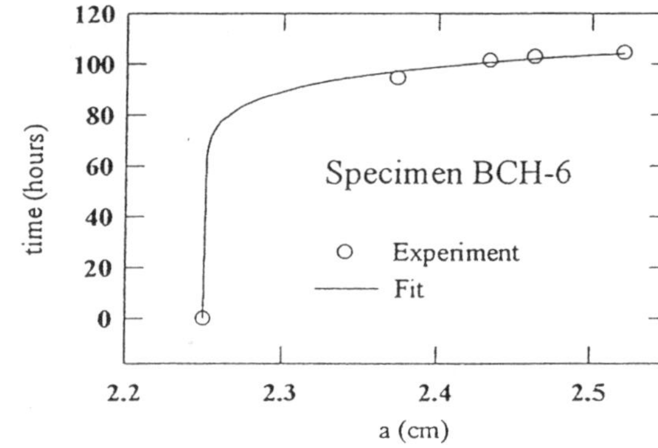

Figure 20 Experimentally determined crack growth history of an aluminum alloy 2519-T87 CT specimen tested at 135°C and the curve fit used to describe the behavior in a finite element study (Hall *et al.*, 1997; reproduced with permission from ASTM).

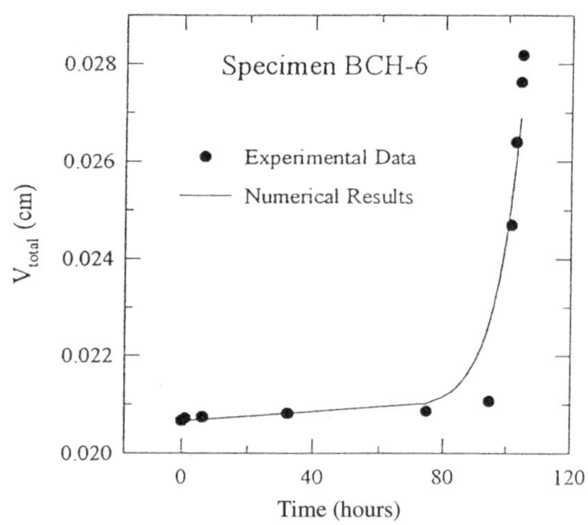

Figure 21 Comparison between the experimentally measured load-line displacement as a function of time and those predicted by the finite element analysis (Hall *et al.*, 1997; reproduced with permission from ASTM).

Interpretation of the finite element results was enhanced using a visualization program to simulate the evolution of the various crack-tip parameters at different stages of crack growth. Figure 22 shows the contour plots of the effective creep strain and effective stress at three stages of crack growth for this specimen. Deflections in all of these plots are magnified by a factor of 60. Color is used to illustrate the creep zone boundaries. The yellow areas mark the boundary of where the creep strains exceeded 0.005. In the stress plots, the yellow areas mark the region where the stress exceeded 310 MPa, and the blue areas where the stress exceeded 103 MPa.

Significant levels of creep deformation are confined to a thin region adjacent to the crack growth path. The characteristic "knob" observed near the initial crack tip is due to the initially elastic crack-tip fields, which produce high creep strain rates while the crack growth is slow due to incubation. As the stresses near the crack tip relax, the driving force for creep deformation decreases, resulting in a corresponding decrease in the creep zone size. Both residual stress and residual creep strain are left in the wake of the growing crack. As the crack grows, the crack-tip fields become more intense, resulting in larger creep zone expansion rates. However, the extent of the creep zone is limited by the accelerating crack growth, since accumulating creep deformation is quickly left in the wake of the growing crack.

Profiles of effective stress along a radial line from the crack tip at $\theta = 90°$ reveal that stresses

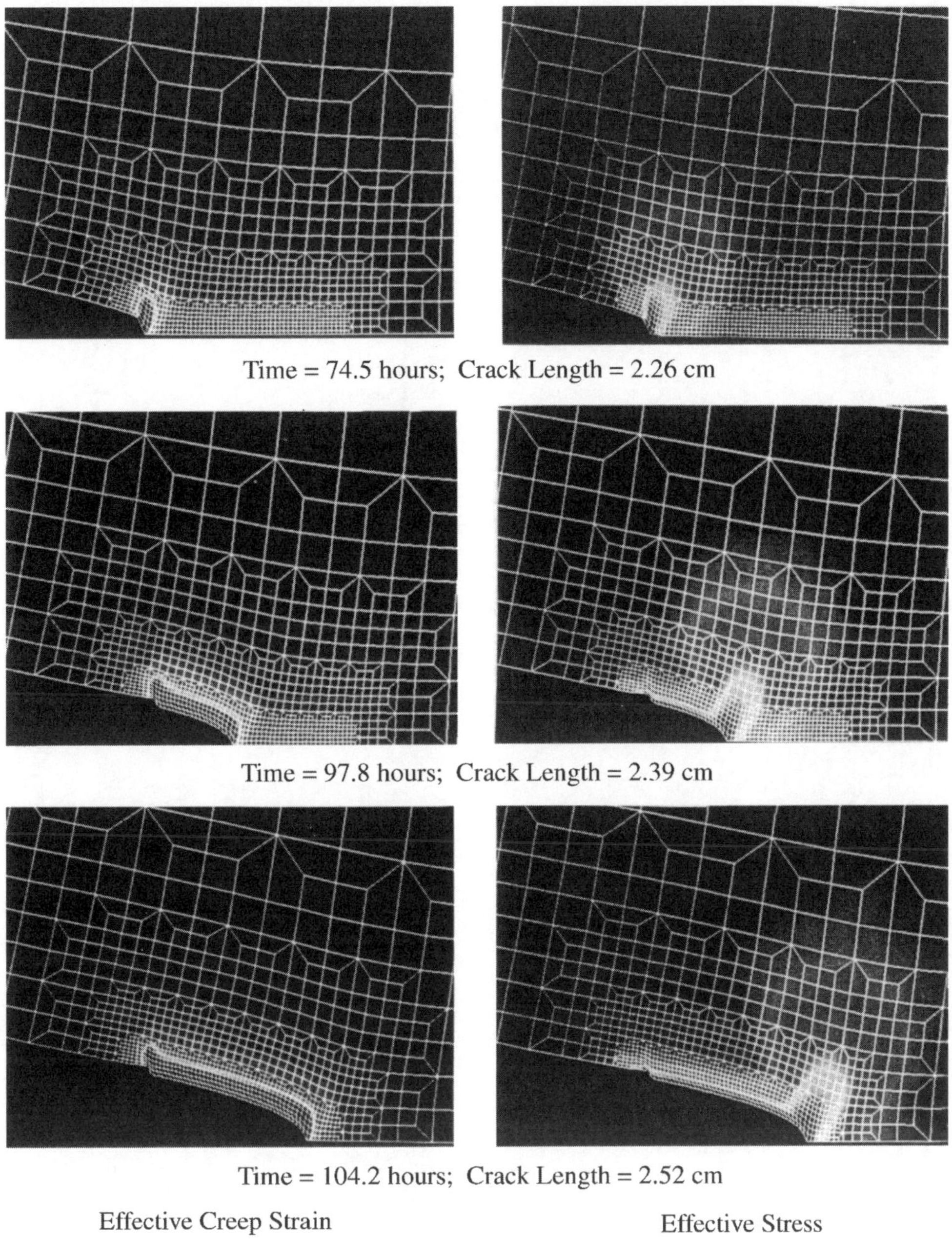

Figure 22 Contour plots of effective stress and creep strain at various stages of crack growth in a finite element simulation of a real test on 2519-T87 aluminum alloy (Hall *et al.*, 1997).

in a region just outside the creep zone scale with $r^{-1/2}$ throughout the crack growth history (Figure 23). Consequently, SSC conditions exist and K describes the crack-tip fields in an annular region around the crack tip. Both the amplitude and the strength of the stress singularity within the creep zone ($r < 0.25$ mm) increase with time. It is not clear whether the amplitude scales with K or with the crack growth rate ($\dot{a}$), since both quantities increase significantly during the test. The strength of the singularity is given by $r^{-1/16}$ just after incubation (after one crack growth increment) and approximately by $r^{-1/12}$ throughout most of the remainder of the test. Similar trends are predicted analytically when a stationary crack-tip field (HRR-type field) gives way to a growing crack-tip field (HR-type field). Thus, the results presented here seem to support the notion of "nested" crack-tip fields, although the particular form of the fields or the character of the nesting cannot be directly compared with theory because of the complex constitutive law employed in this analysis.

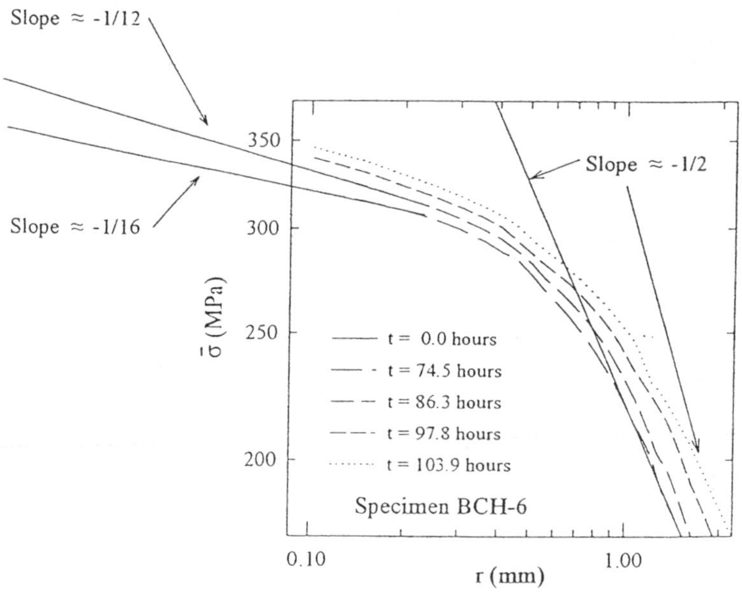

Figure 23 Profiles of effective stress along $\theta=90°$ showing a slope equal to (–1/2) outside the creep zone in the simulation on 2519-T87 aluminum specimen (Hall *et al.*, 1997; reproduced with permission from ASTM).

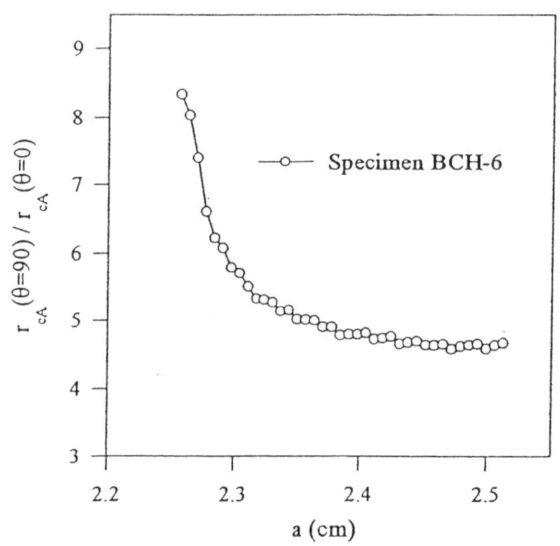

Figure 24 Ratio of the creep zone size at $\theta=90°$ with its size at $\theta=0°$ as check for similitude in the crack-tip deformation processes (Hall *et al.*, 1997; reproduced with permission from ASTM).

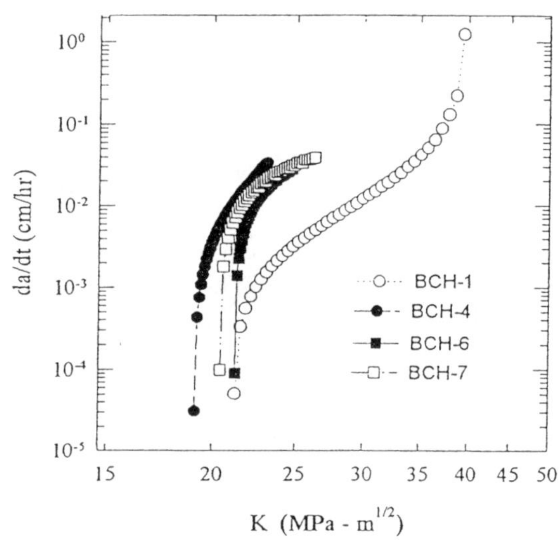

Figure 25 Creep crack growth rate as a function of K for 2519-T87 aluminum alloy (Hall *et al.*, 1997; reproduced with permission from ASTM).

5.04.5.3.3 Correlation of crack growth rate with K

From an analytical viewpoint, successful correlation of crack growth with a fracture parameter requires that the deformation and damage fields surrounding the crack-tip scale with the parameter in a roughly "self-similar" fashion. For example, the creep zone for a stationary crack under small-scale yielding expands in a self-similar fashion such that its shape does not change with time (i.e., r_c $(\theta=90)/r_c$ $(\theta=0°)$ is constant). Figure 24 indicates that the shape of the creep zone for specimen BCH-6 changes significantly during the early stages of crack growth, and later assumes a somewhat constant shape. Thus crack growth for this specimen appears to approach self-similar conditions, following a transient period when the character of the creep zone changes significantly (the other three specimens follow similar trends).

The applicability of SSC conditions, coupled with the existence of similitude, suggest that the later stages of crack growth can be correlated with K (Figure 25). This figure essentially represents smoothed-out experimental data, since K is uniquely determined by the current crack length for a given specimen geometry and loading and is unaffected by the results of the finite element analyses. Specimens BCH-4 and BCH-7 have a thickness of 2.21 cm, while specimen BCH-1 has a thickness of 0.635 cm. Figure 25 reveals good correlation of K with crack growth rates after the initial period of transience for the thicker specimens. Crack growth rates also appear to have a power-law dependence on K for the thinner specimen, although the dependence is shifted relative to the thicker specimens. The higher resistance to creep crack growth demonstrated by the thinner specimen is attributed to reduced constraint, since it is believed that crack-tip damage processes accelerate under increasing triaxiality levels, particularly for creep-brittle materials.

5.04.5.3.4 Transition times for growing cracks

The initial stages for crack growth in aluminum alloy 2519-T87 are characterized by highly transient crack-tip fields (Figure 24) and crack growth rates that do not correlate well with K (Figure 25). These initially transient conditions are expected, since the stationary crack-tip fields (HRR-type fields) that develop during the incubation period are replaced by growing crack-tip fields (possibly HR-type fields) as the crack extends through the initial creep zone. The numerical results indicate that the period of time after which crack growth becomes self-similar and crack growth rates correlate with K corresponds roughly to the time required for the crack to grow through the initial creep zone. Transition times that relate to the passage of the initially transient crack-tip fields are presented in this section.

Owing to nesting of the HR, HRR, and K fields (Hawk and Bassani, 1986), the expansion of the creep zone during the initial stage of crack growth is given by Equation (28) for a material deforming according to an elastic, power-law creep constitutive relation. Differentiating Equation (28) with respect to time shows that $\dot{r}_c$ decreases with $(1/t)^{(n-3)(n-1)}$. Regardless of the crack growth rate, da/dt, a unique reference time, t'_g, will eventually be reached where $\dot{a} = \dot{r}_c$ even when $\dot{a}$ varies with time. The creep zone size at this reference time, denoted by r_g, approximates the distance that the crack must extend to leave the initial creep zone in the wake of the growing crack. The time required for the crack to extend a distance r_g may be considered as a transition time for growing cracks, designated here as t_g.

Examples of the evolution of crack growth rate and creep zone expansion rate are shown in Figure 26 for constant and varying crack growth rates. Notice that for the growing crack is expected to approximately follow the stationary crack solution for $t < t'_g$. However, as the crack tip approaches the reference distance r_g, the creep zone expansion rate is dominated by crack growth effects and approaches $\dot{a}$ at time t_g. The creep-brittle character of the material ensures that $\dot{a} \approx \dot{r}_c$ during the quasi-steady-state crack growth regime which occurs after time t_g, assuming the rate of expansion of the creep zone is measured relative to a fixed position along $\theta = 0°$. The quasi-steady-state crack growth regime is characterized by crack-tip stress fields that change slowly with crack size.

Closed-form expressions for t'_g and t_g can be developed when the crack growth rate is constant. The reference time t'_g is defined as

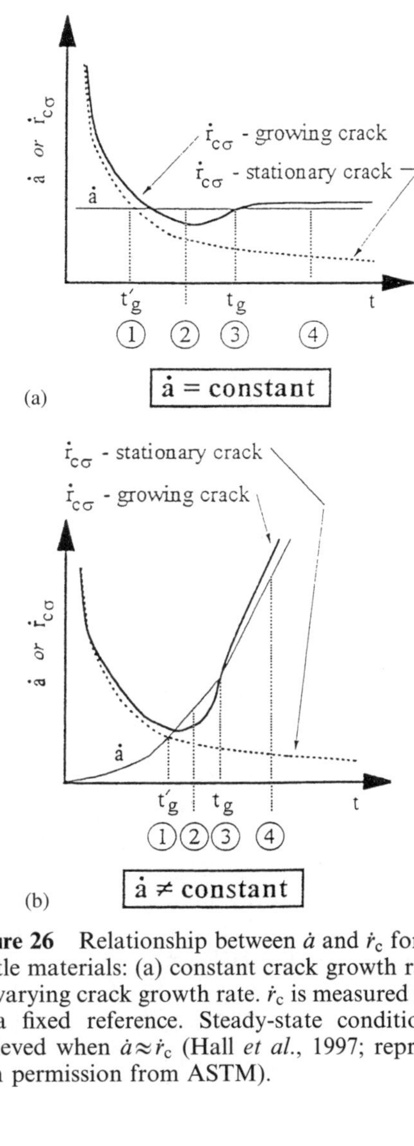

Figure 26 Relationship between $\dot{a}$ and $\dot{r}_c$ for creep-brittle materials: (a) constant crack growth rate and (b) varying crack growth rate. $\dot{r}_c$ is measured relative to a fixed reference. Steady-state conditions are achieved when $\dot{a} \approx \dot{r}_c$ (Hall *et al.*, 1997; reproduced with permission from ASTM).

the time when $\dot{r}_c$ is first equal to $\dot{a}$. Equating $\dot{a}$ with $\dot{r}_c$ and solving for time results in

$$t'_g = \left(\frac{1}{\pi(n-1)}\right)^{(n-1)/(n-3)} \left(\frac{EA(n+1)I_n}{2\pi(1-\nu^2)}\right)^{2/(n-3)} \left(\frac{1-2\nu}{\bar{\sigma}(\theta=0^\circ)}\right)^{2(n+1)/(n-3)} \left(\frac{K^2}{\dot{a}}\right)^{(n-1)/(n-3)} \tag{75}$$

Since the creep zone radius along $\theta = 0^\circ$ is larger than the local amount of crack extension at t'_g, additional crack growth is necessary for the crack to grow out of the influence of the initial creep zone. It follows that $\dot{r}_c = (2/(n-1))r_c/t$. Since $\dot{r}_c = \dot{a}$ and $r_c = r_g$ at time t'_g the initial creep zone size can be written in terms of t_g as $r_g = ((n-1)/2)\dot{a}t'_g$. The transition time for growing cracks (t_g) corresponds to the time required for the crack to extend a distance r_g through the initial creep zone. For a constant crack growth rate, $r_g = \dot{a}t_g$, resulting in:

$$t_g = \left(\frac{n-1}{2}\right)t'_g \tag{76}$$

While Equations (75) and (76) apply to elastic, power-law creeping materials, similar expressions may also be derived for materials that deform by primary creep.

The duration of the transient regime of crack growth can also be estimated by qualitatively examining the finite element results. Figure 24 indicates that the crack has grown through the initial creep zone at a crack length of 2.3 cm or after 84.5 h, since the creep zone radius along $\theta = 90^\circ$ changes slowly with crack length after this time. Thus, the finite element results indicate that $t_g = 87.4$ h is a good estimate of the time required for the onset of quasi-steady-state crack growth. Repeating the analysis for the other specimens and comparing the finite element results with computed values of t_g yields Figure 27. This comparison shows good agreement between t_g and the time required for the crack to grow through the initial creep zone. Examining r_g for specimens BCH-1, BCH-4, BCH-6, and BCH-7 indicates that quasi-steady-state crack growth occurs after 0.89 mm, 0.58 mm, 0.39 mm, and 0.43 mm of crack growth, respectively.

The reference time t_R (Hawk and Bassani, 1986) can also be used to mark the onset of quasi-steady-state crack growth conditions. The formulation of this reference time is based on equating the radius of the creep zone for growing cracks (the HR zone, which depends on K and $\dot{a}$) to the radius of the creep zone for stationary cracks (the HRR zone, which depends on K and t) and solving for the time. When the crack growth rate is constant, the expression for t_R is similar to the expression for t_g above, since both quantities scale with $(K^2/\dot{a})^{(n-1)/(n-3)}$. However, interpretation of t_R is not clear for varying crack growth rates. Moreover, t_T may not exist when crack growth is preceded by an incubation period, since the creep zone predicted from stationary crack mechanics may always be greater than the creep zone predicted from growing crack mechanics.

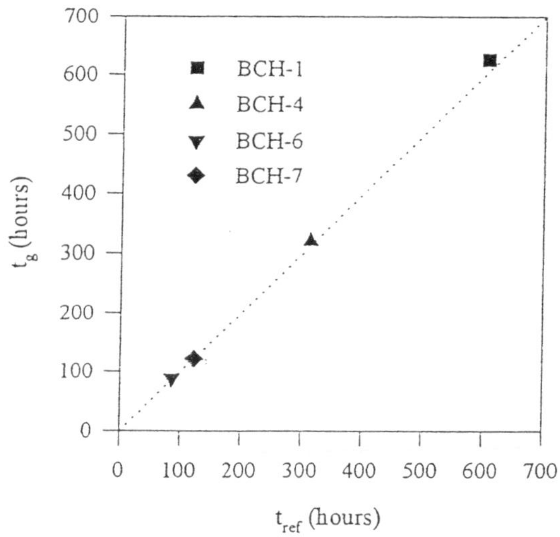

Figure 27 Comparison of t_{ref} and t_g, where t_{ref} is the time for the crack to grow through the initial creep zone obtained from the finite element results (Hall *et al.*, 1997; reproduced with permission from ASTM).

The aim of examining these transition times is to define the region where quasi-steady-state crack growth occurs and correlates well with K. Computing the transition times for all of the aluminum 2519 specimens and plotting only the (K, $\dot{a}$) pairs which occur after the transition time for each specimen results in Figure 28. Notice that the tails which occur during the initial stages of crack growth are removed when compared to Figure 25. Consequently, the transition time t_g presented here provides a useful estimate of the time after which crack growth can be better correlated with K, at least for aluminum 2519. The utility of this approach is that the duration of the highly transient regime of creep-brittle crack growth can be estimated based on material properties and the crack growth history, without detailed information from growing crack finite element analyses.

5.04.6 EFFECTS OF CYCLIC LOADING

In the previous sections we have built the analytical framework for considering time-dependent creep deformation in the crack-tip region, and have used the analyses to define crack-tip parameters for predicting crack growth under the conditions of sustained loading. Cyclic or fatigue loading is quite prominent in several components that are operated at elevated temperature, making crack growth behavior under creep-fatigue conditions quite important. Duty cycles involving combinations of time-varying stresses and periods of sustained stress are typical of several elevated-temperature components. Creep deformation can occur during both the sustained load period and also during the loading portion of the cycle. Some of the common loading waveforms used in the study of creep-fatigue crack growth behavior are shown in Figure 29. The important time parameters characterizing the loading waveforms are t_r = rise time, t_h = hold time, and t_d = decay time, as shown in Figure 29. The amplitude of the loading waveform can be characterized by $\Delta P = P_{max} - P_{min}$, the difference between the maximum and minimum loads, and the load ratio, $R = P_{min}/P_{max}$. The loading frequency is termed as ν.

During a trapezoidal loading waveform, creep deformation can occur at the crack tip during hold time, as well as during the loading portions of the cycle. During loading, creep deformation at the crack tip will also be accompanied by plastic and elastic deformation (instantaneous deformation) and the relative amount of creep and instantaneous deformation depends on the loading rate. For fast loading rates creep can be negligible,

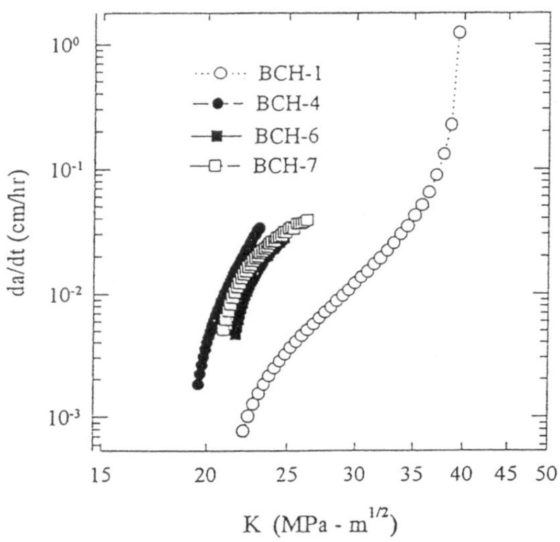

Figure 28 Creep crack growth rate as a function of K for 2519-T87 aluminum after excluding the data up to t_g (Hall *et al.*, 1997; reproduced with permission from ASTM).

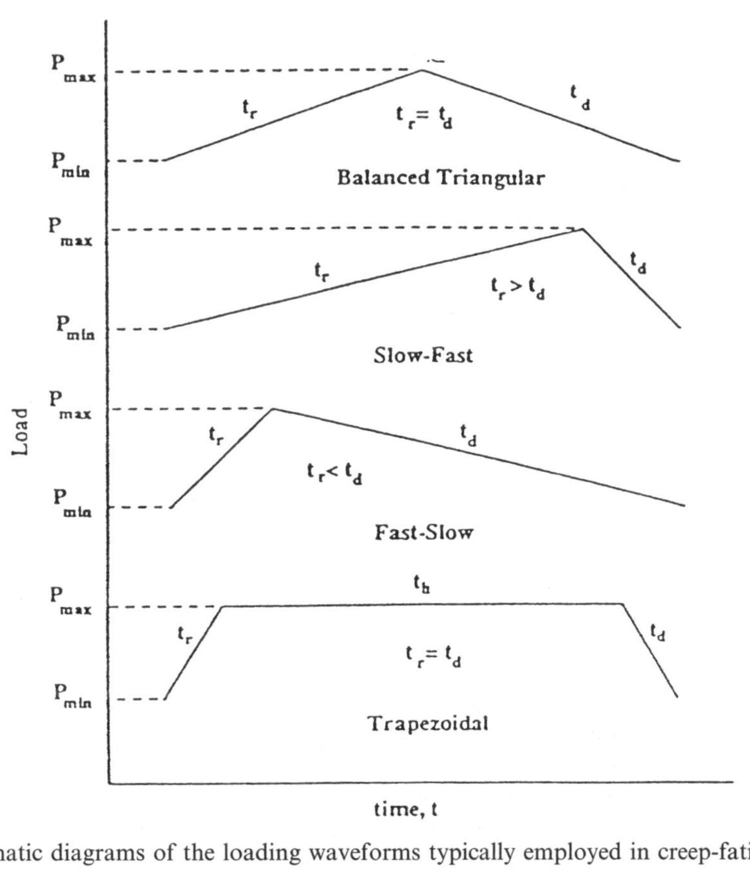

Figure 29 Schematic diagrams of the loading waveforms typically employed in creep-fatigue experiments.

whereas for very slow loading rates creep can be dominant. During the hold period creep deformation will occur, causing the crack-tip stresses to redistribute and resulting in elastic strains. If the hold period is sufficiently long, extensive creep conditions will ultimately prevail. Therefore, to gain a complete understanding of creep-fatigue crack-tip mechanics, solutions must be obtained for a wide range of loading and material deformation conditions. However, analytical solutions for crack-tip stresses are not possible for such complex conditions; thus one must depend on the numerical approach. In this section, the results of some basic numerical studies will first be described. These results will then be used to define crack-tip parameters that are suitable for describing some aspects of creep-fatigue crack growth.

5.04.6.1 Crack-tip Stresses under Creep-fatigue Loading

As a first step toward understanding crack-tip conditions under creep-fatigue loading, Riedel (1981) in his analysis performed in the early 1980s assumed that the material deformed by elastic and power-law creep. He considered triangular loading waveforms and trapezoidal loading waveforms. For a triangular waveform without hold time, the load was increased linearly with time. The peak stress in the crack-tip region was shown to depend on the loading rate. The higher the loading rate, the higher the peak stress, because less time was available for creep strains to develop and redistribute the stresses. Riedel also derived that for a linearly increasing load, the transition from SSC to EC conditions will occur at a time t_{TC} given by

$$t_{TC} = \frac{1+2n}{n+1} \frac{K^2(t_{TC})(1-\nu^2)}{C^*(t_{TC})E} \tag{77}$$

This equation is quite similar to Equation (30), which estimates the transition time for constant load, whereas Equation (77) is for linearly increasing load. In the above equation, $K\,(t_{TC})$ and C^* (t_{TC}) are the values of K and C^* at $t = t_{TC}$, respectively. Also note that the transition time for monotonic loading is larger by a factor of $(1+2n)$ in comparison to the transition time for constant load.

The value of t_{TC} can be used to select loading frequencies during fatigue crack growth testing at high temperatures to maintain dominantly elastic conditions. Alternatively, we can also see that if the loading rate is slow enough, there is little justification for using ΔK for characterizing fatigue crack growth rates at elevated temperatures.

The next type of loading waveform was a trapezoidal loading waveform, in which he varied the loading rate and then held the load constant (Riedel, 1983). Figure 30 summarizes his results, in which the normalized load (P/P_{max}) and the normalized value of the crack-tip stress (σ/σ_{ss}), where σ_{ss} is the fully relaxed, steady-state stress, are plotted against normalized time, t/t_T. If the loading is reasonably rapid, t_T is the transition time as determined by Equation (30). The magnitude of the peak crack-tip stress depends on the loading rate. A slower loading rate will result in a lower peak crack-tip stress. This would imply that, at low loading rates, cycle-dependent crack growth due to fatigue will be suppressed and time-dependent crack growth due to creep may become dominant.

The above analyses were not performed for repeated cycles. During cyclic loading, considerable cyclic deformation occurs at the crack tip during repeated loading and unloading. In order to develop a clearer picture of creep-fatigue interactions, it is necessary to include the effects of cyclic deformation and also the effects of repeated cycling (Yoon *et al.*, 1992; Adefris *et al.*, 1996a, 1996b) for trapezoidal loading waveforms. These studies were carried out on a 1.25Cr–0.5Mo steel and a 1Cr–1Mo–0.25V steel, with the purpose of investigating the applicability of C_t for creep-fatigue crack growth. Thus, the numerical analyses results are presented in the form of variation in C_t as a function of time, following the sudden application of load for several successive cycles.

Detailed finite element analyses were carried out using a code developed by McDowell

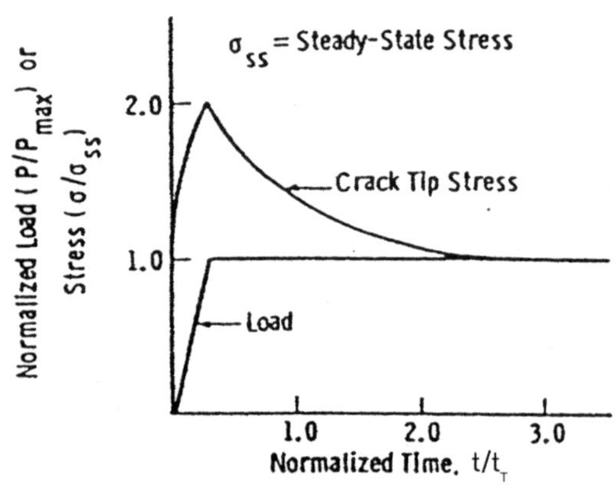

Figure 30 Normalized crack-tip stress as a function of time normalized by transition time, at a fixed distance from the crack tip for fast loading followed by sustained load (Riedel, 1981; reproduced with permission of ASTM).

(1992) that combines the rate-independent state-variable plasticity model applicable to fatigue loading with nonlinear kinematic hardening. The total strain is partitioned into elastic, plastic, and creep. The plasticity and creep deformations are decoupled. The model has the flexibility to include both secondary and primary creep behavior. Both analyses were carried out on 50.8 mm wide CT specimens assuming plane strain conditions.

We will first discuss the results of Yoon *et al.* (1992), which include stable cyclic plasticity behavior and secondary creep in addition to elastic deformation. Figure 31(a) shows a plot of the numerically calculated value of the C_t parameter as a function of normalized time, and Figure 31(b) shows a plot of the C-integral, Equation (31), plotted similarly. In these analyses, the specimens were rapidly loaded to the maximum load level; the load

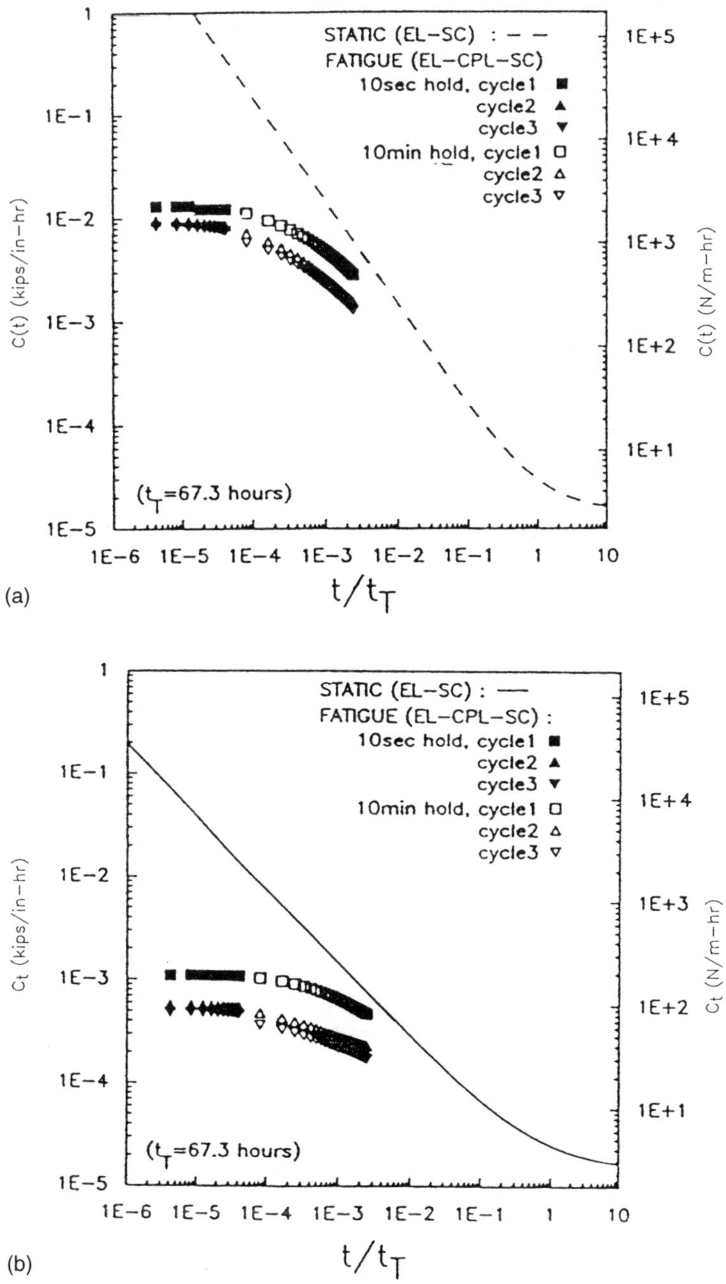

Figure 31 (a) C_t and (b) $C(t)$ as a function of time in fatigue analyses for 10-s and 10-min hold times. The material was a 1.25Cr–0.5Mo steel at 538°C with $A = 1.26 \times 10^{-26}\,\text{MPa}^{-n}\,\text{h}^{-1}$ and $n = 10.3$ (Yoon *et al.*, 1992; reproduced with permission from ASTM).

was held for 10 s in one case and for 10 min in the other, and was then quickly removed. The C_t parameter is calculated using Equations (46) and (48). In terms of the overall trends in the C-integral and C_t, there is not much difference. Since C_t is used as a crack-tip parameter for characterizing crack growth rate, we will be primarily using C_t in the discussion from this point forward to characterize the crack-tip conditions. We observe in Figures 31(a) and (b) that there is a considerable influence of cyclic plasticity in the initial portion of the C_t and C-integral trends. This is evident when the trends for the first cycle are compared with the trend for the elastic-secondary creep trend, which can be analytically predicted using the combination of Equations (47) and (48). There also is a significant difference between C_t values between the first and second cycles. By comparison, the difference between the second and third cycle is marginal, indicating that the crack-tip conditions quickly settle into a repeatable pattern that is followed during each cycle. The initial value of C_t for each cycle does not appear to be significantly influenced by the hold time, because it seems the same for the 10-s and 10-min hold time cases.

Figures 32 and 33, respectively, show the development of the monotonic plastic zone at the end of the hold period of each cycle and the development of the cyclic plastic zone between the consecutive cycles from the 600-second hold time case. The monotonic plastic zones are defined as the locus of points for which the magnitudes of effective elastic and plastic strains are equal. The cyclic plastic zone is similarly defined by the locus of points where the ranges of the elastic and plastic effective strains are equal. The ranges are calculated based on the difference between the strain at the end of a cycle and the beginning of the hold time during the subsequent cycle. Even though the monotonic plastic zone increases cycle by cycle, the cyclic plastic zone size from the second cycle onward is almost constant, much like the C_t and $C(t)$ response. Therefore it is believed that the cyclic plasticity, and not the monotonic plasticity, is governing the initial crack-tip stress field. The effective stress as a function of distance from the crack is shown in Figure 34.

Next, we move to discuss the results of the analyses performed on 1Cr–1Mo–0.25V steel by Adefris *et al.* (1996a, 1996b). In this material, it was found that in order to adequately represent the observed trends, it was necessary to account for plasticity, secondary creep, and also primary creep in the finite element analysis of the specimen. Figure 35 shows a plot of C_t as a function of normalized time that also includes measured values of C_t. We note that the analysis including elastic, plastic, and only secondary creep underpredicts the C_t value considerably. When primary creep is added, the results agree reasonably well with measured values of C_t and also with the

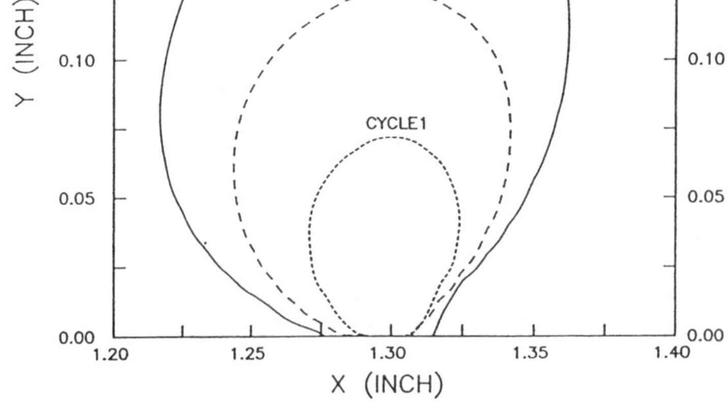

Figure 32 Development of the monotonic plastic zones at the end of the hold period of each cycle in fatigue analysis in 10-min hold time. The crack tip is located at 33 mm (1.3 in) along the x-axis (Yoon *et al.*, 1992; reproduced with permission from ASTM).

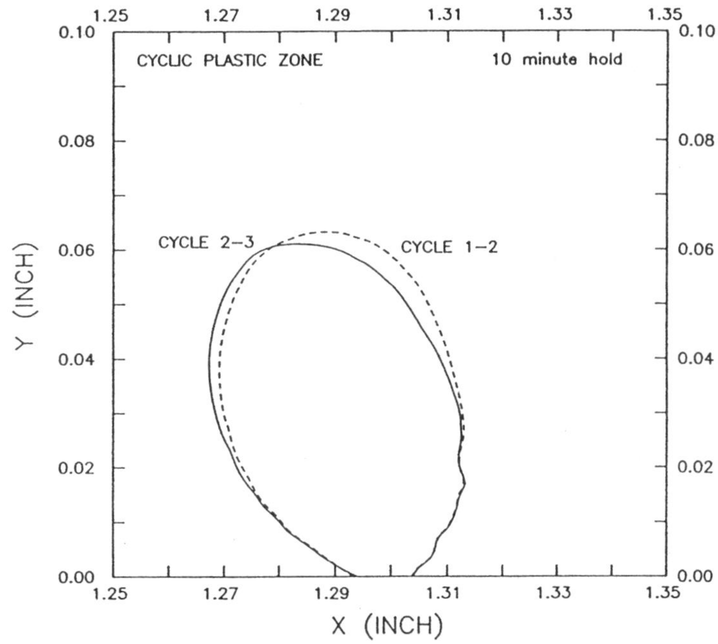

Figure 33 Development of cyclic plastic zones between the end of a cycle and the beginning of the hold period of the following cycle in fatigue analysis with a 10-min hold time, for the same specimen as in Figure 32 (Yoon *et al.*, 1992; reproduced with permission from ASTM).

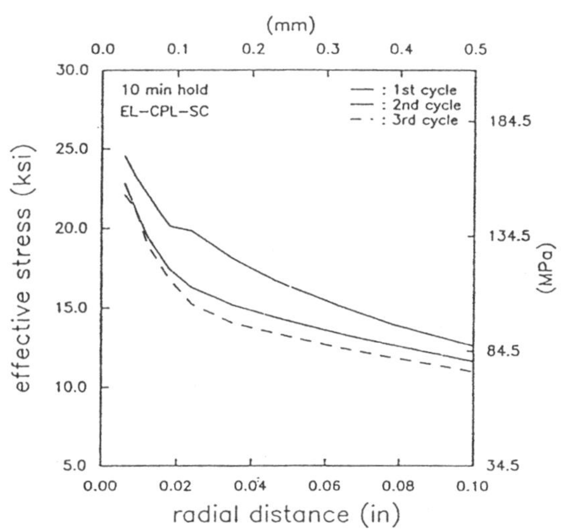

Figure 34 Effective stress distribution along the 94° radial line from the crack tip at the beginning of the hold period during three successive cycles (Yoon *et al.*, 1992; reproduced with permission from ASTM).

analytical values of C_t that include primary and secondary creep but not plasticity. However, the difference between including cyclic plasticity or not is somewhat small. This behavior is quite different from the earlier case, in which the cyclic deformation during loading and unloading was quite significant. Figure 36(a) shows a plot of C_t as a function of normalized time, with the time initialized to coincide with the start of each hold period. There seem to be significant differences between the first, second, and third cycles, with no clear trend toward stabilization from one cycle to the next of the type observed for 1.25Cr–0.5Mo steel shown in Figure 31. In Figure 36(b), we plot the same results as in Figure 36(a), except the time running between cycles is considered. In other words, we set the time at the beginning of the current cycle to equal the time at the end of the previous cycle. The C_t vs. time behavior falls into a single trend. This strongly implies that the loading and unloading events have no lasting influence on the crack-tip stress behavior of the type seen earlier in 1.25Cr–0.5Mo steel. Therefore, the two steels represent very different kinds of response during creep-fatigue loading. The relative sizes of the creep, cyclic plastic zone, and monotonic plastic zones are compared at the end of the third cycle in Figure 37. Clearly, the cyclic plastic zone is the smallest and is embedded in the other two, suggesting the domination of creep over fatigue in this material. Therefore, it is not surprising that the influence of cyclic plasticity is not significant.

Figure 38 attempts to reconcile the two very diverse trends between a 1Cr–1Mo–0.25V steel and a 1.25Cr–0.5Mo steel when subjected to creep-fatigue loading. It schematically represents the comparison between the relative sizes of creep zone and the cyclic plastic zone for two types of material behavior. Figure 38(a)

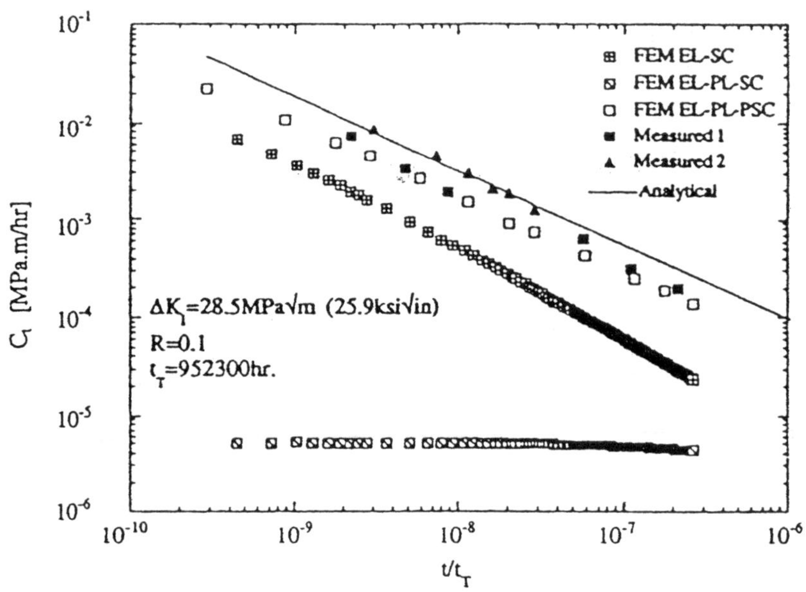

Figure 35 C_t as a function of normalized time for elastic-secondary creep, (EL-SC), elastic–plastic-secondary creep (EL-PL-SC) and elastic–plastic-secondary and primary creep (EL-PL-PSC) finite element analyses for 15-min hold time tests, along with experimental results from two separate tests (Adefris *et al.*, 1996a; reproduced with permission from Blackwell Publishing Ltd).

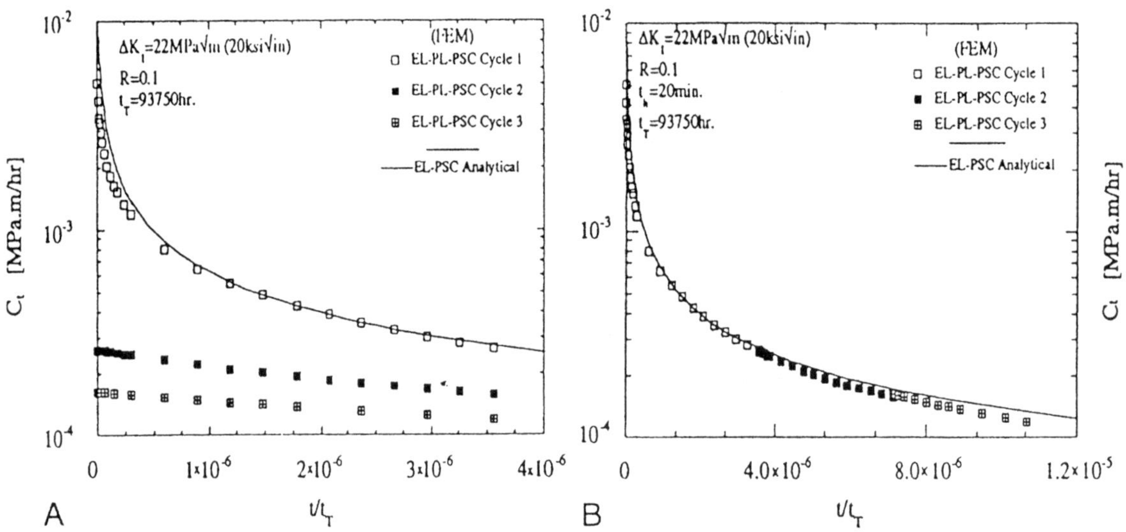

Figure 36 The creep fracture mechanics parameter C_t as a function of normalized time: (a) for the finite element analysis of a CT specimen of 1Cr–1Mo–0.25V at 538°C using elastic–plastic and primary and secondary creep for three successive fatigue cycles. The time is reinitialized at the beginning of each hold time. (b) the same results as in (a) except the times for the second and third cycles are taken cumulatively (source Adefris *et al.*, 1996a; reproduced with permission from Blackwell Publishing Ltd).

represents a material in which the creep zone is larger than the cyclic plastic zone, such as in the case of a 1Cr–1Mo–0.25V steel. This is expected in a material with high cyclic yield strength; therefore not only is the cyclic plastic zone smaller for the same applied ΔK, but the creep rates in the crack-tip region are accelerated due to the high stresses. This condition can also be induced by simply increasing the length of the hold time to increase the creep zone size. Reinstatement of the stress amplitude will not occur from cycle to cycle, as shown schematically in Figure 38(a), where the C_t in each subsequent cycle is shown to start from the value at the end of the previous cycle. Figure 38(b), however, represents a material whose creep zone is completely engulfed by the cyclic plastic zone, as a result of either a short hold time or low cyclic yield strength of the material. Under this condition, significant

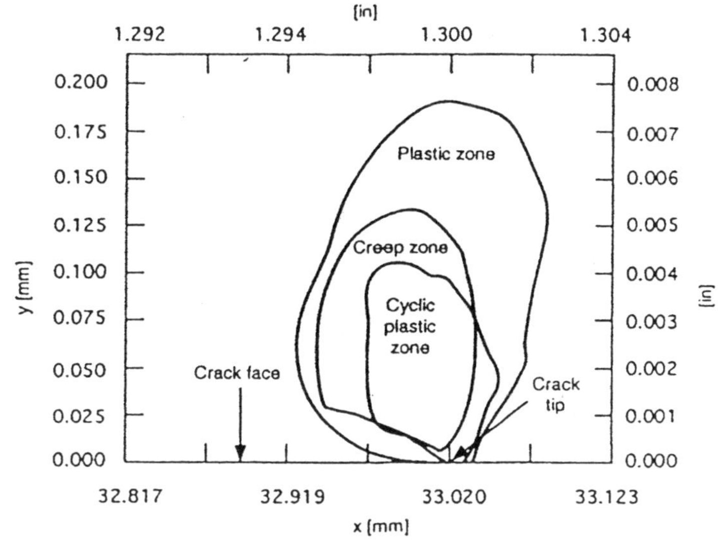

Figure 37 Relative sizes of the cyclic plastic zone, monotonic plastic zone and creep zone at the end of the third cycle of a 20-min hold trapezoidal loading analysis from Figures 35 and 36 (source Adefris *et al.*, 1996a; reproduced with permission from Blackwell Publishing Ltd).

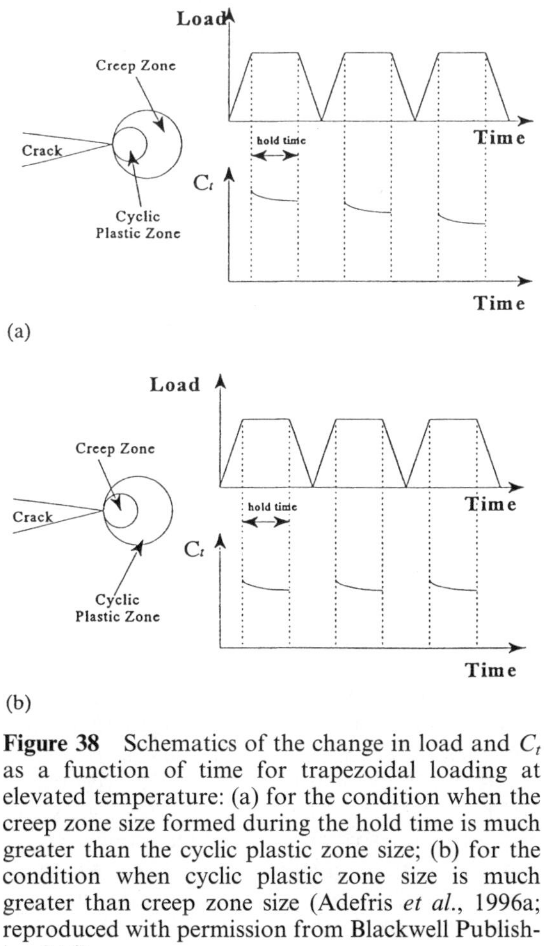

Figure 38 Schematics of the change in load and C_t as a function of time for trapezoidal loading at elevated temperature: (a) for the condition when the creep zone size formed during the hold time is much greater than the cyclic plastic zone size; (b) for the condition when cyclic plastic zone size is much greater than creep zone size (Adefris *et al.*, 1996a; reproduced with permission from Blackwell Publishing Ltd).

reversal of creep deformation occurs and the material will behave as one in which the C_t value is reinstated during each cycle.

5.04.6.2 Crack-tip Parameters in Creep Fatigue

Most successful correlations of time-dependent crack growth during trapezoidal loading waveform at elevated temperature are with time-dependent fracture mechanics parameters. These parameters are based on estimating C^* values using measured load-line deflection rates. The first available evidence of this approach is in the work of Jaske and Begley (1978) and Taira *et al.* (1979). These proposals were for EC conditions. A similar approach based on the $C(t)$-integral was proposed by Saxena *et al.* (1981, 1984). The latter approach also included SSC and TC conditions which are quite commonly encountered in creep fatigue. This approach was later modified so that the time-dependent crack growth rate during the hold time of a fatigue cycle is characterized by the average value of the C_t parameter, $(C_t)_{\text{avg}}$, during the hold time in a given cycle (Saxena and Gieseke, 1987):

$$(C_t)_{\text{avg}} = \frac{1}{t_h} \int_{t(N)}^{t(N)+t_h} C_t \, dt \qquad (78)$$

where $t(N)$ is a function of the number of fatigue cycles, N. For materials in which the crack-tip stresses are completely reinstated and insignificant creep deformation accumulates from cycle to cycle, $t(N)$ can be chosen as 0, such as for 1.25Cr–0.5Mo steels. Alternatively, when no creep reversal due to cyclic plasticity occurs, such as in the case of 1Cr–1Mo–0.25V steel, $t(N)$ is a simple linear function of N which for the Nth cycle will be $(N-1)t_h$. Thus, for materials which exhibit complete stress reversal each cycle:

$$(C_t)_{avg} = \frac{1}{t_h}\int_0^{t_h} C_t\,dt \quad (79a)$$

and for materials that do not exhibit any stress reversal:

$$(C_t)_{avg} = \frac{1}{t_h}\int_{(N-1)t_h}^{Nt_h} C_t\,dt \quad (79b)$$

A method to interpolate between the two limits will be presented in the next section. The ΔJ_c of Taira *et al.* (1979) is defined in the following way,

$$\Delta J_c = \frac{1}{t_h}\int_0^{t_h} C^* dt \quad (80)$$

From Equations (79) and (80), it is clear that $(C_t)_{avg} = \Delta J_c/t_h$ for conditions when $C_t = C^*$. Since this condition is met for EC conditions, we can state that ΔJ_c is a subset of the $(C_t)_{avg}$ parameter. The $(C^*)_{exp}$ parameter (Nishida and Webster, 1990) is based on the measured value of C^*-integral, is also identical to $(C_t)_{avg}$ for EC conditions. For SSC and TC, it is approximately equal to $(C_t)_{avg}$, to within a factor of 2, with the difference varying with the geometry of the specimen. Thus, it seems that the seemingly different approaches proposed by various researchers are not so different after all. The $(C_t)_{avg}$ approach is able to capture the essence of all other parameters. Therefore, in the subsequent discussion, it will be used more frequently to address creep-fatigue crack growth. In the next section, we will focus on methods of determining $(C_t)_{avg}$.

5.04.6.3 Methods of Determining $(C_t)_{avg}$

Methods of determining $(C_t)_{avg}$ include (i) those that are more suitable for test specimens in which both load and load-line deflection behavior with time are measured; and (ii) those that are more suitable for components in which the deflection rates must be analytically predicted. These methods are discussed in that order.

5.04.6.3.1 Methods for determining $(C_t)_{avg}$ in test specimens

$(C_t)_{avg}$ for test specimens can be obtained from the following equation (Yoon *et al.*, 1993):

$$(C_t)_{avg} = \frac{\Delta P \Delta V_c}{BWt_h}\frac{F'}{F} - C^*((F'/F)/\eta_1 - 1) \quad (81)$$

The above equation is a direct adaptation from Equation (51), with the following modifications: $\dot{V}_c$ is replaced by $\Delta V/t_h$, where ΔV is the amount of deflection accumulation during the hold time; ΔP corresponds to the amplitude of the applied load given by P_{max}–P_{min}; C^* is calculated using the load during the hold time, P_{max}. Thus, the value calculated represents the average value of C_t during the hold time. Since the deflection change is measured, the contributions from instantaneous plasticity and primary and secondary creep are already included in that measurement. If primary creep is significant, it should be included in the calculation of C^* as in Equation (63).

5.04.6.3.2 Analytical Methods of Determining $(C_t)_{avg}$

In deriving these equations, we will consider instantaneous cyclic plasticity and secondary creep, assuming plane strain conditions. Primary creep can be added, if necessary, following the methods outlined in Section 5.04.4. We will also separate the discussion for the cases of: (i) complete reversal of creep deformation; (ii) no creep deformation reversal by cyclic plasticity; and (iii) a condition in between the two limiting cases.

(i) $(C_t)_{avg}$ for complete creep reversal

We will first derive an expression for elastic, secondary creep deformation and then modify it to include instantaneous cyclic plasticity. Combining Equations (80), (47), and (48), we can easily show (Saxena and Gieseke, 1987) that

$$(C_t)_{avg} = \frac{2\alpha\beta(1-\nu^2)}{E}F_{cr}(\theta,n)\frac{\Delta K^4}{W} \times \frac{F'}{F}(EA)^{2/(n-1)}t_h^{-(n-3)/(n-1)} + C^* \quad (82)$$

The above equation can be adjusted for instantaneous cyclic plasticity using the following a procedure (Kubo, 1988; Yoon *et al.*, 1992). When load is suddenly applied to the specimen and if we assume that $m = n$ (this may be an oversimplification and will be discussed further), the initial stress distribution must be the same as for an elastic secondary creep material with a creep zone size, r_c, which is proportional to the cyclic plastic zone size, r_p^c. If we then assume that a time, t_{pl}, is needed for the creep zone size to become equal to r_p^c, the crack-tip stress field for elastic–plastic-creeping material at time t is equivalent to that of the elastic, secondary creep material at time $t + t_{pl}$. Thus the $(C_t)_{avg}$ can be obtained by first substituting $t + t_{pl}$ for t in Equation (47), and then substituting the result into Equation (80). The resulting expression for $(C_t)_{avg}$ is given by

$$(C_t)_{avg} = \frac{2\alpha\beta(1-\nu^2)}{E} F_{cr}(\theta, n) \frac{\Delta K^4}{W} \frac{F'}{F} (EA)^{2/(n-1)} \times \left[\frac{(t_h + t_{pl})^{2/(n-1)} - t_{pl}^{2/(n-1)}}{t_h} \right] + C^* \quad (83)$$

The value of t_{pl} can be analytically estimated by relating r_c and r_p^c as follows:

$$r_p^c = \xi_1 r_c |_{t=t_{pl}} \quad (84)$$

If we designate the cyclic plastic exponent by m, and relate r_p^c to mN and the 0.2% cyclic yield strength, σ_0^c, we get,

$$r_p^c = \xi_2 \left(\frac{m'-1}{m'+1}\right) \left(\frac{\Delta K}{\sigma_0^c}\right)^2 \quad (85)$$

Then combining Equations (84), (85) and the equation for r_c, (Equation 28(a)), we get,

$$t_{pl} = \frac{1}{EA} \left[\xi \left(\frac{m'-1}{m'+1}\right) \left(\frac{1}{2\sigma_0^c}\right)^2 \frac{1}{\alpha F_{cr}(90, m')} \right]^{(n-1)/2} \quad (86)$$

where

$$\alpha = \frac{1}{2\pi} \left(\frac{(n+1)^2}{2n\alpha_n^{n+1}} \right)^{2/(n-1)} \quad (87)$$

and

$$\xi = \xi_2 / \xi_1 \quad (88)$$

This equation shows that t_{pl} is not a function of geometry or applied load and is a material constant. The value of ξ_1 is one for $m = n$, but not for $m \neq n$; its value will depend on m and n. For example, for $m = 5.4$ and $n = 8$, $\xi_1 = 1.26$ from comparing the results of the finite element analysis with Equation (84) and $\xi_2 = 0.69$ for $m' = 5.4$, making $\xi = 0.54$. Using the above value of t_{pl}, the $(C_t)_{avg}$ was calculated for specimens in which $(C_t)_{avg}$ was also measured. Figure 39 shows the comparison between the predicted and experimental values of $(C_t)_{avg}$. The comparison appears to be quite satisfactory. However, for other materials, in which $m' \neq n$ and these values are different from 5.4 and 8, respectively, we do not have an analytical method for estimating t_{pl}. Perhaps $\xi = 0.547$ is a good starting point, and the other parameters in Equation (86) are also available. No verification of such an assumption either by finite element or by experiments is available. If experimental estimates of $(C_t)_{avg}$ are available, it is perhaps best to determine the value of t_{pl} by comparing experimental and predicted values of $(C_t)_{avg}$.

One can also see that if SSC conditions dominate, the contribution of C^* will be negligible in comparison to the first term on

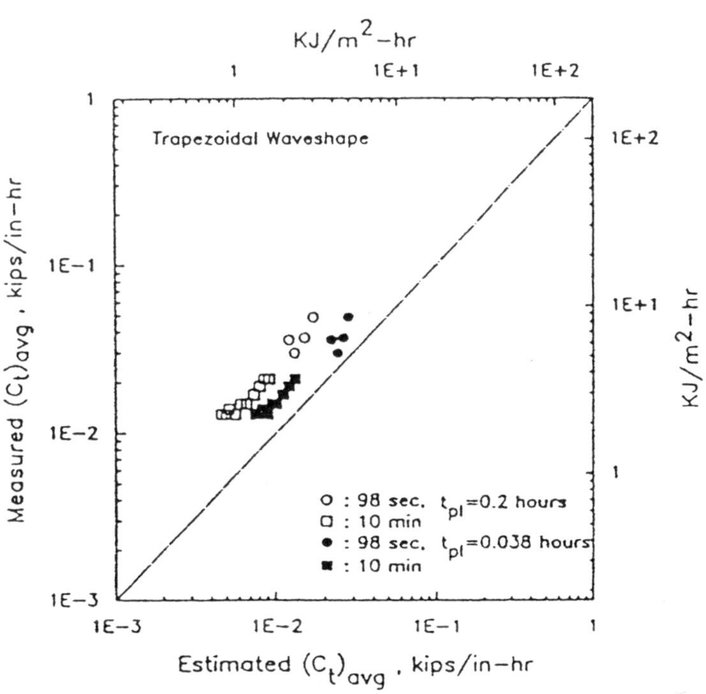

Figure 39 Comparison between the measured and estimated values of $(C_t)_{avg}$ for 1.25Cr–0.5Mo CT specimens subjected to fatigue cycles with 98-s and 10-min hold times (Yoon *et al.*, 1992; reproduced with permission from ASTM).

the right-hand side of Equation (83). Then for a given material and hold time, one can make the argument that $(C_t)_{avg}$ and ΔK are uniquely related. Hence, if da/dN for a constant hold time were plotted as a function of ΔK, a unique correlation is expected. as has been repeatedly shown in the literature. Such a correlation does not contradict the correlation between $(da/dt)_{avg}$ and $(C_t)_{avg}$. However, a distinct trend will be observed for each hold time, as has also been observed. Such equivalence relates this more recent approach with the early approach for characterizing creep-fatigue crack growth.

(ii) $(C_t)_{avg}$ for no creep reversal

The $(C_t)_{avg}$ for this condition can be estimated (Adefris *et al.*, 1996a, 1996b) by

$$(C_t)_{avg} = \frac{4\alpha\beta F_{cr}(\theta, n)}{(n-1)EW}(1-\nu^2)K^4\frac{F'}{F}(EA)^{2/(n-1)} \times (Nt_h)^{-(n-3)/(n-1)} + C^* \tag{89}$$

This equation is identical to Equation (47) except for time that is replaced by Nt_h. Also, note that since t_h is multiplied by N, the result for finite fatigue cycles will be much larger than t_{pl}. Therefore, t_{pl} can be ignored, as was also the case for sustained loading.

(iii) $(C_t)_{avg}$ for partial creep reversal

Grover (1996) and Grover and Saxena (1999) suggested a method by which $(C_t)_{avg}$ can be estimated for situations which are in between complete reversals and no reversal at all. Figure 40 schematically shows the applied load vs. time in a trapezoidal loading waveform and the deflection vs. time behavior. He noted that if ΔV_h is the total hold time accumulated during the hold time and ΔV_R is the displacement range that is reversed, the latter is the difference between the displacement at the end of the hold time and the displacement at the beginning of the hold time during the subsequent cycle. Thus, the two displacements are measured immediately prior to unloading and immediately after the completion of the subsequent loading to obtain ΔV_R. The creep reversal parameter, C_R, is given by

$$C_R = \frac{\Delta V_R}{\Delta V_h} \tag{90}$$

Then the value of $(C_t)_{avg}$ is calculated as

$$(C_t)_{avg} = [1 - C_R]\frac{1}{t_h}\int_{(N-1)t_h}^{Nt_h} C_t\,dt + C_R\frac{1}{t_h}\int_0^{t_h} C_t\,dt \tag{91}$$

As one can easily see, when $C_R = 1$, Equation (91) reduces to Equation (79a) and if $C_R = 0$, it reduces to Equation (79b). Grover (1996) used experimental data to study the variation in C_R as a function of fatigue cycles. An example of this variation is shown in Figure 41. It seems that large variations in C_R occur during the early portion of the test and during the final portion of the test, and in the middle it remains approximately constant. We expect C_R to be a function of the ratio of the cyclic plastic zone size to the creep zone size and the fatigue cycles, N.

$$C_R = f\left(\frac{r_p^c}{r_c}, N\right) \tag{92}$$

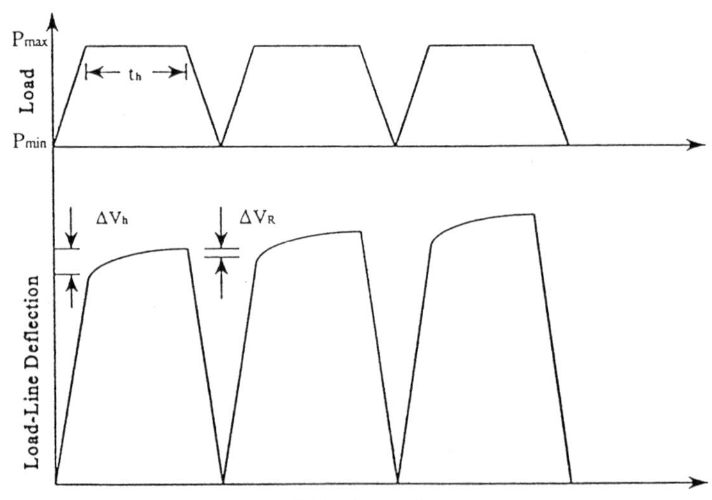

Figure 40 Definition of the creep-reversal parameter, C_R, and the experimental quantities required for calculating C_R (source Grover and Saxena, 1999; reproduced with permission from Blackwell Publishing Ltd).

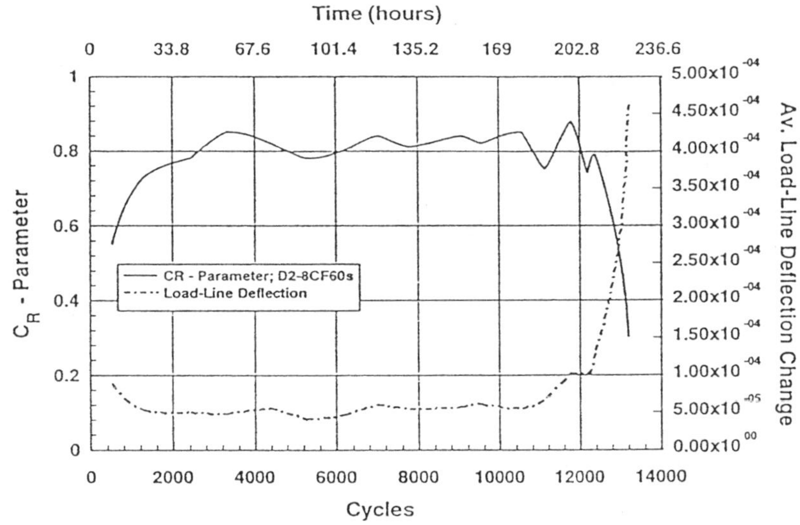

Figure 41 Variation of C_R and the load-line deflection change during hold time during a typical CFCG experiment (source Grover and Saxena, 1999; reproduced with permission from Blackwell Publishing Ltd).

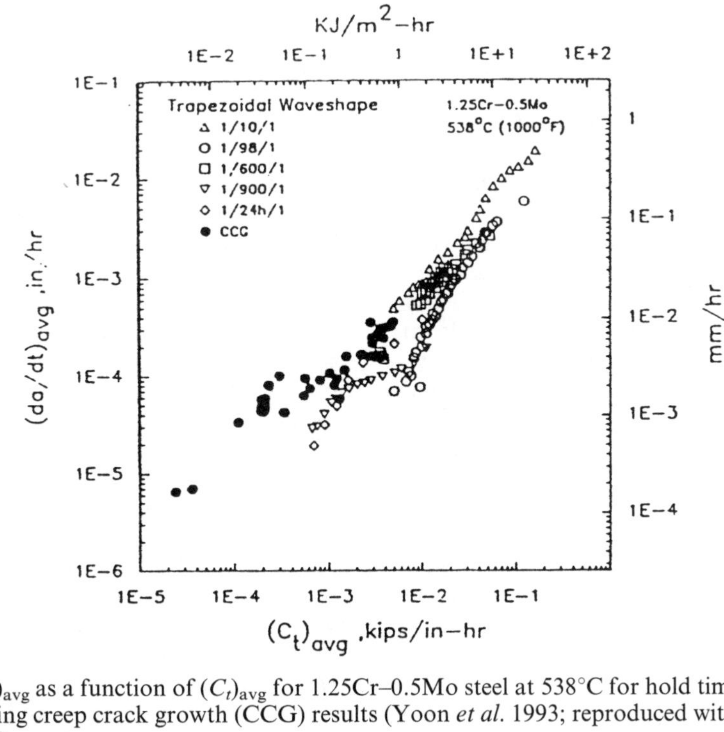

Figure 42 $(da/dt)_{avg}$ as a function of $(C_t)_{avg}$ for 1.25Cr–0.5Mo steel at 538°C for hold times ranging from 10 s to 24 h and including creep crack growth (CCG) results (Yoon *et al.* 1993; reproduced with kind permission of Kluwer Academic).

No forms for the function f have yet been proposed.

5.04.6.4 Creep Fatigue Crack Growth Rate Correlations

Figure 42 shows the correlation between $(da/dt)_{avg}$ and $(C_t)_{avg}$ for a 1.25Cr–0.5Mo steel at 538°C for tests conducted with a variety of hold times ranging from 10 s to 24 h, and also including creep crack growth rate for the same material (Yoon *et al.*, 1993). All data appear to lie in the same scatter band. These data imply that time-dependent crack growth during creep fatigue is identical to creep crack growth. This is a very significant observation for applications of these data. The 10-s hold time tests were completed within 75 h, the 24-h hold time tests required up to 10^4 h for completion, and typical creep crack growth tests require 2,000 h for completion. Thus, the 10-second hold time

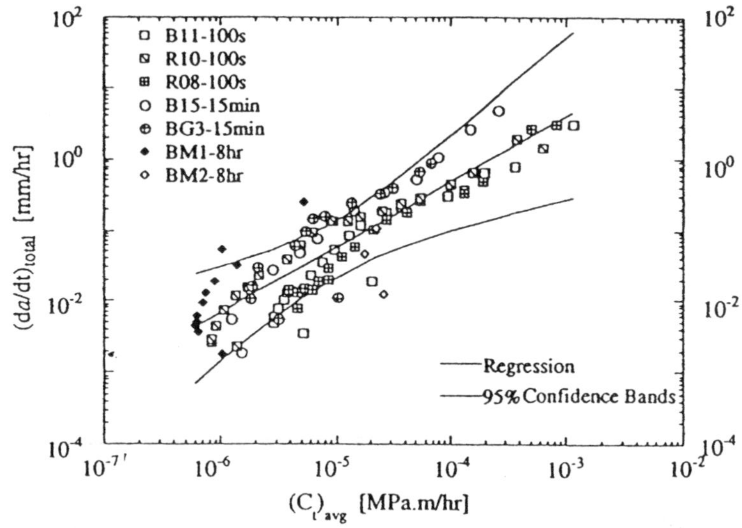

Figure 43 Same as Figure 42 except for 1Cr–1Mo–0.25V steel at 538°C (source Adefris *et al.*, 1996b; reproduced with permission from Blackwell Publishing Ltd).

test can be used to predict the behavior of tests conducted over periods of 2–3 orders of magnitude longer. This conclusion, although very useful, must be used with caution, because it may only hold for certain types of materials such as the creep-ductile materials. Figure 43 shows creep-fatigue crack growth rate data for 1Cr–1Mo–0.25V steel at various hold times. For this material, the agreement between creep-fatigue data at various hold times followed a similar trend. However, the trend for creep crack growth rate was quite distinct from the creep-fatigue trend and was lower than the creep-fatigue trend (Adefris *et al.*, 1996a, 1996b). Yet another study (Grover and Saxena, 1995) showed that in 2.25Cr–1Mo steel at 594°C, the creep-fatigue crack growth rate data and creep crack growth rate data show very comparable trends to the observations on 1.25Cr–0.5Mo steels. Thus, both possibilities exist, if we are considering a variety of materials.

5.04.7 SUMMARY

In this chapter, the crack-tip stress fields have been derived in the presence of creep deformation. Initially, the deformation has been modeled with a power-law relationship between strain-rate and stress, ignoring elastic stress redistribution near the crack tip, primary creep, and also effects due to crack growth. The C^*-integral has been identified for such conditions as the relevant crack-tip parameter for characterizing creep crack growth rates. Methods for estimating C^* in test specimens and in components were described. When the additional effects of stress redistribution and primary creep are considered, the C-integral has been shown to characterize the amplitude of the crack-tip stress field. However, the C-integral cannot be measured at the loading pins under SSC and TC conditions. Therefore, an alternative crack-tip parameter, C_t, was defined. It was shown that under EC condition, C^*-integral, C-integral and C_t are identical. Under SSC and TC, C^* is not defined and C_t and C-integral become distinct from each other. C_t has been shown to uniquely characterize creep crack growth rates for a variety of materials in experimental studies. Methods to incorporate primary creep in the estimation of C_t for test specimens and components were also described.

The effects of crack growth on the crack-tip stress fields were also considered in this artuicle. The behavior of creep-ductile materials was considered, and it was shown that even though C_t and C^* are derived for stationary cracks, they are quite suitable for characterizing creep crack growth, provided the crack growth is slow, such as in creep-ductile materials. The behavior of creep-brittle materials was also considered. It was shown that after an initial transient period, the creep crack growth rates can be correlated with K. The growth of cracks in creep-brittle materials still remains a subject in which considerable further research is needed.

When linear-elastic fracture mechanics (LEFM) conditions can be maintained, such as during high loading frequencies, ΔK is an appropriate crack-tip parameter for characterizing creep-fatigue crack growth rates for a constant loading frequency and waveform. As the temperature rises and the loading frequencies decrease, the validity of ΔK becomes questionable and nonlinear fracture mechanics parameters are needed. Parameters such as $(C_t)_{avg}$ and (ΔJ_c) are more appropriate where ΔK becomes invalid. These parameters are described in this chapter. It was shown that ΔJ_c and $(C_t)_{avg}$ are equivalent under EC conditions, and also that ΔK and $(C_t)_{avg}$ are equivalent under conditions in which the creep deformation during hold time is completely reversed by cyclic plasticity during unloading. Thus, $(C_t)_{avg}$ emerges as the most widely applicable crack-tip parameter for representing hold time effects during creep-fatigue crack growth. Methods for estimating $(C_t)_{avg}$ in specimens and components are described in this chapter. Some examples of data correlation with $(C_t)_{avg}$ were shown.

5.04.8 REFERENCES

N. Adefris, D. L. McDowell and A. Saxena, 1998, An alternative analytical approximation of the C_t parameter. *Fatigue Fract. Eng. Mater. Struct.*, **21**, 375–385.

N. Adefris, A. Saxena and D. L. McDowell, 1996a, Creep-fatigue crack growth behavior in 1Cr–1Mo–0.25V steel: I. Estimation of crack tip parameters. *Fatigue Fract. Eng. Mater. Struct.*, **19**, 387–399.

N. Adefris, A. Saxena and D. L. McDowell, 1996b, Creep-fatigue crack growth behavior in 1Cr–1Mo–0.25V steel. II: Crack growth behavior and models. *Fatigue Fract. Eng. Mater. Struct.*, **19**, 401–411.

J. L. Bassani, D. E. Hawk and A. Saxena, 1989, Evaluation of C_t parameter for characterizing creep crack growth rate in the transient regime. In: "Nonlinear Fracture Mechanics: Time-dependent Fracture Mechanics, ASTM STP 995," American Society for Testing and Materials, Philadelphia, PA, vol. I, pp. 7–29.

J. L. Bassani and F. L. McClintock, 1981, Creep relaxation of stress around a crack tip. *Int. J. Solids Struct.*, **17**, 479–492.

R. Ehlers and H. Riedel, 1981, A finite element analysis of creep deformation in a specimen containing a macrosopic crack. In: "Advances in Fracture Research," ed. D. Francois, ICF-5, Pergamon, Cambridge, UK, vol. 2, pp. 691–698.

Y. Gill, A. Saxena and J. M. Bloom, 1994, "Creep Crack Growth Behavior of an SA-106C Carbon Steel," Unpublished research, Georgia Institute of Technology, Atlanta, GA.

N. L. Goldman and J. W. Hutchinson, 1975, Fully plastic crack problems: the center cracked strip under plane strain. *Int. J. Solids Struct.*, **11**, 575–591.

P. S. Grover, 1996, Creep-fatigue crack growth in Cr–Mo–V base material and weldments. Ph.D. thesis, Georgia Institute of Technology.

P. S. Grover and A. Saxena, 1995, Characterization of creep fatigue crack growth behavior in 2.25Cr–1Mo Steel using $(C_t)_{avg}$. *Int. J. Fract.*, **73**, 273–286.

P. S. Grover and A. Saxena, 1999, Modeling the effect of creep-fatigue interaction on crack growth. *Fatigue Fract. Eng. Mater. Struct.*, **22**, 111–122.

D. E. Hall, 1995, Analysis of crack growth in creep-brittle materials. Ph.D. dissertation, Georgia Institute of Technology.

D. E. Hall, B. C. Hamilton, D. L. McDowell and A. Saxena, 1997, Creep crack growth behavior of aluminum alloy 2519: Part II. Numerical analysis. In: "Fatigue and Fracture Mechanics: 27th Volume, ASTM STP 1296," eds. R. S. Piascik, R. P. Gangloff and A. Saxena, American Society for Testing and Materials, Philadelphia, PA, vol. 27, pp. 19–36.

D. E. Hall, D. L. McDowell and A. Saxena, 1998, Crack tip parameters for creep-brittle crack growth. *Fatigue Fract. Eng. Mater. Struct.*, **21**, 387–401.

D. E. Hawk and J. L. Bassani, 1986, Transient crack growth under creep conditions. *J. Mech. Phys. Sol.*, **34**, 191–212.

C. Y. Hui, 1983, Steady-state crack growth in elastic power law creeping materials. In: "Elastic Plastic Fracture, ASTM STP 803," eds. J. Gudas and C. F. Shih, American Society for Testing and Materials, Philadelphia, PA, vol. 1., pp. 573–593.

C. Y. Hui, 1986, The mechanics of self-similar crack growth in an elastic power-law creeping materials. *Int. J. Solids Struct.*, **22**, 1357–1372.

C. Y. Hui and H. Riedel, 1981, The asymptotic stress and strain field near the tip of a growing crack under creep conditions. *Int. J. Fract.*, **17**, 409–425.

J. W. Hutchinson, 1968, Singular behavior at the end of a tensile crack in a hardening material. *J. Mech. Phys. Sol.*, **16**, 13–131.

C. E. Jaske and J. A. Begley, 1978, An approach to assessing creep/fatigue crack growth. In: "Ductility and Toughness Considerations in Elevated Temperature Service, MPC-ASME-8," ed. M. Prager, American Society for Mechanical Engineers, New York, pp. 391–409.

S. Kubo, 1988, Private communication.

V. Kumar, M. D. German and C. F. Shih, 1981, "An Engineering Approach for Elastic–Plastic Fracture Analysis," EPRI report NP-1931, Electric Power Research Institute, Palo Alto, CA.

J. D. Landes and J. A. Begley, 1976, A fracture mechanics approach to creep crack growth. In: "Mechanics of Crack Growth, ASTM STP 590," American Society for Testing and Material, Philadelphia, pp. 128–148.

C. P. Leung and D. L. McDowell, 1990, Inclusion of primary creep in the estimation of the C_t parameter. *Int. J. Fract.*, **46**, 81–104.

C. Leung, D. L. McDowell and A. Saxena, 1989, A numerical study of nonsteady-state creep at stationary crack tips. In: "Nonlinear Fracture Mechanics: Time-dependent fracture mechanics, ASTM STP 995," eds. A. Saxena, J. D. Landes and J. L. Bassani, American Society for Testing and Materials, Philadelphia, PA, vol. I, pp. 141–158.

D. L. McDowell, 1992, A nonlinear kinematic hardening theory for cyclic thermo-plasticity and thermo-viscoplasticity. *Int. J. Plasticity*, **8**, 695–728.

K. M. Nikbin, G. A. Webster and C. E. Turner, 1976, Relevance of nonlinear fracture mechanics to creep crack growth. In: "Crack and Fracture, ASTM STP 601," American Society for Testing and Materials, Philadelphia, pp. 47–62.

K. Nishida and G. A. Webster, 1990, Interaction between build-up of local and remote damage on creep crack growth. In: "Creep and Fracture of Engineering Materials and Structures," eds. B. Wilshire and R. Evans, Institute of Metals, Metals Park, Ohio, pp. 703–711.

K. Ohji, K. Ogura and S. Kubo, 1979, "Stress–Strain Fields and Modified *J*-Integral in the Vicinity of the Crack Tip under Transient Creep Conditions," Japan Society of Mechanical Engineering, no. 790-13, pp. 18–20 (in Japanese).

J. R. Rice, 1968, A path-independent integral and the approximate analysis of strain concentration by notches and cracks. *J. Appl. Mech., Trans. ASME*, **35**, 379–386.

J. R. Rice and G. F. Rosengren, 1968, Plane strain deformation near a crack tip in a power law hardening material. *J. Mech. Phys. Sol.*, **16**, 1–12.

H. Riedel, 1981, Creep deformation of crack tips in elastic-viscoplastic solids. *J. Mech. Phys. Sol.*, **29**, 35–49.

H. Riedel, 1983, Crack-tip stress fields and crack growth under creep-fatigue conditions. In: "Proceedings of 2nd Symposium on Elastic–Plastic Fracture: Inelastic Crack Analysis, ASTM STP 803, St. Louis, MO," American Society for Testing and Materials, Philadelphia, PA, vol. I, pp. I/505–I/520.

H. Riedel and J. R. Rice, 1980, Tensile cracks in creeping solids. In: "Proceedings of 12th Conference on Fracture Mechanics, ASTM STP 700, St. Louis, MO," ed. P. C. Paris, American Society for Testing and Materials, Philadelphia, PA, pp. 112–130.

H. Riedel and W. Wagner, 1981, The growth of macroscopic cracks in creeping materials. In: "Advances in Fracture Research," Pergamon, Oxford, pp. 683–690.

A. Saxena, 1980, Evaluation of C^* for characterization of creep crack growth behavior in 304 stainless steel. In: "Proceedings of 12th Conference on Fracture Mechanics, ASTM STP 700," ed. P. C. Paris, American Society for Testing and Materials, Philadelphia, PA, pp. 131–151.

A. Saxena, 1986, Creep crack growth under nonsteady-state conditions. In: "Fracture Mechanics: 17th Volume ASTM STP 905," ed. J. H. Underwood, American Society for Testing and Materials, Philadelphia, vol. 17, pp. 185–201.

A. Saxena, 1989, Mechanics and mechanisms of creep crack growth. In: "Fracture Mechanics: Microstructure and micromechanisms," eds. S. V. Nair, *et al.*, ASM International, Metals Park, Ohio, pp. 283–334.

A. Saxena, 1991, Creep crack growth in high temperature ductile materials. *Eng. Fract. Mech.*, **40**, 721–736.

A. Saxena, 1998, "Nonlinear Fracture Mechanics for Engineers," CRC Press, Boca Raton, FL.

A. Saxena and J. L. Bassani, 1984, Time-dependent fatigue crack growth behavior at elevated temperature. In: "Fracture: Interactions of microstructure, mechanisms, and mechanics, TMS-AIME," eds. A. Saxena, J. D. Landes and J. L. Bassani, Warrendale, PA, pp. 357–383.

A. Saxena, H. A. Ernst and J. D. Landes, 1983, Creep crack growth behavior in 316 stainless steel at 514°C. *Int. J. Fract.*, **23**, 245–257.

A. Saxena and B. Gieseke, 1987, Transients in elevated temperature crack growth. In: "Proceedings of MECAMAT: International Seminar on High Temperature Fracture Mechanisms and Mechanics III," EGF-6, pp. 19–36.

A. Saxena, R. S. Williams and T. T. Shih, 1981, A model for representing and predicting the influence of hold time on fatigue crack growth behavior at elevated temperature. In: "Proceedings of 13th Conference on Fracture Mechanics, ASTM STP 743," American Society for Testing and Materials, Philadelphia, PA, pp. 86–99.

A. Saxena, K. Yagi and M. Tabuchi, 1994, Crack growth under small scale and transition creep conditions in creep-ductile materials. In: "Fracture Mechanics: 24th Volume ASTM STP 1207," American Society for Testing and Materials, Philadelphia, vol. 24, pp. 481–497.

S. Taira, R. Ohtani and T. Komatsu, 1979, Application of J-integral to high temperature crack propagation: II. Fatigue crack propagation. *Trans. ASME, J. Eng. Mater. Technol.*, **101**, 162–167.

K. B. Yoon, A. Saxena and P. K. Liaw, 1993, Characterization of creep-fatigue crack growth behavior under trapezoidal waveshape using C_t parameter. *Int. J. Fract.*, **59**, 95–114.

K. B. Yoon, A. Saxena and D. L. McDowell, 1992, Influence of crack tip plasticity on creep-fatigue crack growth. In: "Proceedings of 22nd Symposium on Fracture Mechanics: 22nd Symposium (vol. I) ASTM STP 1131," American Society for Testing and Materials, Philadelphia, PA, vol. I, pp. 367–392.

K. B. Yoon, A. Saxena and D. L. McDowell, 1993, Effect of cyclic overload on the crack growth behavior during hold period at elevated temperature. *Int. J. Fract.*, **59**, 199–211.

Comprehensive Structural Integrity
ISBN (set): 0-08-043749-4

Volume 5; (ISBN: 0-08-044154-8); pp. 201–240

5.05
Creep Crack Growth

G. A. WEBSTER
Imperial College, London, UK

NOMENCLATURE

a, $\dot{a}$	crack depth, crack growth rate
$\dot{a}_i$	crack growth rate of ith element
a_0	initial crack size
$\dot{a}_1$, $\dot{a}_s$	initial, steady-state creep crack growth rate
A	coefficient in plastic strain relation
A_c, D, D_0	coefficients in creep crack growth laws
b	uncracked ligament
B, H, W	specimen, component dimensions
B_n	net thickness between side grooves

B_r coefficient in creep rupture law
c half crack length at surface
C, C_A coefficients in uniaxial creep law
C^*, $C(t)$, C_t creep fracture mechanics parameters
E, E' elastic, effective elastic modulus
$g(n)$ function of n
G elastic strain energy release rate
h nondimensional function of crack size
h_{ref} value of h for use with reference stress
i counter
I_n normalizing factor
J elastic–plastic contour integral around a crack tip
K stress intensity factor
m, ϕ power exponents in creep crack growth laws
n power exponent in uniaxial creep law
N power exponent in plasticity law
P load
P_{LC} limit load of cracked body
p, q constants in multiaxial creep ductility relation
Q constraint term in crack-tip stress equation
r distance ahead of crack tip
$\dot{r}$ rate of approach to crack tip
r_c creep process zone size
r_d crack extension detection limit
s arc length along contour Γ
t, t_r time, rupture time
t_i, t_T incubation, transition time
$t_{r(ref)}$ rupture life at reference stress
t_{ru} rupture life of uncracked plate
T_i traction vector
$\dot{u}_i$ creep displacement rate vector
$\dot{W}_S^*$ creep strain energy rate density
x, y coordinates along and perpendicular to crack
Y nondimensional function of crack and component dimensions
Y_{ref} value of Y for use with reference stress
Γ contour around crack tip
δ, $\dot{\delta}$ crack opening displacement, displacement rate
δ_i incubation crack opening displacement
δ_{ij} Kronecker delta
δ_0 initial crack opening displacement at $t=0$
$\dot{\Delta}$ creep displacement rate
ε, $\dot{\varepsilon}$ creep strain, strain rate
$\dot{\varepsilon}_A$, $\dot{\varepsilon}_s$ average, secondary creep strain rate
ε_{ij}, $\dot{\varepsilon}_{ij}$ creep strain, strain rate tensor
ε_{ref}, $\dot{\varepsilon}_{ref}$ creep strain, strain rate at reference stress
$\dot{\varepsilon}_0$, σ_0 coefficients in uniaxial creep law
$\tilde{\dot{\varepsilon}}_{ij}(\theta, n)$, $\tilde{\sigma}_{ij}(\theta, n)$ normalizing factors
ε_f, ε_f^* uniaxial, multiaxial creep failure strain
ε_{f0}, ε_{f0}^* uniaxial, multiaxial creep failure strain at stress σ_0
ε_Y yield strain
η nondimensional function of cracked specimen geometry
λ ratio of average to secondary creep strain rates
ν Poisson's ratio
ζ function of n in multiaxial ductility relation
θ angle at crack tip
σ, σ_{ij} stress, stress tensor
σ_1, σ_e, σ_m maximum principal, equivalent, mean stress
σ_{ref}, σ_Y reference, yield stress
υ power exponent in creep rupture law
ω creep damage fraction
ω_e, ω_i ligament damage fraction in service exposed, ith element

5.05.1 INTRODUCTION

The failure of engineering components that are subjected to steady loading at elevated temperatures can occur by creep rupture, creep crack growth, or a combination of both processes. Creep rupture is most likely to occur in those components that do not contain local stress concentrations and which are initially defect free. In contrast, failure by crack growth is most likely to take place in components which contain a sharp initial defect or which develop such a defect during use. In other situations crack initiation and growth may take place prior to net section rupture occurring after some amount of crack extension.

With the increasing use and sensitivity of nondestructive inspection techniques resulting in the detection of smaller and smaller cracks, there is a growing need to provide methods of assessing the significance of defects in equipment. This chapter concentrates on the procedures that have been developed for dealing with defects in components that are subjected to creep loading.

Initially the nature of the cracking process is considered. Mechanisms of how crack growth takes place and can be measured are discussed. The parameters used for characterizing the behavior are reviewed. Models are derived for describing the onset of cracking and crack propagation in terms of creep damage accumulation in a process zone ahead of a crack tip.

Figure 1 Typical intergranular creep crack growth fracture surface obtained on type 316H stainless steel at 550°C (source Bettinson, 2002).

The approach chiefly follows that adopted by Webster (1994) and Webster and Ainsworth (1994). Use is made of fracture mechanics, damage mechanics, and limit analysis concepts. Procedures are developed for dealing with crack growth when a steady-state distribution of damage has been generated at the crack tip in materials that deform in secondary creep. Allowance for the importance of state of stress at the crack tip is included. Afterwards, extensions to primary, secondary, and tertiary creep are considered. Emphasis is then placed on the early stages of cracking. Criteria for the onset of cracking are introduced that are based on the attainment of a critical crack opening displacement or a specific amount of damage at a crack tip. Procedures for dealing with a gradual buildup of damage at the crack tip are also examined. The situation of a crack growing into progressively deteriorating material is considered for making remaining life assessments of plant that has been in service for a period of time. Finally the procedures are used, in conjunction with design and failure assessment codes, to illustrate their application to some practical examples.

5.05.2 NATURE OF CRACKING PROCESS

Most failures due to creep that occur over periods of time of practical interest are intergranular. It has been found from detailed inspections that the appearance of a fracture surface is very similar whether failure is by creep rupture or creep crack growth. A picture of a fracture surface when failure was by crack growth is presented in Figure 1. It is evident that it is mainly intergranular. A micrograph taken through the tip of an interrupted creep crack is shown in Figure 2. From this figure it is apparent that crack advance takes place intermittently by the linking up of microcracks and voids that are generated on grain boundaries ahead of the main crack tip, to the crack tip. The implication is that the same mechanisms that control failure by creep rupture also control failure by creep crack growth. Consequently, it may be expected that it should be possible to predict creep crack growth properties from uniaxial creep data. In some instances the crack tip remains relatively sharp during cracking; in others blunting occurs. Mainly, the creep crack growth properties of materials are measured experimentally.

5.05.3 MEASUREMENT OF CREEP CRACK GROWTH

The procedures that are used for measuring creep crack growth properties are similar to those that are employed for measuring fatigue crack growth and fracture toughness. A cracked specimen is subjected to a constant load at elevated temperature, and the crack extension and displacement are measured as a function of time (Webster, 1982). The types of specimen that are used most often are compact tension (CT), single edge notch tension (SENT), single edge notch bend (SENB), center cracked plate (CCP), and double cantilever bend (DCB) test pieces. Different thicknesses B and widths W are employed to represent different component dimensions.

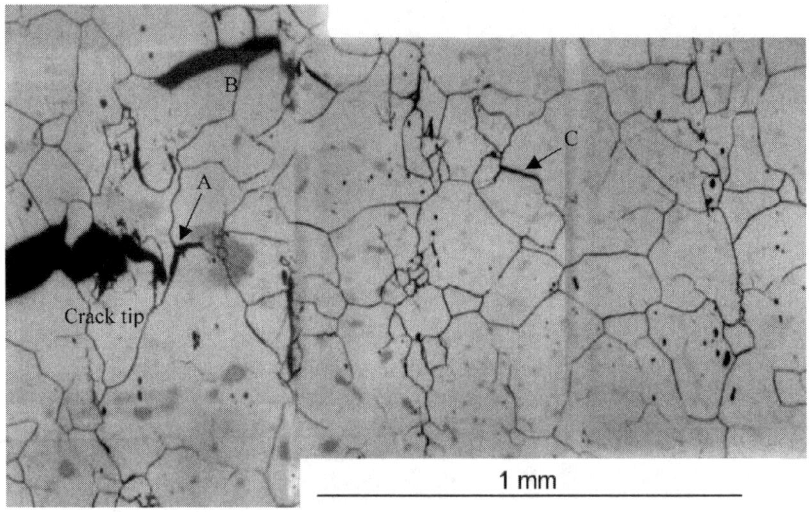

Figure 2 Section through an interrupted creep crack in type 316H stainless steel at 550°C showing microcracking A, B, C, ahead of main crack tip (source Bettinson, 2002).

Frequently side grooves (SGs) are introduced to give a net thickness B_n at the crack plane, to increase constraint, and assist in promoting flat, straight fronted cracks. Recommended procedures for making these measurements on CT specimens are given in ASTM E-1457 (2001). Crack length can be recorded by optical and/or electrical potential methods. Precautions are included for ensuring that reliable data are produced. Also requirements, which must be satisfied, are specified for determining when valid results have been obtained.

Examples of crack growth against time curves that are commonly observed are shown in Figure 3. These curves all indicate an increasing crack propagation rate with increase in time after crack growth has become well established. This is mainly because the magnitude of the stresses generated at the crack tip in these specimens increases with crack extension when they are subjected to a constant load.

The different trends depicted in the early stages of cracking can be attributed to a number of factors. Sometimes a prolonged incubation period (curve A) is observed prior to the onset of crack growth. During this stage damage gradually accumulates at the crack tip whilst crack blunting takes place. Eventually when sufficient damage has been introduced, crack growth initiates. This type of behavior is characteristic of materials which exhibit appreciable creep ductility. Curves B and C show examples of crack growth starting immediately on loading. This situation is most likely to occur in materials of more limited ductility where a sharp crack is preserved during crack propagation. For curve B a progressively rising crack propagation rate is illustrated and for curve C an initial rapid cracking rate is observed which gradually reduces before adopting the common rising trend. The initial rapid cracking rate of curve C can be explained in two ways: by redistribution of stress at the crack tip from its initial elastic state to its steady-state creep distribution; and by primary creep.

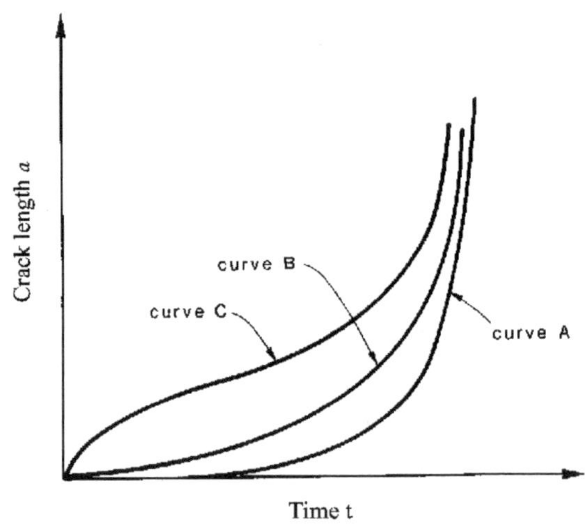

Figure 3 Typical shapes of creep crack growth curves (source Webster and Ainsworth, 1994).

5.05.4 CHARACTERIZATIONS OF CREEP CRACK GROWTH

Historically several parameters have been employed for describing creep crack growth rate (Siverns and Price, 1973; Neate and Siverns, 1973; Floreen, 1975; Ellison and Harper, 1978; Sadananda and Shahinian, 1981; Webster, 1983; Saxena, 1986; Riedel, 1987; Viswanathan, 1989). Both linear and nonlinear fracture mechanics concepts have

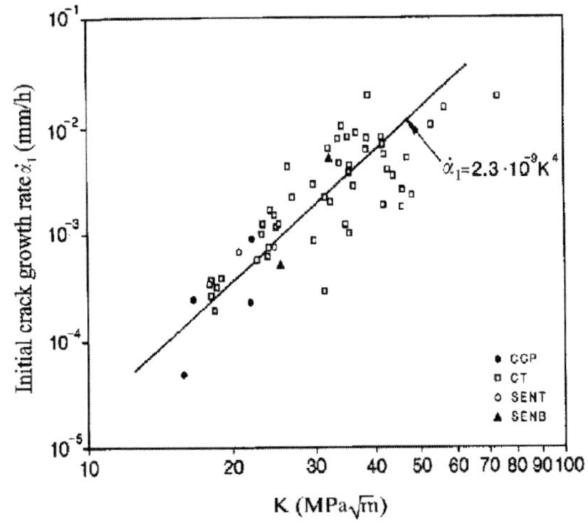

Figure 4 Correlation of initial creep crack growth rate with K for 1% CrMoV steel at 550°C (source Kienzler and Hollstein, 1987).

been used (see also Chapter 5.04). Immediately after loading, the stress distribution ahead of a crack tip in the absence of plastic deformation is elastic and is described by the elastic stress intensity factor K, which in general can be expressed as

$$K = Y\sigma\sqrt{a} \tag{1}$$

where σ is the applied stress, a is the crack size, and Y is a nondimensional function of the crack and component dimensions and also the type of loading. When little creep deformation accompanies fracture, the stress distribution will remain virtually unaltered by creep and K will continue to characterize the stresses around the crack tip. For this "creep brittle" circumstance, some success has been found in correlating creep crack growth rate $\dot{a}$ by an expression of the form

$$\dot{a} = A_c K^m \tag{2}$$

where A_c and m are material constants. It is also possible sometimes to describe the initial crack growth rate $\dot{a}_I$ by K, as illustrated in Figure 4, when cracking occurs before stress redistribution has had time to take place. In this case, typically m in Equation (2) is found to be close to the stress sensitivity of creep n in the uniaxial creep law

$$\dot{\varepsilon} - \dot{\varepsilon}_0\left(\frac{\sigma}{\sigma_0}\right)^n = C\sigma^n \tag{3}$$

where $\dot{\varepsilon}$ is creep strain rate, and $\dot{\varepsilon}_0$, σ_0, and C are material constants.

When substantial creep deformation accompanies fracture, stress redistribution will occur at the crack tip, as shown in Figure 5, and K will no longer adequately describe the local stress state. Eventually when creep strains dominate, a steady-state stress distribution is obtained which can be characterized by the creep fracture mechanics parameter C^*. This parameter is the creep equivalent of the J contour integral (Rice, 1968) used to describe the stress state ahead of a crack tip in the presence of extensive plasticity. It can be evaluated for a contour Γ surrounding a crack tip in a two-dimensional (2D) body of unit thickness from

$$C^* = \int_\Gamma W_s^*\,\mathrm{d}y - T_i\frac{\partial \dot{u}_i}{\partial x}\mathrm{d}s \tag{4}$$

where W_s^* is the creep strain energy rate density given by

$$W_s^* = \int \sigma_{ij}\,\mathrm{d}\dot{\varepsilon}_{ij} \tag{5}$$

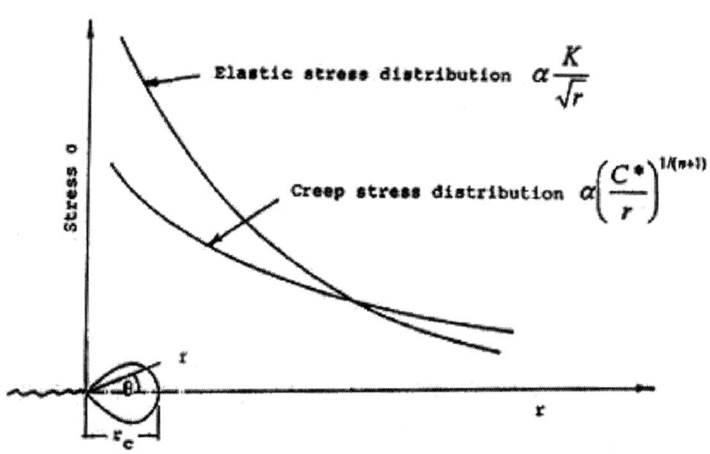

Figure 5 Creep process zone at a crack tip subjected to elastic and creep stress distributions (source Webster, 1994).

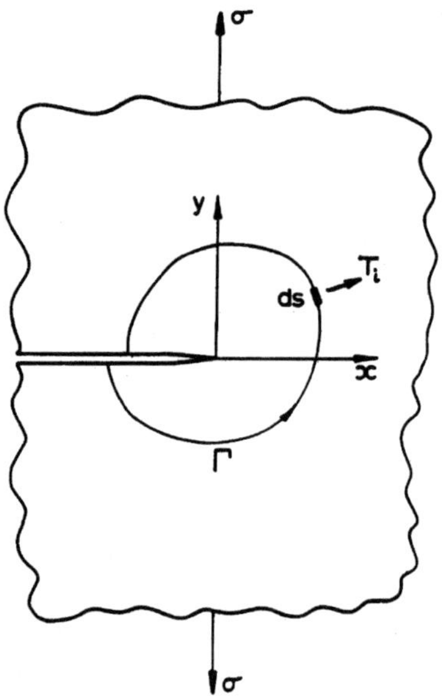

Figure 6 Integration contour around a crack tip (source Webster and Ainsworth, 1994).

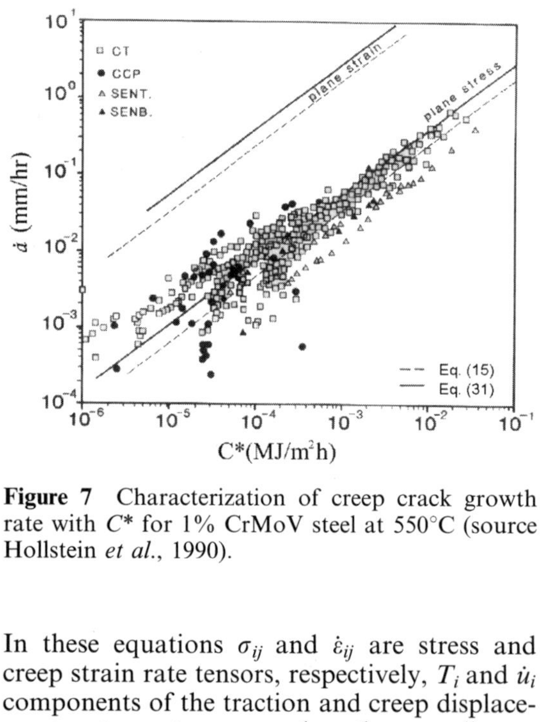

Figure 7 Characterization of creep crack growth rate with C^* for 1% CrMoV steel at 550°C (source Hollstein *et al.*, 1990).

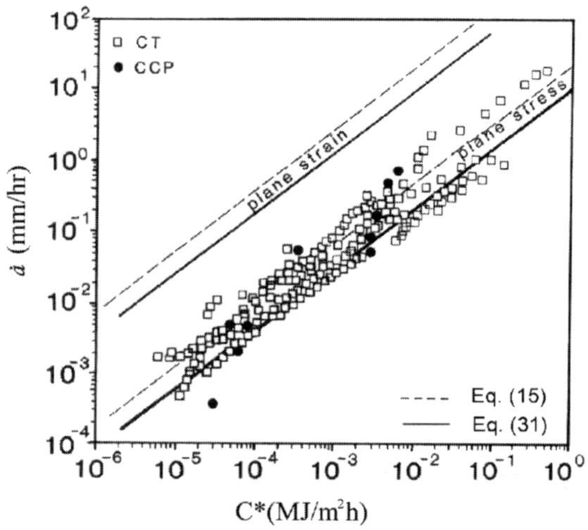

Figure 8 Dependence of creep crack growth rate on C^* for alloy 800 H at 800°C (source Hollstein *et al.*, 1990).

In these equations σ_{ij} and $\dot{\varepsilon}_{ij}$ are stress and creep strain rate tensors, respectively, T_i and $\dot{u}_i$ components of the traction and creep displacement rate vectors, x and y the coordinates along and perpendicular to the crack tip, and s is arc length along Γ as shown in Figure 6. When steady-state conditions have been achieved, C^* becomes path independent and can be used to characterize the stress and creep strain rate fields around a crack tip where creep strains dominate. In this circumstance it is usually found (Nikbin *et al.*, 1984, 1986) that creep crack growth can be described, as indicated in Figures 7 and 8, by an expression of the form

$$\dot{a} = D_0 C^{*\phi} \tag{6}$$

where D_0 is a material constant and ϕ is slightly less than 1.

Whilst stress redistribution due to creep is taking place at a crack tip, C^* is not path independent and it is more appropriate to describe the local stress field in terms of the parameters $C(t)$ or C_t (Riedel and Rice, 1980; Bassani *et al.*, 1987). $C(t)$ is evaluated from Equations (4) and (5) like C^* but for a contour that has shrunk to the crack tip. Prior to stress redistribution $C(t) > C^*$, but as steady state is reached $C(t) \rightarrow C^*$. A conservative estimate of the time t_T to reach this steady state is given by

$$t_T = \frac{G}{(n+1)C^*} \tag{7}$$

where G is elastic strain energy release rate.

The term C_t has been developed by Saxena (1998) for dealing with small-scale creep. It makes allowance for the generation of a creep process zone r_c at a crack tip as indicated in Figure 5. Like $C(t)$ initially $C_t > C^*$, and as steady-state creep conditions are reached at a crack tip $C_t \rightarrow C^*$. The time for this to be achieved is similar to that given by Equation (7).

The available information suggests that, despite the introduction of the above fracture mechanics parameters, creep crack growth rate can be interpreted most often in terms of C^* by an expression of the form of Equation (6). This is because in most practical situations stress redistribution is complete relatively early in the cracking process. The checks imposed in

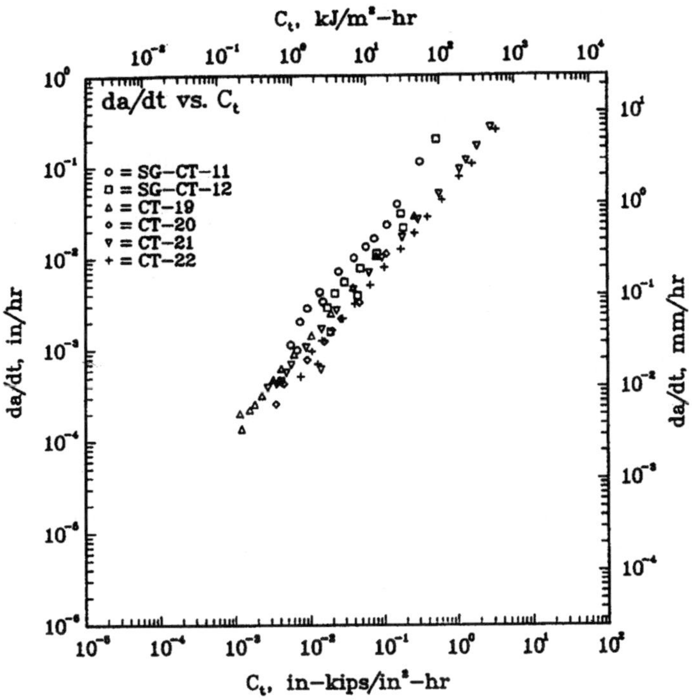

Figure 9 Correlation of creep crack growth rate with C_t for 1% CrMoV steel at 593°C (source Saxena *et al.*, 1992) (different symbols correspond to different tests).

ASTM E1457 (2001) for obtaining valid data are aimed at ensuring that only results after stress redistribution is complete are used to determine D_0 and ϕ in Equation (6). In principle, prior to steady-state conditions having been reached at a crack tip, it should be possible to describe the cracking rate by an expression of the form of Equation (6) but with C^* replaced by $C(t)$ or C_t (see Chapter 5.04 for further discussion of these terms). An illustration of a successful correlation of creep crack growth data with C_t is depicted in Figure 9.

5.05.5 MODELS OF CRACKING PROCESS

Models for describing the initiation of creep crack growth and estimating crack growth rates are useful for identifying the parameters that are expected to control these properties. They are also useful for indicating the quantities that should be measured to characterize their behavior. In addition, they may allow the behavior to be predicted purely from the uniaxial creep strain and rupture properties of a material. Although the models involve simplifying assumptions, often satisfactory assessments of component integrity can be made by using this approach when detailed crack initiation and propagation data are not available. Allowance for progressive material deterioration can also be included, when appropriate, for making residual life estimates.

It has been demonstrated that the stress distribution around a crack tip can be described by C^* when creep strains dominate there. In this circumstance the stress field at (r, θ) ahead of the crack tip (as shown in Figure 5) is given by

$$\frac{\sigma_{ij}}{\sigma_0} = \left[\frac{C^*}{I_n \sigma_0 \dot{\varepsilon}_0 r}\right]^{1/(n+1)} \tilde{\sigma}_{ij}(\theta, n) + Q\delta_{ij} + \cdots \quad (8)$$

for a material which creeps according to Equation (3) where in this expression: I_n is a factor which is sensitive to n and state of stress and is given approximately for plane-stress and plane-strain conditions by, respectively,

$$I_n = 7.2\left(0.12 + \frac{1}{n}\right)^{1/2} - \frac{2.9}{n} \quad (9)$$

and

$$I_n = 10.3\left(0.13 + \frac{1}{n}\right)^{1/2} - \frac{4.6}{n}, \quad (10)$$

the term $\tilde{\sigma}_{ij}(\theta, n)$ is a nondimensional function of n and θ which is chosen in such a way that its maximum equivalent value is unity, δ_{ij} is the Kronecker delta, and Q represents a hydrostatic stress component ahead of the crack tip. In addition to being sensitive to n and θ, Q also depends on the applied loading conditions, crack geometry and test piece or component dimensions. It is approximately independent of distance ahead of the crack tip (O'Dowd and Shih, 1994). For highly constrained conditions at a crack tip for which the HRR (Hutchinson, 1968; Rice and Rosengren, 1968) stress distribution is valid, $Q=0$. For progressively decreasing constraint at a crack tip, Q becomes increasingly negative (O'Dowd and Shih, 1994). The role of Q, in addition to C^*, in describing the mechanism of creep crack growth has been considered by Budden and Ainsworth (1999) and Bettinson *et al.* (2000a, 2000b). It provides an extra term for dealing with the influence of different degrees of constraint at a crack tip. However close to a crack tip, the first term in Equation (8) dominates. In this circumstance, the creep strain rate field ahead of the crack tip becomes

$$\frac{\dot{\varepsilon}_{ij}}{\dot{\varepsilon}_0} = \left[\frac{C^*}{I_n \sigma_0 \dot{\varepsilon}_0 r}\right]^{n/(n+1)} \tilde{\dot{\varepsilon}}_{ij}(\theta, n) \qquad (11)$$

where $\tilde{\dot{\varepsilon}}_{ij}(\theta, n)$ is another nondimensional function of n and θ with a maximum equivalent value of 1.

It is possible to employ Equations (8) and (11) to make predictions of creep crack initiation and growth from uniaxial creep deformation and rupture data. Several models have been described in the literature (Webster, 1994; Webster and Ainsworth, 1994). These will be reviewed here.

5.05.6 STEADY-STATE CRACK GROWTH

Initially the situation when a steady-state distribution of damage has been produced at a crack tip is examined (Nikbin *et al.*, 1983, 1984, 1986).

5.05.6.1 Ductility Exhaustion Criterion

Consider Figure 5 which shows a creep process zone ahead of a crack which is assumed to be extending by creep at constant speed. It is postulated that this zone encompasses the region over which creep damage is accumulating ahead of the crack tip. Material will first experience damage when it enters the creep zone at $r=r_c$ at time $t=0$ and will have accumulated creep strain ε_{ij} over a period of time t by the time it is at a distance r from the crack tip such that

$$\varepsilon_{ij} = \int_0^t \dot{\varepsilon}_{ij}\, \mathrm{d}t \qquad (12)$$

The available microstructural observations of Figure 2 indicate that voiding and microcracking can extend up to several grain sizes ahead of the main crack tip and that crack advance takes place in a series of small steps by the progressive linking of the main crack to this damage. The process can therefore be regarded as quasi-static. Provided the zone is sufficiently small, Equation (11) can be taken to represent the distribution of creep strain rate ahead of the extending crack so that substitution into Equation (12) gives

$$\varepsilon_{ij} = \int_{r_c}^{r} \dot{\varepsilon}_0 \left(\frac{C^*}{I_n \sigma_0 \dot{\varepsilon}_0 r}\right)^{n/(n+1)} \tilde{\dot{\varepsilon}}_{ij}(\theta, n) \frac{\mathrm{d}t}{\mathrm{d}r}\, \mathrm{d}r \qquad (13)$$

Integrating for a constant growth rate at constant C^* by substituting $\dot{r} = -\dot{a}$, because an element ahead of the crack is approaching the crack tip in the opposite direction to the crack extension, results in

$$\varepsilon_{ij} = (n+1)\dot{\varepsilon}_0 \left(\frac{C^*}{I_n \sigma_0 \dot{\varepsilon}_0}\right)^{n/(n+1)} \frac{\tilde{\dot{\varepsilon}}_{ij}(\theta, n)}{\dot{a}} \times \left(r_c^{1/(n+1)} - r^{1/(n+1)}\right) \qquad (14)$$

It is possible to propose that fracture occurs at the crack tip when the available creep ductility of the material is exhausted there. This will happen first at the value of θ at which the equivalent value of $\tilde{\dot{\varepsilon}}_{ij}(\theta, n)$ reaches its maximum magnitude of unity. Therefore, substituting $\varepsilon_{ij} = \varepsilon_f^*$ at this value of θ at $r=0$ into Equation (14) gives

$$\dot{a} = \frac{(n+1)\dot{\varepsilon}_0}{\varepsilon_f^*} \left(\frac{C^*}{I_n \sigma_0 \dot{\varepsilon}_0}\right)^{n/(n+1)} r_c^{1/(n+1)} \qquad (15)$$

where ε_f^* is the creep ductility appropriate to the state of stress at the crack tip. From Equation (3), Equation (15) can be re-expressed as

$$\dot{a} = \frac{(n+1)}{\varepsilon_f^*} \left(\frac{C^*}{I_n}\right)^{n/(n+1)} (C r_c)^{1/(n+1)} \qquad (16)$$

Both Equations (15) and (16) have the same form as Equation (6) which was previously determined experimentally. The model implies that in the experimental expression

$$\phi = \frac{n}{n+1} \qquad (17)$$

and

$$D_0 = (n+1)\frac{\dot{\varepsilon}_0}{\varepsilon_f^*} \left(\frac{1}{I_n \sigma_0 \dot{\varepsilon}_0}\right)^{n/(n+1)} r_c^{1/(n+1)} \qquad (18a)$$

or its equivalent value

$$D_0 = \frac{(n+1)}{\varepsilon_f^*}\left(\frac{Cr_c}{I_n^n}\right)^{1/(n+1)} \tag{18b}$$

For large values of n, ϕ will be a fraction close to unity in agreement with most test data. Similarly, since r_c is raised to a small fraction in Equations (18a) and (18b), relative insensitivity to the size of the creep damage zone is predicted. Equation (18b) also shows relative insensitivity to C. This term gives the temperature dependence of creep deformation in Equation (3). Consequently, Equation (18b) indicates that D_0 and, therefore, creep crack growth rate should be only weakly dependent on temperature at a given value of C^* for no change in creep ductility.

In contrast, Equations (15) and (16) indicate that crack growth rate should be inversely proportional to the creep ductility ε_f^* appropriate to the state of stress at a crack tip. Bounds on this value for plane-stress and plane-strain conditions can be obtained from models of the influence of hydrostatic stress σ_m on void growth because of the microstructural evidence (Figure 2) which suggests that creep crack growth occurs by void growth and microcracking ahead of the main crack tip. Several models have been proposed. These allow the ratio $\varepsilon_f^*/\varepsilon_f$ to be calculated for various states of stress where ε_f is the uniaxial creep ductility of a material. All the models predict that void growth rate increases with the triaxial stress ratio σ_m/σ_e, where σ_e is the equivalent stress. From the work of Rice and Tracey (1969), who considered the behavior of a spherical void in a rigid plastic material, Bettinson (2002) has shown that

$$\frac{\varepsilon_f^*}{\varepsilon_f} = \frac{0.521}{\sinh \dfrac{3\sigma_m}{2\sigma_e}} \tag{19}$$

This approach was extended by Budiansky *et al.* (1982) for a power-law creeping material obeying Equation (3) to produce

$$\frac{\varepsilon_f^*}{\varepsilon_f} = \left[\frac{(n+\zeta)}{(3n(\sigma_m/\sigma_e)+\zeta)}\right]^n \tag{20}$$

where $\zeta = 2(n-1)(n+0.432)$ although the expression is strictly not valid for low triaxialities. A further development by Cocks and Ashby (1980) for an array of voids on a grain boundary assuming that the time to initiate voids is negligible gives

$$\frac{\varepsilon_f^*}{\varepsilon_f} = \frac{\sinh[(2/3)[(n-0.5)/(n+0.5)]]}{\sinh[2[(n-0.5)/(n+0.5)](\sigma_m/\sigma_e)]} \tag{21}$$

Finally, Spindler (1994) and Spindler *et al.* (2001) have produced an empirical expression specifically for stainless steels to account for void nucleation and growth of the form

$$\frac{\varepsilon_f^*}{\varepsilon_f} = \exp\left[p\left(1-\frac{\sigma_1}{\sigma_e}\right)\right]\exp\left[q\left(0.5-\left(\frac{3\sigma_m}{2\sigma_e}\right)\right)\right] \tag{22}$$

where σ_1 is the maximum principal stress, and p and q are material-dependent parameters. Other expressions have been developed for incorporating the effects of diffusion. However, in all cases for plane-stress conditions ε_f^* can be taken to be close to ε_f. For plane strain, the situation is more complicated. Some guide can be provided by considering the stresses associated with a Prandtl field for which $\sigma_m/\sigma_e = 2.4$. In this case the various models give $\varepsilon_f/\varepsilon_f^*$ ranging between about 20 and 80 depending on the choice of n, p, and q, as appropriate. This suggests that a typical value for ε_f^* for plane-strain conditions may be about $\varepsilon_f/50$. For other states of stress between plane stress and plane strain, an intermediate value for ε_f^* between ε_f and $\varepsilon_f/50$ will be expected. Experimental determinations may be obtained from tests on circumferentially notched bars of different notch acuities that generate different multiaxial stress states in the notch throat region (Webster *et al.*, 1992, 2001). Also when the second term in Equation (8) is significant, some account of the influence of Q on the stress state at a crack tip may be required (Budden and Ainsworth, 1999; Bettinson *et al.*, 2000a, 2000b).

5.05.6.2 Predictions Based on Rupture Data

In general, an element of material will undergo primary, secondary, and tertiary creep, as illustrated in Figure 10, as it traverses the process zone. Also, often only creep rupture data are available. Normally Equation (3) is employed to describe secondary creep rate $\dot{\varepsilon}_s$. This is satisfactory for making an assessment of creep crack growth when this stage of creep dominates. When primary and tertiary creep are appreciable, some allowance for these stages is needed. This can be achieved in an approximate manner as follows. Provided the uniaxial creep failure strain ε_f is known, an average creep strain rate $\dot{\varepsilon}_A$ can be defined as shown in Figure 10:

$$\dot{\varepsilon}_A = \varepsilon_f/t_r \tag{23}$$

which enables Equation (3) to be rewritten as

$$\dot{\varepsilon}_A = \dot{\varepsilon}_0\left(\frac{\sigma}{\sigma_0}\right)^n = C_A\sigma^n \tag{24}$$

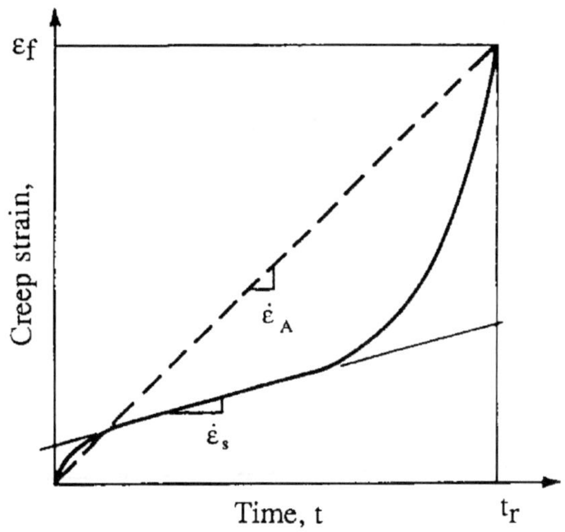

Figure 10 Simplification of primary, secondary, and tertiary creep to an average creep strain rate $\dot{\varepsilon}_A$ (source Webster and Ainsworth, 1994).

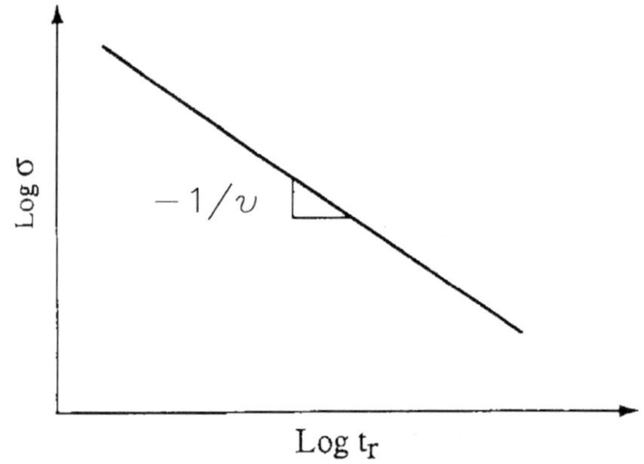

Figure 11 Typical stress rupture diagram (source Webster and Ainsworth, 1994).

where the constants are now obtained from rupture data rather than secondary creep properties.

The same procedure can now be repeated as was carried out in the previous section to obtain crack growth rate. The only modification is that the constants used for describing creep strain rate are taken from Equation (24) rather than Equation (3). In this circumstance, creep crack growth rate is again given by Equation (16) but with C replaced by C_A. Since in this Equation C_A (or C) is raised to a small fractional power, relative insensitivity of cracking rate to this term is anticipated. Consequently almost the same predictions of crack growth rate will be expected at the same value of C^* irrespective of whether they are based on secondary creep or creep rupture data.

The previous analyses have been based on the assumption that creep failure strain ε_f is constant. However, it is frequently dependent on stress. For this situation, the stress sensitivity of rupture life will be different to that of the average creep rate. Even so, it is often possible to approximate stress rupture properties by the relation shown in Figure 11. In this case

$$t_r = \frac{\varepsilon_{f0}}{\dot{\varepsilon}_0}\left(\frac{\sigma_0}{\sigma}\right)^{\upsilon} = \frac{B_r}{\sigma^{\upsilon}} \tag{25}$$

so that creep ductility becomes

$$\varepsilon_f = \varepsilon_{f0}\left(\frac{\sigma}{\sigma_0}\right)^{n-\upsilon} = C_A B_r \sigma^{n-\upsilon} \tag{26}$$

where ε_{f0} is the uniaxial creep failure strain at stress, σ_0, and B_r is a material constant. For $n > \upsilon$, ductility decreases with decrease in stress, and when $n = \upsilon$ a constant failure strain is predicted.

When creep failure strain is not constant, a slight modification to the derivation of the creep crack growth law is required. This can be achieved by application of the life fraction rule (Webster and Ainsworth, 1994) to give fracture at the crack tip when

$$\int \frac{dt}{t_r} = 1 \tag{27}$$

at $r = 0$. Therefore substituting Equations (25) and (8), assuming that the Q term can be ignored, into Equation (27) gives a criterion for fracture at the crack tip as

$$1 = \int_{r_c}^{0} \frac{\dot{\varepsilon}_0}{\varepsilon_{f0}^*}\left(\frac{C^*}{I_n \sigma_0 \dot{\varepsilon}_0 r}\right)^{\upsilon/(n+1)} \left[\tilde{\sigma}_{ij}(\theta, n)\right]^{\upsilon} \frac{dt}{dr} dr \tag{28}$$

where ε_{f0} has been replaced by ε_{f0}^* to allow for the influence of state of stress on creep ductility. Integrating this expression for a constant cracking rate at constant C^* for $\dot{r} = -\dot{a}$, as before, and inserting the maximum value of $\tilde{\sigma}_{ij}(\theta, n) = 1$ for equivalent stress-controlled behavior gives

$$\dot{a} = \left(\frac{n+1}{n+1-\upsilon}\right)\frac{\dot{\varepsilon}_0}{\varepsilon_{f0}^*}\left(\frac{C^*}{I_n \sigma_0 \dot{\varepsilon}_0}\right)^{\upsilon/(n+1)} r_c^{(n+1-\upsilon)/(n+1)} \tag{29}$$

By using Equations (24) and (25), this equation can be rewritten in the form

$$\dot{a} = \left(\frac{n+1}{n+1-\upsilon}\right)\frac{\varepsilon_{f0}}{\varepsilon_{f0}^*}\frac{1}{B_r C_A}\left(\frac{C^*}{I_n}\right)^{\upsilon/(n+1)} \times (C_A r_c)^{(n+1-\upsilon)/(n+1)} \tag{30}$$

In deriving this expression, allowance has been made for rupture taking place at a time $t = (\varepsilon_{f0}^*/\varepsilon_{f0}) t_r$ for multiaxial stress conditions.

Both Equations (29) and (30) have the same form as the experimental relation (Equation (6)) if $\upsilon/(n+1)=\phi$.

In general, ε^*_{f0} will be less than its corresponding uniaxial value ε_{f0}. It will be expected to vary with stress state in the same manner as ε_f, so that for plane-stress conditions it can be taken to equal ε_{f0} and for plane-strain conditions to equal $\varepsilon_{f0}/50$. For a constant failure strain, when $n=\upsilon$, Equations (29) and (30) reduce, respectively, to Equations (15) and (16) as required, but with C replaced by C_A.

It is recognized that the zone over which creep damage develops ahead of a crack tip is likely to increase with crack extension (Riedel, 1987; Saxena, 1998). However, since r_c in the crack growth rate expressions is raised to a small fractional power, crack growth rate is relatively insensitive to the magnitude of r_c. Choice of a constant value for r_c will not, therefore, affect predicted crack propagation rates appreciably. Equations (29) and (30) allow all stages of creep and a variable ductility to be incorporated into the creep crack growth models in a simple but realistic approximate fashion.

Most practical engineering alloys undergo primary, secondary, and tertiary creep prior to fracture. Also their creep properties often do not exhibit the simple power-law stress dependence illustrated in Equation (3) and Figure 11. It is possible to incorporate more accurate constitutive relations into the creep crack growth models but then it is not possible to derive analytical expressions and numerical integration is required. In addition, accurate descriptions of the creep behavior of materials are not always available. When this is the case, the procedure of this section allows estimates of crack propagation rates to be made from a knowledge of the creep rupture properties of materials only.

5.05.6.3 Comparison with Experimental Data

The creep crack growth laws derived in Sections 5.05.6.1 and 5.05.6.2 will now be compared with experimental crack propagation data that have been taken from a number of sources. The experimental results were collected using the procedures discussed in ASTM E1457 (2001). The appropriate parameters used for describing the creep properties of the materials investigated are listed in Table 1. The crack growth results were obtained on a selection of test piece geometries including CT, DCB, SENT, SENB, and CCP specimens of various thicknesses. In some instances SGs were employed, in others they were not. In making the calculations it has been assumed that the creep process zone size is equal to the material grain size, although in view of the preceding discussion it could have been taken to be several times larger without significantly affecting the predictions.

The predictions of Equation (15), or the equivalent form Equation (16), for those materials exhibiting a constant creep ductility assuming both plane-stress and plane-strain conditions are shown in Figures 7, 8 and 12–14. It is clear that close correspondence with the plane-stress estimates is achieved for the relatively ductile low alloy steels (Figures 7 and 12) and alloy 800H (Figure 8). This suggests that very thick sections would be needed to achieve plane-strain conditions in these materials.

With the nickel base superalloy AP1 (Figure 13), behavior between the plane-stress and plane-strain limits is obtained. This material showed less creep ductility, and the specimens contained deep SGs, so that crack growth rates closer to the plane-strain bounds would be anticipated.

Further influence of the effect of specimen size is indicated in Figures 14 and 15 which

Table 1 Uniaxial tensile creep properties (Webster and Ainsworth, 1994).

Material	*Temperature* (°C)	n	σ_0[a] (MPa)	υ	ε_f	*Grain size* (μm)
(a) Materials with a constant creep ductility						
1% CrMoV steel	550	6.5	1146		0.15	22
1% CrMoV steel	538	13.4	624		1.6[b]	50
Alloy 800 H	800	6.5	170		0.3	40
21/4% CrMo steel	538	9	340		0.45	20
AP1	700	15	1060		0.15	30
(b) Materials with variable creep ductility						
RR 58	150	35	361	32	0.09[c]	150
1/2% CrMoV steel	565	17	483	11	0.19[c]	250

[a] Assumes $\dot{\varepsilon}_0 = 1/h$. [b] True failure strain calculated from reduction in area at fracture. [c] Correspond to ε_{f0} values.

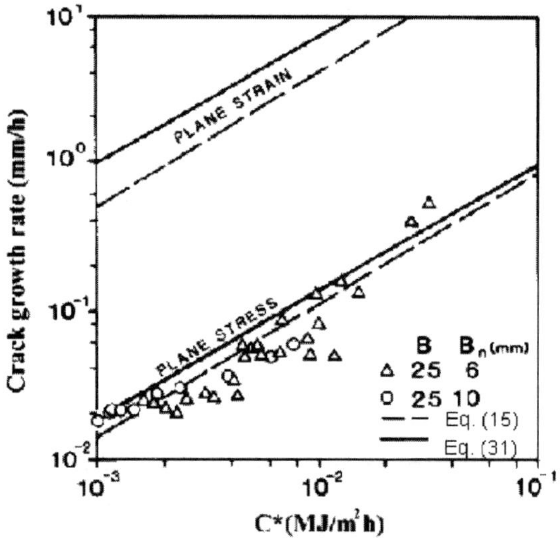

Figure 12 Creep crack growth characteristics of $2\frac{1}{4}$% CrMo steel measured on CT specimens at 538°C (source Nikbin *et al.*, 1986) compared with predictions.

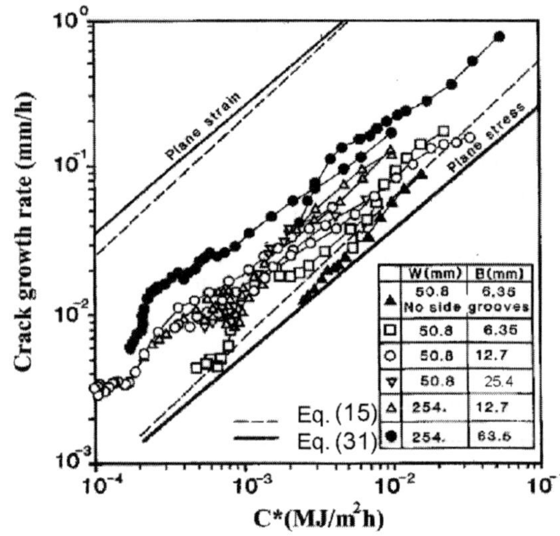

Figure 14 Creep crack growth properties of different sizes of CT specimen of 1% CrMoV steel at 538°C (source Tabuchi *et al.*, 1990) with $(B_n/B) = 0.75$ compared with predictions.

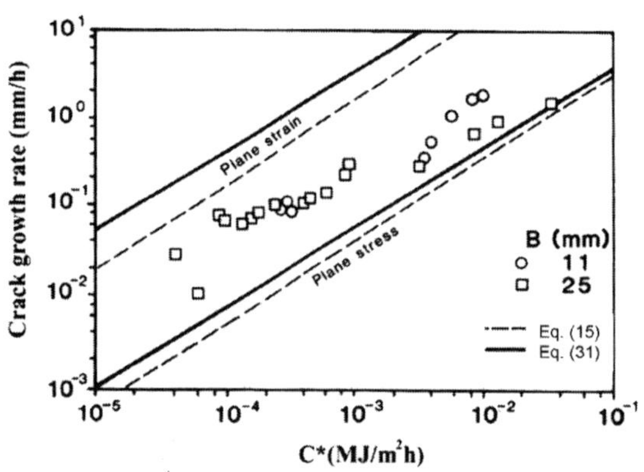

Figure 13 Creep crack growth behavior of nickel base superalloy AP1 determined on CT specimens at 700°C (source Nikbin *et al.*, 1984) compared with predictions.

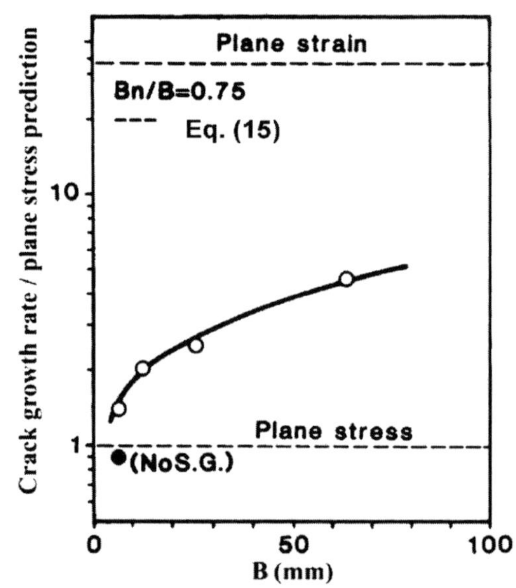

Figure 15 Dependence of creep crack growth rate on thickness for 1% CrMoV steel specimens tested in Figure 14 (source Tabuchi *et al.*, 1990).

show crack growth data on specimens having thicknesses ranging between 6.35 mm and 63.5 mm. Also included in Figure 14 are the predictions of Equation (15). It can be seen that all the results lie between the plane-stress and plane-strain bounds and that crack propagation rate increases with increasing specimen size and thickness. This observation is consistent with 3D (as opposed to 2D) finite element calculations of Q in Equation (8) which predict an increase in constraint with depth from a free surface (Bettinson, 2002; Bettinson *et al.*, 2002). It would seem that as full plane-strain conditions are approached, cracking rate tends to a value approaching 10 times the plane-stress rate implying that choice for plane-strain conditions of a creep failure strain of 1/50th of the uniaxial ductility is overly conservative in this case.

Additional crack growth data are shown in Figures 16 and 17 for two alloys which exhibited a decrease in creep ductility with decreasing stress. In this case $n > \upsilon$ and comparison has been made with the variable ductility model (Equation (29)). Both these alloys demonstrate only a limited creep failure strain, and closest agreement is obtained with the plane-strain predictions. Similar trends have been reported elsewhere (Nikbin *et al.*,

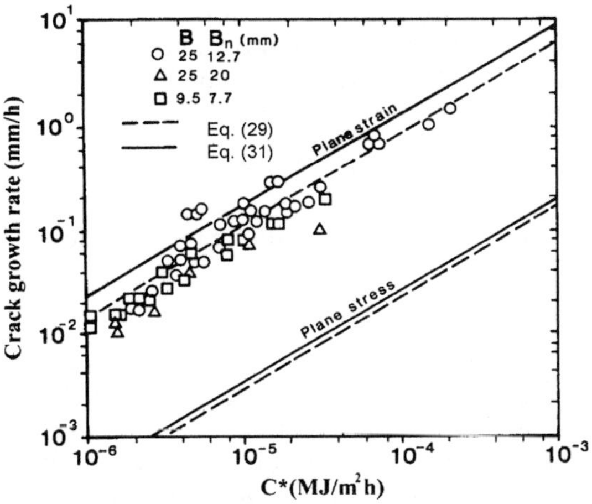

Figure 16 Creep crack growth behavior of aluminum alloy RR 58 obtained on DCB specimens at 150°C (source Nikbin *et al.*, 1983).

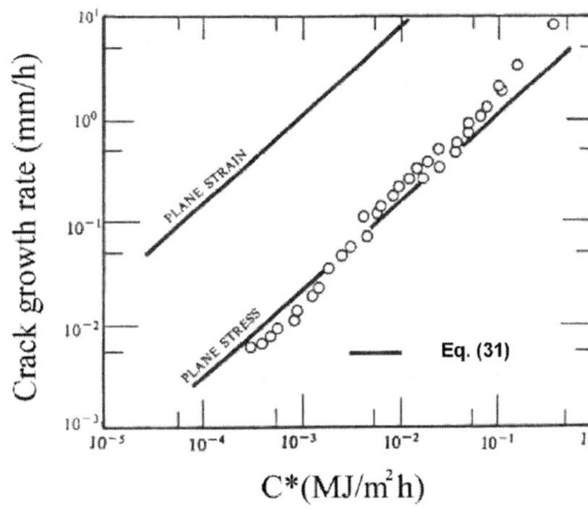

Figure 18 Creep crack growth properties of 0.16% C steel at 500°C measured on circumferentially cracked cylinders of 11 mm internal and 14 mm external diameter (source Nikbin *et al.*, 1986).

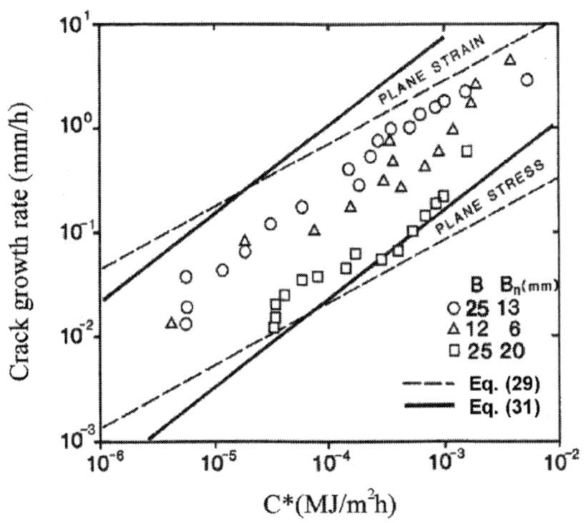

Figure 17 Crack propagation characteristics of 0.5% CrMoV steel obtained on DCB specimens at 565°C (source Nikbin *et al.*, 1983).

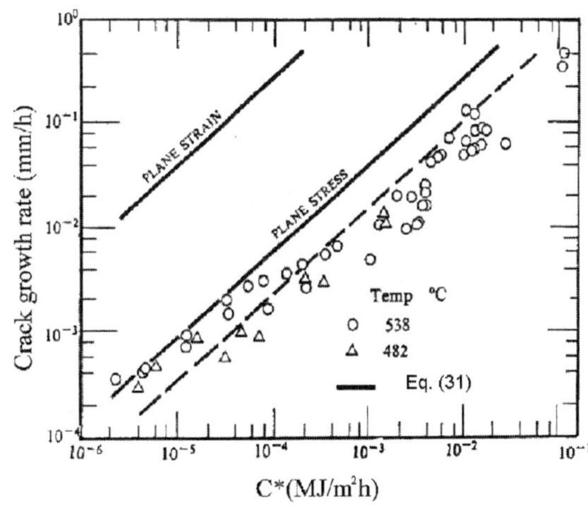

Figure 19 Creep crack growth properties determined on nonside-grooved CT specimens of A 470 class 8 steel (source Nikbin *et al.*, 1986).

1986; Tan *et al.*, 2001) on a wider range of materials.

5.05.6.4 Universal Creep Crack Growth Law

It can be demonstrated (Nikbin *et al.*, 1986), by substitution of appropriate values for the parameters in Equations (15) and (29) for a broad spread of materials, that the power dependence of C^* varies only over the range 0.7–1.0 and that crack growth rates can be predicted approximately within a factor of ~2 by

$$\dot{a} = \frac{3C^{*0.85}}{\varepsilon_f^*} \qquad (31)$$

with $\dot{a}$ in mm h^{-1}, ε_f^* as a fraction, and C^* in MJ m^{-2} h^{-1}. Application of this formula to the data presented previously, which indicated a constant creep ductility, is shown in Figures 7, 8 and 12–14. It is clear that reasonable agreement is achieved. Additional correlations are presented in Figures 16 and 17 for the materials which demonstrated a variable creep failure strain and in Figures 18–20 for a plain carbon steel, A 470 class 8 steel, and lead. The ductilities used in Equation (31) for making the predictions are listed in Table 2. Some flexibility in choice is possible for materials with a variable ductility. The values listed for these materials correspond with the middle of the experimentally measured range. Satisfactory agreement is again obtained.

Figure 19 demonstrates the relative insensitivity of cracking rate to temperature when characterized in terms of C^*. This behavior has been seen also for other materials (Riedel,

1987) and is to be expected from the form of Equations (16) and (31) unless there is a significant influence of temperature on the creep ductility term ε_f^*.

It is apparent from the correlations obtained with Equation (31) that this simple approximate expression can be used to provide satisfactory estimates of crack propagation rates in a broad range of materials. The chief parameters governing behavior at a given value of C^* are creep ductility and degree of constraint. For plane-stress conditions, Equation (31) can be written as

$$\dot{a}\varepsilon_f = 3C^{*0.85} \tag{32}$$

and for plane-strain conditions it becomes

$$\dot{a}\varepsilon_f = 150^{*0.85} \tag{33}$$

These expressions are shown plotted in Figure 21 with $\dot{a}\varepsilon_f$ as ordinate. The shaded area represents the spread of all the experimental results recorded here and other data presented previously (Nikbin *et al.*, 1984, 1986). It can be seen that the two equations approximately span the data. Equation (32) gives an approximate upper bound to plane-stress crack growth rates and Equation (33) a corresponding prediction

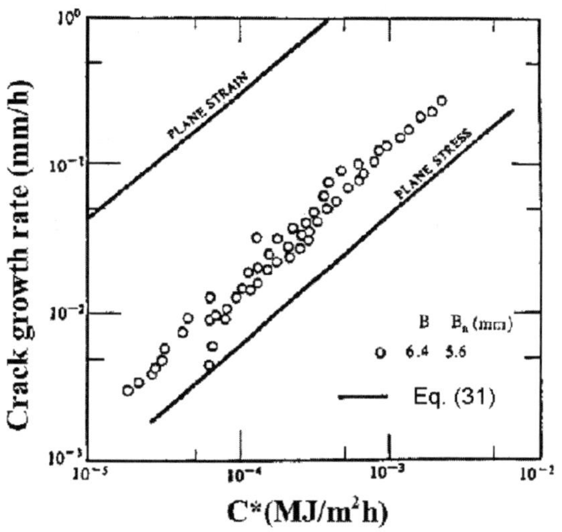

Figure 20 Creep crack growth properties determined on CT specimens of lead at 20°C (source Nikbin *et al.*, 1986).

Table 2 Creep ductilities used in universal crack growth law (Webster and Ainsworth, 1994).

Material	*Temperature* (°C)	*Ductility*
RR 58	150	0.05
0.5% CrMoV steel	565	0.05
0.16% C steel	500	0.38
A 470 class 8	538	0.18
A 470 class 8	482	0.18
Lead	20	0.20

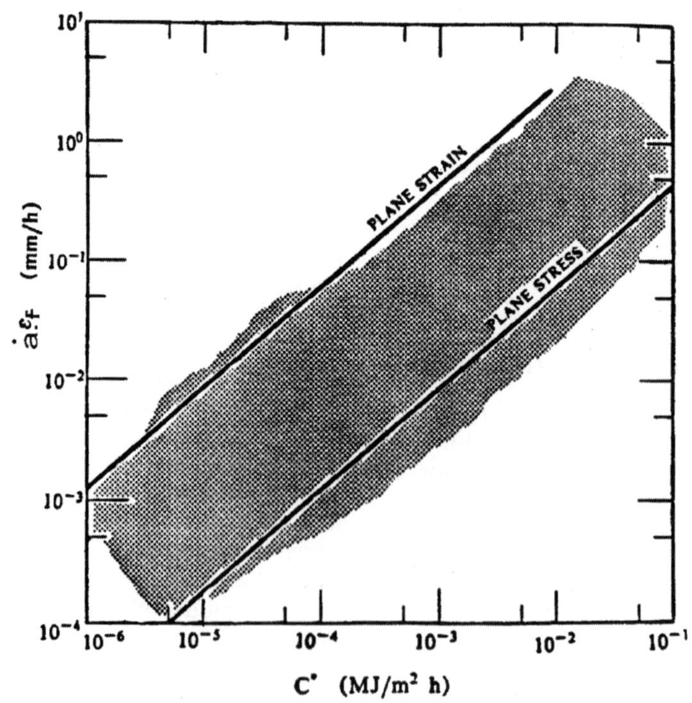

Figure 21 Material-independent engineering creep crack growth assessment diagram (source Webster and Ainsworth, 1994).

for plane-strain behavior. The experimental results which lie below the line representing Equation (32) were obtained on ductile materials under plane-stress conditions. These results can also be brought within the bounds if reduction in area is used as a measure of ε_f for these circumstances, as illustrated by the dashed line in Figure 19. However use of uniaxial failure strain, rather than reduction in area, introduces an element of conservatism into the estimates. Figure 21 has been plotted in such a way that it is independent of material properties. It is proposed that it can be used as a material-independent engineering creep crack growth assessment diagram to give approximate bounds to plane-stress and plane-strain behavior when more detailed information of the crack growth characteristics of materials is not available. It has been developed from physical models of the cracking process and is well supported by the wide spread of experimental results examined. Figure 21 can therefore be regarded as representing an approximate universal creep crack growth law.

It should be noted that Equations (15), (16), (29) and (30) and Figure 21 do not suggest that crack growth rates are 50 times greater under plane-strain conditions than when plane-stress exists for the same external applied loads. This is because C^* will be different in each case. Typically for $\phi = 0.85$ which corresponds to $n = 5.67$, C^* for plane-strain situations is expected to be ~ 0.5 times that for plane stress. When allowance is made for a change in the value of I_n from about ~ 3.4 for plane stress to 4.9 for plane strain, this gives cracking rates from Equations (15), (16), (29) or (30) for the same external applied loads of ~ 20 times greater for plane-strain conditions than for plane-stress situations (Webster and Ainsworth, 1994).

5.05.6.5 Crack Propagation into Damaged Material

So far no allowance has been made in the crack growth models for a crack propagating into damaged material. It has been assumed always that the crack grows into virgin material. With prolonged use, however, it may be expected that progressive material deterioration may take place in components. This situation will now be examined.

Consider Figure 22. This figure shows the local region at a crack tip where damage accumulates to produce crack growth and a region ahead where material deterioration takes place in the uncracked ligament. The influence of ligament damage is to cause less

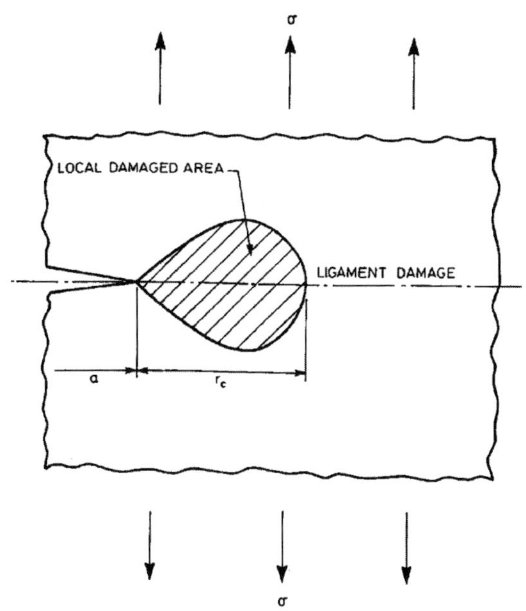

Figure 22 Regions of local and ligament damage ahead of a crack tip (source Webster and Ainsworth, 1994).

ductility to be available to be exhausted in the process zone at the crack tip. The amount of damage accumulated in the ligament can be determined by using the life fraction rule as in Section 5.05.6.2. With this approach the fraction of damage ω suffered in the ligament up to the present time is given by

$$\omega = \int_0^t \frac{dt}{t_r} \tag{34}$$

where t_r is the rupture life of the uncracked ligament. The remaining fraction of damage available to be used up in the local region at the crack tip is now only $(1-\omega)$. To allow for this deterioration, D_0 in Equation (6) must be replaced by a variable D given by

$$D = \frac{D_0}{1-\omega} \tag{35}$$

so that Equation (6) becomes

$$\dot{a} = \frac{D_0 C^{*\phi}}{1-\omega} \tag{36}$$

where D_0 and ϕ are obtained from any of the steady-state crack growth expressions derived in Sections 5.05.6.1 and 5.05.6.2. For example when rupture data are employed to determine crack growth rates, D from Equation (29) becomes

$$D = \left(\frac{1}{1-\omega}\right)\left(\frac{n+1}{n+1-\upsilon}\right)\frac{\dot{\varepsilon}_0}{\varepsilon^*_{f0}}\left(\frac{1}{I_n\sigma_0\dot{\varepsilon}_0}\right)^{\upsilon/(n+1)} \times r_c^{(n+1-\upsilon)/(n+1)} \tag{37}$$

Consequently, Equation (36) can be used to estimate creep crack growth involving ligament damage interaction. In evaluating this expression it is appropriate to use reference stress concepts to determine ω (Nishida *et al.*, 1989; Penny and Marriott, 1995). The definition of reference stress σ_{ref}, in terms of load P, is

$$\sigma_{\text{ref}} = \frac{P}{P_{\text{LC}}}\sigma_{\text{Y}} \tag{38}$$

where P_{LC} is the collapse load of the uncracked ligament for a material of yield stress σ_{Y}. Further discussion on how it is evaluated is given in Section 5.05.10. However, in general, it will be necessary to allow for a change in its value with crack extension when estimating t_{r} in Equation (34).

Nishida *et al.* (1989) and Webster and Ainsworth (1994) have examined the influence of ligament damage on creep crack growth in an infinite and a finite width CCP. An example of the results obtained for a finite width plate with $n = \upsilon = 5$, in Equation (37), is shown in Figure 23. The calculations were made for $r_c/W = 0.001$ which represents typical component dimensions and for different relative initial crack sizes a_0/W. The results have been normalized with respect to the rupture life t_{ru} of the uncracked plate. It is apparent that the inclusion of ligament deterioration only has a significant effect on lifetimes for plane-stress conditions. It can also be seen that the magnitude of the effect becomes more pronounced with decrease in initial crack size a_0 consistent with more time being available for ligament damage to take place. Crack growth is too rapid under plane-strain loading for appreciable deterioration to develop in the uncracked ligament.

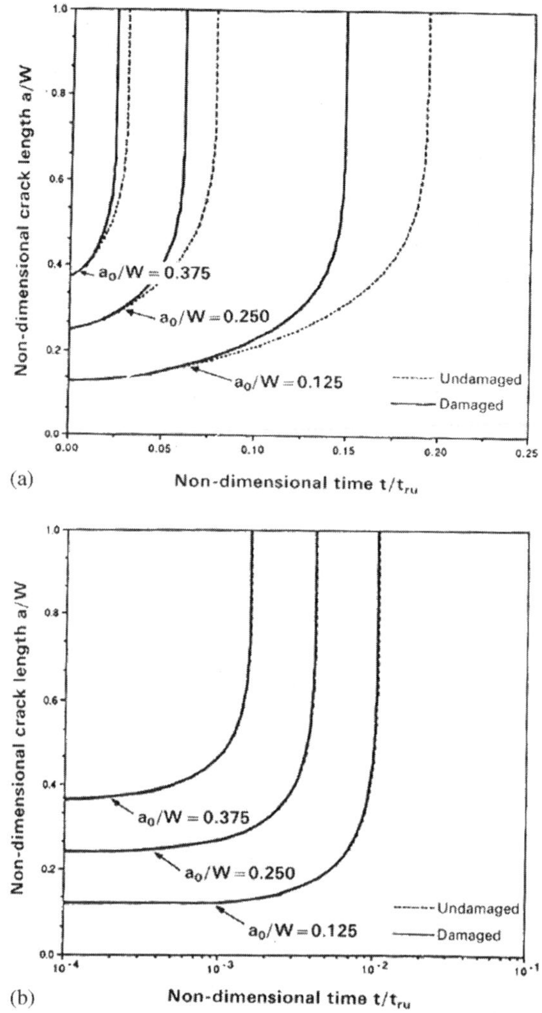

Figure 23 Effect of initial crack size and ligament damage on creep crack growth in finite width CCP for $n = \upsilon = 5$ for: (a) plane-stress and (b) plane-strain conditions (source Webster and Ainsworth, 1994).

5.05.6.6 Residual Life Assessment

Residual life assessments of components that have been in service for a period of time can involve the use of nondestructive inspection methods, reanalysis of the component based on actual operating experience, and/or removal of material for subsequent testing.

Removal of material is a destructive procedure but it does allow the remaining properties of service-exposed material to be measured and for these properties to be inserted into the appropriate expressions for determining crack growth rates. No change in the method of calculation is required except for the use of current material properties in the equations. Where sufficient material is available, creep crack growth tests can be performed on service-exposed material to obtain D_0 and ϕ.

A successful application of this approach to the prediction of creep crack growth in service-exposed type 340H stainless steel is shown in Figure 24. This figure indicates that a significant reduction in the resistance of the material to creep crack growth has occurred due to service exposure. This can be attributed, in this instance, to the aging which has taken place causing deterioration in the uniaxial creep strength and ductility of the material. However, when the actual creep properties of the stainless steel in the service-exposed condition are inserted in Equation (15), a satisfactory prediction of creep crack growth behavior is obtained.

When data on service-exposed material are not available and cannot be produced, remanent life assessments of plant must be based on virgin material properties and some estimate of the proportion of life that has already been

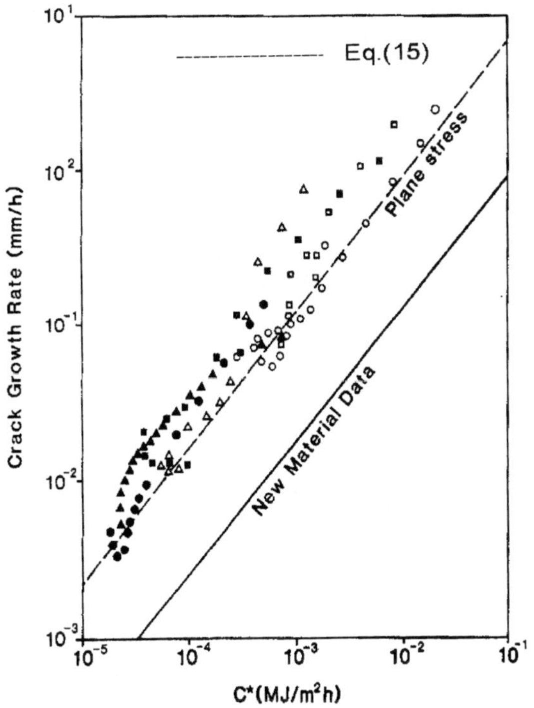

Figure 24 Influence of service exposure on creep crack propagation behavior of type 304H stainless steel at 760°C (source Buchheim *et al.*, 1989) (different symbols represent different tests).

used up (Smith and Webster, 1983). The accuracy with which predictions of lifetimes can be made with this method will depend on the reliability of the information that is available concerning the previous operating conditions of the plant and whether the material undergoes overaging effects.

An estimate of the creep crack propagation properties of service-exposed material can be obtained by adopting the approach described in the previous section for a crack growing into previously damaged material. With this method, the instantaneous cracking rate is determined from Equation (36) from a knowledge of the fraction of damage that has been incurred in the uncracked ligament ahead of the crack. For new plant this fraction ω is taken to be zero at the start of life and to gradually accumulate with use according to a ductility exhaustion criterion or Equation (34). When making a residual life assessment, account must be taken of the fact that some fraction of damage ω_e has already been incurred because of the previous operational history, so that from Equation (36) the relevant crack growth law for service-exposed material becomes

$$\dot{a} = \frac{D_0}{1-\omega_e} C^{*\phi} \tag{39}$$

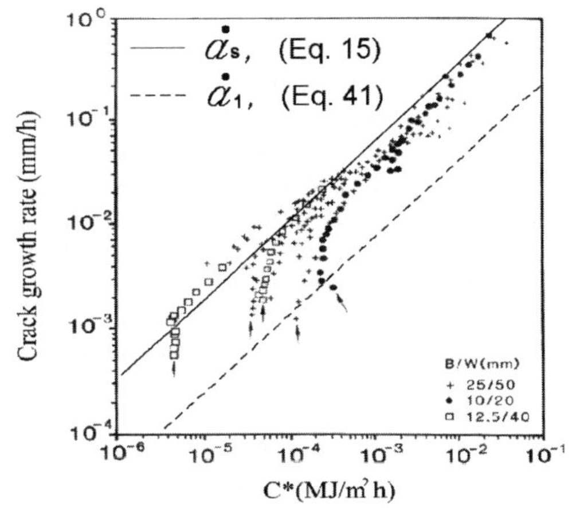

Figure 25 Creep crack growth in several sizes of CT specimen of 1% CrMoV steel at 550°C showing "tails" (source Webster, 1992b).

A residual life assessment, based on crack growth, can therefore be made by integrating the original Equation (36) over future anticipated operating conditions starting with $\omega = \omega_e$ instead of $\omega = 0$ for virgin material.

It should be borne in mind, when adopting this procedure, that no provision is included in Equation (39) for aging effects during service exposure which may alter the uniaxial creep properties of a material. It is assumed that these remain constant with use and that all that service exposure does is gradually exhaust the available damage fraction ω.

5.05.7 TRANSIENT CRACK GROWTH

During the early stages of crack growth, crack propagation rates are often less (Taira *et al.*, 1979; Webster, 1992a, 1992b) than predicted by the steady-state models discussed in the previous section. An example is shown in Figure 25 where the arrows indicate the beginning of each test. It is apparent that each test exhibits a "tail" and it is not until crack growth is well established that all the data superimpose onto one line. This early stage of cracking is referred to as the transient period of crack growth. During this period, cracking rates from different tests that are subjected to different loading conditions do not superimpose. The steady-state analysis developed earlier will now be extended to predict crack growth during this transient period (Nishida and Webster, 1990; Webster, 1992a, 1992b).

The models considered so far have postulated a process zone at the crack tip in which a steady-state distribution of creep damage has accumulated. Under this circumstance a gradient of damage exists in the process zone. Material which has just entered the process zone will have experienced little damage whereas material near the crack tip will have almost exhausted its available ductility. However, this situation does not exist immediately after loading as no damage will have had time to accumulate anywhere at the crack tip.

The influence of a buildup of damage in the process zone at a crack tip on crack growth rate can be examined by considering Figure 26. Once a steady-state distribution of damage has been formed in this region, little extra strain is required to break an element dr ahead of the crack tip since it will be almost broken before the crack reaches it. This situation will not exist in element 1 on first loading as it will not have suffered any creep strain at this stage. Under this circumstance, the time taken dt for element 1 to fail is obtained by substituting Equation (25) into Equation (8). Assuming that the first term dominates in Equation (8), and taking the maximum equivalent value of $\tilde{\sigma}_{ij}(\theta, n)$ as unity, gives

$$\mathrm{d}t = \frac{\varepsilon_{f0}^*}{\dot{\varepsilon}_0}\left(\frac{I_n \sigma_0 \dot{\varepsilon}_0 \,\mathrm{d}r}{C^*}\right)^{\upsilon/(n+1)} \tag{40}$$

where ε_{f0}^* has been introduced to allow for the effects of state of stress on creep failure strain and t_r has been replaced by dt. After failure the crack will have advanced dr, so that Equation (40) can be used to calculate an initial crack growth rate

$$\dot{a}_1 = \frac{\mathrm{d}r}{\mathrm{d}t} = \frac{\dot{\varepsilon}_0}{\varepsilon_{f0}^*}\left(\frac{C^*}{I_n \sigma_0 \dot{\varepsilon}_0}\right)^{\upsilon/(n+1)} (\mathrm{d}r)^{(n+1-\upsilon)/(n+1)} \tag{41}$$

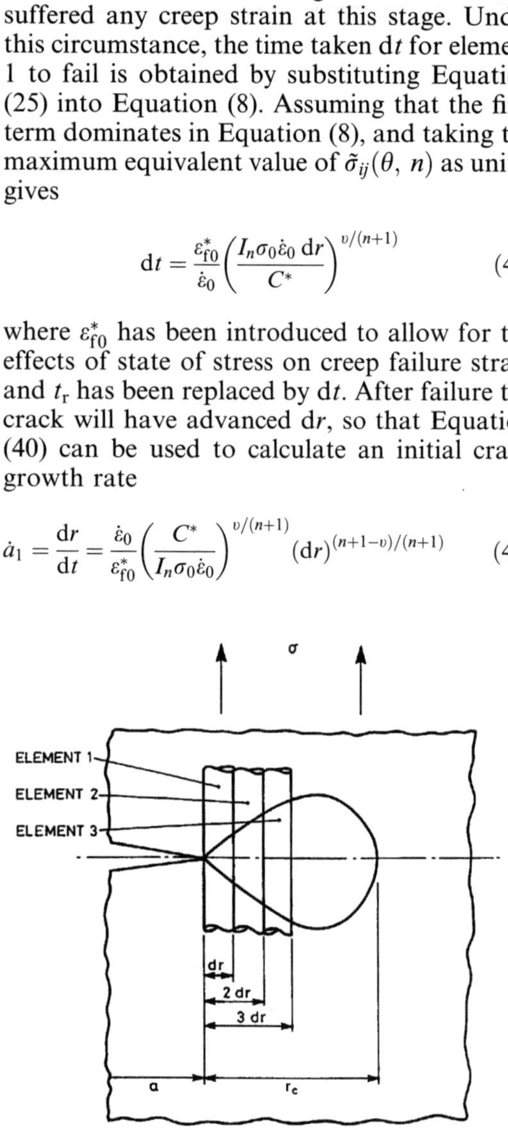

Figure 26 Elements in creep process zone in which damage builds up initially (source Nishida and Webster, 1990).

which can be expressed from the steady-state crack growth Equation (29) as

$$\dot{a}_1 = \left(\frac{n+1-\upsilon}{n+1}\right)\left(\frac{\mathrm{d}r}{r_c}\right)^{(n+1-\upsilon)/(n+1)} \dot{a}_s \tag{42}$$

where $\dot{a}_s$ is the crack propagation rate when a steady-state distribution of damage has been reached at the crack tip. The element dr can be chosen to be a suitable fraction of r_c. However, since $n \gg 1$ *and* $n \approx \upsilon$, dr/r_c will be raised to a small fractional power so that

$$\dot{a}_1 \approx \frac{n+1-\upsilon}{n+1}\dot{a}_s \tag{43}$$

For a constant failure ductility when $n = \upsilon$, Equation (43) reduces to

$$\dot{a}_1 \approx \left(\frac{1}{n+1}\right)\dot{a}_s \tag{44}$$

Therefore, for most materials the initial crack growth rate will be expected to be approximately an order of magnitude less than that predicted from a steady-state analysis. This is consistent with most observations as illustrated in Figure 25 for a low alloy steel.

With each crack advance dr, each successive element in Figure 26 will progressively accumulate more damage prior to fracture. For example after element 1 has failed, element 2 distance 2dr from the original crack tip will have accumulated strain and used up some fraction of its life. This can be defined in terms of the life fraction or ductility exhaustion criteria. Taking the life fraction rule, Equation (34), and defining the fraction of life used up in element 2 after element 1 has broken as ω_2, it can be shown by substituting 2dr into Equation (40) that

$$\omega_2 = \left(\frac{1}{2}\right)^{\upsilon/(n+1)} \tag{45}$$

Similarly, after element 2 has broken, element 3 will have used up a life fraction:

$$\omega_3 = \left(\frac{1}{3}\right)^{\upsilon/(n+1)} + \left(\frac{1}{2}\right)^{\upsilon/(n+1)} - \left(\frac{1}{2}\right)^{2\upsilon/(n+1)} \tag{46}$$

and so on for successive elements. Consequently, for the ith element when damage ω_i has been incurred in it, the crack growth rate $\dot{a}_i$ will be

$$\dot{a}_i = \left(\frac{1}{1-\omega_i}\right)\frac{\dot{\varepsilon}_0}{\varepsilon_{f0}^*}\left(\frac{C^*}{I_n \sigma_0 \dot{\varepsilon}_0}\right)^{\upsilon(n+1)} (\mathrm{d}r)^{(n+1-\upsilon)/(n+1)} \tag{47}$$

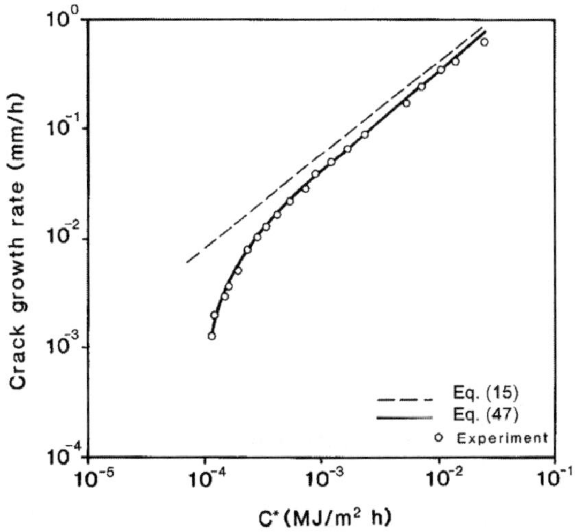

Figure 27 Prediction of transient crack growth in "tail" region of test taken from Figure 25 (source Webster and Ainsworth, 1994).

where

$$\omega_i = \int \frac{\mathrm{d}t}{t_\mathrm{r}} = \left(\frac{1}{i}\right)^{\upsilon/(n+1)} + (1-\omega_2)\left(\frac{1}{i-1}\right)^{\upsilon/(n+1)} + \cdots + (1-\omega_{i-1})\left(\frac{1}{2}\right)^{\upsilon/(n+1)} \quad (48)$$

for $i > 1$. In general, in evaluating Equation (47) it will be necessary to allow for an increase in C^* with crack advance.

An illustration of the application of this analysis to the prediction of crack growth in one of the tests presented in Figure 25 is shown in Figure 27. Similar correlations have been achieved for other materials (Webster, 1992b). The procedure is relatively insensitive to the choice of magnitude for dr provided it is kept less than r_c.

It is apparent that this procedure can be employed to provide a satisfactory description of the early stages of cracking. It is most relevant to materials which deform predominantly in secondary creep and to situations where crack growth does not occur until stress redistribution is complete and can be described in terms of C^*.

Other shapes of "tail" to those seen in Figures 25 and 27 can be observed as indicated diagrammatically in Figure 28. These correspond approximately to the creep crack growth curve C shown in Figure 3. Several causes are possible. Curve I is most likely to be exhibited by a material which starts to crack before stress redistribution is complete at the crack tip. In this case crack growth rate should be plotted against C_t or $C(t)$ (Saxena, 1998). Since both these terms are greater than C^* prior to complete stress redistribution, plotting the results against either of these parameters will cause the "tail" to move to the right with a tendency for it to superimpose on the steady-state portion. The reason why the cracking rate decreases initially is because it is occurring in a region at the crack tip where stress relaxation is taking place during the redistribution process and C_t and $C(t)$ are reducing.

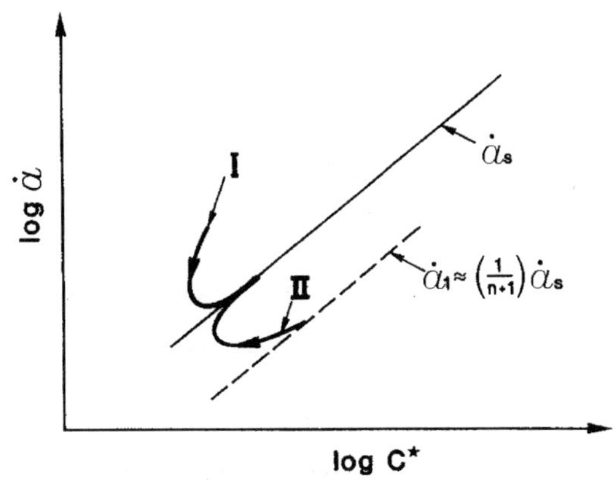

Figure 28 Alternative shapes of "tail" that have been observed experimentally (source Webster and Ainsworth, 1994).

A curve like II in Figure 28 can be attributed to primary creep effects whilst damage builds up at the crack tip. In this case the transient analysis presented in this section is still valid provided stress redistribution at the crack tip is complete and $t > t_\mathrm{T}$ in Equation (7). During primary creep, C^* initially decreases whilst creep rate is decreasing and the cracking rate gradually approaches that for a steady-state distribution of damage. This leads to the hook in the "tail." However, the initial crack growth rate will still be given by Equation (41) and then subsequently by Equation (47) so that a curve starting on the line representing the initial cracking rate $\dot{a}_1$, and finishing on the line representing the steady-state crack propagation rate $\dot{a}_\mathrm{s}$, will be obtained as indicated in Figures 28 and 29.

5.05.8 INCUBATION PERIOD

Materials accumulate damage at different rates depending on their creep properties. With ductile materials an incubation period, prior to the onset of crack extension, may exist whilst damage develops at the crack tip. An example of such a situation is shown in Figure 30 where the incubation period occupies ~40% of life (Holdworth, 1994). Also, prior to crack extension, crack blunting may occur. To some extent, the duration of an incubation period will depend on the sensitivity of the equipment used to detect crack growth. This may be

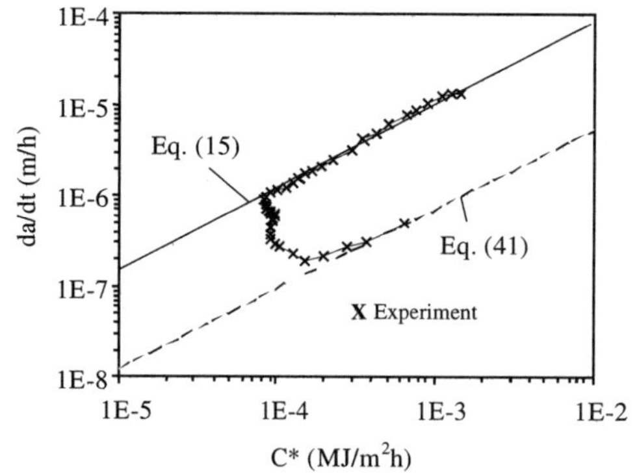

Figure 29 "Tail" in type 316H stainless steel exhibiting primary creep at 550°C (source Bettinson, 2002).

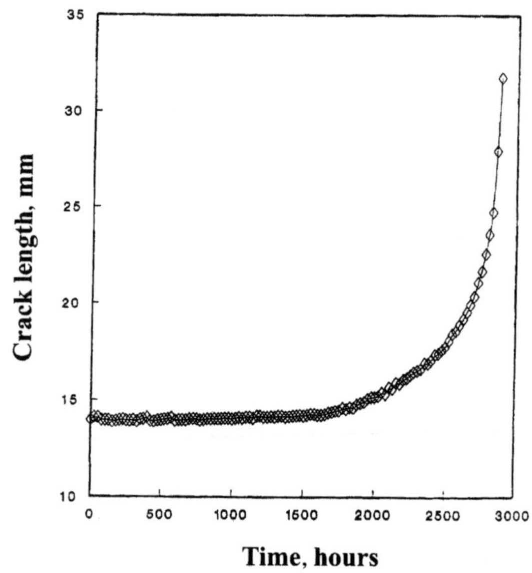

Figure 30 Example of an incubation period prior to the onset of creep crack growth in an SENT specimen of 2.25% CrMo steel weld metal at 565°C (source Holdworth, 1994).

different in laboratory specimens to components. For fracture assessments of the onset of tearing, an engineering definition of crack initiation is often taken as 0.2 mm of crack extension (ASTM 1820-99a, 2000) and this criterion can also be used for determining the onset of creep crack growth (ASTM E1457, 2001). Two methods of estimating incubation periods will now be considered.

5.05.8.1 Prediction Based on Damage Accumulation

The same concept can be employed (Austin and Webster, 1992) to calculate an incubation period t_i prior to the onset of cracking as was used to determine transient crack growth rates in Section 5.05.7. With this approach, if r_d is the minimum extent of cracking that can be detected, then

$$t_i = \int_0^{r_d} \frac{dr}{\dot{a}} \tag{49}$$

where $\dot{a}$ will increase from its initial value $\dot{a}_1$ to $\dot{a}_s$ with crack extension. Provided r_d is sufficiently small, C^* can be taken as approximately constant with crack growth so that a lower bound estimate of t_i can be obtained by substituting $\dot{a} = \dot{a}_s$ from Equation (29) to give

$$t_i = \frac{r_d}{\dot{a}_s} = \left(\frac{n+1-\upsilon}{n+1}\right)\frac{\varepsilon_{f0}^*}{\dot{\varepsilon}_0}\left(\frac{I_n\sigma_0\dot{\varepsilon}_0 r_d}{C^*}\right)^{\upsilon/(n+1)} \tag{50a}$$

which can be re-expressed using Equation (30) as

$$t_i = \left(\frac{n+1-\upsilon}{n+1}\right)\frac{\varepsilon_{f0}^*}{\varepsilon_{f0}}B_r\left(\frac{I_n C_A r_d}{C^*}\right)^{\upsilon/(n+1)} \tag{50b}$$

Some indication of an upper bound can be determined by taking $\dot{a} = \dot{a}_1$ so that from Equation (43), t_i becomes

$$t_i = \frac{\varepsilon_{f0}^*}{\dot{\varepsilon}_0}\left(\frac{I_n\sigma_0\dot{\varepsilon}_0 r_d}{C^*}\right)^{\upsilon/(n+1)} \tag{51a}$$

$$t_i = \frac{\varepsilon_{f0}^*}{\varepsilon_{f0}}\cdot B_r\left(\frac{I_n C_A r_d}{C^*}\right)^{\upsilon/(n+1)} \tag{51b}$$

When creep failure strain is constant, $n = \upsilon$ and $B_r = \varepsilon_f/C_A$ so that some simplification of Equations (50) and (51) is possible. For example, Equation (50b) becomes

$$t_i = \left(\frac{1}{n+1}\right)\varepsilon_f^* r_d\left(\frac{I_n}{C^*}\right)^{n/(n+1)}\left(\frac{1}{C_A r_d}\right)^{1/(n+1)} \tag{51c}$$

This equation demonstrates that at constant C^* the incubation period is almost insensitive to C_A and therefore temperature for a given value of crack extension resolution r_d.

Alternative estimates of an incubation period can also be obtained from the approximate universal creep crack growth law (Equation (31)). When this law is employed, the lower bound becomes

$$t_i = \frac{r_d\varepsilon_f^*}{3C^{*0.85}} \tag{52}$$

and the upper bound

$$t_i = \frac{2.2 r_d\varepsilon_f^*}{C^{*0.85}} \tag{53}$$

where r_d is in mm and t_i in h. An illustration of the application of Equations (50)–(53) to

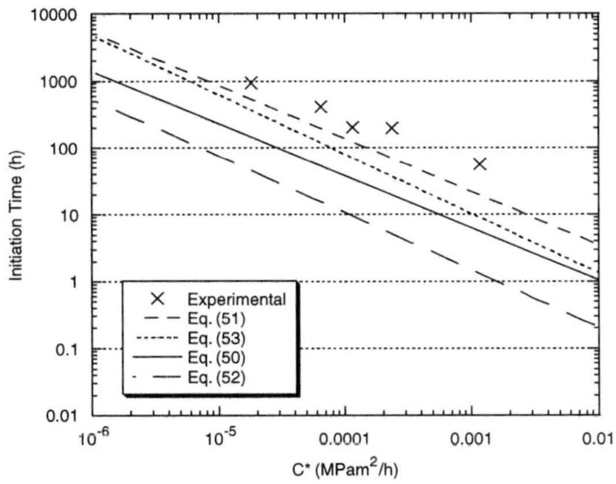

Figure 31 Comparison between predicted and experimental incubation periods for 1% CrMoV steel at 550°C (source Austin and Webster, 1992).

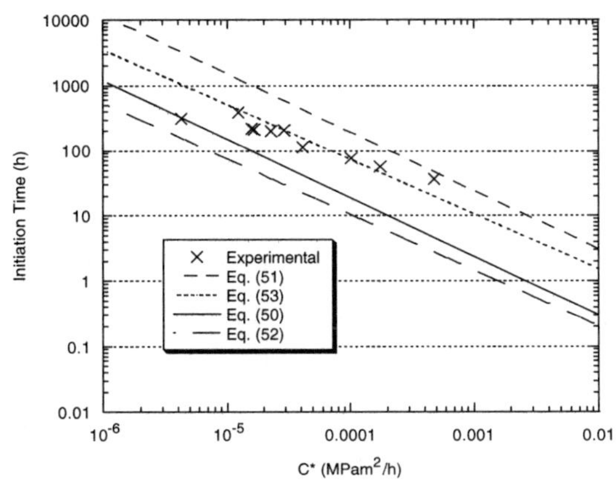

Figure 32 Comparison between predicted and experimental incubation periods of type 316L stainless steel at 600°C (source Webster, 1994).

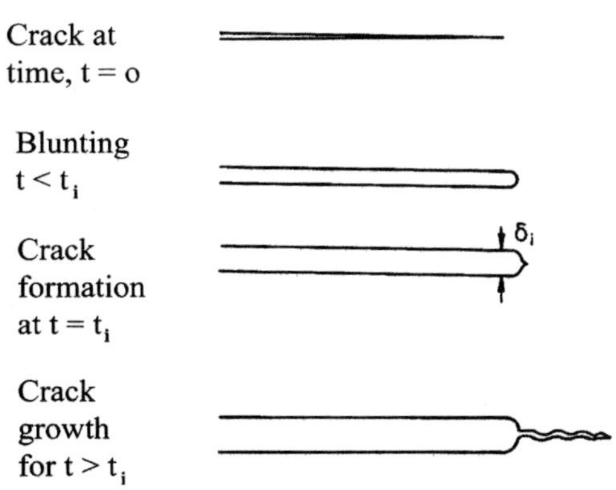

Figure 33 Development of crack blunting and crack initiation (source Webster and Ainsworth, 1994).

experimental results on a low alloy steel and a stainless steel (Piques *et al.*, 1991) are shown in Figures 31 and 32, respectively. The calculations were made for $r_d = 0.05\,\text{mm}$ to correspond approximately with the material grain sizes. It is evident that the predictions of incubation periods based on upper- and lower-bound estimates span the data or provide conservative determinations. Alternatively, choice of lower-bound estimates derived from a value of $r_d = 0.2\,\text{mm}$, to correspond with the definition of fracture initiation in conventional fracture toughness tests, would also have given satisfactory predictions. It is apparent, therefore, that either criterion can be adopted to provide satisfactory estimates of incubation periods.

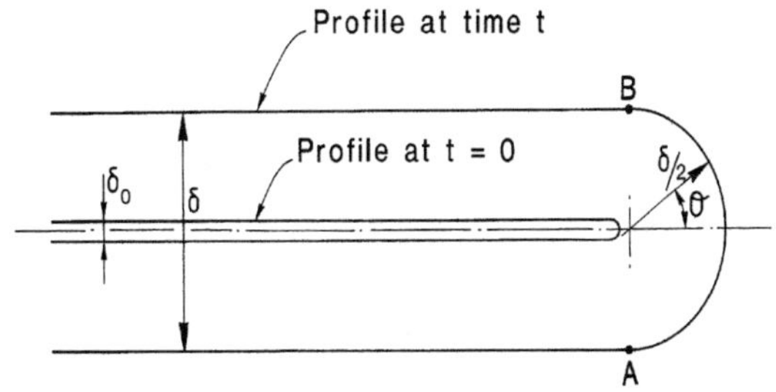

Figure 34 Model of the blunting process (source Webster and Ainsworth, 1994).

5.05.8.2 Determination from Crack Blunting

In ductile material substantial crack-tip blunting may accompany creep crack growth. The mechanism by which this takes place is shown in Figure 33. The blunting continues until the ductility of the material ahead of the blunted crack is exhausted and further creep crack growth takes place. It is argued that this occurs when a critical crack opening displacement δ_i is attained (Ainsworth, 1982, 1986).

The process of crack blunting leads to a notch profile which may be smooth or may have sharp corners depending on the local microstructure. However, the blunted shape may be represented with sufficient accuracy as a semi-circle of radius equal to one-half the crack opening displacement δ as depicted in Figure 34. The rate of crack opening $\dot{\delta}$ may be evaluated in terms of the creep fracture mechanics parameter $C(t)$ discussed earlier. By taking the integration path Γ along the notch surface from point A to point B in Figure 34, it may be shown that (Webster and Ainsworth, 1994)

$$C(t) = \sigma_0 \varepsilon_0^{-1/n} \delta \left(\frac{\dot{\delta}}{\delta} \right)^{1+1/n} g(n) \qquad (54)$$

where $g(n)$ is a known function of n. Integration of this equation from its initial crack opening displacement δ_0 at time $t = 0$ to δ_i gives the incubation period.

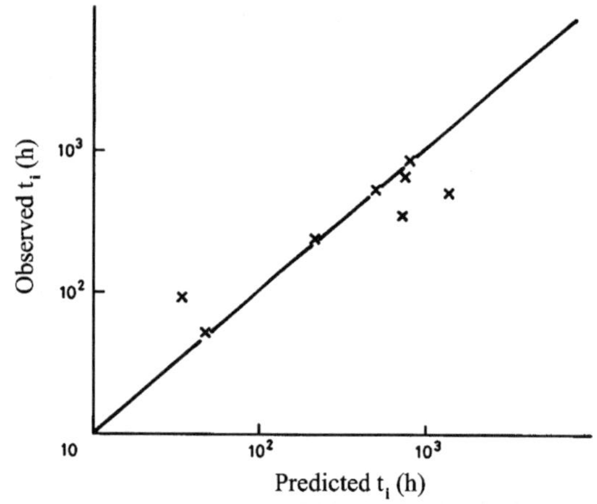

Figure 35 Comparison between observed and predicted incubation periods for 0.5% CrMoV pipe steel at 565°C (source Neate, 1986).

For ductile materials normally steady-state creep conditions are established early in an incubation period. Consequently, $C(t)$ can be replaced by C^* in Equation (54). Similarly δ_0 can usually be neglected so that integration of the expression produces

$$t_i \approx \frac{1}{\dot{\varepsilon}_0}\left(\frac{\sigma_0\dot{\varepsilon}_0\delta_i}{C^*}\right)^{n/(n+1)} = \left(\frac{1}{C}\right)^{1/(n+1)}\left(\frac{\delta_i}{C^*}\right)^{n/(n+1)} \tag{55}$$

when allowance is made for $g(n)$ and secondary creep behavior is assumed. If primary creep is significant, C^* will be a function of time in the integration to obtain t_i.

For a constant critical crack opening displacement, the form of Equation (55) is such that it is relatively independent of temperature at a constant value of C^*. This is because the temperature-dependent term C, in the uniaxial creep law (Equation (3)), in it is raised to a small fractional power. This is consistent with the predictions of the temperature dependence of the incubation period obtained in Section 5.05.8.1 and of creep crack growth in Section 5.05.6.

The validity of using the critical crack opening displacement approach for determining incubation periods can be demonstrated by making comparisons with experimental data (Taylor and Batte, 1980; Neate, 1986). From experiments, the initiation of cracking can be obtained metallographically or from electrical potential measurements. An example of a comparison is shown in Figure 35 which indicates satisfactory agreement.

5.05.9 METHODS OF ESTIMATING C^*

Estimates of C^* can be obtained by numerical, limit analysis, and experimental methods.

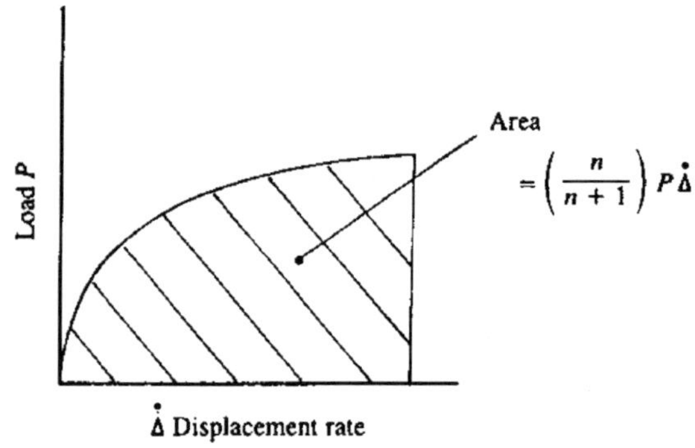

Figure 36 Load–creep displacement rate diagram for a specimen containing a crack of length a (source Webster, 1994).

The experimental procedure is most applicable to laboratory specimens that are used for measuring the creep crack growth properties of materials. The other two methods are needed when calculating values of C^* for components.

5.05.9.1 Experimental

When a steady-state creep stress distribution has been achieved ahead of a crack, C^* can be obtained by an analogous procedure to that recommended in ASTM E1820-99a (2000) for measuring the contour integral J in fracture toughness testing. The method is given in ASTM E1457 (2001). It makes use of Figure 36, which shows a load-creep displacement rate diagram for a test piece containing a crack of length a. With this approach, C^* is given from the shaded area as

$$C^* = \frac{(\text{Area})\eta}{B_n b} \tag{56}$$

where b is the uncracked ligament, η depends on the geometry of the cracked specimen, and B_n is the net thickness between SGs when these are employed to produce a straight-fronted crack. For a specimen subjected to load P and experiencing a creep displacement rate $\dot{\Delta}$, it can be shown (Webster, 1992a) that when the material deforms according to Equation (3), Equation (56) can be expressed as

$$C^* = \frac{P\dot{\Delta}}{B_n b}\left(\frac{n}{n+1}\right)\eta \tag{57}$$

From ASTM E1820-99a (2000), ASTM E1457 (2001), and limit analysis (Nikbin *et al.*, 1986) for CT specimens of width W, $\eta = 2 + 0.52b/W$, for bend specimens $\eta = 2.0$ and for tension specimens $\eta = 1.0$.

Although Equation (57) has been derived for secondary creep, it is applicable to primary and tertiary creep as well, provided these stages can also be characterized by the same stress sensitivity n. Equation (57) has the attraction of allowing accurate estimates of C^* to be obtained without detailed knowledge of a material's creep properties, so long as $\dot{\Delta}$ is measured during an experiment.

5.05.9.2 Numerical

Numerical solutions for C^* are only available for a restricted range of crack geometries and loading conditions. For an infinite width CCP subjected to a remote tensile stress σ, and containing a crack of length $2a$, C^* is given for a material deforming according to Equation (3) by He and Hutchinson (1983) as

$$C^* = \pi\sqrt{n}\sigma_0\dot{\varepsilon}_0\left(\frac{\sigma_e}{\sigma_0}\right)^{n+1} a \tag{58}$$

where $\sigma_e = \sigma$ for planes-stress conditions and $(\sqrt{3}/2)\sigma$ for plane-strain loading. In general, C^* can be written as

$$C^* = h\sigma\dot{\varepsilon}a \tag{59}$$

where h is a nondimensional function of crack size a, component dimensions, state of stress, and n. Also, as $\dot{\varepsilon}$ is instantaneous creep strain rate, Equation (59) can be employed to incorporate primary, secondary, and tertiary creep when calculating C^*. When secondary creep prevails, substitution of Equation (3) into Equation (57) gives

$$C^* = h\sigma_0\dot{\varepsilon}_0\left(\frac{\sigma}{\sigma_0}\right)^{n+1} a = hC\sigma^{n+1}a \tag{60}$$

Estimates of h can be obtained from solutions for J for power-law hardening materials which deform plastically in uniaxial tension according to the expression

$$\varepsilon = \varepsilon_Y\left(\frac{\sigma}{\sigma_Y}\right)^N = A\sigma^N \tag{61}$$

where N is a constant and σ_Y and ε_Y are yield stress and yield strain, respectively. This equation allows J to be written as

$$J = h\sigma_Y\varepsilon_Y\left(\frac{\sigma}{\sigma_Y}\right)^{N+1} a = hA\sigma^{N+1}a \tag{62}$$

The definition for C^* in Equations (4) and (5) is exactly the same as that for J except that creep strain rates and displacement rates replace plastic strains and displacements. Consequently, from the similarity of Equations (3) and (61) when $N = n$ the same stress distribution is obtained ahead of crack in a creeping material as in a work-hardening material and h will be the same in both Equations (60) and (62). Consequently C^* can be determined from J using

$$C^* = \frac{\dot{\varepsilon}_0}{\varepsilon_Y}\left(\frac{\sigma_Y}{\sigma_0}\right)^n J = \left(\frac{C}{A}\right)J \tag{63}$$

Equation (63) enables estimates of C^* to be obtained when solutions for J only are available (Kumar *et al.*, 1981; Saxena, 1998).

5.05.9.3 Limit Analysis

When solutions for C^* are not available by other means, limit analysis methods can be employed to produce approximate estimates. With this approach a reference stress σ_{ref} is defined to characterize the stress distribution in the uncracked ligament of a cracked body as given by Equation (38). Since σ_{ref} is proportional to the applied stress σ, it is possible to rewrite Equations (1) and (59), respectively, as

$$K^2 = Y_{ref}^2\sigma_{ref}^2 a \tag{64}$$

and

$$C^* = h_{ref}\sigma_{ref}\dot{\varepsilon}_{ref}a \tag{65}$$

where $\dot{\varepsilon}_{ref}$ is the creep strain rate at σ_{ref} and Y_{ref} and h_{ref} are appropriate nondimensional functions to replace Y and h. When Equation (65) is used to estimate C^*, it has been found (Miller and Ainsworth, 1989), for a wide range of crack geometries and loading conditions, that h_{ref} is almost independent of n. Consequently it can be obtained from solutions with $n = 1$. This is equivalent to an elastic solution for which

$$J = G = K^2/E' \tag{66}$$

where G is elastic strain energy release rate and E' is the elastic modulus E for plane stress and $E/(1-v^2)$ for plane strain with v as Poisson's ratio. By comparing Equations (64)–(66), it is evident that for h_{ref} to be independent of n it must satisfy the relation

$$h_{ref} = Y_{ref}^2 \tag{67}$$

so that substitution into Equation (65) gives

$$C^* = \sigma_{ref}\dot{\varepsilon}_{ref}\left(\frac{K}{\sigma_{ref}}\right)^2 \tag{68}$$

Since solutions of K and σ_{ref} are available for a wide range of cracked components (Rooke and Cartwright, 1976; Murakami, 1987; Tada *et al.*, 2000; Miller, 1988), Equation (68) is a convenient expression for estimating C^* when other solutions are not available. It is also applicable to all stages of creep and is not restricted to secondary creep behavior.

When stress rupture information only is available, some modification to Equation (68) is required. In this case, $\dot{\varepsilon}_{ref}$ is replaced by $\varepsilon_f/t_{r(ref)}$, where $t_{r(ref)}$ is the rupture life at σ_{ref}, so that

$$C^* = \frac{\sigma_{ref}\varepsilon_f}{t_{r(ref)}}\left(\frac{K}{\sigma_{ref}}\right)^2 \tag{69}$$

Equation (69) is, therefore, an alternative expression to Equation (68) for calculating C^* in terms of K and σ_{ref}. When $\dot{\varepsilon}_{ref}$ can be described by Equation (3) or Equation (24), Equations (68) and (69) can be rewritten, respectively, as

$$C^* = CK^2\sigma_{ref}^{(n-1)} \tag{70}$$

and

$$C^* = C_A K^2\sigma_{ref}^{(n-1)} \tag{71}$$

The use of Equation (69) requires knowledge of ε_f which may not always be available when creep rupture data are provided. In this circumstance, some guide to cracking rate can be obtained by substituting Equation (69) into the universal creep crack growth law (Equation (31)) to give

$$\dot{a} = \frac{3}{\varepsilon_f^*}\left[\frac{\varepsilon_f K^2}{\sigma_{ref} t_{r(ref)}}\right]^{0.85} \tag{72}$$

for $\dot{a}$ in mm h^{-1}, K in MPa m$^{1/2}$, σ_{ref} in MPa, and $t_{r(ref)}$ in h.

For plane-stress conditions, this equation reduces to

$$\dot{a} = \frac{3}{\varepsilon_f^{0.15}}\left[\frac{K^2}{\sigma_{ref} t_{r(ref)}}\right]^{0.85} \tag{73}$$

This expression is relatively insensitive to failure ductility and a conservative prediction of crack growth rate for most circumstances can be obtained by taking $\varepsilon_f \approx 0.03$ to give

$$\dot{a} = 5\left[\frac{K^2}{\sigma_{ref} t_{r(ref)}}\right]^{0.85} \tag{74}$$

The corresponding equation for plane strain becomes

$$\dot{a} = 250\left[\frac{K^2}{\sigma_{ref} t_{r(ref)}}\right]^{0.85} \tag{75}$$

A similar approach can be applied for calculating incubation periods from Equations (52) and (53). Again, taking $\varepsilon_f \approx 0.03$ with $\varepsilon_f^* \approx \varepsilon_f/15$ for reasonably constrained conditions and $r_d = 0.2$ mm, for consistency with ASTM E1457 (2001), gives a lower-bound estimate of t_i from Equation (52) of

$$t_i = 0.0025\left[\frac{\sigma_{ref} t_{r(ref)}}{K^2}\right]^{0.85} \tag{76}$$

Equations (74)–(76) are alternative forms to Equations (32), (33) and (52) when creep failure strain is not known. These equations have been incorporated into the British Standards defect assessment procedures BSPD 6539 (1994) and BS 7910 (1999) for estimating creep crack growth and incubation periods in components when experimental crack propagation, incubation period, and creep failure strain data are not available.

5.05.10 PRACTICAL APPLICATIONS

Several models have been presented for describing creep crack growth. They all result in an expression of the form of Equation (6) where the terms D_0 and ϕ can be determined experimentally or calculated from uniaxial creep strain or stress rupture data. It is likely that the most accurate predictions of cracking in components will be obtained when experimental values of D_0 and ϕ are available. This is provided that these values have been determined on specimens that experience the same degree of constraint local to the crack tip as exists in the component. In this case, the use of experimental determinations of D_0 avoids having to establish an accurate estimate of multiaxial creep ductility for insertion into the crack growth laws based on uniaxial creep data.

Another requirement for obtaining reliable predictions of creep crack propagation is to have accurate determinations of C^*. Although the widest range of solutions for C^* is available from limit analysis methods, different formulas for reference stress may be quoted in different design and defect assessment codes (see, e.g., AFCEN, 1985; BS 7910, 1999; R5, 1999; R6, 2000). Consequently, although the procedures in the codes are based on the same principles, different predictions of cracking rate can be obtained, dependent on choice of expression for σ_{ref}. This choice can be avoided by using numerical finite element estimates of C^*, but again the reliability of the calculations depends on how accurately the materials data are replicated in the computations.

Examples of applying the different procedures to some practical situations will now be considered.

5.05.10.1 Pressure Vessel

Experimental data have been reported (Coleman *et al.*, 1977) for the growth of a circumferential crack of initial depth 30 mm in a normalized and tempered 0.5% CrMoV steel which was tested under a constant internal pressure of 62.5 MPa at a temperature of 565°C. The cylinder had a wall thickness of 60 mm. The results are shown in Figure 37 together with several predictions. In all cases the predictions were based on experimental determinations of D_0 and ϕ for a crack growing into a progressively deteriorating ligament. In addition, C^* was estimated from finite element calculations (Nishida and Webster, 1990). In the figure, lines 1 and 3 assume a steady-state distribution of damage in the process zone at the crack tip; lines 2 and 4 allow for a buildup of damage in this region to give a transient period of cracking. It is apparent that the buildup of damage during the transient stage results in a slower initial cracking rate with an effective incubation period of ~500 h.

Lines 1 and 2 were calculated on the assumption of secondary creep behavior. Lines 2 and 4 show the influence of allowing for tertiary creep with a creep strain rate law of the form

$$\dot{\varepsilon} = \dot{\varepsilon}_s \left(1 - \frac{t}{t_r}\right)^{(1/\lambda)-1} \tag{77}$$

where $\dot{\varepsilon}_s$ is the initial minimum (secondary) creep strain rate given by Equation (3) and $\lambda = \dot{\varepsilon}_A/\dot{\varepsilon}_s$. It is evident that the inclusion of tertiary creep has very little effect on failure times. This is because of the acceleration in cracking caused by the accumulation of damage in the uncracked ligament as failure is approached. The most accurate prediction is achieved when transient cracking, ligament damage and tertiary creep are all modeled.

5.05.10.2 Plate

In this example, the case of an austenitic stainless steel plate subjected to combined tension and bending loading and containing a semi-elliptical surface defect is examined. The overall dimensions of the plate and the loading conditions imposed are shown in Figure 38. Further details of how the testing was carried out are given in Poussard and Moulin (1998). The loading involved creep-fatigue cycling. A full analysis of the results, using a cumulative damage creep-fatigue crack growth law (Winstone *et al.*, 1985), is included in Webster *et al.* (1998). Here only the creep component of cracking is considered to examine the significance of using different formulas for reference stress taken from the codes AFCEN, Appendix 16 (1985), BS 7910 (1999), and R6 (2000). A significant variation in this term is found as indicated in Table 3. This is because both R6 and BS 7910 base σ_{ref} on "local" collapse of the uncracked ligament ahead of the crack, whereas A16 (AFCEN Appendix 16) employs "global" collapse of the entire cracked cross-section. Analysis (Goodall and Webster, 2001) and finite element calculations (Lei, 2001) are tending to the view that basing σ_{ref} on a "global" collapse mechanism is more realistic for a plate containing a semi-elliptical surface defect that is subjected to combined tension and bending loading.

The codes do not consider a transient period of cracking involving a buildup of damage at a

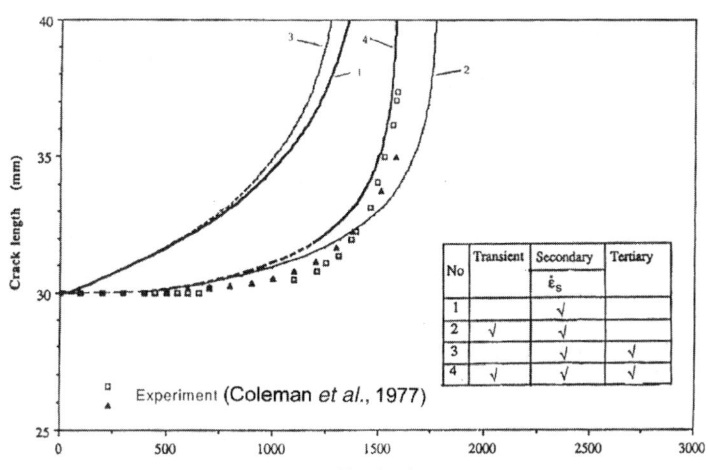

Figure 37 Influence of transient cracking and tertiary creep on predicted crack growth in 0.5% CrMoV steel pressure vessel undergoing ligament damage (source Nishida and Webster, 1990).

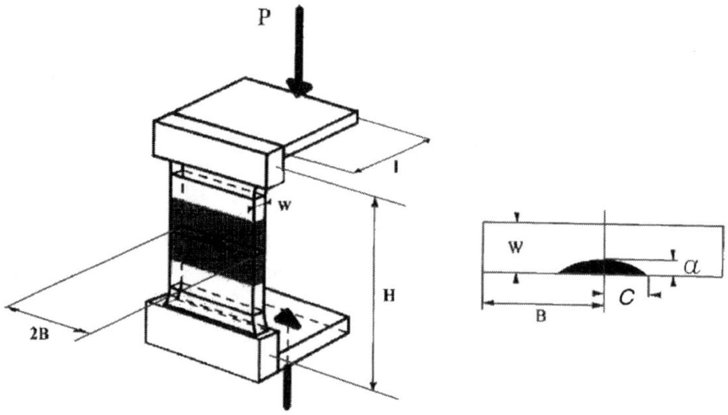

Figure 38 Geometry of plate containing a semi-elliptical surface crack showing method of load application (source Webster *et al.*, 1998).

Table 3 Predictions of incubation periods based on different definitions of reference stress (Webster *et al.*, 1998): A16, 93.3 MPa; R6, 153.4 MPa; and BS7910, 148.2 MPa.

	Incubation period (h)			
r_d (μm)	*Experiment*	*A16*	*R6*	*BS7910*
50	35	214	12.0	14.5
200	140	857	48	58

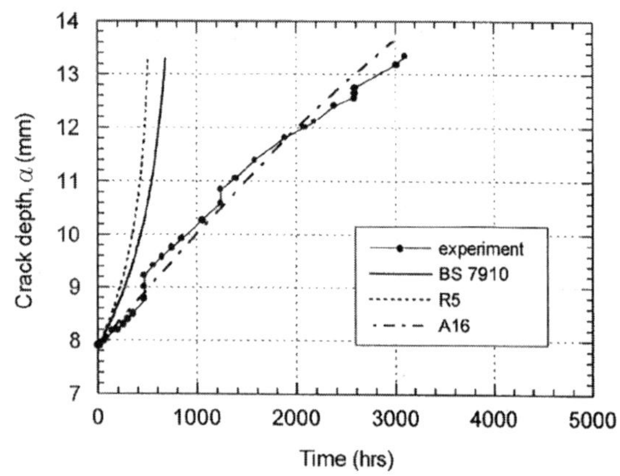

Figure 39 Comparison of experimental crack growth at deepest point in plate with creep crack growth predictions made by different assessment methods (source Webster *et al.*, 1998).

crack tip. Instead they approximate this behavior to an incubation period followed by steady-state crack growth.

Table 3 compares predictions of incubation periods obtained using the code determinations of σ_{ref} with experimental measurements for two crack detection limits r_d of 0.05 mm and 0.2 mm. It is evident that a wide spread is obtained between the predictions. This is because of the use of different values of σ_{ref} in Equations (70) and (71) for calculating C^* and, therefore, t_i in Equations (50)–(53). In all cases the reference stress is raised to a high power in determining C^* for typical values of creep stress index n. This causes a significant variation in predicted incubation periods. A similar comparison is indicated in Figure 39 between experimentally measured and predicted crack extensions. Again a wide range in predictions is obtained because of the high sensitivity of C^* to reference stress in Equation (6).

These findings demonstrate the importance of having accurate estimates of σ_{ref} when making predictions of incubation periods and crack growth in components when the limit analysis approach to calculating C^* is adopted.

5.05.10.3 Pipes

Pipes that are subjected to internal pressure loading and which contain axial semi-elliptical defects are further examples of when numerical solutions for C^* are not readily available and limit analysis procedures must be applied. In this case, all the codes use the same expression due to Raju and Newman (1982) for calculating K but a selection of formulas from Folias (1965), Kiefner *et al.* (1973), Miller (1988), and Sattari-Far (1994) for calculating σ_{ref}. Some of these formulas are based on "global" collapse and others on "local" collapse estimates of reference stress. By definition estimates based on "global" collapse mechanisms are always less than (or equal to) those determined from a "local" collapse mechanism because the collapse load P_{LC} in Equation (38) for "global" collapse is greater than (or equal to) that for "local" collapse (Goodall and Webster, 2001).

Examples of the range of predictions of crack extension that are obtained on a pressurized straight pipe of a 9% Cr steel (designated P91) are shown in Figure 40 (Wasmer *et al.*, 2002). As expected, solutions based on a

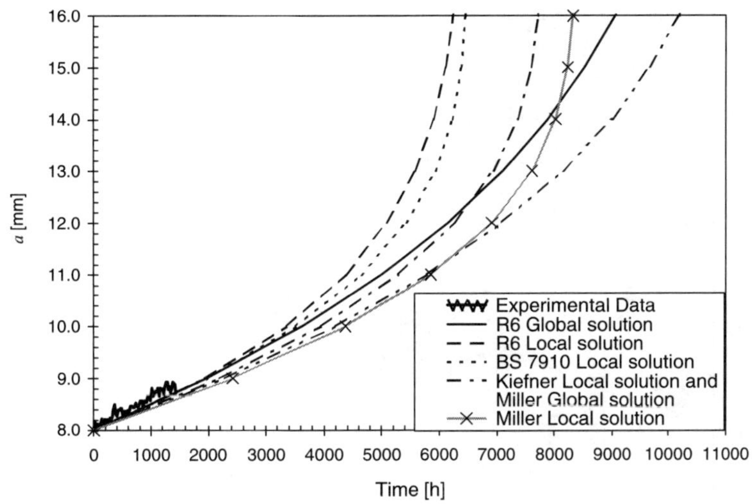

Figure 40 Comparison of experimental crack growth in straight pipe of 9% Cr steel at 625°C with predictions based on different definitions of σ_{ref} (source Wasmer *et al.*, 2002).

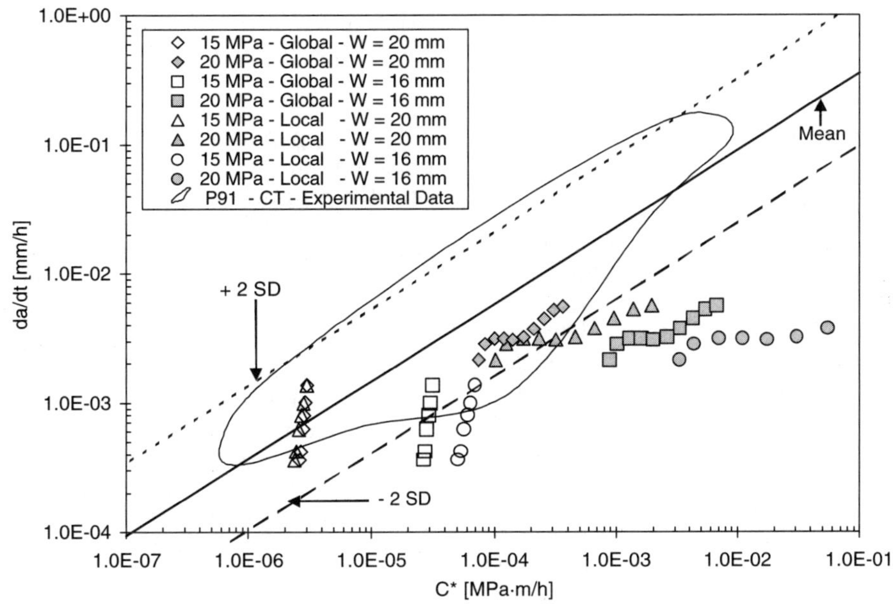

Figure 41 Crack growth from axial semi-elliptical defect Y in pressurized bent pipe of 9% Cr steel at 625°C based on R6 (2000) "global" and "local" σ_{ref} solutions compared with CT data (source Wasmer *et al.*, 2003).

"local" collapse mechanism result in the fastest accumulation of cracking.

A similar result is shown in Figure 41 for tests on pressurized bent pipes which also contained a series of axial semi-elliptical defects of different dimensions. The pipes had a nominal wall thickness W of 20 mm and were subjected to hoop stresses between 11 MPa and 20 MPa. The figure compares experimental data obtained from one defect (labeled Y) with experimental results recorded on CT specimens of the same material. The pipe data have been plotted against values of C^* determined from "global" and "local" collapse estimates of σ_{ref} using both local and average pipe wall thickness. It is clear that use of "local" collapse determinations of σ_{ref} give high values of C^* which cause the bent pipe data to lie outside the scatter band for the CT tests. This deviation is amplified when calculations of σ_{ref} are based, in addition, on local wall thickness (to compensate for thinning during the bending operation) rather than nominal wall thickness. Figure 42 demonstrates that close agreement is achieved with the CT data for all the defects examined when calculations of C^* are restricted to "global" collapse estimates of σ_{ref} and nominal wall thickness to represent the behavior of the entire pipe cross-section.

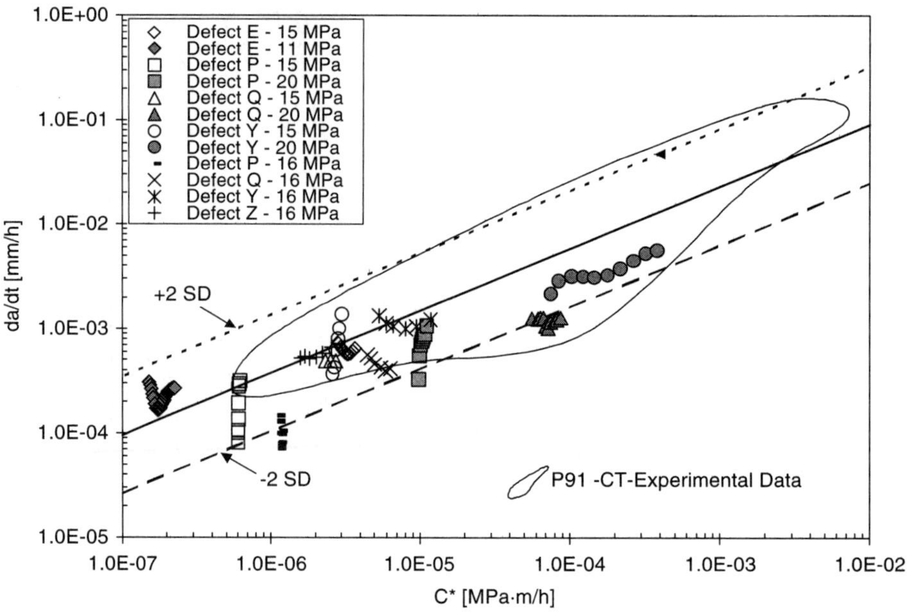

Figure 42 Crack growth from several defects in pressurized bent pipe of 9% Cr steel at 625°C based on R6 (2000) 'global' σ_{ref} solutions for nominal pipe-wall thickness compared with CT data (source Wasmer *et al.*, 2003).

5.05.10.4 Observations

It has been shown from these practical applications that the most accurate predictions of creep cracking are obtained when a transient analysis, using experimentally determined values of D_0 and ϕ in conjunction with numerical determinations of C^*, is employed. This procedure, however, requires numerical integration to perform the transient analysis. The code defect assessment methods replace the transient analysis by a simpler calculation of an incubation period followed by steady-state crack growth. Also when numerical solutions for C^* are not available, they recommend that it is calculated from handbook formulas of K and σ_{ref} as discussed in Section 5.05.9.3. There is close agreement in these handbooks on formulas for K but not σ_{ref}. This is because of the range of "global" and "local" collapse mechanisms proposed. As mentioned earlier, normally use of a reference stress based on a "local" collapse mechanism will provide the most conservative assessment of cracking. This can be envisaged by rewriting Equations (16) and (50) from Equations (68) and (69) as

$$\dot{a} = \frac{(n+1)}{\varepsilon_{\mathrm{f}}^*}\dot{\varepsilon}_{\mathrm{ref}}\left(\frac{K^2}{I_n\sigma_{\mathrm{ref}}^2}\right)^{n/(n+1)} r_{\mathrm{c}}^{1/(n+1)} \qquad (78)$$

and

$$t_{\mathrm{i}} = \left(\frac{n+1-\upsilon}{n+1}\right)\frac{\varepsilon_{\mathrm{f0}}^*}{\varepsilon_{\mathrm{f0}}}t_{\mathrm{r(ref)}}\left(\frac{I_n r_{\mathrm{d}}\sigma_{\mathrm{ref}}^2}{K^2}\right)^{\upsilon/(n+1)} \qquad (79)$$

It is apparent from these equations that cracking rate and the incubation period are directly proportional to the creep strain rate and rupture life calculated at the reference stress, respectively. For $n \gg 1$, this will result in a wide range in predictions for small changes in σ_{ref} as demonstrated. These equations also indicate the importance of having accurate creep strain rate and stress rupture data for making reliable predictions. When primary and tertiary creep deformation are significant, use of average creep strain rate rather than secondary creep rate will result in creep crack growth rates that are elevated in the ratio $\dot{\varepsilon}_{\mathrm{A}}/\dot{\varepsilon}_{\mathrm{S}}$ for the same values of K and σ_{ref}.

5.05.11 SUMMARY

The behavior of cracks in components that are subjected to creep has been examined. It has been shown how creep cracking occurs by the linking of damage ahead of the main crack to the crack tip. The stress and strain rate distributions ahead of a crack tip have been interpreted using fracture mechanics concepts. It has been demonstrated how the distributions can be described by the creep fracture mechanics parameter C^* once these have become stabilized. Models based on uniaxial creep data have been presented for predicting incubation periods and crack propagation rates in terms of exhaustion of the available ductility in a process zone at the crack tip. Allowance has been made for a buildup of damage in the process zone initially which results in a transient period of cracking prior to the onset of steady-state growth. Consideration has also been given to the influence of state of stress at the crack tip and for the effect of progressive material deterioration in the ligament ahead of a crack. The development of material deterioration has been extended to enable residual life assessments to be made of components that have already experienced a period of creep loading.

Comparisons have been made with a wide spread of experimental data and universal laws have been proposed for predicting incubation periods and crack growth rates, approximately, when only creep ductility and stress rupture data are available. It has been shown how limit analysis can be employed for obtaining approximate estimates of C^* in terms of stress intensity factor K and reference stress when numerical finite element solutions are not available. The principles outlined have been incorporated into a number of failure assessment codes. Application of these to a number of practical situations has demonstrated that a range of predictions can be obtained depending on whether a "global" or "local" collapse mechanism is employed for determining the reference stress. It has been established that the most accurate predictions are achieved when a transient analysis is adopted and when experimental crack growth data and numerical solutions for C^* are available.

5.05.12 REFERENCES

AFCEN, 1985, "Design and Construction Rules for Mechanical Components of FBR Nuclear Islands," RCC-MR, Appendix 16, AFCEN, Paris.

R. A. Ainsworth, 1982, The initiation of creep crack growth. *Int. J. Solids Struct.*, **18**, 873–881.

R. A. Ainsworth, 1986, Approximate blunting solutions for tensile cracks. In: "Applied Solid Mechanics—1," eds. A. S. Tooth and J. Spence, Elsevier, London, pp. 59–72.

ASTM E 1457-00, 2001, Standard test method for measurement of creep crack growth rates in metals. *ASTM*, **03-01**, 936–950.

ASTM E 1820-99a, 2000, Standard test method for J_{IC}, a measure of fracture toughness. *ASTM*, **03-01**, 1000–1033.

T. S. P. Austin and G. A. Webster, 1992, Prediction of creep crack growth incubation periods. *Fatigue Fract. Eng. Mater. Struct.*, **15**, 1081–1090.

J. L. Bassani, D. E. Hawk and A. Saxena, 1987, Evaluations of the C_t parameter for characterising creep crack growth rates in the transient regime. In: "Nonlinear Fracture Mechanics, ASTM STP 995," eds. A. Saxena, J. D. Landes and J. L. Bassani, ASTM, Philadelphia, pp. 7–26.

A. D. Bettinson, 2002, The influence of constraint on the creep crack growth of 316H stainless steel. Ph.D. thesis, University of London.

A. D. Bettinson, K. M. Nikbin, N. P. O'Dowd and G. A. Webster, 2000a, The influence of constraint on creep crack growth in 316H stainless steel. In: "Structural Integrity in the 21st Century—The Lifetime of Plant, Structures and Components: Evaluation, Design, Extension and Management," ed. J. H. Edwards, EMAS, Cambridge, pp. 149–159.

A. D. Bettinson, N. P. O'Dowd, K. M. Nikbin and G. A. Webster, 2002, Experimental investigation of constraint effects on creep crack growth. In: "ASME Pressure Vessels and Piping Conference," ASME, vol. 434, pp. 143–150.

A. D. Bettinson, N. P. O'Dowd and G. A. Webster, 2000b, Two parameter characterisation of crack tip fields under creep conditions. In: "Proc. IUTAM Symp. on Creep in Structures, Nagoya, Japan," eds. S. Murakami and N. Ohno, Kluwer, Netherlands, pp. 95–104.

BSPD 6539:1994, 1994, "Guide to Methods for the Assessment of the Influence of Crack Growth on the Significance of Defects in Components Operating at High Temperatures," BSI, London.

BS 7910:1999, 1999, "Guide on Methods for Assessing the Acceptability of Flaws in Fusion Welded Structures," BSI, London.

G. M. Buchheim, C. Becht, K. M. Nikbin, V. Dimopulos, G. A. Webster and D. J. Smith, 1989, Influence of aging on high temperature creep crack growth in type 304H stainless steel. In: "Nonlinear Fracture Mechanics: Volume I—Time-dependent Fracture, ASTM STP 995," eds. A. Saxena, J. D. Landes and J. L. Bassani, pp. 153–172.

P. J. Budden and R. A. Ainsworth, 1999, The effect of constraint on facture assessments. *Int. J. Fract.*, **97**, 237–247.

B. Budiansky, J. W. Hutchinson and S. Slutsly, 1982, Void growth and collapse in viscous solids. In: "Mechanics of Solids," eds. H. G. Hokins and M. J. Sewell, Pergamon, Oxford.

A. C. F. Cocks and M. F. Ashby, 1980, Intergranular fracture during power-law creep under multi-axial stress. *Mater. Sci.*, **14**, 395–402.

M. C. Coleman, A. T. Price and J. A. Williams, 1977, Crack growth in pressure vessels under creep conditions. In: "Fracture," ed. D. M. Taplin, Pergamon, Oxford, pp. 649–662.

E. G. Ellison and M. P. Harper, 1978, Creep behaviour of components containing cracks—a critical review. *J. Strain Anal.*, **13**, 35–51.

S. Floreen, 1975, The creep fracture of wrought nickel base alloys by a fracture mechanics approach. *Met. Trans. A*, **6A**, 1741–1749.

E. S. Folias, 1965, 'An axial crack in a pressurized cylindrical shell', *Int. J. Fract. Mech.*, **1**, 104–113.

I. W. Goodall and G. A. Webster, 2001, Theoretical determination of reference stress for partially penetrating flaws in plates. *Int. J. Pres. Ves. Piping*, **78**, 687–695.

M. Y. He and J. W. Hutchinson, 1983, Elastic-plastic fracture. In: "Bounds on Fully Plastic Crack Problems for Infinite Bodies in Elastic–Plastic Fracture, ASTM STP 803," eds. C. F. Shih and J. P. Gudas, ASTM, Philadelphia, vol. 1, pp. 277–290.

S. R. Holdworth, 1994, Private communication.

T. Hollstein, G.A. Webster, F. Djavanroodi, 1990, Creep crack growth in a 1% CrMoV steel and a 32% Ni 20% Cr alloy. In: "Fracture Behaviour and Design of Materials and Structures, EGF," 8, ed. D. Firrao, EMAS, Halesowen, vol. III, pp. 1249–1256.

J. W. Hutchinson, 1968, Plastic stress and strain fields at a crack tip. *J. Mech. Phys. Solids*, **16**, 337–347.

J. F. Kiefner, W. A. Maxey, A. Eiber and A. R. Duffy, 1973, Failure stress levels of flaws in pressurized cylinders. In: "Progress in Flow Growth and Fracture Toughness Testing, ASTM STP 536," pp. 461–481.

R. Kienzler and T. Hollstein, 1987, Experimental and numerical investigations of creep crack growth. In: "Creep and Fracture of Engineering Materials and

Structures," eds. B. Wilshire and R. W. Evans, Inst. Metals, London, pp. 563–576.

V. Kumar, M. D. German and C. F. Shih, 1981, "An Engineering Approach for Elastic–Plastic Fracture," EPRI report, NP 1931.

Y. Lei, 2001, *J*-integral and limit load analysis of a semi-elliptical surface crack in a plate under combined tensile and bending load. British Energy Generation Ltd. Report E/REP/ATEC/0015/GEN/01.

A. G. Miller, 1988, Review of limit loads of structures containing defects. *Int. J. Pres. Ves. Piping*, **32**, 197–327.

A. G. Miller and R. A. Ainsworth, 1989, Consistency of numerical results for power-law hardening materials and the accuracy of the reference stress approximation for. *J. Eng. Fract. Mech.*, **32**, 233–247.

Y. Murakami, 1987, "Stress Intensity Factor Handbook," vols. 1 and 2, Pergamon, Oxford.

G. J. Neate, 1986, Creep crack growth in 1/2Cr–Mo–V steel at 838 K. I: Behaviour at a constant load. *Mater. Sci. Eng.*, **82**, 59–76.

G. J. Neate and M. J. Siverns, 1973, The application of fracture mechanics to creep crack growth. In: "Int. Conf. on Creep and Fracture in Elevated Temperature Applications," Inst. Mech. Eng., London.

K. M. Nikbin, D. J. Smith and G. A. Webster, 1983, Influence of creep ductility and state of stress on creep crack growth. In: "Advances in Life Prediction Methods," eds. D. A. Woodford and J. R. Whitehead, ASME, New York, pp. 249–258.

K. M. Nikbin, D. J. Smith and G. A. Webster, 1984, Prediction of creep crack growth from uni-axial creep data. *Proc. Roy. Soc. A*, **396**, 183–197.

K. M. Nikbin, D. J. Smith and G. A. Webster, 1986, An engineering approach to the prediction of creep crack growth. *J. Eng. Mater. Technol., Trans ASME*, **108**, 186–191.

K. Nishida, K. M. Nikbin and G. A. Webster, 1989, Influence of net section damage on creep crack growth. *J. Strain Anal.*, **24**(2), 75–82.

K. Nishida and G. A. Webster, 1990, Interaction between build-up of local and remote damage on creep crack growth. In: "Creep and Fracture of Engineering Materials and Structures," eds. B. Wilshire and R. W. Evans, Inst. Metals, London, pp. 703–713.

N. P. O'Dowd and C. F. Shih, 1994, "Two-parameter Fracture Mechanics: Theory and Applications, ASTM STP 1207," eds. J. D. Landes, D. E. McCabe and J. A. M. Boulet, ASTM, Philadelphia, pp. 21–47.

R. K. Penny and D. L. Marriott, 1995, "Design for Creep," Chapman and Hall, London.

R. Piques, E. Molinie and A. Pineau, 1991, Comparison between two assessment methods for defects in the creep range. *Fatigue Fract. Eng. Mater. Struct.*, **14**, 871–885.

C. Poussard and D. Moulin, 1998, Creep-fatigue in austenitic stainless steel centre cracked plate at 650°C: Part 1. Experimental study and interpretation. *Mater. High Temp.*, **15**, 131–138.

R5, 1999, "Assessment Procedure for the High Temperature Response of Structures Containing Defects," Issue 2, British Energy Generation.

R6, 2000, "Assessment of the Integrity of Structures Containing Defects," Revision 3, British Energy Generation.

I. S. Raju and J. C. Newman, 1982, Stress intensity factors for internal and external surface cracks in cylindrical vessels. *J. Pres. Ves. Technol.*, **104**, 293–298.

J. R. Rice, 1968, A path independent integral and the approximate analysis of strain concentrations by notches and cracks. *J. Appl. Mech.*, **35**, 379–386.

J. R. Rice and G. F. Rosengren, 1968, Plane strain deformation near a crack tip in a power law hardening material. *J. Mech. Phys. Solids*, **16**, 1–12.

J. R. Rice and D. M. Tracey, 1969, On the ductile enlargement of voids in triaxial stress fields. *J. Mech. Phys. Solids*, **17**, 201–217.

H. Riedel, 1987, "Fracture at High Temperatures," Springer, Berlin.

H. Riedel and J. R. Rice, 1980, Tensile cracks in creeping solids. In: "Fracture Mechanics 12th Conf., ASTM STP 700," ASTM, Philadelphia, pp. 112–130.

D. P. Rooke and D. J. Cartwright, 1976, "Compendium of Stress Intensity Factors," HMSO, London.

K. Sadananda and P. Shahinian, 1981, Creep-fatigue crack growth. In: "Cavities and Cracks," ed. J. H. Gittus, Applied Science, London, pp. 109–196.

I. Sattari-Far, 1994, Finite element analysis of limit loads for surface cracks in plate. *Int. J. Pres. Ves. Piping*, **57**, 237–243.

A. Saxena, 1986, Creep crack growth under non-steady state conditions. In: "Fracture Mechanics, ASTM STP 905," ASTM, Philadelphia, pp. 185–201.

A. Saxena, 1998, "Non-linear Fracture Mechanics for Engineers," CRC Press, Boca Raton, USA.

A. Saxena, T. Hollstein, G. A. Webster and T. Yokobori, 1992, Intercomparison of creep crack growth data. In: "Harmonisation of Testing Practice in High Temperature Materials," eds. M. S. Loveday and T. B. Gibbons, Elsevier, London, pp. 295–330.

M. J. Siverns and A. T. Price, 1973, Crack propagation under creep conditions in a quenched $2\frac{1}{4}$ CrMoV steel. *Int. J. Fract.*, **9**, 199–207.

D. J. Smith and G. A. Webster, 1983, Estimates of the C^* parameter for crack growth in creeping materials. In: "Elastic–Plastic Fracture, ASTM STP 803," eds. C. F. Shih and J. P. Gudas, ASTM, Philadelphia, vol. 1, pp. 654–674.

M. W. Spindler, 1994, "The Multi-axial Creep of Austenitic Stainless Steels," Nuclear Electric report TIGM/REP/0014/94.

M. W. Spindler, R. Hales and R. P. Skelton, 2001, The multi-axial creep ductility of an ex-service type 316H stainless steel. In: "9th Int. Conf. on Creep and Fracture of Eng. Mater. and Struct., April 2001, Swansea," Inst. Materials, London, pp. 679–688.

M. Tabuchi, K. Yagi, K. Kubo and C. Tanaka, 1990, Effect of geometrical constraint on creep crack growth behaviour of 1 Cr MoV turbine rotor steel. In: "Creep and Fracture of Engineering Materials and Structures," eds. B. Wilshire and R. W. Evans, Inst. Metals, London, pp. 729–738.

H. Tada, P. C. Paris and G. R. Irwin, 2000, "The Stress Analysis of Cracks Handbook," 3rd edn., ASME, New York.

S. Taira, R. Ohtani and T. Kitamura, 1979, Application of *J*-integral to high-temperature crack propagation: Part 1. Creep crack propagation. *J. Eng. Mater. Technol. Trans. ASME*, **101**, 154–161.

M. Tan, N. J. C. Celard, K. M. Nikbin and G. A. Webster, 2001, Comparison of creep crack initiation and growth in four steels tested in HIDA. *Int. J. Pres. Ves. Piping*, **78**, 737–747.

E. Taylor and A. D. Batte, 1980, Creep crack formation in a 1% CrMoV rotor forging steel. In: "Int. Conf. on Engineering Aspects of Creep," Inst. Mech. Eng., Sheffield, London, paper C216/80.

R. Viswanathan, 1989, "Damage Mechanisms and Life Assessment of High Temperature Components," ASM International, Metals Park, Ohio.

K. Wasmer, K. M. Nikbin and G. A. Webster, 2002, "A sensitivity study of creep crack growth in pipes," ASME Pres. Ves. Piping Conf., Vancouver, Aug. 2002. ASME-PVP, vol. 438, pp. 17–24.

K. Wasmer, K. M. Nikbin and G. A. Webster, 2003, "Creep crack initiation and growth in thick section steel pipes under internal pressure," Int. J. Pres. Ves. Piping, to be published.

G. A. Webster, 1982, Methods of measuring crack growth at elevated temperatures. In: "Measurement of High Temperature Mechanical Properties of Materials," eds. M. S. Loveday, M. F. Day and B. F. Dyson, HMSO, London, pp. 255–277.

G. A. Webster, 1983, Crack growth at high temperature. In: "Engineering Approaches to High Temperature Design," eds. B. Wilshire and D. R. J. Owen, Pineridge Press, Swansea, pp. 1–56.

G. A. Webster, 1992a, Lifetime estimates of cracked high temperature components. *Int. J. Pres. Ves. Piping*, **50**, 133–145.

G. A. Webster, 1992b, Methods of estimating C^*. *Mater. High Temp.*, **10**, 74–78.

G. A. Webster, 1994, Fracture mechanics in the creep range. *J. Strain Anal.*, **29**, 215–223.

G. A. Webster and R. A. Ainsworth, 1994, "High Temperature Components Life Assessment," Chapman and Hall, London.

G. A. Webster, P. F. Aplin, B. J. Cane, B. F. Dyson and M. S. Loveday, 1992, A code of practice for notched bar creep rupture tests: procedures and interpretation of data for design. In: "Harmonisation of Testing Practice for High Temperature Materials," eds. M. S. Loveday and T. B. Gibbons, Elsevier, London, pp. 295–330.

G. A. Webster, S. R. Holdsworth, M. S. Loveday, I. J. Perrin and H. Purper, 2001, A code of practice for conducting notched bar creep rupture tests and for interpreting the data. ESIS P10-02, Issue 2a, Nov. 2001.

G. A. Webster, K. M. Nikbin, M. R. Chorlton, N. J. C. Celard and M. A. Ober, 1998, Comparison of high temperature defect assessment methods. *Mater. High Temp.*, **15**, 295–304.

M. R. Winstone, K. M. Nikbin and G. A. Webster, 1985, Modes of failure under creep/fatigue loading of a nickel-based superalloy. *J. Mater. Sci.*, **20**, 2471–2476.

Comprehensive Structural Integrity
ISBN (set): 0-08-043749-4

Volume 5; (ISBN: 0-08-044154-8); pp. 241–271

5.06
Creep-fatigue Crack Growth

S. KUBO
Osaka University, Japan

and

T. KITAMURA
Kyoto University, Japan

NOMENCLATURE

a	crack length
A	area of crack surface
b	ligament length
B	thickness of plate
C, C_c	constants in $\mathrm{d}a/\mathrm{d}t$–J^* correlation
C_f	constant in $\mathrm{d}a/\mathrm{d}N$–ΔJ_f correlation
$C(t)$	J^* near crack tip under small-scale creep

C_t experimentally determined J^* value
C_h^* C^* or J^* under primary and hardening creep laws
da/dN crack propagation rate per cycle
da/dt crack propagation rate per time
E Young's modulus
F fracture driving force of grain boundary for crack initiation
J the J-integral
J^* modified J-integral, C^*-integral, creep J-integral
J_{av}^* averaged creep J-integral
J_{nt}^* value of J^* near crack tip
J_{steady}^* steady-state value of J^*
J_{el} elastic J-integral
K stress intensity factor
K_F fracture driving force of grain boundary for crack propagation
m stress exponent for plastic strain
m' stress exponent in cyclic stress strain curve
m_f exponent in da/dN–ΔJ_f correlation
M_J correction factor for crack shape and boundary condition
n creep exponent
N number of cycles
N_f number of cycles to failure
p Poisson's ratio
P load
P_L limit load
r distance form crack tip
R fracture resistance of grain boundary
S_T traction boundary
t time duration after load application
t_H tensile hold time
t_{pl} plastic correction time for small-scale creep effect
t_{tr} time measure of transition from small-scale creep to large-scale creep
T_i traction
u_i displacement vector
$\dot{u}_i$ displacement rate vector
U work done by load
V load point displacement
$\dot{V}$ load point displacement rate
W strain energy density
W^* strain energy rate density
x_i Cartesian coordinate
Γ path for the J-integral
δ crack mouth opening displacement
$\dot{\delta}$ crack mouth opening displacement rate
ΔJ_c creep J-integral range
ΔJ_f J-integral range, cyclic J-integral, fatigue J-integral range
$(\Delta J_f)_{th}$ threshold J-integral range
ΔK stress intensity factor range
ΔK_{eff} effective stress intensity factor range
$\Delta\varepsilon$ strain range
$\Delta\varepsilon_t$ total strain range
$\Delta\sigma$ stress range
ε strain
ε_c material constant in Norton type creep law
ε_{ct} strain near crack tip
$\dot{\varepsilon}_e$ equivalent strain rate corresponding to σ_e
ε_{el} elastic strain
ε_{pl} plastic strain
ε_{ij} strain tensor
$\dot{\varepsilon}_{ij}$ strain rate tensor
$\dot{\varepsilon}_{net}$ net section strain rate corresponding to σ_{net}
$\dot{\varepsilon}_{nom}$ net section strain rate corresponding to σ_{nom}
$\dot{\varepsilon}_{ref}$ equivalent strain rate corresponding to σ_{ref}
ε_Y yield strain
η eta factor
θ angular coordinate
ν loading frequency
Π^* creep potential
σ stress
σ_c material constant in Norton type creep law
σ_e equivalent stress
σ_{ij} stress tensor
σ_{net} net section stress
σ_{nom} nominal stress
σ_{ref} reference stress
σ_Y yield stress
τ time of tension in one cycle
Ω volume of body
$\dot{\phi}$ angular displacement rate

5.06.1 INTRODUCTION

Due to increasingly severe economical and environmental requirements for plants and their components operating at high temperatures, their operating temperatures have increased considerably. Many of the plants and their components operate in the ranges where creep deformation is a significant concern, and are also subjected to nonsteady loading. To evaluate their safety, creep effects are essential and have to be taken into account in addition to fatigue effects. Since cracking is observed before fracture, the quantitative approach for the evaluation can be constructed by applying fracture mechanics, which was found useful in evaluating crack initiation and ductile crack growth at room temperature.

Application of fracture mechanics to high temperature crack problems started from the characterization of high-cycle fatigue crack growth using the stress intensity factor K. For the evaluation of creep crack growth, K was used during the early stage of research on this topic (Siverns and Price, 1970). Then C^*-integral, which was also called the modified J-integral J^*, and the creep J-integral, was proposed by several research groups (Ohji *et al.*, 1974a, 1974b, 1976; Landes and Begley, 1976; Nikbin *et al.*, 1976). J^* was found superior to K for the correlation of creep crack growth. These results and much related research contributed to the quantitative characterization of the behavior of cracks under creep and fatigue conditions (Saxena, 1993, 1998; Webster and Ainsworth, 1994).

This chapter is devoted to the description of stress, strain, and strain rate fields near the crack tip under fatigue and creep conditions, the evaluation of fracture mechanics parameters including J^*, and growth laws of macroscopic cracks and small cracks, which are essential for the assessment of life and integrity of structures of components at high temperatures.

5.06.2 STRESS AND STRAIN FIELDS NEAR A CRACK TIP AND FRACTURE MECHANICS PARAMETERS

5.06.2.1 Elastic–Plastic Stress and Strain Fields (see also Chapter 2.03)

Consider an ideally sharp two-dimensional (2D) crack whose surface is stress free. For simplicity only the mode I (opening mode) crack is considered, since mode I crack is relevant to most failure of structures and their components for high temperature use. As is well known from the linear elastic fracture mechanics, the stress near the crack tip is expressed in terms of the stress intensity factor K, when the effects of plasticity and creep are not taken into account:

$$\sigma_{ij} = K\frac{f_{ij}(\theta)}{\sqrt{2\pi r}} + T\delta_{1i}\delta_{1j} + O(r^{1/2}) \qquad (1)$$

where r and θ are polar coordinates whose origin is located at the crack tip, and $f_{ij}(\theta)$ is the eigen function which expresses the angular distribution of singular stress. T is called T-stress which is parallel to the crack face and δ_{ij} denotes the Kronecker delta. The first term is singular and large compared with higher nonsingular terms for small r. Yielding occurs near the crack tip for most materials, and plastic strain prevails in the near-tip region. As far as the plastic zone size is well-contained by the region where the elastic singular stress field prevails, the stress and plastic strain fields are controlled by the stress intensity factor K.

This situation is called small-scale yielding. Consider a material whose stress σ–strain ε relation beyond yield is expressed by a power-law:

$$\varepsilon = \varepsilon_Y\left(\frac{\sigma}{\sigma_Y}\right)^m \qquad (2)$$

where σ_Y, ε_Y, and m denote the yield stress, yield strain, and stress exponent, respectively. For the material, the stress and strain near the crack tip are expressed as

$$\sigma_{ij} = \sigma_Y\left[\frac{K^2}{E'I(m)\varepsilon_Y\sigma_Y r}\right]^{1/(m+1)}\tilde{\sigma}_{ij}(\theta) \qquad (3)$$

$$\varepsilon_{ij} = \varepsilon_Y\left[\frac{K^2}{E'I(m)\varepsilon_Y\sigma_Y r}\right]^{m/(m+1)}\tilde{\varepsilon}_{ij}(\theta) \qquad (4)$$

where $\tilde{\sigma}_{ij}(\theta)$, $\tilde{\varepsilon}_{ij}(\theta)$, and $I(m)$ are nondimensional stress and strain distribution functions of θ and a constant, which are dependent on the stress exponent m and whether the condition of plane stress or plane strain applies. E' is given in terms of Young's modulus E and Poisson's ratio p as

$$E' = E \qquad \text{for plane stress} \qquad (5)$$

$$E' = E/(1-p^2) \qquad \text{for plane strain} \qquad (6)$$

The singular stress and strain fields of Equations (3) and (4) are referred to as the Hutchinson–Rice–Rosengren (HRR) field after Hutchinson (1968) and Rice and Rosengren (1968), who first reported the plastic singular stress and strain fields.

Under repeated fatigue loading the stress and strain and residual plastic strain are controlled by the maximum and the minimum values of the stress intensity factor K as far as the small-scale-yielding condition prevails. Thus the stress intensity range ΔK and stress ratio can be used for characterizing the near-tip stress and strain ranges. When residual strain distribution due to crack growth exists, the effective stress intensity range ΔK_{eff} is replaced for K in Equations (3) and (4) for evaluating the stress range and strain range near the crack tip considering crack closure.

When the plastic zone is not well contained in the region where the elastic singular fields prevail, the stress intensity factor no more expresses the near-tip singular fields. The J-integral (Rice, 1968) is proposed for one-parameter representation of the plastic singular stress and strain fields under the large-scale yielding as well as under the small-scale yielding.

When the Cartesian coordinates x_1x_2 (xy) are taken in such a way that crack face is parallel to the $x_1(x)$-direction as in Figure 1, the J-integral is defined by the following line

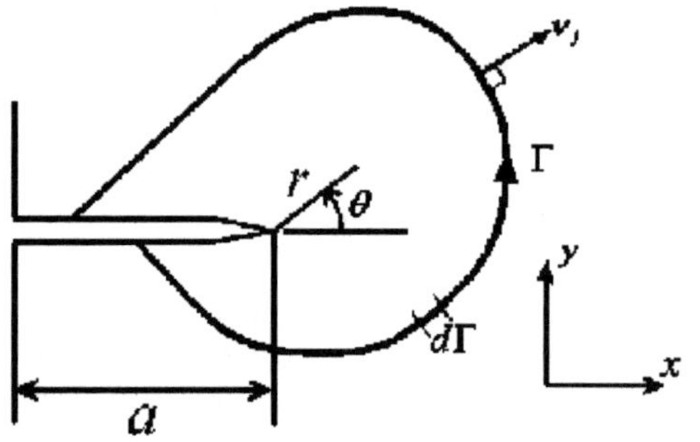

Figure 1 Integration path Γ in the definition of J and J^* integrals.

integral for the counterclockwise path Γ enclosing the crack tip:

$$J = \int_\Gamma \left(W\,dx_2 - T_i \frac{\partial u_i}{\partial x_1} d\Gamma \right) \quad (7)$$

where $\partial u_i/\partial x_1$ is the gradient of displacement u_i in the $x_1(x)$-direction, T_i is traction acting on Γ, and dΓ denotes the line segment. W denotes the strain energy density, which is defined as

$$W = \int_0^{\varepsilon_{ij}} \sigma_{ij}\, d\varepsilon_{ij} \quad (8)$$

The J-integral is independent of the selection of path Γ. Thanks to the path independence, the J-integral can be evaluated for a path taken far from the crack tip. Alternatively, when the path is taken near the crack tip the J-integral is given by the near-tip stress and strain fields. Thus the J-integral can be a parameter representing the singular stress and strain fields of the HRR type near the crack tip as

$$\sigma_{ij} = \sigma_Y \left[\frac{J}{(I(m)\varepsilon_Y \sigma_Y r)} \right]^{1/(m+1)} \tilde{\sigma}_{ij}(\theta) \quad (9)$$

$$\varepsilon_{ij} = \varepsilon_Y \left[\frac{J}{(I(m)\varepsilon_Y \sigma_Y r)} \right]^{m/(m+1)} \tilde{\varepsilon}_{ij}(\theta) \quad (10)$$

For the small-scale yielding, J is equal to the elastic J-integral given as

$$J = J_{el} = \frac{K^2}{E'} \quad (11)$$

Then Equations (9) and (10) are consistent with Equations (3) and (4) for the small-scale yielding. For fatigue loading under the large-scale-yielding condition, where the plastic zone is not contained by the elastic singular region near the crack tip, the stress intensity factor range ΔK is no more valid for expressing the stress and strain ranges. The J-integral was extended to J-integral range ΔJ_f for characterizing the near-tip stress and strain ranges, and experimental methods were proposed for evaluating ΔJ_f (Dowling and Begley, 1976).

A mathematical expression of ΔJ_f was also proposed (Kubo *et al.*, 1989; Anderson, 1995) for the case when the cyclic stress $\Delta\sigma$–strain $\Delta\varepsilon$ relation beyond yield is stable and is expressed as

$$\Delta\varepsilon = \varepsilon'_Y \left(\frac{\Delta\sigma}{\sigma'_Y} \right)^{m'} \quad (12)$$

where ε'_Y, σ'_Y, and m' are material constants. In this case the J-integral range ΔJ_f can be defined by the following equation, and is used for characterizing the near-tip stress and strain range as

$$\Delta J_f = \int_\Gamma \left(\Delta W\, dx_2 - \Delta T_i \frac{\partial \Delta u_i}{\partial x_1} d\Gamma \right) \quad (13)$$

where ΔT_i and Δu_i are increments of traction and displacement evaluated from their values at the minimum load or crack opening load. The strain energy density ΔW in this expression is defined as

$$\Delta W = \int_0^{\Delta\varepsilon_{ij}} \Delta\sigma_{ij}\, d\Delta\varepsilon_{ij} \quad (14)$$

The stress and strain ranges near the crack tip are expressed in terms of ΔJ_f as

$$\Delta\sigma_{ij} = \sigma'_Y \left[\frac{\Delta J_f}{(I(m')\,\varepsilon'_Y \sigma'_Y r)} \right]^{1/(m'+1)} \tilde{\sigma}_{ij}(\theta) \quad (15)$$

$$\Delta\varepsilon_{ij} = \varepsilon'_Y \left[\frac{\Delta J_f}{I(m')\,\varepsilon'_Y \sigma'_Y r} \right]^{m'/(m'+1)} \tilde{\varepsilon}_{ij}(\theta) \quad (16)$$

Thus the stress and strain ranges can be represented solely by the J-integral range ΔJ_f.

5.06.2.2 Stress and Strain Fields under Creep Condition (see also Chapter 5.04)

The stress and strain under creep condition are given by applying the analogy between the nonlinear elasticity and the creep. Assume the Norton type steady-state creep, for which the creep strain rate is expressed by a power function of stress as

$$\dot{\varepsilon} = \dot{\varepsilon}_c \left(\frac{\sigma}{\sigma_c} \right)^n \quad (17)$$

Here $\dot{\varepsilon}_c$ and σ_c are material constants and n is the creep exponent. By applying the analogy, the creep solution for the creep law (Equation (17)) is readily obtained by replacing the strain rate and displacement rate together with material constants $\dot{\varepsilon}_c$, σ_c, and n, for the strain and displacement together with material constants ε_Y, σ_Y, and m in the corresponding solution under the nonlinear stress–strain relation of Equation (2). An application of the analogy to the J-integral for the nonlinear elastic material yielded the following energy rate integral, called the C^* integral and the modified J-integral J^*, for the creeping material (Ohji *et al.*, 1974a, 1974b, 1976; Landes and Begley, 1976; Nikbin *et al.*, 1976):

$$J^* = \int_\Gamma \left(W^* \, dx_2 - T_i \frac{\partial \dot{u}_i}{\partial x_1} d\Gamma \right) \tag{18}$$

where $\dot{u}_i$ denotes the displacement rate, and W^* is the strain energy rate density given as

$$W^* = \int_0^{\dot{\varepsilon}_{ij}} \sigma_{ij} \, d\dot{\varepsilon}_{ij} \tag{19}$$

For the power law creeping material the stress and strain rate in the vicinity of the crack tip are expressed in terms of J^* as

$$\sigma_{ij} = \sigma_c \left[\frac{J^*}{I(n)\dot{\varepsilon}_c \sigma_c r} \right]^{1/(n+1)} \tilde{\sigma}_{ij}(\theta) \tag{20}$$

$$\dot{\varepsilon}_{ij} = \dot{\varepsilon}_c \left[\frac{J^*}{I(n)\dot{\varepsilon}_c \sigma_c r} \right]^{n/(n+1)} \tilde{\varepsilon}_{ij}(\theta) \tag{21}$$

Thus one-parameter representation of the stress and strain rate fields near the creep crack tip can be made using J^*.

5.06.2.3 Acceleration of J^* due to Small-scale-creep Effect

In Section 5.06.2.2 the stress and strain rate fields are discussed for the case when the creep strain is predominant compared with the elastic strain. The elastic strain is significant compared with the creep strain for small time duration after loading.

Consider that the total strain rate is composed of the elastic strain rate and the Norton type creep strain rate as

$$\dot{\varepsilon} = \dot{\varepsilon}^{(el)} + \dot{\varepsilon}^{(cr)} = \frac{\dot{\sigma}}{E} + \dot{\varepsilon}_c \left(\frac{\sigma}{\sigma_c} \right)^n \tag{22}$$

The creep zone, where the creep strain dominates over the elastic strain, is small and increases with time as shown in Figure 2. When the creep zone is well contained by a region where the linear elastic singularity dominates, the value of J^* near the crack tip, J^*_{nt}, is much larger than the steady-state value J^*_{steady} which is attained when the creep zone is fully developed (Ohji *et al.*, 1979, 1980; Riedel, 1978; Riedel and Rice, 1980). This situation is called small-scale-creep (SCC) condition (Figure 2(a)). It was shown that the behavior of J^*_{nt} is approximated by the following equation using the approximate path independence of the J-integral, as is shown in Section 5.06.7 (Appendix A) (Ohji *et al.*, 1979, 1980):

$$J^*_{nt} = \frac{J_{el}}{(n+1)t} \tag{23}$$

In this equation the elastic J-integral J_{el} can be evaluated from the stress intensity factor K using Equation (11), with t denoting the time duration after loading. J^*_{nt} near the crack tip is also

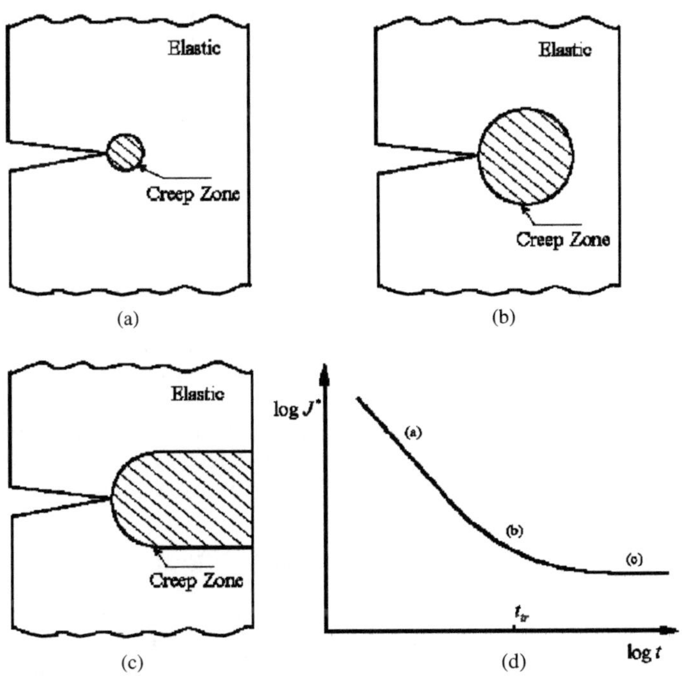

Figure 2 Transition from small-scale creep to large-scale creep in a cracked body showing the relative size of the creep zone in comparison with the size of the crack and the uncracked ligament. (a) Small-scale creep, (b) transition, (c) large-scale creep, and (d) the definition of transition time t_{tr}.

denoted by $C(t)$. The value of J^*_{nt} is very large for very small t and decreases with t. The validity of Equation (23) is confirmed by finite element calculations (Ohji *et al.*, 1979, 1980; Bassani and McClintock, 1981; Biner and Wilkinson, 1985). When the creep zone becomes large, the contribution of the elastic strain decreases and J^* approaches its steady state value J^*_{steady}:

$$J^* = J^*_{steady} \tag{24}$$

This situation is called large-scale-creep (LSC) condition (Figure 2(c)). The time measure of the transition from the SSC to the LSC is obtained by equating Equations (23) and (24) (Figure 2(d)), and is given as

$$t_{tr} = \frac{J_{el}}{(n+1)J^*_{steady}} \tag{25}$$

For approximating the behavior of J^*_{nt} from the SSC to the LSC, the following equation is proposed (Ehlers and Riedel, 1981; Riedel, 1987):

$$J^*_{nt} = \frac{J_{el}}{(n+1)t} + J^*_{steady} \tag{26}$$

Materials which undergo creep crack growth accompanied by small amounts of creep deformation are called creep-brittle, while those which undergo creep crack growth accompanied by large amounts of deformation are called creep-ductile.

Plasticity has an effect equivalent to eliminate the initial stage of very high J^*_{nt} for a period of t_{pl}, and J^*_{nt} can be modified as (Ohji *et al.*, 1979, 1980)

$$J^*_{nt} = \frac{J_{el}}{(n+1)(t+t_{pl})} \tag{27}$$

Under repeated fatigue loading the elastic straining effect due to SSC is recovered in every cycle of loading. The effect of the residual strains introduced by fatigue loadings on J^*_{nt} was also discussed (Ohji *et al.*, 1985; Joch and Ainsworth, 1992).

When the crack propagation rate da/dt is high, the elastic strain rate due to crack advance is significant. For $n \leq 3$, singular stress and strain fields of the order of $r^{-1/2}$ prevail. For $n > 3$, singular stress and strain fields, which are different from the HRR fields and the linear elastic singular fields and are called the HR singular fields after Hui and Riedel (1981), prevail in the very vicinity of the growing crack tip. This singularity is written as

$$\sigma_{ij} \propto \left(\frac{da/dt}{r}\right)^{1/(n-1)} \tag{28}$$

$$\dot{\varepsilon}_{ij} \propto \left(\frac{da/dt}{r}\right)^{n/(n-1)} \tag{29}$$

As is seen in Equations (28) and (29) the HR singularity is stronger than the HRR singularity. In usual situations of creep-crack growth the size of the region where HR singularity prevails is extremely small, and therefore the effect of the HR singularity is not significant.

5.06.2.4 Acceleration of J^* due to Dynamic Recovery

In addition to the SSC effect the dynamic recovery may increase the value of J^*. The creep strain rate under the steady-state creep condition is minimum and much lower than that under primary creep. Under varying loads the strain rate may be increased due to dynamic recovery and reinitiation of primary creep. This results in a considerable increase in the value of J^*, which may not be ignored in actual situations which undergo nonsteady loading.

It was shown that the HRR type singular stress and strain rate fields prevail in the vicinity of crack tip even in the presence of creep recovery and hardening (Kubo, 1983), the intensity of the singular fields being enhanced by the change in loads. The J^* was extended to primary and hardening creep laws (Ohji *et al.*, 1975; Riedel, 1981, 1983), for which J^* was denoted by C^*_h also. Combined effects of SSC and dynamic recovery were discussed (Kubo, 1981; Riedel, 1983, 1987).

5.06.3 METHODS OF EVALUATION OF FRACTURE MECHANICS PARAMETERS

5.06.3.1 Evaluation of Stress Intensity Factor

Equations for evaluating the stress intensity factor K are available in the literature. Many numerical studies were conducted on the evaluation of K. Many handbooks of the stress intensity factor compiling these results are available.

For evaluating K for geometries and situations other than found in the handbooks, the finite element method, the boundary element method, and other numerical schemes can be applied to obtain a numerical solution of the stress intensity factor K.

5.06.3.2 Evaluation of J-integral Range

Referring to the experimental evaluation method of the J-integral, the experimental evaluation of the J-integral range ΔJ_f was proposed. For center-cracked panel under tension (M(T)) (Figure 3(a)) and double-edge-cracked panel under tension (DE(T))

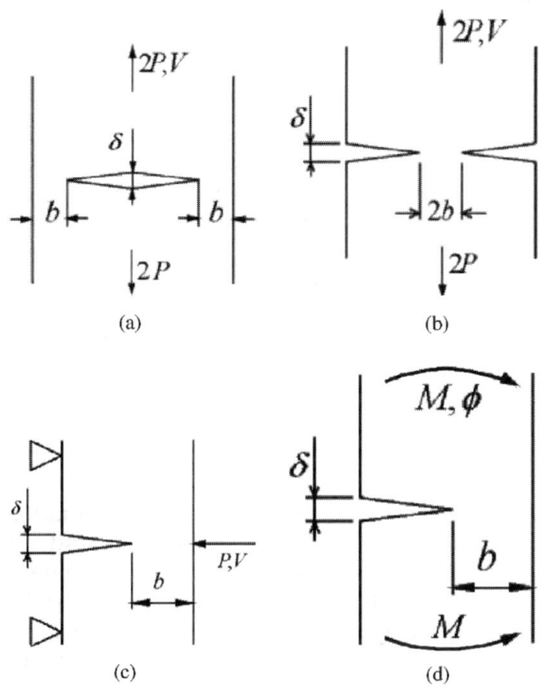

Figure 3 Geometries of common deeply-cracked specimens. (a) Center-cracked panel under tension (M(T)), (b) double-edge-cracked panel under tension (DE(T)), (c) single-edge-cracked panel under three-point bending, and (d) single-edge-cracked panel under pure bending (SE(B)).

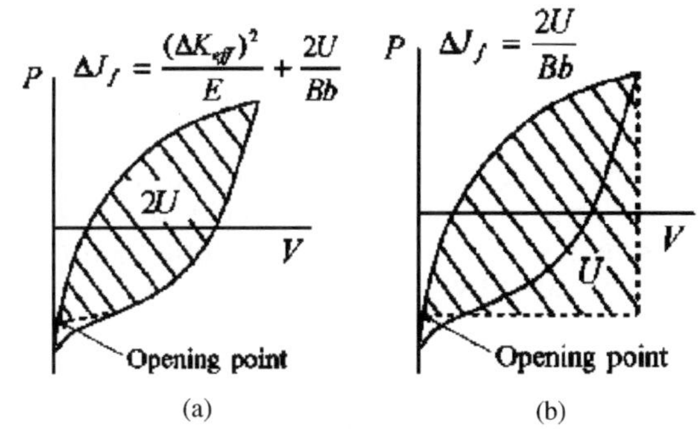

Figure 4 Experimental evaluation of ΔJ_f for deeply cracked specimens. (a) Center-cracked panel under tension (M(T)) and double-edge-cracked panel under tension (DE(T)) and (b) single-edge-cracked panel under bending.

(Figure 3(b)), the value of ΔJ_f is evaluated as

$$\Delta J_f = \frac{\Delta K_{\text{eff}}^2}{E} + \frac{2U}{Bb} \tag{30}$$

where U denotes an area in the load–displacement diagram above crack opening point shown in Figure 4(a). For single-edge-cracked panel under bending (SE(B)) (Figures 3(c) and (d)), ΔJ_f is given as

$$\Delta J_f = \frac{2U}{Bb} \tag{31}$$

where B and b are thickness and ligament length, and U denotes work done by load above crack opening point shown in Figure 4(b). When the cyclic stress–strain relation is available, the finite element method can be applied to evaluate ΔJ_f from the contour integral given by Equation (13).

5.06.3.3 Evaluation of J^*

Some but very limited equations for J^* under steady-state LSC conditions are available in the literature when the analogy between creep stress analysis and nonlinear elastic stress analysis is applied. For a small crack of length $2a$ in an infinite 2D body subjected to remote tension σ_∞, J^* is evaluated by

$$J^* = g_1(n)a\sigma_\infty\dot{\varepsilon}_\infty \tag{32}$$

Here $\dot{\varepsilon}_\infty$ denotes the creep strain rate which corresponds to σ_∞. The function $g_1(n)$ is approximated for plane stress by (Shih and Hutchinson, 1976; Ohji *et al.*, 1981)

$$g_1(n) = 3.85\,n^{1/2}\left(1 - \frac{1}{n}\right) + \frac{\pi}{n} \quad \text{for plane stress} \tag{33}$$

$$g_1(n) = \pi(\pi n/2)^{(n-1)/(2n-1)} \quad \text{for plane stress} \tag{34}$$

For a plane strain crack subjected to remote normal tension σ_∞ and transverse remote stress σ_T, $g_1(n)$ is approximated as (He and Hutchinson, 1981)

$$J^* = (3/4)\pi n^{1/2} a\,\sigma_e\dot{\varepsilon}_e\left(\frac{\sigma_\infty}{\sigma_e}\right)^2 \quad \text{for plane strain} \tag{35}$$

where σ_e is the equivalent stress calculated using σ_∞ and σ_T, $\dot{\varepsilon}_e$ being the strain rate corresponding to σ_e.

For a deeply double-edge-cracked panel with a ligament length of $2b$, J^* is approximately given by (Ohji *et al.*, 1981)

$$J^* = g_2(n)b\,\sigma_{\text{net}}\,\dot{\varepsilon}_{\text{net}} \tag{36}$$

Here σ_{net} and $\dot{\varepsilon}_{\text{net}}$ are net section stress and the corresponding strain rate, and an approximate equation for $g_2(n)$ is given for plane stress as

$$g_2(n) = \frac{4\pi}{\sqrt{n}(\pi^2 - 4)} \tag{37}$$

For a penny-shaped crack of radius a embedded in an infinite body subjected to remote tension σ_∞ normal to the crack face and σ_T parallel to the crack, J^* is approximated as

(He and Hutchinson, 1981)

$$J^* = g_3(n) a\, \sigma_e \dot{\varepsilon}_e \left(\frac{\sigma_\infty}{\sigma_e}\right)^2 \tag{38}$$

where σ_e is again the equivalent stress calculated using σ_∞ and σ_T, $\dot{\varepsilon}_e$ being the corresponding strain rate. The function $g_3(n)$ is approximated by

$$g_3(n) = \frac{6(1 + 3/n)^{-1/2}}{\pi} \tag{39}$$

The reference stress method was proposed for evaluating J^* (Webster and Ainsworth, 1994). The reference stress σ_{ref} is defined in terms of load P, and limit load P_L for yield stress σ_Y as

$$\sigma_{ref} = P\sigma_Y / P_L \tag{40}$$

In the reference stress method geometrical factor is assumed to be given by that of K^2 or the energy release rate of linear elastic material. Then J^* is given as

$$J^* = \sigma_{ref} \dot{\varepsilon}_{ref} \left(\frac{K}{\sigma_{ref}}\right)^2 \tag{41}$$

Here $\dot{\varepsilon}_{ref}$ is the strain rate corresponding to σ_{ref}.

In the engineering approach (Kumar *et al.*, 1981) J values are calculated numerically for specific configurations. The J value was tabulated in a nondimensional form. The results give expressions of J^* in the following form for the Norton type creep law of Equation (17):

$$J^* = a\sigma_{nom}\, \dot{\varepsilon}_{nom} h(n) \tag{42}$$

where σ_{nom} and $\dot{\varepsilon}_{nom}$ are nominal stress and the corresponding nominal strain rate, $h(n)$ is the so-called hardening function, which is nondimensional and depends on geometrical parameters. The value of J^* can be evaluated experimentally using displacement rate measured at certain points (Kubo, 1975; Ohji *et al.*, 1978; Smith and Webster, 1983; Ellison and Musicco, 1984).

The creep potential Π^* can be defined as

$$\Pi^* = \int_\Omega W^*\, dV - \int_{S_T} T_i \dot{u}_i\, dS \tag{43}$$

The first integral in Equation (43) is evaluated over volume Ω of the body concerned, and the second integral is evaluated over the boundary S_T, where traction is specified. By applying the analogy between J and J^*, J^* is interpreted as the release rate of Π^* with increase in area of crack surface A:

$$J^* = -d\Pi^*/dA \tag{44}$$

From Equation (43) Π^* is equal to the area under load (P)–load point displacement rate ($\dot{V}$) relation under displacement control, as shown in Figure 5. Then J^* can be graphically

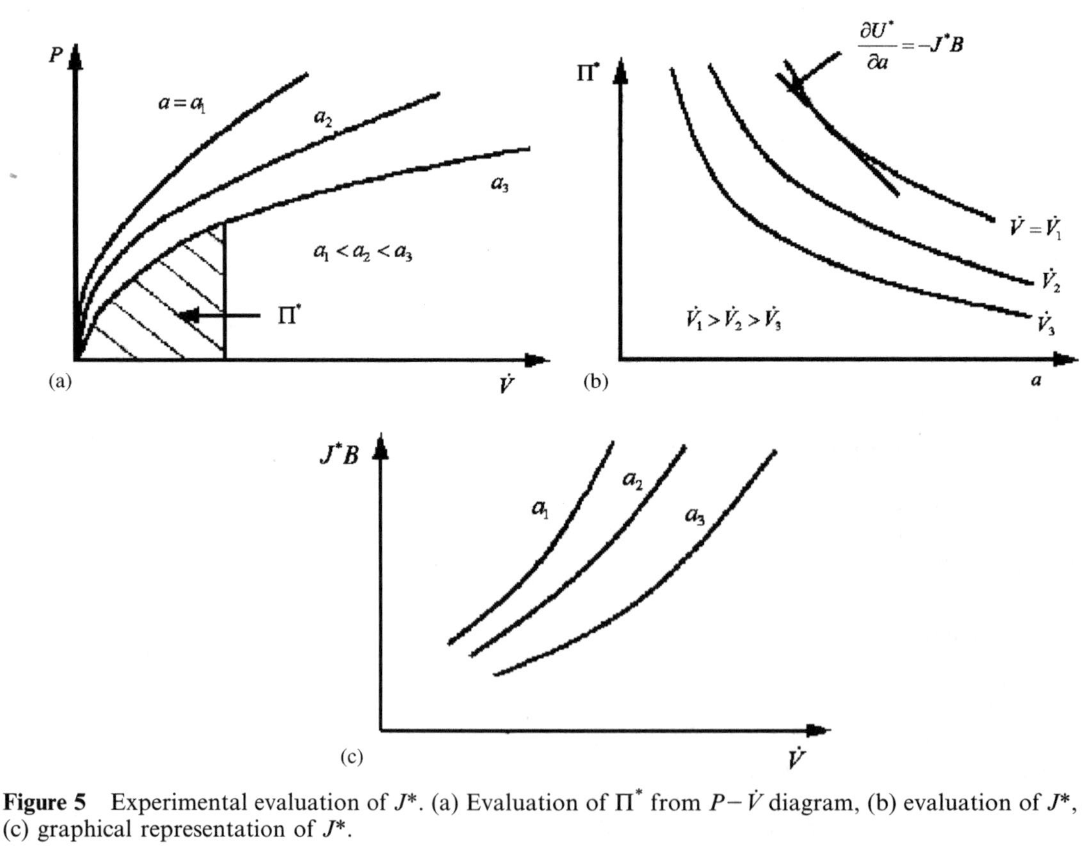

Figure 5 Experimental evaluation of J^*. (a) Evaluation of Π^* from $P-\dot{V}$ diagram, (b) evaluation of J^*, and (c) graphical representation of J^*.

determined using gradient obtained from the curve. Using the evaluation of Π^* from $P-\dot{V}$ relation, alternative expressions of J^* can be reduced from Equation (44) as

$$J^* = \frac{1}{B}\int_0^P \left(\frac{\partial \dot{V}}{\partial a}\right)_P \mathrm{d}P \tag{45}$$

$$J^* = -\frac{1}{B}\int_0^{\dot{V}} \left(\frac{\partial P}{\partial a}\right)_{\dot{V}} \mathrm{d}\dot{V} \tag{46}$$

where B denotes the thickness of plate or the length of crack front. For a deeply double-edge-cracked plate, the geometrical parameter affecting $\dot{V}$ is only the ligament length b. Then $\dot{V}$ can be given as

$$\dot{V} = C_1\left(\frac{P}{Bb}\right)^n b \tag{47}$$

with C_1 denoting a constant. By introducing Equation (47) into Equation (45) the following equation is obtained

$$\begin{aligned} J^* &= \frac{n-1}{n+1}\frac{\dot{V}P}{Bb} \\ &= \frac{n-1}{n+1}\dot{V}\sigma_{\text{net}} \end{aligned} \tag{48}$$

where σ_{net} denotes the net section stress. From this equation J^* can be readily evaluated from load and displacement rate. This equation is generalized as

$$J^* = \eta\left(\frac{n}{n+1}\right)\dot{V}\sigma_{\text{nom}} \tag{49}$$

where σ_{nom} is a nominal stress and η is a so-called eta factor, which depends slightly on specimen geometry. It was found that crack mouth opening displacement rate $\dot{\delta}$ can be replaced for $\dot{V}$ in evaluating J^* (Ohji *et al.*, 1978):

$$J^* = \eta\left(\frac{n}{n+1}\right)\dot{\delta}\sigma_{\text{nom}} \tag{50}$$

This equation using $\dot{\delta}$ gives a reasonable estimate of J^* for shallow cracks as well as deep cracks. Equations for J^* evaluation using displacement rate are summarized in Section 5.06.8 (Appendix B).

Equations (44)–(46), or the equations using displacement rates, are used for J^* value in the transient condition from the small-scale creep to large-scale creep. The experimentally determined J^* value under the transient condition is denoted by C_t (Saxena, 1986; Bassani *et al.*, 1989; Leung and McDowell, 1990). The load point displacement rate under the transient condition can be estimated from the expansion of the creep zone as

$$\dot{V} \propto t^{-(n-3)/(n-1)} \tag{51}$$

Finite element calculation showed that the J^* value evaluated using the load point displacement rate was close to the J^* integral for a path taken along the contour of specimen, not that near the crack tip (Ohji *et al.*, 1979, 1980; Leung and McDowell, 1990). By introducing Equation (51) into Equation (48) C_t is expected to be proportional to $t^{-(n-3)/(n-1)}$, not to t^{-1} found in the behavior of near-tip value J^*_{nt} given in Equation (23).

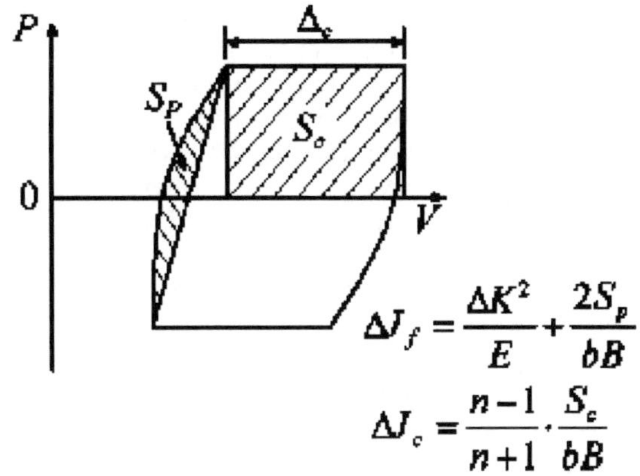

Figure 6 Experimental evaluation of ΔJ_f and ΔJ_c for deeply center-cracked panel.

5.06.3.4 Simultaneous Evaluation of J and J^*

By applying the evaluation methods of ΔJ_f and J^* described in the foregoing, these can be evaluated in fatigue loadings. ΔJ_f is sometimes called fatigue J-integral range. In comparison with the fatigue J-integral range ΔJ_f, the time integration of J^* in one loading cycle can be called creep J-integral range, and may be used for evaluating time-dependent crack propagation:

$$\Delta J_c = \int_0^\tau J^* \mathrm{d}t \tag{52}$$

Then referring to Equations (30) and (48) ΔJ_f and ΔJ_c can be experimentally evaluated from area S_p and S_c in load displacement diagram as in Figure 6, for deeply center-cracked panel (Figure 3(a)) subjected to trapezoidal load with hold time:

$$\Delta J_f = \frac{\Delta K^2}{E} + \frac{2S_p}{Bb} \tag{53}$$

$$\Delta J_c = \frac{n-1}{n+1}\frac{S_c}{Bb} \tag{54}$$

5.06.4 MACROCRACK PROPAGATION LAWS

The characteristic behavior of macrocrack propagation in high-temperature fatigue stems from the time-dependent deformation, namely "creep." We will focus on the creep effect on the macrocrack propagation in fatigue on the basis of the fracture mechanics concept. The propagation behavior is classified mainly into two types (Solomon and Coffin, 1973; Taira *et al.*, 1979a; Saxena and Bassani, 1984):

(i) Cycle-dependent fatigue where no or little creep deformation takes place in the fatigue cycle.

(ii) Time-dependent fatigue where the creep deformation dominates.

We begin to discuss the governing fracture mechanics parameter in the two types.

5.06.4.1 Cycle-dependent Fatigue

In room temperature fatigue of metallic material, a crack propagates due to blunting and resharpening of the crack tip in each cycle. Thus, the period (or frequency) of a stress (or strain) cycle has little effect on the crack propagation rate (cycle-dependent crack propagation). The propagation rate is correlated well with the stress intensity factor range, ΔK, (or effective stress intensity factor range, ΔK_{eff}) in the elastic fatigue (high-cycle fatigue) (Paris and Erdogan, 1963; Elber, 1971) with the cyclic J-integral, ΔJ_{f}, in the elastic/plastic fatigue (low-cycle fatigue) (Dowling and Begley, 1974). Even at high temperature, a crack propagates by the same mechanism if the creep deformation is negligible and the propagation shows cycle-dependent behavior. Figure 7(a) (Ohtani and Kitamura, 1988) shows the relationship between the crack propagation rate (crack extension per unit cycle), $\mathrm{d}a/\mathrm{d}N$, and ΔK in the cycle-dependent fatigue conditions of low carbon steel at the temperature of 673 K with high frequency of stress cycle, where no or little creep deformation takes place in each cycle. $\mathrm{d}a/\mathrm{d}N$ has a fairly good correlation at lower stress conditions (small scale yielding: SSY), while it departs from the correlation at higher ones. The crack opening stress does not explain the acceleration of propagation rate because there remains the deviation in the $\mathrm{d}a/\mathrm{d}N$–ΔK_{eff} relation shown by the dotted line in the figure. This is attributed to the effect of plasticity (large scale yielding: LSY), so that the crack propagation rate is governed by the cyclic J-integral (fatigue J-integral), ΔJ_{f}, as shown in Figure 7(b). The relationship except the threshold region is formulated as

$$\mathrm{d}a/\mathrm{d}N = C_{\mathrm{f}}\Delta J_{\mathrm{f}}^{m_{\mathrm{f}}} \tag{55}$$

where C_{f} and m_{f} are the material constants.

Figure 8 shows the experimental results of heat resisting alloys and steels in cycle-dependent fatigue obtained by Ohtani *et al.* (1988). The $\mathrm{d}a/\mathrm{d}N$–ΔJ_{f}, relation shows little dependence on the temperature and the test material, and the magnitude of m_{f} is ~ 1.5. It is well known that the compressive creep deformation influences the life of smooth specimen in high temperature fatigue (strain (or stress) waveform effect) (Halford *et al.*, 1973). However, it affects little on the macrocrack propagation behavior as shown in Figure 9 (Ohtani *et al.*, 1982), where $\mathrm{d}a/\mathrm{d}N$ in the cycle-dependent fatigue without compressive hold (p–p type) is compared with that in the one with it (p–c type) on a 304 stainless steel at the temperature of 923 K. In other words, the crack propagation is governed by the tensile half cycle.

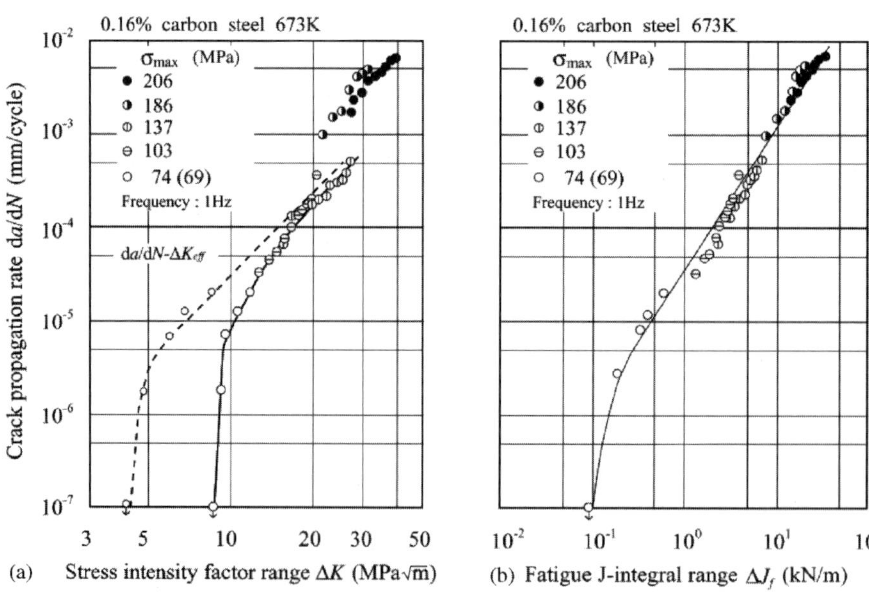

Figure 7 Relationship between $\mathrm{d}a/\mathrm{d}N$ and elastic stress intensity factor (a), and between $\mathrm{d}a/\mathrm{d}N$ and fatigue J-integral (b) range in cycle-dependent fatigue.

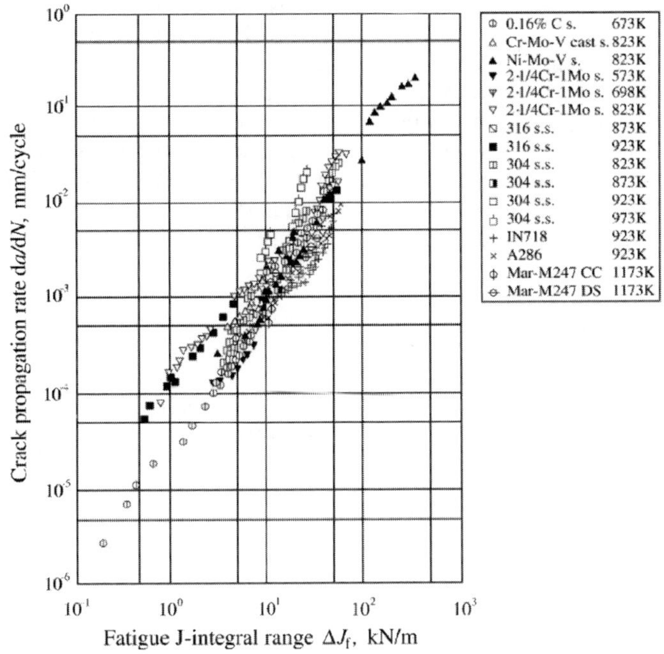

Figure 8 Relationship between crack propagation rate and fatigue *J*-integral range in cycle-dependent fatigue on several kinds of steels and alloys.

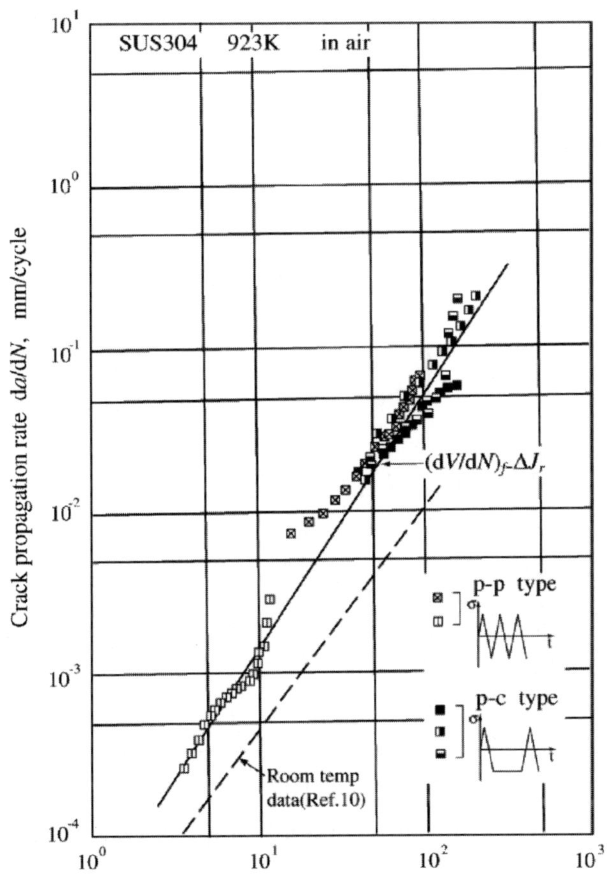

Figure 9 Effect of compressive creep on crack propagation behavior in cycle-dependent fatigue.

5.06.4.2 Time-dependent Fatigue

Figures 10(a) and (b) (Ohtani *et al.*, 1988) show the relationships between the crack propagation rate (crack extension per unit time), da/dt, and the strain rate near the crack tip, $d\varepsilon_{ct}/dt$, and $d\varepsilon_{ct}/dt$ and creep J-integral (C^*-parameter or modified J-integral), J^*, in monotonic creep of a low carbon steel at the temperature of 673 K, respectively. The strain rate is measured *in situ* by taking photographs of the elongated nickel grids, which are secured on the specimen surface by baking before the test. The figures elucidate that the creep crack propagation is controlled by the strain rate near the crack tip, which is correlated with the creep J-integral. Thus, da/dt is governed by J^* as shown in Figure 10(c) and is nearly proportional to J^*.

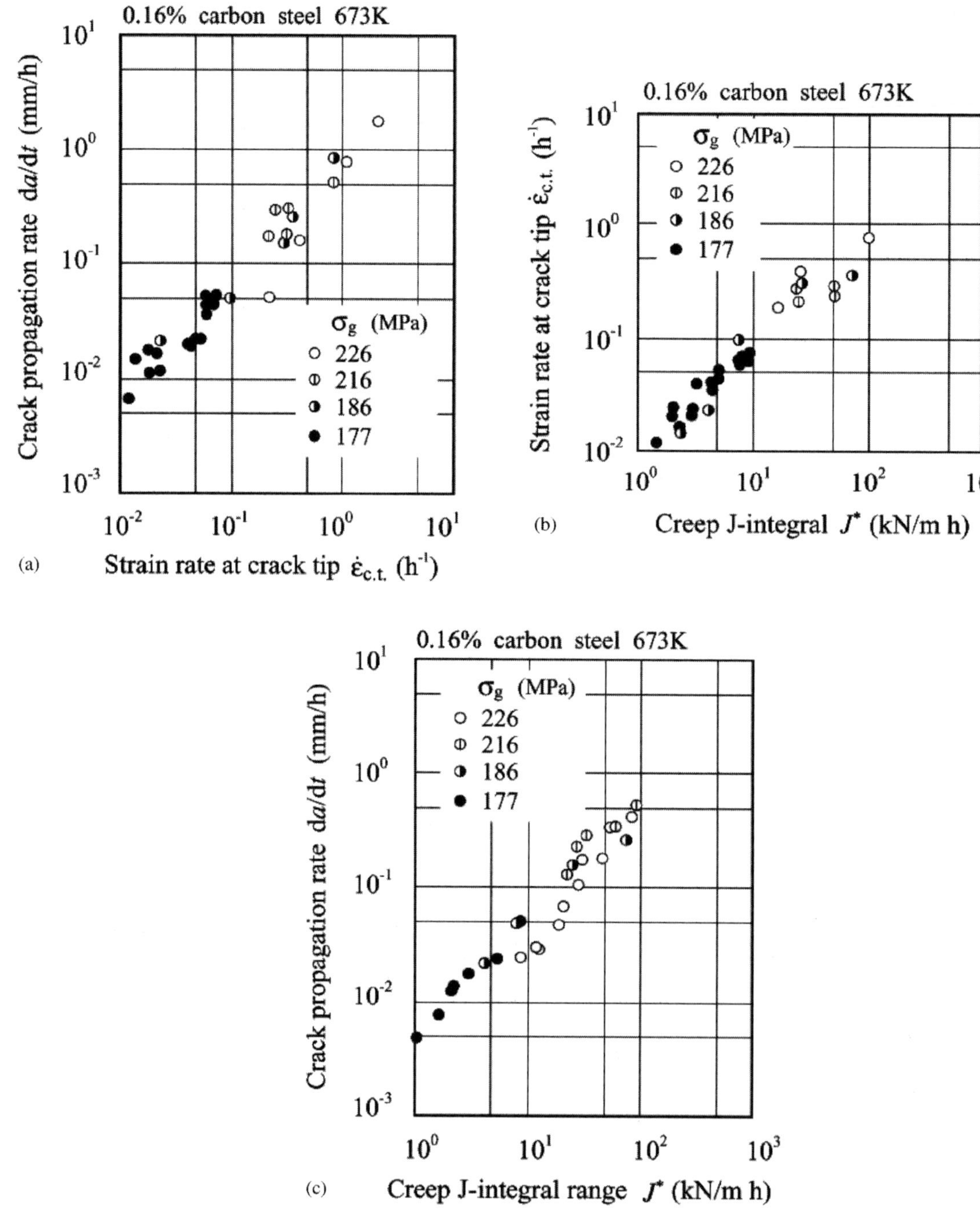

Figure 10 (a) Relationship between crack propagation rate and strain rate near the crack tip in monotonic creep. (b) Relationship between strain rate near the crack tip and creep J-integral in monotonic creep. (c) Relationship between crack propagation rate and creep J-integral in monotonic creep.

The relation is formulated as

$$da/dt = CJ^{*n/(n+1)} \approx C_c J^* \quad (56)$$

where n is the creep exponent, and C_c is the material constant (Ohji *et al.*, 1974a; Landes and Begley, 1976; Nikbin *et al.*, 1976; Harper and Ellison, 1977; Koterazawa and Mori, 1977; Sadananda and Shahinian, 1977; Taira *et al.* 1979b).

Figure 11 (Ohtani and Kitamura, 1988) shows the da/dt–$d\varepsilon_{ct}/dt$ relationship in cyclic creep (trapezoidal stress waveform with the stress holds of 1 min) of the low carbon steel at 673 K. Here, da/dt is the average crack propagation rate during the tensile hold, $(da/dN)/t_H$ (t_H: tensile hold period). The strain rate is measured *in situ* during the tensile hold period by the photographs of the elongated nickel grids. The broken lines indicate the scatter of da/dt–$d\varepsilon_{ct}/dt$ relationship in the monotonic creep shown in Figure 10(a). Figure 11 signifies that the cyclic load influences little on the creep crack propagation mechanism. This suggests the applicability of the fracture mechanics concept in the static creep on the time-dependent crack propagation in low-cycle fatigue at high temperature.

It is not easy to evaluate the crack propagation rate, da/dt, and the creep J-integral, J^*, at a moment during the loading cycle of time-dependent fatigue. The crack propagation usually is measured as the extension during the cycle, da/dN, using the crack propagation curve (the relationship between the crack length and the number of stress (strain) cycles). Considering

$$da/dN = \int_0^\tau (da/dt)\, dt \quad (57)$$

the creep J-integral range, ΔJ_c, is defined by (Taira *et al.*, 1979b, 1979c; Yoon *et al.*, 1992; Saxena, 1993)

$$\Delta J_c = \int_0^\tau J^*\, dt \quad (58)$$

where τ is the time of tension in one cycle. Then, Equation (56) is transformed into

$$da/dN = C_c \Delta J_c \quad (59)$$

which is the governing equation of creep crack propagation in cyclic loading conditions. It should be noted that ΔJ_c is in the same dimension as ΔJ_f. This is beneficial in the discussion of crack propagation behavior under the interaction between creep and fatigue.

Figure 12 (Ohtani *et al.*, 1982) shows the experimental result on the cyclic creep of a 304 stainless steel with the tension–compression holds (c–c type) and with the tension one (c–p type) at the temperature of 923 K. da/dN is correlated well with ΔJ_c as formulated by Equation (58), and the compression hold scarcely influences the relationship. The crack propagates only in the tensile half cycle, so that the J-integral (range) is the governing parameter of the crack propagation in the time-dependent fatigue. However, if one examines the figure in detail, the cyclic creep shows a little slower propagation rate, da/dt (da/dN), for a same magnitude of J^* (ΔJ_c) than the monotonic creep. Although the detailed investigation has not been carried out yet, it might be attributed to the healing of creep damage during the compressive period.

Figures 13(a) and (b) (Ohtani *et al.*, 1988) show the relationship between da/dt and J^* in the monotonic creep and the one between da/dN and ΔJ_c in time-dependent fatigue for

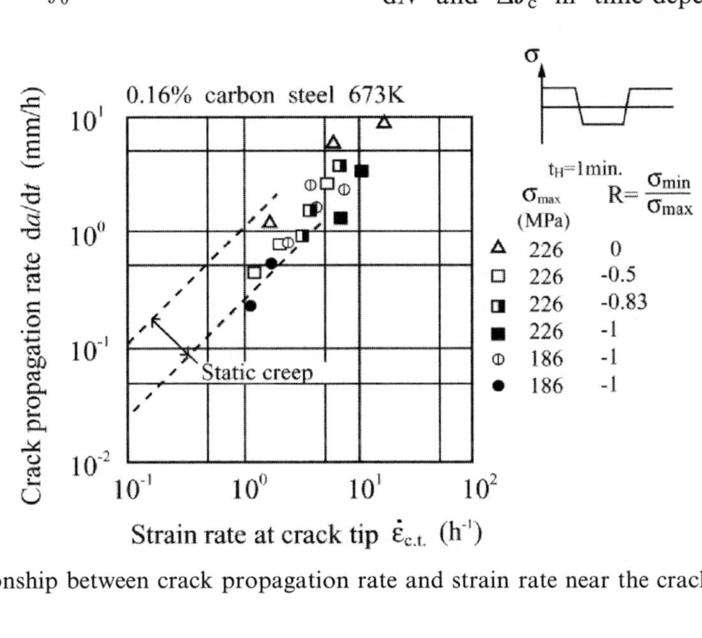

Figure 11 Relationship between crack propagation rate and strain rate near the crack tip in cyclic creep.

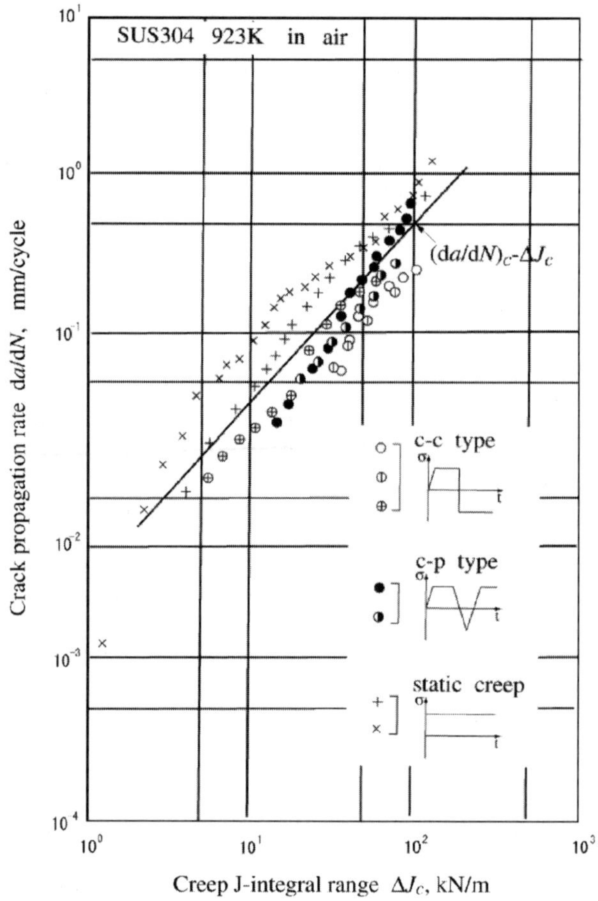

Figure 12 Effect of compressive creep on crack propagation behavior in time-dependent fatigue.

several alloys and steels that we have obtained, respectively. Though the relationship shows little dependence on the material, the magnitude of C_c, which represents the toughness for the creep cracking, is smaller with the increase of creep ductility of material. The temperature also affects little on the relationships.

5.06.4.3 Acceleration of Creep-crack Propagation due to Cyclic Loading

As discussed in Section 5.06.2.3, the magnitude of J^* ($C^*(t)$-integral or $J^*(t)$) increases with the cyclic loading. Then, the acceleration of J^* due to the primary creep (Taira *et al.*, 1979a; Ohtani *et al.*, 1981) and the small-scale creep (Ohji *et al.*, 1980; Riedel and Rice, 1980) have to be taken into account in the evaluation of ΔJ_c. The values of J^* and ΔJ_c in Figures 12 and 13(b) are evaluated by

$$J^* = \frac{n-1}{n+1}\sigma_{net}\dot{\delta} \quad (60)$$

for center-cracked cylindrical (tube with thin wall) specimen and center-cracked plate specimen which include the acceleration effect (Ohji *et al.*, 1980) through the increase of $\dot{\delta}$ in the time-dependent fatigue. Here, σ_{net} and $\dot{\delta}$ are the net section stress and the crack center opening displacement rate, respectively. In this section, we will carefully examine the acceleration effect on the crack propagation in detail.

Dynamic recovery of microstructure during the compressive or unloading process brings about the primary creep in the tensile reloading process. The transient acceleration due to the decrease in the long-range internal stress (or back stress) causes the increase of not only the strain rate near the crack tip, $d\varepsilon_{ct}/dt$, but also the crack propagation rate, da/dt. Figures 14(a) and (b) (Taira *et al.*, 1980; Ohtani and Kitamura, 1988) show experimental results of $d\varepsilon_{ct}/dt$ and da/dt in the cyclic creep (see Figure 11) with the different compressive hold stress. As the compressive hold stress is larger (the recovery effect is stronger), the acceleration of $d\varepsilon_{ct}/dt$ as well as da/dt is more remarkable. The resemblance of the dependence on the compressive stress and the half crack length suggests that the acceleration in da/dt can be

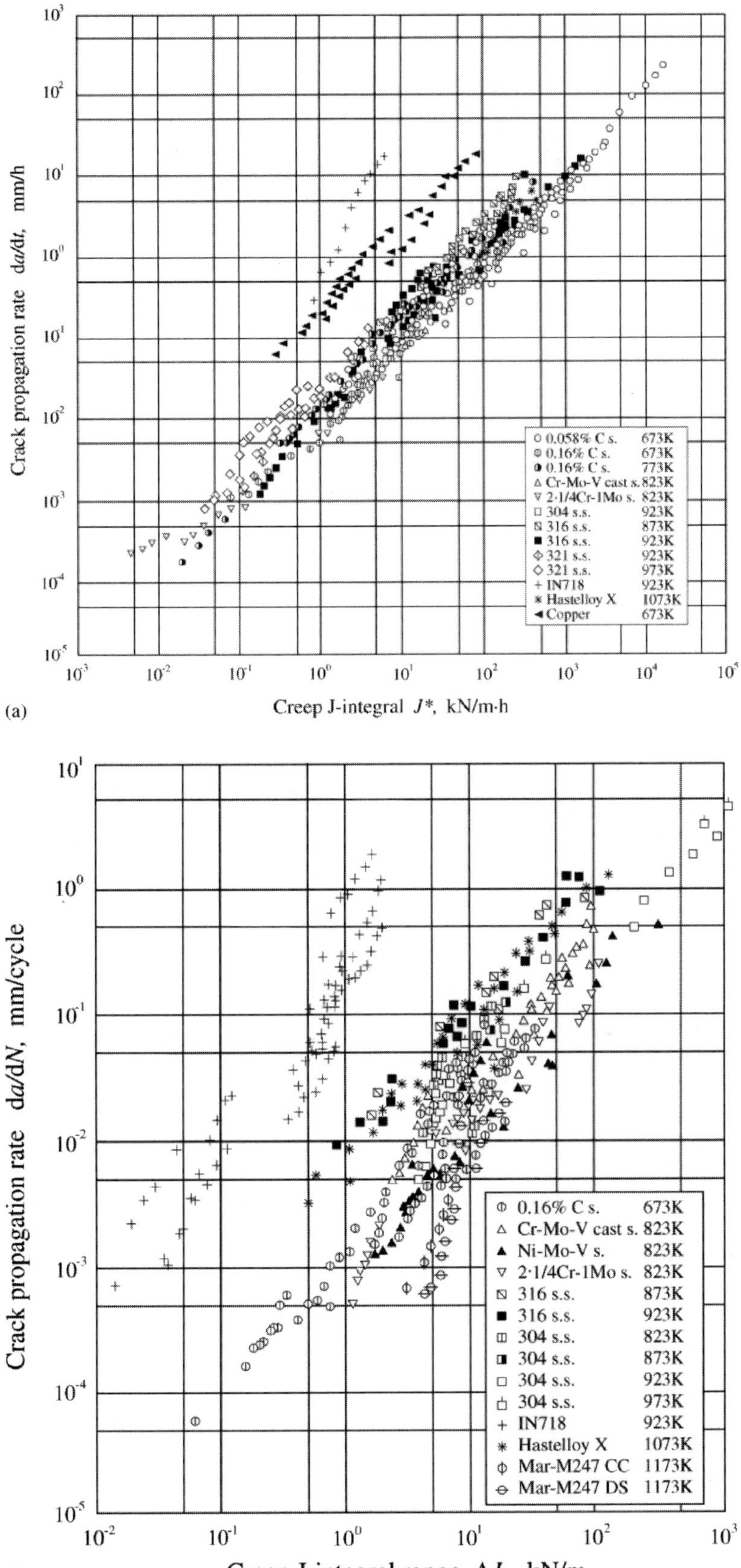

Figure 13 Relationship between $\mathrm{d}a/\mathrm{d}N$ and creep J-integral in monotonic creep (a), and between $\mathrm{d}a/\mathrm{d}N$ and creep J-integral range in time-dependent fatigue (b) on several kinds of steels and alloys.

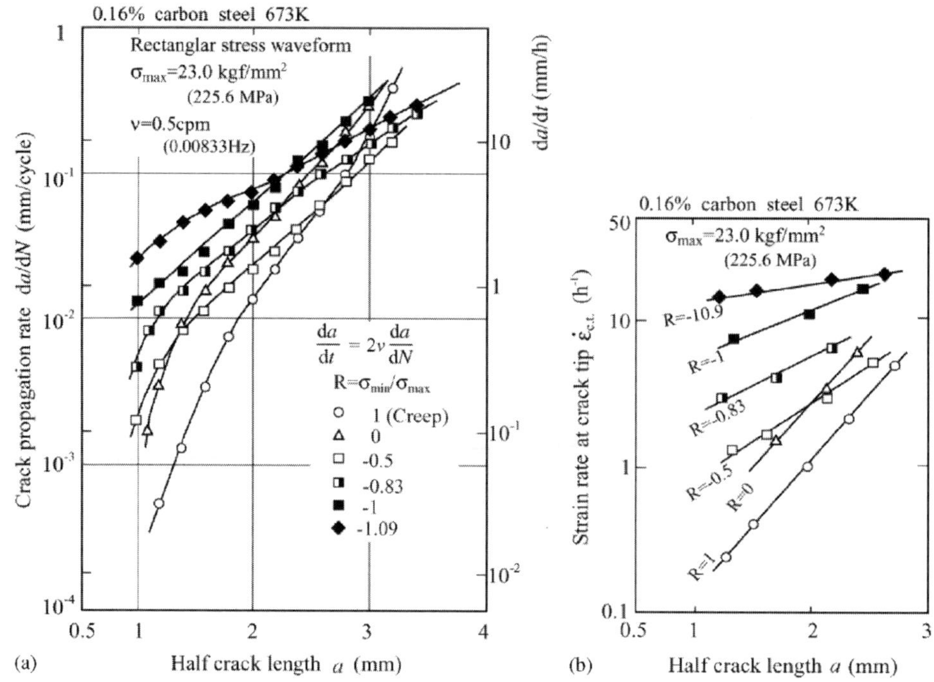

Figure 14 Effect of compressive stress hold on crack propagation rate (a), and strain rate near the crack tip (b) in time-dependent fatigue.

attributed to the primary creep. Thus, da/dN in a ductile material is correlated fairly well with ΔJ_c, which takes into account the acceleration of J^* due to the primary creep.

Another factor, which markedly accelerates the creep strain rate in the vicinity of the crack tip, is the transition from the small scale creep (SSC) to the large scale creep (LSC) under rapid loading without a large plastic zone followed by a sustained creep load. Attention must be paid to the fact that $d\varepsilon_{ct}/dt$ is governed by not only the elastic stress intensity factor, K, but also the time passed after the loading. This is because the creep zone in the vicinity of the crack tip expands as the time passes. Figure 15 (Kuwabara *et al.*, 1988) shows the experimental result of a superalloy in the time-dependent fatigue. The averaged crack propagation rate, $(da/dN)/t_H$, is controlled well with the averaged creep J-integral,

$$J^*_{av} = \Delta J_c / t_H \tag{61}$$

(Figure 15(b)), while it has a poor correlation with K (Figure 15(a)). Here, the magnitude of ΔJ_c is evaluated by substituting

$$J^* = \frac{K^2}{(n+1)(t+t_p)} \tag{62}$$

into Equation (58), where t is the time from the loading in every cycle and t_p represents the plastic effect at the loading. If the expansion of the creep zone is constrained during the stress hold, the transition from SSC to LSC does not take place. The whole stress state of the body remains in SSC condition so that the strain rate (stress) field in the vicinity of the crack tip is prescribed by only the elastic stress intensity factor, K. This means that da/dt in not only the monotonic creep but also in the time-dependent fatigue is correlated with K. Figure 16 (Ohtani *et al.*, 1993) shows an example of the creep cracking under the SSC condition. The tested material is a carbon-fiber-reinforced polymer (PEEK/AS4), and the creep deformation of matrix near the crack tip is constrained by the rigid fibers in the monotonic creep and the time-dependent fatigue at the temperature of 473 K. It is clear from the figure that da/dt is correlated well with K.

5.06.4.4 Creep–Fatigue Interaction

Figure 17(a) (Kitamura and Ohtani, 1986) shows the effect of the period of stress cycle on the crack propagation rate, da/dN, obtained from the fatigue tests on a 1Cr–1Mo–1/4V steel at the temperature of 823 K. There are distinct three regions:

(i) pure cycle-dependent region where da/dN is constant,

(ii) interaction region, and

(iii) pure time-dependent region where da/dN is proportional to v (da/dt is constant).

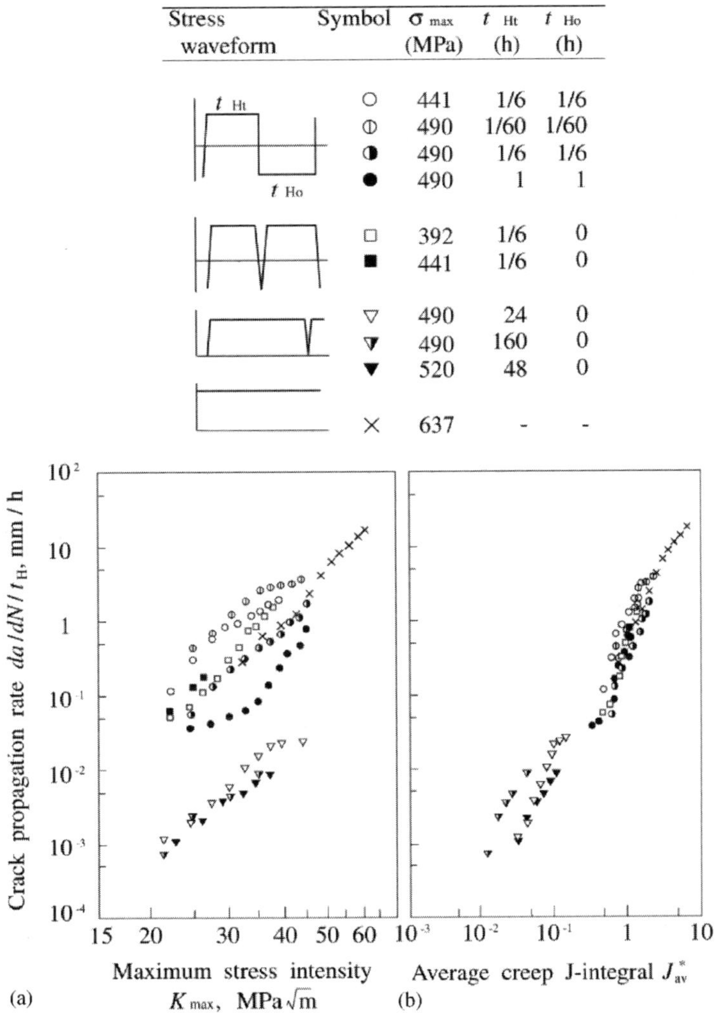

Figure 15 Relationship between $\mathrm{d}a/\mathrm{d}N$ and elastic stress intensity factor (a), and between $\mathrm{d}a/\mathrm{d}N$ and creep J-integral (b) under the transition from SSC to LSC.

As discussed in Sections 5.06.4.1 and 5.06.4.2, the crack propagation rates in the regions (A) and (C), $(\mathrm{d}a/\mathrm{d}N)_\mathrm{f}$ and $(\mathrm{d}a/\mathrm{d}N)_\mathrm{c}$, are governed by ΔJ_f and ΔJ_c, respectively. The region (C) expands (the (A) region shrinks) with increasing temperature because of activation in the creep deformation. In the intermediate frequency (the region (B)), $\mathrm{d}a/\mathrm{d}N$ observed is faster than the linear summation of $(\mathrm{d}a/\mathrm{d}N)_\mathrm{f}$ and $(\mathrm{d}a/\mathrm{d}N)_\mathrm{c}$ extrapolated from (A) and (C). Figure 17(b) shows the dependence of ΔJ_f and ΔJ_c on the cycle period for the crack propagation in Figure 17(a). It is found that the discrepancy of ΔJ_c from the pure time-dependent line with the slope 1 (ascending line) appears in the short period (high frequency) region. This means J^* is not constant except in the region (C). The acceleration of J^* is mainly attributed to the primary creep (the major cause in this case). Figure 17(b) reveals also that the discrepancy of ΔJ_f from the pure cycle-dependent line (horizontal line) arises in the long period. This comes from the change in the crack-tip plastic strain, which is accelerated by the interacted creep strain. The figures signify that the interaction between creep deformation and elastic-plastic deformation influences the crack propagation in the high temperature fatigue through the interaction between the fracture mechanics parameters.

Figure 18(a) shows the relationship between $\mathrm{d}a/\mathrm{d}N$ and ΔJ_c under the constant ΔJ_f ($=10\,\mathrm{kN\,m^{-1}}$) condition picked up from the experimental results in the high temperature fatigue of 1Cr–1Mo–1/4V steel (Figure 17). In the figure, $\mathrm{d}a/\mathrm{d}N$ must be proportional to ΔJ_c in the time-dependent fatigue (Equation (59)), while $\mathrm{d}a/\mathrm{d}N$ must be constant in the cycle-dependent one (Equation (55)). The figure indicates two distinct regimes and there is no intermediate one. Including the crack propagation in the (B) region in Figure 17(a), the propagation rate in the high temperature fatigue is dominated by one of the J-integral ranges. In the other words, Equations (55) or (59) govern the crack propagation in creep fatigue.

Figure 18(b) shows the da/dN–ΔJ_f relationship in the cycle-dependent fatigue and the da/dN–ΔJ_c one in the time-dependent fatigue. It is clear from the figure that the creep-crack propagation is faster than the fatigue one at the same magnitude of J-integral range. Accordingly, the transition condition from the cycle-dependent to the time-dependent crack propagation is prescribed by

$$\Delta J_c = \frac{C_f}{C_c} \Delta J_f^{m_f} \tag{63}$$

on the basis of Equations (55) and (59), except the region near the threshold of cycle-dependent crack propagation. Under the condition of $\Delta J_f < (\Delta J_f)_{th}$, the crack does not propagate

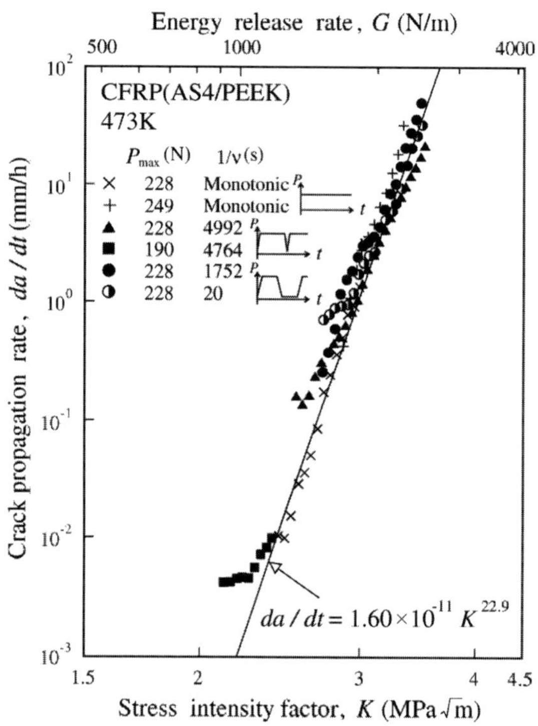

Figure 16 Relationship between crack propagation rate and elastic stress intensity factor on CFRP in monotonic creep and time-dependent fatigue.

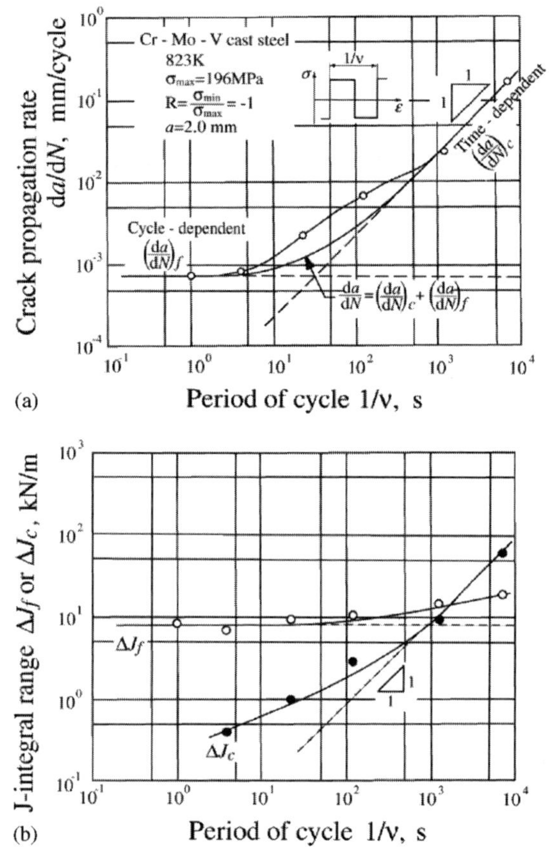

Figure 17 Effect of period of stress cycle on (a) crack propagation rate and (b) J-integral ranges.

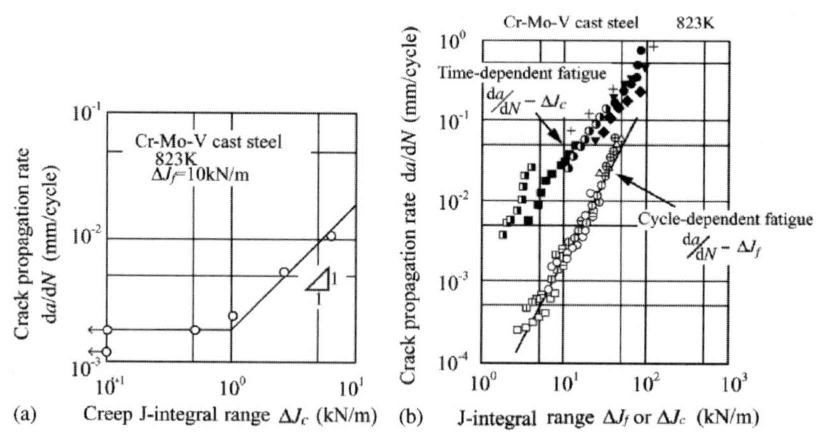

Figure 18 (a) Relationship between crack propagation rate and creep J-integral under constant fatigue J-integral range. (b) Comparison between da/dN–ΔJ_f in cycle-dependent fatigue and da/dN–ΔJ_c in time-dependent fatigue.

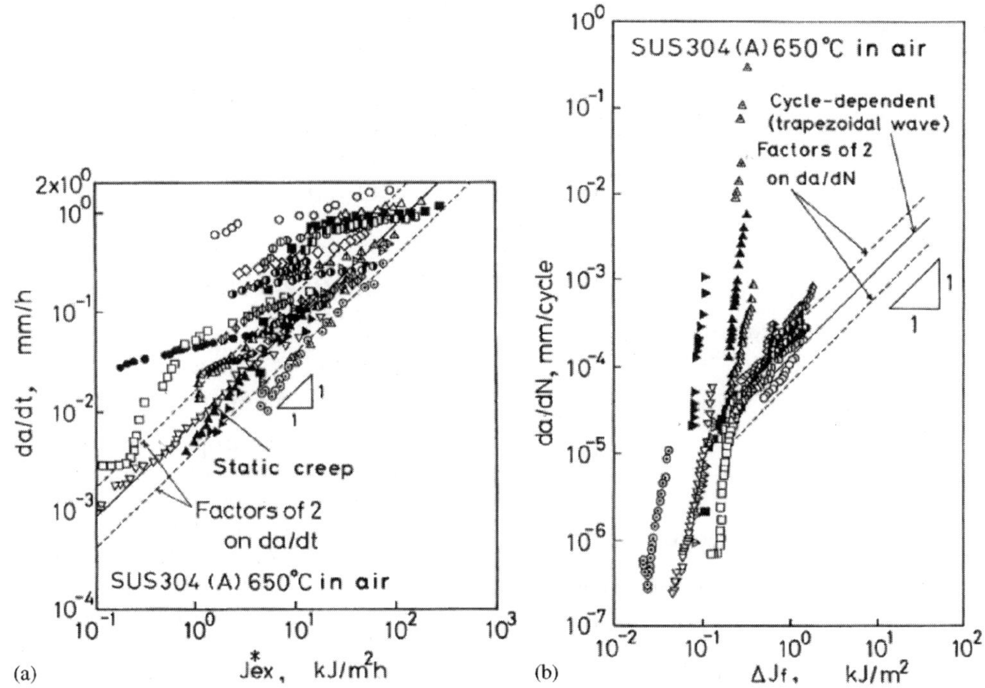

R	σ_{max0} (*MPa*)	ν (*Hz*)	*Symbol*
0·0	137	4·0	○
	137	1·1	⦶
	137	0·30	◑
	137	0·080	●
0·3	137	1·1	◇
0·5	137	1·1	□
0·7	137	1·1	▽
0·0	170	0·080	◇ (with vertical line)
0·3	170		◆ (half-filled)
0·5	170	6·0	△
	170	1·6	△ (with vertical line)
	170	0·080	◭
	170	0·008 0	▲
	170	0·001 0	△ (with dot)
0·57	170	5·0	◧
0·63	170	10·0	■
0·7	170	6·0	▷ (with line)
	170	0·080	▶
0·85	170	10·0	⊙

Figure 19 Crack propagation behavior near the threshold.

if no creep deformation takes place (Figure 7). However, the crack propagates due to creep even under the fatigue threshold as shown in Figure 19 (Ohji and Kubo, 1988). Thus, the transient condition is schematically illustrated in Figure 20.

5.06.4.5 Characterization of Crack Propagation at High Temperature

The strength of metallic materials at high temperatures has been labeled phenomenologically as monotonic creep, dynamic creep,

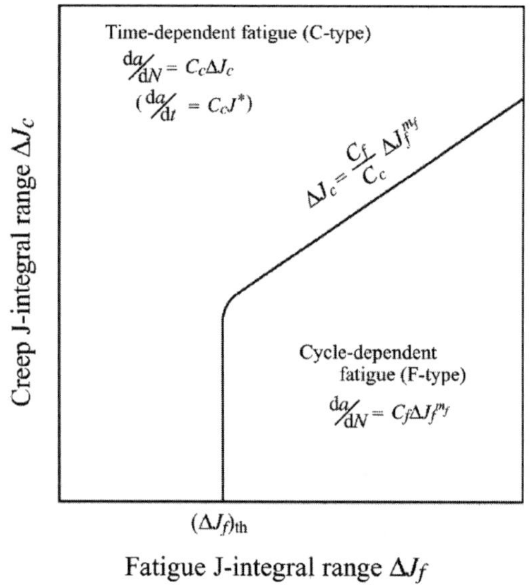

Figure 20 Schematic diagram of transient condition.

cyclic creep, low-cycle fatigue, thermal fatigue, etc., according to the waveform of stress (or strain) cycle and temperature. On the basis of the fracture mechanics laws discussed above, the crack propagation in the phenomenologically labeled condition is classified into two types, namely a cycle-dependent one (F-type) and a time-dependent one (C-type). Figure 21 summarizes the phenomenon in terms of the crack propagation. Here it should be noted that the crack propagation is governed by not the waveform of stress (strain) and temperature but the magnitude of J-integral ranges, ΔJ_f and ΔJ_c.

Because the J-integral can be defined on an unloading condition for an elastic body (although it cannot for a plastic body), the creep J-integral can be valid on a decreasing load condition (relaxation). Figure 22 (Ohtani and Kitamura, 1988) shows the da/dt–J^* relationship in the stress relaxation compared with that in the sustain load (monotonic creep) condition. This points out that the J-integral approach can be applied in the strain-hold period during creep fatigue.

The da/dN–ΔJ_f relationship in the cycle-dependent fatigue is not sensitive to the temperature as shown in Figure 8. The crack propagation rate in the time-dependent fatigue shows a strong dependence on the temperature, and its activation energy is almost equal to that of creep rate. Since the activation energy of creep J-integral is almost the same as for creep itself, the da/dt–J^* (or da/dN–ΔJ_c) depends little on the temperature. Thus, the fundamental laws (Equations (55), (59), and (63)) are applicable to the thermal fatigue if the temperature range is not large. Figure 23 (Ohtani and Kitamura, 1988) shows the crack propagation results under specific stress–temperature conditions simulating the thermal fatigue. The creep effect is stronger than the fatigue effect under the in-phase type (tension at high temperature) and vice versa in out-of phase type (compression at high temperature).

5.06.5 PROPAGATION OF SMALL CRACKS

5.06.5.1 Effect of Microstructure on Crack Propagation

The macrocrack propagation test is conducted under the time-dependent fatigue using a center-cracked plate specimen of an oxide (Y_2O_3) dispersion strengthened nickel base superalloy, MA754, which has coarse and elongated grains (Figure 24(a), Kitamura *et al.*, 1993). Figures 24(b) and (c) show the cracking morphology and the crack propagation curve, respectively. They indicate the retardation of transverse propagation by the blocking, kinking, or branching at longitudinal grain boundaries. Alternatively, the crack propagates swiftly in the grain. The alternating period of acceleration/deceleration synchronizes with the transverse grain diameter. The relationship between da/dN and ΔJ_c is shown in Figure 25. Although da/dN is almost proportional to ΔJ_c, there are large amounts of scatter due to the effect of microstructure. This suggests that the fracture mechanics laws apply on the average propagation behavior.

It should be noted that the microstructure has strong influence on the propagation of a crack, which is smaller than a few grain diameters (see also Chapter 4.05). Since the cracks initiate at an early stage of life in low-cycle fatigue in general, it is important to understand the characteristics of small cracks.

5.06.5.2 Morphology of Small Cracks on Smooth Surfaces

The strain-controlled fatigue tests are conducted under a fast–fast strain waveform (cycle-dependent fatigue: F-type) and a slow–fast wave-form (time-dependent fatigue: C-type) at the temperature of 923 K in a vacuum using a cylindrical specimen of a 304 stainless steel with a smooth surface. The test is interrupted at regular intervals and the cracking morphology is observed by means of an optical microscope. In particular, the number of small cracks and their length are measured carefully.

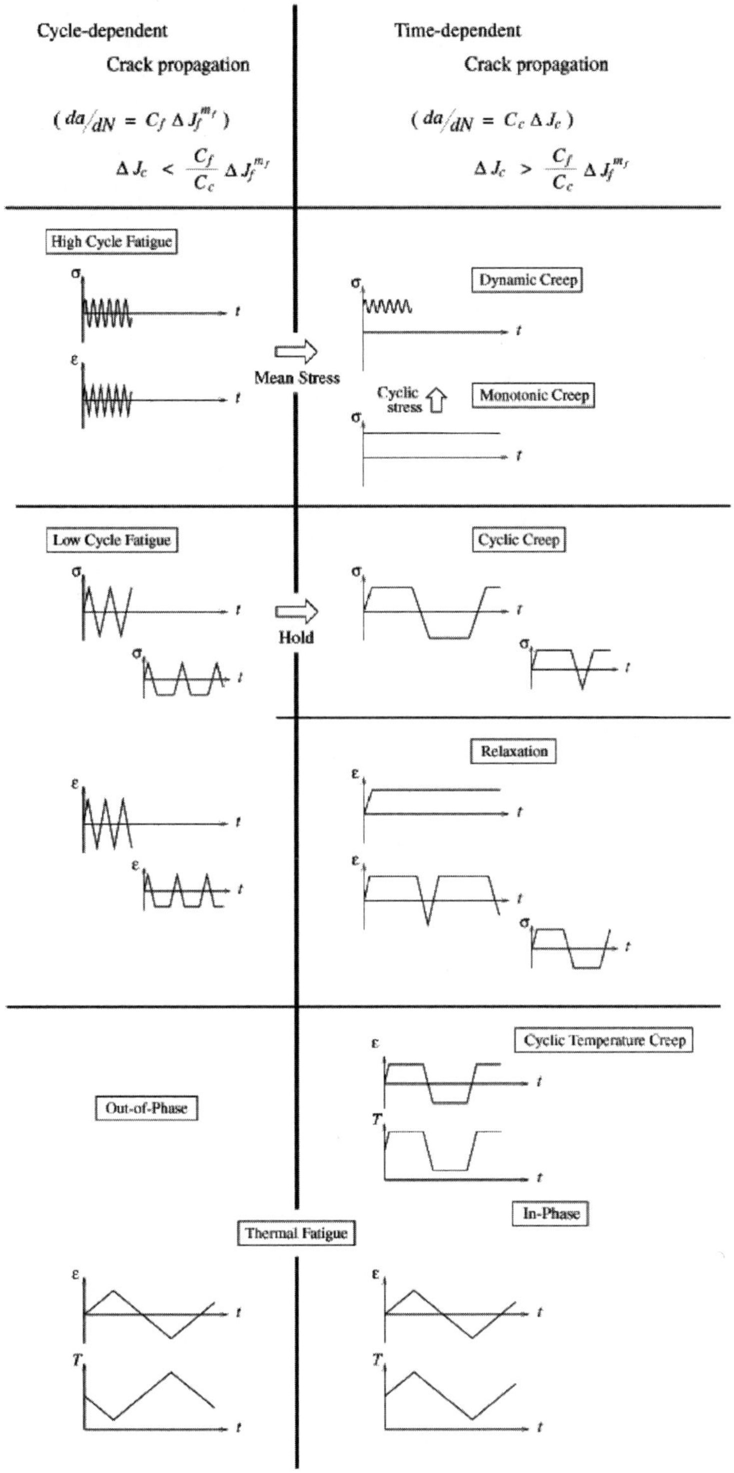

Figure 21 Characterization of high-temperature strength in terms of crack propagation.

Figure 26(a) shows the morphology of small cracks initiated and grown on the surface in the cycle-dependent low-cycle fatigue of $\Delta\varepsilon_t = 1\%$ ($\Delta\varepsilon_t$: total strain range) (Ohtani and Kitamura, 1994). The cracks begin to initiate at less than 5% of the life ($N_f = 3{,}300$ cycles). The crack density, which is the number of cracks per unit surface area, increases with the increase of the strain cycles (Figure 27). The growth and coalescence of several small cracks form the main crack, which dominates the fatigue life of the specimen. Figure 26(b), which is a micrograph of an intergranular crack that initiated, shows that the crack initiates at the

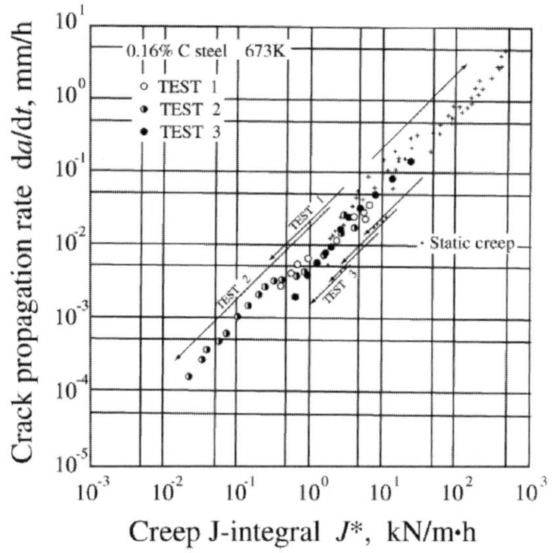

Figure 22 Time-dependent crack propagation under stress relaxation.

intersection of a grain boundary and slip bands. This suggests that the crystallographic slips govern the crack initiation, and the mechanism is similar to that in room temperature fatigue.

Figure 28(a) (Ohtani and Kitamura, 1994) shows the failure process in the time-dependent fatigue of $\Delta\varepsilon_t = 1\%$. The life ($N_f$ = 388 cycles) is much shorter than in the cycle-dependent fatigue under the same magnitude of $\Delta\varepsilon_t$. Again, the small cracks begin to initiate at the very early stage of life. The crack density is much higher than that in the cycle-dependent one and reaches 50 cracks mm^{-2} at the middle stage (Figure 27). The decrease at the late stage is due to the frequent crack coalescences because of the high density. Thus, the propagation and coalescence of small cracks dominate the failure process in low-cycle fatigue at high temperature. Although the cracks are also intergranular in the time-dependent fatigue as shown in Figure 28(b), the morphology is different from the one in the cycle-dependent

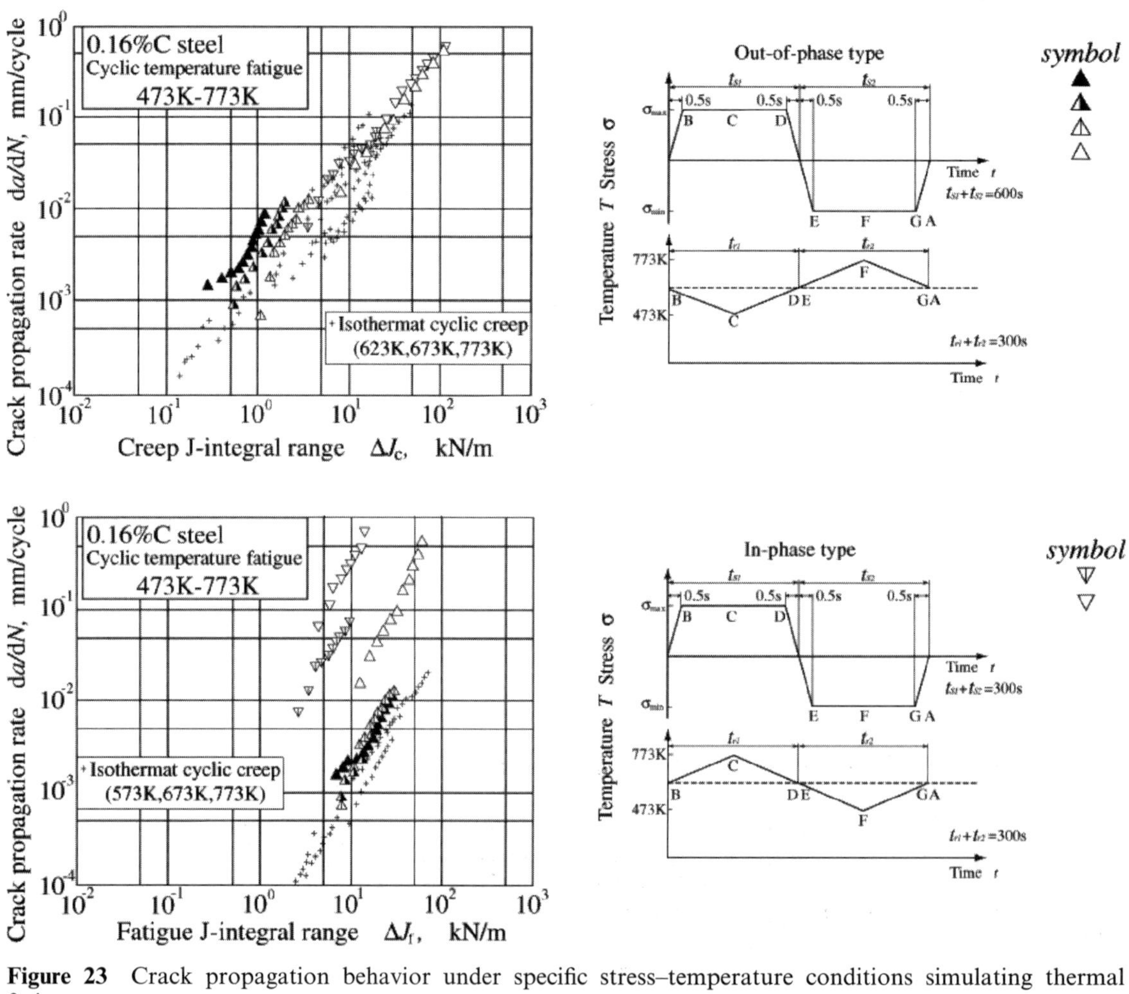

Figure 23 Crack propagation behavior under specific stress–temperature conditions simulating thermal fatigue.

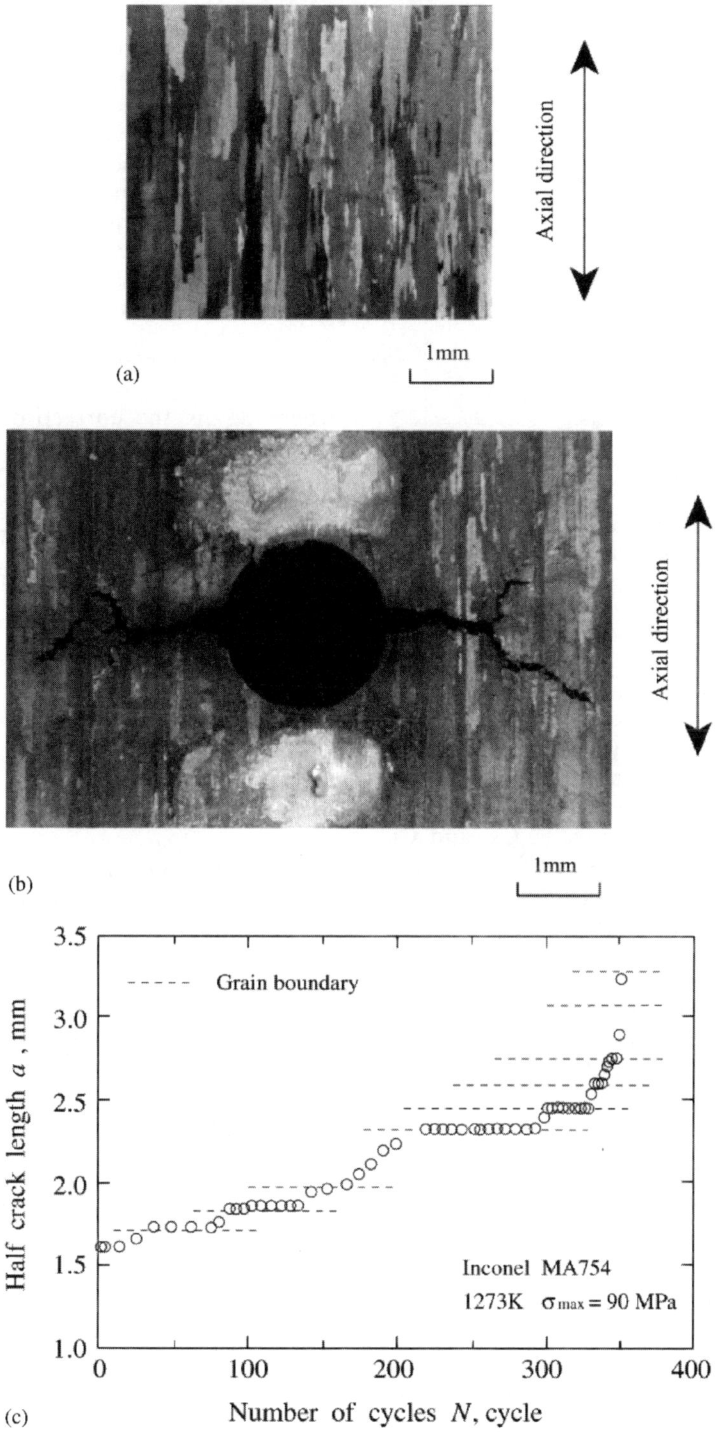

Figure 24 (a) Microstructure of ODS superalloy, MA754, (b) cracking morphology, and (c) crack propagation curve.

fatigue. The array of cavities is found along the crack, suggesting the preferential behavior of grain boundary sliding at the surface of material. Here, the observation on the cross-section (Figure 29(a)) reveals that the cracks appear only on the surface in this case (Cs-type cracking). Moreover, *in situ* observation (Figure 30) reveals that the Cs-type crack progresses facet-by-facet.

Under a creep-fatigue at higher temperature in slower tensile strain rate, $d\varepsilon_t/dt$, many cracks initiate inside of specimen due to the cavities caused by diffusion creep, and they form a main crack (Figure 29(b): Ci-type cracking)

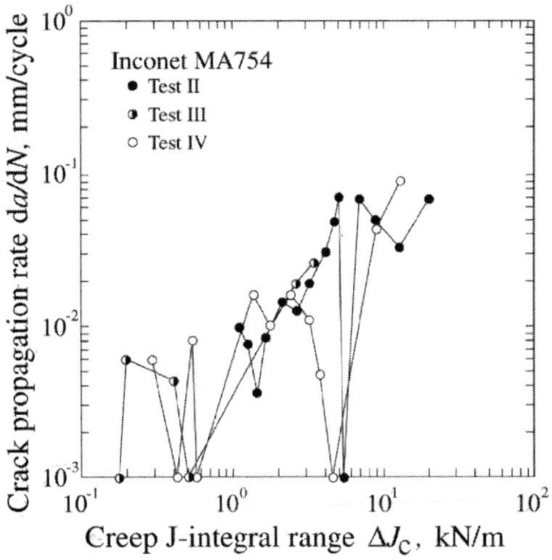

Figure 25 Relationship between crack propagation rate and creep J-integral range in time-dependent fatigue of MA754.

owing to the coalescences (Zhou *et al.*, 1995). Figure 31 shows the inner crack density on a cross-section of a specimen under the creep-fatigue with 10^{-4}–1% s^{-1} (tensile strain rate—compressive strain rate) at the temperature of 1,073 K in air.

The three types of cracking, F, Cs, and Ci, are tabulated in Figure 32(a), and the modes are summarized as a map (Figure 32(b)) in terms of the test conditions such as tensile strain rate, compressive strain rate, and temperature (Zhou *et al.*, 1995). Figure 33 shows the fatigue life, which strongly depends on the cracking morphology. The F-type gives the longest life, while the Ci-type gives the shortest.

5.06.5.3 Propagation Behavior of Small Cracks

From the micrographs on the specimen surface (e.g., F-type: Figure 26(a) and Cs-type: Figure 34), the change in the crack length is examined carefully for all small cracks that appeared in the observed area. The propagation rate of each crack, da/dN, in F-type and Cs-type is evaluated by the propagation curve (the relationship between the crack length and the number of cycles).

Figures 35(a) and (b) (Ohtani and Kitamura, 1990, 1994) show the propagation (growth) rates of small cracks in the F-type and Cs-type, respectively. Since the crack-propagation rate has remarkable scatter, the upper bound, lower bound, and mode are shown in the figure. The time-dependent crack propagation shows a faster rate than the cycle-dependent one. The scatter in the propagation rate in both types is over several orders at the crack length shorter than 20 μm. The upper bound is almost independent of the crack length, though the lower bound is accelerated for the longer crack. The crack propagation rate tends to converge as the crack becomes longer.

For a semicircular surface crack under a uniform tensile remote stress, σ, of elastic–plastic body in an infinite plate, the magnitude of J-integral is given by

$$J = M_J\left(\pi\sigma\varepsilon_e + f(n_p)\sigma\varepsilon_p\right)a \tag{64}$$

where M_J is the correction factor for crack shape and boundary condition ($M_J = 0.51$ for a semicircular surface crack), ε_e is the elastic remote strain, ε_p is the plastic remote strain, n_p is the reciprocal of strain hardening exponent, and $f(n_p)$ is the function of n_p. Substituting the fatigue J-integral range, which is evaluated from Equation (64) in a tensile half cycle, into Equation (55),

$$da/dN \propto a^{m_f} \tag{65}$$

is derived readily. By a similar procedure,

$$da/dN \propto a \tag{66}$$

is obtained from the macrocrack propagation law in the time-dependent fatigue. Here, the creep J-integral, J^*, for the small crack in an infinite plate of ductile creep body under LSC is in the form

$$J^* = M_J f(n)\sigma\dot{\varepsilon}a \tag{67}$$

The solid line indicated by "macrocrack propagation law" in Figures 35(a) and (b) points out the relationship obtained from the fracture mechanics laws of 304 stainless steel at 923 K. The mode–crack propagation rates coincide with the extrapolated line of the fracture mechanics law in the cycle-dependent fatigue as well as the time-dependent one.

Figure 36 (Ohtani and Kitamura, 1990, 1994) shows the effect of strain range on the propagation rate of small cracks. The average propagation rate coincides with the extrapolation of macrocrack propagation law. This points out that the average behavior of small-crack propagation obeys the fracture mechanics law for the macrocrack though the microstructure brings about a large scatter on it.

5.06.5.4 Monte Carlo Simulation

Because the microstructurally small cracks show a remarkable scatter in the initiation times, the location, and the propagation rate,

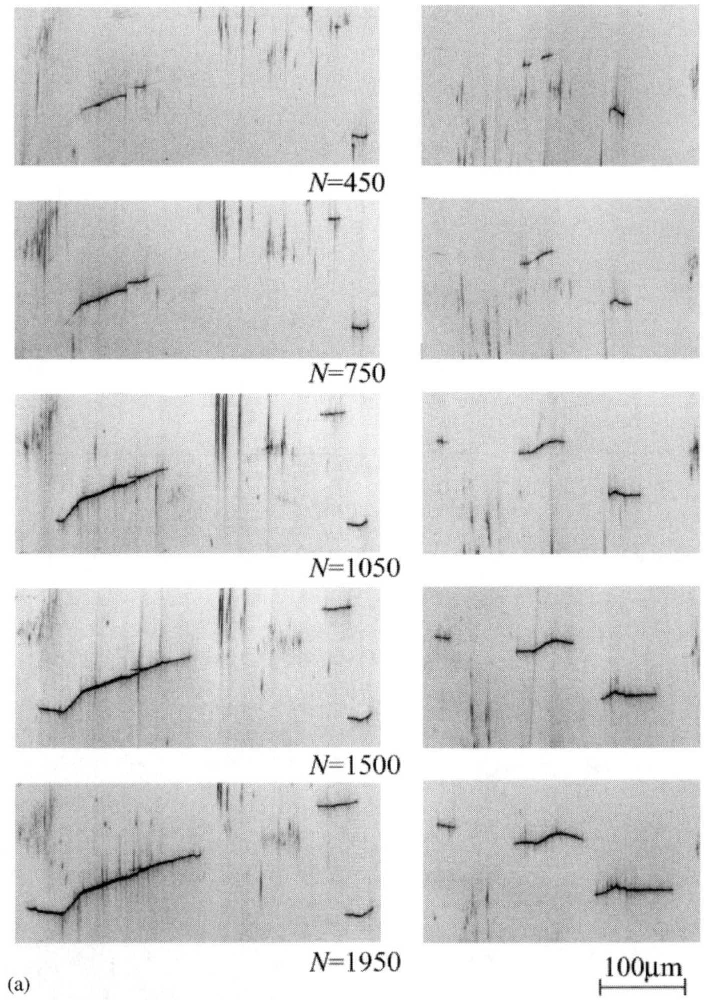

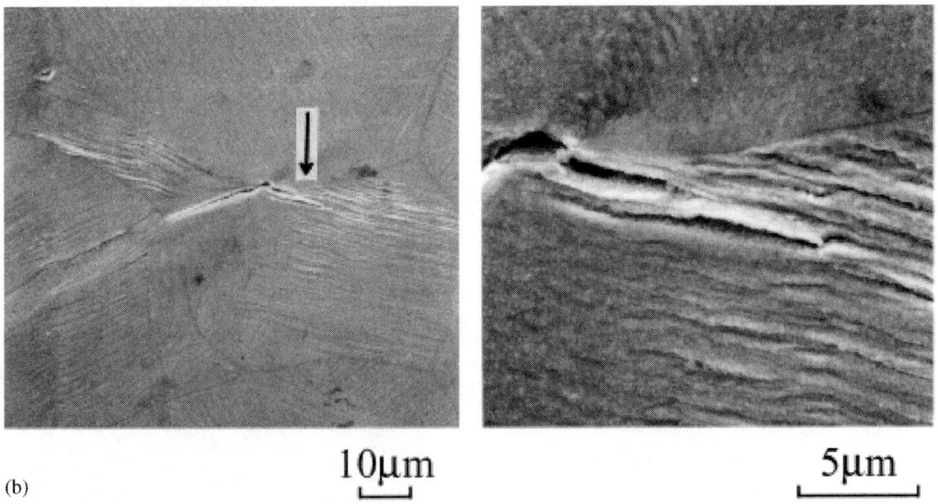

Figure 26 (a) Morphology of small cracks initiated on specimen surface in cycle-dependent fatigue and (b) magnified view of small crack in cycle-dependent fatigue.

the random characteristics should be analyzed on the basis of a stochastic concept. The Monte Carlo simulation is one of the powerful tools which can handle the stochastic nature (Ohtani and Kitamura, 1987). However, the characteristic has not yet been well understood in detail because of the difficulty of experimental observation and the complexity of microstruc-

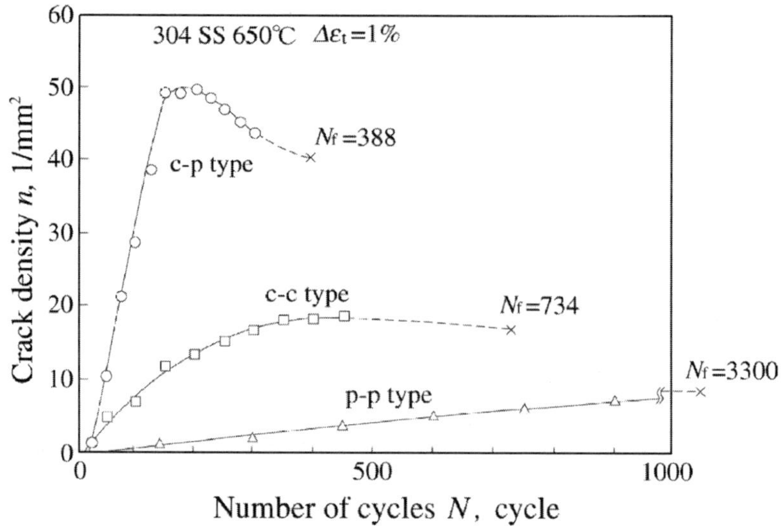

Figure 27 Density of small cracks initiated on specimen surface.

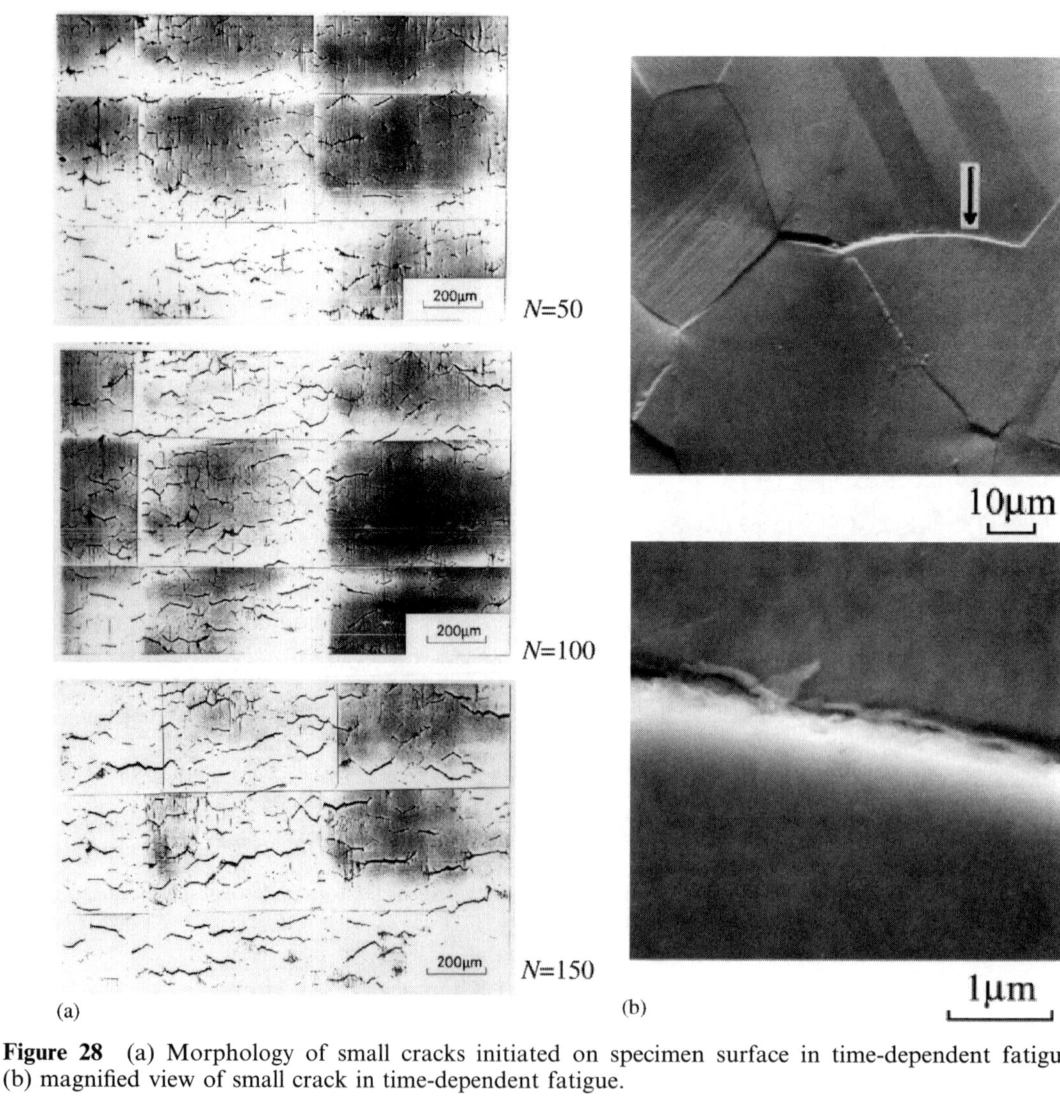

Figure 28 (a) Morphology of small cracks initiated on specimen surface in time-dependent fatigue and (b) magnified view of small crack in time-dependent fatigue.

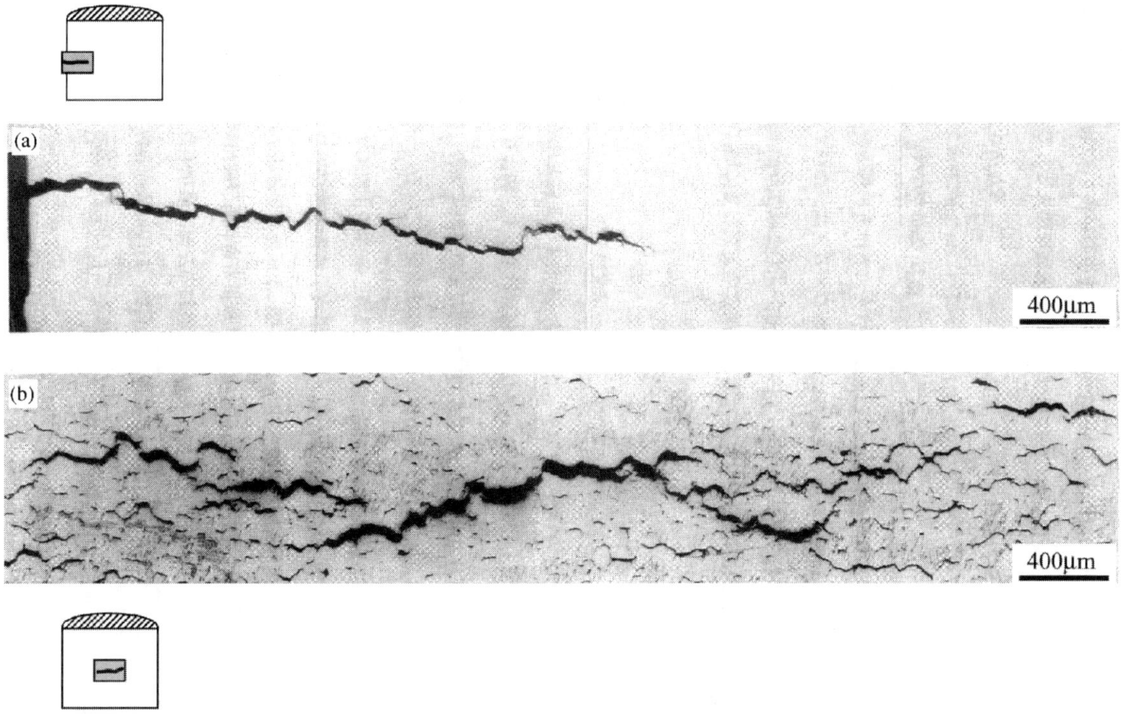

Figure 29 (a) Cs-type cracking and (b) Ci-type cracking on cross-section of specimen.

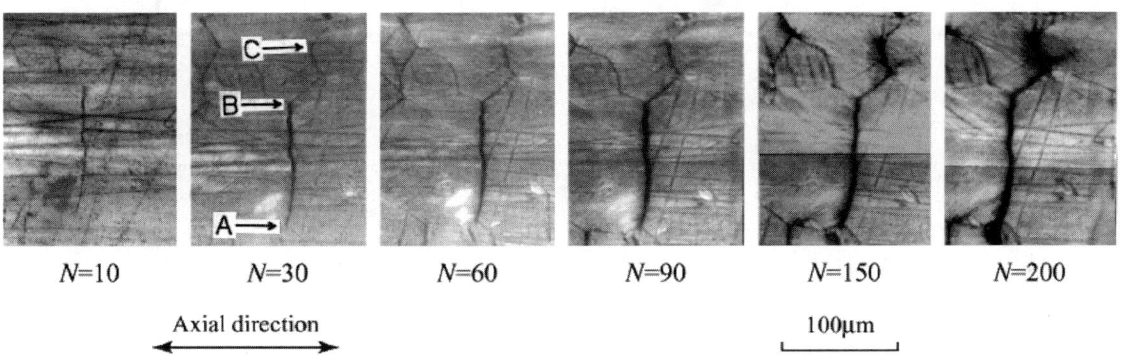

Figure 30 *In situ* observation on process of small crack propagation in time-dependent fatigue.

tural effects. In order to get a key to analysis, a simplified model for the simulation of initiation and propagation of intergranular small cracks in the creep-fatigue is developed by one of authors. The model and numerical simulation method (Kitamura *et al.*, 1992) is shown in this section as a handhold for future investigation.

Assumptions of simulations are summarized below.

(i) Crack initiation.

(a) Each grain boundary has its own magnitude of resistance; R. (R_i: the resistance at the ith grain boundary.) R is a stochastic parameter depending on the microstructure of the grain boundary.

(b) When the grain boundary is subjected to the creep-fatigue, the driving force for the

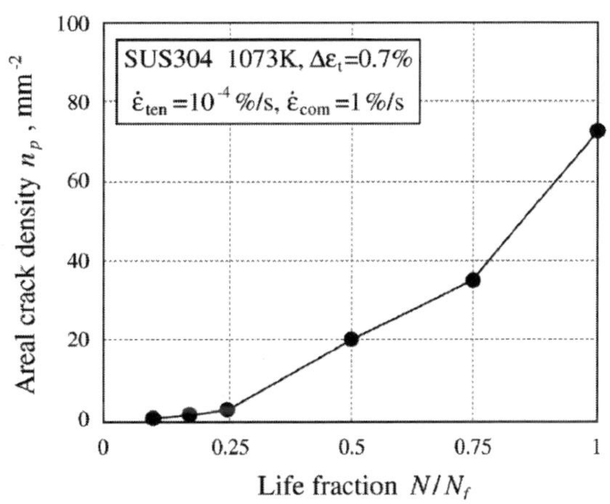

Figure 31 Density of inner cracks (Ci-type) on cross-section of specimen.

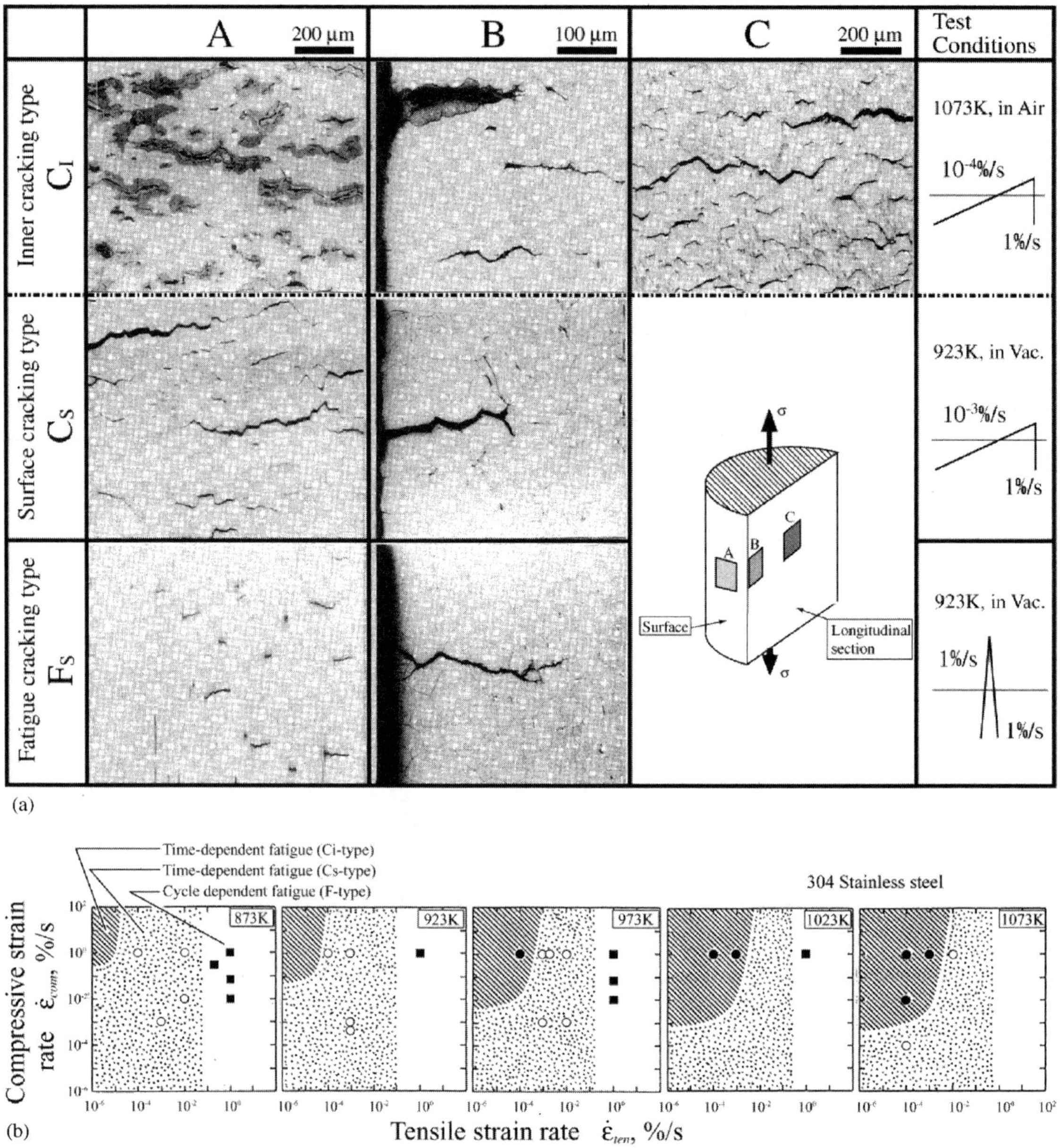

Figure 32 (a) Creep-fatigue cracks classified into three types and (b) creep-fatigue fracture mode map.

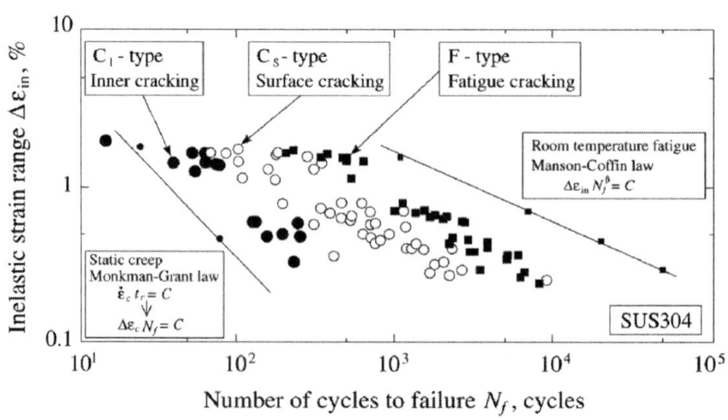

Figure 33 Effect of fracture mode on creep-fatigue life.

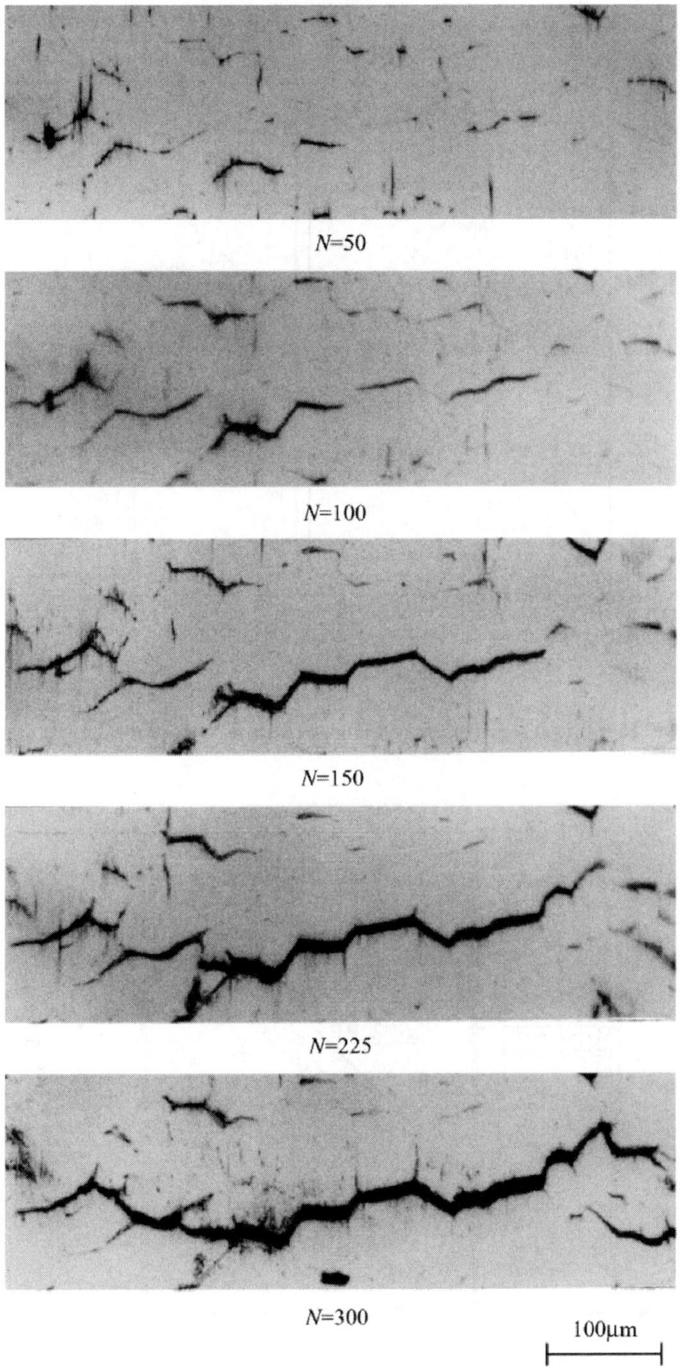

Figure 34 Crack propagation in Cs-type.

crack initiation, F, is applied on the grain boundary in every strain cycle. (F_i: the driving force at the ith grain boundary for crack initiation.) F is a stochastic parameter depending on the load, angle of grain boundary to the stress axis, etc.

(c) The resistance reduces by F in every cycle.

(d) When the residual resistance in the ith grain boundary, $R_i - F_i N$, reaches zero, it fails and brings about an intergranular crack with a one grain boundary length.

(ii) Crack propagation.

(e) Once a crack appears, the driving force for cracking at the neighboring grain boundary, K_F, is accelerated by the stress intensity. K_F is a stochastic fracture mechanics parameter, which depends on the crack length, angle to the stress axis, etc. (K_{Fi}: the driving force at the grain boundary for crack propagation.)

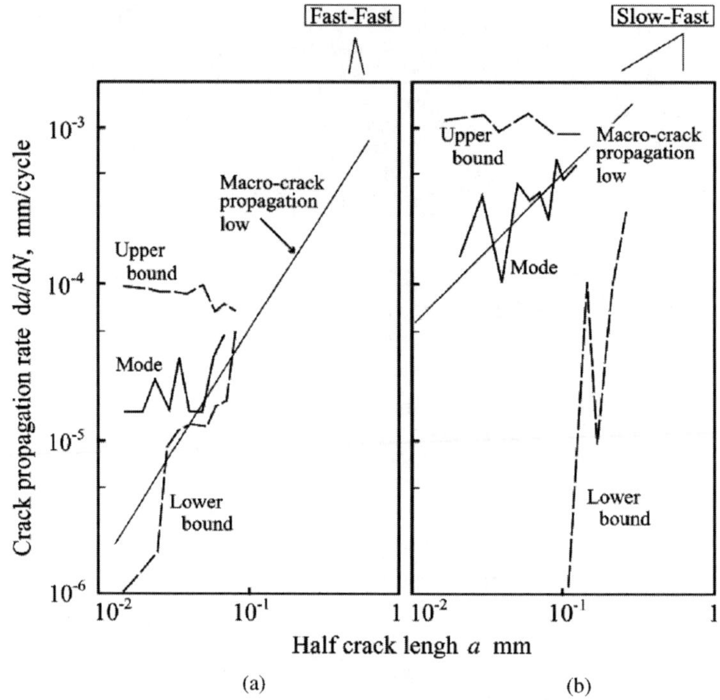

Figure 35 Propagation rate of (a) F-type crack and (b) Cs-type crack.

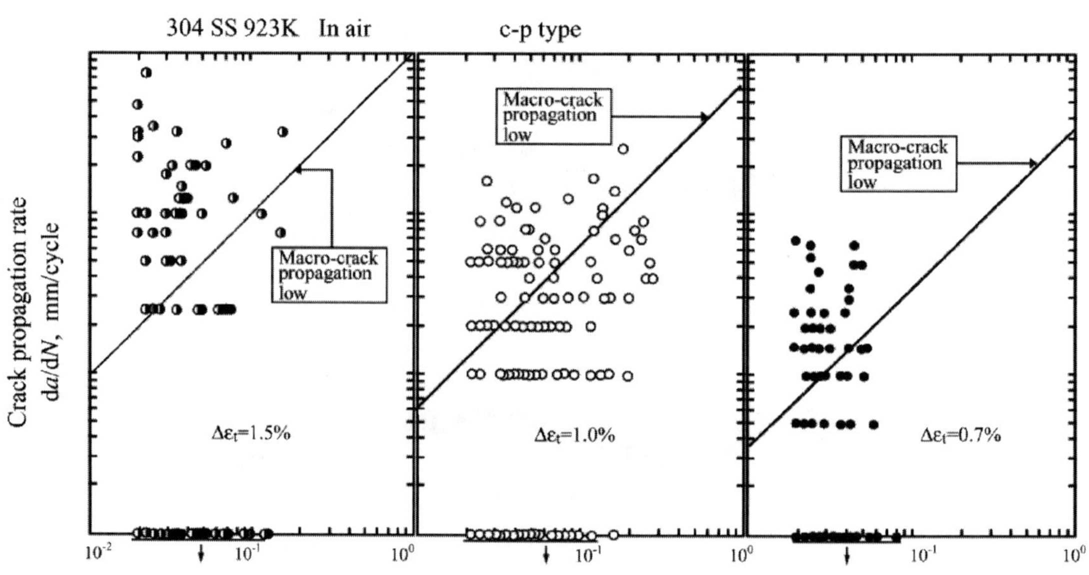

Figure 36 Effect of strain range on propagation rate of small cracks.

(f) The resistance at the grain boundary reduces by K_F in every cycle after a crack initiates at the neighboring grain boundary.

(g) When the residual resistance at the neighboring grain boundary reaches zero, it fails and the crack propagates in one grain boundary length.

The process of crack initiation and propagation is numerically simulated by a Monte Carlo method, in which the stochastic variables are determined using the random numbers generated by a computer.

Figure 37 (Tada *et al.*, 1996) shows the 2D model for inner cracking (Ci-type) simulation, where the grain boundary facets in the material are represented by the projection on a plane perpendicular to the stress axis and their stratification. In the Cs-type cracking, it is impossible to observe continuously the initiation and propagation process of inner small

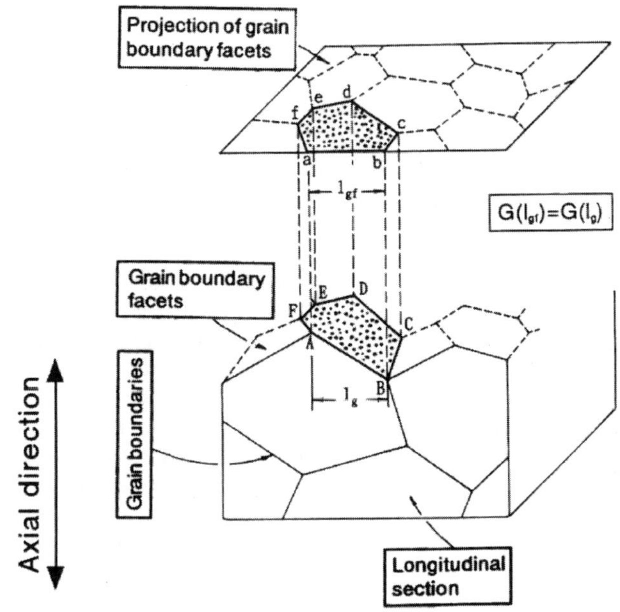

Figure 37 Model for Monte Carlo simulation on inner cracking.

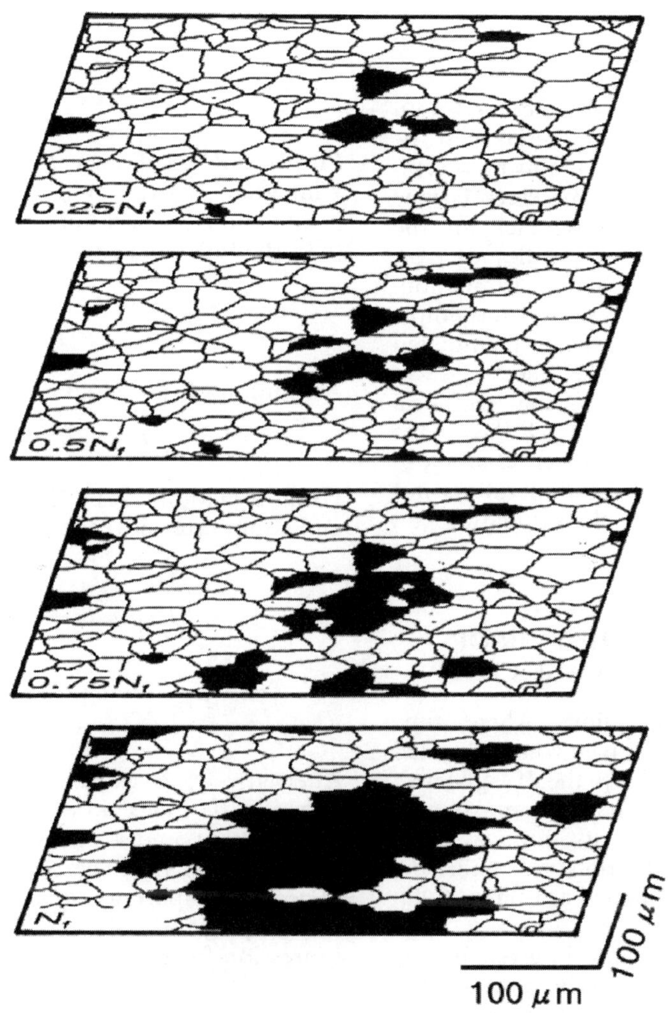

Figure 38 Visual expression of inner crack propagation.

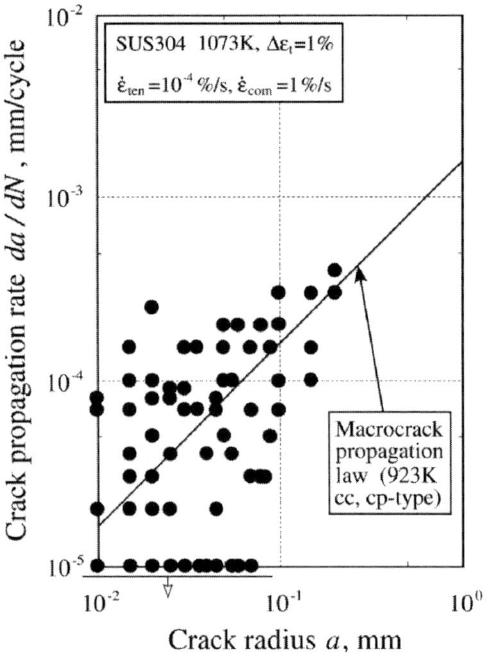

Figure 39 Crack propagation rate obtained by the Monte Carlo simulation.

cracks. The crack density and the crack length distribution inside of the material can be measured at several stages of life using several specimens subjected to the same creep-fatigue condition. Adjusting the parameter on the basis of the observable data by trial-and-error, the process is restored by the simulation. Figure 38 (Tada *et al.*, 1996) shows an example of the visual expression of a part of simulated cracks in the Cs-type. The facet-by-facet cracking dominates the spatial and temporal distribution of cracks. Figure 39 shows the crack propagation rates of inner cracks obtained. Although the propagation rate is not experimentally observable, the simulation can rationally infer it.

5.06.6 APPENDIX A: BEHAVIOR OF J^* UNDER SMALL-SCALE CREEP

Under the SSC conditions where time duration is small and the creep zone is located in the vicinity of the crack tip, the J^* value may be very high due to elastic straining by the surrounding elastic singularity dominated region. An approximate equation of J^* under the small-scale creep was obtained as follows (Ohji *et al.*, 1979, 1980).

When the total strain rate is given by the summation of the elastic strain rate and the Norton type creep strain rate as in Equation (22), the strain at time t is given as

$$\varepsilon = \varepsilon_{\rm el} + \int_0^t \dot{\varepsilon}_{\rm c} \cdot [\sigma(T)/\sigma_{\rm c}]^n \, {\rm d}T = \sigma(t)/E + \dot{\varepsilon}_{\rm c}\sigma_{\rm c}^{-n}F(t)[\sigma(t)]^n \tag{68}$$

where $F(t)$ is defined by

$$F(t) = \int_0^t \left(\frac{\sigma(T)}{\sigma(t)}\right)^n {\rm d}T \tag{69}$$

Since the second term in the right-hand side of Equation (68) is small in the region far from the crack tip, $F(t)$ evaluated in the vicinity of the crack tip may be used throughout the body without deteriorating the accuracy of the approximation. In this case, Equation (68) can be regarded as a nonlinear stress–strain relationship. Then referring to the solution to nonlinear elastic materials, the HRR singular stress strain fields prevail in the near-tip region:

$$\sigma_{ij} = \sigma_{\rm c}\left(\frac{J}{I(n)\dot{\varepsilon}_{\rm c}\sigma_{\rm c}F(t)r}\right)^{1/(n+1)} \tilde{\sigma}_{ij}(\theta) \tag{70}$$

$$\varepsilon_{ij} = \dot{\varepsilon}_{\rm c}F(t)\left(\frac{J}{I(n)\dot{\varepsilon}_{\rm c}\sigma_{\rm c}F(t)r}\right)^{n/(n+1)} \tilde{\varepsilon}_{ij}(\theta) \tag{71}$$

For the SSC conditions, J is equal to the elastic J integral, $J_{\rm el}$ given by Equation (11).

Substituting Equation (70) into Equation (69), the following equation is deduced:

$$F(t) = \int_0^t \left(\frac{F(t)}{F(T)}\right)^{n/(n+1)} {\rm d}T \tag{72}$$

A solution of this equation is expressed as

$$F(t) = (n+1)t \tag{73}$$

Using Equations (70), (71), and (73), the stress and strain rates near the crack tip are given as

$$\sigma_{ij} = \sigma_{\rm c}\left(\frac{J_{\rm el}}{I(n)\dot{\varepsilon}_{\rm c}\sigma_{\rm c}(n+1)tr}\right)^{1/(n+1)} \tilde{\sigma}_{ij}(\theta) \tag{74}$$

$$\dot{\varepsilon}_{ij} = \dot{\varepsilon}_c\left(\frac{J_{\rm el}}{I(n)\dot{\varepsilon}_{\rm c}\sigma_{\rm c}(n+1)tr}\right)^{n/(n+1)} \tilde{\varepsilon}_{ij}(\theta) \tag{75}$$

Alternatively, J^* can be defined in the region very near the crack tip where the creep strain is predominant under the SSC conditions. The near-tip fields are given by Equations (20) and (21). By comparing Equations (20) and (21) with Equations (74) and (75) the following expression for the near-tip J^*, $J^*_{\rm nt}$, is obtained.

$$J^*_{\rm nt} = \frac{J_{\rm el}}{(n+1)t} \tag{76}$$

5.06.7 APPENDIX B: ESTIMATES OF J^* USING DISPLACEMENT RATE FOR DEEPLY CRACKED SPECIMENS

Estimates of J^* under the LSC conditions can be made using load point displacement rate $\dot{V}$, angular displacement rate $\dot{\phi}$, and crack mouth opening rate $\dot{\delta}$, as follows (Kubo, 1975):

(i) Center-cracked panel under tension (M(T)) (Figure 3(a)) and double-edge-cracked panel under tension (DE(T)) (Figure 3(b))

$$J^* = \frac{n-1}{n+1}\frac{\dot{V}P}{Bb} = \frac{n-1}{n+1}\dot{V}\sigma_{\text{net}} \quad (77)$$

$$J^* = \frac{n-1}{n+1}\frac{\dot{\delta}P}{Bb} = \frac{n-1}{n+1}\dot{\delta}\sigma_{\text{net}} \quad (78)$$

(ii) Single-edge-cracked panel in bending (SE(B)) (Figures 3(c) and (d))

$$J^* = \frac{2n}{(n+1)}\frac{\dot{V}P}{Bb} \quad (79)$$

$$J^* = [2n/(n+1)](\dot{\phi}M/Bb) \quad (80)$$

(iii) Circumferentially cracked bar under tension (Figure 40(a))

$$J^* = \frac{2n-1}{n+1}\frac{\dot{V}P}{2\pi b^2} \quad (81)$$

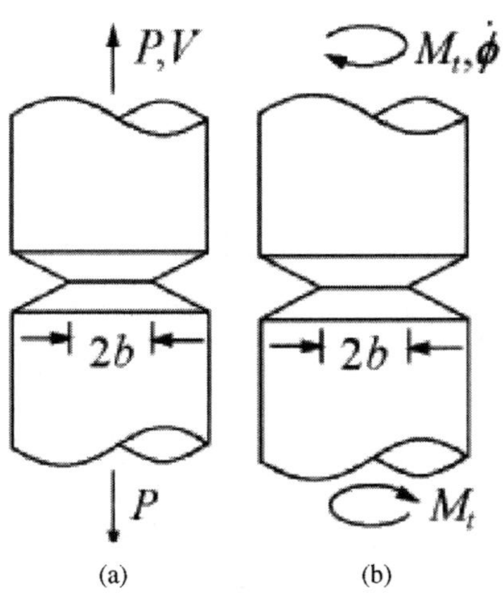

Figure 40 Experimental evaluation of ΔJ_{f} for deeply cracked specimens. Circumferentially cracked bar under (a) tension and (b) torsion.

(iv) Circumferentially cracked bar under torsion (Figure 40(b))

$$J^* = \frac{3n}{n+1}\frac{\dot{\phi}M_{\text{t}}}{2\pi b^2} \quad (82)$$

5.06.8 REFERENCES

T. L. Anderson, 1995, Appendix 10: Application of the J contour integral to cyclic loading. In: "Fracture Mechanics: Fundamentals and Applications, 2nd edn," CRC Press, pp. 559–563.

J. L. Bassani, D. E. Hawk and A. Saxena, 1989, Evaluation of the C_t parameter for characterizing creep crack growth rate in the transient regime. In: "Nonlinear Fracture Mechanics: Volume I Time-Dependent Fracture," ASTM STP 995, eds. A. Saxena, J. D. Landes and J. L. Bassani, American Society for Testing and Materials, Philadelphia, pp. 7–29.

J. L. Bassani and F. A. McClintock, 1981, Creep relaxation of stress around a crack tip. *Int. J. Solids Struct.*, **17**, 479–492.

S. B. Biner and D. S. Wilkinson, 1985, On the stress and strain fields ahead of a stationary crack in creeping solids. *Eng. Fract. Mech.*, **21**, 315–328.

N. E. Dowling and J. A. Begley, 1974, "Fatigue Crack Growth during Plasticity and the J-integral," Westinghouse Research Lab., Tech. Document Summary, No. 74-1E7-CREEP-P1.

N. E. Dowling and J. A. Begley, 1976, Fatigue crack growth during gross plasticity and the J-integral. In: "Mechanics of Crack Growth," ASTM STP 590, eds. J. R. Rice and P. C. Paris, American Society for Testing and Materials, Philadelphia, pp. 82–103.

R. Ehlers and H. Riedel, 1981, A finite element analysis of creep deformation in a specimen containing a macroscopic crack. In: "Advances in Fracture Research (Fracture 81)," eds. D. Francios *et al.*, vol. 2, pp. 691–698.

W. Elber, 1971, The significance of fatigue crack closure. In: "ASTM STP 486," American Society for Testing and Materials, Philadelphia, pp. 230–242.

E. G. Ellison and G. G. Musicco, 1984, Notch-rupture and crack-growth behavior under creep conditions. In: "Subcritical Crack Growth Due to Fatigue, Stress Corrosion and Creep," ed. L. H. Larsson, Elsevier Applied Science, London, pp. 403–448.

G. R. Halford, M. H. Hirschberg and S. S. Manson, 1973, Temperature effects on the strainrange partitioning approach for creep fatigue analysis. In: "ASTM STP 520," American Society for Testing and Materials, Philadelphia, pp. 658–667.

M. P. Harper and E. G. Ellison, 1977, The use of the C^* parameter in predicting creep crack propagation rates. *J. Strain Analysis*, **12**, 167–179.

M. Y. He and J. W. Hutchinson, 1981, The penny-shaped crack and the plane strain crack in an infinite body of power-law material. *J. Appl. Mech.*, **48**, 830–840.

C. Y. Hui and H. Riedel, 1981, The asymptotic stress and strain field near the tip of a growing crack under creep conditions. *Int. J. Fract.*, **17**, 409–425.

J. W. Hutchinson, 1968, Singular behaviour at the end of a tensile crack in a hardening material. *J. Mech. Phys. Solids*, **16**, 13–31.

J. Joch and R. A. Ainsworth, 1992, The effect of geometry on development of creep singular fields for defects under step-load controlled loading. *Fatigue Fract. Eng. Mater. Struct.*, **15**, 229–240.

T. Kitamura, L. J. Ghosn and R. Ohtani, 1992, Stochastic modeling of crack initiation and short-crack growth under creep and creep-fatigue conditions. *J. Appl. Mech.*, **59**, S35–S42.

T. Kitamura and R. Ohtani, 1986, Crack propagation under creep-fatigue interaction condition. *Trans. JSME, Ser. A*, **52**, 1816–1823.

T. Kitamura, R. Ohtani, H. Miki and M. Tsutsumi, 1993, Initiation and propagation of creep-fatigue cracks in an ODS superalloy inconel MA754. In: "Asian Pacific Conference on Fracture and Strength '93," pp. 151–156.

R. Koterazawa and T. Mori, 1977, Applicability of fracture mechanics parameters to crack propagation under creep condition. *J. Eng. Mater. Technol.*, **99**, 298–305.

S. Kubo, 1975, Study on mechanics of creep crack initiation and growth. Ph.D. thesis, Osaka University.

S. Kubo, 1981, "Combined Effect of Creep-recovery-hardening and Elastic Straining on the Stress and Strain-rate Fields near a Crack Tip in Creeping Materials," Brown University report, MRL E-133, pp. 1–35.

S. Kubo, 1983, Effects of creep recovery and hardening on the stress and strain-rate fields near a crack tip in creeping materials. In: "Elastic–Plastic Fracture: 2nd Symposium, Volume I—Inelastic Crack Analysis," ASTM STP 803, eds. C. F. Shih and J. P. Gudas, American Society for Testing and Materials, Philadelphia, pp. I-594–I-614.

S. Kubo, T. Yafuso, M. Nohara, T. Ishimaru and K. Ohji, 1989, Investigation on path-integral expression of the *J*-integral range using numerical simulations of fatigue crack growth. *JSME Int. J.*, **32**, 237–244.

V. Kumar, M. D. German and C. F. Shih, 1981, "An Engineering Approach for Elastic–Plastic Fracture Analysis," NP-1931, Electric Power Research Institute.

K. Kuwabara, A. Nitta, T. Kitamura and T. Ogata, 1988, Effect of small-scale creep on crack initiation and propagation under cyclic loading. In: "ASTM STP 924," American Society for Testing and Materials, Philadelphia, pp. 41–59.

J. D. Landes and J. A. Begley, 1976, A fracture mechanics approach to creep crack growth. In: "Mechanics of Crack Growth," ASTM STP 590, eds. J. R. Rice and P. C. Paris, American Society for Testing and Materials, Philadelphia, pp. 128–148.

C.-P. Leung and D. L. McDowell, 1990, Inclusion of primary creep in the estimation of ct parameter. *Int. J. Fract.*, **46**, 81–104.

K. M. Nikbin, G. A. Webster and C. E. Turner, 1976, Relevance of nonlinear fracture mechanics to creep cracking. In: "Cracks and Fracture," ASTM STP 601, eds. J. L. Swedlow and M. L. Williams, American Society for Testing and Materials, Philadelphia, pp. 47–62.

K. Ohji and S. Kubo, 1988, Fracture mechanics evaluation of crack growth behavior under creep and creep-fatigue conditions. *Current Japan. Mater. Res.*, **3**, 91–113.

K. Ohji, S. Kubo, T. Shibata and K. Ogura, 1985, The effect of interaction of elastic and creep strains on the modified J-integral under cyclic load conditions. *Trans. Japan Soc. Mech. Eng.*, **51**, 90–96.

K. Ohji, K. Ogura and S. Kubo, 1974a, Application of the J-integral to creep crack problems. *Preprint Japan Soc. Mech. Eng.*, **740**(11), 207–210.

K. Ohji, K. Ogura and S. Kubo, 1974b, Creep stress analysis of a notched body under longitudinal shear and its application to crack propagation problems. In: "Proceedings of the 1974 Symposium on the Mechanical Behavior of Materials, 1974," pp. 455–466.

K. Ohji, K. Ogura and S. Kubo, 1975, Path-independent integral (modified *J*-integral) in transient creep conditions. *Preprint Japan Soc. Mech. Eng.*, **750**(13), 65–68.

K. Ohji, K. Ogura and S. Kubo, 1976, Mechanics of creep crack propagation under longitudinal shear and its applications. *Trans. Japan Soc. Mech. Eng.*, **40**, 350–358.

K. Ohji, K. Ogura and S. Kubo, 1978, Estimates of *J*-integral in the general yielding range and its application to creep crack problems. *Trans. Japan Soc. Mech. Eng.*, **44**, 1831–1837.

K. Ohji, K. Ogura and S. Kubo, 1979, Stress and strain fields and modified *J*-integral near a crack tip in initial transient conditions. *Preprint Japan Soc. Mech. Eng.*, **790**(13), 18–20.

K. Ohji, K. Ogura and S. Kubo, 1980, Stress field and modified *J*-integral near a crack tip under condition of confined creep deformation. *J. Soc. Mater. Sci., Japan*, **29**, 465–471.

K. Ohji, K. Ogura and S. Kubo, 1981, A simple method for estimation of *J*-integral and modified *J*-integral. *Trans. Japan Soc. Mech. Eng.*, **47**, 400–405.

R. Ohtani and T. Kitamura, 1987, Numerical simulation of microstructurally-short-crack propagation in creep. *JSME Int. J.*, **30**, 1741–1749.

R. Ohtani and T. Kitamura, 1988, Characterization of high temperature strength of metals based on the mechanics of crack propagation. *Current Japan. Mater. Res.*, **3**, 65–90.

R. Ohtani and T. Kitamura, 1990, Initiation and propagation of microstructurally small cracks under creep-fatigue condition. In: "Proceedings of the 4th International Conference on the Creep and Fracture of Engineering Materials and Structures," The Institute of Metals, pp. 791–802.

R. Ohtani and T. Kitamura, 1994, "Handbook of Fatigue Crack," Elsevier, vol. 2, pp. 1347–1383.

R. Ohtani, T. Kitamura, A. Nitta and K. Kuwabara, 1988, High-temperature low cycle fatigue crack propagation and life laws of smooth specimens derived from the crack propagation laws. In: "ASTM STP 942," American Society for Testing and Materials, Philadelphia, pp. 1163–1180.

R. Ohtani, T. Kitamura, Y. Uematsu and M. Hojo, 1993, Mode I delamination of carbon fiber reinforced thermoplastic polymer under static and cyclic creep at elevated temperatures. In: "ASTM STP 1174," American Society for Testing and Materials, Philadelphia, pp. 23–38.

R. Ohtani, T. Kitamura and K. Yamada, 1981, A nonlinear fracture mechanics approach to crack propagation in the creep-fatigue interaction range. In: "Fracture Mechanics of Ductile and Tough Materials and Its Applications to Energy Related Structures," Martinus Nijhoff Publishers, pp. 263–270.

R. Ohtani, K. Yamada, T. Kashiwagi and H. Matsubara, 1982, Crack propagation of high-temperature low-cycle fatigue in 304 stainless steel. *Trans. JSME, Ser. A*, **48**, 1378–1390.

P. Paris and F. Erdogan, 1963, A critical analysis of crack propagation laws. *J. Basic Eng.*, **85**, 528–534.

J. R. Rice, 1968, A path independent integral and the approximate analysis of strain concentration by notches and cracks. *J. Appl. Mech.*, **35**, 379–386.

J. R. Rice and G. F. Rosengren, 1968, Plane strain deformation near a crack tip in a power-law hardening material. *J. Mech. Phys. Solids*, **16**, 1–12.

H. Riedel, 1978, Cracks loaded in anti-plane shear under creep conditions. *Zeitschrift fuer Metallkunde*, **69**, 755–760.

H. Riedel, 1981, Creep deformation at crack tips in elastic-viscoplastic solids. *J. Mech. Phys. Solids*, **29**, 35–49.

H. Riedel, 1983, Crack-tip stress fields and crack growth under creep-fatigue conditions. In: "Elastic–Plastic Fracture: Second Symposium, Volume I—Inelastic Crack Analysis," eds. C. F. Shih and J. P. Gudas, American Society for Testing and Materials, Philadelphia, pp. I-505–I-520.

H. Riedel, 1987, "Fracture at High Temperatures," Springer, Berlin.

H. Riedel and J. R. Rice, 1980, Tensile cracks in creeping solids. In: "Fracture Mechanics: 12th Conference," ed. P. C. Paris, American Society for Testing and Materials, Philadelphia, pp. 112–130.

K. Sadananda and P. Shahinian, 1977, Creep crack growth in alloy 718. *Metal. Trans. A*, **8A**, 439–449.

A. Saxena, 1986, Creep crack growth under non-steady-state conditions. In: "Fracture Mechanics: 17th Volume," eds. J. H. Underwood *et al.*, American Society for Testing and Materials, Philadelphia, pp. 185–201.

A. Saxena, 1993, Fracture mechanics approaches for characterizing creep-fatigue crack growth. *JSME Int. J., Ser. A*, **36**, 1–20.

A. Saxena, 1998, "Nonlinear Fracture Mechanics for Engineers," CRC Press.

A. Saxena and J. L. Bassani, 1984, Time-dependent fatigue crack growth behavior at elevated temperature in fracture. *TMS-AIME*, **357**, 357–383.

C. F. Shih and J. W. Hutchinson, 1976, Fully plastic solutions and large scale yielding estimates for plane stress crack problems. *J. Eng. Mater. Technol.*, **98**, 289–295.

M. J. Siverns and A. T. Price, 1970, Crack growth under creep conditions. *Nature*, **228**, 760–761.

D. J. Smith and G. A. Webster, 1983, Estimates of the C^* parameter for crack growth in creeping materials. In: "Elastic–Plastic Fracture: Second Symposium, Volume I—Inelastic Crack Analysis," eds. C. F. Shih and J. P. Gudas, American Society for Testing and Materials, Philadelphia, pp. I-654–I-674.

H. D. Solomon and L. F. Coffin, Jr., 1973, Effect of frequency and environment on fatigue crack growth in A286 at 1100F. In: "ASTM STP 520," American Society for Testing and Materials, Philadelphia, pp. 112–122.

N. Tada, W. Zhou, T. Kitamura and R. Ohtani, 1996, Analysis of the intergranular cracking process inside polycrystalline heat-resistant materials under creep-fatigue conditions. In: "ASTM STP 1297," American Society for Testing and Materials, Philadelphia, pp. 87–101.

S. Taira, R. Ohtani and T. Kitamura, 1979a, Application of *J*-integral to high-temperature crack propagation: Part i. Creep crack propagation. *J. Eng. Mater. Technol.*, **101**, 154–161.

S. Taira, R. Ohtani, T. Kitamura and K. Yamada, 1979b, *J*-integral approach to crack propagation under combined creep and fatigue condition. *JSMS, Japan*, **28**, 414–420.

S. Taira, R. Ohtani and T. Komatsu, 1979c, Application of *J*-integral to high-temperature crack propagation: Part ii. Fatigue crack propagation. *J. Eng. Mater. Technol.*, **101**, 162–167.

S. Taira, R. Ohtani, T. Yonekura, M. Osada and T. Kitamura, 1980, Crack propagation in time-dependent fatigue at high temperature. *Trans. JSME*, **46**, 861–869.

G. A. Webster and R. A. Ainsworth, 1994, "High Temperature Component Life Assessment," Chapman & Hall, London.

K. B. Yoon, A. Saxena and D. L. McDowell, 1992, Influence of crack-tip cyclic plasticity on creep-fatigue crack growth. In: "ASTM STP 1131," American Society for Testing and Materials, Philadelphia, pp. 367–392.

W. Zhou, R. Ohtani, T. Kitamura, N. Tada and A. Kosaka, 1995, Creep-fatigue intergranular fracture of inner cracking type in type 304 stainless steel. *JSMS, Japan*, **44**, 78–83.

Comprehensive Structural Integrity
ISBN (set): 0-08-043749-4
Volume 5; (ISBN: 0-08-044154-8); pp. 273–307

5.07
Remnant Creep Life Prediction in Ferritic Materials

D. J. GOOCH
GenExp Ltd., Swindon, UK

NOMENCLATURE

a, b, m, v	constants
A, A^*	cavitation parameters
A_f, A_f^*	cavitation parameters at failure
B, C	constants
D_0	diffusion constant
f	cavity volume fraction
G	shear modulus
G'	Goto hardness parameter
H	hardness
H_∞	equilibrium hardness
k_A, k_B	constants
L_r	reference length
n	stress exponent of uniaxial creep rate
N	number density of cavitated grain boundary facets
P	Larson Miller parameter
r_i, r_o	inner and outer radii
R	gas constant
t	time
t_r	rupture life
t_{rem}	remaining life
t_{ser}	elapsed service life
T_q	torque
α, β, η	constants
γ^*	shear strain
ε	strain
ε_r	uniaxial rupture strain
ε_r^*	multiaxial rupture ductility
ε_s	Monkman Grant constant
θ	twist (radians)
λ	mean interparticle spacing
Λ	ratio of secondary to tertiary creep strain
ξ	Avrami function
ρ	Riedel damage parameter
σ	stress
σ_0	threshold stress
$\sigma_1, \sigma_2, \sigma_3$	principal stresses
σ_m	hydrostatic stress
σ_{rep}	representative stress
σ_{vm}	Von Mises equivalent stress
τ^*	shear stress
Φ	nucleation and growth function
ω	generalized damage function
ΔH	diffusion activation enthalpy

5.07.1 INTRODUCTION

Thick section high temperature, high pressure components in large-scale steam generating power plants are largely manufactured from creep resistant low alloy steels. These components are required to operate for many years under severe conditions of temperature and stress. The design of high temperature pressure parts is governed by national and international standards and allowable stresses are generally based upon mean 1×10^5 h creep rupture data with appropriate safety factors to take account of data scatter and departure from design conditions. Designs must be conservative and are based upon laboratory generated creep rupture data. For example, the safety factor applied to the 1×10^5 h rupture stresses for a life of 1×10^5 h for a straight pipe is generally 1.5, with higher factors being applied for longer design lives. (It is rare that creep rupture data for durations longer than 1×10^5 h are available.) Additional factors are also applied in specific cases where there are particular design aspects that need to be taken into consideration. Examples include pipe bends, where the bending process may have resulted in wall thinning, and steam headers, where the stress raising effect of multiple pipes penetrating the header wall has to be allowed for.

Some high temperature but nonpressure containing components are not covered by national design standards because of the proprietary nature of their designs. In particular steam turbine rotors and blades are generally designed to company standards. Furthermore, these components may be designed using creep strain rather than creep rupture data if dimensional stability is critical.

The rupture stress variability for a specific grade of steel is generally considered to be ±20%, this being a result of variations in detailed chemical composition and heat treatment allowed within the specification. Consequently, at the end of its nominal design life, a component may in practice be perfectly fit for many years of safe further operation. In the 1960s, when many of the current large power plants were constructed, nominal component design lives of 1×10^5 h were the norm, corresponding to only 14 years of operation at a typical availability level of 85%. (Design lives are generally greater than this nowadays with 2.5×10^5 h being common for high temperature rotors.) The economic life of a power station is much greater than 14 years, so a strategy had to be formulated to facilitate decisions as to whether components would have to be replaced or could be safely operated beyond their design lives. This led to the expenditure of a great deal of effort in the 1980s and early 1990s in developing science based remnant life assessment procedures.

The components of concern are fabricated predominantly from low alloy ferritic steels used in the quenched and tempered or normalized and tempered condition. The microstructures are, although heavily tempered, in a metastable condition and it is this feature which both increases the complexity of assessments and, paradoxically, offers a potential means for quantifying service exposure and remaining life. The driving force for microstructural changes in service is very low, with the result that changes will be very slow and therefore potentially useful over the lifespan of a power plant. Table 1 (Bhadeshia *et al.*, 1998) illustrates the stored energy of typical microstructures relative to the equilibrium or paraequilibrium states. In the case of a power plant steel containing M_2X on entering service, the reduction in energy on transition from M_2X to $M_{23}C_6$ during service is very small compared with a transition from bainite to ferrite.

Although the term "remnant life assessment" is generally taken to refer to creep life assessment, it must be remembered that other degradation processes, which may not determine design life, may become the critical processes beyond design life. Consequently, any remnant life assessment program must consider all other possible mechanisms. These will include fatigue, creep-fatigue, loss of toughness through embrittlement, corrosion, stress corrosion, etc. These processes are not covered in this chapter, except where there is a direct interaction, but are covered in other chapters or volumes.

Development of remnant life assessment techniques continues but at a lower activity level because changes in the power generation industry—notably privatization and the introduction of large gas turbine power plants—have increased uncertainty over required plant lives. However, it is now recognized that life assessment techniques will also be needed for the new gas turbine plants and a short section on this application is included at the end of this chapter.

5.07.2 GENERAL METHODOLOGY

Planning cycles are dependent upon the ability to predict the future performance and requirements of the plant. In particular, plant life extension requires a predictive capability for establishing the safe and economic life for each component. The care and maintenance requirements for each major item need formal documentation which informs station staff about the actions to be taken at predetermined intervals in terms of plant monitoring, inspection and determination of remaining life. In

Table 1 Stored energy as a function of microstructure, defined relative to microstructure for which stored energy is arbitrarily set to zero.

Case		*Stored energy* ($J\,mol^{-1}$)
Phase mixture in Fe–0.2C–1.5Mn wt.% at 300 K		
1	Ferrite, graphite, and cementite	0
2	Ferrite and cementite	70
3	Paraequilibrium ferrite and paraequilibrium cementite	385
4	Bainite and paraequilibrium cementite	785
5	Martensite	1,214
Phase mixture in Fe-0.1C–2.2Cr–1Mo wt.% at 873 K		
6	Ferrite and paraequilibrium cementite → ferrite and $M_{23}C_6$ after extensive service	0
7	Ferrite and paraequilibrium cementite → ferrite and M_2X on entering service	63

Source: Bhadeshia *et al.* (1998).

Table 2 Components with potential for life extension programs.

Component	*Lead time* (Years)	*Outage* (Weeks)
Boiler drums and headers	2	4–9
High pressure pipework	1	9–15
Pipework valves	1	2
Turbine valve chests	2	6–9
Turbine casings	3	12
Turbine rotors	3	9

Source: Gooch (2001).

developing a methodology for mechanical plant the following factors need to be considered for each component:

(i) safety aspects;
(ii) operating conditions;
(iii) materials of construction;
(iv) operational regime (base load, two shifting, etc.);
(v) potential failure mechanisms (creep, fatigue, corrosion, etc.);
(vi) current failure statistics;
(vii) difficulties and costs of repair;
(viii) lead times and costs of component replacement (see Table 2); and
(ix) outage times for replacement.

Consideration of these factors leads to the identification of the following key components for development of remnant life assessment procedures and techniques:

(i) boiler headers,
(ii) turbine valve chests,
(iii) high temperature turbine rotors,
(iv) main and reheat steam lines, and
(v) superheater and reheater tubing.

5.07.2.1 The Three Stage Approach

In the United Kingdom, the former Central Electricity Generating Board (CEGB) promoted the concept of progressive assessment procedures for major components, which consisted of a number of stages of increasing complexity and decreasing conservatism (Gooch and Townsend, 1987; Gooch, 1988a; Gooch *et al.*, 1989a). This philosophy was subsequently adopted by other organizations (e.g., EPRI, 1986). The general form of these assessment procedures is as follows:

Stage 1. This is essentially a recalculation of the design life taking into account the actual operating conditions. It will involve:

(i) measurement of steam temperatures,
(ii) actual operating hours and numbers of hot and cold starts,
(iii) minimum materials property data, and
(iv) design dimensions.

Stage 2. This stage generally involves access to the component for condition monitoring procedures, which may be performed *in situ*. It may also involve more refined theoretical assessment calculations than stage 1. These may include:

(i) installation of additional thermocouples for metal temperature measurements,
(ii) refined stress analysis,
(iii) direct measurement of strain, and
(iv) use of less conservative materials property data.

Stage 3. This usually requires the removal of samples from the component and may in some cases necessitate the component to be removed from the unit (e.g., turbine rotors). These techniques include those that require technical specialist input and may not yet be considered routine. Stage 3 may involve:

(i) postexposure creep testing,
(ii) creep cavitation assessment, and
(iii) microstructure degradation characterization.

All stages are accompanied by standard nondestructive examination (NDE) techniques, which are applied at intervals determined by plant fault statistics and by current assessments of remaining life. The techniques used in stages 2 and 3 assessments are largely research based and are at various stages of development. These will be described in more detail in later sections.

5.07.2.2 Components and Materials of Construction

The key components for remnant life assessment procedures have been identified above and are listed in Table 3, together with materials of construction; Table 4 gives the composition ranges of these materials. Until the 1980s the dominant materials in use were the low alloy ferritic steels with chromium levels up to ~2.5 wt.%. These were used because they were cheap, easy to cast and forge in thick sections, generally readily weldable, and had good creep strength, with operating temperatures up to 565°C attainable. These are the classes of steels on which most remnant life assessment development work has been conducted.

Austenitic stainless steels, which have superior creep resistance, are used extensively in boilers as superheater tubes but the main issues in that application are concerned with corrosion and are not covered in this chapter. Austenitic steels were also used for main steam pipework and other thick section components in a handful of plants in the 1960s in order to pursue steam temperatures up to, and even above, 600°C. However, operation of these plants caused considerable difficulties due to

Table 3 Components requiring life assessment procedures.

Component	*Materials of construction*[a]
Boiler drums and headers	CMn, 1CrMo, 2CrMo, (1/2)CrMoV, 10CrMoVNb
Boiler tubes	CMn, 1CrMo, 2CrMo, 9CrMo, 10CrMoVNb
High pressure pipework	1CrMo, 2CrMo, (1/2)CrMoV, 12CrMoV, 10CrMoVNb
Pipework valves	1CrMo, 2CrMo, (1/2)CrMoV, 10CrMoVNb
Turbine valve chests	2CrMo, (1/2)CrMoV, 1CrMoV, 10CrMoVNb
Turbine casings	2CrMo, 1CrMoV, 10CrMoVNb
Turbine rotors	1CrMoV, 12CrMoV, 10CrMoVNb
High temperature bolts	1CrMoVNbB, 12CrMoVNb, Nimonic 80A

[a] Excluding austenitic steels.

thermal fatigue and thermal ratcheting of the austenitic components. This arises because of the poor thermal conductivity and high thermal expansion coefficients of austenitic steels, which in turn lead to the development of high transient stresses during thermal cycling. Consequently, these plants were either down-rated or decommissioned and there has been no call for the development of detailed remnant life assessment procedures for them. Nevertheless, the general principles developed for the ferritic steels will be equally valid.

The pursuit of higher steam temperatures has also led to the development of a range of new ferritic steels with 9–12 wt.% chromium for use at steam temperatures up to 600°C. These have tempered martensitic structures and are now entering service in thick section components such as steam pipes, valve chests, turbine casings, and turbine rotors (Strang and Gooch, 1997). The remnant life assessment techniques developed for the lower alloy ferritic steels are equally applicable to these newer grades although, in the case of microstructural based techniques, specific quantitative data will have to be generated to enable plant application in due course.

5.07.3 DESIGN REASSESSMENT

5.07.3.1 Materials Data

Materials data used for design purposes are approved by the relevant national (e.g., the American Society of Mechanical Engineers, ASME) or international body (e.g., the International Standardization Organization, ISO) or by individual companies in the case of noncodified materials. Data are continually

Table 4 Typical composition ranges (wt.%) of ferritic power plant steels.[a]

Material	*Cr*	*Mo*	*Ni*	*C*	*Si*	*Mn*	*Others*
CMn				0.30 max.	0.50 max.	0.40 min.	
(1/2)CrMoV	0.30/0.60	0.40/0.70		0.08/0.18	0.10/0.40	0.40/0.70	V: 0.22/0.35
1CrMo	0.70/1.20	0.40/0.65		0.10/0.20	0.40 max.	0.40/0.70	
1CrMoV	0.98/1.15	0.65/1.08	0.60/0.76	0.24/0.31	0.17/0.27	0.74/0.81	
1CrMoVTiB (D1055)	0.90/1.30	0.85/1.10		0.15/0.25	0.35 max.	0.75 max.	V: 0.60/0.80, Ti: 0.05/0.20, B: 0.001/0.01
2CrMo	2.20/2.50	0.90/1.20		0.08/0.15	0.50 max.	0.40/0.70	
9CrMo	7.50/1.00	0.80/1.20		0.18 max.	1.00 max.	0.30/0.60	
Modified 9CrMo (P91)	8.00/9.50	0.85/1.05	0.40 max.	0.08/0.12	0.20/0.50	0.30/0.60	V: 0.18/0.25, Nb: 0.06/0.10, N: 0.03/0.07
10CrMoVNb (EM12)	8.50/10.50	1.70/2.30	0.030 max.	0.15 max.	0.20/0.65	0.80/1.30	V: 0.20–0.40, Nb: 0.30/0.55
12CrMoV (W)	10.00/12.50	0.80/1.20	0.30/0.80	0.17/0.23	0.50 max.	1.00 max.	V: 0.25/0.35, W: 0.70 max.
12CrMoVNb (Z20)	10.00/12.00	0.70/1.10	1.00 max.	0.18/0.25	0.20/0.60	0.60/1.00	V: 0.20/0.40, Nb: 0.25/0.55

[a] The above compositions are indicative; there are many variants for different component applications.

reviewed by these bodies and the design values updated as necessary. In the cases of the well-established steels these reviews generally result in little or no change in the approved values. However, for the newer grades there may be significant revisions, either upward or downward, as longer term data are accumulated. In general, the data are analyzed using a polynomial function of (log stress) and a time–temperature parameter such as the Larson–Miller or Manson–Haferd parameters. The use of these and other parameters has been reviewed by Viswanathan (1989). Care must be taken to define allowable extrapolation limits to ensure that short term, high-temperature test data are not used to predict long term, lower-temperature results as it is known that reductions in ductility at long durations can result in the over prediction of rupture lives from shorter term tests.

5.07.3.1.1 *The influence of ductility*

Creep rupture life is a function both of creep resistance and rupture ductility. The same life may be obtained from a steel with high creep resistance and low ductility as from one with lower creep resistance but higher ductility. Both types of behavior may be observed within one grade of steel and it is important to be able to distinguish between them when performing an assessment. A high strength, low ductility steel may, e.g., for example, be more susceptible to damage by creep cavitation, particularly at stress raising features such as corners or changes in section. An example of this difference in behavior within a single steel grade occurs in the 1CrMoV class of rotor forgings where the long-term rupture ductility may be as high as 30% or as low as 2% (Figure 1). This variation is due to a wide variety of factors but the reasons for the low ductilities are thought to include: (i) high temperature austenitizing temperatures resulting in large grain sizes and high creep strength, (ii) low tempering temperatures, and (iii) high levels of impurity elements such as P, As, Sn, and Sb. Figure 1 is an illustration of the widely observed behavior mode in which initially high short-term ductility gives way to low ductility at long times—the transition time being widely variable.

When one examines national and international standards for creep resistant materials and structures there is very little mention of creep ductility—the standards relying on the simple concept of rupture strength. There is occasionally an implicit recognition of the importance of ductility through the inclusion of data from tests on notched bars and the use of the concept of "notch weakening and notch strengthening" behavior. However, there are little quantitative ductility data.

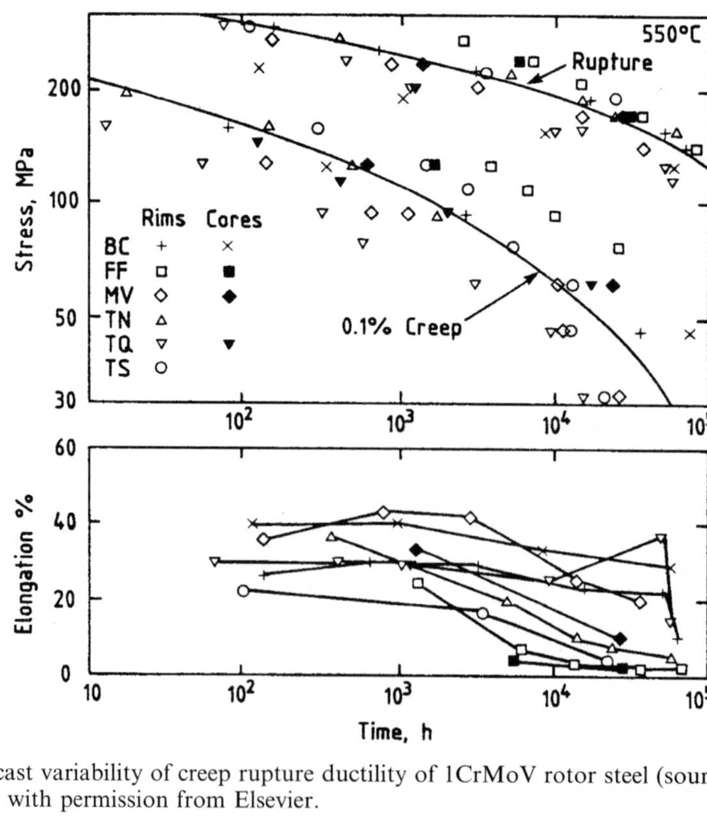

Figure 1 Cast to cast variability of creep rupture ductility of 1CrMoV rotor steel (source Batte and Gooch, 1985). Reproduced with permission from Elsevier.

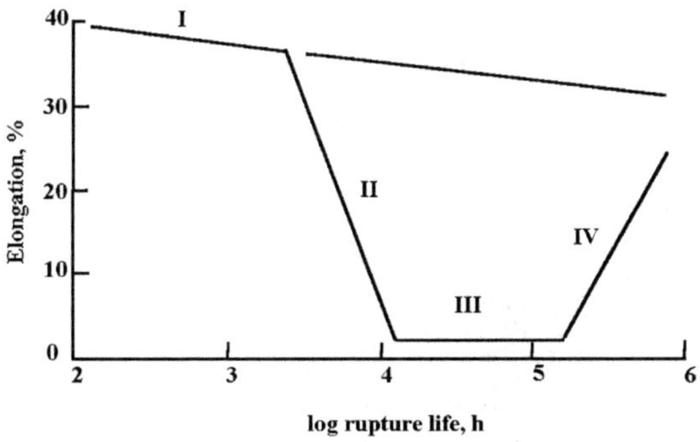

Figure 2 Schematic representation of the creep rupture ductility regimes (source Gooch, 1991).

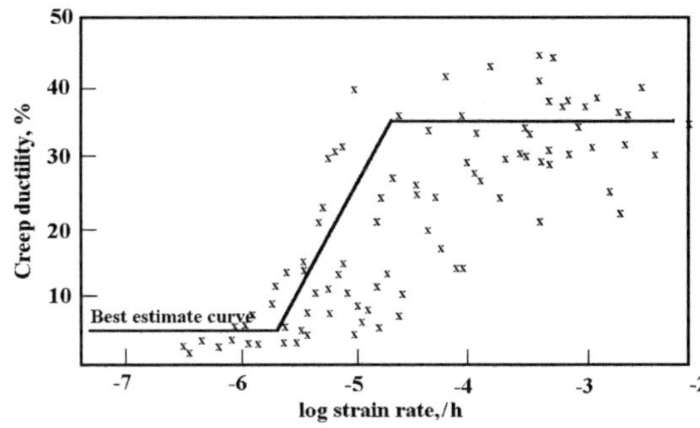

Figure 3 Creep ductility data for a range of 1CrMoV rotor steels with best fit line for one specific cast (source Gooch, 1991).

In many cases creep fracture strain *per se* is of little importance, with dimensional stability during service being more significant. This is certainly the case for most heavy section components manufactured from high ductility (>5%) steels. However, problems may arise when a steel has been designed to maximize creep strength, sometimes resulting in a loss of rupture ductility. Many of the vanadium bearing steels, such as the 1CrMoV steel cited above, fall into this category. Failures within design lifetime have occurred in high temperature fasteners, valves, and pipework systems because of inadequate creep ductility.

Many remnant life assessment procedures rely on creep ductility arguments either implicitly or explicitly. Creep cavitation assessment procedures, discussed later, are an implicit recognition that creep ductility is the life determining factor, as are creep fatigue assessment procedures which invoke ductility exhaustion arguments (e.g., Wareing *et al.*, 1986; Ainsworth, 2001).

The ability to predict long-term rupture ductility from short-term tests would represent a major advance. However, the variation of ductility with test duration is a complex function of time and temperature and is also dependent upon the microstructure and composition of the steel concerned. Figure 2 shows schematically the various ductility regimes, which may be summarized as follows (Gooch, 1991):

Stage I. Ductile rupture by the formation of discrete voids that may be either inter- or intragranular. The ductility tends to decrease with increasing test duration as the density of intergranular voids increases but remains typically >10%.

Stage II. A transition region in which ductility drops sharply due to the increasing incidence of grain boundary cavitation, although still accompanied by relatively high levels of matrix deformation.

Stage III. The region of minimum ductility, which may be as low as 1% in extreme cases. Failure is by the nucleation and subsequent diffusive growth of grain boundary cavities. This is the regime, which has received most attention in terms of mechanistic study.

Stage IV. Overaging of the microstructure lowers the rate of cavity nucleation and/or growth, leading to a progressive recovery of ductility.

At test durations of practical interest creep resistant steels may exhibit one or more of these stages. Ferrite/pearlite C-Mn steels tend to exhibit stage I behavior whereas bainitic steels strengthened by fine precipitates may show all four stages. Figure 3 shows an empirical analysis of rupture ductility data for 1CrMoV rotor steels, which shows the transition through stages I, II, and III.

5.07.3.1.2 Chemical composition effects

Within a particular steel grade there is a permitted range of chemical compositions which will contribute to the position of an individual heat within the rupture strength scatter band. The influence of the composition will be only one of many but it may be possible in some circumstances to draw tentative inferences as to the possible position of the cast in the scatter band from the composition. For example,

(i) The presence of all major elements at levels in the middle of their permitted ranges may give some confidence that the material should not exhibit lower bound rupture strength.

(ii) Levels of a precipitation hardening or solution-strengthening element at the bottom of the permitted range could indicate the possibility of below average strength.

(iii) High levels of Ni in 9–12 wt.% Cr steels may result in low rupture strength (Strang and Vodárek, 1997).

However, there are many complex interactions between elements which are not yet fully understood and it is potentially dangerous to place too much reliance on compositional observations.

A development which shows great promise both for the design and assessment of creep resistant steels is that of neural network analysis (Mackay, 1997; Brun *et al.*, 2001). This is a nonlinear method of data analysis in which the network is "trained" on a set of weighted input and output data. The training process involves a search for the optimum nonlinear relationship (based on multiple hyperbolic tangents) between the input and output data, together with a series of "weights" for the input variables. This is "tested" against a second set of input and output data. The training is then repeated with adjusted weights in order to minimize the error of the "test" predictions. The process is computer intensive but, once the network is trained, estimation of the outputs for any given input is rapid. For creep strength predictions the input parameters will include all compositional variables including appropriate functions which are believed to be physically relevant (e.g., V:C ratio); the output parameter will be the rupture strength. A large database is initially required but the technique is then able to predict how rupture strength varies with compositional changes and could become a valuable remnant life assessment tool.

5.07.3.1.3 The effect of oxidation

Creep rupture data quoted in standards for ferritic steels are generally obtained using test pieces typically 6–10 mm in diameter. Data are available for times in excess of 1×10^5 h, at which duration substantial loss of material by oxidation occurs. Application of creep rupture data obtained from standard specimens to thick section components may therefore be thought to underestimate component life. Application of "oxidation correction factors" to design creep rupture therefore offers a potential means of extending operating life.

In practice, the effect of oxidation on creep rupture duration will depend upon the properties of the oxide. At one extreme the oxide may spall throughout the test and therefore have no load bearing capacity. In this case, the rupture strength of a standard specimen tested in air would significantly underestimate the true strength of a thick section component. At the other extreme, if the oxide were to remain fully adherent and have a creep resistance greater than the metal, it could actually increase the relative life of the standard section specimen. It has generally been assumed that the true situation approximates to the former and a number of theoretical studies have been performed on that basis. Hart (unpublished work, 1985) calculated oxidation correction factors using a life fraction approach originally proposed by Moles *et al.* (1980). Dyson and Osgerby (1987) derived modified strain–time relationships by taking into account the effect of specimen deformation in enhancing oxidation rates through the formation of microcracks, but still assumed the creep strength of the oxide to have no influence on the behavior of the metal core. Cane and Manning (unpublished work, 1981) took into account the load bearing capacity of the oxide by deriving a computer model requiring a numerical solution which assumed that the oxide was load bearing up to a critical fracture strain. This model predicted strengthening behavior at small specimen diameters and rapid oxidation rates.

In addition to oxidation rate data, the Dyson and Osgerby (1987) and the Cane and Manning models require the input of critical parameters of oxide creep resistance and fracture strain, for which there are little data. The Hart model requires knowledge of the life fraction constant, which is not necessarily unity under increasing stress conditions.

Little quantitative information is available on the creep resistance of iron oxides, particularly with alloying additions. However, the creep resistance of unalloyed Fe_3O_4 at temperatures of 550–700°C is comparable to that of the low alloy steels (Cane and Manning, unpublished work, 1981) and it would be expected that the addition of alloying elements would increase it. However, oxides have lower ductility than steels and are prone to cracking during creep testing. At first through-scale cracks are formed, followed by delamination at the oxide–metal interface. Finally the cracks are healed by access of air to the metal surface. A laminated oxide appearance, frequently observed in creep specimens, is taken as evidence of repeated oxide fracture and healing. Oxidation is generally considered to be governed by parabolic kinetics as a consequence of diffusion distances increasing as oxidation progresses. However, if frequent spalling of the oxide occurs then the kinetics may appear to be linear.

The phases formed during oxidation in air of low alloy steels are based upon Fe_2O_3 (hematite), Fe_3O_4 (magnetite), $(Fe,Cr,Mo,V)_3O_4$ (complex spinel), and FeO (wustite), forming in different proportions depending on the test temperature. At creep testing temperatures of 550–600°C the usual structure is an inner

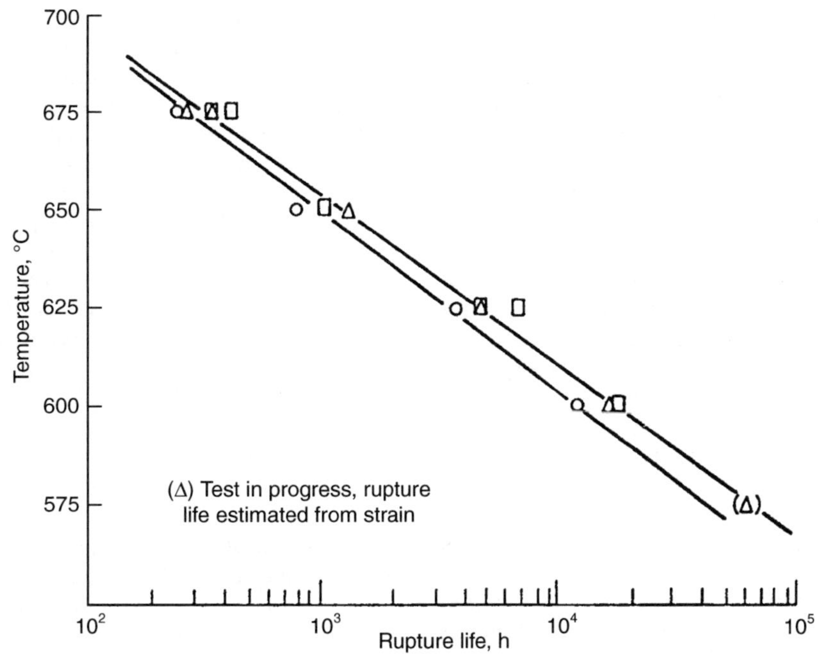

Figure 4 The effect of oxidation on the rupture endurance at 70 MPa of (1/2)Cr(1/2)Mo(1/4)V steel showing the effect of specimen size and test environment (source Gooch, 1988b). □ 5 mm, Inert atmosphere, ○ 5 mm, air, and △ 25 mm, air.

magnetite layer surrounded by a thin outer layer of haematite. A common observation is that, during test interruptions and on test termination, the outer red layer of haematite readily spalls leaving a very adherent hard inner layer of magnetite. This lends credence to the suggestion that oxidation could actually increase rupture life in some circumstances.

Figure 4 shows creep rupture data for a CrMoV steel tested at 70 MPa and temperatures of 575–675°C using specimens of 5 mm and 25 mm diameter in air and inert atmospheres. The data demonstrate a strengthening effect for large specimens and for specimens tested in inert atmosphere. The measured effect for the large specimens corresponds to an oxide correction factor of ~ 1.2, which is somewhat lower than model predictions. Other unpublished data confirm the suggestion that strengthening effects can occur for smaller specimens at long test duration.

5.07.3.2 Operating Conditions

The actual operating conditions of a component may vary from the original design conditions either constantly or variably as a consequence of deviation from design parameters. Clearly a knowledge of operating temperatures and stresses, both historically and for anticipated future operation is required for an accurate life assessment. This information is generally applied using either the life fraction rule or the strain fraction rule or a mixture of the two (Viswanathan, 1989).

The life fraction rule states

$$\sum t_i/t_{ri} = 1 \quad (1)$$

The strain fraction rule states that

$$\sum \varepsilon_i/\varepsilon_{ri} = 1 \quad (2)$$

where t_r and ε_r are the times and strains to rupture.

The life fraction rule has been found to work well for small changes in stress and temperature (e.g., Goldhoff and Woodford, 1982). However, the validity of the rule for stress changes has been questioned. Hart (1980) suggested a microstructural reason why this might be the case. He pointed out that, whereas both temperature and time of exposure result in the degradation of microstructure through thermally activated processes, this is not the case for stress. Fortunately, it is usually temperature rather than stress which is subject to variation during operation.

5.07.3.3 Operating Temperature

Deviations from design temperature may have a profound influence on the operating life of a component because of the thermally activated nature of the creep process, which results in a high sensitivity to small changes in temperature. For example, an increase in operating temperature from 565°C to 575°C for a CrMoV steel results in a halving of the creep rupture life. Consequently, overtemperature operation for even a short time may have a significant influence on component life.

The operating temperature of power plant components is primarily a function of the steam temperature which, in components downstream of the boiler, is well regulated. In these areas the major variations occur during start-up and procedures are defined to ensure that components do not develop unacceptable stresses due to thermal transients. However, in boiler components operating conditions are more variable, both spatially and temporally. Superheater and reheater headers have historically been items of particular concern. These are thick-section cylindrical components pierced by numerous smaller pipes along the length and around the circumference. These smaller pipes transport steam generated in different regions of the boiler, which may consequently be at different temperatures, perhaps up to 25°C higher than the mean steam temperature. The installation of thermocouples along headers enables the determination of the metal temperature distribution and, if appropriate, the re-routing of the feeder pipes to even out temperature variations.

5.07.3.4 Reference Stress Techniques

Application of the life fraction rule requires a knowledge of the representative rupture stress of a component. This is generally determined by an inverse application of the original design rules which, for pipes and pressurized tubes, are based on the mean diameter hoop stress formula. This implicitly assumes the initial elastic stress distribution and takes no account of the redistribution of stress that takes place as a consequence of creep. This approach can be pessimistic in that failure under pressure loading can be correlated with a "reference stress," which may be up to 20% lower than the mean diameter stress.

The reference stress for a component is defined as that stress which would fail a simple tensile specimen in the same time as the component. Depending on the complexity of the component under examination, reference stresses may be determined either by analytical techniques or by the use of experimental model test programs. In the initial elastic state the maximum principal stress (σ_1) in a pressurized cylinder, which is in the hoop direction, occurs at the inner surface with the stress decreasing through the wall. Conversely, the Von Mises octahedral shear stress (σ_{vm}), which is generally assumed to control creep deformation, is lowest at the inner wall and rises towards the outer surface. As creep occurs, the stresses redistribute resulting in steady state through wall distributions which are functions of the stress sensitivity of the creep rate. However, it is found that there are unique positions within the wall where the stresses are essentially invariant and these are known as skeletal points (Figure 5). It is generally considered that a suitable reference stress for deformation, as opposed to rupture, is the Von Mises equivalent stress at its skeletal point:

$$\sigma_{vm} = [(\sigma_1 - \sigma_2)^2 + (\sigma_2 - \sigma_3)^2 + (\sigma_3 - \sigma_1)^2]^{1/2}/\sqrt{2} \qquad (3)$$

However, it does not necessarily follow that rupture may be predicted using this stress since rupture may be a function of equivalent, maximum principal and hydrostatic (σ_m) stresses.

$$\sigma_m = \tfrac{1}{3}(\sigma_1 + \sigma_2 + \sigma_3) \qquad (4)$$

Various functional forms of a representative stress for the control of rupture have been proposed, for each of which a skeletal point, and hence a reference stress, may be defined. These include the following (Webster *et al.*, 2001):

$$\sigma_{rep} = \alpha\sigma_1 + (1-\alpha)\sigma_{vm} \quad (0 \leq \alpha \leq 1) \qquad (5)$$

$$\sigma_{rep} = 3\beta\sigma_m + (1-\beta)\sigma_{vm} \quad (0 \leq \beta < 1) \qquad (6)$$

$$\sigma_{rep} = \alpha\sigma_1 + 3\beta\sigma_m + (1-\alpha-\beta)\sigma_{vm} \quad (0 \leq \alpha+\beta \leq 1) \qquad (7)$$

The complexity of the rupture process of a component dictates a need for model test programs to validate analytical predictions, even for relatively simple geometries. In order to reproduce the correct failure mechanisms these tests should ideally be performed under creep conditions as close to those experienced in service as possible. Where creep rupture is controlled primarily by the equivalent stress, however, reference stresses may be obtained by performing short-term limit load tests on model components under noncreep conditions. The result of a short-term test on an internally pressurized pipe-to-pipe intersection is shown in Figure 6.

It is also possible to extend the reference stress approach to anisothermal temperature distributions by determining a reference temperature to be used in conjunction with an

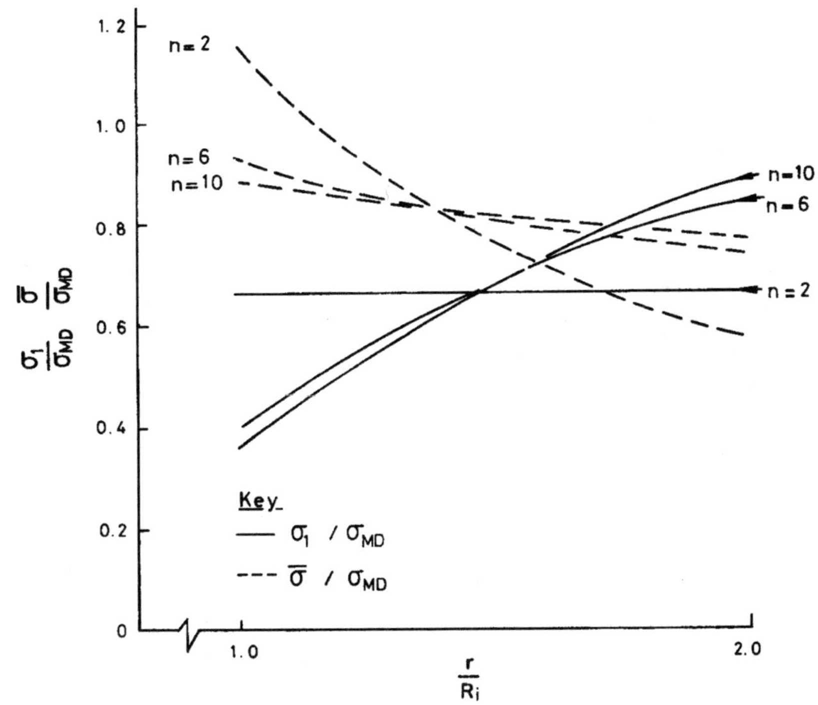

Figure 5 Variation of the maximum principal tensile stress and the equivalent stress with radius for a pressurized thick tube, $R_o/R_i = 2$ (source Goodall, 1986; reproduced with permission from Elsevier).

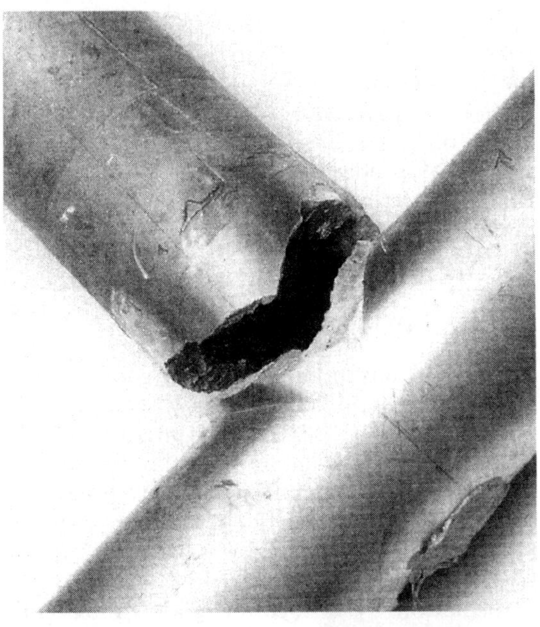

Figure 6 The result of a short-term test on an internally pressurized model header fabricated from boiler tubing (source Gooch, 1988b).

isothermal reference stress (Fielding, unpublished work, 1986). This has been achieved by adapting standard isothermal limit load analyses for cylinder-to-cylinder intersections to anisothermal conditions. Application to a stub penetration in a header only then requires a knowledge of the metal temperatures of the stub and the surrounding header material.

5.07.3.5 Finite Element Analysis and Continuum Damage Mechanics

For complex components, where neither analytical nor experimental reference stresses are available, computer finite element analysis FEA is used (e.g., Goodall *et al.*, 1975). The component is "broken down" into an aggregate of finite elements (FE) with prescribed properties. The computer analysis evaluates the response of the component as a whole and enables the stresses at any given point to be determined. A detailed picture of the distribution within the component is produced. In its simplest form the analysis may only be of elastic stresses but knowledge of the steady state creep properties of the material enables a steady state or time dependent analysis to be performed (Figure 7).

The development of computational continuum damage mechanics now allows FEA to be performed using physically based constitutive equations to describe the material behavior. This enables the full time dependent behavior of a structure to be modeled, including, by the input of ductility values, the transition from generalized damage to discrete crack growth. Increasing speed and data

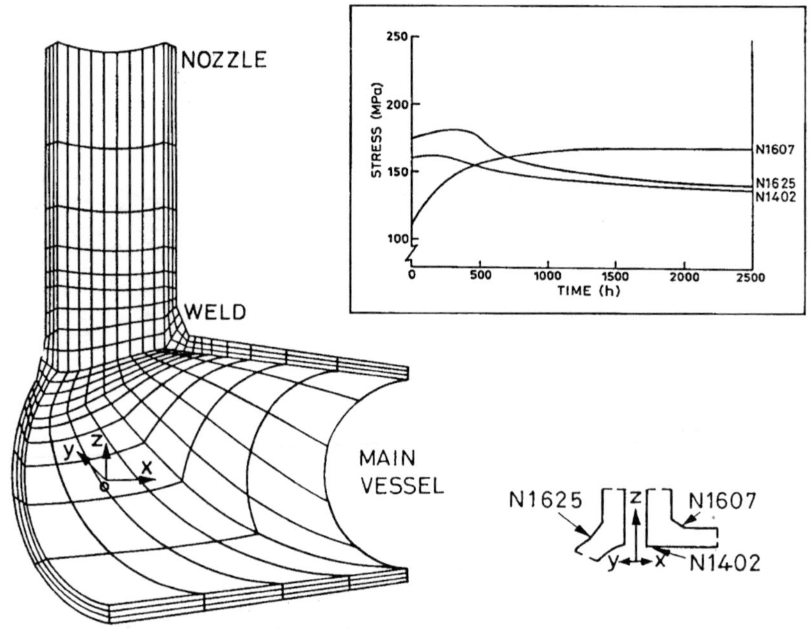

Figure 7 A FE mesh for a header geometry showing typical results of a time dependent analysis for critical locations (source Gooch, 1988b).

storage capacity of computer workstations will rapidly reduce the time required for such computations. However, the procedure is critically dependent upon the availability of validated multiaxial constitutive equations for deformation and damage over the mechanistic regimes encountered (Hayhurst, 1996).

5.07.3.6 System Stresses

In addition to the stresses arising from steam pressure in boiler parts there are also stresses arising from the self-weight of components and from thermal expansion during the transition from ambient to operating temperature. These stresses are referred to as system stresses and are particularly important in heavy section pipework systems. When these systems are fabricated, allowance is made for the thermal expansion during heating to operating temperature by subjecting the pipes to "cold pull" prior to the production of "closure" welds. The intention is to ensure that the sum of the stresses in the direction of the pipe axis due to (i) internal pressure and (ii) self-weight and thermal expansion does not exceed the design stress, which is based on the hoop stress in the pipe. (The axial stress due to pressure alone is nominally half that of the hoop stress.) Pipes are supported by hangers incorporating deflection measuring devices to facilitate monitoring of movements during service. However, maintaining stress free systems over extended lifetimes is a difficult task and system stresses are thought to have contributed to a number of failures, particular in welds. Bends are also considered to be vulnerable to the presence of excessive system stresses. Attempts at measurement of these stresses, necessarily at operating temperature, have met with limited success and they are not incorporated quantitatively in life assessment calculations. Strategies are instead based on good maintenance practice and on the use of nondestructive examination regimes derived from failure statistics to maintain plant integrity.

5.07.3.7 Multiaxial Stresses

Mention has been made above of the fact that creep rupture of low alloy steels may be governed either by the maximum principal stress, the equivalent stress, the hydrostatic stress, or a combination of the three. The subject has received a great deal of attention and a more detailed explanation is given elsewhere in this series. Briefly however, creep deformation by dislocation glide or by grain boundary sliding is a shear stress controlled process and is hence controlled by the octahedral shear stress, or equivalent stress. Creep cavitation, however, is caused by diffusional flows of vacancies controlled by the maximum principal stress. The controlling stresses are therefore the maximum principal and the hydrostatic stresses, since the latter is required for the volume increase accompanying cavity formation. The practical situation is of course not as clear-cut as this since each mechanism has its own stress and temperature dependence, which results in overlap between the different mechanisms. Furthermore, although creep cavitation requires a vacancy flux driven by the maximum principal stress, under certain conditions of temperature and stress cavity growth is limited by the ability of the surrounding material to deform to accommodate the growth—this ability being governed by the equivalent stress. The various regimes of temperature and stress in which the various mechanisms dominate have been documented in the mechanism maps of Ashby (e.g., Frost and Ashby, 1982). For remnant life assessment purposes the main consideration is that, under uniaxial stress conditions, processes that are controlled by the equivalent stress dominate at high stresses and low temperatures while the maximum principal stress is increasingly important at combinations of higher temperature and lower stress. Increased stress multiaxiality favors cavitation, leading to low ductility failure, and has to be given particularly careful consideration at features such as intersections, changes in section, and other stress concentrating features.

5.07.4 NONDESTRUCTIVE EVALUATION

5.07.4.1 Creep Cracks and Creep Fatigue Cracks

The final evidence for the development of creep damage during service is the initiation and growth of discrete cracks either by creep processes alone or by creep fatigue interactions. These may form either as single crack at stress concentrating features or as the final phase of more generalized damage. In the case of the former, it is often possible to extend life by local repair procedures involving grinding out of the crack and replacement of the metal by welding, often accompanied by reprofiling to remove the stress concentration. One example of this situation is the growth of circumferential creep fatigue cracks from sharp changes in section in high-temperature rotors. Another example is the growth of creep cracks in the heat affected zones (HAZs) of welds in heavy section steam pipes. In this case, the formation of the cracks may be exacerbated by the presence of localized regions of weak or low ductility microstructure, which may be modified during the repair.

If, however, the cracking is a consequence of more widespread creep damage, then it will generally appear as multiple cracks on either the macro- or microscopic level. Examples of this form of end of life damage may occur in the weld metal of heavy section welds and in also pipe bends.

5.07.4.1.1 Detection

Many nondestructive examination techniques exist for the detection of discrete cracking, as opposed to general creep damage. These are routinely applied at inspection intervals throughout the life of a component and form an indispensable part of a life assessment program. The frequency of examination and the percentage coverage will be a function of the age of the component and past experience of cracking or damage of the component type.

Crack detection techniques include:

Radiography. This is used for detecting internal defects in, e.g., pipe walls. A radiation source is placed inside the pipe and film is wrapped around the outside. Radiography is normally only used during manufacture or construction because of access and safety issues. It is not suitable for detecting planar defects such as cracks.

Fluorescent die penetration. This is used for detecting surface cracking. This technique is particularly useful in situations where direct access is limited but where a fluorescent liquid can be poured and the surface subsequently examined remotely using fiber optics.

Magnetic particle inspection. This is used for surface crack detection, particularly in weldments. A direct current passed through the component creates positive and negative poles on either side of a crack. The surface is painted white and iron oxide powder sprinkled on, forming lines along the crack.

Eddy current detection. It is routinely used as an inspection tool for detection of surface or subsurface defects. Eddy current patterns due to the defects are detected using excitation/detection probes.

Ultrasonic examination. It is probably the most versatile of the inspection techniques and used routinely as an inspection tool. In angle beam mode, ultrasonic examination can detect crack like defects down to less than 1 mm in size and can be used to size cracks down to less than 3 mm. Using compression probes it can also be used to measure pipe or tube wall thickness, which is valuable when only nominal design dimensions are available. It is not, however, suitable for the detection of surface or near surface defects. A further restriction is that "noise" generated in large grain size materials such as castings or austenitic steels limits its resolution. Ultrasonic examination is the source of most of the crack size data used in defect analysis.

5.07.4.1.2 Analysis

It is not the purpose of this chapter to deal in depth with the analysis of creep and creep fatigue crack initiation and growth, this being covered elsewhere in the series (also see Webster and Ainsworth, 1994). However, it must be emphasized that these are an essential part of any remnant life assessment program. Moreover, the effect of service exposure on creep strength and particularly ductility must be taken into account in any defect analysis.

Creep crack growth assessment procedures (Ainsworth *et al.*, 1987; Ainsworth, 2001) utilize experimental crack growth rate data correlated with the C^* parameter, which, for a given specimen geometry and applied load, is essentially directly proportional to the displacement rate. Consequently, the dominant factor distinguishing between materials when using C^* correlations is the rate of crack growth for a given displacement, i.e., the ductility (Figure 8). It was shown earlier that a single steel may exhibit widely variable creep ductility depending upon the test conditions, with high ductility being characteristic of short-term tests and ductilities decreasing as the test duration increases. A short-term creep crack growth test is therefore likely to show lower crack growth rates for a given value of C^* than a long-term test and may lead to nonconservative predictions of service crack growth rates. Preaging, without stress, of test material in an attempt to simulate the microstructural degradation which occurs in service will not reproduce the creep damage that is ultimately responsible for low ductility behavior. Indeed the opposite is likely to occur, in that the material will be softened but not damaged, thus leading to even higher ductility behavior in subsequent testing. It is, therefore, essential that laboratory creep crack growth tests are performed in the same mechanistic regime that applies to the service conditions.

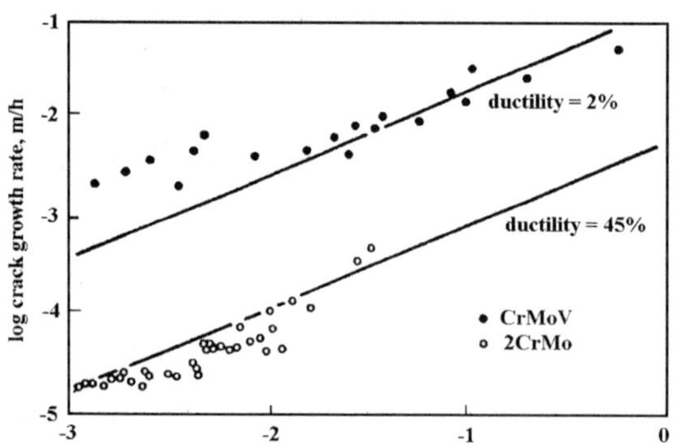

Figure 8 Observed and predicted creep crack growth rates for tempered 2CrMo (high ductility) and untempered (1/2)Cr(1/2)Mo(1/4)V (low ductility) steels (source Gooch, 1991).

5.07.4.2 Creep Cavitation

Creep cavitation is the precursor to the creep cracking that ultimately results in creep failure. In some steels grain boundary cavitation develops early in life, progressing steadily through stages of growth and linkage into microcracks before reaching the failure point. This behavior is generally characteristic of low ductility steels, the cavities forming preferentially on grain boundaries orientated normal to the maximum principal stress on particles such as sulfides or carbides (Middleton, 1981; Cane and Middleton, 1981). Conversely, higher ductility materials may not exhibit detectable cavitation until a late stage in life and the cavities or voids may form intragranularly or on grain boundaries experiencing the maximum shear stress. In either case the detection of cavities as early as possible is essential and this has become one of the major tools in any life management program.

5.07.4.2.1 Detection

A repeatable, nonsubjective, method of detection and measurement is required for the quantitative assessment of the extent of cavitation present in a plant component. It follows that this requires a consistent technique for metallographic preparation, suitable for use either in the laboratory or in the more onerous conditions of access and cleanliness that exist on site. The technique must also be, as far as possible, operator independent and is attained through operator qualification procedures.

As in the laboratory, cavitation on plant is revealed by metallographic polishing and etching techniques. However, whereas in the laboratory samples will generally be examined directly under an optical or scanning electron microscope, this is difficult, though not impossible on plant. Consequently plastic replicas are usually taken of the etched surface, for later examination in the laboratory, again by optical or scanning electron microscopy. Replicating materials are available commercially, usually in the form of cellulose acetate sheets of 35–250 μm thick. These are softened by spraying with or immersing in solvent and are then laid onto the area under examination with minimal pressure and no lateral smoothing. When dry, which may take between 5 min and 45 min depending on the thickness, the sheets are peeled from the component and flattened between glass slides for removal to the laboratory for later examination. The replica is coated by sputtering or vacuum deposition of carbon, aluminum, gold, or platinum to improve contrast. A suitable dye in the acetate sheet or solvent will also enhance contrast but is not as effective as coating. A good quality replica should show the same amount of detail as the original surface (Figure 9). Neubauer (1984) demonstrated that cavities as small as 0.1 μm could be detected by scanning electron microscopy.

Site preparation requires the optimum combination of repeatability, resolution, and economy that is consistent with the quantitative assessment procedure to be used. Two options are available—the fully manual route and that using electropolishing for the later stages. In either case it is essential that the procedure is fully documented for quality assurance purposes. In the fully manual route successive stages of mechanical polishing are followed by multiple fine polishing and etching. This is because surface layer "smearing" may limit cavity resolution if a single polish and etch is used. Conversely, if a sample is over-etched, cavity sizes may be accentuated or, in the extreme, rows of discrete cavities may assume the appearance of microcracks.

A fully manual procedure is time consuming, however, particularly in plant conditions, and may impact upon the required plant outage time with unacceptable economic consequences. The use of electropolishing for the final preparation stages offers a more rapid alternative. The quality of the surface produced by electropolishing may be inferior to that produced by the manual technique but may be adequate for those situations where the detection of heavy cavitation or microcracking is sufficient. If electropolishing is to be used it is advisable to validate it against the fully manual route for the level of cavitation detection required for the specific steel and microstructure of concern.

Other techniques for cavity detection which have been investigated, but not yet developed to the point of routine plant application, include the following (Eggeler and Hurst, 1994; Nakashiro *et al.*, 1996):

Ultrasound velocity attenuation. In this variation of the procedure used for crack detection, the velocity of the ultrasonic wave is attenuated by the presence of cavities, which change the elastic modulus. The attenuation is used as a measure of the cavitation. Trials have been performed on prior damaged creep samples and the technique has been shown to be more sensitive to cavitation detection than conventional ultrasonic techniques, which rely on amplitude rather than velocity attenuation. However, the effect is small (typically $<1\%$) compared with scatter from other sources.

Ultrasound frequency analysis. Nakashiro *et al.* (1996) suggested a method based on analysis of the high frequency noise, which precedes the first backwall echo. The noise spectrum from a defect-free sample is used for calibration purposes. A noise parameter is defined which is the integration of the frequency/decibel spectrum component due to the presence of defects. It is claimed to be very sensitive to defects and has been used to characterize "type IV" cavitation damage in the HAZ of 2Cr1Mo welds.

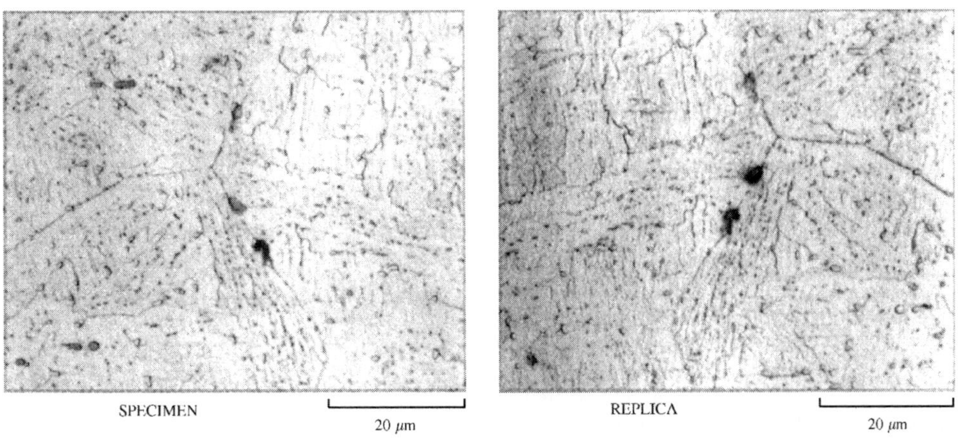

Figure 9 Comparison between metal surface and acetate replica using optical microscopy showing creep cavities on grain boundaries in 1CrMo steel tested at 575°C, 43 MPa for 2×10^4 h. Note the mirror inversion of the replica (source Gooch, 1988b).

Electrical resistance. The electrical resistance of the steel will increase with the presence of cavities and this may have potential for near surface cavity detection.

Micromagnetic parameters. These have been investigated and shown not to correlate with cavitation, the major influences being other microstructural changes that occur during creep, such as dislocation structure and carbide size and distribution.

Density. The presence of cavitiation reduces the density of a material and in principle could be used as a measure of cavitation severity. However, this would in reality be a destructive procedure and would require significant volumes of damaged material.In all of the above a major problem is that microstructural changes other than cavitation, such as carbide coarsening, will also have an effect on the physical property concerned.

5.07.4.2.2 Analysis

The severity of creep cavitation and its effect on remaining life has been studied extensively and both qualitative and quantitative procedures have been variously developed or proposed. The semiquantitative procedure proposed by Neubauer (1984) is the best known and remains the basis for most procedures in use today. The degree of cavitation is described by five "damage parameters" ranging from "undamaged" through to "macrocracked." The damage parameters are then related to the stage that the material has reached on a classic primary–secondary–tertiary creep curve and, on the basis of this, recommendations for future plant actions are made (Table 5). Classification is relatively fast to perform since it is qualitative and not quantitative but the definitions are interpretations of schematic figures rather than formal rules. Various utilities have adopted this general approach, with modifications based upon their own plant experiences. Similar procedures are incorporated in European guidelines for in service damage assessment (Auerkari *et al.*, 1996; VGB, 1992). The categories may be subdivided in order to give a greater range of options. Furthermore, the relationship between damage category and creep life is, in practice, highly dependent upon the particular steel and microstructure concerned and on the nature of the component involved. For example, creep cavitation detected in an isolated region of susceptible microstructure arising from the welding process may be treated differently to the same level of cavitation spread throughout a component. Similarly cavitation observed in a region of a component with a sharp stress gradient would be treated differently to cavitation in a uniformly stressed component. Qualitative classification is also useful as a methodology for comparing results from different research groups. For example, Wu *et al.* (1993) developed a detailed classification which can readily be reduced to the Neubauer categories (Table 6).

Quantitative assessment parameters which have been proposed include number density, area fraction, and volume fraction. Empirical relationships have been suggested for various materials between the *number of cavities per unit area* on a metallographic section and the life fraction consumed. A major difficulty with this approach is that, as creep rupture is neared, the number density of cavities may actually start to decrease as the cavities link up to form microcracks, leading to the possibility of nonconservative assessments. The results will be highly dependent on the sensitivity of the cavity detection and imaging procedures used, e.g., optical or scanning electron microscopy. "Round robin" exercises have suggested that cavity detection limit is a major factor in accounting for interlaboratory variations.

Fully quantitative assessments, where each cavity is individually counted and measured are extremely difficult for similar reasons. Such assessments have been attempted, e.g., when information about cavity growth rates is desired. Comparisons of *mean cavity size* are hindered by continuous nucleation and highly

Table 5 Cavitation damage parameters (after Neubauer, 1984).

Damage rating	*Description*	*Recommended action*[a]
Undamaged	No creep damage observed	Normal 3 year inspections
A: Isolated	Isolated cavities on grain boundaries	Normal 3 year inspections
B: Oriented	Damaged boundaries aligned normal to the maximum stress axis	Reduce inspection intervals to 1–2 years
C: Microcracked	Coalescence of cavities causes the separation of some grain boundaries	Repair or replace part. Continued service for $\sim$6 months may be possible
D: Macrocracked	Cracks many boundaries long	Immediate repair or replacement

[a] Recommended actions related to a normal service life of 1×10^5 h.

Table 6 Damage ratings proposed by Wu *et al.* (1993).

Rating	*Description of damage*
1	No cavities
Isolated cavities	
2B	1–100 cavities mm^{-2}
2C	100–1,000 cavities mm^{-2}
2D	> 1,000 cavities mm^{-2}
Strings of cavities	
3B	Few strings and isolated cavities 2B
3C	Few strings and isolated cavities 2C or 2D
3D	Numerous strings and isolated cavities up to 2D
Microcracks < 0.1 mm	
4A	Reheat cracks[a]
4B	Few microcracks and damage 2B or 3B
4C	Few microcracks and damage 2C, 2D, 3C, or 3D
4D	Numerous microcracks and damage 2C, 2D, 3C, or 3D
Macrocracks > 0.1 mm	
5A	Reheat cracks[a]
5B	One macrocrack (< 1 mm) and damage 2B, 3B, or 4B
5C	Macrocracks (< 5 mm) and damage 2C, 3C, or 4C
5D	Macrocracks (< 5 mm) and damage 2D, 3D, or 4D, or creep crack (> 5mm)

[a] Reheat cracks occur in the coarse grained HAZs of, usually, low alloy ferritic steel weldments strengthened by fine intragranular precipitates such as vanadium carbide.

sophisticated analysis of the data is needed to separate out nucleation and growth. Measurements of *size distribution* will also be prone to error unless there is stringent control of metallographic preparation.

Several techniques have been used for determining the *area fraction or volume fraction* of material occupied by cavities. The volume fraction may first be determined from a full quantitative study. However, this is a difficult and time-consuming procedure and is subject to large errors for the reasons described above. Alternatively, the area fraction may be determined from point counting and image analysis techniques and then converted into volume fraction. This is an attractive option because of the potential for automation but still suffers from preparation error. In principle volume fraction can be used to calculate the lower bound strain that a component has experienced using equations of the forms:

$$\varepsilon = f/\Phi \tag{8}$$

where Φ depends upon nucleation and growth kinetics (Cane and Needham, unpublished work 1983, cited in Gooch *et al.*, 1989b; Auerkari and Vierros, 1987) or

$$\varepsilon = \varepsilon_r (f/f_{crit})^{1/q} \tag{9}$$

where q is a material dependent constant ($1 < q < 2$) and f_{crit} is the critical value of f at failure and is dependent upon test conditions.

This strain-based approach does not, however, lead to a conservative assessment procedure since strain accumulating other by cavitation is not taken into account. This of particular significance in the case of a ductile material in which cavitation may not occur until the later stages of life.

The need for a quantitative assessment parameter which was as independent of preparation procedure as possible but which could still be related mechanistically rather than empirically to creep life led to the development of the *"A" parameter* (Cane and Shammas, 1984; Shammas, 1987). This is defined as the number fraction of grain boundaries that are damaged by cavitation or cracking (measured in traverses parallel to the direction of the maximum principal stress). It has the major advantage that it does not require the measurement of actual numbers of cavities nor area or volume fraction and is thus relatively insensitive to preparation and imaging techniques. It is particularly applicable to replicas taken from plant. However, it does still require the reliable identification of grain boundaries and a consistent size limit for detection of cavities. The theory that relates the A parameter to creep life assumes that:

(i) cavities are growing by constrained cavity growth in which the growth rate is determined by deformation of the surrounding matrix and

(ii) that the observation of even a single cavity on a grain boundary indicates that the whole boundary no longer carries any load. This might be a reasonable assumption since it is well known from scanning electron microscopy of fracture surfaces that once cavities are visible optically a boundary is severely damaged on a fine scale.

According to the Kachanov (1958) continuum damage model a creep curve can be described by the following coupled equations:

$$d\varepsilon/dt = Bt^{-m}[\sigma/(1-\omega)]^{\eta} \quad (10)$$

$$d\omega/dt = Ct^{-m}\sigma^{\nu}/(1-\omega)^{\eta} \quad (11)$$

where B, C, m, η, and ν are constants, ε is the creep strain and ω is a generalized damage parameter, in this case describing cavitation damage, which varies from 0 in the undamaged state to 1 at failure. In the simplified form often used, primary creep is neglected, i.e. $m = 0$, and damage is assumed to be strain controlled, i.e. $\eta = n$, the stress exponent of the uniaxial creep rate.

The use of this simplified form then leads to the following relationship between the A parameter and the consumed life fraction:

$$(1 - t/t_r) = (1 - A)^{n\Lambda/(\Lambda-1)} \quad (12)$$

where Λ is the ratio of secondary to tertiary creep strain.

Figure 10 shows data plotted for a range of low alloy steels. However, a difficulty with the approach is the determination of a value of Λ for different materials and stress states. Even with appropriate values it has often been found that the approach is excessively conservative when compared with experimental data or that the data scatter is too great to permit use of the model relationship. For example, the magnitude of the A parameter at failure, A_f, varies

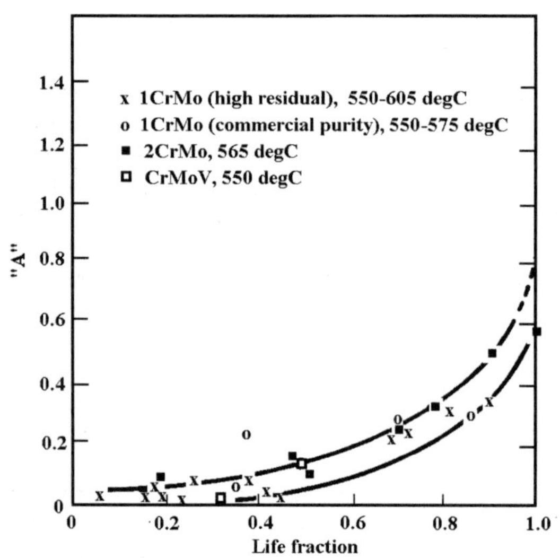

Figure 10 Variation of the "A" parameter with consumed life fraction for ferritic steels (source Viswanathan, 1989).

considerably with material, ductility, grain size, temperature, and stress. In 1CrMoV steels A_f increases from $\sim$0.02 to 0.20 as the tensile creep ductility decreases from 33% to 2% (Strang *et al.*, 1996). This variation is observed in multicast data and also in data from a single cast when ductility decreases with time and temperature.

Consequently, various modifications of the process have been proposed to facilitate practical use in life assessments. For example, a linear lower bound approach (Shammas, 1987) has been proposed which would require validation for each material or microstructure. An example of this approach is shown for 1CrMo steel in Figure 11. The predictive equation then takes the form:

$$t_{rem} = t_{ser}[a/(A+b) - 1] \quad (13)$$

where t_{rem} and t_{ser} are the remaining and service lives, respectively, and a and b are experimentally determined constants which will be material and condition dependent. Viswanathan (1989) showed similar data and noted that, in this case, there was no systematic difference between commercial purity and high purity steels. He also noted that there was no systematic difference between A parameter measurements made by interrupting a single creep test and taking surface measurements or by using a series of specimens and destructively examining them. This is an important observation for data collection purposes.

Riedel (1989) suggested that the apparent overconservatism of the A parameter in some circumstances may be due to its inability to take account of the increasing effect of the interaction of damaged grain boundaries with neighboring damaged boundaries as damage progressed. He proposed the use of damage parameter, ρ (Hutchinson, 1983) which characterizes the constitutive response of a creeping solid in the tertiary stage,

$$\rho = (n+1)Nd^3/2(1+3/n)^{1/2} \quad (14)$$

where N is the number density of facets undergoing constrained cavitation and d is their diameter. He then derived a relationship between ρ and life fraction using continuum damage arguments as with the A parameter,

$$t/t_f = \Lambda\rho^{\Lambda-1} - (\Lambda-1)\rho^{\Lambda} \quad (15)$$

and the following relationship between ρ and the A parameter:

$$\rho = 0.6(n+1)A/(1+3/n)^{1/2} \quad (16)$$

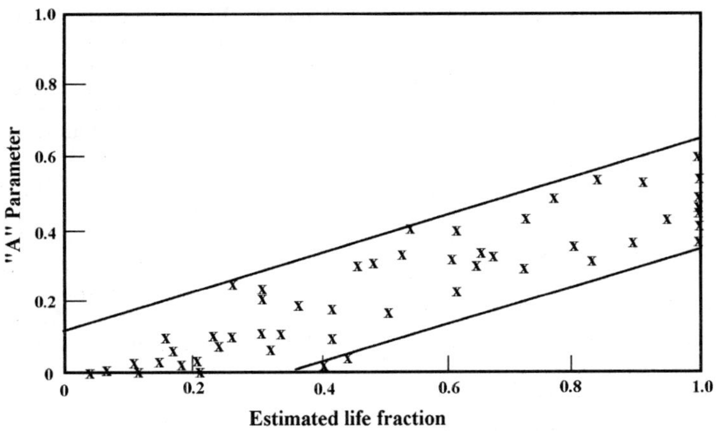

Figure 11 Linear correlation between "A" parameter and life fraction for two casts of coarse grained 1CrMo steel (source Shammas, 1987).

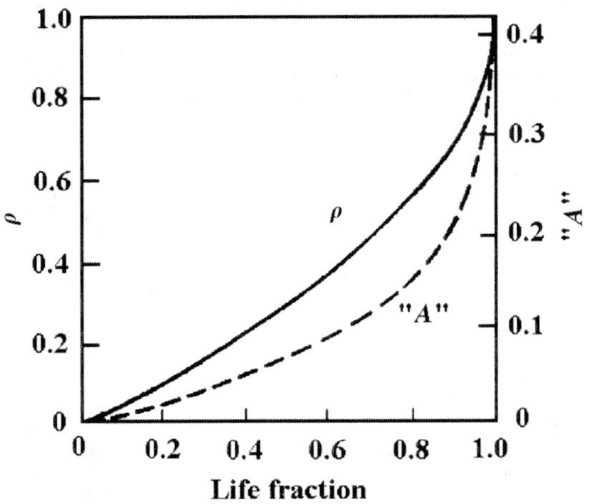

Figure 12 Comparison of life predictions for 1CrMo steel made from the Riedel model with those from an unmodified "A" parameter, using $n = 5$, $\Lambda = 2$ (source Riedel, 1989).

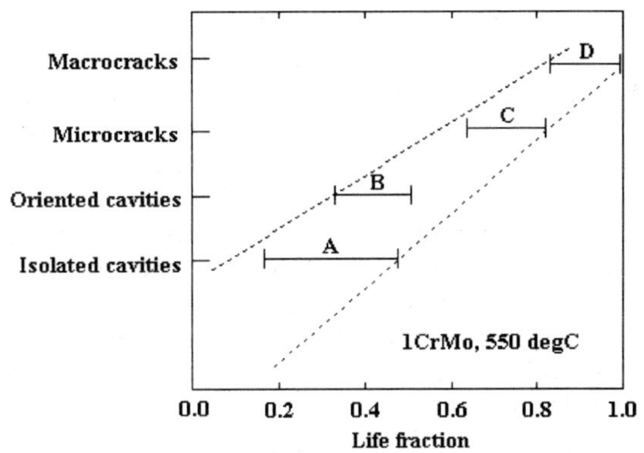

Figure 13 Correlation between cavity classification and consumed life fraction for high purity and commercial purity 1CrMo steels tested at 550°C (source Shammas, 1987).

where the numerical factor depends on the grain shape.

Finally Riedel suggested that ρ could be measured experimentally by a simple modification of the A parameter which weighted damaged boundaries according to the number of interacting damaged boundaries. Thus, a boundary interacting with one other damaged boundary would be counted twice and so on. Figure 12 shows a comparison between predictions using ρ and A, from which it is apparent that the overconservatism of A is indeed reduced.

Eggeler (1991) proposed a further refinement, termed the A^* parameter, similar in concept to ρ, but in which the interaction field with damaged grain boundaries was extended to a box centered on each damaged boundary and extending $\pm 5d$ perpendicular to and $\pm 1d$ in the direction of the measurement line. The theoretical basis for this lies in an analysis of constrained cavity growth by Chockshi (1987). Eggeler claimed better correlation of experimental results on 2Cr1Mo steel with A^* than with A or ρ.

Despite the considerable effort expended on developing quantitative models for relating cavitation to consumed creep life it remains the case that, for practical purposes, data scatter generally precludes the use of the models in isolation—empirical correlations based on multicast data being generally used. In recognition of this Shammas (1987) proposed correlating the Neubauer damage classifications directly with life fraction and this approach has been adopted by others (Figure 13). The potential advantage of this approach is that it could enable longer term forward planning than the original reinspection recommendation strategy. However, the correlations will again be material specific as previously discussed.

5.07.5 MICROSTRUCTURE ANALYSIS

The high creep resistance of the ferritic steels used in power plant is primarily due to the formation of a fine dispersion of alloy carbides, which are stable or metastable under service conditions. There is also a strong contribution from solid solution strengthening by substitutional atoms, which becomes more important at long service exposures as the carbide structure degrades. The initial alloy carbide distribution is formed during the manufacturing heat treatment sequence. This is an austenitizing treatment followed by either quenching or air cooling to form microstructures ranging from martensite through bainite to ferrite and pearlite. These are then tempered at ~100–200°C above the service temperature, during which treatment the fine alloy carbides are formed.

Although the alloy carbides produced may be stable or metastable at the service temperature the structure is not necessarily the equilibrium structure. Further phase changes may occur in which the carbides produced by tempering may transform or dissolve and other carbides may be precipitated. In addition, the size and distribution of the carbides may change as a result of Ostwald ripening, the process by which particle distributions coarsen in order to minimize the sum of volumetric and surface energy. These changes in microstructure occur extremely slowly at service temperatures—this feature enabling the steels to be used for the extreme lengths of time required. Monitoring the changes offers potential means for assessing the loss of creep strength due to service exposure and consequently the remaining creep life.

Qualitative assessments of microstructural changes have been used for this purpose for many years. Toft and Mardsen (1961) used optical microscopy to relate the spheroidization of pearlite or bainite in carbon and low alloy boiler steels to service history. Subsequently, Carruthers and Day (1968) and Coade (1985) have correlated the stages of breakdown identified by Toft and Mardsen (1961) with various time–temperature parameters. Dobrzanski *et al.* (1996) established atlases of microstructure for use in remaining life assessment of 1CrMo steels (Table 7). However, it must be recognized that such correlations must be specific to particular service conditions since the effects of temperature and stress or strain are not separated out.

The possibility of using the increase of the bainite lath size with creep strain as a life monitor has also been investigated. Figure 14 shows results for a 2CrMo steel, which suggest a correlation. However, the approach has not been developed to the point of practical application, largely because of the lack of sensitivity at strains of practical interest and the need to use transmission electron microscopy.

The development of quantitative electron optical and X-ray microanalytical techniques enabled the development of more quantitative procedures. In most cases it is relatively simple to extract small samples of metal from components for analysis without compromising future performance. Where this is not practicable it may be possible to use plastic replicas, similar to those used for cavitation assessment, to extract carbides from the

Table 7 Microstructural changes as a function of life consumed for 1CrMo steel.[a]

t/t_f	*Microstructure observed optically*
0	Ferrite plus bainite and/or pearlite M_3C and M_2C
0.2	Ferrite and partially disintegrated bainite/pearlite M_3C, $M_2C\uparrow$ and $M_{23}C_6\uparrow$
0.3–0.4	Ferrite and completely disintegrated bainite/pearlite $M_3C\downarrow$, $M_2C\downarrow$, $M_{23}C_6$ and $M_6C\uparrow$
0.4–0.6	Ferrite and coarse carbides $M_2C\downarrow$, $M_{23}C_6$ and M_6C Small number of irregularly spaced isolated cavities
0.6–0.8	As above but more cavities Many cavities coalesced or about to do so
0.8–0.9	As above but crack-like coalescence of cavities
0.9–1.0	Macroscopic cracking (removed from service)

Source: Bhadeshia *et al.* (1998) (after Dobrzanski *et al.*, 1996).
[a] Arrows indicate whether phase is increasing or decreasing.

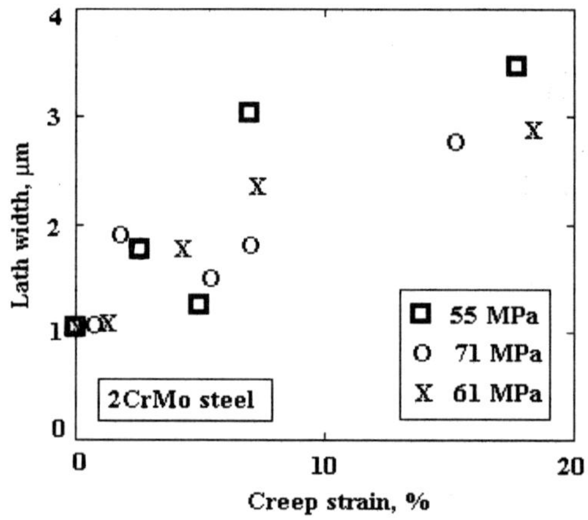

Figure 14 Lath size in 2CrMo steel as a function of secondary creep strain. Strain measurements are ±0.01 and lath sizes are ±25 μm (source Lonsdale and Flewitt, 1979).

component surface (Viswanathan, 1989). For this purpose the component surface is heavily etched following the normal surface preparation procedure. The surface is then washed with alcohol, dried, and a dilute solution of 1% polyvinyl formal in chloroform is applied. This deposits a fine film which is allowed to harden and then backed with softened cellulose acetate foil, which is subsequently removed for examination. In the laboratory the foil is coated with carbon on the extraction side and is then dissolved away leaving the extracted carbides embedded in the carbon film. This may then be examined in, e.g., the scanning transmission electron microscope (STEM), which enables both microstructural and chemical studies of the precipitates. The plastic extraction replica technique, however, suffers from some limitations, notably inconsistency in the extraction of very fine carbides such as VC and Mo_2C and, as yet, is not widely used in the field.

5.07.5.1 Particle Spacing and Coarsening

Under typical service conditions creep strain accumulation, and hence life exhaustion, is governed by the coarsening, due to thermal exposure, of the fine carbide distributions. Consideration of the carbide coarsening kinetics and the movement of dislocations between the carbides enabled Askins *et al.* (1988) to develop a computer model, which directly related carbide distribution to creep strain or remaining life. In service application small samples are removed from the component and carbon extraction replicas are made from them. The carbide distribution is determined by transmission electron microscopy using advanced image analysis techniques. Different parameters characterizing the carbide distribution were investigated but the mean interparticle spacing was selected because of its relative simplicity and because of its mechanistic relevance.

The model assumes that a threshold stress, σ_0, is needed to enable dislocations to move between particles:

$$\sigma_0 = \alpha' \mu b / \lambda \tag{17}$$

where α' is a constant, μ is the shear modulus, b is the Burgers vector, and λ is the mean interparticle spacing.

The creep rate then becomes

$$d\varepsilon/dt = B(\sigma - \alpha' \mu b / \lambda)^n \tag{18}$$

where B is the constant containing the temperature dependence, $B = B_0 \exp(k_A T)$.

Askins *et al.* (1988) performed interparticle spacing measurements on samples of 1CrMo steel which had been subjected to accelerated aging treatments and found that the kinetics could be described by a cube root growth law (Figure 15):

$$\lambda_t^3 = \lambda_0^3 + C_0 \exp(\beta T) t \tag{19}$$

where λ_t is the interparticle spacing at time t, λ_0 is the spacing at $t = 0$, T is the temperature in K, and C_0 and β are constants.

Equations (17)–(19) and experimentally determined values of the constants enabled a series of strain versus time curves to be constructed. The model was validated by

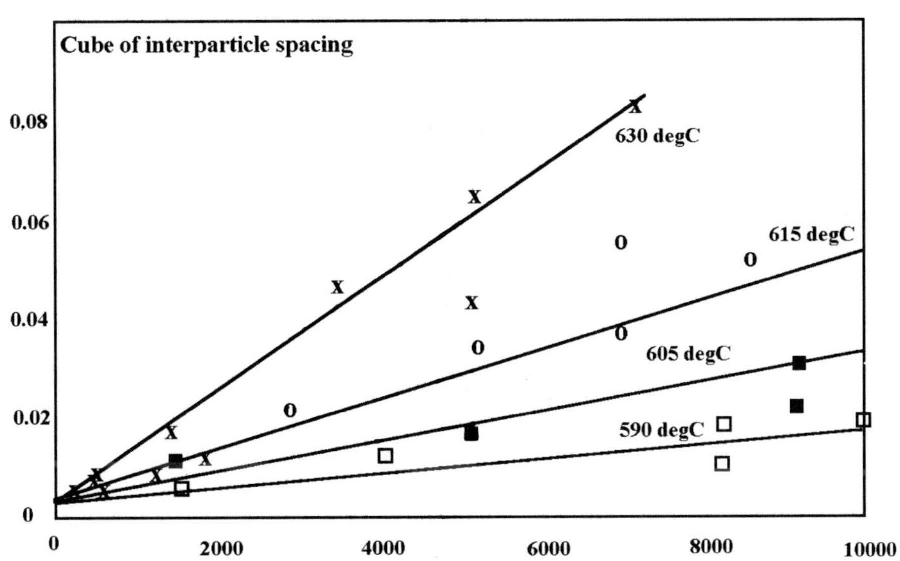

Figure 15 Carbide coarsening kinetics for a 1CrMo steel (souce Askins *et al.*, 1988).

interrupting creep tests, performed at 590°C, 42 MPa, for the measurement of interparticle spacing and for the manufacture of miniature creep specimens for continued testing. These were then tested at the original stress level but at elevated temperatures between 605°C and 650°C. A comparison between the actual creep curves and those predicted on the basis of interparticle spacing measurements on a specimen which had been interrupted at 8,222 h at a strain of 1.2% shows that there is good agreement between the predictions and the experimental results (Figure 16). However, the cast-to-cast variability of the carbide coarsening kinetics and the solid solution strengthening element of the creep strength are such that cast specific data for each of these are required to calibrate the computer model. Furthermore the determination of the carbide spacing is experimentally difficult and subject to segregation effects. In order to produce cast specific coarsening data it is necessary either to:

(i) take material from at least three areas of the component operating at different temperatures or

(ii) produce data using archive material or material extracted from the component.

Even if time permits such a strategy, the application of the coarsening data to predict creep behavior requires knowledge or assumptions of the values of the constants n, α', B_0, and k_A in the model equations. It is thus unlikely that the coarsening model in its fullest form will find significant service application. A simpler alternative approach would be to use the carbide coarsening data alone as a means of monitoring the mean effective historical operating temperature in the same manner as other microstructural features described later.

5.07.5.2 Hardness

It is well established that the hardness of low alloy ferritic steels operating in the creep range is a function of time, temperature, and stress. The relative simplicity of making hardness measurements on service components has promoted interest in using these for remnant life assessment purposes, either as a means of determining or confirming operating temperature, as a qualitative indicator of material condition, or in the extreme, as a means of direct estimation of remaining creep life. Changes in hardness occur due to coarsening of carbide particles, recovery, and recrystallization. All creep resistant power plant steels are severely tempered before they enter service and are thus almost always beyond the point where secondary hardening is expected. Hardness can therefore be expected to decrease monotonically during service in a manner that may be described by an Avrami type equation.

$$\xi = 1 - \exp(-k_B t^n) \tag{20}$$

where t is the time, k_B and n are constants, and ξ is given by

$$\xi(t) = [H_0 - H(t)]/[H_0 - H_\infty] \tag{21}$$

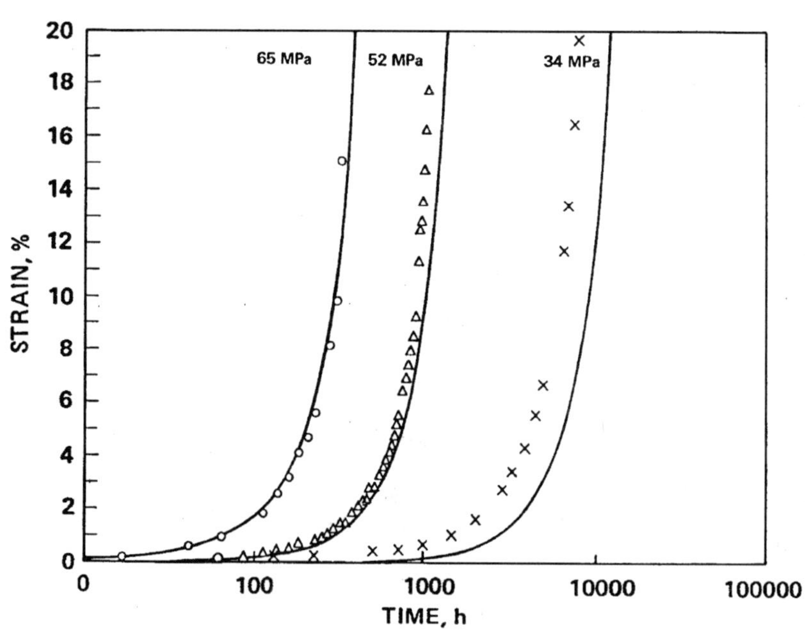

Figure 16 Comparison of predicted (lines) and observed (symbols) creep behavior of precrept 1CrMo, using the interparticle spacing model (source Askins, 1988).

where H_0 is the initial hardness, H_∞ is hardness at the end of useful life, and $H(t)$ is hardness at time t. The initial hardness may not always be available but can sometimes be estimated from a cold part of the component.

In practice the Larson Miller parameter (P) is more often used for correlations; e.g., Gooch *et al.* (1993) found the following relationship for a 1CrMoV rotor steel,

$$H(t,T)/H_0 = 1.605 \times 10^{-3}P - 4.962 \times 10^{-8}P^2 - 11.99 \qquad (22)$$

where $P = T(16.2 + \log t)$, T is in K, and t is in hours.

The data from which this equation is derived are shown in Figure 17, the apparent peak in hardness at low values of P being due to the exceptional occurrence of secondary hardening. The spread of data corresponds to 95% confidence limits of $\pm 15°C$ at exposure conditions equivalent to 1×10^5 h at 550°C. If multiheat data are treated in the same way, the confidence limits become unrealistically large (Figure 18), although use of the mean lines may provide a semiquantitative estimate of the effective operating temperature.

Published data suggest that values of H_∞ for power plant steels are likely to be in the range 120–150 HV, the main contributors being the intrinsic strength of iron and solid solution

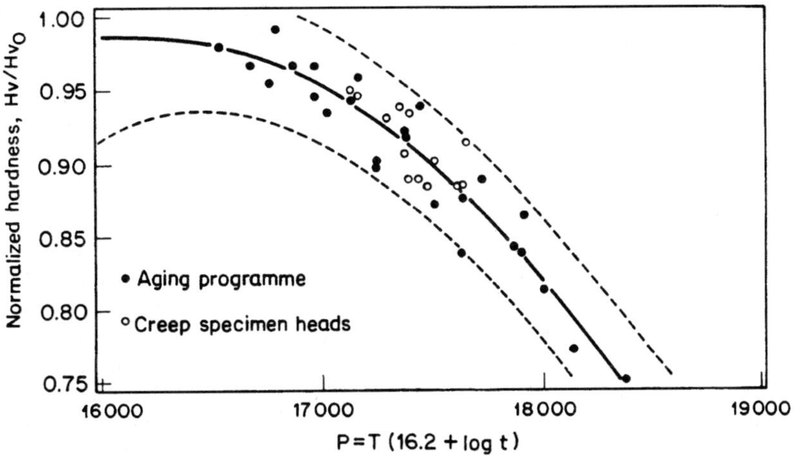

Figure 17 Hardness variation, normalized to initial value, with Larson-Miller parameter of constant 16.2, for a 1CrMoV rotor steel; showing 95% confidence limits (source Gooch *et al.*, 1993; reproduced by permission of the Institute of Materials).

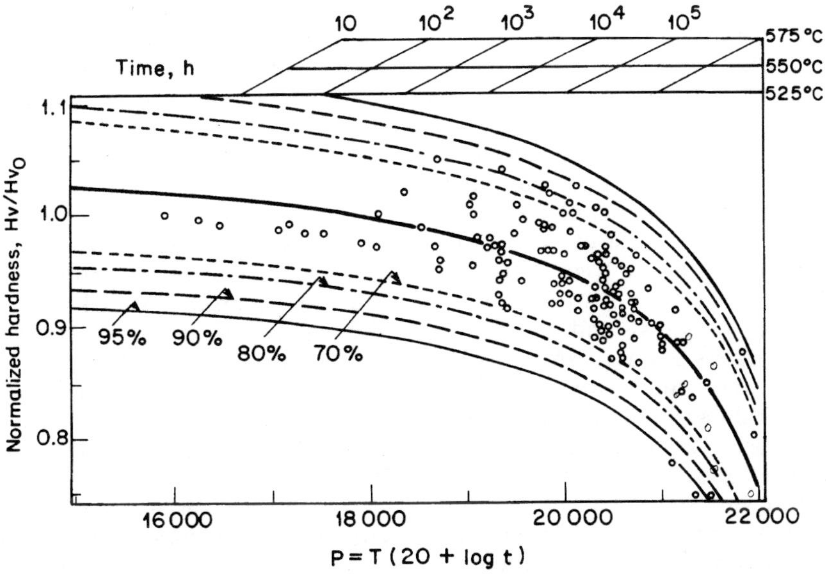

Figure 18 Hardness variation, normalized to initial value, with Larson-Miller parameter of constant 20, for a range of 1CrMoV rotor steels (source Gooch *et al.*, 1993; reproduced by permission of the Institute of Materials).

strengthening. The starting hardness for low alloy steels is likely to be in the range $H_0 = 160-240$ HV, the lower value corresponding to normalized and tempered boiler steels and the upper value to quenched and tempered rotor steels.

Kimura *et al.* (1986) correlated hardness with creep rupture strength by destructive testing of ex-service rotors and developed a relationship which they claimed could be used empirically for direct estimation of remaining creep life by hardness measurement alone. However, this was for a specific class of rotor steels for which they had extensive data and does not suggest a universal correlation.

The effect of composition on softening rates was demonstrated by Kusumoto *et al.* (2001). They investigated the softening of a range of 2CrMo weld metals and determined the effects of C, Si, Mn, Cr, and Mo by regression analysis, assuming each to be independent. They found that C, Si, Mn, and Mo each retarded softening whereas Cr appeared to accelerate it. However, there are some uncertainties about the details of the analysis since Cr is generally thought to retard softening and carbon to increase it. Inspection of the data actually shows that the two weld metals that softened the fastest were those with the highest carbon levels.

For steam headers Cane *et al.* (1985) suggested that the experimental difficulties of the carbide spacing assessment technique could be overcome by using hardness as an empirical measure of the carbide spacing. A direct life estimation model was proposed in which cast specific data were obtained for calibration purposes by accelerated aging in the laboratory. Figure 19 shows an example of an empirical relationship between hardness and interparticle spacing following creep exposure of a 1CrMo steel.

Although it is well established that softening rates are increased by the application of a stress, the quantitative data are conflicting and the question as to whether stress or strain control the mechanism is unresolved. Batte and Gooch (1985) and Roberts and Strang (1987) reported the effect of stress and the latter suggested that hardness losses of more than 15% in service were indicative of either significant creep strain accumulation or operation for extended periods above the maximum design temperature. Gooch *et al.* (1993) studied the hardness of creep specimens of a large number of 1CrMoV rotor steels and found that the reduction in hardness of the gauge length was always greater than that of the head of the test piece. Moreover the gauge length softening was proportional to that of the head, irrespective of stress, temperature or test duration, being described by the relationship (Figure 20):

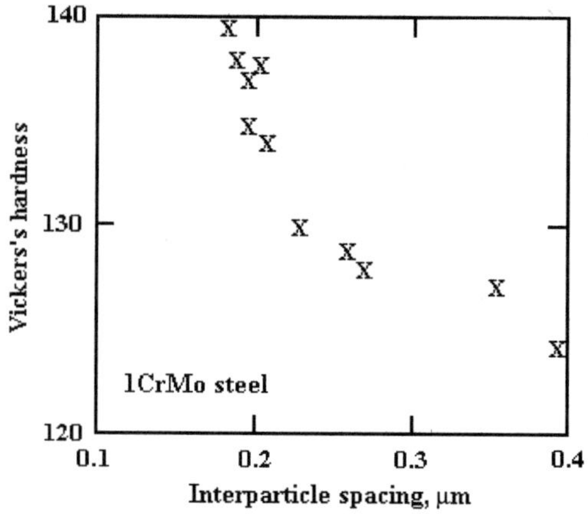

Figure 19 Hardness, as a function of nearest neighbor spacing of carbides, in a 1CrMo steel after creep testing at 630°C for various times (source Askins and Menzies, 1985).

$$H(t)_{\text{gauge}} - H_0 = 1.234[H(t)_{\text{head}} - H_0] - 14.55 \quad (23)$$

The fact that the correlation line does not go through the origin was attributed to the secondary hardening observed for some of the alloys examined. It was hypothesized that a possible mechanism for the enhanced softening might be the establishment, above a threshold stress of 62 MPa (the lowest test stress), of a dislocation network promoting pipe diffusion, thus enhancing carbide coarsening rates.

Goto (1984) found that the hardness of 1CrMoV rotor steels could be related to a function G', which was essentially the logarithm of the Larson Miller parameter, with an additional stress softening term. There was however a critical value of stress (108 MPa) below which there was no additional softening.

$$G' = \log P + 0.000217(\sigma - 108) \quad (24)$$

Contrary to the work of Gooch *et al.* (1993) this relationship suggests a strong dependence of softening on the magnitude of the applied stress. Conversely, Hayakawa *et al.* (1996) investigated the effect of stress on hardness changes in 2CrMo weld metals and found that the additional softening due to stress could be described in terms of a stress independent Larson Miller relationship over the stress range 20–100 MPa.

The observation of stress or strain enhanced softening of hardness is consistent with other work on the tempering of ferrite, bainite, and martensite. Strain tempering, in which the steel is first deformed then tempered, leads to large increases in strength with some loss in ductility

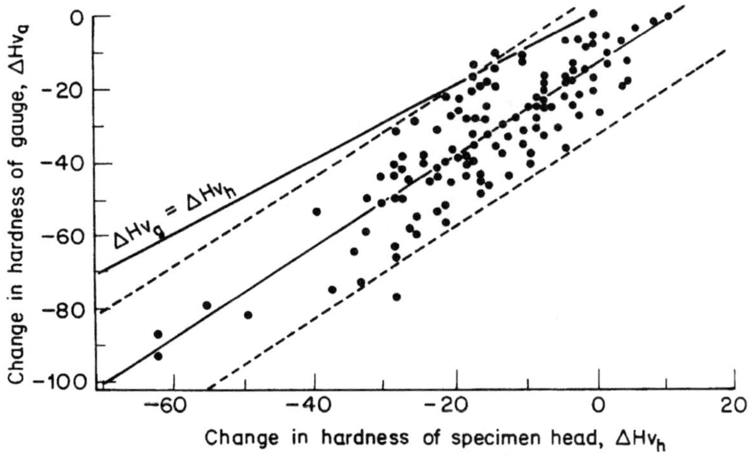

Figure 20 Variation of creep test piece gauge softening with head softening for a range of 1CrMoV rotor steels (source Gooch *et al.*, 1993; reproduced by permission of the Institute of Materials).

and toughness. The tempering causes a significant rise in hardness by the precipitation of very fine alloy carbides on deformation-induced defects, particularly dislocations (Kalish *et al.*, 1965). Precipitation rarely occurs in the matrix in the absence of the dislocations (Honeycombe and Bhadeshia, 1995). The increase in precipitation varies in proportion with the plastic strain (e.g., Jonas and Weiss, 1979). The recovery and recrystallization that occurs when bainite is tempered is also accelerated by creep deformation (creep tempering) (Ridal and Quarrell, 1962; Murphy and Branch, 1969). Ridal and Quarrell showed that, in Mo containing bainitic steels, Mo_2C is replaced more rapidly by the more stable M_6C carbide when the tempering occurs during creep deformation. The acceleration is greater for lower temperatures, higher creep strain rates, and for steels where the kinetics of the Mo_2C to $M_{23}C_6$ reaction are slow in the absence of creep. This is consistent with more rapid diffusion and nucleation rates, attributable to increased dislocation densities. Conversely, Ridal and Quarrell also found that the M_7C_3 to $M_{23}C_6$ reaction during the tempering of Fe–Cr–C martensitic alloys may be slightly retarded by creep deformation. This would suggest that dislocations stabilize the M_7C_3.

For all techniques based on hardness measurements, the extent to which the various claims may be realized in practice is dependent upon many factors, including:

(i) variations of aging response and creep strength between heats;

(ii) variations in hardness within heats due to segregation and heat treatment;

(iii) scatter on individual hardness measurements associated with the use of portable measuring equipment; and

(iv) surface condition due to machining, oxidation, decarburization, etc.

Figure 21 shows the variation of hardness with creep life fraction predicted for a single heat of 1CrMo from the carbide spacing model (Askins, 1988). It is clear from this that, at temperatures and life fractions of practical interest, hardness changes with time are so slow as to be unsuitable for direct use in quantitative life assessments, bearing in mind the limitations discussed above. It must be concluded, therefore, that the use of hardness measurements is primarily to provide a qualitative indication of whether a component has experienced excessive temperatures during operation.

The situation is further complicated in weldments since the hardness is likely to be highly variable both between and within individual microstructural regions, i.e., weld metal, HAZ, and parent material, even when the welds are made with matching composition weld metal. The potential location of failure is then difficult to identify since creep ductility, creep strength, and creep strain may vary with position. The weld metal may be of lower intrinsic ductility than the parent material because of:

(i) the larger inclusion content;

(ii) the presence of regions of large grain size such as partially transformed columnar grains; or

(iii) trace levels of creep-embrittling species such as As, Sn, Sb, and P.

The weld metal may also be either softer or harder than the parent depending upon the composition and postweld heat treatment. Nominally matching composition weld metals will generally have lower carbon contents, and potentially lower hardness, than the parent to reduce susceptibility to weld cracking.

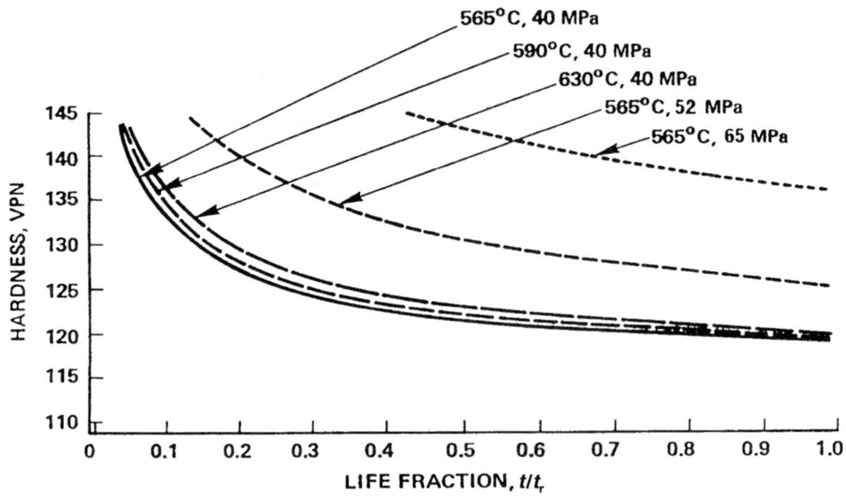

Figure 21 Variation of hardness with creep life fraction predicted from a carbide spacing model for a 1CrMo steel (source Askins, 1988).

However, the cooling rate may be faster than that of the original component leading to harder transformation products. Another possible scenario is that the weld metal composition may be deliberately chosen to have a lower creep strength in order to avoid cracking of the weld metal or the HAZ of the parent during welding. This is generally the case when welding CrMoV steels, a 2CrMo weld metal being used. Although of lower creep strength, the 2CrMo metal has a higher hardness because the higher chromium level increases the hardenability so that the structure is bainite rather than the ferrite of the normalized and tempered CrMoV. It cannot therefore be assumed that failure will always occur in the softest part of the joint.

The HAZ of the parent material also has a highly inhomogeneous microstructure with the potential for narrow bands of either very large or very fine grains due to the different time–temperature history during welding. Conventional wisdom is that a fine grain size should result in a high ambient temperature hardness. However, where the fine grain size is a consequence of a temperature cycle in which full austenitization occurs but the alloy carbides do not fully dissolve, the structure of the fine grains may be ferrite as opposed to the harder bainite which forms from higher temperature cycles. Thus, local variations in time–temperature history may result in extreme differences in grain size and carbide distribution, which may be the principal determinators of local hardness and creep strength although not necessarily overall ductility or rupture life.

Finally, local regions of multiaxial stress are produced as a result of the presence of (i) narrow bands of weak material in the HAZ and (ii) the interface between weld metal and HAZ. Increasing triaxiality reduces creep deformation rates but increases creep cavitation and the conflict between these two effects may result in either reduced or increased rupture life. Reduced overall joint ductility because of this is a contributory factor in "Type IV" cracking in fine grained regions of low alloy ferritic weld HAZ (Gooch and Kimmins, 1987). Conversely, the supporting effect of a strong parent material is the reason why it is often possible to specify a weld metal of lower creep strength if there is a concern over ductility.

Hardness may, therefore, only be used as a general indication of the state of the microstructure and only then if there are ample specific data for the materials and detailed welding processes concerned. This is well illustrated by the classic case of creep failures of low alloy ferritic welds, which may be in the weld metal or coarse or fine HAZs depending on chemical composition, weld process details, weld geometry or stress, or residual welding stress level.

5.07.5.3 Phase Evolution

As previously discussed, low alloy ferritic power plant steels are generally used in the quenched and tempered or normalized and tempered condition. This results in a metastable microstructure, which evolves slowly on service exposure and offers a potential means for quantifying the temperature of exposure. The average effective operating temperature is frequently the missing link in a life assessment program.

Table 8 Equilibrium mole fractions of precipitate phases at 565°C predicted using MTDATA and SGTE.

Steel	M_2X	M_3C	M_7C_3	$M_{23}C_6$	M_6C	*Laves*	*NbC*	*NbN*	*VN*
(1/2)CrMoV	0.0053			0.0247					
1CrMoV	0.0089			0.0412					
2CrMo				0.0335					
9CrlMo				0.0222					
Modified 9CrMo				0.0222				0.0009	0.0030
12CrMoV				0.0443					
12CrMoVW				0.0444		0.0007			
12CrMoVNb				0.0318			0.0006	0.0029	0.0018

Source: Bhadeshia *et al.* (1998) with permission.

The results of thermodynamic calculations which give the phase fractions in equilibrium at 565°C are given in Table 8 (Bhadeshia *et al.*, 1998) for the common power plant steels. The calculations were performed using the MTDATA computer program and the Scientific Group Thermodata Europe (SGTE) database, taking into account the carbide phases and Laves phase listed. (MTDATA has been developed by the Thermochemistry Group of the National Physical Laboratory, Teddington.) The chemical elements considered are carbon, silicon, manganese, chromium, nickel, molybdenum, vanadium, and niobium; nitrogen is not included but is known to be important in forming MX particles. An M_5C_2 phase has also been identified in 1CrMo steels (Mann *et al.*, 1995) but this is not included in the analysis.

The table is interesting for the absence of some commonly observed phases, such as M_2X and M_6C in 2CrMo and M_7C_3 in 12CrMoV, which implies that these are in fact only metastable phases. M_6C in particular has often been considered to form after $M_{23}C_6$ and therefore to be the more stable carbide.

However, these equilibrium structures are approached through complex sequences of (i) distinct phase changes and (ii) compositional changes within phases. Consequently, potential indicators of effective operating temperature include phase composition and phase proportions.

5.07.5.4 Phase Compositions

Cementite that forms in association with bainite or martensite enriches with elements such as chromium during service. Measurement of this, and of other elements such as Mn and Mo, has been used to estimate the effective service temperature of 1CrMo and other low alloy steels to an accuracy of ± 12°C (Carruthers and Collins, 1983). Cementite precipitated from martensite or bainite is thought to grow by a displacive transformation mechanism in which the substitutional lattice is displaced into the new structure at a rate controlled by the diffusion of carbon (Chance and Ridley, 1981). Cementite thus precipitates in a few seconds during the tempering of martensite.

Low alloy steels used in the normalized and tempered condition will frequently contain pearlite. The cementite component of the pearlite grows by a diffusion mechanism, which allows atoms to partition to the most energetically favorable phase. Consequently, the cementite within pearlite always has a chemical composition that is closer to equilibrium. For example, the initial Cr level in pearlitic cementite is higher than in cementite formed from bainite (Bhadeshia *et al.*, 1998). However, this is not a practical problem for long aging times since the degree of enrichment in pearlitic cementite is generally small relative to the difference between the formation and equilibrium values.

Irrespective of its origin, cementite will tend towards its equilibrium composition when exposed for long periods of time, which generally results in increases in the levels of the substitutional elements. Bhadeshia (1989) has shown, by considering diffusion fluxes across the interface of a thin plate particle, that the rate of change is proportional to $t^{1/2}$ rather than the more usually assumed $t^{1/3}$. He also showed that the rate of diffusion within the cementite was such that the concentration distribution of the substitutional elements within it are likely to be uniform. This is important for practical purposes since it means that microanalysis techniques which average the concentration in the whole particle may be used. It is often not possible in practice to distinguish between the two exponents. However, values of the diffusion coefficient obtained graphically are only correct in the case of parabolic ($t^{1/2}$) kinetics (Thomson and Bhadeshia, 1994). This supports the physical significance of a $t^{1/2}$ relationship, which can then be extrapolated with more confidence than a $t^{1/3}$.

Figure 22 shows experimental results for enrichment of cementite in a bainitic 1CrMo steel on aging at 550°C after initial tempering at 730°C (Afrouz *et al.*, 1983). The data are consistent with parabolic kinetics and also illustrate the different rates of enrichment of the substitutional elements, which reflect the different solubilities in cementite as well as the different diffusion rates. It is apparent that the most useful element to monitor for practical purposes is Cr.

It should be noted that solubility in cementite varies with temperature, becoming lower with increasing temperature and also that the approach to equilibrium is faster at higher temperatures. These factors will influence the temperature–time window over which the technique will find practical application. Other factors that must be taken into account include:

(i) particle size;

(ii) solute saturation and formation of other phases; and

(iii) chemical composition, in particular carbon.

Figure 23 shows that smaller particles enrich more rapidly; this is a consequence of the fact that they are smaller reservoirs for solute. Consequently, the prior thermal history of the steel, which influences particle size, will also affect enrichment rates. Furthermore, at long exposure times, the diffusion fields of individual particles will start to overlap resulting in a phenomenon known as soft impingement, which is not taken into account in the analytical derivation of the parabolic growth law. This process has the effect of slowing down enrichment rates so that there is a deviation from the parabolic law as the equilibrium state is approached. Bhadeshia (1989) has computed enrichment rates for the same 1CrMo steel cited in Figure 23. The results (Figure 24) show that Mn and Mo saturate more rapidly than Cr. Ultimately of course, the cementite, which is not the long term equilibrium phase (Table 8), will be replaced by other phases, in particular $M_{23}C_6$.

Detailed chemical composition will also affect enrichment rates; with carbon having a particular strong influence. First, an increase in carbon level is accompanied by a decrease in the equilibrium chromium level in the cementite that will slow down the enrichment rate.

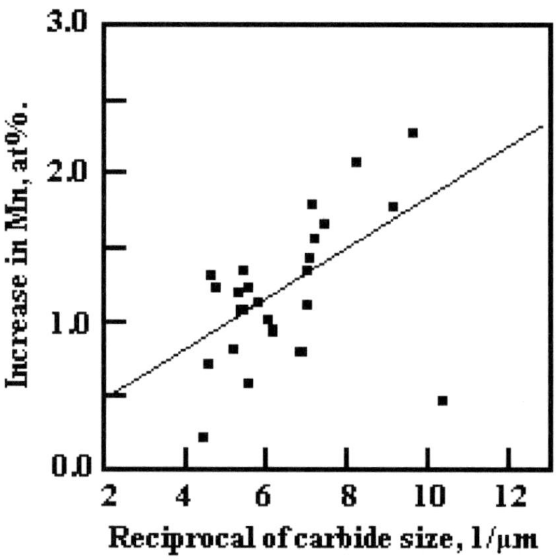

Figure 23 Particle size dependence of Mn enrichment of cementite in 1CrMo after aging at 565°C for 4 weeks (Thomson and Bhadeshia, 1994; reproduced by permission of the Institute of Materials).

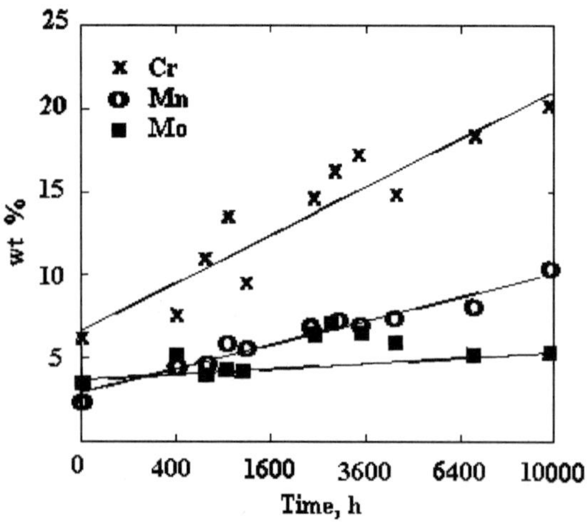

Figure 22 Measured changes in the composition of cementite particles during aging at 550°C of a 1CrMo steel heat treated to give a tempered bainitic structure (source Afrouz *et al.*, 1983).

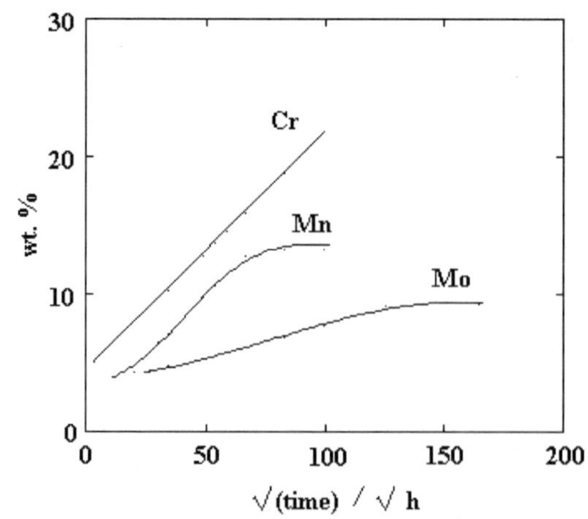

Figure 24 Calculated changes in cementite composition in 1CrMo on aging at 550°C (Bhadeshia, 1989; reproduced by permission of the Institute of Materials).

Second, an increase in carbon results in an increase in the amount of cementite present, which is achieved partly through an increase in particle size. This will also result in a reduction in enrichment rate both directly by the particle size effect and indirectly by accelerating the onset of soft impingement.

In addition to the effect of the major alloying elements it is known that some impurity elements accelerate the carbide coarsening process. For example, Pilling *et al.* (1982) demonstrated that *P* increases the microstructural degradation rate of 2CrMo and speculated that this might be due to cooperative diffusion of Mo and P. It is quite conceivable that similar effects may influence enrichment rates.

Most work on carbide solute enrichment has centered upon cementite, since this carbide is present initially in most low alloy steels, is of a size amenable to analysis, and remains present for much of a typical service life. However, in the high alloy 9–12% Cr steels, which are increasingly used, particularly in the new generation of power plants with advanced steam conditions, the kinetics of cementite enrichment and alloy carbide precipitation are much more rapid. In fact, cementite is rarely present after a typical tempering or stress relief treatment (Figure 25). Attention therefore turns to the alloy carbides. The difficulty here is that the alloy carbides, of which $M_{23}C_6$ is the most dominant, generally form by diffusion alone and consequently with chemical compositions close to equilibrium. However, as the carbides form initially during the tempering treatment, some changes may occur during subsequent service at a lower temperature. Strang and Vodárek (2000) found that in a 12CrMoVNb steel, the Cr level in $M_{23}C_6$ increased from 60 wt.% to 70 wt.% on long term aging at 550°C or 600°C. Moreover, by normalizing against the decreasing Fe content they proposed that the Cr : Fe ratio could be used as a time–temperature indicator. Similarly, they found that the Cr : V ratio in the M_2X precipitate changed by a factor of more than 2 on long-term exposure.

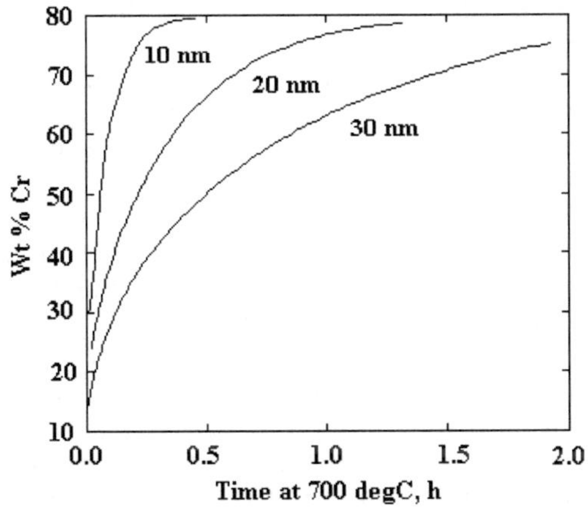

Figure 25 Rapid saturation of cementite in 12CrMoV on ageing at 700°C (Thomson and Bhadeshia, 1992; reproduced by permission of the Institute of Materials).

5.07.5.5 Phase Proportions

An alternative approach is to follow the evolution of the various phases during service through the proportions of the phases present, either qualitatively or quantitatively. Dobrzanski and Hernas (1996) related the remaining safe residual life of service exposed 1CrMo steel to the "main" and subsidiary carbides present and related these to equivalent times and temperatures of exposure, described in terms of a Larson Miller type parameter. Creep rupture curves were constructed for different carbide distributions. The rupture strength of heavily exposed material ($>5\times10^5$ equivalent hours at 550°C, predominantly M_6C with some $M_{23}C_6$ and Mo_2C) was found to be approximately half that of new material (mainly M_3C with some Mo_2C).

Robson and Bhadeshia (1996, 1997) modeled the phase evolution of a 10CrMoV during tempering at 600°C. For times of practical interest, the amounts of the alloy carbides remain essentially constant. However, of interest as a possible exposure time or temperature monitor is the development of the intermetallic Laves phase (Figure 26), which forms in this class of steel, typically, between 100 h and 1000 h at 600°C and at longer times at lower temperatures.

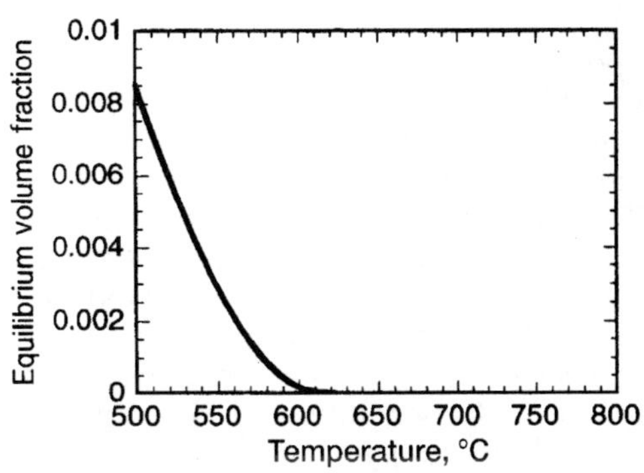

Figure 26 Variation of equilibrium volume fraction of Laves phase with temperature in 10CrMoV steel, calculated using MTDATA (Robson and Bhadeshia, 1997; reproduced by permission of the Institute of Materials).

5.07.6 DIRECT MEASUREMENT OF CREEP PROPERTIES

5.07.6.1 Strain Rate Measurements on Plant

Strain or dimensional measurements on plant are made for a variety of reasons. At one extreme they may be used as dimensional checks to ensure that strain accumulation is within the limits required by various codes or does not exceed allowable clearance tolerances. At the other extreme they may be made for direct life estimation purposes. The accuracy required will depend upon the purpose of the measurements. For example, routine measurements on headers are made using "pi" tape or bow-gauge micrometers whereas other procedures which have been proposed for more accurate measurements include photographic (photogrammetry) (Clayton, 1989), or replication (Eggeler *et al.*, 1989) techniques. Each of these involves the measurement at successive outages of reference marks on the component. In the case of the replication technique these consist of a grid of scribed lines, protected during service by a layer of nickel which is removed chemically prior to replication. The photogrammetric technique consists of stereo pair photography of a grid of small holes on the component. A specialized device is required for this, consisting of a frame containing two cameras, accurately mounted and calibrated. The developed film is measured in the laboratory, using a specialized table and projection equipment. Image analysis is used to perform a statistical assessment of the strain distribution from the grid.

Each of these techniques can only be used when the component is off-line, thus severely limiting the amount of data available for analysis. Continuous monitoring using *in situ* measuring devices would be preferable but is difficult because of the high temperatures involved. Two types of capacitance measuring strain gauge have been developed for this purpose.

(i) "Planer" strain gauges were developed by the former Central Electricity Research Laboratories (CERL) in the UK (Noltingk *et al.*, 1972) and consist of two arches, with capacitance plates between them. The ends of the arches are spot welded to the component so that strains in the component over an effective gauge length of ~20 mm are translated into changes in the separation of the capacitance plates. These gauges can operate up to approximately 700°C.

(ii) "SJB" sensors (Parker, 1998) consist of two parts which are separately welded to the component, thus allowing the gauge length to be varied. The first part is a reference target and the second is a guard ring capacitance transducer focused on the target. This system has the advantage that the transducer can be reset thus easing installation and enabling a wider displacement range to be covered.

In order to estimate remaining component life from strain measurements a suitable parametric model which adequately describes the creep behavior is required. A model derived by Cane (1982) from the earlier work by Kachanov (1958) and Robotnov (1969) gives the following expressions that can be used to give direct life estimates:

$$\varepsilon = \varepsilon_r[1 - (1 - t/t_r)^{\varepsilon_s/\varepsilon_r}] \tag{25}$$

$$\varepsilon_t = \varepsilon_s(t/t_r)(1 - t/t_r)^{\varepsilon_s/\varepsilon_r - 1} \tag{26}$$

where t_r is the rupture life, ε_r is the rupture strain, and $\varepsilon_s = t_r(d\varepsilon/dt)$, the Monkman Grant constant.

The service strain rate, either from direct measurement or from extrapolation of post-exposure testing measurements, can be substituted in these equations to predict the remaining life. This is illustrated in Figure 27, where the strain rate and remaining life fraction are given for various values of ε_r and ε_s. In general, the remaining life is insensitive to ε_r for most of the life. In principle, ε_s may be obtained from existing data but in practice the cast-to-cast and experimental scatter is large. In the absence of cast-specific data use of conservative values results in unacceptably pessimistic estimates so that in practice plant strain measurements are generally analyzed using less sophisticated approaches.

In the UK, a number of pipework systems were fitted with creep measurement points (Carmichael, unpublished work 1984, cited by Gooch and Townsend, 1987) and analysis of the data from these suggested total lives in excess of 2.5×10^4 h if the hoop strain at the outer diameter were limited to 1%. These predictions were greater than those based on temperature measurement and inverse application of the design codes, thus illustrating the value of direct strain measurement.

In high temperature steam turbine rotors the positions of maximum creep strain accumulation may be the rotor bore, the blade root fixings, or the disk balance holes, depending on the specific design. 1CrMoV rotor steels exhibit extreme heat-to-heat variability with regard to rupture ductility, values of less than 2% being recorded for some heats. For this reason, a criterion used for rotor assessments is the attainment of a specified strain, which may be as low as 0.5% in some cases, depending on the pedigree of the particular forging. Many of the larger rotors in the UK were provided with creep measurement points, either on the outer diameter or on the disk faces, and datum measurements were taken before the rotors entered service. Measurements in service were then made using diametral or chordal bow gauge micrometers. The diametral strain at the point of measurement can then be converted into bore strain by applying the calculated ratio of the steady-state creep strain rates at the bore and the measurement position. Bore creep strains calculated in this manner for one family of rotors are shown in Figure 28 and compared with the values predicted on the basis of average material properties (Carlton *et al.*, 1987). The measured strains are in reasonable agreement with the predicted values and show

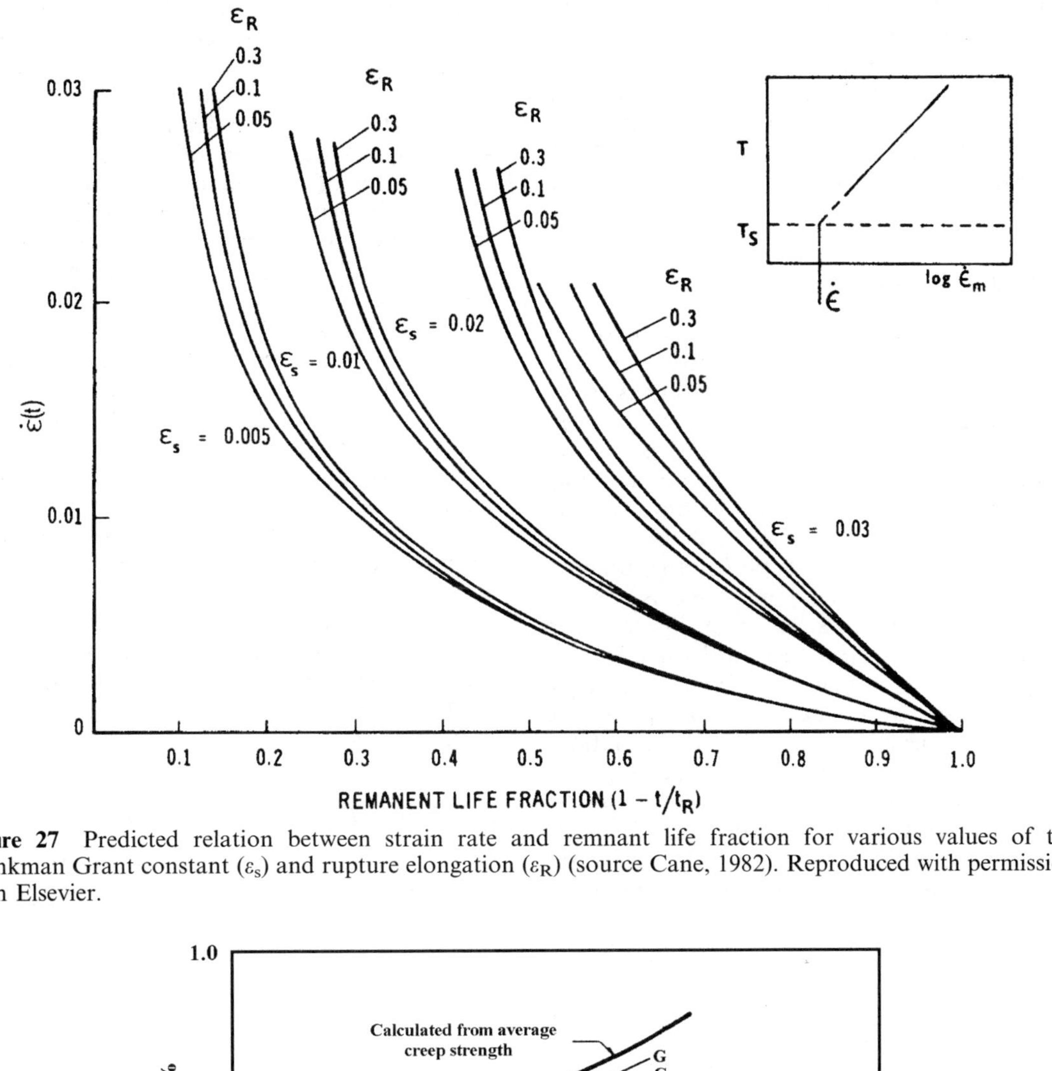

Figure 27 Predicted relation between strain rate and remnant life fraction for various values of the Monkman Grant constant (ε_s) and rupture elongation (ε_R) (source Cane, 1982). Reproduced with permission from Elsevier.

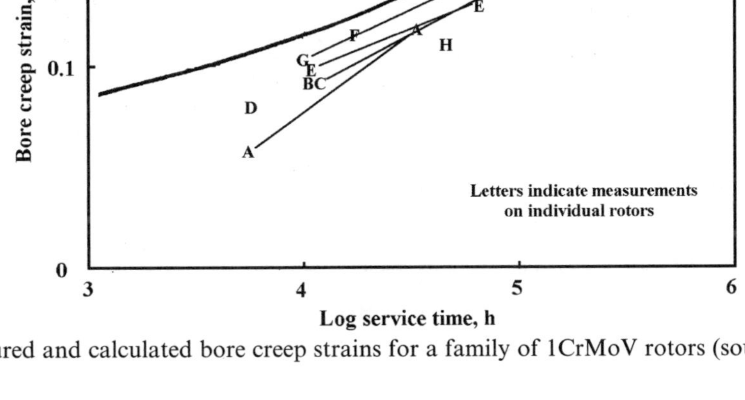

Figure 28 Measured and calculated bore creep strains for a family of 1CrMoV rotors (source Carlton *et al.*, 1987).

consistent trends with exposure time. This demonstrates that consistent, useable information can be obtained by simple measurement techniques. For this particular group of rotors the measured values are below those predicted but this is not always the case. The low measured strains could, e.g., be due to stronger than average material or operation below design temperature.

Techniques are not available for the measurement of strain at rotor blade root fixings or balance holes. These are generally assessed on the basis of calculated strain accumulation (balance holes) or rupture life (blade root fixings) in conjunction with standard NDE techniques and replication for cavitation detection.

5.07.6.2 Creep Testing

The most reliable method of life prediction is that of direct creep testing of material removed from service. There are many practical issues associated with this simple statement, however. One of the most important of these is the ability to obtain sufficient material from representative areas of the component in question to enable a realistic test program to be conducted without prejudicing the integrity of the component. A second issue is how to obtain reliable information on long-term creep properties from the necessarily short-term creep programs required to enable strategic decisions to be made. This involves the complex subject of constitutive equations for creep data representation that has occupied the minds of successive generations of researchers. This topic is dealt with in some depth elsewhere in the series. However, the major requirement is that the test program must either be performed in the same creep mechanism regime as that experienced in service or must involve extrapolation procedures which take into account changes in mechanism. The most obvious change in mechanism between short- and long-term tests is that from failure by ductile void growth to failure by intergranular cavitation. As discussed earlier this change is influenced not only by test duration but also by test temperature and by multiaxial stress state. However, there are also more subtle changes in mechanism such as changes in the control mechanism for cavity growth. These changes can result both in differences in the stress sensitivity of cavity growth rates and in the contribution of cavitation to the accumulated creep strain, with a consequent variation in the stress exponent for the overall creep rate. The three control regimes in which cavities are believed to grow are illustrated in Figure 29. These are (Cane, 1981):

(i) diffusion controlled growth in which cavities grow by grain boundary vacancy diffusion at a rate proportional to the maximum principal stress;

(ii) continuum growth where cavities grow under the action of the hydrostatic stress at a rate proportional to the Von Mises effective creep deformation rate and a multiaxiality factor; and

(iii) constrained growth where the diffusional growth rate is limited by the ability of the surrounding matrix to accommodate the dilation; growth then follows the creep deformation kinetics.

5.07.6.2.1 Isostress and isotemperature testing

In order to obtain life predictions within an acceptable timescale, accelerated creep testing

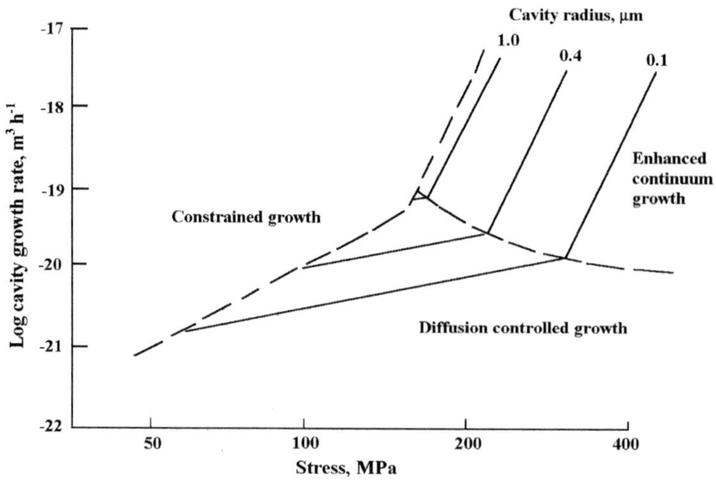

Figure 29 Cavity growth mechanisms for a 2CrMo steel of grain size 150 μm (source Needham and Cane, 1983).

techniques need to be employed. The two alternatives available are to increase stress or temperature. Increasing stress has the advantage that, if standard size specimens are available, tests may be performed in air rather than in inert atmosphere. However, increases in stress generally result in changes in the mechanism for creep deformation and/or rupture such that extrapolation of short-term data to service stress levels may become nonconservative. Nevertheless, tests of this nature may be used to position the specific cast of steel under study within the design data band for that grade of steel. This avoids the assumption of minimum creep rupture properties in life assessments and in many cases this is sufficient to underwrite the operation of the component to the required extended life.

When a direct estimate of rupture life is required accelerated temperature "isostress" testing in inert atmosphere is used. Temperature acceleration is less likely to result in changes in creep mechanism because the structural changes influencing creep deformation, such as carbide coarsening, are accelerated to a similar degree to the creep process, both being diffusion controlled. In order to obtain sufficient confidence in the extrapolation of creep rupture data to service temperatures, a statistical assessment of the data is performed. It is generally considered that a minimum of five and preferably seven tests at different temperatures are required for most purposes. Moreover it is advisable that the chosen temperatures do not approach the tempering temperature of the steel in order to avoid excessive structural degradation. For extrapolation purposes creep rupture data may be plotted against either T or $1/T$. In practice it is usual to plot against T since this reduces the possibility of a nonconservative life prediction. When time is available it is advisable to conduct a concurrent series of isotemperature tests with minimal stress acceleration in order to minimize the degree of data extrapolation required.

Figure 30 illustrates the statistical analysis of a set of isostress results for miniature specimens of a 1CrMoV rotor steel and emphasizes the importance of appreciating the significance of the error bands, which may be substantial, particularly at the higher test temperatures. The steel in question would have been originally tempered between 680°C and 710°C.

A linear relationship between test temperature and logarithm of rupture life is generally assumed for extrapolation purposes and this is usually found to be the case. Moreover, it is also found that slopes of plots for tests conducted at different stress levels are usually similar. However, it is important to understand that these assumptions are not universally true. For example, Steen and de Witte (1987) suggested that for bainitic 2CrMo steel the extrapolation may only hold true for material that has received sufficient service exposure to stabilize the microstructure, and may in some circumstances be nonconservative for virgin material. Conversely, other work has suggested that linear temperature–log stress extrapolations of heavily creep damaged (1/2) Cr (1/2) Mo (1/4) V may be nonconservative due to changes in creep ductility and creep curve shape at the higher temperatures. For these reasons test programs must be designed to use minimum temperatures and obtain maximum test duration consistent with plant requirements. Ideally, test program duration should be only one order of magnitude lower than the anticipated remaining life.

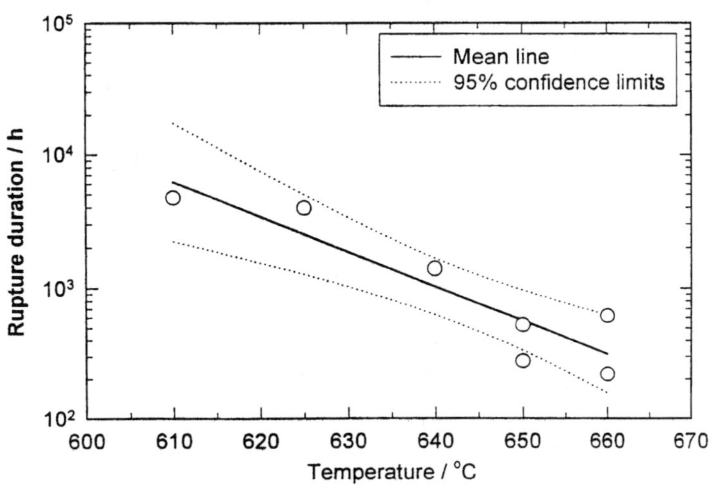

Figure 30 Isostress creep rupture test results for miniature specimens of 1CrMoV rotor steel (source Beech *et al.*, 1995; reproduced by permission of the Institute of Materials).

A further consideration is the position in the component from which the samples are taken. If the requirement is simply to place the material within the data band then the sample material should be taken from a region seeing minimum temperature and stress, so that the steel is as close to the virgin state as possible. Care must be taken, however, to ensure that the sampled material is representative of that of the critical regions. For example, the microstructure at the "cold" end of a rotor forging may not be characteristic of that at the "hot" end due to differences in segregation, forging severity, or cooling rate. Conversely, if a direct life estimate is required then samples should, in principle, be taken from the critical region. However, this is frequently not possible either because of inaccessibility or difficulty of subsequent repair. Thus, postexposure test programs will in general test material from a noncritical region, in terms of both stress and temperature, and will require the use of a life fraction rule, together with extrapolated data, to predict the remnant creep life of the critical region. Therefore, the aim should be to remove samples from those regions where the combination of stress and temperature predicts the greatest remaining creep life subject to the above considerations. This approach will minimize the reliance on knowledge of the value of the consumed life fraction, which can have a large scatter associated with an uncertain service history (Askins *et al.*, 1985).

5.07.6.2.2 *Miniature specimen testing*

It has already been stated that a minimum of five and preferably seven tests should be conducted in an isostress creep rupture program. It is often not possible to obtain sufficient material from a component for such a test program using conventional test pieces which typically each require a cylinder of material 12–16 mm in diameter and up to 100 mm long. Consequently, procedures have been developed to produce and test creep rupture specimens with gauge lengths as small as 10 mm and gauge diameters as low as 2 mm. These specimens are necessarily tested in inert atmosphere and special-test chambers have been designed to ensure the accurate measurement of the small applied loads (Figure 31).

In order to validate the use of miniature specimens Beech *et al.* (1995) conducted

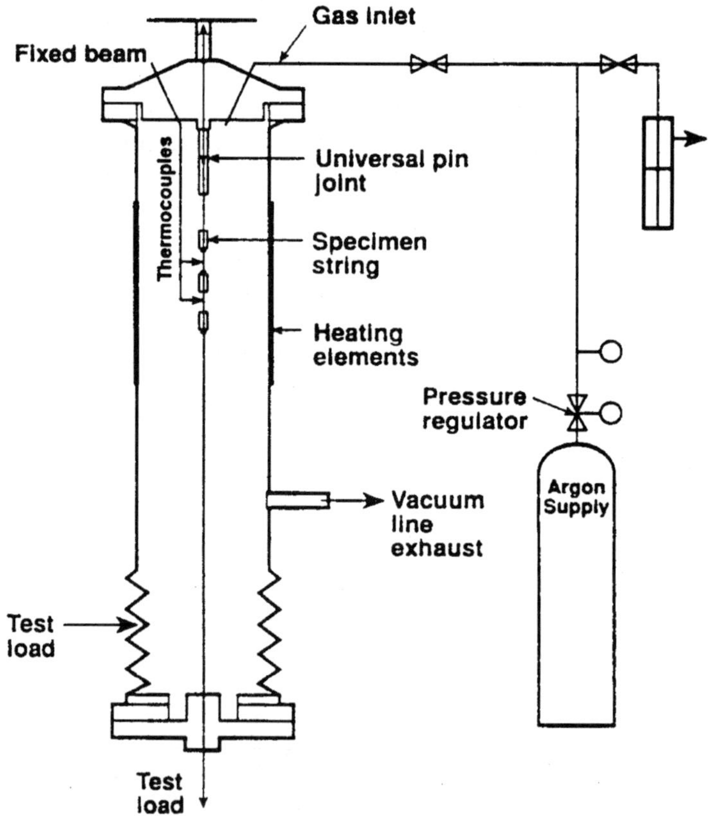

Figure 31 Schematic drawing of argon atmosphere testing chambers for miniature specimens (source Beech *et al.*, 1995; reproduced by permission of the Institute of Materials).

isostress creep rupture tests on 1CrMoV rotor material using standard and miniature test pieces of gauge diameters of 9 mm and 2 mm respectively. The tests on the standard specimens were conducted in air whereas the miniature specimens were tested in dry argon. The results (Figure 32) showed linear correlations between temperature and the logarithm of rupture life for both standard and miniature specimens, with similar slopes to the plots. However, the miniature specimens lasted typically 25% longer than the standard specimens. This was attributed to oxidation effects on the standard specimens whereby grain boundary oxide penetration led to earlier failure.

The same workers compared predictions of rupture life from isostress tests on five 1CrMoV rotor steels with the lives predicted from the long-term isothermal test data available from extensive test programs on the rotors. Close agreement was found between the two sets of data at predicted lives out to more than 10^6 h (Figure 33).

Finally, miniature specimens were also extracted from long-term creep specimens of a range of 1CrMoV rotors of different creep strengths. The results confirmed the same form of linear relationship as seen with tests on virgin material with lines of similar slope but separated according to the rotor creep strength. This was claimed to provide additional confidence in the extrapolation technique over the range of material strengths and ductilities similar to that seen in production forgings.

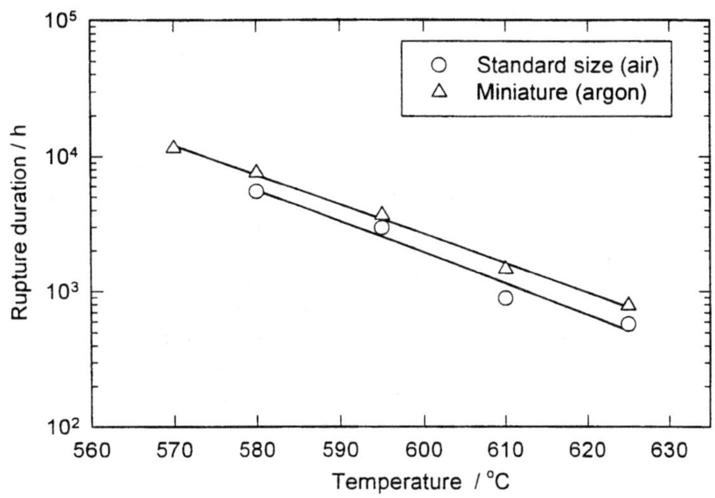

Figure 32 Comparison of miniature and standard specimen isostress rupture data for a 1CrMoV steel at 154 MPa (source Beech *et al.*, 1995; reproduced by permission of the Institute of Materials).

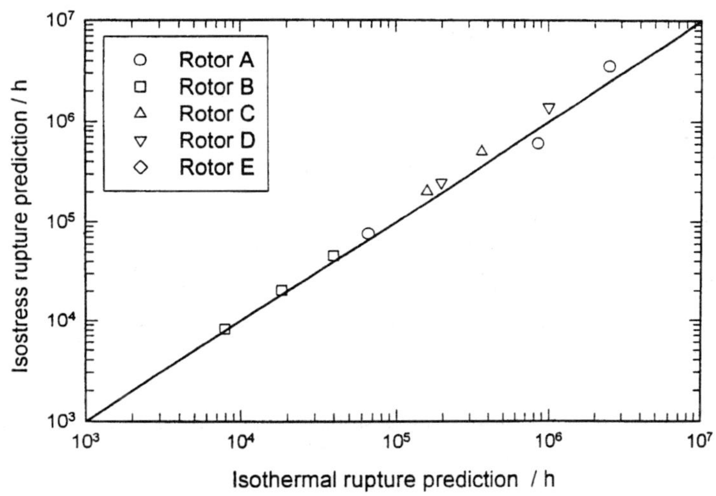

Figure 33 Comparison of isothermal and isostress rupture data for 1CrMoV steels (source Beech *et al.*, 1995; reproduced by permission of the Institute of Materials).

5.07.6.2.3 Cyclic loading

Power plants that were originally designed for base load operation are now required to operate under cyclic conditions in order to match electricity demand. This means that components are now operating under onerous cyclic stress and/or temperature conditions. For example, steam reheat pipelines may be subjected during shutdown to several hours at zero or partial pressure while the thermal insulation results in a high temperature being maintained. Similarly, during startup, the reheat tubing in some designs may receive radiant heat with little or no steam pressure. These conditions can lead to creep under variable load or to recovery/aging processes while the material is at high temperature but not under load. These recovery processes may lead to a reduction in "on-load" creep life as a consequence of the thermal aging or repeated primary creep. The use of strain and time hardening methodologies and of the life fraction rule have been shown to be significantly in error under these conditions (e.g., Gibbons *et al*, 1973).

Cane (unpublished work, 1984), proposed an extension of the isostress procedure to investigate the effect of load cycling. In this methodology a series of accelerated temperature cyclic creep test programs to 1% strain is conducted under conditions chosen to give a range of cycles to 1% strain. The data for each individual frequency are then extrapolated to give the time to 1% strain at service conditions. A second correlation is then performed in which the service times to 1% strain are plotted against the number of cycles and the data extrapolated to the anticipated number of cycles in service. In this way relatively short-term tests may be used to predict the rate of creep strain accumulation under cyclic load regimes similar to those experienced in service.

Gooch and Askins (1989) performed a series of square wave-form (40 MPa/10 MPa) cyclic creep tests on (1/2) Cr (1/2) Mo (1/4) V steel using loading periods chosen to give between 20 and 500 cycles to 1% strain; the results of which are shown in Figure 34. The effect of load cycling is to reduce the on load time to 1% strain, with the influence being marginally greater for the higher cyclic frequencies. However, the effects are small and may be insignificant at service conditions.

5.07.6.3 Multiaxial Testing

Few design codes make specific reference to creep ductility under multiaxial conditions, reliance being on uniaxial rupture data, sometimes supplemented by information on whether a material is "notch strengthening" or "notch weakening" in a uniaxial notched bar rupture test. The safety factors incorporated in the design codes are generally sufficient to counter any uncertainties which may exist, although there are many instances of premature failure attributed either to poor creep ductility or multiaxial stresses. For a life extension

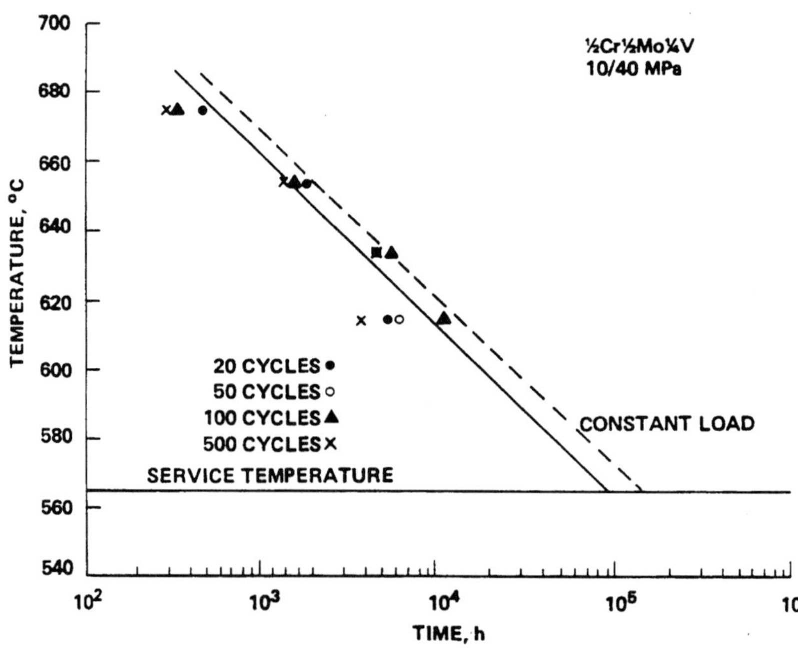

Figure 34 Uniaxial cyclic load creep strain data (time to 1% strain) for a (1/2)Cr(1/2)Mo(1/4)V steel (source Gooch and Askins, 1989).

program, the explicit purpose of which is to take advantage of the design safety factors, it is therefore often necessary to determine the response of the steel to multiaxial situations. There are two possible approaches: (i) by determining the multiaxial stress rupture criterion (MSRC) and then applying this analytically to the component or (ii) by testing model components. This section is concerned with the former, model component testing having been discussed in the section on reference stresses. The major multiaxial test procedures available are as follows (Gooch and How, 1986):

(i) uniaxial circumferentially notched bar testing;

(ii) internally pressurized tube testing, with or without applied end load; and

(iii) torsion testing, with or without applied end load.

5.07.6.3.1 Notched bar testing

Notched bar testing is by far the most widely used of these procedures because of its simplicity, low cost, and minimal material requirements. Historically the most common form of notch used has been that with a "V" section which results in a high elastic stress concentration factor (K_t) at the root of the notch. Examples of values of for K_t are 3.9 (BS 3550 pt 1) and 4.5 (DIN 50118). Materials are described as notch strengthening or notch weakening on the basis of comparison of the plain bar uniaxial life with the life of a notched specimen of the same net section stress. If a material is notch strengthening in both long- and short-term tests then this may be sufficient information since it implies that the dominant stress controlling rupture is the equivalent stress, which is lower than the net section stress due to the constraint imposed by the notch. However, it is frequently observed that there is a transition from notch strengthening to notch weakening at extended test durations. In this case, it may become necessary to conduct a more quantitative determination of the MSRC.

The Bridgman or semicircular notch geometry has the advantage of a more uniform stress field than the V notch and a less severe triaxiality. This results in a more uniform distribution of creep damage across the net section than the V notch specimen, which tends to fail by a discrete crack initiation and growth process in the more brittle materials. The semicircular notch therefore leads itself more readily to metallographic interpretation. In principle this notch geometry should also enable the direct measurement of rupture strain under multiaxial conditions. For example, Hayhurst and Webster (1986) computed the variation of the equivalent strain with time and radial distance and showed that this may be related to the tangential strain at the notch surface. The latter may be measured through the use of tangential extensometers. However, these procedures are experimentally and theoretically complex and are complicated further by the nonuniformity of the strain across the specimen throat. An empirical approach is to use a composite plain/notched specimen in which the plain part of the specimen is designed so that the uniaxial stress on it is the same as the Von Mises skeletal point stress across the notch. If it is accepted that the deformation of the notch is controlled by this stress then the plain section strain at the point of rupture of the notch may be used as a practical estimation of the notch rupture strain. Gooch and Maguire (1988) used this approach to estimate the multiaxial ductility of a 1CrMoV rotor steel and found that the elongation of the plain sided gauge length at the point of rupture of the notch was on average 43% of the uniaxial rupture elongation. They also found a similar relationship between reductions of area at rupture for the plain and notched specimens. A number of different theoretical relationships have been proposed to describe the relationship between the multiaxial rupture ductility (ε_r^*) and the uniaxial rupture ductility (ε_r). These include the following:

$$\varepsilon_r^*/\varepsilon_r = 2^{(1-3\sigma_m/\sigma_{vm})} \qquad \text{(Manjoine, 1982)} \qquad (27)$$

$$\varepsilon_r^*/\varepsilon_r = (\sigma_{vm}/\sigma_1)^{n/2} \qquad \text{(Cane, 1979)} \qquad (28)$$

$$\varepsilon_r^*/\varepsilon_r = 1.65\exp(-3\sigma_m/2\sigma_{vm}) \qquad \text{(Rice and Tracey, 1969)} \qquad (29)$$

where n is the stress exponent of the uniaxial creep rate, σ_{vm} is the Von Mises equivalent stress, σ_m is the hydrostatic stress, and σ_1 is the maximum principal stress.

For the notch geometry used by Gooch and Maguire (1988) these relationships give values for $\varepsilon_r^*/\varepsilon_r$ between 0.4 and 0.6, consistent with the experimental results but not sufficient to distinguish between the proposed relationships.

Modification of the Bridgman notch by locating a semicircular notch root at the bottom of a parallel sided notch enables a variety of stress states to be investigated by varying the notch acuity and hence the MSRC to be determined. Table 9 (Webster *et al.*, 2001) shows how the various skeletal point stresses vary with notch acuity. From these it is clear

Table 9 Normalized skeletal point stresses for a Bridgman notch ($D/d = \sqrt{2}$).[a]

Notch acuity ratio (d/r)	Normalized stresses		
	$\sigma^*_{vm}/\sigma_{net}$	σ^*_m/σ_{net}	σ^*_1/σ_{net}
1	0.86	0.42	1.02
2	0.81	0.49	1.04
3	0.77	0.54	1.05
4.82	0.72	0.61	1.08
10	0.66	0.73	1.14
20	0.63	0.83	1.22
30	0.61	0.87	1.28
35	0.61	0.89	1.29
50	0.58	0.91	1.34

Source: Webster *et al.* (2001).
[a] D = gross section diameter, d = net section diameter, r = notch root radius.

that distinction between equivalent stress and maximum principal stress or hydrostatic stress control is straight forward. However, given the likely scatter in data in a practical test program, distinction between maximum principal stress control and hydrostatic control is not so easy.

5.07.6.3.2 *Internal pressure testing of tubes*

Internal pressure testing of tubes may be used to: (i) determine empirically the reference stress for rupture, (ii) establish the failure mode of a component, particularly in the case of welded tubes or pipes, or (iii) determine the MSRC for a material. Test pieces are usually produced from tubes of appropriate dimensions. However, for a remnant life program on a thick section component it may be possible to manufacture tubular specimens as well as uniaxial specimens from the component wall (e.g., Browne, 1986).

In a tube subjected to internal pressure alone, the maximum principal stress is in the hoop direction and is typically twice the stress in the axial direction. Consequently the failure mode, whether it be maximum principal stress or equivalent stress controlled, is invariably by axial splitting. However, the stress state may be varied either by changes in wall thickness or by the application of an additional end load. Increasing the relative wall thickness results in an increase in the maximum principal stress (i.e., the hoop stress) relative to the Von Mises equivalent stress (Browne, 1986). Consequently, thick walled vessels are more likely to fail by low ductility cavitation than thin tubes. The further application of an end load enables the hoop and axial stresses to be varied resulting in a test procedure in which all of the principal stresses may be varied. For example, Figure 35 shows how the through wall hoop and axial stress distribution varies as the end load is increased. It is instructive to note that as the end load is increased the position of maximum hoop stress moves from the outer surface to the inner surface.

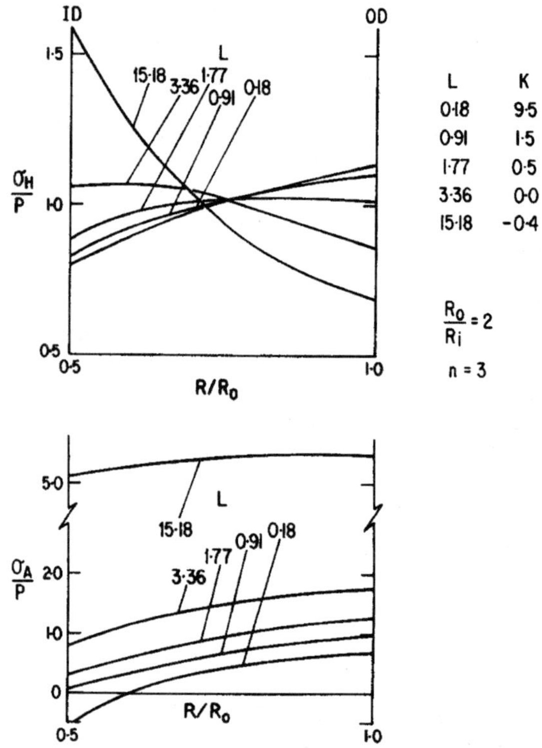

Figure 35 Stationary state hoop and axial stresses for various applied end loads and hoop:axial strain rate ratios in a thick walled pressurized tube (source Cane, 1986). Reproduced with permission from Elsevier.

The ability to vary independently the hoop and axial stresses is particularly useful for study of the failure mode of tubes containing circumferential welds. The failure mode of such components is frequently by a circumferential failure through a weak microstructural region in the weld. In extreme cases this may occur under internal pressure alone, despite the fact that the axial stress is only half the hoop stress. In less extreme cases the progressive application of an end load may result in a change in failure mode from axial to circumferential cracking, thus enabling an axial "reference" stress for the welded tube to be determined.

A full mathematical description of the steady state stress system of an end loaded internally pressurized tube has been given by Finnie (1960) and a code of practice for testing these components has been published by How *et al.* (1992).

5.07.6.3.3 *Torsion testing of tubes*

The torsion test is sometimes used to simulate the stress state in components subjected predominantly to shear stresses in service. In particular the torsion test may be preferred to reliance on equivalent stress correlations from uniaxial test pieces when there is an issue of material anisotropy or when large rotational strains are possible. In applying data from a torsion test it is important to identify shear creep strains with a representative shear stress. This aspect has been considered by Rees *et al.* (1992) in a proposed code of practice for torsional creep testing of tubular test pieces. They show that an approximate skeletal point exists and that the shear strain (γ^*) and shear stress (τ^*) at the skeletal point may be described by the following relationships:

$$\gamma^* = r^*\theta/L_r \qquad (30)$$

$$\tau^* = 3T_q[1/\{1-(r_i/r_o)^3\}]/2\pi r_o^3 \qquad (31)$$

where r^* is the skeletal point radius, θ is the measured twist in radians, L_r is a reference length, T_q is the applied torque, and r_i and r_o are the inner and outer radii, respectively.

As with internal pressure testing of tubes, the torsion test may be further refined by the application of an end load (Trampczynski and Kowalewski, 1986).

A material for which rupture is controlled by a MSRC would be expected to show a longer rupture life for the same equivalent stress in a torsion test compared with a uniaxial test because of the lower value of the maximum principal stress in the former. This is borne out by the results of torsion, uniaxial and Bridgman notch tests on a 2CrMo steel shown in Figure 36.

5.07.6.4 "Small Punch" Testing

The requirement to obtain creep data from very small samples extracted from service exposed components has led to the development of the "small punch" test. This utilizes thin disks of material, typically 10 mm in diameter and 0.5 mm thick, and was originally developed to test material extracted from rotor bores using a remote sampling device, which uses a hemispherical shaped cutter to remove a thin slice of material (Parker *et al.*, 1989).

In this test the disk shaped sample is located in a stainless steel die and clamped around the edges by a stainless steel sheath. The punch has a hemispherical tip, typically manufactured in Nimonic 80A or a ceramic, and is pushed, either under constant load or at a constant strain rate (CSR), through the specimen into a circular hole located in the die below the specimen. Testing may be either in air or in a controlled atmosphere. Parker *et al.* (1998) have demonstrated that the displacement curves for a 1CrMoV rotor steel obtained under constant load conditions are similar in shape and repeatability to those obtained in a standard uniaxial creep test. In addition, the temperature dependence is similar leading to the possibility of extrapolative procedures for life determination from accelerated temperature testing. However, a major limitation of the test procedure is the complex nature of the stress state in the disk sample. Parker *et al.* (1998) used a reference stress type approach to determine the load to be applied to the sample, the load being determined by relating the "life" of a disk sample to the life of a standard uniaxial creep rupture test on a well characterized material.

Tettamanti and Crudelli (1998) have attempted to quantify the relationship between the applied load and the equivalent uniaxial

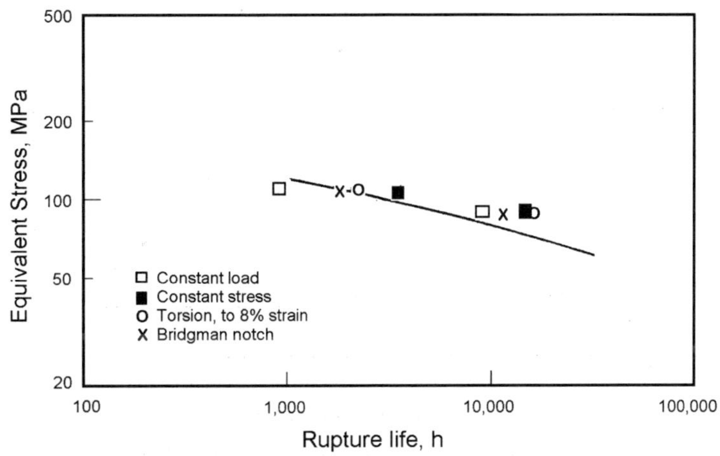

Figure 36 Uniaxial and multiaxial creep rupture data for a 2CrMo steel at 575°C (source Gooch, 1992).

stress both through a balance of forces method and through a method derived from relationships developed for drawing of thin sheets. For the specimen geometry used (8 mm diameter, 0.5 mm thickness, die hole diameter 4 mm, punch radius 2 mm) they estimated values for the ratio of the applied load (N) to the equivalent uniaxial stress (MPa) between ~1.9 and 2.1. Subsequently, using a value of 2.0, they demonstrated a good correlation between accelerated temperature small punch rupture and standard uniaxial rupture results for a 12CrMoV steel (Figure 37). Milička and Doběs (2001), who used a similar specimen size but a smaller punch (2.5 mm radius) found a value of 2.5 for the applied load to uniaxial stress ratio. Zhi (2001) has described procedures whereby deflection curves may be translated to a uniaxial creep curves by either (i) matching deflection curves to uniaxial creep curves held in databases, using numerical optimization procedures or (ii) calculation of the strains in the outer fibers of the disk sample using geometrical relationships or, more rigorously, using steady-state FEA.

It appears, therefore, that the small punch test shows promise for use in life assessment programs although further validation work is required. The need for "calibration" of the technique against uniaxial data also indicates a requirement for standardization of test procedures, with dimensions, tolerances, surface finishes, and test rig alignment being of particular importance.

5.07.6.5 CSR Tensile Testing

The accumulation of creep rate data for life assessment programs by conventional creep testing requires multiple specimens and test programs extending to several thousand hours. In order to reduce the time required to perform a creep life assessment test program Steen and de Witte (1987) proposed the use of slow strain rate (typically 10^{-5}–$10^{-8}\,s^{-1}$) CSR tests. This requires the use of a closed loop control system since the CSR must be measured over the specimen gauge and corrected for elastic strain.

In the proposed approach the tests are continued only to the point at which the load achieves a constant value. It is then assumed that the corresponding stress equates to the internal stress (σ_i) in a load controlled creep test in which the stationary creep rate is equal to the imposed strain rate in the CSR test. The test data are then interpreted according to an appropriate constitutive equation using the internal stress concept. Steen and de Witte (1987) proposed the use of the following equation:

$$d\varepsilon/dt = (D_o Gb/kT)\{(\sigma_a - \sigma_i)/G\}^3 \exp(-\Delta H/RT) \qquad (32)$$

where D_o is the pre-exponential diffusion coefficient, ΔH is the diffusion activation enthalpy, G is the shear modulus, b is the Burgers vector, σ_a is the applied stress, and R is the gas constant.

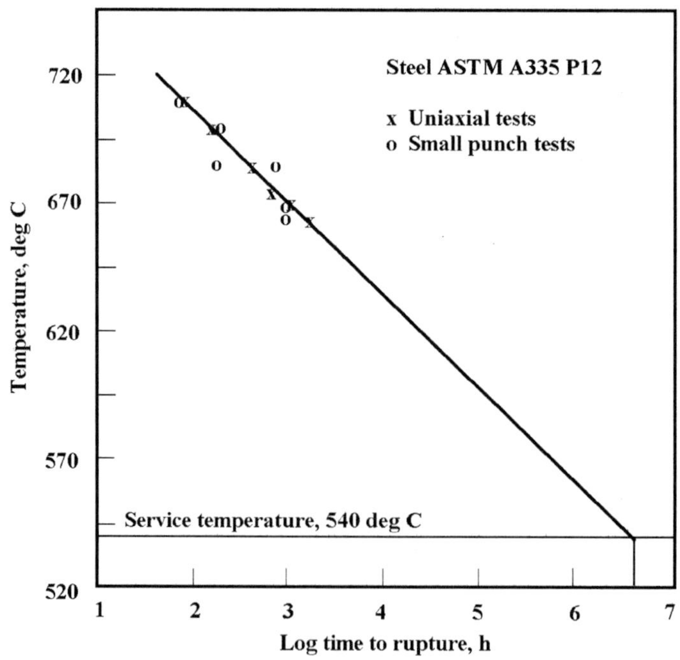

Figure 37 Remnant life estimation from small punch ($F_0 = 74$ N) and uniaxial "isostress rupture" tests ($\sigma = 37$ MPa) on a 12CrMoV steel (source Tettamanti and Crudeli, 1998).

Steen and de Witte (1987) found linear relationships between $(d\varepsilon/dt)^{1/3}$ and σ_i, implying a proportionality between σ_a and σ_i. However, there were three distinct regimes of linearity, which were taken to be different mechanistic regimes, across the boundaries of which extrapolation was not allowed. The observed proportionality between σ_a and σ_i was rationalized by consideration of the component parts of the internal stress, namely that due to second phase particles, σ_p, and that due to the matrix σ_m.

The methodology by which extrapolated values for σ_a at service strain rate levels are obtained is not altogether clear from the published papers since it requires knowledge of the proportionality constant between σ_a and σ_i. This would seem to require either knowledge of or experimental determination of the prestress term in the constitutive equation. One possible means by which the relationship between σ_a and σ_i could be determined would be by switching the tensile test from strain control to load control once the constant load state had been reached. However, this may be experimentally difficult because of possible transients in the test machine response. Nevertheless, the authors use extrapolated σ_a and strain rate values to estimate service creep rate and hence rupture life; good agreement with predictions made from isostress accelerated temperature creep rupture tests is claimed.

A difficulty with this methodology is that its is intuitively difficult to differentiate between the steady state stress in a CSR test and the stress at the same strain rate in the secondary creep regime in a constant load test. Osgerby and Dyson (1993) attempted to resolve this by using a physically based model of primary, secondary, and tertiary creep to predict CSR behavior from uniaxial creep data. They found that the stress at a given minimum creep rate is virtually identical to the peak stress in the corresponding CSR test. (The model did not actually predict a constant stress region.) Whilst acknowledging that this could be used in a life assessment procedure they pointed out that a rapid creep test could generate the same data in a shorter time than the CSR test using less expensive test equipment. At least as of early 2000s the CSR test appears to have been used very little for life assessment purposes.

5.07.6.6 Stress Relaxation Testing

The stress relaxation test also offers the possibility of acquiring data over several decades in creep rate from a single test lasting less than 100 h. In a stress relaxation test the total (i.e., elastic plus inelastic) strain on the specimen is held constant in a closed loop machine and the stress allowed to relax. Knowledge of the elastic modulus allows the elastic and inelastic strain rates at any point in the test to be calculated; and a stress versus "creep" strain rate curve to be constructed.

Woodford (1993, 1996) has suggested that the stress relaxation test may be used to characterize the creep properties of a material at a fixed point in time, allowing rapid evaluation of the consequences of microstructural evolution rather than attempting to incorporate the changes in a conventional long-term creep test. In the case of materials that undergo thermal degradation this would of course require the availability of material in different microstructural states for an evaluation of the long-term properties. However, it could provide a "fitness for purpose" characterization, for continued operation of service exposed material.

Quantitative use of the data generated for direct life estimation, however, poses a number of issues, not the least of these being that the technique does not address the issue of creep rupture, which must be considered separately. Woodford suggests that this may actually be an advantage by emphasizing the need to decouple creep strength and fracture resistance in relation to service performance. He suggests that the technique may be used to predict the time to a given level of strain, within the approximate limits of the initial strain applied in the stress relaxation tests (typically $<1\%$). This would be particularly useful for components where the service life is limited by strain tolerances rather than by rupture. It could also be used for evaluation of materials for design purposes, again when strain limitation is the design criterion.

A major limitation of the technique for life assessment, however, remains the inability to simulate continued thermal degradation during ongoing service. For this reason Woodford suggests that the technique may be most useful for alloys with a low sensitivity to thermal mechanical history such as the cast nickel based superalloys. It seems unlikely that it will find significant application for low alloy steels.

5.07.7 STEAM GENERATING POWER PLANT APPLICATIONS

The initial stage in a life assessment program is to confirm the absence of gross defects or service embrittlement, which might limit safe operation before creep damage becomes important. If defects are absent or can be repaired, those regions that are likely to be at the most advanced stages of deterioration are then subject to a detailed life assessment procedure. The critical regions in the major high temperature components in fossil fueled power plants are as follows.

5.07.7.1 High Temperature Steam Headers

Steam headers are critical components because they are subject to large variations in temperature and pressure during operation and because failure has potentially life threatening consequences. In addition, it is extremely expensive and time consuming to replace them (Townsend, 1991). It is significant that the original concept of the three-stage approach was developed in the UK for the life assessment of headers in order to comply with statutory requirements for the operation of pressure vessels. This followed the explosive failure of a header in the 1960s.

Boiler headers are built to many different designs. However, they usually comprise a large diameter cylindrical pressure vessel into which smaller diameter tubes deliver steam. Consequently they may involve butt welds, longitudinal seam welds and end cap welds between thick section materials as well as branch and stub welds. Creep damage, in the form of cavitation or cracking, generally occurs first in the vicinity of these weldments. Cracks may also form in the parent material ligaments between stub penetrations as a result of creep fatigue interactions generated by thermal stresses during rapid transients (Viswanathan, 1989; Cane, 1996). These damage modes are detected by regular inspections based on service experience. Weld cracking is usually repairable by grinding and rewelding. However, ligament cracking may necessitate component replacement, accompanied by redesign and/or new materials of construction. Headers originally fabricated from low alloy steels may be replaced by components made from higher creep strength material such as a modified 9Cr steel. The stronger material enables the wall thickness of the header to be reduced, thus lowering the thermal stresses developed during transient operation.

In addition to these macroscopic damage modes, headers accumulate microscopic or submicroscopic creep damage throughout operation. This has to be assessed if the component is to be used beyond its original design life. In general, the critical steady state stresses are at a maximum at outer surfaces of cylindrical pressure vessels so that access is not a major problem. Direct observation, replication, and material sampling are then options for damage assessment by microstructural techniques. Techniques are available for removing samples for accelerated creep testing without compromising the integrity of the component (Gooch and Townsend, 1987). It is possible, e.g., to trepan from a thick section header wall a circular plug of material of sufficient size to extract near standard size creep specimens. The hole in the wall is then repaired by welding in place a "blind" stub. If this is not possible then a smaller sample of material, sufficient for extracting miniature creep specimens, may be removed using a "chain drilling" technique (Figure 38). However, the critical region is not always accessible. For example, damage in welds generally initiates subsurface due to microstructural variations and the position of maximum stress at stub penetrations is at the bore. In these cases stress analysis, service experience, and ultrasonic examination become the primary assessment tools.

5.07.7.2 High and Intermediate Pressure Steam Turbine Rotors

High temperature rotors vary in design from manufacturer to manufacturer. However, the critical regions for the accumulation of creep damage are the bore (if one is present), the blade root fixings, balance holes, and stress concentrating features, such as grooves on the outer surface (Carlton *et al.*, 1987).

The position of maximum steady-state strain accumulation in many designs is the rotor bore at the inlet stages. The potential failure mode here is the initiation and growth of intergranular creep cracks to a size which could result in brittle fracture, either at cold startup, due to centrifugal and thermal stresses, or during overspeed tests. Ultrasonic and eddy current techniques are used to detect macroscopic defects but the direct examination of the bore surface for microscopic damage is particularly difficult. Remotely controlled devices are, however, now available to prepare and examine surfaces *in situ* and also to remove samples for examination and testing (Parker *et al.*, 1989). If creep damage is detected then it may be possible to remove this by enlarging the rotor bore, providing of course that stress analysis confirms that this is acceptable for future operation. A potential complication is that the region of maximum triaxial stress, which may be the area most susceptible to creep cavitation, is actually subsurface from the rotor bore (Gooch, 1991).

Some rotors have bands or pips machined into the outer diameter or disks for the

Figure 38 Chain drilling jig for removal of material for manufacture of miniature creep specimens from a steam pipe (source Gooch, 1988b).

measurement of creep strain during service, using datum measurements taken prior to service. The remaining safe life can then be estimated using critical strain data determined from experimental programs.

Possible failure mechanisms in blade root fixings (Carlton *et al.*, 1987) are creep crack initiation and growth or shear failure due to excessive steady-state creep strain accumulation. Blade root entry fixings may be axial or circumferential. Axial entry fixings can be examined directly or by the taking of plastic replicas without removing the blades. Circumferential entry fixings are less accessible but fortunately are generally less highly stressed.

The critical region on some designs may be at the balance holes in the disks or at the transition between disk and body. These regions are generally accessible for direct examination. It is not often practicable to remove sufficient material from critical areas to produce specimens for direct determination of remaining life by accelerated creep testing. In this case samples may be prepared from "cold" regions with the aim of positioning the original creep properties of the specific forging within the scatter band assumed by the design procedure.

5.07.7.3 High Temperature and Pressure Pipework Systems

Welds in high pressure pipework systems are, as in headers, the areas most susceptible to creep damage. However, whereas in headers the stresses are generally determined by the pressure and thermal stresses alone, which are known or can be calculated, pipework systems are also subject to highly variable and largely unquantified external forces. These arise from the self-weight of the pipes and from thermal expansion effects. In principle the erection procedures and the support systems should ensure that these are minimized. However, practical experience has shown this not always to be the case. In addition, the microstructural variations within a weldment prevent the quantitative application of microstructure based assessment techniques.

This means that life management procedures for welds have remained largely experience based despite a great deal of effort expended in trying to devise quantitative prediction procedures (Gooch *et al.*, 1989b). The major examination techniques are ultrasonic and magnetic particle crack detection. However, qualitative assessment of the severity of creep cavitation using surface replication is also used for weld life management.

Operation beyond design life requires that the condition of the parent material also be determined. Obtaining material for this should not be a problem but the number of heats and individual forgings and castings in a typical system presents difficulties. Strain monitoring of straight sections either by measuring creep pips at outages or by continuous monitoring using high temperature capacitance gauges has probably the most technical potential but also suffers from heat-to-heat variation.

In practice straight lengths of pipe are seldom life limiting; more critical items include T-pieces and, in particular, bends. The latter are more highly stressed than straight pipes because of wall thinning and ovality produced during the bending process as well as the greater potential for stresses generated by system forces. Design codes make allowance for these effects by the inclusion of an additional safety factor, with the consequence that the straight pipes from which bends are produced are often overdesigned. Plant experience in continental Europe has confirmed the analytical prediction that creep cracks initiate at outer surfaces. However, there remains a debate about the profile of creep damage below the surface; e.g., Viswanathan (1988) has suggested that subsurface cavitation may be more severe than that detected at the surface. (This is certainly observed in weld heat affected zones where the geometry of the weld may result in planes of vulnerable microstructure being oriented perpendicular to the axial stress field subsurface but not at the surface itself.)

Methods which have been examined for quantification of cavitation damage in pipes and welds include the usual metallographic procedures as well as ultrasonic, magnetic, and resistivity based techniques. Surface replication results must clearly be interpreted with caution if the damage is most severe below the surface.

5.07.7.4 Valve Chests and Turbine Casings

These are large complex components, which are usually castings but may sometimes be closed-die forgings. The more complicated components may also consist of two or more castings or forgings welded together. Thermal fatigue cracking at inside surfaces is the most common mode of parent material degradation encountered; procedures for assessing this are described elsewhere. There have been instances where poor design has led to creep damage in internal ligaments of valve chests but, in general, remnant life assessment of chests and casings has not been as great an issue as for other components. The greater concern has been with the welds joining valve chests to pipework systems.

5.07.7.5 High Temperature Bolts

Creep resistant bolts are used in critical locations such as joints between (i) upper and lower halves of turbine casings and (ii) valve casings and covers. The failure of such bolts could have very serious consequences and there has been at least one instance of a valve cover being ejected through the roof of a power plant. Bolt failures generally occur through crack initiation in the first engaged thread (Townsend, 1995). Detection of these cracks *in situ* is particularly difficult because of the access issues. Bolts operate under nominally constant strain rather than constant load conditions; consequently, the ability to maintain a steam tight joint is determined by the stress relaxation performance of the bolt (Bolton, 1995). Although physically based models for predicting stress relaxation behavior in bolting alloys have been proposed (e.g., Osgerby and Dyson, 1995), life management of high temperature bolts is usually effected through pragmatic formalized procedures which define conservative replacement intervals in terms of operating hours and tightening events. Procedural documents also place requirements on: (i) bolt tightening, removal and replacement practices, in order to avoid over tightening (Jones, 1995) and (ii) thermal insulation in order to avoid excessive temperature gradients which may result in strain concentration and bending moments at the first engaged thread.

The most commonly used materials are bainitic or martensitic ferritic steels of the 1CrMoVTiB or 12CrMoVNb types. However, in situations where the creep strength of these is not adequate the nickel based superalloy Nimonic 80A may be used. (Creep resistant austenitic steels are not suitable because of their high thermal expansion coefficients.) A characteristic of Nimonic 80A, in common with other superalloys, which has to be taken into account in operating procedures, is the phenomenon of "negative creep." This is a consequence of short- and long-range ordering of the Ni_2Cr lattice superstructure during exposure between 450°C and 550°C which results in an increase in strength, a loss of ductility, and a decrease in volume of up to 0.16% (Nath *et al.*, 1995). Different tightening regimes must therefore be specified for Nimonic 80A bolts in order to guard against brittle failures during service.

5.07.7.6 Weldments

Requirements for weld assessment procedures have been identified throughout this chapter and have been addressed as they have arisen. As weldments are by far the most common location for creep failures in thick walled components much effort has been devoted to the modeling of their behavior under creep conditions and major advances in understanding have resulted. In particular, the use of ever more powerful computers to perform time dependent FEA using continuum damage constitutive equations is now permitting predictions of weld failure locations and times (Hyde *et al.*, 1998). However, formal design and assessment codes rely on more established procedures such as semiempirical weld strength reduction or efficiency factors, and reference stress techniques (Budden and Goodall, 1998).

Despite all this, the complexities of microstructure and stress distribution mean that weld life management has remained primarily inspection based, with inspection regimes being optimized through plant experience and through the analytical activities.

5.07.8 GAS TURBINE HOT GAS PATH APPLICATIONS

Remnant life assessment procedures were originally developed to enable operation beyond the end of nominal design life of steam generating power plants. However, developments in gas turbine technology and the availability of large reservoirs of natural gas has led to the emergence of large industrial gas turbines as the power generating technology of choice in many countries.

Technological advances have been rapid and competition for orders strong with the result that new or modified designs enter the market before the previous generation has been proven in long-term service. Consequently, availability and reliability during the early years of operation has not always lived up to expectation. Life assessment procedures for gas turbine hot gas path components may therefore be more appropriate for predicting whether a component will be able to achieve rather than exceed its expected life. The principles of assessment are the same as for a steam generating power plant and will not be repeated here. However, opportunities for material sampling are fewer and the potential efficacy lower because of component costs and variations in operating temperature and stress within individual components. Consequently, the emphasis is on visual inspection, *in situ* nondestructive examination and analytical techniques.

The major driving force for technology improvement has been the desire for ever higher operating temperatures and hence cycle efficiency. The figure usually quoted is the turbine inlet temperature, which already exceeds 1300°C in the latest designs and is likely to exceed 1400°C in the near future. Clearly, metals cannot operate at these temperatures and machines incorporate complex cooling systems to keep metal temperatures to acceptable levels.

The main components in the hot gas path comprise the combustor chambers, transition pieces between combustor and turbine, turbine rotor, turbine vanes (stationary airfoils), and turbine blades (rotating airfoils). The most critical items are probably the first stage blades, which see the extremes of temperature and stress, where it has been reported (Nath, 2000) that typical equivalent operating hours (EOH) of only 3,000–2.4×10^4 EOH are being achieved in comparison with expected lives of 4.8×10^4 EOH. (The precise quantification of EOH varies between manufacturers but is in general the sum of the steady-state operating hours and "equivalent hours" assigned to each cyclic operation (e.g., normal start, high temperature trip, low temperature trip, etc.).) There are also factors to account for the use of substitute fuels that influence erosion and corrosion behavior.

The metals used in the hot gas path are cast or forged cobalt or nickel based superalloys. For example the first stage blades are made from cast nickel based alloys, which allows the incorporation of complex air cooling channels. In order to permit the highest possible metal temperatures (>900°C) equiaxed grained materials are being superseded by directionally solidified and single crystal alloys. Table 10 lists the compositions of the most commonly used hot gas path alloys.

The main mechanisms of alloy degradation are as follows:

Oxidation. A single alloy composition cannot in general optimize both mechanical properties and oxidation resistance. Critical components are therefore covered with oxidation resistant metallic coatings, typically 0.2 mm thick, of the MCrAlY type, where M is Ni and/or Cr. The oxidation resistant phase in these coatings is the Al rich beta phase, depletion of which during service exposure may result in a requirement for stripping and recoating at some point during life. Erosion and foreign object impact damage may also lead to recoating. Simms *et al.* (2000) have described a procedure for modeling the corrosion and erosion behavior and for predicting the remnant life of corrosion resistant coatings.

Low cycle fatigue (LCF) and thermomechanical fatigue (TMF). The consequence of the onerous stress and temperature cycles is that these are mechanisms of major importance, particularly for turbine blades and vanes. Analysis is complex and requires the

Table 10 Typical industrial gas turbine blade and vane alloy compositions (wt.%).[a]

Alloy	*Ni*	*Co*	*Cr*	*Mo*	*W*	*Nb*	*Ta*	*Al*	*Ti*	*Zr*	*B*	*C*	*Others*
IN738LC	Rem	8.5	16.0	1.7	2.6	0.9	1.7	3.4	3.4	0.05	0.011	0.11	
IN939	Rem	19.0	22.4		2.0	1.0	1.4	1.9	3.7	0.10	0.01	0.15	
Udimet 520	Rem	12.0	19.0	6.0	1.0			2.0	3.0		0.005	0.05	
FSX414	10.5	Rem	29.5		7.0						0.012	0.25	
ECY768	10.0	Rem	22.5		7.0		3.5	0.15	0.2	0.05	0.01max	0.60	
EA/DS GTD111	Rem	9.5	14.0	1.5	3.8		2.8	3.0	4.9		0.01	0.10	
DS CM247LC	Rem	9.2	8.4	0.6	9.5		3.2	5.5	0.8	0.01	0.015	0.08	Hf 1.5
SC Rene N4	Rem	8.0	10.0	2.0	6.0	0.5	5.0	4.2	3.5				Hf 0.2
SC Rene N5	Rem	8.0	7.0	2.0	5.0		6.0	6.2					Re 3.0, Hf 0.2
SC PWA1483	Rem	9.0	12.0	1.9	3.9		5.0	3.5	4.1			0.07	
SC CMSX4	Rem	9.0	6.5	0.6	6.0		6.5	5.6	1.0				Re 3.0, Hf 0.1

[a] EA, equiaxed; DS, directionally solidified; SC, single crystal.

performance of aerothermal analyses of the turbine to determine gas temperature distributions over the airfoil surface, together with thermal and stress analysis of the airfoil itself (Hepworth *et al.*, 2000). This information is then used in an appropriate damage summation procedure (e.g., Ainsworth, 2001) to determine crack initiation times and growth rates. Other influences may also have to be factored into the analysis. For example, oxidation of uncoated components may accelerate crack initiation by the oxidation of carbides on grain boundaries intersecting the surface. If components are coated then a crack in the coating may propagate through the interdiffusion zone into the substrate below (MCrAlY coatings are bonded to the base material by a diffusion treatment). Experimental data on coated materials are therefore required.

Creep. High temperature components in industrial gas turbines are typically designed for a creep life of 1×10^5 h steady-state operation. In practice, the equivalent operating hours concept means that actual operating hours are substantially lower than this and it is unusual for creep *per se* to be life limiting. Nevertheless, estimates have to be made of the remaining creep life as part of any life assessment program. A rigorous assessment requires aerothermal, thermal, and stress analyses as for creep fatigue damage.

Creep rupture (Wood, 1995) or stress relaxation tests (Woodford and Daleo, 2000) may be used in principle for estimating remaining creep lives from miniature specimens extracted from sacrificial components (if available). However, interpretation of the results is limited by: (i) blade to blade variations, (ii) positional variations in stress and temperature, (iii) specimen size effects, and (iv) data scatter.

Creep cavitation occurs in nickel base superalloys and attempts have been made to relate this semiquantitatively (Lindblom, 1985; Wortmann, 1985; Karlsson *et al.*, 1995) to consumed life fraction. However, cavitation in these alloys occurs at a relatively late stage of rupture life and is of little practical use in situations where life is strain limited. Other potential techniques, for creep damage assessment in superalloys, which have been investigated but not found practical application include X-ray line broadening (Wood and Raynor, 1996), small angle neutron diffraction (Krautwasser *et al.*, 1992) and electron beam back scattering (Quested *et al.*, 1988).

A common requirement for assessment of the above degradation mechanisms is data on actual metal temperatures during operation. Some machines have been equipped by their operators with optical pyrometers for measuring blade metal temperatures but access, cost, and data interpretation issues may restrict the number of installations. Temperature estimation through microstructure change has therefore been one experimental area where there has been significant effort.

The microstructure of nickel based superalloys is characterized by a distribution of cuboidal gamma prime (γ') precipitates in a gamma (γ) matrix and it is this distribution which is responsible for the creep strength of the alloys. Correlations between γ' particle size and operating time and temperature for specific alloys have been compiled by plant operators and manufacturers but the accuracy of the technique is probably only ± 25°C (Aurrecoechea and Brentnall, 1990), which is insufficient for most purposes. A particular issue is that the knowledge of the γ' size prior to service is required and this information is not generally available. Another complication is that in directionally solidified and single crystal materials the γ' particles form into "rafts" that are preferentially oriented perpendicular to the applied stress. Other parameters which have been proposed for effective temperature estimation include (i) the width of the interdiffusion zone between the MCrAlY coating and the base metal (Ellison *et al.*, 1998) and (ii) oxide thickness measurements (Swaminathan *et al.*, 1997).

Ceramic thermal barrier coatings (TBCs) may be applied to critical components to reduce metal temperatures to levels acceptable for creep performance. In the past these were only applied to stationary parts because of concerns over long-term integrity on rotating blades. However, the technology subsequently advanced to the point where TBCs could be applied to blades in the newer machines. Retrospective application to reduce metal temperatures is also becoming a remedial option where service lives are being limited by excessive temperatures.

5.07.9 CONCLUDING REMARKS

Remnant creep life assessment and consequently plant life extension cannot be defined simply in terms of individual procedures; rather it is a formalized strategy requiring the application of a wide range of techniques by experienced metallurgists and engineers. These techniques include nondestructive examination, stress analysis, defect assessment, creep data analysis, microstructural analysis, creep and creep crack growth testing, low cycle fatigue testing, and others. Not all have been covered in this chapter since many are described elsewhere in the series—the emphasis having been on strategy and on those techniques associated with the detection and assessment of generalized creep damage. It is hoped that these have been covered sufficiently to give the reader sufficient understanding of the subject to devise programs for component life assessment. However, the coverage has necessarily been relatively broad rather than deep and the reader is referred to the reference list for more detail. In particular, the text books by Viswanathan (1989) and Webster and Ainsworth (1994) are recommended.

5.07.10 REFERENCES

A. Afrouz, M. J. Collins and R. J. Pilkington, 1983, Microstructural examination of 1Cr-0.5Mo steel during creep. *Met. Technol.*, **10**, 461–463.

R. A. Ainsworth (ed.), 2001, "R5, An Assessment Procedure for the High Temperature Response of Structures, Issue 2, Revision 2," BEGL, Barnwood.

R. A. Ainsworth, G. G. Chell, M. C. Coleman, I. W. Goodall, D. J. Gooch, J. R. Haigh, S. T. Kimmins and G. J. Neate, 1987, CEGB assessment procedure for defects in operating plant in the creep range. *Fatigue Fract. Eng. Mater. Struct.*, **10**, 115–127.

M. C. Askins, 1988, Creep life assessment by measurement of interparticle spacing. In: "Proceedings of the International Conference on Life Extension and Assessment, The Hague," Netherlands Institute voor Lastechniek, The Netherlands.

M. C. Askins, D. Barraclough, C. W. King and D. Lonsdale, 1985, "A Guide to Current Procedures for Post Exposure Creep Testing of Samples Removed from Thick Walled, High Temperature Pressure Components," CEGB report OED/STB(S)/87/0022/R, Central Electricity Generating Board, Gravesend, UK.

M. Askins and K. Menzies, 1985, Unpublished work cited in Gooch and Townsend, 1987.

M. C. Askins, *et al.*, 1988, "Remaining Life of Boiler Pressure Parts—Base Material Model," Report RP2253-1, vol. 3, Electric Power Research Institute, Palo Alto, CA.

P. Auerkari, J. Salonern and K. Borgreen, 1996, Towards unified evaluation of replica inspections for power plant applications. In: "Proceedings of the 6th International Conference on Creep and Fatigue," Institute of Mechanical Engineering, London, pp. 501–510.

P. Auerkari and P. Vierros, 1987, In: "Proceedings of the 3rd International Conference on Creep and Fracture of Engineering Materials and Structures, Swansea, 1987," Institute of Metals, London, pp. 905–928.

J. M. Aurrecoechea and W. D. Brentnall, 1990, Operating temperature estimation and life assessment of gas turbine airfoils. In: "Proceedings of a Conference on Life Assessment and Repair Technology for Combustion Turbine Hot Section Components," ASM International, Ohio.

A. D. Batte and D. J. Gooch, 1985, Metallurgical aspects of remanent life assessment of high temperature rotor forgings. In: "Proceedings of the EPRI Conference on Turbogenerator Rotors for Fossil Plants, Raleigh, 1984," Pergamon, pp. 3.77–3.98.

S. M. Beech, D. J. Gooch and A. Strang, 1995, Remanent life assessment of 1CrMoV HP and IP steam turbine rotors. In: "Proceedings of the 3rd International Charles Parsons Turbine Conference, Newcastle upon Tyne, 1995," Institute of Materials, London, pp. 277–291.

H. K. D. H. Bhadeshia, 1989, Theoretical analysis of changes in cementite composition during tempering of bainite. *Mater. Sci. Technol.*, **5**, 131–137.

H. K. D. H. Bhadeshia, A. Strang and D. J. Gooch, 1998, Ferritic power plant steels: remanent life assessment and approach to equilibrium. *Int. Mater. Rev.*, **43**(2), 45–69.

J. Bolton, 1995, Design considerations for high temperature bolting. In: "Performance of Bolting Materials in High Temperature Plant Applications," ed. A. Strang, Institute of Materials, London, pp. 1–14.

R. J. Browne, 1986, Creep rupture testing of tubular model components. In: "Techniques for Multiaxial Creep Testing," eds. D. J. Gooch and I. M. How, Elsevier Applied Science, London, pp. 311–332.

F. Brun, T. Yoshida, J. D. Robson, V. Narayan, H. K. D. H. Bhadeshia and D. J. C. Mackay, 2001, Theoretical design of creep resistant steels. *Mater. Sci. Technol.*, to be published.

P. J. Budden and I. W. Goodall, 1998, Assessment procedures and design codes. In: "Integrity of High Temperature Welds," Institute of Mechanical Engineers, London, pp. 309–322.

B. J. Cane, 1979, Inter-relationship between creep deformation and creep rupture in 2(1/4)CrlMo steel. *Met. Sci.*, **13**, 287–294.

B. J. Cane, 1981, Mechanistic control regimes for intergranular cavity growth in 2(1/4)CrlMo steel under various stresses and stress states. *Met. Sci.*, **15**, 302–310.

B. J. Cane, 1982, Remaining creep life estimation by strain assessment on plant. *Int. J. Pres. Ves. Piping*, **10**, 11–30.

B. J. Cane, 1986, Stress state distributions in thick-walled pressurized tubes under creep loading. In: "Techniques for Multiaxial Creep Testing," eds. D. J. Gooch and

I. M. How, Elsevier Applied Science, London, pp. 245–266.

B. J. Cane, 1996, Surveillance and control of damage in elevated temperature pressure parts in power and process plants. In: "Proceedings of the 6th International Conference on Creep and Fatigue," Institute of Mechanical Engineers, London, pp. 473–489.

B. J. Cane, P. F. Aplin and J. M. Brear, 1985, A mechanistic approach to remanent creep life assessment of low alloy ferritic components based on hardness measurements. *J. Pres. Ves. Tech.*, **107**(3), 295–300.

B. J. Cane and C. J. Middleton, 1981, Intergranular creep-cavity formation in low-alloy bainitic steels. *Met. Sci.*, **15**, 295–301.

B. J. Cane and M. S. Shammas, 1984, "A Method for Remanent Life Estimation by Quantitative Assessment of Creep Cavitation on Plant," CEGB report TPRD/L/2645/N84, Central Electricity Generating Board, Leatherhead, UK.

R. G. Carlton, D. J. Gooch, E. M. Hawkes and H. D. Williams, 1987, The CEGB approach to the determination of the remanent life of high temperature turbine rotors. In: "Proceedings of an International Conference on Refurbishment and Life Extension of Steam Plant, London, 1987," Institute of Mechanical Engineers, London, pp. 45–52.

R. B. Carruthers and M. J. Collins, 1983, *Met. Sci.*, **17**, 107–110.

R. B. Carruthers and R. V. Day, 1968, CEGB report SSD/NE/R138, Central Electricity Generating Board, Harrogate, UK.

J. Chance and N. Ridley, 1981, *Metall. Trans. A*, **21A**, 1205–1213.

A. H. Chockshi, 1987, *Mater. Sci. Technol.*, **3**, 656.

R. Clayton, 1989, Photogrammetry in the range 300 mm to 3000 mm: techniques and applications in the electricity supply industry. *Photogrammetric Record*, **13**(73), 27–62.

R. Coade, 1985, "Temperature Determination Based on Microstructural Changes Occurring in 1%Cr–0.5%Mo Steel," SECV report SO/85/87, State Electricity Commission of Victoria, Australia.

J. Dobrzanski and A. Hernas, 1996, Relationship between microstructure and residual life-time of low alloy CrMo steels. In: "Proceedings of the 6th International Conference on Creep and Fatigue," Institute of Mechanical Engineers, London, pp. 451–461.

J. Dobrzanski, P. Milinskp and J. Woszczek, 1996, In: "Proceedings of the 9th International Symposium on Creep Resistant Metallic Materials, Hradec nad Moravici, Czech Republic, 1996," Vitkovice, pp. 152–163.

B. F. Dyson and S. Osgerby, 1987, Modeling creep corrosion interactions in nickel-base superalloys. *Mater. Sci. Technol.*, **3**, 545–553.

G. Eggeler, 1991, Microstructural parameters for creep damage quantification. *Acta Metall. Mater.*, **39**(2), 221–231.

G. F. Eggeler and R. C. Hurst, 1994, Development and validation of models for the lifetime assessment of steam power plant components. In: "Proceedings of the International Conference on Materials for Advanced Power Engineering, Liege, 1994," Kluwer Academic Publishers, Amsterdam, pp. 109–132.

G. Eggeler, C. Millward and B. J. Cane, 1989, A remanent life assessment technique based on local strain measurement using replication. In: "Proceedings of an International Conference on Fossil Power Plant Rehabilitation, Cincinnati, 1989," ASM International, Ohio, pp. 13–17.

K. A. Ellison, J. A. Daleo and D. H. Boone, 1998, Metallurgical temperature estimates based on interdiffusion between CoCrAlY overlay coatings and a directionally solidified nickel based superalloy substrate. In: "Proceedings of the 6th Liege Conference on Materials for Advanced Power Engineering, Liege, 1998," Forschungszentrum, Jülich, Germany, pp. 1523–1534.

EPRI, 1986, "Generic Guidelines for the Life Extension of Fossil Fuel Power Plants," EPRI report CS-4778, Electric Power Research Institute, Palo Alto, CA.

I. Finnie, 1960, Steady state creep of a thick-walled cylinder under combined axial load and internal pressure. *J. Basic Eng.*, **1960**(Sept), 689–694.

H. J. Frost and M. F. Ashby, 1982, "Deformation Mechanism Maps: The Plasticity and Creep of Metals and Ceramics," Pergamon, Oxford, UK.

T. B. Gibbons, J. S. Herd, L. N. McCartney and D. McLean, 1973, Creep under varying load. *Met. Sci. J.*, **7**, 196–204.

R. M. Goldhoff and D. A. Woodford, 1982, "The evaluation of creep damage in a CrMoV steel, STP 515," American Society for Testing and Materials, pp. 89–106.

D. J. Gooch, 1988, Plant life extension—remanent life assessment of high temperature components. *Power Eng. J.*, **2**(6), 323–331.

D. J. Gooch, 1988, How much longer—remanent life assessment of high temperature components. In: "CEGB Research, May, 1988," Central Electricity Generating Board, Leatherhead, UK, pp. 60–67.

D. J. Gooch, 1991, Creep ductility considerations in design and assessment. In: "Proceedings of the Conference on Rupture Ductility of Creep Resistant Steels, York, 1990," Institute of Metals, London, pp. 302–312.

D. J Gooch, 1992, "Creep Damage Accumulation under Multiaxial Stress Conditions, Final Report, COST 501(2), Project 5UK2," Commission of the European Communities, Brussels, Belgium.

D. J. Gooch, 2001, Creep design and life extension of large-scale plant. In: "Encyclopedia of Materials: Science and Technology," eds. K. H. J. Buschow, R. W. Cahn, M. C. Flemings, B. Ilschner, E. J. Kramer and S. Mahajan, Pergamon, Oxford.

D. J. Gooch and M. C. Askins, 1989, "Effect of Load Cycling on Creep Life Prediction of Ferritic Steels, Final Report, COST 505 Project UK22," Commission of the European Communities, Brussels, Belgium and CEGB Report ESTD/L/0003/R89, Central Electricity Generating Board, Leatherhead, UK.

D. J. Gooch and I. M. How (eds.), 1986, "Techniques for Multiaxial Creep Testing," Elsevier, London.

D. J. Gooch, J. A. W. Hunter, W. J. Stevens and R. G. Wilson, 1989a, CEGB approach to the life assessment of fossil fuel power plant. In: "Proceedings of the AIM Conference, Liege, 1989,"

D. J. Gooch and S. T. Kimmins, 1987, Type IV cracking in 0.5Cr0.5Mo0.25V/2.25Cr1Mo weldments. In: "Proceedings of the 3rd International Conference on Creep and Fracture of Engineering Materials and Structures, Swansea, 1987," Institute of Metals, London, pp. 689–703.

D. J. Gooch and J. M. Maguire, 1988, Creep cavitation assessment in 1CrMoV rotor steels. In: "Proceedings of the 2nd Parsons Turbogenerator Conference, Cambridge, 1988," Institute of Metals, London, pp. 129–135.

D. J. Gooch, M. S. Shammas, M. C. Coleman, S. J. Brett and R. A. Stevens, 1989b, Creep cavitation assessment using field replication. In: "Proceedings of the Conference on Fossil Power Plant Rehabilitation, Cincinnati, 1989," ASM International, Ohio, 1989, Paper 8901-004.

D. J. Gooch, A. Strang and S. M. Beech, 1993, Hardness evaluation for remanent life assessment of 1CrMoV rotors. In: "Proceedings of the 5th International Conference on Creep and Fracture of Engineering Materials and Structures, Swansea, 1993," Institute of Materials, London, pp. 743–764.

D. J. Gooch and R. D. Townsend, 1987, CEGB remanent life assessment procedures. In: "Proceedings of the EPRI

Conference on Life Assessment and Extension of Fossil Fuel Plants, Washington, DC, 1986," Pergamon, Oxford, pp. 927–943.

I. W. Goodall, 1986, Multiaxial data requirements for structural integrity assessments in creep. In: "Techniques for Multiaxial Creep Testing," eds. D. J. Gooch and I. M. How, Elsevier, London, pp. 3–27.

I. W. Goodall, R. D. H. Cockcroft and E. J. Chubb, 1975, An approximate description of the creep rupture of structures. *Int. J. Mech. Sci.*, **17**, 351–360.

T. Goto, 1984, In: "Proceedings of the 2nd International Conference on Creep and Fracture of Engineering Materials and Structures, Swansea," Institute of Metals, London, pp. 1135–1146.

R. V. Hart, 1980, The effect of prior creep on the rupture properties of power plant alloys and the use of such properties to determine remaining creep life. In: "Proceedings of the International Congress on the Engineering Aspects of Creep, Sheffield, 1980," Institute of Mechanical Engineers, London, pp. 207–216.

H. Hayakawa, A. Kanaya, A. Koduru, H. Watanabe, M. Kubota, H. Deguchi and K. Ohta, 1996, Developments of creep life assessment technique for boiler parts of fossil-fired power plants. In: "Proceedings of the 6th International Conference on Creep and Fatigue," Institute of Mechanical Engineers, London, pp. 463–472.

D. R. Hayhurst, 1996, High temperature design and life assessment of structures using continuum damage mechanics. In: "Proceedings of the 6th International Conference on Creep and Fatigue," Institute of Mechanical Engineers, London, pp. 399–410.

D. R. Hayhurst and G. A. Webster, 1986, An overview of studies on stress state effects during creep of circumferentially notched bars. In: "Techniques for Multiaxial Creep Testing," eds. D. J. Gooch and I. M. How, Elsevier, London, pp. 137–175.

J. K. Hepworth, J. E. Fackrell, L. W. Pinder and J. D. Wilson, 2000, A modeling approach to gas turbine life management. In: "Life Assessment of Hot Section Gas Turbine Components, Edinburgh, 1999," Institute of Materials, London, pp. 281–291.

R. W. K. Honeycombe and H. K. D. H. Bhadeshia, 1995, "Steel Microstructures and Properties," Edward Arnold, London.

I. M. How, R. J. Browne, M. C. Coleman, I. H. Craig, W. H. Ham, R. C. Hurst and P. C. Meecham, 1992, A code of practice for internal pressure testing of tubular components at elevated temperatures. In: "Harmonization of Testing Practice for High Temperature Materials," eds. M. S. Loveday and T. B. Gibbons, Elsevier, London, pp. 375–400.

J. W. Hutchinson, 1983, Constitutive behavior and crack tip fields for materials undergoing creep-constrained grain boundary cavitation. *Acta Metall.*, **31**, 1079–1088.

T. Hyde, A. Tang and W. Sun, 1998, Analytical and computational stress analysis of welded components under creep conditions. In: "Integrity of High Temperature Welds," Institute of Mechanical Engineers, London, pp. 285–308.

J. J. Jonas and I. Weiss, 1979, *Met. Sci.*, **13**, 238–245.

G. T. Jones, 1995, Review of HT bolt tightening, removal and replacement procedures. In: "Performance of Bolting Materials in High Temperature Plant Applications," ed. A. Strang, Institute of Materials, London, pp. 331–342.

L. M. Kachanov, 1958, Time of the rupture process under creep conditions. *Izv. Akad. Nauk. SSSR Otd. Tekh. Nauk*, **8**, 26–31.

D. Kalish, S. A. Kulin and M. Cohen, 1965, *J. Met.*, **17**(2), 157–164.

S.-A. Karlsson, C. Persson and P.-O. Persson, 1995, Metallographic approach to turbine blade lifetime prediction. *Mater. Manufact. Process.*, **10**(5), 939–953.

K. Kimura, K. Fujiyama, F. Ishii and M. Murmatsu, 1986, Effect of material degradation on creep properties and life estimation. In: "Proceedings of the International Conference on Creep, Tokyo, 1986," Japanese Society of Mechanical Engineers, pp. 343–348.

P. Krautwasser, M. Widera, H. Schuster and H. Nickel, 1992, Determination of creep strain in alloy IN100 by small angle neutron diffraction scattering. In: "Superalloys 1992," The Mineral, Metals and Materials Society, pp. 665–674.

J. Kusumoto, Y. Gotoh and T. Kobayashi, 2001, Development of creep life assessment for boiler parts of fossil power plants. In: "Proceedings of the 9th International Conference on Creep and Fracture of Engineering Materials and Structures, Swansea, 2001," Institute of Materials, London, pp. 579–588.

Y. Lindblom, 1985, Refurbishing superalloy components for gas turbines. *Mater. Sci. Technol.*, **1**, 636–641.

R. Lonsdale and P. E. J. Flewitt, 1979, *Mater. Sci. Eng.*, **39**, 217–229.

D. J. C. Mackay, 1997, Bayesian non-linear modeling with neural networks. In: "Mathematical Modeling of Weld Phenomena 3," eds. H. Cerjak and H. K. D. H. Bhadeshia, Institute of Materials, London, pp. 359–387.

M. J. Manjoine, 1982, Creep-rupture behaviour of weldments. *Weld. J. Res. Suppl.*, **31**, 50s–57s.

S. D. Mann, D. G. McCulloch and B. C. Muddle, 1995, Identification of M_5C_2 carbides in ex-service 1Cr-0.5Mo steels. *Metall. Mater. Trans. A.*, **26A**, 509–520.

C. J. Middleton, 1981, Reheat cavity nucleation and nucleation control in bainitic creep-resisting low-alloy steels: Roles of manganese sulphide, residual, and sulphur-stabilizing elements. *Met. Sci.*, **15**, 154–167.

K. Milička and F. Doběs, 2001, Comparison of small punch tests performed on exposed power plant part. In: "Proceedings of Baltica V, Condition and Life Management for Power Plants, Helsinki, 2001," Technical Research Centre of Finland, Espoo, Finland, vol. 1, pp. 383–393.

M. D. C. Moles, D. V. Leemans and H. J. Westwood, 1980, Residual life estimation of boiler tubing in thermal power plants. In: "Proceedings of the International Conference on Engineering Aspects of Creep, Sheffield, 1980," Institute of Mechanical Engineers, London, vol. 2, pp. 241–247.

M. C. Murphy and G. D. Branch, 1969, *J. Iron and Steel Inst.*, **207**, 1347–1364.

M. Nakashiro, H. Yoneyama, S. Shibata, H. Umaki and N. Izawa, 1996, "Proceedings of the 9th International Symposium on Creep Resistant Metallic Materials, Hradec nad Moravici, Czech Republic, 1996," Vitkovice, Czech Republic, pp. 183–192.

B. Nath, 2000, Life management issues for operators. In: "Life Assessment of Hot Section Gas Turbine Components, Edinburgh, 1999," Institute of Materials, London, pp. 35–47.

B. Nath, K. H. Mayer, S. M. Beech and R. W. Vanstone, 1995, Recent developments in Alloy 80A for high temperature bolting applications. In: "Performance of Bolting Materials in High Temperature Plant Applications," ed. A. Strang, Institute of Materials, London, pp. 306–317.

N. G. Needham and B. J. Cane, 1983, Creep strain and rupture predictions by cavitational assessment in 2(1/4)CrlMo steel weldments. In: "Advances in Life prediction Methods," ed. D. A. Woodford, ASME, New York, pp. 65–73.

B. Neubauer, 1984, Remaining life estimation for high temperature materials under creep load by replicas. *Nucl. Technol.*, **66**, 308–312.

B. E. Noltingk, F. A. McLachlan, C. K. V. Owen and P. C. O'Neill, 1972, High-stability capacitance strain gauge for use at extreme temperatures. *Proc. IEE.*, **119**(7), 897–903.

S. Osgerby and B. F. Dyson, 1993, Constant strain rate testing: prediction of stress–strain curves from constant load creep data. In: "Proceedings of the 5th International Creep Conference on Creep and Fracture of Engineering Materials and Structures, Swansea, 1993," Institute of Metals, London, pp. 53–61.

S. Osgerby and B. F. Dyson, 1995, Physically based modeling of stress relaxation in superalloys and ferritic steels. In: "Performance of Bolting Materials in High Temperature Plant Applications," ed. A. Strang, Institute of Materials, London, pp. 362–373.

J. D. Parker, 1998, Recent advances in condition assessment of components based on strain monitoring. In: "Proceedings of Baltica IV, Plant Maintenance for Managing Life and Performance, Helsinki, 1998," Technical Research Centre of Finland, Espoo, Finland, vol. 1, pp. 239–250.

J. D. Parker, A. McMinn and J. Foulds, 1989, Material sampling for the assessment of component integrity. In: "Proceedings of a Conference on Life Assessment and Life Extension of Power Plant Components," ed. T. V. Narayanan, American Society of Mechanical Engineers, New York, pp. 223–230.

J. D. Parker, G. C. Stratford, N. Shaw, G. Spink and H. Metcalfe, 1998, The application of miniature disc testing for the assessment of creep damage in CrMoV rotor steel. In: "Proceedings of Baltica IV, Plant Maintenance for Managing Life and Performance, Helsinki, 1998," Technical Research Centre of Finland, Espoo, Finland, vol. 2, pp. 477–488.

J. Pilling, N. Ridley and D. J. Gooch, 1982, The effect of phosphorus on creep in 2.25CrlMo steels. *Acta Metall.*, **30**, 1587–1595.

P. N. Quested, P. J. Henderson and M. McLean, 1988, Observations of deformation and fracture heterogeneities in a nickel based superalloy using electron back scattering patterns. *Acta Metall.*, **36**, 2743–2752.

D. W. A. Rees, M. W. Brown, T. Hyde, R. D. Lohr, C. J. Morrison and M. Shammas, 1992, A code of practice for torsional creep testing of tubular test pieces at elevated temperatures. In: "Harmonization of Testing Practice for High Temperature Materials," eds. M. S. Loveday and T. B. Gibbons, Elsevier, London, pp. 331–361.

J. R. Rice and D. M. Tracey, 1969, On the ductile enlargement of voids in triaxial stress fields. *J. Mech. Phys. Solids*, **17**, 201–207.

K. A. Ridal and A. G. Quarrell, 1962, *J. Iron and Steel Inst.*, **200**, 366–373.

H. Riedel, 1989, Life prediction methods for constrained grain boundary cavitation. *Int. J. Pres. Ves. Piping*, **39**, 119–134.

B. W. Roberts, A. Strang, 1987, In: "Refurbishment and Life Extension of Steam Plant, London," Institute of Mechanical Engineers, London, pp. 205–213.

Y. N. Robotnov, 1969, "Creep Problems in Structural Members," North-Holland, Amsterdam.

J. D. Robson and H. K. D. H. Bhadeshia, 1996, "Proceedings of the 9th International Symposium on Creep Resistant Metallic Materials, Hradec nad Moravici, Czech Republic, 1996," Vitkovice, Czech Republic, pp. 83–91.

J. D. Robson and H. K. D. H. Bhadeshia, 1997, Modeling precipitation sequences in power plant steels. *Mater. Sci. Technol.*, **13**, 631–644.

M. S. Shammas, 1987, Remanent life assessment of ferritic weld heat affected zones by a metallographic measurement of cavitation damage—the "A" parameter. In: "Proceedings of the International Conference on Refurbishment and Life Extension of Steam Plant, London, 1987," Institute of Mechanical Engineers, London, pp. 289–299.

N. J. Simms, J. E. Oakey and J. R. Nicholls, 2000, Predicting the remanent life of corrosion resistant coatings. In: "Life Assessment of Hot Section Gas Turbine Components, Edinburgh, 1999," Institute of Materials, London, pp. 271–279.

M. Steen and M. de Witte, 1987, Impact of structural instability on the extrapolation of short-term creep test results. In: "Proceedings of the 3rd International Conference on Creep and Fracture of Engineering Materials and Structures, Swansea, 1987," Institute of Metals, London, pp. 773–802.

A. Strang, S. M. Beech and D. J. Gooch, 1996, The potential of creep cavitation measurements for the assessment of the remnant life of 1CrMoV steam turbine rotors. In: "Proceedings of the 9th International Symposium on Creep Resistant Metallic Materials, Hradec nad Moravici, Czech Republic, 1996," Vitkovice, Czech Republic.

A. Strang and D. J. Gooch (eds.), 1997, "Microstructural Development And Stability In High Chromium Ferritic Power Plant Steels," Institute of Materials, London.

A. Strang and V. Vodárek, 1997, Precipitation processes in martensitic 12CrMoVNb steels during high temperature creep. In: "Microstructural Development and Stability in High Chromium Ferritic Power Plant Steels," eds. A. Strang and D. J. Gooch, Institute of Materials, London, pp. 31–52.

A. Strang and V. Vodárek, 2000, Modeling and experimental verification of minor phase composition changes in creep resistant 12CrMoVNb steels. In: "Proceedings of the Fifth International Charles Parsons Turbine Conference, Cambridge, 2000," Institute of Materials, London, pp. 572–589.

V. P. Swaminathan, J. M. Allen and G. L. Touchton, 1997, Temperature estimation and life prediction of turbine blades using post service oxidation measurements. *Trans. ASME*, **119**, 922–929 (Proceedings of the 1996 International Gas Turbine and Aeroengine Congress & Exhibition. ASME, 1996).

S. Tettamanti and R. Crudelli, 1998, Small punch creep test: A promising methodology for high temperature plant components life evaluation. In: "Proceedings of Baltica IV, Plant Maintenance for Managing Life and Performance, Helsinki, 1998," Technical Research Centre of Finland, Espoo, Finland, vol. 2, pp. 501–510.

R. C. Thomson and H. K. D. H. Bhadeshia, 1992, Carbide precipitation in 12CrMoV power plant steel. *Metall. Trans. A.*, **23A**, 1171–1179.

R. C. Thomson and H. K. D. H. Bhadeshia, 1994, Changes in chemical composition of carbides in 2.25Cr–1Mo power plant steel. *Mater. Sci. Technol.*, **10**, 193–203; 205–208.

L. H. Toft and R. A. Mardsen, 1961, "Special Report 70, Structural Processes in Creep," Iron and Steel Institute.

R. D. Townsend, 1991, Historical perspective: a review of service problems. In: "Rupture Ductility of Creep Resistant Steels," ed. A. Strang, Institute of Metals, London, pp. 1–16.

R. D. Townsend, 1995, Performance of high temperature bolting in power plant. In: "Performance of Bolting Materials in High Temperature Plant Applications," ed. A. Strang, Institute of Materials, London, pp. 15–39.

W. Trampczynski and Z. Kowalewski, 1986, A tension-torsion testing technique. In: "Techniques for Multiaxial Creep Testing," eds. D. J. Gooch and I. M. How, Elsevier, London, pp. 79–92.

VGB, 1992, "Guideline for the Assessment of Microstructure and Damage Development of Creep Exposed Materials for Pipes and Boiler Components, Report VGB-TW 507," VGB, Essen, Germany.

R. Viswanathan, 1988, Residual life techniques for plant life extension. *Mater. Sci. Eng. A.*, **A103**, 131–139.

R. Viswanathan, 1989, "Damage Mechanisms and Life assessment of High Temperature Components," ASM International, Ohio.

J. Wareing, I. Bretherton and V. B. Livesey, 1986, Life prediction for elevated temperature components subjected to cyclic deformation. In: "Proceedings of the International Conference on Fatigue of Engineering Materials and Structures, Sheffield, 1986," Institute of Mechanical Engineers, London, Paper C269/86, pp. 147–163.

G. A. Webster and R. A. Ainsworth, 1994, "High Temperature Component Life Assessment," Chapman and Hall, London.

G. A. Webster, S. R. Holdsworth, M. S. Loveday, I. Perrin and H. Purper, 2001, "A Code of Practice for Conducting Notched Bar Creep Rupture Tests and for Interpreting the Data," European Structural Integrity Society Report ESIS TC11, also *Fatigue Fract. Eng. Mat.*, to be published.

M. I. Wood, 1995, The assessment of service induced degradation of nickel based superalloy gas turbine blades. *Mater. Manufacturing Processes*, **10**, 903–924.

M. I. Wood and D. Raynor, 1996, Condition assessment techniques for degraded gas turbine super-alloy materials. *Int. J. Pres. Ves. Piping*, **66**, 341–350.

D. A. Woodford, 1993, Test methods for accelerated development, design and life assessment of high temperature materials. *Mater. Des.*, **14**(2), 231–242.

D. A. Woodford, 1996, Creep design analysis for superalloys based on stress relaxation testing. In: "Proceedings of the 6th International Conference on Creep and Fatigue," Institute of Mechanical Engineers, London, Paper C494/090/96, pp. 61–70.

D. A. Woodford and J. A. Daleo, 2000, Stress relaxation as a basis for blade creep life assessment. In: "Life Assessment of Hot Section Gas Turbine Components, Edinburgh, 1999," Institute of Materials, London, pp. 293–309.

J. Wortmann, 1985, Improving reliability and lifetime of rejuvenated turbine blades. *Mater. Sci. Technol.*, **1**, 644–650.

R. Wu, J. Storesund and R. Sandström, 1993, Influence of postweld heat treatment on creep properties of 1Cr0.5Mo welded joints. *Mater. Sci. Technol.*, **9**, 773–780.

L. Y. Zhi, 2001, An analytical approach for interpreting creep curves from small punch tests. In: "Proceedings of the Ninth International Conference on Creep and Fracture of Engineering Materials and Structures, Swansea, 2001," Institute of Materials, London, pp. 589–600.

Comprehensive Structural Integrity
ISBN (set): 0-08-043749-4

Volume 5; (ISBN: 0-08-044154-8); pp. 309–359

5.08
Creep Resistance in Nonferritic Materials

H. HARADA

National Institute for Materials Science, Tsukuba, Japan

5.08.1 INTRODUCTION

Advancements in creep-resistant high-temperature materials have contributed greatly to improving the thermal efficiency of jet engines, and land- and sea-based gas turbines. Or, conversely, the need for larger (more powerful) and efficient gas turbines has provided, and continues to provide, major impetus for developments in advanced heat-resistant materials. The metal temperatures in modern gas turbine disks are in the range of 650–700°C and in the combustors and the first stage blades and vanes they can reach as high as 1,000°C. These temperatures are beyond the range of heat-resistant ferritic and austenitic steels, which have maximum temperature capabilities of 600°C and have exclusively served the steam turbine industry for several decades. The composition, physical metallurgy, and the processing techniques for Ni-based superalloys are quite distinct from those of heat-resistant ferritic and austenitic steels and offer unique technical challenges. A discussion of the creep and creep-fatigue and thermal-fatigue properties of the ferritic and austenitic steels can be found in Chapters 5.02–5.06 of this volume.

This chapter is dedicated only to the discussion of Ni-based superalloys, which are extensively used in modern gas turbines, and to other advanced materials such as intermetallics, some high-melting-point alloys, ceramics and high-temperature composites that are being considered for applications in future gas turbines. The physical metallurgy of these materials is enormously complex and is covered in detail in previous publications (Sims *et al.*, 1987). The objective of this chapter is to provide a historical perspective on the development of these materials and to discuss their temperature capabilities in relation to applications to gas turbines, and to provide a brief account of the technological advances in the processing techniques that have led to their fruition. This chapter also provides the technological motivation for studying the detailed science behind the phenomenon of creep and its interactions with fatigue and thermal-fatigue (described in the previous chapters of the this volume) and collectively determine the life of high-temperature components.

5.08.2 ADVANCEMENTS IN SUPERALLOYS

Several Ni-based superalloys are used in high-temperature components (e.g., combustors and high-pressure regions of gas turbines such as first-stage blades and vanes) that determine the power and efficiency of jet engines and industrial gas turbines. The microstructure of these alloys consists of solid-solution-strengthened matrix known as γ and intermetallics precipitates of the type $Ni_3(Ti, Al)$ termed as γ'. Table 1 lists the materials used in various components of the jet engine, whereas Table 2 provides the chemical compositions of the various Ni-based superalloys used in gas turbines. In keeping with the objectives of this volume, we focus our attention on materials and parts of components of the gas turbines that experience the extreme temperatures and stresses and consequently creep deformation and rupture, as well as the creep—fatigue and creep—thermal-fatigue interactions become a major consideration in the structural integrity analysis. Because of different temperature requirements, we first describe materials that are currently used in first-stage blades and vanes and then the materials for turbine disks.

5.08.2.1 Ni-based Alloys for First-stage Turbine Blades and Vanes

Wrought alloy, invented in 1941, was the first base superalloy applied to the hot sections of gas turbines, which was followed by the introduction of conventionally cast (CC), directionally solidified (DS), and single crystal (SC) alloys. Figure 1 shows the CC, DS, and SC blades. The microstructures associated with the time line of development of superalloys is illustrated in Figure 2 Sims, *et al.* (1987). We find that a steady increase in the volume fraction of γ' precipitates has occurred with improvements in technology. Since γ' is an intermetallic phase with excellent temperature-resistant properties, the temperature capability of these alloys has also steadily improved with time. Figure 3 illustrates the typical microstructure of an SC alloy with cubical γ' precipitates and the high volume fraction that these precipitates occupy in the structure. Since their insertion into service, the alloy composition of SC superalloys is being continuously improved. For instance, researchers are focusing on the development of fourth-generation SC alloys containing precious metal elements, such as ruthenium (Ru). Figure 4 illustrates the evolution in the temperature capability of superalloys (Harada 2000). The target temperature capability indicated in Figure 4

Table 1 Materials for use as parts in aeroengines and their manufacturing methods.

Components	*Parts*	*Materials*	*Fabrications*
	Casing	Ti alloys, Al alloys	Wrought
	Frame	Ti alloys	Cast
Fun	Disk	Ti alloys	Wrought
	Blade	Ti alloys	Precision cast
	Vane	Ti alloys, Al alloys	Precision cast
Compressor	Casing	Ti alloys, Ni superalloys	Wrought
	Disk	Ti alloys, Ni superalloys	Wrought
	Blade	Ti alloys, Ni superalloys	Precision cast
	Vane	Ti alloys, Ni superalloys	Precision cast
Combuster	Casing	Ni superalloys	Wrought, Cast
After burner	Liner	Co alloys, Ni superalloys	Sheet
High-pressure turbine	Casing	Ni superalloys	Wrought, Cast
Low-pressure turbine	Disk	Ni superalloys	Wrought
	Blade	Ni superalloys	Precision cast
	Vane	Ni superalloys, Co alloys	Precision cast
	Shaft	Special steel, Ni superalloys	Wrought
Accesory	Housing	Al alloys	Cast
Bearing		Special steel	Wrought

Table 2 Chemical compositions of typical Ni-based superalloys for use as turbine blades.

	Chemical composition (wt.%)														
Alloy name	*Co*	*Cr*	*Mo*	*W*	*Al*	*Ti*	*Nb*	*Ta*	*Hf*	*Re*	*C*	*B*	*Zr*	*Others*	*Notes*
CC															
IN 738	8.5	16	1.7	2.6	3.4	3.4		1.7			0.17	0.01	0.1		
IN 792	9	12.4	1.9	3.8	3.1	4.5		3.9			0.12	0.02	0.2		
Rene′80	9.5	14	4	4	3	5					0.17	0.015	0.03		
MarM24	10	8.5	0.7	10	5.6	1		3			0.16	0.015	0.04		
TM-321	8.2	8.1		12.6	5	0.8		4.7			0.11	0.01	0.05		
DS															
GTD111	9.5	14	1.5	3.8	3	4.9		2.8			0.1	0.01			1st
CM247L	9	8	0.5	10	5.6	0.7		3.2	1.4		0.07	0.015	0.01		1st
TMD-5	9.5	5.8	1.9	13.7	4.6	0.9		3.3	1.4		0.07	0.015	0.015		1st
PWA142	12	6.5	1.7	6.5	6			4	1.5	3	0.1	0.015	0.03		2nd
CM186L	9	6	0.5	8.4	5.7	0.7		3.4		3	0.07	0.015	0.005		2nd
TMD-103	12	3	2	6	6			6	0.1	5	0.07	0.015			3rd
SC															
PWA148	5	10		4	5	1.5		12							1st
Rene′N4	8	9	2	6	3.7	4.2	0.5	4							1st
CMSX-2	4.6	8	0.6	8	5.6	1		9							1st
MC2	5	8	2	8	5	1.5		6							1st
MDSC	4.5	10	0.7	6	5.4	2		5.4		0.1					1st
TMS-26	8.2	5.6	1.9	10.9	5.1			7.7							2nd
PWA-148	10	5	2	6	5.6			9		3					2nd
Rene′N5	8	7	2	5	6.2			7	0.2	3					2nd
CMSX-4	9	6.5	0.6	6	5.6	1		6.5	0.1	3					2nd
TMS-82+	7.8	5	3.4	8.7	5.2	0.5		4.4	0.1	2.4					2nd
YH 61	1	7.1	0.8	8.8	5.1		0.8	8.9	0.25	1.4	0.07	0.02			2nd
Rene′N6	12.5	4.2	1.4	6	5.75			7.2	0.15	5.4	0.05	0.004		0.01Y	3rd
CMSX-10	3	2	0.4	5	5.7	0.2	0.1	8	0.03	6					3rd
TMS-75	12	3	2	6	6			6	0.1	5					3rd
OD															
MA6000	2	15	2	4	4.5	2.5		2			0.05	0.01	0.15	$1.1Y_2O$	
TMO-20	8.7	4.3	1.5	11.6	5.5	1.1		6			0.05	0.01	0.05	$1.1Y_2O$	

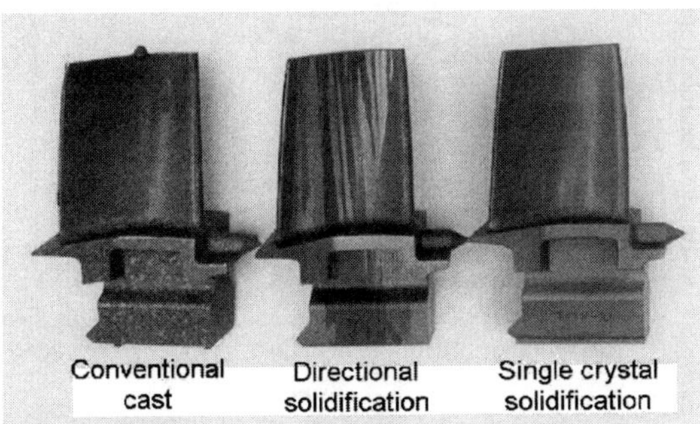

Figure 1 Photograph of the turbine blades made by CC, DS, and SC solidification method, respectively.

represents part of the 1999–2003 objectives of a Japanese project "High Temperature Materials 21 Project" led by National Institute for Materials Science in collaboration with gas turbine industries. One notices the almost 400°C increase in the temperature capability of these materials since the 1940s. The following discussion traces through the most significant developments in the technology of Ni-based alloys.

5.08.2.1.1 A shift from wrought alloys to cast alloys

In the 1940s, manufacturers started using wrought alloys, such as Nimonic alloys, for turbine blades and turbine vanes for jet engines. The first step taken to strengthen wrought alloys was to increase the Al and Ti content, which lead to γ' formation, to provide additional precipitate strengthening. Subsequently, they were strengthened by further enhancing γ' precipitation by increasing the Mo, W, and Nb content. However, difficulties related to forging arose as the alloy strength reached the level of Udimet700/720, temperature capability of 900°C for a creep life of 1,000 h at a stress of 137 MPa (see Figure 4). As a solution to this problem, vacuum melting technique was introduced in the mid-1950s. Using this technique, Ni-based superalloys with high reactive element content (Al, Ti, etc.) can be cast as conventionally cast hollow blades and vanes. This promoted further strengthening of alloys through increased γ' precipitation and increased solid-solution

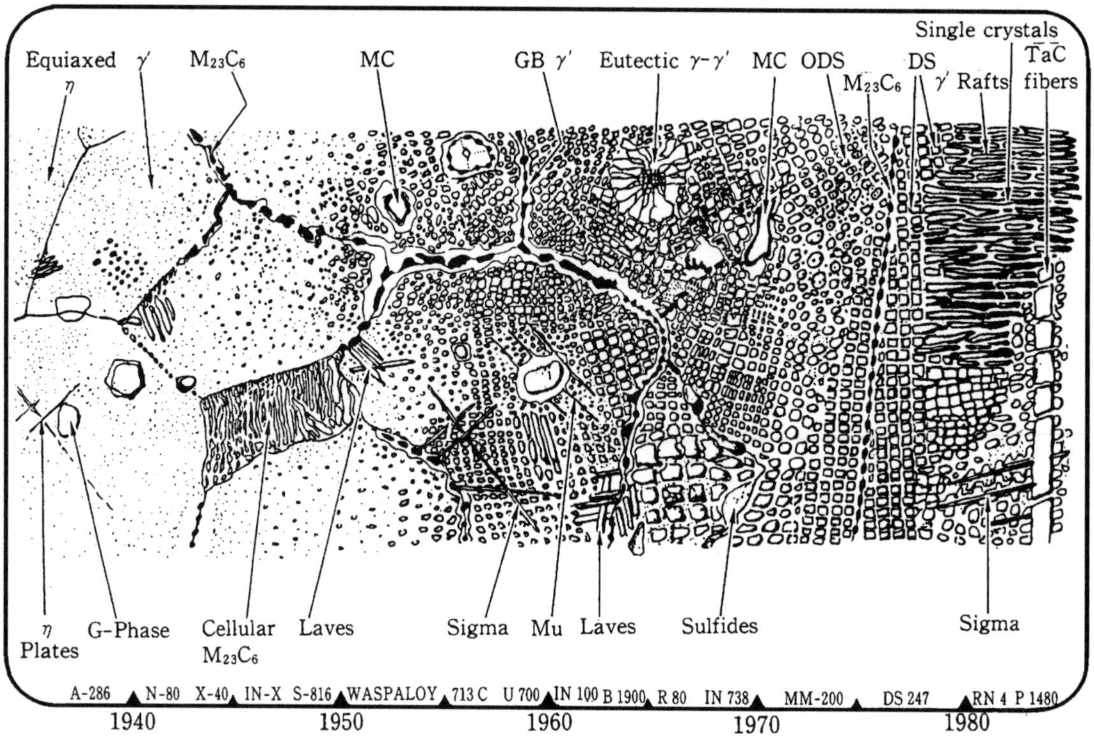

Figure 2 Panoramic and historical view of the microstructures for Ni-based superalloys.

Figure 3 Typical microstructure of Ni-based SC superalloys. The γ' phase is based on Ni, whereas the γ' phase is based on Ni_3Al compound. Ni-based SC superalloys are regarded as SC alloys although they have a two-phase structure. Because both phases in Ni-based superalloys have same f.c.c. crystal structures and similar lattice parameters, the lattices of each crystal are continuously connected.

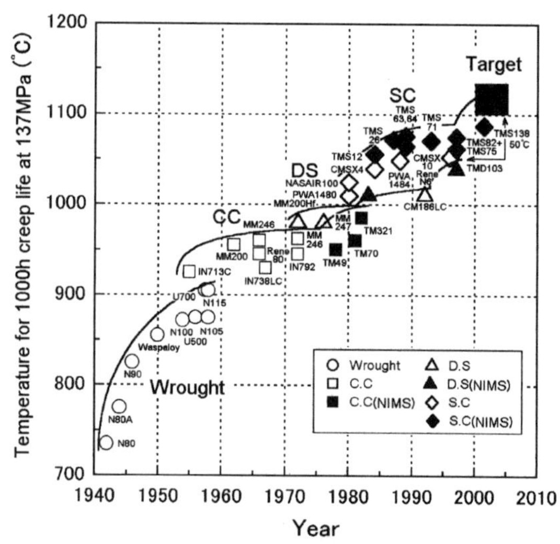

Figure 4 Evolution of creep rupture temperature capability of Ni-based superalloys under 1,100°C and 137 MPa conditions.

strengthening elements. This new processing technique motivated the development of new alloys such as MarM200, IN738, Rene'80, and MarM247, which are the typical CC alloys available now but developed in the 1960s.

5.08.2.1.2 Development of DS alloys

Directional solidification process, illustrated schematically in Figure 5, became popular in the US in the 1970s and greatly improved the creep strength, ductility, and fatigue properties in the longitudinal direction of the turbine blades. Thus, the application of this process has contributed greatly to the enhancement of power, reliability, and availability of jet engines. One of the first DS alloys developed during this period is MarM200Hf, which is the combination of MarM200 and hafnium (Hf). Hf contributes to the prevention of

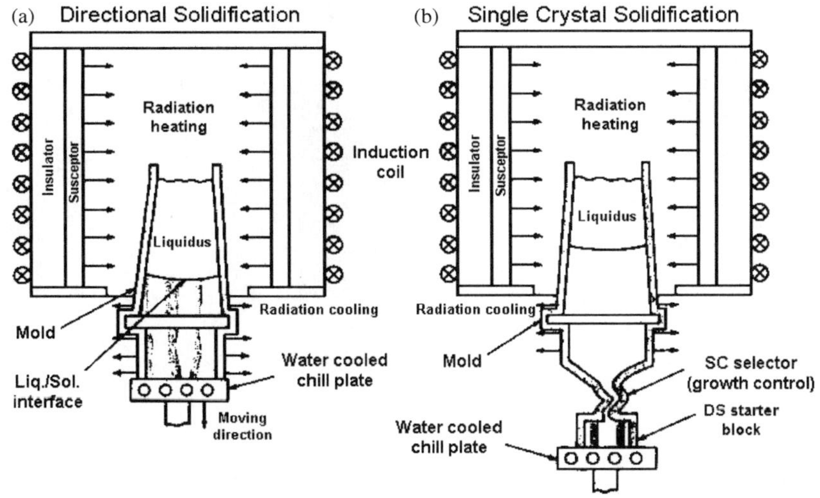

Figure 5 Schematics of the DS and the SC solidification processes.

intergranular cracking during directional solidification. Similar improvements were made on many other typical CC alloys so that they could be directionally solidified. Some of the DS alloys widely used today are Rene'80DS and GTD111DS, which are characterized by high Cr content and good hot corrosion resistance, and MarM247DS, which has good high-temperature strength and oxidation resistance.

TMD-5 was developed in the late 1970s to mid-1980s in Japan for application in the first-stage turbine blades of a pilot plant and was tested for its strength. TMD-5 was the strongest among the first-generation DS alloys, with a strength equivalent to that of the second-generation DS alloys. Mitsubishi Heavy Industries was the first manufacturer in Japan to use DS Ni-based alloys (Mitsubishi's original MGA1400, MGA2400, etc.) in the blades and vanes of a 1500-class large gas turbine, which was put in service in 1999 Kawai *et al.* (1997). The use of DS MGA1400 shown in Figure 6 in the first and second stages of blades improved the temperature capability of the turbine by 50°C, compared to IN738LC (CC material).

Second-generation DS alloys contain 3 wt.% Re, which enhances solid-solution strengthening of the γ phase, and have creep strength equivalent to that of first-generation SC alloys. CM186C and Rene'142 are typical DS alloys widely used in jet engines. TMD-103 (Kobayashi *et al.* 1998), developed in Japan, is the first third-generation DS alloy in the world, which was developed by converting a third-generation SC alloy into a DS material by adding grain-boundary strengthening elements such as carbon (C), boron (B), etc. This alloy has a creep strength equivalent to that of a second-generation SC alloy, CMSX-4. Figure 7 illustrates the creep strength of DS alloys in comparison to IN738LC (CC material).

Figure 6 DS blade cast with MGA1400 alloy for use as first-stage turbine blade in 1,500°C land-based gas turbine.

The production cost of DS alloys is relatively low because DS alloys have higher casting yield than single crystals. Therefore, DS alloys are likely to be used widely in turbine vanes or in last stages of larger blades, where casting defect is more likely to occur if the SC casting process is used.

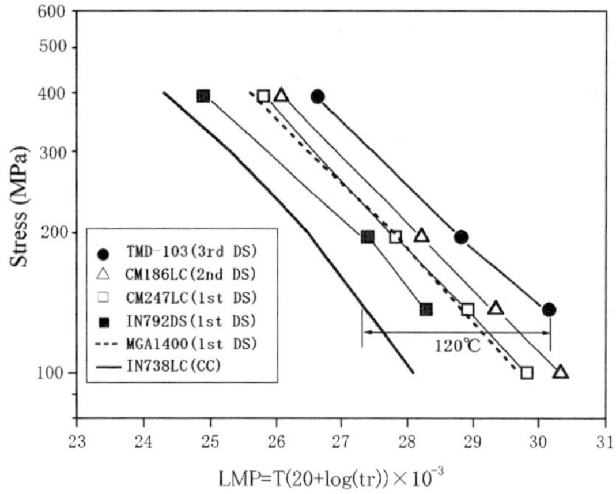

Figure 7 Creep rupture strengths of various DS alloys, where LMP is the Larson–Miller parameter, T is temperature in K, and t_r is creep rupture life in h.

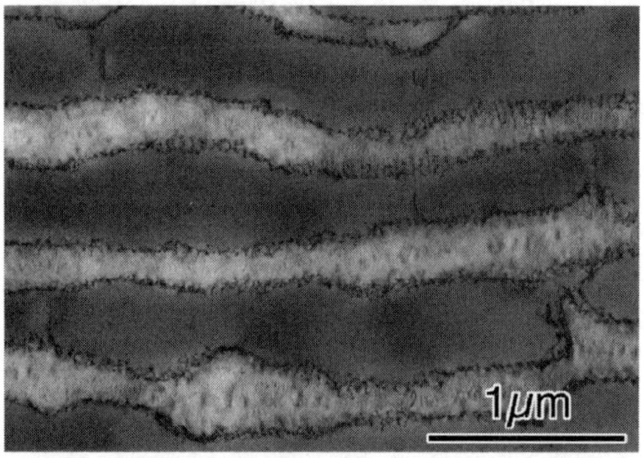

Figure 8 Micrograph of the rafted structure that appeared during a creep rupture test in Ni-based SC superalloy.

5.08.2.1.3 Development of SC alloys

In 1980, a group of researchers developed an SC alloy called PWA1480, which does not contain grain-boundary strengthening elements, such as C, B, and Zr. Removal of grain-boundary strengthening elements can significantly raise the incipient melting point, and enables solution treatment at temperatures as high as 1,300°C. Thus, PWA1480 with 70 vol.% γ' phase can be perfectly solution-treated, and the solution-treated phase can be ideally re-precipitated. As shown in Figure 8, when PWA1480 is heat treated, raft structure, which is a continuous connection of precipitates, is formed. The raft structure prevents changes in dislocation arrangement and contributes greatly to increasing creep strength of the alloy at temperatures exceeding 1,000°C.

In 1984, ReneN5 and CMSX-4, which are typical second-generation SC alloys that contain 3 wt.% Re, were developed. In Japan, the alloy development program called "Advanced Alloys with Controlled Crystalline Structures 1981–88" (Yamasaki, 1991) developed new superalloys by optimizing the composition and structure of SC alloys. These alloys, which do not contain Re, were named TMS-12 and TMS-26, and were among the most promising SC superalloys ever developed. Furthermore, extensive research was conducted during the project to develop a molding technique for SC superalloys.

In the 1990s, research and development of third-generation SC alloys have progressed. Newly developed alloys, e.g., Rene'N6, CMSX-10, and TMS-75, have been released. Rolls Royce applied CMSX-10 and its improved version CMSX-10+ under the names of RR3000 and RR3010 to the uncooled intermediate-pressure turbine blades in their most recent commercial jet (Trent 800) engines. TMS-75, which is one of the third-generation SC alloys developed in Japan, has relatively good microstructural stability and can be heat treated easily. Japan's Toshiba Corporation is bench-testing TMS-75 in the 1300-class 15MW gas turbine.

The development of the fourth-generation alloys, which contain precious metal elements, is underway. In Japan, "High Temperature Materials 21 Project" successfully produced several fourth-generation alloys. One such alloy, called TMS-138, was developed by adding a mixture of strengthening elements and Ru (or iridium (Ir)) to an SC alloy. The improved strength of fourth-generation alloys is achieved by an elaborate, stable dislocation network on the rafted interface of the γ and γ' phases, which works as a barrier that prevents changes in dislocation arrangement from occurring Zhang *et al.* (2003). Figure 9 presents the electron microphotograph of the interfacial dislocation network. "Environmentally

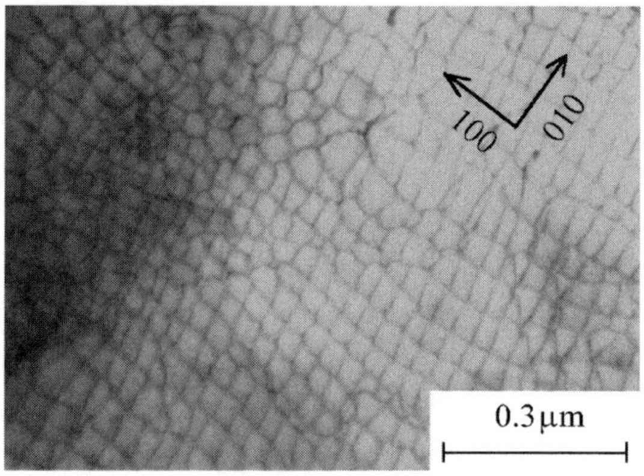

Figure 9 Micrograph of fine dislocation network observed at γ–γ' phase boundary in fourth-generation SC superalloy.

Compatible Propulsion System for Next-Generation Supersonic Transport (ESPR) Research Program," is conducting applied research of fourth-generation alloys as the materials for turbine blades. The European and US jet engine manufacturers are also evaluating fourth-generation alloys. Since the first generation, temperature capability of SC alloys has improved by 70°C. Researchers involved in material design believe that fourth-generation alloys can contribute greatly to increasing temperature capability.

5.08.2.1.4 Application of SC alloys to gas turbines for large power generators

Although SC alloys are used in various jet engines, their application to large gas turbine blades for combined duty cycle for electricity-generating facility has been limited due to difficulties involved in the casting process. Nevertheless, ABB Corporation and Siemens seem to be using an SC alloy equivalent to CMSX-4 in their newest turbine models. In Europe, an extensive material property database for CMSX-4 was developed through a project called COST-501. In 1998, COST-522 project was launched with an aim to develop a 1450-class power generation turbine systems with 60% thermal efficiency (LHV) Allen *et al.* (1998). In the US, "Advanced Turbine System (ATS) Project," funded by Department of Energy (DOE), was launched to promote application of SC alloys. As part of the project, GE and other gas turbine industries applied a second-generation alloy called Rene'75 in the first stage of the blades of a 1450-class gas turbine.

In Japan, TMS-82+ (Hino *et al.*, 1999) was developed through "High Temperature Materials 21 Project." This alloy contains relatively low amount of Re (2.5 wt.%) and its strength in the high-temperature low-stress region exceeds that of third-generation alloys. Verification test of TMS-82+ and TMS-75 was conducted using a 1300-class 15MW experimental turbine. Figure 10, which shows a large blade, illustrates the result of the castability test conducted for TMS-82. The following SC alloys are being developed as of early 2002: YH61 (Tamaki *et al.*, 1998), which is an SC alloy that contains only 1.4 wt.% Re and has excellent tolerance to new crystals and low-angle boundary, good castability, and good high-temperature strength; and MDSC-7M (Mitsuhashi *et al.*, 1990), which is an SC alloy for larger blades and contains 10 wt.% Cr to ensure hot corrosion resistance. They are likely to find applications in the gas turbines in the future.

Figure 10 SC mockup blade cast with TMS-82+ alloy developed in Japan, for use as first-stage turbine blade in a 200 MW-class land-based gas turbine.

The 1450–1500°C-class combined cycle power generation system can contribute greatly to reducing CO_2 emission from fuel consumption. In order to develop more advanced, super-efficient combined cycle power generation systems with higher operating temperature and improved thermal efficiency, researchers and manufactures should find a way to cast high-performance Ni-based SC alloys into larger blades.

5.08.2.1.5 Oxide-dispersion-strengthened superalloys

As shown in Figure 11, oxide-dispersion-strengthened (ODS) superalloys are developed through the process of mechanical alloying, extrusion, and directional re-crystallization. ODS alloys have excellent creep strength at temperatures exceeding 1,000°C because they are dispersion-hardened by using stable oxides such as Y_2O_3. MA series (MA6000, etc.) and PM series (PM3000, etc.) are the ODS alloys available commercially. In Japan, ODS alloys called TMO-2 and TMO-20 were developed through "Advanced Alloys with Controlled Crystalline Structures Development Project" (Kawasaki, 1992). The creep strength of these alloys exceeds that of SC superalloys in all temperature and stress regions; these properties were achieved by strengthening the base alloy depending upon the material design method. Figure 12 compares the creep properties of ODS superalloys, TMO-2 and TMO-20, with those of SC superalloys. They show a clear and

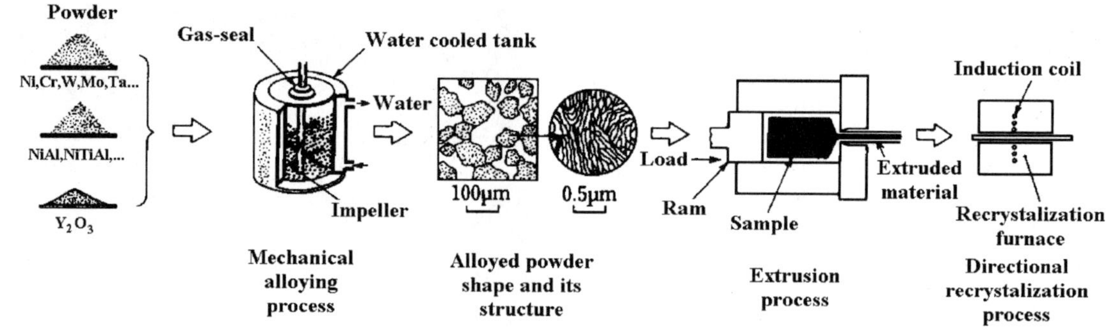

Figure 11 Schematics of fabrication process for the ODS superalloys.

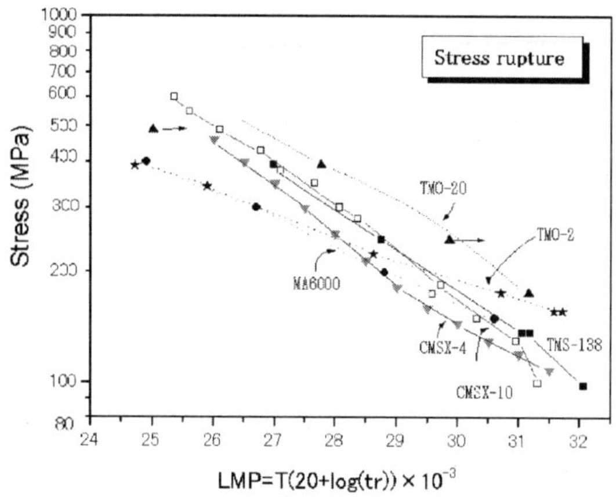

Figure 12 Creep rupture strength of the ODS superalloys compared with SC superalloys.

significant advantage in creep strength over the common SC alloys.

ODS superalloys that are ductile enough to be forged or rolled into sheet, such as MA754, have good oxidation resistance and are suitable high-temperature materials for combustors. However, the above-mentioned TMO-2 and TMO-20 lack such ductility (1–3%) and are not suitable for manufacturing air-cooled blades; therefore, their application to turbine blades has not been materialized yet. Nevertheless, they are much more ductile than ceramics or intermetallic alloys and can contribute greatly to improving gas temperature if applied in small uncooled gas turbines or to jet engine blades and vanes. However, for TMO-2 and TMO-20 to be utilized in gas turbines and jet engines, the production cost must be lowered and approaches should be taken to ensure stabilization and homogenization of material quality across different rods and within the billet.

5.08.2.2 Wrought and Powder Metallurgy Alloys for Turbine Disks

Wrought alloys and powder metallurgy (P/M) alloys are typical alloys used for manufacturing turbine disks. Inconel 718 is a wrought alloy created by utilizing the precipitation-hardening of the Ni_3Nb-type γ'' phase and has widely been used for the manufacturing of disks. However, many manufactures today are shifting to an ordinary superalloy called Udimet 720, which is precipitation hardened with Ni_3Al-type γ' phase and is more stable compared to Ni_3Nb at higher temperatures. Udimet 720Li, which is created by reducing the micro elements (e.g., C) of Udimet 720, is specially popular.

The use of P/M materials allows segregation-free processing and near net shaping by superplastic forging or high-temperature forging. However, P/M materials require rigid quality control during high-temperature processing, in order to prevent foreign matters from mixing in, which increases the production cost. One of the strongest P/M materials available in the market is AF115, which contains relatively large amount of W as a strengthening element. In recent years, more researchers and manufactures are paying attention to damage-tolerant designs, which optimize resistance to crack growth in structures. N18 and Rene 88 DT alloys, which are one of the P/M materials developed based on the damage-tolerant design, are widely in use. In Japan, a P/M superplastic alloy, called TMP-3, was developed through "Advanced Alloys with Controlled Crystalline Structures Development Project." Subsequently, a dual-property disk was developed, which contains the properties of TMP-3 and AF-115. The composition of typical disk alloys is shown in Table 3. Figure 13 illustrates the advancements in the strength of superalloys for disks (Hattori, 1997).

In order to meet the increasing demand for higher-temperature turbines, superalloys for turbine disks should have good creep resistance at temperatures exceeding 700°C. Most of the disks used in Japan today are imported, because high-tonnage forging presses with 5×10^4 t capacity, which are required for the

Table 3 Chemical compositions of typical Ni-based superalloys for use as turbine disks.

	Chemical composition (wt.%)													
Alloy name	*Co*	*Cr*	*Mo*	*W*	*Al*	*Ti*	*Nb*	*Ta*	*Hf*	*Re*	*C*	*B*	*Zr*	*Others*
P/M														
MERL76	18.6	12.4	3.2		5.0	4.3	1.4		0.4		0.03	0.02	0.06	
AF115	15	10.5	2.8	5.9	3.8	3.9	1.8		0.8		0.05	0.02	0.05	
N18	15.7	11.5	6.5		4.4	4.4			0.45		0.02	0.02	0.03	
Wrought														
Inconel 718		19	3		0.5	0.9	5.1				0.04			18.5Fe, 0.2Mn, 0.2Si
WASPALOY	9.5	5.8	1.9	13.7	4.6	0.9		3.3	1.4		0.07	0.015	0.015	
U 720Li	8.7	4.3	1.5	11.6	5.5	1.1		6			0.05	0.01	0.05	

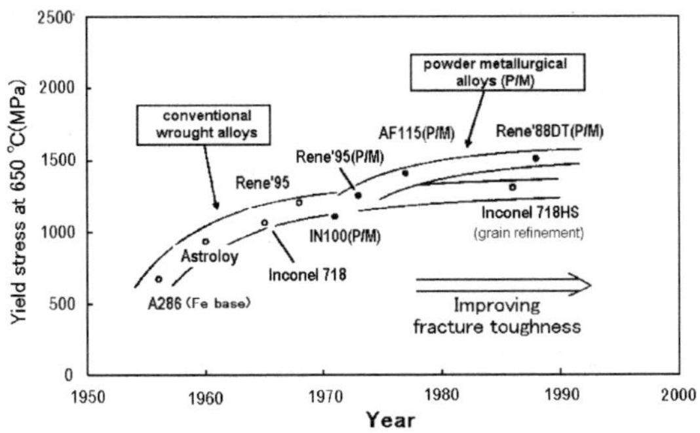

Figure 13 Evolution of high-temperature strength of superalloys for turbine disk.

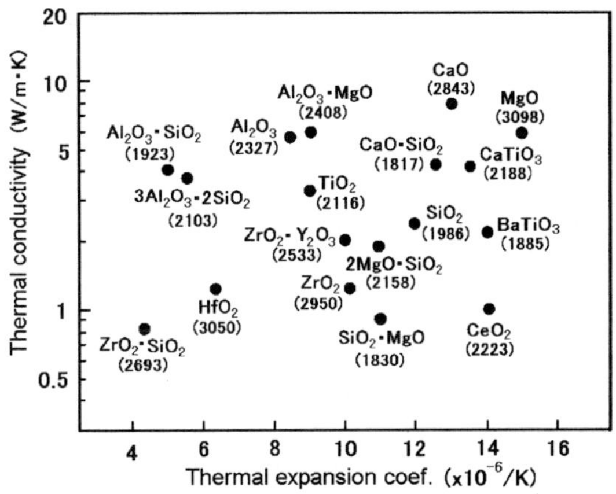

Figure 14 Thermal expansion coefficients and thermal conductivities of typical oxide ceramics.

manufacturing of disks, are not currently available in the country.

5.08.3 ADVANCEMENTS IN THE COATINGS TECHNOLOGY

Ni-based superalloys used in gas turbines in combuster and blade applications are treated with corrosion-resistant coating, oxidation-resistant coating, and thermal barrier coating (TBC) to provide further protection from heat and to reduce the metal temperature. The evolution of various deposition techniques for coatings is described by Yoshiba (1997).

Some of the widely used coatings are chromium/aluminum diffusion coating, Pt-Al coating, and NiCoCrALY thermal spray coating. The coating that can best withstand high gas temperatures is the TBC. The most commonly applied TBC material is yttria-stabilized zirconia (YSZ), which is generally deposited on the bond coat (alloy layer) surface by thermal spraying and electron beam physical vapor deposition (EBPVD) processes. This technique, when used with internal cooling of blades and vanes, creates a large temperature gradient under the coating layer and reduces the metal temperature. EBPVD process allows ceramics to grow vertically on the alloy surface as a group of mutually independent fine columnar grains. This coating can alleviate the difference between thermal expansion of the alloy (bond) layer and that of the ceramic layer.

After prolonged use, however, the properties of EBPVD coating begin to diminish gradually because of a sintering that occurs on the ceramic surface, which can cause a coating failure. As a solution to this problem, researchers are developing zirconia oxide, which has slower sintering kinetics. Furthermore, research and development of totally new oxide coatings is also in progress. Figure 14 illustrates the thermal properties of typical oxide ceramics including the thermal expansion coefficient and thermal conductivity (Yoshiba, 1998). Ideal oxide ceramics are supposed to have higher thermal expansion coefficient (closer to that of the metal underneath) and lower thermal conductivity to further reduce the metal temperatures. Furthermore, in order to prevent damage in the bond coat, introduction of diffusion barrier layer is being considered. The effect of coating on the fatigue crack initiation and propagation is of concern to designers and has been investigated and discussed in Chapter 5.02 of this volume.

5.08.4 FUTURISTIC MATERIALS FOR GAS TURBINE APPLICATIONS

5.08.4.1 Intermetallic Alloys

Intermetallic alloys generally have poor room-temperature ductility. However, titanium–aluminum (TiAl) alloys exhibit a certain level of room-temperature ductility and are characterized by light weight and high specific strength. Currently, the TiAl alloy is the best candidate material for jet engines and is being tested in the second-stage blades of low-pressure turbines. The ground test conducted using the actual jet engine was reported to be successful. However, although TiAl alloy can contribute greatly to the goal of lighter-weight jet engines, it is still too expensive for actual application.

In Japan, the possibility of applying TiAl alloy to low-pressure turbine blades and stationary components has been discussed in the ESPR Research Program. Japan Society for the Promotion of Science Related Institutes and Universities initiated a project called "Application of Advanced High Temperature Material (Heat-resistant Intermetallic Alloy) 1996–2000" and conducted extensive research on TiAl alloys, including the use of directional solidification method to control crystal orientation and lamellar structure of intermetallic alloys (Yamaguchi *et al.*, 2001). Furthermore, this project successfully achieved high creep strength in TiAl alloy without compromising room-temperature tensile ductility.

Automobile manufactures in Japan are already using TiAl alloy in turbo charger rotors. Compared to rotor blades made of ceramics, which tend to be thick due to their limited toughness, those made of TiAl alloy have the following advantage: because TiAl alloy can be cast as thin hollow blades, it has become possible to increase the number of rotor blades, thus enhancing the turbine efficiency.

5.08.4.2 High-melting Point Alloys

A high-melting point alloy is known as a candidate material for uncooled turbine blades. Among the best-known high-melting point alloy is niobium (Nb), which melts at 2,468°C. Although Nb exhibits good ductility and its specific gravity (8.58) is equivalent to that of a superalloy, its application has been difficult due to poor oxidation resistance. However, alloying Nb with silicon (Si) at 1,200°C can produce an alloy with a composite structure consisting of Nb solid solution and Nb silicide (Nb_5Si_3), which has an oxidation resistance equivalent to that of an Ni-based superalloy (Jackson *et al.*, 1996). In Japan, an ongoing project called "Advanced Processing Technique for High Melting Point Metallic Materials" is conducting research and development of various Nb alloys and is reported to have greatly improved the high-temperature strength of Nb alloys (Ma *et al.*, 1999). Figure 15 illustrates the strength of Nb-silicide alloys that have been strengthened by adding Mo and W. The goal of this project is to achieve a certain level of strength in Nb alloys without compromising the oxidation-resistant properties.

Meanwhile, researchers in Japan are developing high-melting point superalloys by generating the γ/γ' structure observed in Ni-based superalloys in the base metals with high melting points, Figure 16 (Yamabe *et al.*, 1998).

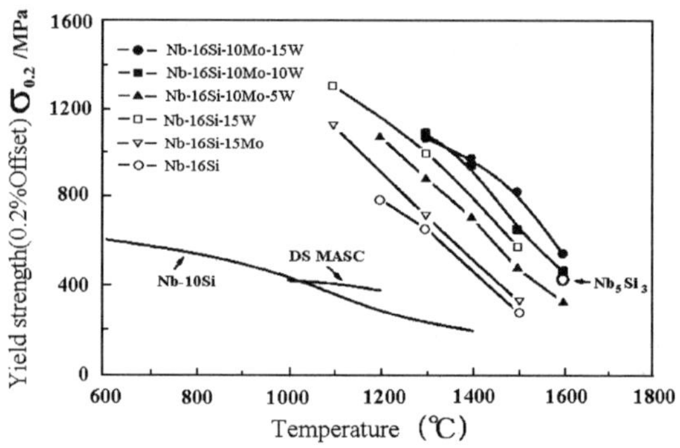

Figure 15 Yield strength of the Nb-silicide refractory superalloys strengthened with Mo and W.

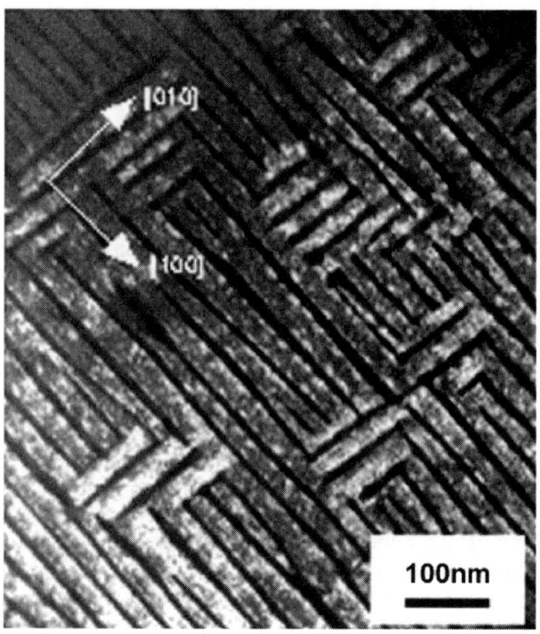

Figure 16 Microstructure of iridium- (Ir-)based refractory superalloy.

For instance, an alloy developed using iridium (2,447°C melting point) as a base metal exhibited 100 h of creep life in the creep test conducted at 1,800°C under a stress of 137 MPa. Furthermore, it has far greater oxidation resistance compared to conventional high-melting-point metals and alloys. Researchers are trying to establish new material design for this alloy, which can be used to lower the specific gravity (by ~20%) and reduce production cost, such as by mixing it with Ni-based superalloys.

Properties of chrome (Cr) alloys are receiving much attention in recent years. For instance, an Fe—(50–60 wt.%) Cr alloy produced by ultrahigh-purity melting process exhibits strength equivalent to Inconel 718 and has excellent ductility and workability at room temperature (Yano and Abiko, 2000). Cr–W and Cr–Re alloys, which are produced through ordinary melting process using materials that have industrially acceptable levels of purity, are reported to have strength equivalent to Ni-based SC superalloys at high temperatures exceeding 1,100°C. Furthermore, Cr–Re alloy can be machined at room temperature. However, ingress of nitrogen during elevated-temperature exposure in air is known to harden and embrittle Cr alloys. Researchers are currently exploring solutions to this problem.

5.08.4.3 Ceramics

Over a million turbo charger rotors for automobiles have been manufactured using silicon nitride (Si_3N_4), which can be produced by selective laser sintering. In "Ceramic Gas Turbine (CGS) Project (1985–2000)," Si_3N_4 was used in the rotor of a small gas turbine, which resulted in greatly increased operational temperature and efficiency. Currently, a project entitled "Industrial Co-generation: Hybrid Gas Turbine (1999–2003)," is conducting research on increasing the thermal efficiency of gas turbines by using Si_3N_4 material in uncooled turbine blades. In addition, there is a plan to run an endurance test on blades made of Si_3N_4 material using an 8 MW-class gas turbine operating at 1,250°C (Karasawa, 2001). Thus, the application of Si_3N_4 in gas turbines is still a viable goal and is being explored by several research groups.

"High Temperature 21 Project" is conducting a research to improve the high-temperature strength of Si_3N_4 by crystallizing the grain-boundary glass phase (Hirosaki, 2001). Its ultimate goal is to achieve 1,000 h of creep rapture life at the temperature of 1,500°C under the stress of 137 MPa. Silicon nitride, combined with a mixture of ytterbium (Yb) (or lutetium (Lu)) and some other elements and then gas-pressure-sintered to crystallize the grain boundary phase, exhibits excellent creep strength at 1,400°C, as seen in Figure 17. For practical widespread use, toughness, long-term oxidation resistance, and aqueous corrosion resistance of these materials must be improved.

Oxide ceramics are another class of ceramic materials that are being considered for gas turbine applications. Compared to Si_3N_4, these materials have better oxidation resistance. A ceramic eutectic called melt growth composite (MGC) (Kazuhisa, 1999), which was released in 1997, is receiving much attention from researchers. MGC maintains consistent strength at temperatures ranging from room temperature to 1,800°C and, like metals, exhibits plastic deformation due to dislocation motion at high temperatures. It has good oxidation resistance and its form and structure remain stable during 1,000 h of exposure to air at 1,700°C. Figure 18 compares the reported strength of MGC with that of a sintered Si_3N_4 ceramic with crystallized grain boundary

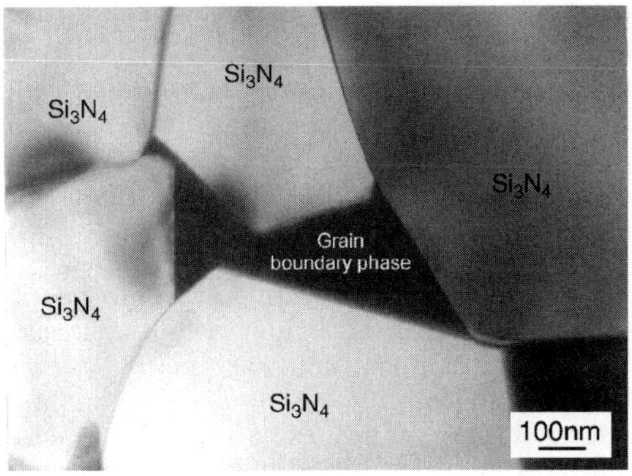

Figure 17 Microstructure of the silicon nitride with lutetium (Lu), which is sintered for crystallization of grain boundary phase.

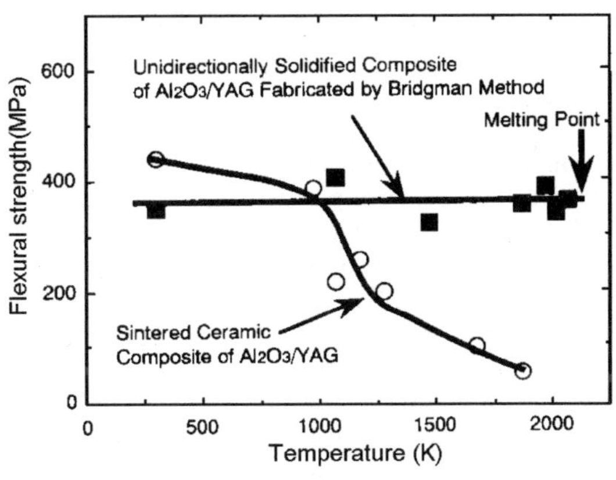

Figure 18 Flexural strengths of MGC and sintered ceramic composite of Al_2O_3/YAG.

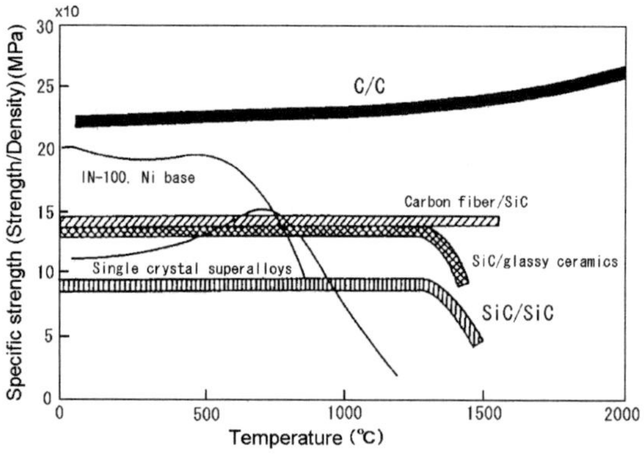

Figure 19 Specific strengths of various composites, with those of CC and SC superalloys as reference.

Table 4 Fracture toughness of several composites and ceramics.

CMCs, ceramics	*Flexural strength* (MPa)	*Fracture toughness, K_{Ic}* ($MPa\,m^{1/2}$)
SiC/LAS	830	17
LAS	200	2
SiC_W/Al_2O_3	800	8.7
Al_2O_3	550	4–5
ZrO_2(TZP)	954	11.6
SiC/SiC	193	30
SiC (Pressureless sintered)	402	3.4
SiC_W/Si_3N_4	800	15.6
CVD-SiC/RBSN	620	
HIP'ed Si_3N_4	863	5

phase. A project called "MGC Super Efficient Gas Turbine System (2001–2005)" is conducting product development research of MGC, including the production of prototype vane for a 1,700°C-class uncooled gas turbine.

Future widespread application of ceramic materials depends on whether the room temperature toughness and thermal shock resistance of ceramic materials can be improved (see also Chapters 2.08 and 4.14). Furthermore, gas turbines can be modified and designed in a way that is appropriate for the properties of ceramics.

5.08.4.4 Composite Materials

Composite materials can contribute greatly to the realization of lighter-weight materials and higher-temperature capability for gas turbine applications. Al-based, Ti-based, and intermetallic composites are typical metal matrix composites (MMCs) with lighter weight and higher specific strength (Figure 19). Ti-based composite materials have the potential to replace Ti-based alloys as a materials for jet engine fan blades and compressor components. The potential of Ti-based composite materials is being explored in ESPR Research Program. However, in order for the use of Ti-based composites to become practical, the production cost of Ti-based composite materials must be reduced and a complete material database required for designing must be made available.

As shown in Table 4 (Tanaka, 2000), ceramic matrix composites (CMCs) have significantly greater fracture resistance compared to ordinary ceramics. Known as light-weight super-high-temperature materials, they are excellent candidate materials for combustor liners, high-pressure turbine vanes and shrouds. CMCs developed in Japan contain high-quality SiC fibers, known as Nicalon and Tirano. Although Japan's technology in CMCs development is very advanced, CMCs are still too expensive for practical applications. Research is also needed to enhance environmental degradation resistance of the coatings and to develop a technique for controlling the fiber–matrix interface properties.

Carbon–carbon (C/C) composite is a light-weight high-temperature material which increases its strength in extremely high temperatures up to 2,000°C. However, for these materials to find widespread use, the oxidation resistance must be improved through the development of reliable coating technique.

Latest development in the application research of composite materials include Ni-based SC alloys, TiAl intermetallic alloys, and other advanced materials that were reported in "International Symposium on Environmentally Compatible Propulsion System for Next-Generation Supersonic Transport."

5.08.5 REFERENCES

D. H. Allen *et al.*, 1998, "Materials for Power Engineering: Proceedings of Conference," ed. J. Lecomte-Beckers *et al.*, p. 1825.

H. Harada, 2000, *J. Gas Turbine Soc. Japan*, **28**(4), 278.

H. Hattori, 1997, *J. Gas Turbine Soc. Japan*, **25**(98), 71.

T. Hino, Y. Ishiwata, Y. Yoshioka, K. Nagata, T. Kobayashi, Y. Koizumi, H. Harada and T. Yamagata, 1999, "Gas Turbine Congress: Proceedings of International Conference, Kobe, November 14–19, 1999," p. 169.

N. Hirosaki, 2001, *J. Gas Turbine Soc. Japan*, **29**(3), 203.

M. R. Jackson, B. P. Bewlay, R. G. Rowe, D. W. Skelly and H. A. Lipsitt, 1996, *JOM*, p. 48.

H. Karasawa, 2001, Ceramic Data Book 2001 (separate volume), Industries and Products, vol. 29, p. 83.

H. Kawai, *et al.*, 1997, *Mater. Japan*, **36**, 643.

Y. Kawasaki, 1992, *Metal*, May, 9.

Y. Kazuhisa, 1999, *J. Gas Turbine Soc. Japan*, **27**(2), 65.

T. Kobayashi, Y. Koizumi, H. Harada, T. Yamagata, A. Tamura and S. Nitta, 1998, "Materials for Power Engineering: Proceedings of Conference," eds. J. Lecomte-Beckers *et al.*, p. 1079.

C. L. Ma, A. Kasama, Y. Tan, H. Tanaka, R. Tanaka, Y. Mishima and S. Haneda, 1999, "Reports of the 123rd Meeting of JSPS High Temperature Materials Committee," vol. 40(3), p. 349.
A. Mitsuhashi *et al.*, 1990, "High Temperature Materials for Power Engineering: Proceedings of Conference, Liege, Belgium, September 24–27, 1990," p. 1301.
H. Tamaki, A. Yoshinari, A. Okayama and S. Nakamura, 1998, "Materials for Power Engineering," ed. J. Lecomte-Beckers *et al.*, p. 1099.
R. Tanaka, 2000, "High Temperature Corrosion and Protection, Hokkaido, Japan, September 17–22, 2000," p. 31.
C. T. Sims *et al.*, 1987, "Superalloys," Wiley, p. 12.
Y. Yamabe, Mitarai, Y. Ro, T. Maruko and H. Harada, 1998, *Metall. Mater. Trans. A*, **29**, 537.
M. Yamaguchi *et al.*, 2001, "Report of the 123rd Meeting of High Temperature Materials Committee, November 2001," as part of Japan Society for the Promotion of Science (JSPS) Research for the Future Program.
M. Yamasaki, 1991, "A Handbook for Advanced Alloying Technique for Crystallization Control," Japan Standard Association, Japan.
K. Yano and K. Abiko, 2000, *Mater. Trans. Japan Inst. Metals (JIM)*, **41**, 122.
M. Yoshiba, 1997, *J. Gas Turbine Soc. Japan*, **25**(98), 81.
M. Yoshiba, 1998, *J. Gas Turbine Soc. Japan*, **26**(101), 63.
J. S. Zhang, T. Murakumo, H. Harada and Y. Koizumi, 2003, Dependence of creep strength on the interfacial dislocations in a fourth generation SC superalloy TMS-138. *Scr. Mater.*, **48**(3), 287–293.

Comprehensive Structural Integrity
ISBN (set): 0-08-043749-4
Volume 5; (ISBN: 0-08-044154-8); pp. 361–373

Volume Subject Index

The index is in letter-by-letter order, whereby hyphens and spaces within index headings are ignored in the alphabetization (e.g. acid-base precedes acid casein). Terms in parenthesis are excluded from the initial alphabetization. In line with normal materials science practice, compound names are not inverted but are filed under substituent prefixes.

The index is arranged in set-out style, with a maximum of three levels of heading. Location references refer to the page number. Major discussion of a subject is indicated by bold page numbers.